FOUNDATIONS OF COLLOID SCIENCE

FOUNDATIONS OF COLLOID SCIENCE

Volume I

ROBERT J. HUNTER
University of Sydney

Written in collaboration with

Lee R. White *University of Melbourne*
Leonard R. Fisher *CSIRO, Sydney*
Norman Parker *CSIRO, Sydney*
Richard M. Pashley *Australian National University*
Donald H. Napper *University of Sydney*
Richard W. O'Brien *University of NSW, Sydney*
John Ralston *SAIT, Adelaide*
Franz Grieser *University of Melbourne*

CLARENDON PRESS · OXFORD

Oxford University Press, Walton Street, Oxford OX2 6DP

Oxford New York Toronto
Delhi Bombay Calcutta Madras Karachi
Petaling Jaya Singapore Hong Kong Tokyo
Nairobi Dar es Salaam Cape Town
Melbourne Auckland

and associated companies in
Berlin Ibadan

Oxford is a trade mark of Oxford University Press

Published in the United States
by Oxford University Press, New York

First published 1987
Reprinted 1989 (with corrections), 1991

British Library Cataloguing in Publication Data
Hunter, Robert J.
Foundations of colloid science.
Vol. 1
1. Colloids
I. Title II. White, Lee R.
541.3'451 QD549
ISBN 0-19-855187-8 (Pbk)

Library of Congress Cataloging in Publication Data
Hunter, Robert J. (Robert John), 1933–
Foundations of colloid science.
Bibliography: p.
Includes index.
1. Colloids. I. White, Lee R. II. Title.
QD549.H94 1986 541.3'451 85-28902
ISBN 0-19-855187-8 (Pbk)

Printed in Northern Ireland by The Universities Press (Belfast) Ltd

PREFACE

Over the past forty years or so, colloid science has undergone something of a revolution, transforming itself from little more than a collection of qualitative observations of the macroscopic behaviour of some complex systems into a discipline with a solid theoretical foundation. It can now boast a set of concepts which, if used judiciously, can go a long way towards providing an understanding of the many strange and interesting behaviour patterns exhibited by colloidal systems. Almost every technique and theoretical procedure of modern physics and chemistry has been and is being applied to the study of colloids so that even the specialist colloid chemist finds it difficult to remain *au fait* with the many ramifications of the subject. How much more difficult is it then, for those many scientists and engineers in industry, biology, mineral processing and agriculture, who find themselves confronted with problems involving colloidal systems and who have only their undergraduate studies of physical chemistry or physics to help them?

In May 1982 I organized an intensive course for graduate students and industrial scientists on the principles of colloid and surface chemistry. As a result of that experience, my colleagues and I decided that there was a need for a textbook which would assume a knowledge of elementary physical chemistry but no prior knowledge of colloid science, and take the subject through to a point where the reader could tackle the research literature with reasonable confidence. Hopefully, such a treatment would provide some grasp of the many areas which now form the basis of colloid science: electrostatics, thermodynamics, hydrodynamics, statistical mechanics, and rheology to name a few.

A number of groups of eminent colloid scientists have embarked on such a project over the past thirty years, since the classic texts on colloid science by Alexander and Johnson (Oxford University Press (1949)) and by Kruyt (Elsevier (1952)) appeared. Most of these efforts have, however, been abandoned, and the few books which have appeared do not seem to us to have met the challenge at the appropriate level. That is hardly surprising for there are few people who can claim expertise across such a wide area of fundamental science, whilst retaining some knowledge of the practical applications, and some ability to communicate that knowledge in a straightforward way.

The facile approach to such a problem is to ask a number of experts to write chapters in their special areas and to assemble the results after some judicious editing. Unfortunately, the result usually lacks the unity, the sense of purpose, and the integrity of a book by a single author. To reconcile the conflicting aims of authority and coherence we have opted for a single author, myself, who would be assisted by a number of willing colleagues. Their job was to make a preliminary assembly of material, and to follow the evolution of that material through several revisions until there emerged a version which satisfied their standards of rigour, but was integrated into a single philosophical perspective (mine) on the subject. Only our readers will be able to judge how successful we have been in this enterprise. I know that I have benefited enormously from the effort required to come to grips with all the material presented here. I also know that the book is a much better one than I could possibly have written myself, even with the usual informal help from my friends.

For some chapters (1, 2, 3, and 6) I took full responsibility and for some I was able to rely very heavily on the material initially provided for me (by Lee White (Chapter 4) and Don Napper (Chapter 8) in particular). In other cases, the initial material went through a number of more or less complete rewrites until it fitted into the overall scheme. Through the trauma of seeing their pearls of wisdom scattered, sometimes dropped and reassembled, my collaborators showed unfailing good humour; they were always prepared to sacrifice their time and energy in an effort to 'get it right'. Apart from supplying the initial versions (Len Fisher and Norman Parker (Chapter 5), Richard Pashley (Chapter 7), Richard O'Brien (Chapter 9), and John Ralston and Franz Grieser (Chapter 10)) all these and more (including Derek Chan) were involved in vetting the various developments. Then in 1983 and 1984 the preliminary version was tested with a graduate class in the University of Sydney and further modifications made. The task of establishing a coherent view was greatly assisted by the frequent contacts which have occurred over the years between members of the colloid fraternity in Australia.

Even with all this it must be said that the degree of difficulty of the material varies somewhat between chapters. Some areas, like hydrodynamics, if they are to be treated at the level required to understand the problems (let alone the solutions) require such a powerful mathematical apparatus (tensors and vector calculus) that we were forced to compromise a little on the level of detail we could present. In other areas, like micellar solutions, the techniques applied are more familiar to the physical chemist and we could develop the subject to a fairly sophisticated level. In Chapter 8 (on the effect of polymers on colloid behaviour) the problem was of a different kind: we decided that there was little point in trying to make colloid scientists also polymer scientists, so we opted there for a fairly qualitative (though conceptually demanding) approach. In each case, our primary concern has been to promote an *understanding* of the areas we have tackled, by making each step in the argument accessible. Once again, I thank my collaborators for their patience in trying to make it all understandable to me so that I could try to pass on that understanding to the reader.

I should also like to thank the other people and organizations who made this possible: the University of Sydney for granting me study leave in the summer of 1982/3 and again in 1984/5; Canterbury University in Christchurch, New Zealand, which looked after me in that summer of 1982/3 and, particularly, Ms Sue Bennett (of the Department of Chemical Engineering there) who typed much of the initial draft and whose enthusiasm and unfailing good humour were a positive inspiration. Also Mrs Carol Neville who typed the rest of the manuscript and Mr John Kent for his careful preparation of the diagrams.

The completed manuscript was read for Oxford University Press by Professor Douglas Everett, FRS, and his colleagues at the University of Bristol; their many comments provided us with a further welcome opportunity to correct some errors and misconceptions and to remove some ambiguities. I must also thank Oxford University Press for the care they have taken in producing the final text.

We had hoped the entire work would occupy a single volume about the size of this one. Alas, that proved to be too optimistic so we have put into this one all the material we would expect could be tackled in a one year graduate course in colloid science. This volume is, therefore, essentially complete in itself. Volume II will build on the material developed here and will look to particular applications and more specialized topics.

Sydney R.J.H.
January 1985

Solutions manual

A manual containing solutions to all the problems in Volume I is available from:

Professor R. J. Hunter
School of Chemistry
The University of Sydney
Sydney 2006
Australia

CONTENTS

1

CHARACTERIZATION OF COLLOIDAL DISPERSIONS

1.1 Nature of the colloidal state

When one substance dissolves in another to form a true solution, the ultimate particles of the solute are of molecular dimensions: at most a few molecules may be joined together to form an associated species, such as occurs when benzoic acid dimerizes to form the molecule $(C_6H_5COOH)_2$ in benzene. The radius of the solute molecule in these cases is seldom more than a nanometre and usually rather less. Solute and solvent molecules are of comparable size and we normally assume that the solute molecules are, on average, dispersed uniformly through the (continuous) solvent. There is an important class of materials, however, in which the kinetic units that are dispersed through the solvent are very much larger in size than the molecules of the solvent. Such systems are called *colloidal dispersions* and they may arise in a variety of ways.

If a substance, A, is insoluble in substance B it will usually be possible to break A down into very small particles that can be distributed more or less uniformly through the substance B. Substance A is then called the *disperse phase* and substance B, the *dispersion medium.* In general, A and B may be either solids, liquids or gases so the dispersion should be regarded as a *state of matter,* accessible to any substance A, given the appropriate temperature and pressure and a means of producing and maintaining small discrete lumps of A distributed throughout B. If A is a solid, the particles may be produced by crushing and grinding a macroscopic piece of pure A, by growing small crystals of A by some chemical reaction, or by controlled crystallization of A from some solvent. Just how one goes about distributing A through the medium B (which might be a solid, a liquid or a gas) and maintaining the discrete nature of the A particles (i.e. preventing *aggregation*) forms a considerable part of the theory and practice of colloid science.

The lower limit of size for dispersions of this kind is around 1 nm. Smaller particles would ultimately become indistinguishable from true solutions. The upper limit is normally set at a radius of 1 μm but there is no clear distinction between the behaviour of particles of 1 μm and the somewhat larger particles often encountered in emulsions, in mineral separation processes, and in ceramic engineering.

There are, of course, some molecules that are individually larger than 1 nm in size. These 'macromolecules' can often be uniformly dispersed through a fluid medium and they then form a colloidal solution or dispersion. Proteins, polysaccharides (like starch) and many synthetic polymers fall into this category. It was just such a substance (a naturally occurring gum) that suggested the name 'colloid' (from the Greek word for glue) to the pioneer investigator in the field, Thomas Graham, in the 1860s. The field of polymer science has now developed into an entirely separate discipline with a vast specialized literature of its own. We will, therefore, treat only those few parts of it that are of most significance in colloid science; this is done chiefly in Chapter 8 but references to polymer and protein behaviour are scattered through the text.

A third class of colloidal dispersions arises when a number of molecules of normal size associate together to form an aggregate. Soap molecules, for example, if they are at a sufficiently high concentration in a suitable solvent, can associate together to form *micelles.* These structures are of colloidal dimensions and the resulting system is referred to as an *association colloid.* (The term *colloidal electrolyte* is also used for ionic soap and detergent systems).

The distinguishing feature of all colloidal systems is that the area of contact between the disperse particles and the dispersion medium is relatively large. The energy associated with creating and maintaining that interface is significant, so the study of interfacial (or surface) chemistry is an integral part of the study of colloids. We will, however, limit ourselves to those aspects of surface chemistry of direct relevance to colloids. It is, incidentally, not necessary that all of the dimensions of the disperse particle be very small for the system to be of interest to colloid scientists. Some important colloidal systems have particles that can readily be seen with the aid of an ordinary microscope, such as textile fibres or the cellulose fibres used in making paper. It is sufficient if one of the characteristic dimensions of the particle (in this case the diameter) falls in the colloid size range. The gas bubbles that make up a foam are usually larger than colloid size but in that case the thin lamellae of liquid between the bubbles have a thickness of the order of colloidal dimensions. They, therefore, also fall within the purview of colloid science.

Most of the examples quoted above refer to dispersions of solids in liquids and this is the area with which we will be principally concerned. We will, however, also consider dispersions of liquids in liquids (emulsions: Chapter 16, Vol. II) and gases in liquids (foams). Dispersions in which the continuous (dispersion) phase is solid are becoming increasingly important in the field of material science where many new composites are being developed, but this area remains outside the scope of the present volume (see, for example, Evans and Langdon 1976). Likewise, dispersions of solids and liquids in gases will not be treated in

Table 1.1
The various types of colloidal dispersion with some common examples. The nomenclature is adapted from Ostwald (1907)

Disperse phase	Dispersion medium	Notation	Technical name	Examples
Solid	Gas	S/G	Aerosol	Smoke
Liquid	Gas	L/G	Aerosol	Hairspray, mist, fog
Solid	Liquid	S/L	sol or dispersion	Printing ink, paint
Liquid	Liquid	L/L	Emulsion	Milk, mayonnaise
Gas	Liquid	G/L	Foam	Fire-extinguisher foam
Solid	Solid	S/S	Solid dispersion	Ruby glass (Au in glass); some alloys
Liquid	Solid	L/S	Solid emulsion	Bituminous road paving; ice cream
Gas	Solid	G/S	Solid Foam	Insulating foam.

any detail. Such *aerosols* as they are called, are certainly colloidal systems but the theoretical concepts required to describe their behaviour, although similar in some respects, differ quite significantly from those with a liquid dispersion medium. There are, in any case, a number of recent works concerned with the solid–gas and liquid–gas interface (see references at the end of this chapter).

Table 1.1 lists the various types of colloidal system with common examples of each. Note that all possible combinations can be realized except that of gas in gas.

Exercises. 1.1.1 Give further examples of each of the colloidal systems listed in Table 1.1 and state their approximate composition.

1.1.2 Starting with a cube of solid, 1 cm along each edge, what is the total surface area when the solid is subdivided into cubes 10^{-4} cm on each edge? Repeat the calculation for 10^{-5} cm and 10^{-6} cm cubes. Calculate the surface energy per particle in each case, assuming the surface energy is 70 mJ m^{-2} and compare this to thermal energy (kT) at room temperature (25 °C). What is the total surface energy for each system?

1.1.3 Show that the surface area per unit mass, A_m, of particles of density ρ is given by: $A_m = k'/\rho r$, where r is some characteristic dimension, and $k' = 3$ for spheres, 2 for thin cylindrical discs and long rods, and 4 for long square prisms. (A_m is called the specific surface area and typical values for a colloidal material fall in the range 1–10^3 m^2 g^{-1}.

1.1.4 Chemical bonding energies are commonly of the order 100 kJ mol^{-1}. Show that surface energies of particles will approach this value for sizes of about 1 nm. (Assume the surface energy is 0.1 J m^{-2} and take reasonable values for the density and molar mass).

1.1.5 A mineral oxide of density 2.8 g cm^{-3} is broken up into colloid sized particles in a ball mill. Calculate the total surface area of the crushed material (m^2 g^{-1}) when the average particle radius is 5×10^{-4} cm. What is the total area when the particle radius is 500 Å?

1.2 Technological and biological significance of colloidal dispersions

Almost all of the ancient and modern craft industries draw much of their technical expertise from colloid science. In paper-making both the cellulose fibre used as the meshwork and the clay used as a filler to improve opacity and produce a shiny texture are colloidal. The inks used in ball point pens, in xerography, and in high-speed printing presses each owe their special properties to their colloidal character, as do also the many varieties of paints and cosmetics.

Ceramic products from expensive china to building bricks are made from clay/water sols and modern colloid techniques are currently being used to develop a new generation of very tough (fracture resistant) ceramic materials for use in rocket nose cones, car engines and in medical prostheses (Kuhn 1963; Evans and Langdon 1976).

Colloid science is important in extracting oil from geological deposits, in converting it to petroleum and in making rubber tyres as well as in mineral extraction. The aerosols for dispensing domestic products like shaving cream and deodorants have their agricultural counterparts in the sprays used for dispensing weedicides and insecticides. On the debit side, the same techniques are used for making defoliants, and the gels and dispersions used in flame throwers, napalm, and riot control gases.

Apart from the widely recognized colloidal nature of protein and polysaccharide solutions there are many other biological systems that have been studied by the methods of colloid science. The flow properties of blood are best understood in terms of its being a colloidal dispersion of (deformable) flat plates (the red corpuscles) in a liquid (Goldsmith and Mason 1975). The flow properties of faecal material must sometimes be modified by colloid chemical techniques to avoid unpleasant physiological consequences. The synovial fluids that lubricate the joints and bearing surfaces in the body owe their remarkable properties to their colloidal character. Also the adhesion between cells and the interaction between antigens and antibodies is currently being treated by the same mathematical theory that applies to the coagulation of colloidal particles. The greater part of the food processing, preserving, and packaging industry rests heavily on colloid chemistry, and agricultural scientists require a knowledge of the colloidal properties of soils in order to induce optimum plant growth.

Medical practitioners have used a colloidal dispersion of gold (*potable gold*) since the Middle Ages for treating a variety of ailments. Modern colloidal microcapsule techniques allow controlled release of a drug (Gardner 1976) and, in some cases, accurate targeting onto a particular organ. More routine applications of colloid chemical principles crop up in the preparation of emulsions and suspensions that must remain homogeneous for long periods on the shelf (or at least be readily redispersed on shaking). Aqueous emulsions of perfluoro-hydrocarbons have recently been developed as (temporary) blood substitutes (Riess and Le Blanc 1978)).

Apart from its contributions to engineering, agriculture, biology, and medicine, colloid science also has an important role to play in reducing the harmful effects of technological development. Many pollution problems are due to the presence of unwanted colloidal materials, and their removal (from air or waterways) calls for the application of colloid chemical techniques. The specific adsorptive properties of colloids can also be used to remove, to concentrate and possibly to recover industrial products (especially metal ions) from air and water.

1.3 Classification of colloids

Freundlich, in his classical text on the subject (1926), suggested that colloidal dispersions could be divided into two classes, called *lyophilic* (solvent loving) and *lyophobic* (solvent hating) respectively, depending on the ease with which the system could be redispersed if once it was allowed to dry out. As with most such dichotomies, further study has revealed a complete range of intermediate types, but it is still useful to distinguish between the extremes. Alexander and Johnson (1949) list their properties as shown in Table 1.2. Kruyt (1952) uses the same classification and refers to them also as reversible and irreversible systems, respectively. This terminology expresses more clearly the real nature of the distinction because the ultimate test of whether a system is lyophilic is to determine whether the dispersion process occurs spontaneously when the solvent is added to the colloid. In the grey area between the two extremes lie systems that can exhibit both forms of behaviour. For example, the clay mineral montmorillonite (section 1.5.5) will disperse spontaneously in water if its negative charge is neutralized by strongly hydrated cations (e.g. Li^+) but not if the cation is poorly hydrated (e.g. Cs^+) or highly charged (Ca^{2+}).

One contributing factor to the difference in behaviour between reversible (lyophilic) and irreversible (lyophobic) systems is the extent to which the dispersion medium is able to interact with the atoms of the suspended particle. If it can come into contact with all or most of those atoms then solvation energy will be important and the colloid should be lyophilic (reversible) in some suitable solvent. If it is prevented, by the structure of the suspended particles (i.e. the disperse phase) from coming into contact with any but a small fraction of the atoms of those particles then the colloid will almost certainly be lyophobic (i.e. irreversible) in its behaviour, even if the surface atoms interact strongly with the solvent. An exception is the case of some microemulsions (section 1.5.4), which appear to be reversible (in the sense that they form 'spontaneously' with a minimum of mechanical energy input) even though the size of the emulsion drops may be ~20 nm.

When the dispersion medium is water, the terms hydrophilic and hydrophobic are used; the great majority of the present work is devoted to the examination of hydrophobic sols.

The lyophilic colloid solution is thermodynamically stable since there is a reduction in the Gibbs free energy when the 'solute' is dispersed. The strong interaction between 'solute' and solvent usually supplies sufficient energy to break up the disperse phase ($\Delta H < 0$) and there is often an increase in entropy as well; any reduction in solvent entropy due to the interaction with 'solute' is usually more than compensated by the entropy

Table 1.2

Lyophilic	*Lyophobic*
1. High concentrations of disperse phase frequently stable.	1. Only low concentrations of disperse phase stable.[a]
2. Unaffected by small amounts of electrolytes. 'Salted out' by large amounts.	2. Very easily precipitated by electrolytes.
3. Stable to prolonged dialysis.[b]	3. Unstable on prolonged dialysis[c] (due to removal of the small amount of electrolyte necessary for stabilization).
4. Residue after dessication will take up dispersion medium spontaneously.	4. Irreversibly coagulated on dessication.
5. Coagulation gives a gel or jelly.	5. Coagulation gives definite granules.[d]
6. Usually give a weak Tyndall beam.[e]	6. Very marked light scattering and Tyndall beam.
7. Surface tension generally lower than dispersion medium.	7. Surface tension not affected.
8. Viscosity frequently much higher than that of medium.	8. Viscosity only slightly increased.[f]

[a] This is no longer true, especially if one allows the possibility of an adsorbed stabilizing layer of lyophilic material (steric stabilization).
[b] Dialysis refers to a membrane filtration technique for separating colloidal particles from small molecules or ions (see section 1.7.1(b) below).
[c] Note that this is not true of lyophobic sols with dissociable ionic surface groups (COOH, $-SO_3OH$, $-NH_3^+$) attached.
[d] Except for concentrated systems.
[e] When a light beam passes through a colloidal sol it is visible from the side because the light is scattered by the particles. This is the Tyndall effect (see Exercise 1.5.1 and section 2.3.).
[f] This is true only for dilute, *stable* sols with more or less spherical particles.

increase of the 'solute'. For the lyophobic colloid, the Gibbs free energy *increases* when the disperse phase is distributed through the dispersion medium so that it is a minimum when the disperse phase remains in the form of a single lump. A lyophobic colloid can, therefore, only be dispersed if its surface is treated in some way that causes a strong repulsion to exist between the particles. In this way the particles can be prevented from aggregating (or *coagulating*) for long periods, although it must be emphasized that they are still thermodynamically *unstable* and the barrier to coagulation is merely a kinetic one†. Given enough time they will ultimately form an aggregate.

There is a well developed theory to describe the interaction between particles of a lyophobic colloid but the behaviour of lyophilic colloids is more difficult to describe. The reason for this is that all of the forces involved in lyophobic systems are also important for lyophilic systems but in addition, for the lyophilics, there are very strong specific solvent effects that are difficult to predict. We will, therefore, spend quite some time developing the theory of lyophobic colloids, choosing as typical examples the silver halide, clay mineral, metal oxide, and polymer latex sols, since these have been well characterized and much studied. The principles that emerge from those studies will then be applied to a number of more practical systems.

1.4 Some thermodynamic considerations

1.4.1 *The phase rule*

The references made above to the surface area of the disperse material suggest that it is being treated as a separate phase. On the other hand, a homogeneous solution of a protein in water would normally be treated as a one-phase system. How, then, can one reconcile this one-phase or two-phase treatment of colloidal systems with the requirements of Gibbs' phase rule? The problem is discussed in some detail by Kruyt (1952, p. 11) using the generalized form of the phase rule:

$$\mathcal{V} + \mathcal{C} = f + \mathcal{P} \tag{1.4.1}$$

where $\mathcal{V}$ is the number of independent variables, and $\mathcal{C}$, f, and $\mathcal{P}$ are the numbers of components, degrees of freedom and phases respectively. The variables, $\mathcal{V}$, include temperature and pressure (the usual two) and any other quantities that might influence the thermodynamic behaviour of the system (e.g. magnetic field strength, angular velocity etc). Each of the $\mathcal{C}$ components has a chemical potential ($\mu_1, \ldots, \mu_{\mathcal{C}}$), which will be the same in every phase at equilibrium and will determine its activity in that phase. The total number of variables is, therefore, $\mathcal{V} + \mathcal{C}$. Since there is a Gibbs–Duhem relation between the variables in each of the $\mathcal{P}$ phases, the number of degrees of freedom, f, is $(\mathcal{V} + \mathcal{C} - \mathcal{P})$, as stated in eqn (1.4.1)‡.

† Strictly speaking one should distinguish the aggregate of particles from the bulk solid. The dispersion is *metastable* with respect to the aggregate which is metastable with respect to the bulk solid.

‡ Defay *et al.* (1966) give a more general analysis which reduces to eqn (1.4.1) when there are no chemical reactions and each surface in the system is a single surface phase. We will return to this more detailed treatment in section 5.4.4.

Kruyt shows that one can choose to treat a colloidal system as one- or two-phase, whichever is most convenient. The usual lyophobic colloid is best treated as a two-phase system since the disperse material has a negligible effect on the chemical potential of the dispersion medium. If one were to treat it as a one-phase system one would also have to acknowledge that the chemical potential of the sol is not effectively a variable so that $\mathscr{C}$ in eqn (1.4.1) is also reduced by one and the number of degrees of freedom remains unchanged.

A lyophilic sol, on the other hand, is best treated as a single-phase system because the disperse phase does have a significant effect on the properties of the dispersion medium and this introduces an extra degree of freedom into the system. (See also Hall and Pethica 1967).

The difference in behaviour of the two types of system can be illustrated by considering the vapour phase of the dispersion medium as a function of temperature. For the lyophobic system, the vapour pressure is essentially unaffected by the concentration of the colloid whereas the lyophilic colloid will lower the solvent vapour pressure at any temperature, to an extent dependent on its concentration.

A more complete resolution of this problem can be given using the theory of the thermodynamics of small systems. A short account of that theory and its application to micellar solutions is given by Hall and Pethica (1967). In that case the single-phase description is much more satisfactory than the two-phase description.

1.4.2 *Particle size distribution*

Kruyt (1952) also shows, by a general thermodynamic argument, that the degree of dispersion, λ, of a lyophobic sol is not an independent thermodynamic variable. λ is defined by the relation

$$\lambda = (\partial \mathscr{A} / \partial v) \tag{1.4.2}$$

where $\mathscr{A}$ is the particle surface area and v is the particle volume. It can be shown that if the system is in equilibrium (as one would expect for droplets of liquid in a vapour) then all of the particles must be of the same size, so λ is constant. Furthermore, this is an *unstable* equilibrium (unless the volume of the container is very small). William Thomson (later dubbed Lord Kelvin) showed that the vapour pressure, p, of a spherical liquid droplet of radius r_e, is related to the normal equilibrium vapour pressure over a flat liquid surface (p_∞) by the equation:

$$\ln p / p_\infty = \frac{2\bar{V}_L \gamma}{RTr_e} \tag{1.4.3}$$

where $\bar{V}_L$ is the molar volume of the liquid, γ is the surface energy (or surface tension) and T is the temperature. (We will derive equation

(1.4.3) in Chapter 5.) Note that the equilibrium vapour pressure of small drops is higher than that of larger drops so the large ones grow at the expense of the small ones. The equilibrium radius, r_e, is unstable because the slightest fluctuation from it produces a particle that must either continue to grow at the expense of the others ($r > r_e$) or continue to evaporate ($r < r_e$). An analogous relation can be written for the solubility of solid particles in a liquid (Ostwald 1907):

$$\ln \frac{C(r)}{C_\infty} = \frac{2\bar{V}_{SL}\gamma}{RTr} \tag{1.4.4}$$

where $C(r)$ is the solution concentration in equilibrium with particles of radius r and C_∞ is the bulk solubility. Again the large particles grow at the expense of the small ones, *provided the solid is sufficiently soluble in the liquid to allow equilibrium to be established* in a reasonable time.

There are many colloidal situations, however, where the particle solubility is so low that equilibrium is never established and a quasi-equilibrium exists in which a variety of particle sizes can co-exist. The variable, λ, can in this case be ignored. Likewise, in cases where the surface energy is zero, or nearly so, (as in microemulsions) it would be possible for many different particle sizes to co-exist at equilibrium, although in practice it turns out that such systems are usually quite monodisperse (Overbeek 1982), so λ is again constant.

Exercises. 1.4.1 How small must a droplet of water be in order for it to exhibit a vapour pressure 25 per cent higher than the normal value at 373 K? (Take $\gamma = 72\ \text{mJ m}^{-2}$.)

1.4.2 The surface energy of the solid/solution interface is typically about $1\ \text{J m}^{-2}$. Estimate the size at which a typical solid (molar mass 100, density $2\times10^3\ \text{kg m}^{-3}$) would show a 10 per cent rise in solubility. [This is one of the methods used to estimate the interfacial energy of solids].

1.5 Some typical colloidal dispersions

1.5.1 *Preparation of colloidal dispersions*

The general methods of preparing colloidal dispersions have been known for a long time and are adequately discussed in the older literature (Svedberg 1928; Weiser 1933; Alexander and Johnson 1949).

We will describe these only briefly and then proceed to the more recent developments in which careful control of the growth process has led to the production of dispersions in which the particles all have almost the same size and shape. Such systems are ideal for testing aspects of the theory of dispersions, they have some interesting properties in their own right, and they may even offer some special advantages in certain technological processes.

We will also examine some naturally occurring colloidal systems that offer special advantages for testing various theoretical analyses. It will come as no surprise to learn that the theoretical description of the equilibrium, kinetic, and transport properties of colloidal systems is almost always confined to certain simple geometric shapes: usually the sphere or infinite flat plate but sometimes the cylindrical rod or disc and more rarely the spheroid. To adequately test the validity of these descriptions one must have available colloidal dispersions in which the particle shapes conform as accurately as possible to these simple geometric types and that has only recently become possible. Only by improving our understanding of such model systems can we hope to improve our descriptions of the behaviour of real colloidal systems.

Svedberg (1928) divides the preparation of colloidal dispersions into two categories: dispersion and condensation, of which the latter is probably more important for fine material. In the dispersion methods, a sample of bulk material is broken down to colloidal dimensions by some kind of mechanical process. The most direct method is by grinding in a *colloid mill.* This device subjects a coarse suspension of particles to a very high shear field by forcing it into a narrow gap between two surfaces that are rotating rapidly with respect to one another. The particles are then torn apart by the shearing process and a colloidal dispersion results, provided that the solution contains a suitable *dispersing agent* to prevent the small particles from aggregating together (see section 1.5.3 below). A similar effect can be achieved, especially with liquid-in-liquid dispersions (emulsions), by subjecting a mixture of the two phases to a very high frequency sound wave (~20 kHz). This process, known as *ultrasonication,* also requires the presence of a dispersing agent if a stable sol is to result.

Sols can also be formed by passing an electric arc between two wires placed under the surface of a liquid (Bredig's procedure). This is also an example of a dispersion method in that some of the sol almost certainly results from pieces of metal being torn from the surface of the wire; there is, however, some condensation from the vapour also involved.

Condensation methods are much more numerous and more diverse. They may involve dissolution and reprecipitation, condensation from the vapour or chemical reaction. In the first category is the formation of a solid paraffin sol in water. This can be done by dissolving paraffin wax in ethanol and pouring a little into a large volume of boiling water. The ethanol rapidly boils off leaving an opalescent dispersion of the paraffin. The second type is exemplified by the spontaneous formation of a mist or fog from a supersaturated vapour; provided the degree of supersaturation is sufficiently high (Svedberg suggests a vapour pressure more than eight times the equilibrium value) the formation of many droplets of very small size (<1 μm) is assured.

The chemical methods may involve reduction, oxidation or double decomposition. Metal sols· can be produced by reduction (gold by reducing chlorauric acid, $HAuCl_4$, with hydrogen peroxide or red phosphorus; silver from silver nitrate and ferrous citrate). The particles are usually very small (≪100 nm) and with some recipes are so well stabilized that they behave as though they were lyophilic (Frens and Overbeek 1969).

Oxidation can be used to produce sulphur sols either from hydrogen sulphide:

$$2H_2S + O_2 \rightarrow 2S + 2H_2O$$

or from thiosulphate solutions:

$$S_2O_3^{2-} + H_2O \rightarrow S + SO_4^{2-} + 2H^+ + 2e^-.$$

The latter reaction occurs at a well defined rate, determined by the pH and $S_2O_3^{2-}$ concentration and is the basis of one of the well known 'clock reactions' for demonstrating kinetic concepts in elementary chemistry courses. It has recently been studied in detail by Johnston and McAmish (1973), who suggest that the rate-determining step is probably:

$$HS_2O_3^- + S_2O_3^{2-} \rightarrow S_2 + HSO_3^- + SO_3^{2-}.$$

We will return to this reaction shortly to consider the formation of monodisperse sulphur sols.

Many inorganic compounds that are insoluble in water can be induced to form colloidal dispersions if they are formed by mixing fairly concentrated reagents, especially in the presence of a dispersing agent. Thus barium sulphate sols can be prepared by mixing $Ba(SCN)_2$ and $(NH_4)_2SO_4$ in the presence of a little potassium citrate to act as dispersant. Arsenious sulphide sols are formed by bubbling H_2S through a solution of As_2O_3 while silver halide sols are readily prepared by mixing silver nitrate and alkali halide solutions. Indeed, it is often necessary in gravimetric analysis procedures to treat a precipitate (e.g. $BaSO_4$) in a special way to prevent the formation of a colloidal sol or to aggregate a sol, once formed, in order to filter it effectively.

There are other types of chemical reaction that are useful in certain cases but of limited applicability. Photodecomposition is, for example, important in the formation of silver particles from AgBr in photographic processes. Hydrolysis is also an important technique for preparing sols of the transition metal oxides and hydroxides; it has been very effectively exploited by Matijevic for the preparation of highly *monodisperse systems* (i.e. systems of uniform particle size), the subject to which we will now address ourselves.

1.5.2 *Monodisperse sols*

Overbeek (1981) points out that monodisperse (or homodisperse or isodisperse) systems have played an important part in the development of our understanding of colloidal sols and, more importantly, have allowed colloid science to make essential contributions to our understanding of the behaviour of matter. In the next chapter we will see how Perrin used them to establish the experimental basis of the kinetic theory of matter (and indeed the very existence of molecules) and the value of the Avogadro constant. Likewise, the proof, by Svedberg, that proteins were well-defined molecules with precise molar masses was essential to the development of modern biochemistry.

The key to formation of monodisperse sols is illustrated in Fig. 1.5.1. The chemical reaction by which the sol material is being formed must proceed at a suitable rate (and this can be controlled by temperature and concentration conditions). Precipitation does not occur as soon as the concentration of the product exceeds its saturation (i.e. maximum equilibrium) concentration in the solution. Rather, it is necessary for a certain level of supersaturation to be reached before there are formed the nuclei on which crystal growth can subsequently occur. Conditions must be arranged so that nucleation occurs in a single short burst so that all subsequent deposition occurs on those initial nuclei. The nuclei grow rapidly at first and it is this that causes the concentration to fall below the nucleation level. The formation of new material must be maintained at a

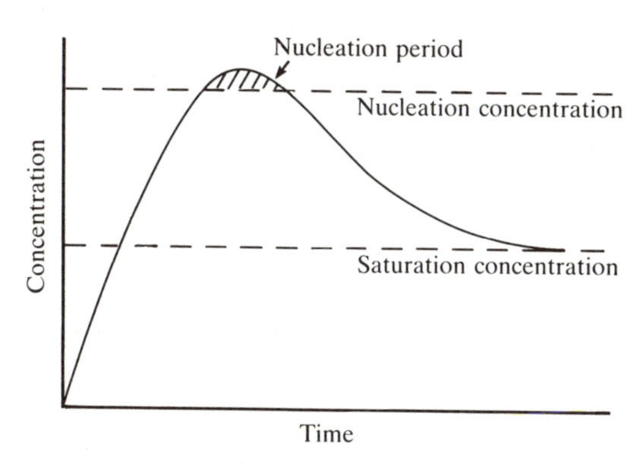

Fig. 1.5.1. Illustrating the production of a monodisperse sol by confining the formation of nuclei to a very short period, so that the particle number remains constant and all grow together to the same size. (After Overbeek 1981.)

rate that keeps the solution concentration between the horizontal broken lines so that no new nuclei can form.

Zaiser and La Mer (1948) used this technique to produce highly monodisperse sulphur sols from dilute (~0.003 M) thiosulphate solution in acid (see Exercise 1.5.2 below). Reiss and La Mer (1950) subsequently showed that so long as the rate-determining step is diffusion of material to the growing surface (and not the incorporation step) the rate of change of surface area with time is the same for all the particles (though it may vary with time). Overbeek (1981) shows that for such a system the particle size distribution must become more narrow with increasing time. He also shows that, even if it is the incorporation step which is rate determining, the particle size distribution will narrow with time if either (i) the rate of incorporation is the same for all particles or (ii) the rate of incorporation is proportional to the surface area of the particles. Only in the less likely event of the incorporation rate being proportional to particle volume does the size distribution fail to become sharper with time, though even then it remains constant.

Instead of relying on the natural formation of nuclei, an alternative approach is to bring the concentration to the region between saturation and nucleation and then add some very small seed crystals of the desired product or some other material of the same crystal habit. The small differences in size between these initial seeds are rapidly smoothed out as the crystals grow. Zsigmondy used this method very effectively to produce monodisperse gold sols from very fine (~3 nm) gold seed particles obtained by reduction of a gold salt with red phosphorus (the Faraday sol).

Matijevic and his co-workers have prepared monodisperse samples of a wide variety of transition metal oxides and hydroxides using a controlled hydrolysis technique, usually under fairly acid conditions, in the presence of certain complexing ions, like sulphate and phosphate (see Exercise 1.5.9). The precursor in these cases is probably a basic complex species like $M_a(OH)_b(SO_4)_c^{(za-b-2c)+}$ (where z is the oxidation number of M), from which the nuclei form. Because these metals have a marked propensity for forming polynuclear complexes of varying molar mass, the exact mechanism of nucleus formation and growth will probably never be fully known since the solution composition depends markedly on the pH and concentration of the various species present. Nevertheless some attempts are being made to reach an understanding of these systems (Matijevic and Bell 1973). Some examples of the sols grown by Matijevic and his co-workers are shown in Figs 1.5.2 and 3. It should be noted that the very regular character of these particles may be misleading in some cases: these micrographs would not reveal whether the particles are porous. Adsorption studies suggest that some of them are (Fuerstenau 1982, personal communication).

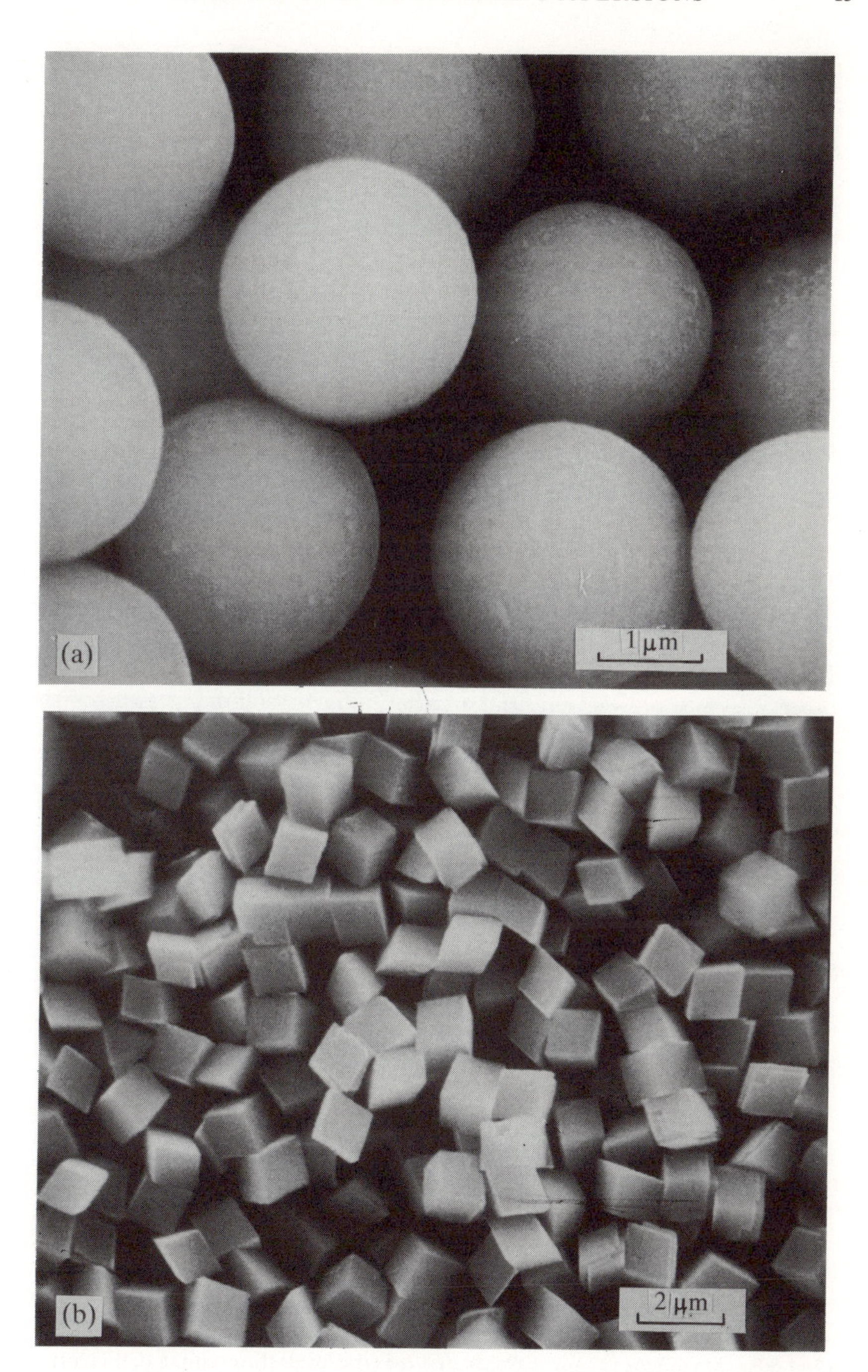

FIG. 1.5.2. Monodisperse inorganic colloids. (a) Zinc sulphide (sphalerite); (b) cadmium carbonate. (Photo courtesy of Prof. Egon Matijevic, Clarkson University, New York.)

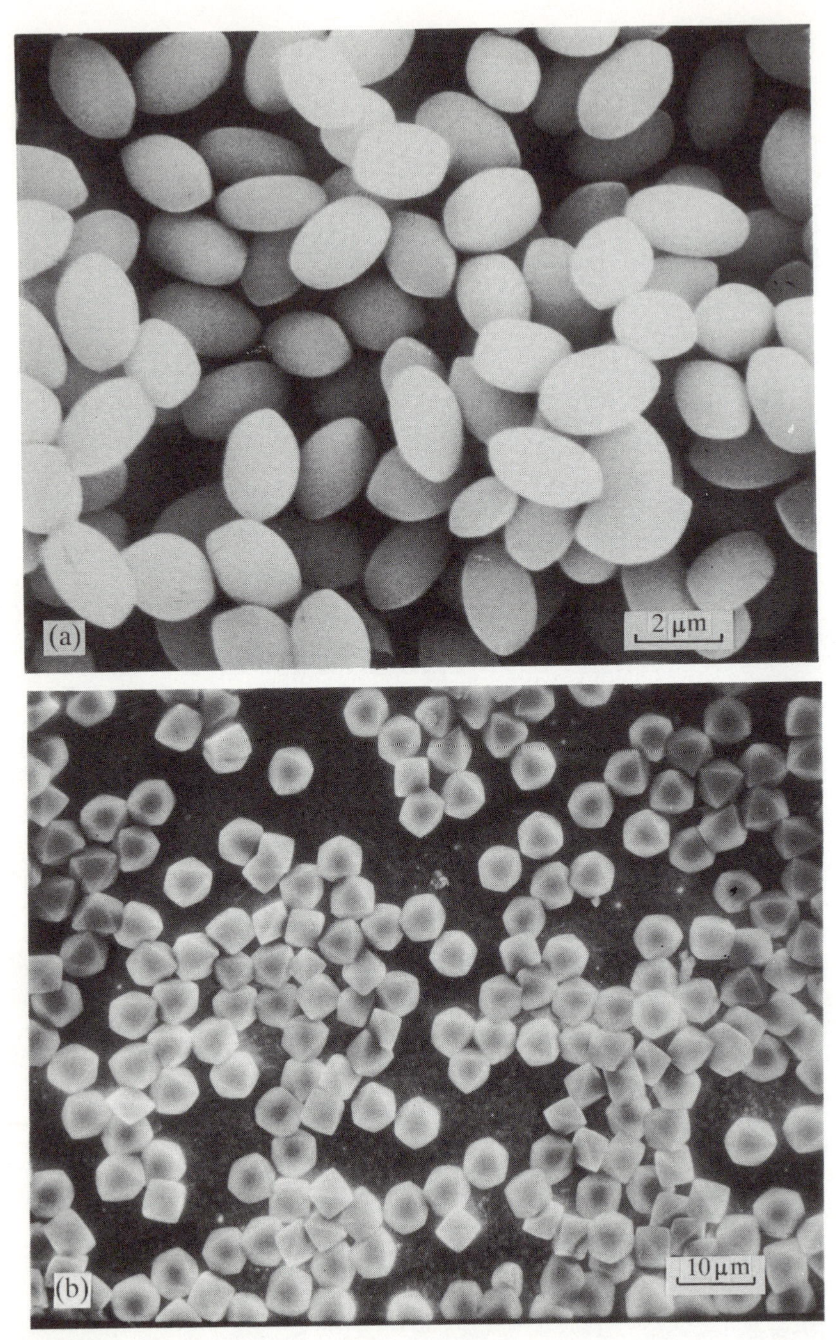

FIG. 1.5.3. Monodisperse iron colloids. (a) α-Ferric oxide (haematite) ($1 \times 2\,\mu m$); (b) basic ferric sulphate (alunite) ($5\,\mu m$ dia.). (Courtesy of Prof. E. Matijevic.)

Another important colloidal material, from the commercial point of view, is silicon dioxide (silica), which is sold as Ludox, Aerosil, etc. for use as a catalytic support, as a rubber reinforcing agent, a filler for paint and for more specialized applications such as antireflecting coatings and encapsulating compounds for electronic components. Silica sols can also be used for making synthetic opal since they form the basis of natural opal. Monodisperse sols can be prepared from silicic acid (see e.g. Iler 1979, p. 312) or by hydrolysis of ethyl orthosilicate (Stöber *et al.* 1968) and their highly spherical appearance suggests that the solid is an amorphous inorganic polymer in this case.

Organic polymers can also be prepared as monodisperse spheres in water by the method of emulsion polymerization; these dispersions are milky in appearance and are called *latices* (by analogy with the natural rubber latex). Polystyrene and poly(methyl methacrylate) (PMMA) latices have been widely used as models in colloid chemical studies because they are easy to prepare as monodisperse spheres (see Liu and Krieger 1978). More recently a range of monodisperse latices with varying densities of negatively and positively charged groups on their surfaces has been prepared by Homola and James (1977). These are often referred to as *amphoteric latices* but should more properly be called *zwitterionic* by analogy with the behaviour of proteins in solutions of varying pH. A note of caution should be sounded here. The various monodisperse latex preparations that have been used as models to test colloid chemical theories are usually assumed to consist of smooth spheres with occasional electrically charged (negative) groups firmly embedded in the surface. Napper and Hunter (1975) draw attention to studies that suggest that in most of these systems there are indications that the surface is not perfectly smooth. Even the polystyrene latex, which on solubility grounds might be expected to form a smooth spherical surface (to minimize the area of contact with the water), shows some evidence of surface roughness (McDonogh and Hunter (1983). The zwitterionic latices, with many charged groups in the surface, are even more likely to show such effects. (See also Healy *et al.* 1978).

1.5.3 *Association colloids*

The term *soap* is applied to the sodium or potassium salts of long-chain fatty acids that are but one example of a general class of substances called *amphiphiles*†. These are substances whose molecules consist of two

† They are sometimes called *amphipathic* molecules, which refers to their ambivalence about what they hate rather than what they like—a bit like the optimist (pessimist) with the half full (empty) wine glass.

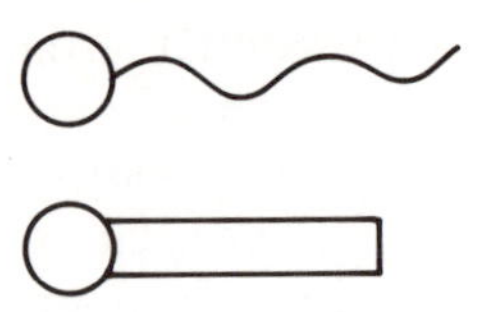

FIG. 1.5.4. Conventional representation of a surfactant molecule as a flexible tail or rod (the hydrocarbon chain) and a head group. The chain is, of course, not infinitely flexible but is limited by the C–C–C bond angle (109°28′). Restricted rotation can occur except at double bonds. The cross-sectional area of the paraffin chain is about 0.2 nm^2 when fully extended and this is comparable to the head group size for $-OH$ and $-NH_2$ but smaller than $-SO_4^-$.

well-defined regions: one which is oil-soluble (lipophilic, oleophilic or hydrophobic) and one which is water-soluble (hydrophilic) (Fig. 1.5.4). The hydrophobic part is non-polar and usually consists of aliphatic or aromatic hydrocarbon residues. The hydrophobic character is not much affected by introducing halogens and similar groups (Laughlin 1981). The hydrophilic part consists of polar groups which can interact strongly with water (especially hydroxyl, carboxyl, and ionic groups).

The fatty acid soaps are typical examples:

$CH_3(CH_2)_nCOO^-Na^+$
sodium stearatc ($n = 16$)
sodium palmitate ($n = 14$)

$CH_3(CH_2)_nCH{=}CH(CH_2)_nCOO^-Na^+$
sodium oleate ($n = 7$ and *cis*)

as are also the common anionic detergents:

$CH_3(CH_2)_{11}O.SO_2O^-Na^+$
sodium lauryl sulphate
(sodium dodecylsulphate or SDS)

R⟨◯⟩$SO_2O^-Na^+$
Alkylbenzene sulphonate
(sodium salt)
R is a (preferably linear) alkyl chain ($\sim C_{12}$)

The most significant characteristic of this type of amphiphile is the tendency to adsorb very strongly at the interface between air and water (Fig. 1.5.5); in both cases the hydrophobic part of the molecule can escape from the aqueous environment whilst the hydrophilic *head group* can remain immersed in the water. Such substances are said to be strongly *surface-active* because they lower the surface (or interfacial) tension, γ. They therefore make the formation of new surface easier and are widely used as *dispersing agents*. Commercial surface active agents (or *surfactants*) are used for a variety of purposes: as cleaning agents (detergents), colloid chemical stabilizers, and wetting agents. We will

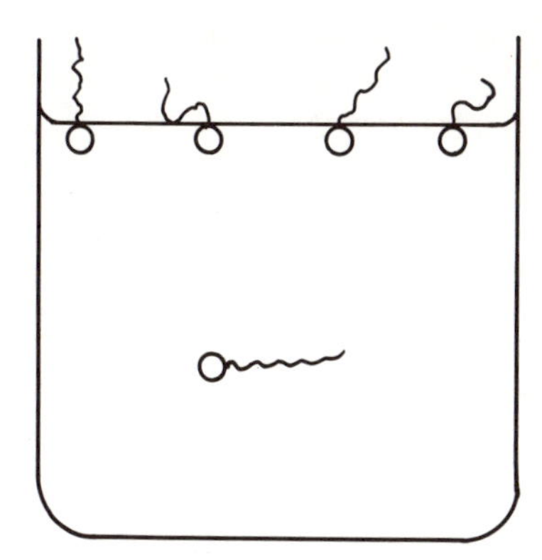

FIG. 1.5.5. Schematic arrangement of amphiphile (soap or detergent) molecules at low concentration in water. (Note that counterions are not shown.)

return to these matters in Chapters 5 and 10. For the moment we are concerned with another of their properties.

At very low concentrations ($<10^{-4}$ M, say) many surface active agents are soluble in water to form simple solutions; if they are ionic, like the fatty acid soaps or the alkyl sulphate detergents they will be dissociated as weak or strong electrolytes. Some of the molecules will also be preferentially adsorbed at the surfaces of the solution (i.e. the air–solution interface if there is one, and at the walls of the container) (Fig. 1.5.5). As the concentration rises this adsorption becomes stronger until saturation is reached when the molecules are packed close together with strong lateral interactions occurring between the hydrophobic chains, which tend to stick up out of the water (Fig. 1.5.6).

Another aggregation process, which often occurs at about the same surfactant concentration, is the formation of *micelles*. These are structures in which the hydrophobic portions of the surfactant molecule associate together to form regions from which the solvent, water, is excluded. The hydrophilic head groups remain on the outer surface to maximize their interaction with the water and the oppositely charged ions (called *counterions*). A significant fraction of the counterions remain strongly bound to the head groups so that the lateral repulsive force between those groups is greatly reduced. The precise structure of the micelle depends upon the temperature and concentration but also on the details of the molecular structure: size of head group, length and number of hydrocarbon chains, presence of branches, double bonds or aromatic rings etc. We will deal with those matters in Chapter 10. For the moment we restrict attention to the simplest amphiphiles: those with a polar (hydrophilic) head group at one end and one or two straight hydrocarbon chains attached. This includes the simple soaps and detergents and some

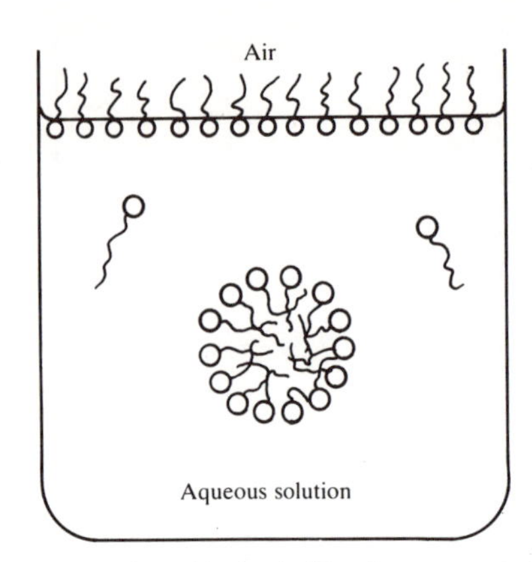

FIG. 1.5.6. The situation above the critical micellization concentration (c.m.c.). Note that adsorption will also occur on the walls of the vessel. The arrangement there is not shown because it is more problematical and is, in any case, dependent on how hydrophilic the surface of the vessel is.

natural lipids. These substances form micelles of colloidal size and, as noted in section 1.1, are called *association colloids* or, more rarely, colloidal electrolytes.

The concentration at which micelles first form in the solution is called the *critical micellization concentration* (c.m.c.). It is marked by quite sharp changes in slope when various transport and equilibrium properties (like conductivity and surface tension) are plotted against concentration.

The initial suggestions on micellar structure by Hartley and by McBain have been refined in recent years by the work of Stigter (1967) and many others, and we will discuss the current models in more detail in Chapter 10. Suffice it to say at this stage that the long chain fatty acid soaps and simple detergents like sodium dodecylsulphate ($CH_3(CH_2)_{11}.O.SO_2.O^-Na^+$) initially form micelles that are spherical in shape and have a fairly well defined aggregation number (~50 molecules for sodium stearate) (Fig. 1.5.7). They are, therefore, monodisperse.

As the surfactant concentration is increased above the c.m.c., the initially spherical micelles become more distorted in shape, forming cylindrical rods or flattened discs (Fig. 1.5.8). Ultimately, at high ratios of soap to water they form *liquid crystals* and other so-called 'mesomorphic phases', a discussion of which would take us beyond the scope of this book (see e.g. Gray and Winsor 1974; and Ekwall *et al.* 1972). Under other circumstances amphiphilic substances can form two dimensional membranes, or *bilayers* to separate two aqueous regions, very similar to a biological membrane. If the bilayer is continuous and encloses an aqueous region the result is a *vesicle*. (Fig. 1.5.8). These types of

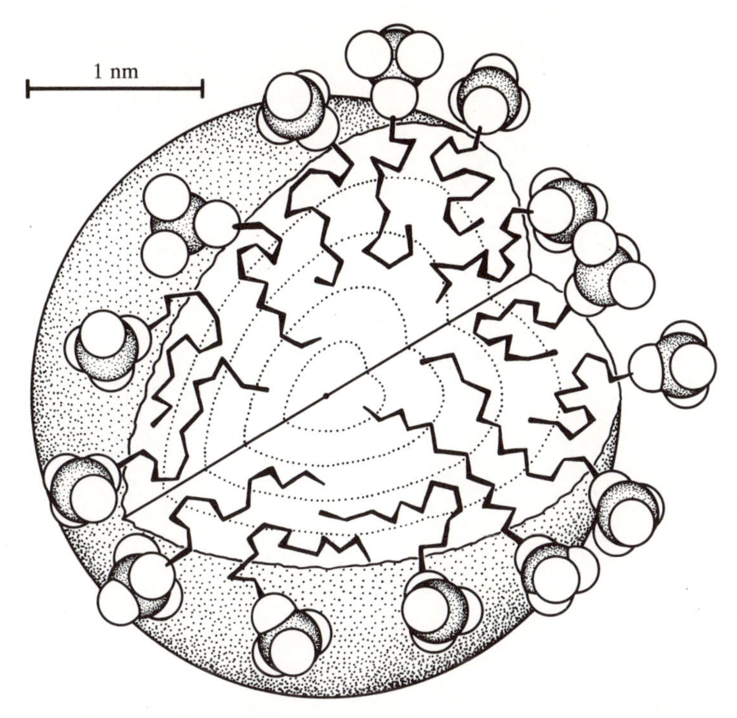

FIG. 1.5.7. A sodium dodecyl sulphate micelle. The more detailed picture, which emerges from a statistical mechanical calculation of the likely structure. (To be discussed in more detail in section 10.10.) (Drawn by Dr J. N. Israelachvili from calculations by Dr D. W. R. Gruen, Australian National University.)

structure, including the liquid crystal, provide possible models for investigating colloid chemical behaviour under controlled conditions.

1.5.4 *Emulsions*

Emulsions, like solid dispersions in liquids, can be formed by either condensation or dispersion methods. The use of mechanical dispersion methods is more common in this case because the energies involved are generally smaller. Apart from the ultrasonication technique mentioned above (section 1.5.1), high-speed stirring or shaking of a two-phase liquid mixture can often induce emulsification, especially if a dispersing agent is present. The resulting droplets are, of course, spherical, provided that the interfacial tension (i.e. surface energy) is positive and sufficiently large.

Many two-phase systems are able to undergo *spontaneous emulsification,* especially if a third component is present. This term

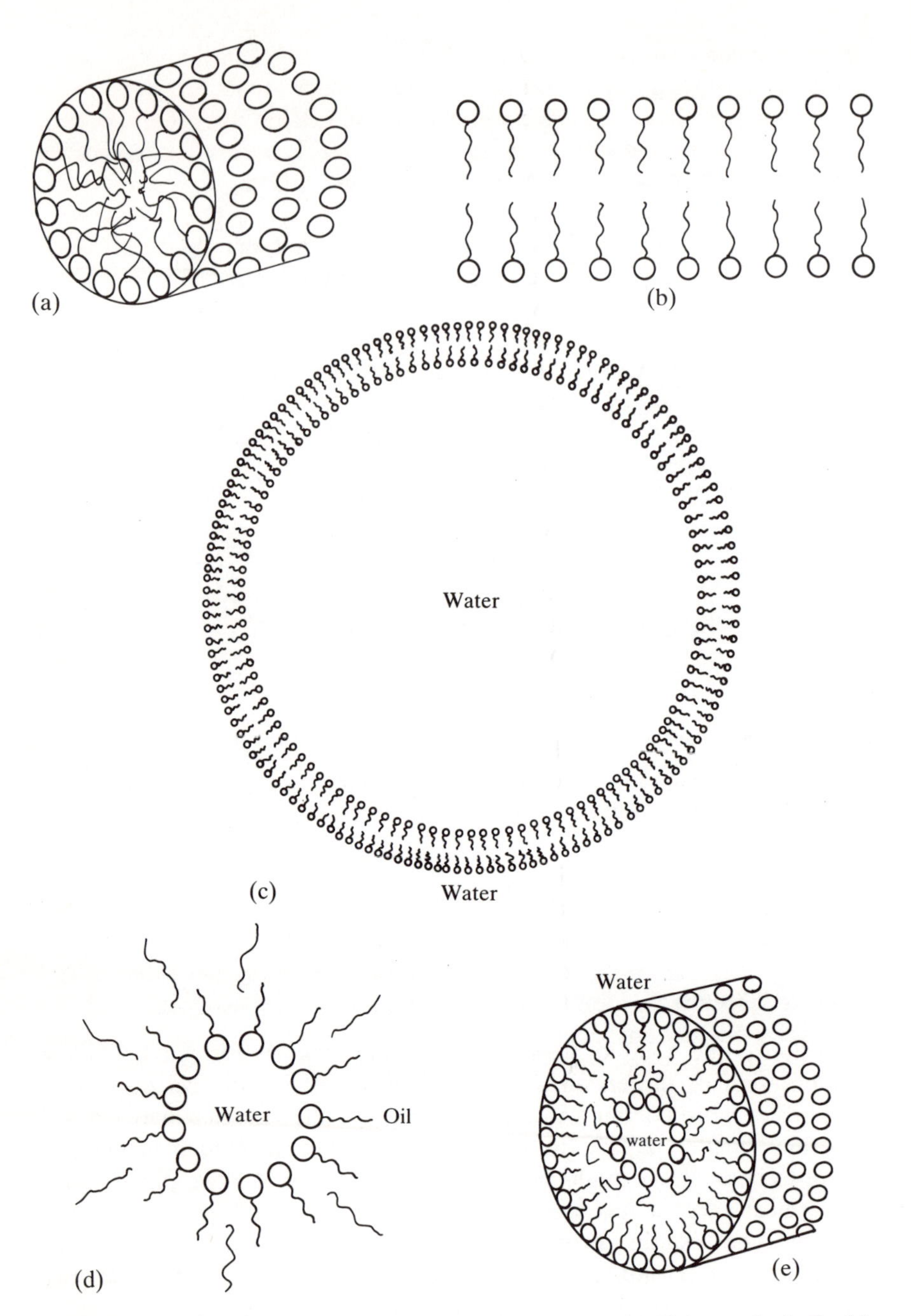

FIG. 1.5.8. (a) Cylindrical micelle; (b) bilayer; (c) bilayer vesicle; (d) inverted micelle; (e) microtubule. [These diagrams are highly schematic. The chains are in all cases expected to adopt a conformation very near to that of the normal liquid state.]

applies strictly only to those systems in which no mechanical energy at all is required, though it is sometimes applied to systems that are simply easy to emulsify. In some cases, the spontaneous emulsification occurs because of the presence of a surface active agent, which lowers the interfacial tension essentially to zero. Negligible energy is then required for the formation of the emulsion. A special case of this sort is the formation of 'microemulsions', in which the droplet size is very small ($\sim$10 nm). Such systems have attracted a good deal of attention recently because of their technological significance. Although the surface tension is essentially zero, the microemulsion droplets are spherical and almost monodisperse, (Overbeek 1980). We will discuss the reason for this later (Chapter 16). Suffice it to note at this stage that the oily interior of a micelle can be used to take up more oil (a process called *solubilization,* which is important in detergency). There is then no sharp dividing line between an oil-swollen micelle (Fig. 1.5.9) and a microemulsion of oil in water (Fig. 1.5.10), nor, for that matter between a microemulsion and a normal emulsion (Fig. 1.5.11).

In some cases of spontaneous emulsification, the interfacial tension (or energy) remains positive but the energy necessary for emulsification is supplied by the redistribution of a solute between the two phases. When, for example, a solution of 15 per cent ethanol in toluene is saturated with water and the mixture is placed in contact with pure water, a very vigorous activity occurs at the interface and an emulsion of water droplets forms in the toluene, as well as an emulsion of toluene in the water. Davies and Rideal (1963) give a detailed account of this phenomenon (with photographs), from which it appears that the mechanism of emulsification is best described as 'diffusion and stranding'. Toluene and alcohol diffuse simultaneously into the aqueous phase and as the soluble alcohol diffuses ahead the insoluble toluene is left stranded in the aqueous phase as small droplets. A similar process can leave water droplets suspended in the toluene. The vigorous turbulent motion of the interface evidently has less effect on the emulsification process in this case than one would expect, though it may play a role in other systems.

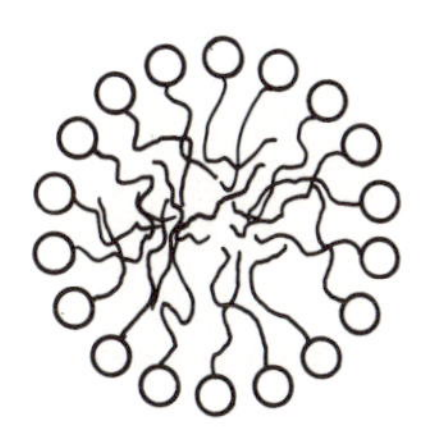

Fig. 1.5.9. A surfactant micelle swollen by the presence of some solubilized oil. The molecules of oil are all in intimate contact with surfactant hydrocarbon chains.

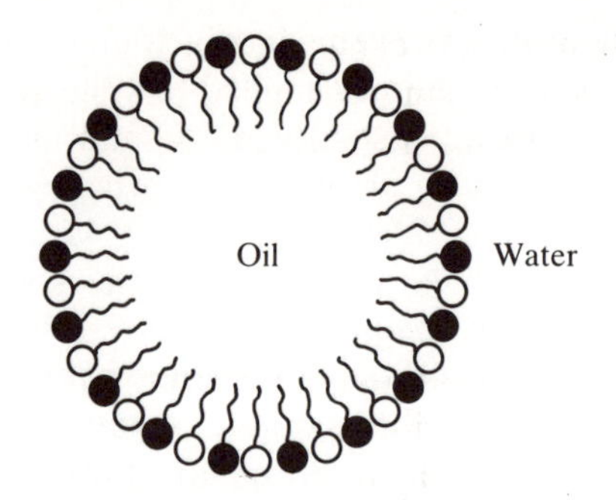

FIG. 1.5.10. An oil in water microemulsion. A separate pure oil region (containing no surfactant) is assumed to be present in the interior. These systems are usually generated by using two surfactants—often an ionic detergent together with a neutral dipolar compound of similar chain length (called a *co-surfactant*).

Provided the interfacial tension or energy, γ, is sufficiently high, the emulsion droplets must be spherical and this should make emulsion systems ideal candidates as model systems for testing colloid theories. Unfortunately, they have some drawbacks: the droplet size distribution is often rather wide, the kinetic behaviour of the droplets is sometimes affected by the fact that the interior is fluid and therefore potentially

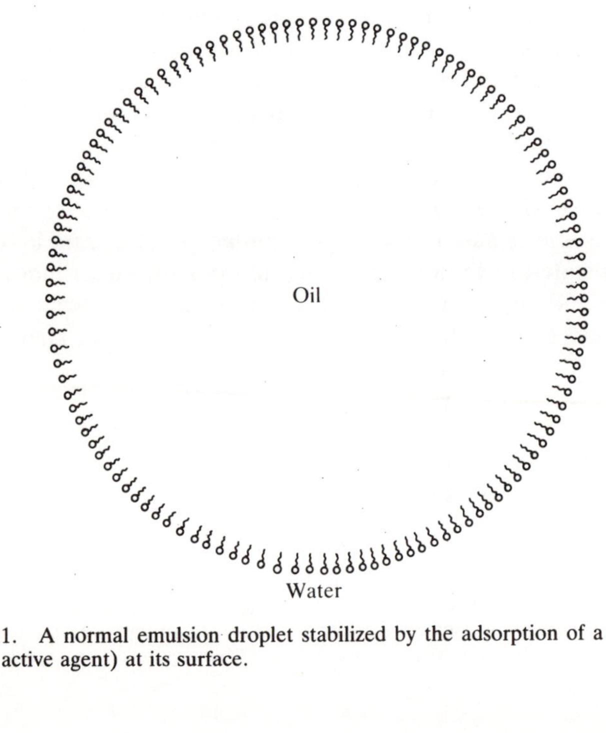

FIG. 1.5.11. A normal emulsion droplet stabilized by the adsorption of a surfactant (i.e. surface active agent) at its surface.

mobile and the possibility of coalescence leading to a change in droplet size is a further complication. The microemulsions referred to earlier, being spherical and almost monodisperse offer special advantages as model systems.

1.5.5 *Clay minerals*

The inorganic fraction of soils and most natural sediments consists almost entirely of silica and the various silicates. The term *clay* is used in soil science and agriculture to mean any material of particle size less than 2 μm but the term *clay mineral* refers to specific groups of silicate minerals. Some clay minerals have long been used in the ceramic industry because their plate-like crystal habit and ability to bond to one another when heated to high temperatures makes them suitable for making bricks, earthenware, and pottery (including china). Clays are also used extensively as fillers in making paper, paint, and rubber tyres, among other things. Indeed, so extensive is the use of these materials in industry that clay minerals rank second only to oil in terms of tonnages used.

We are concerned here with the special property of silicon, when bonded in a certain way with oxygen, to form extensive flat plates or sheets†. When combined with similar flat sheets of an aluminium oxide they can produce layered crystals, which in favourable cases can be cleaved to yield surfaces that are believed to be atomically smooth and flat over relatively large distances (of the order of a few square millimetres). Such smooth solid surfaces have made it possible in recent years to make measurements of the forces between solid particles with an accuracy and reliability not previously possible (see Chapter 7).

Space does not permit a detailed discussion of the structure of the layer silicates. The reader is referred to the texts by Grim (1953) or van Olphen (1977) or to Chapter 25 in Alexander and Johnson (1949). We will consider only the basic structures (talc, pyrophyllite, and kaolinite) and the ways in which the first two are modified to yield vermiculite, mica, and montmorillonite, since these are the systems that have been most extensively used as colloid chemical models.

The basic silicon–oxygen unit is a tetrahedron with four oxygens surrounding the central silicon. The bonds are approximately 50 per cent ionic and 50 per cent covalent in character and in the clay minerals of interest to us the tetrahedra are linked to form hexagonal rings (Fig. 1.5.12). This pattern can be repeated indefinitely in two

† Crystallographers refer to these as layers rather than sheets. We will use 'sheets' to avoid confusion with the electrical double layer formed when the particles are dispersed in water (section 2.5).

FIG. 1.5.12. Arrangement of silica tetrahedra in hexagonal rings to form a layer. Only the oxygens are visible; a silicon sits at the centre of each tetrahedral arrangement of oxygens. The apical oxygens (top layer) are shared with the adjoining alumina layer.

dimensions to form the sheet. Aluminium in combination with oxygen (and hydroxyls) forms an octahedron with the aluminium at the centre and again these octahedra can be linked to form a more closely packed two-dimensional sheet (Fig. 1.5.13).

In the kaolinite crystal a sheet of alumina octahedra sits on top of a sheet of silica tetrahedra with the apical oxygen atoms from the silica being shared with the aluminium atoms of the upper layer (Fig. 1.5.14(a)). This 'ideal' structure would be completely uncharged and the perfect kaolinite crystal would be built up by laying these double sheets one on top of another. The bonding of one double sheet to the next occurs partly through van der Waals forces and partly through hydrogen bonds from the OHs of the octahedral sheet to the oxygens of the next silica sheet (Fig. 1.5.14(b)). These bonds are so numerous that only rather drastic treatments are able to prize open the structure. The crystals are normally hexagonally shaped discs with an axial ratio of the order of 10:1 (Fig. 1.5.14(c)) and they seldom grow to sizes of more than a few

FIG. 1.5.13. A crude model of the structure of a single sheet of a 2:1 layer-lattice alumino-silicate. Note the hexagonal array of holes, which is repeated on the bottom side. The counterions which balance the crystal charge can, in some cases (e.g. K^+) sit in those holes.

micrometres. The real crystals always carry a negative charge on the basal surfaces (i.e. the larger flat surfaces). It cannot be attributed to dissociation of the few Al–OH groups (since they would produce both positive and negative charges). It is generally assumed to be due to substitution of aluminium for silicon in the tetrahedral layer with a consequent imbalance of negative charge. The edges of the crystal, where imperfections necessarily occur because of bond breakage, carry a positive charge at low pH and this decreases to zero as the pH is raised to about 7 (Schofield and Samson 1954). The charge is generally considered to be due to dissociation from the aluminium octahedra:

$$>\text{Al–OH} \rightarrow >\text{Al}^+ + \text{OH}^-,$$

a process that is readily reversed at high pH.

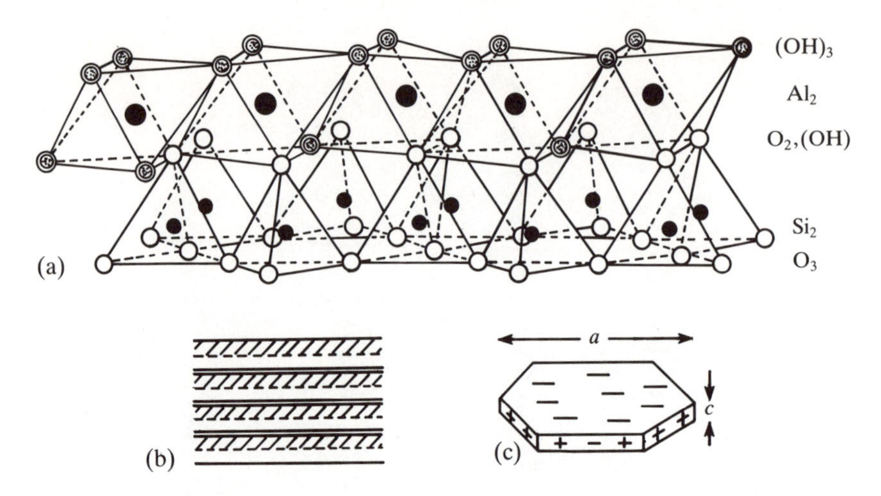

FIG. 1.5.14. (a) A diagrammatic sketch of the 'ideal' kaolin layer $(Al(OH)_2)_2.O.(SiO_2)_2$. One hydroxyl ion is situated within the hexagonal ring of apical, tetrahedral oxygens and there are three others in the uppermost plane of the octahedral sheet. The two sheets combined make up the kaolin layer.
(b) Simplified schematic diagram of the kaolinite structure. Note that the upper and lower cleavage surfaces in the perfect crystal are quite different. A typical crystal would have about 100 or so layers.
(c) A typical kaolinite crystal of aspect ratio (a/c) about 10. Note the negative charges on the basal planes (perpendicular to the c-axis of the crystal) and positive charges around the edges. The latter are eliminated at pHs above about 7.

Kaolinite is called a 1:1 non-swelling dioctahedral clay because (i) it has one silica layer to one alumina layer and (ii) the double sheets do not separate from one another under any normal conditions. The term *dioctahedral* refers to the fact that only two out of every three possible sites in the octahedral layer are occupied by aluminium ions.

The other clay minerals of interest to us are all of the 2:1 type (i.e. two sheets of silica to one of alumina or two of silica to one of magnesium oxide). The two parent materials are *pyrophyllite* (with alumina in the central layer) and *talc* (with magnesia in the central layer) (Fig. 1.5.15(a)). Pyrophyllite is again a dioctahedral whilst talc is a *tri*octahedral mineral, since all three possible sites must be occupied by Mg^{2+} to obtain charge balance.

One important difference between the 1:1 and the 2:1 minerals is that in the 2:1 minerals there is no possibility of hydrogen bonding between successive triple sheets. The basal oxygen planes can interact with each other only by way of van der Waals forces. They are, therefore, very easily cleaved along this plane.

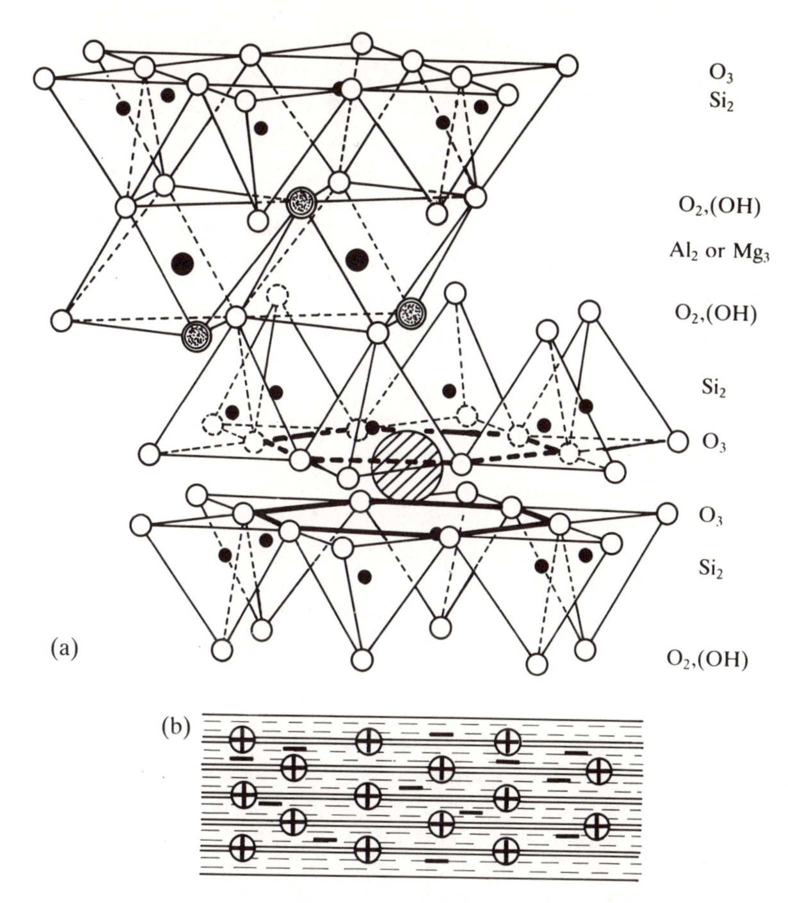

FIG. 1.5.15. (a) A sketch of the ideal 2 : 1 layer silicate. The trioctahedral mineral talc $[(MgO)_2.Mg(OH)_2.(SiO_2)_4]$ has all three octahedral sites occupied by Mg^{2+}. In pyrophyllite $[(AlO(OH))_2(SiO_2)_4]$ only two of the three octahedral sites are occupied by Al^{3+} and as a result the layer is dioctahedral. The large circle shows preferred position of the counterions required to balance the crystal charge after isomorphous replacement. (b) Schematic diagram of the white mica structure. The potassium ions are shown ⊕ and they balance negative charges in the silica layers caused by the substitution of about a quarter of the silicon ions by aluminium. The 'ideal' formula is $K^+[(AlO(OH))_2(AlSi_3O_8)]^-$.

The parent minerals pyrophyllite and talc are not of much interest to us but if they are modified by the substitution of some of the tetrahedral silicon by aluminium, they develop very interesting new properties. Replacing one quarter of the silicons by aluminium in pyrophyllite generates *muscovite* or *white mica* (Fig. 1.5.15(b)). This confers a very large negative charge on each sheet and that charge must be balanced; in muscovite it is balanced by the presence of potassium ions, which can fit snugly in the hexagonal hole of the silica sheet shown by the heavier lines in Fig. 1.5.15(b). This greatly strengthens the bonding between each triple sheet so that mica does not tend to expand in water. A similar replacement of silicon by aluminium in the talc structure generates another form of mica called *phlogopite.* A further substitution of some of the octahedral magnesium in phlogopite by other divalent metal (usually Fe^{2+}) ions produces another mica called *biotite.*

Mica crystals can be very large (many centimetres across) and can be readily cleaved in air or vacuum. Good specimens when carefully cleaved can yield the macroscopic atomically smooth surfaces referred to earlier.

Montmorillonite can also be related to the pyrophyllite structure by the substitution of approximately one in six of the aluminium ions in the octahedral layers by magnesium or other divalent ions. Again this generates a negative charge throughout each triple sheet and this must be compensated by the presence of cations in the inter-layer region (Fig. 1.5.16). The material called *Wyoming bentonite,* which finds many uses (as a filler, a catalytic support and in drilling muds), consists of a mixture of montmorillonite with a related material called *beidellite* in which aluminium ion is isomorphously substituted for silicon in the tetrahedral layers.

When dry montmorillonite is placed in a moist atmosphere, it is able to take up water vapour by adsorbing it between the triple-sheets (that is, in the *interlayer region*) and the shape of the water vapour adsorption isotherm suggests that about four layers of water can be taken up into this region. The same behaviour occurs when the clay is immersed in concentrated salt solution (~1 M) (Norrish, 1954). The spontaneous hydration is presumably associated with the presence of the cations, since

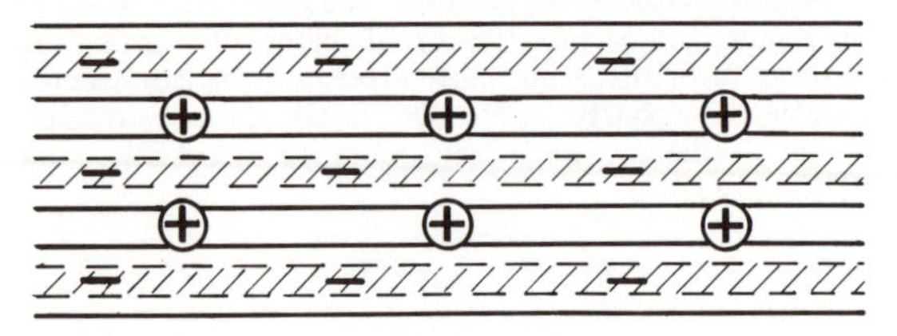

FIG. 1.5.16. Schematic diagram of the montmorillonite structure. It is very similar to that of mica (Fig. 1.5.15(b)) but the negative charges are now in the central (octahedral) layer and are fewer in number (about one in six aluminiums is replaced by Mg^{2+} to give $Na_{0.33}[(Al_{1.67}Mg_{0.33})(O(OH))_2(SiO_2)_4]$. The lower charge density, larger distance between the positive and negative charges, and the poor fit of the counterion (when hydrated) make it possible for montmorillonite to expand readily when placed in water.

the parent material (pyrophyllite), which has no such ions, is quite hydrophobic.

In salt solutions, another important phenomenon occurs and one which has been studied in clay mineral systems for over a century. This is the process of *cation exchange,* which can occur between ions in solution and the ions in the interlayer region (i.e. those on the basal cleavage planes). The process is made easier if the electrolyte concentration is lowered because the triple-sheets (or platelets) can then separate from one another to allow easier access to the adsorbed cations. The separation of the platelets occurs more readily if the interlayer cations are monovalent and strongly hydrated (Na^+ or Li^+ for example) because the inter-platelet repulsion is stronger in that case, as we will show in Chapter 7. The number of cations adsorbed on the clay (the *cation exchange capacity*) is an important characteristic of the material and is a direct measure of the degree of isomorphous substitution that has occurred in the pyrophyllite structure. A typical value for montmorillonite would be about one mole of univalent charge per kilogram of clay (see Exercise 1.5.7). All clay minerals that have crystal structures with unbalanced charge exhibit ion exchange behaviour (e.g. kaolinite, vermiculite), which is of particular importance in determining the retention and availability of plant nutrients in the soil. Indeed, the strong affinity of the potassium ion for certain clay minerals is, to a large extent, responsible for the relative dominance of the sodium ion in sea-water.

The final clay mineral we wish to discuss is *vermiculite.* Its structure is derived from that of talc by the substitution of about one in six of the tetrahedral silicons by aluminium. The balancing (exchangeable) cation is often magnesium but it may be replaced by other divalent or monovalent ions. Vermiculite, like mica, can form large sheets that show a pronounced cleavage parallel to the plane of the sheets.

Apart from the exchange of simple cations, there has been a very large amount of work done on the adsorption of more complex ions and molecules onto montmorillonite and vermiculite or their *intercalation* into the interlayer regions (see e.g. Weiss 1963).

Before leaving the subject of clay minerals it should be emphasized that the chemical compositions given above apply to the ideal (or idealized) crystals. The real materials, as they occur in nature, seldom conform to these idealizations and that must be taken into account in interpreting their behaviour. It should also be mentioned that there are other geometries available in naturally occurring clay minerals (e.g. the thin cylindrical rods of attapulgite; van Olphen 1977, p. 7), which have so far not been much exploited for the testing of physical models of colloidal systems (though, see Buscall 1982).

Exercises. 1.5.1 Prepare a paraffin wax sol in water by the method described above. Use about 5–10 ml ethanol saturated with paraffin wax and pour it into about 600 ml of boiling distilled water in a 1 L beaker. (Careful: the alcohol boils off very vigorously.) Use this sol to examine the general statements made in the text. Note, for example, the Tyndall effect when a light beam is passed through the sol. Put a drop on a microscope slide and examine it in dark field illumination.

1.5.2 Make up a monodisperse sulphur sol by treating a dilute thiosulphate solution with acid (refer to the work of La Mer (Reiss and La Mer 1950; Zaiser and La Mer 1948)). Can you produce the 'chemical sunset' effect? This is caused by the light scattering from particles that are gradually increasing in size.

1.5.3 The formation of micelles can be treated as an equilibrium process between monomer molecules and an n-mer:

$$nS \rightleftharpoons S_n$$

How could one determine experimentally whether the entropy change on micellization is positive or negative?

1.5.4 Estimate the radius of the spherical hydrocarbon core of a sodium stearate micelle assuming an aggregation number of 50 and a density for liquid hydrocarbon of $0.9\,g\,cm^{-3}$. Why do you think a co-surfactant is useful in producing microemulsions?

1.5.5 A kaolinite sample is found to have a surface area of $15\,m^2\,g^{-1}$. Assuming that the particles are cylindrical discs of density $2.8\,g\,cm^{-3}$ and aspect ratio 10:1, estimate the average radius. (This is called the area average size (section 3.3)). How many double (Al/Si) sheets are involved in this average crystal?

1.5.6 Israelachvili and Adams (1978) in their direct measurements of the forces between two mica sheets quote the following composition (by mass) for one of their micas:

SiO_2	Al_2O_3	FeO	MgO	Na_2O	K_2O	H_2O (assumed)
45.94	37.29	0.96	0.56	0.93	10.14	4.19

Compare this with the theoretical structure and suggest a likely arrangement. (Assume that divalent ions can substitute for Al^{3+} and Al^{3+} can substitute for silicon to create sufficient charge to neutralize the Na^+ and K^+). [Hint: convert the composition to a molar basis first.]

1.5.7 The thickness of montmorillonite platelets is 0.97 nm. Calculate the maximum expected specific surface area of a montmorillonite sample assuming its density is $2.8\,g\,cm^{-3}$. If this material has a cation exchange capacity (c.e.c.) of $1\,mol\,kg^{-1}$ of monovalent ion, how far apart, on average, are the exchangeable cations? Repeat the calculation for the kaolinite sample in Exercise 1.5.5 assuming the c.e.c. is $50\,mmol\,kg^{-1}$ and the ions are confined to the basal planes.

1.5.8 What is the relationship between the charges per unit cell in white mica, vermiculite, and montmorillonite?

1.5.9 Prepare a bibliography of the papers by Matijevic and his co-workers on the production of monodisperse sols. (Hint: use *Chemical Abstracts* and the author index to the *Journal of Colloid and Interface Science*). Note the particle shape and composition in each case.

1.6 Surface contamination

One of the most striking features of colloidal systems, especially lyophobic dispersions, is that their properties can be very radically influenced by the presence of quite small quantities of certain substances (particularly surface active agents, multivalent ions or some polymeric materials). In technological situations it is always difficult, if not impossible to deal with perfectly clean systems (i.e. ones of known composition). Nevertheless, even there it frequently happens that a suitable product is only obtained when troublesome impurities are rigorously excluded. To do effective experimental work at a more fundamental level it is absolutely vital that the colloid surfaces be uniform and reproducible and highly desirable that they be well characterized and free of impurities (i.e. clean).

There are three separate impurity problems: (i) the composition of the colloid particles themselves, (ii) the possibility of (slow) chemical reactions occurring at the surface to gradually alter their nature and (iii) the possibility of adsorption of some adventitious material (ions, surface active agents etc.) from the surrounding solutions at some stages during preparation, treatment and storage. It goes without saying that once the system is purified one should exercise the normal purification procedures on all solutions that will come into contact with the colloid. In particular, special care is needed in the preparation of all distilled water to ensure it is free from surface active impurities as well as ions (see section 1.6.4 below). It may also be necessary, in some cases, to take precautions to clean the laboratory air. Special cabinets that provide a laminar flow of filtered air are now available commercially for this purpose.

1.6.1 *Particle composition*

Purity of composition of colloidal particles is obviously a problem with natural materials. Whether of geological or biological origin one can only search for the most promising source of 'pure' material and test it with a barrage of physical and chemical analytical techniques. For synthetic materials the problems are a little different. The classical synthetic colloids are made up of fairly simple molecules (metals, halides, sulphides, etc) and even the more recently studied systems (like the polymer latices) would seem to be readily characterizable. The problem is that, although the bulk composition may be readily ascertained at least for the inorganic sols, the detailed composition of the surface regions may be much more difficult to determine.There is evidence, for example, that when one attempts to prepare pure monodisperse samples of silver iodide, the surface properties become rather variable from sample to sample and the variability between particles in a single sample increases (Ottewill and Woodbridge 1964). The possible structure of the interfacial region in these Ag I preparations is discussed at some length by Sparnaay (1972) (see section 6.7.2). Not enough work has yet been done on the monodisperse metal oxides prepared by Matijevic and his collaborators but it would not be surprising if they turned out to have some residual surface groups derived from the complexing agents (sulphate and phosphate) used in their preparation; as noted earlier they may also be porous to some extent. Fortunately, there are now available a wide range of physical analytical techniques that can be used to examine the surface of solids (see e.g. Parfitt and Rochester 1976). There are also many aspects of colloid chemical theory and behaviour that can be tested using materials whose surface composition need not be completely specifiable so long as it is reproducible.

1.6.2 *Chemical reaction at the surface*

The most common problem here is surface oxidation of, for example, sulphides by dissolved oxygen. Rigorous exclusion of oxygen during a particular experiment is comparatively easy to achieve with the normal dry-box techniques of inorganic chemistry. It is much more difficult to prevent oxidation over long periods of storage. The usual solution is to either prepare the suspension afresh each time, or to develop suitable etching or washing procedures to remove the surface oxidized layer. If the layer is an oxide it may dissolve in strong acid, but the oxidation product of metal sulphides is often thiosulfate (or other thionates), which deposits free sulphur on acid treatment. There are no general rules but suitable etching or washing procedures can usually be devised once the surface reaction product is identified.

1.6.3 *Adventitious impurities*

By far the most pervasive problem is that of impurities derived from the preparation or storage procedures. Grinding processes may introduce multivalent metal ions but these are relatively easily removed by ion exchange: washing the colloid with a fairly concentrated solution (~1 M) of a simple electrolyte like KNO_3 at pH 3 will normally suffice.

Much more stubborn problems are caused by the adsorption of polymeric contaminants: organic polymers from the membranes used for filtration and dialysis (sections 1.7.1(b) and (c)) and inorganic oligomers† of silica derived from glass containers. The organic polymers are

† Oligomer: a low molar mass polymer consisting of a few monomer units.

particularly troublesome in the preparation of polymer latices because they have a strong affinity for the organic surface. The polymer latex is prepared by emulsion polymerization and it is then necessary to remove the small molecules and ions (especially surfactant molecules) remaining in the reaction mixture; this is usually done by dialysis (section 1.7.1(b)) using an organic membrane. Maintaining the pH above 7 retards the acid hydrolysis of the membrane but even this precaution may not be sufficient to prevent some polymeric hydrolysis products from contaminating the colloid. Harding and Healy (1982) claim that cleaner latex surfaces are obtained after several centrifugation/washing cycles, or by filtration through an ultrafilter membrane (section 1.7.1(c)). Even these membranes are, however, suspect. Some commercial membranes contain large quantities of cationic surfactant, which should first be removed by washing with copious quantities of hot distilled water.

Chemists have traditionally regarded glass as an ideal material for the storage and handling of most chemicals. Although borosilicate glass is fairly inert towards most chemicals (except fluoride) it does react with OH^-. If alkaline solutions are allowed to remain in contact with glass, even for quite short periods, (say overnight) there is sufficient dissolution of soluble silicate species (either as monosilicate or an oligomer) to cause significant contamination of a colloidal material (Furlong *et al.* 1981). In slightly acid solutions the process is much slower but is still discernible over several months, especially on oxide surfaces. The problem can be reduced by dry storage or, if the sample must be stored in suspension, using a high volume fraction of sol, so that the surface area is very large and the effect of the contaminant is minimized. Plastic containers are not, generally speaking, a satisfactory substitute since the low molecular mass materials used as plasticisers in most moulded products are able to diffuse out of the plastic and also act as contaminants. Polyethylene and poly(tetrafluoroethylene) (PTFE or Teflon) are more satisfactory than other plastics in this respect.

With these cautionary notes in mind we will return to an examination of the usual procedures for cleaning or 'purifying' a colloidal sol, after first discussing briefly the preparation of 'surface clean' water.

1.6.4 *Surface clean water*

The preparation of water free from inorganic ions (low conductivity water) is normally accomplished by distillation (Petrick *et al.* 1981) or ion exchange (Hughes *et al.* 1971). For colloid and surface chemical purposes it is rather more important that the water also be free of trace quantities of surface active agents and for that reason ion exchange methods are not regarded favourably. Depending on the quality of the exchange resin, its age and the treatment it has suffered (particularly temperature fluctuations) the strong possibility exists that the exchange column will introduce more problems, by way of surface active contaminants, than it removes as ions.

Purification by distillation is, at first sight, a straightforward process, particularly when the electrical conductance requirements are not too stringent (as is usually the case). There are, however, some sources of difficulty that make surface clean water supplies a significant problem in colloid laboratories: first, the large volume of clean water required (necessitating continuous distillation) and the possibility that some surfactants could be steam volatile and, therefore, not removed by simple distillation. Contaminants can also be transferred to the receiver as aerosols in the vapour stream.

The second problem is attacked by distilling from an alkaline solution

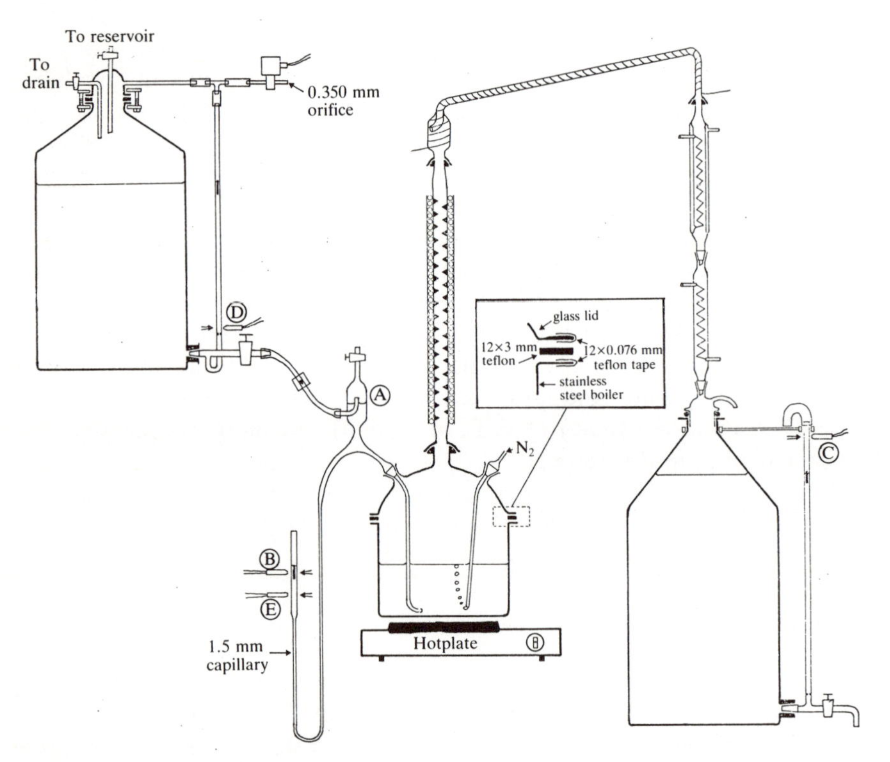

FIG. 1.6.1. Distillation apparatus designed for operation over extended periods from alkaline permanganate solution. Main elements are a stainless steel still pot and film breaker with automatic cut off devices controlled by liquid level. Starting with water of $\gamma \approx 60\ \mathrm{mN\ m^{-1}}$ it produces water of $\gamma \geqslant 71.9\ \mathrm{mN\ m^{-1}}$. (After Lacey *et al.* 1985).

of potassium permanganate (~0.003 M $KMnO_4$ in 0.025 M NaOH) after a preliminary simple distillation to remove most of the impurities. The main steam volatile surfactants are expected to be alcohols and this procedure is expected to oxidize them to carboxylic acids or CO_2, which are retained by the alkaline solution.

In many laboratories, the second distillation is carried out in an all-glass still (see Franks 1961) but there are advantages to be gained from using stainless steel for the still-pot since it does not appear to be so readily attacked by the alkaline permanganate solution and is much better able to withstand the occasional 'bumping' that occurs during distillation. A schematic diagram of a still (with appropriate safeguards for continuous operation) is shown in Fig. 1.6.1; it has been in successful operation in the author's laboratory for some years. It does not provide complete protection against the possibility of an aerosol containing an impurity being carried across with the vapour but that has not proved a problem in our experience.

1.7 Purification procedures

Colloid purification procedures depend upon the fact that impurities adsorbed on the particle surface are normally in equilibrium with the surrounding solution. Even if this is not so, purification by the following techniques is possible provided that the impurity is not completely irreversibly adsorbed. The procedures depend on either:

(a) successive dilution of the impurity in the solution phase; or

(b) competitive adsorption of the impurity onto another surface.

In successive dilution methods the particles are brought into contact with a large volume of liquid and allowed to equilibrate; the two are then separated and the process is repeated with a fresh sample of liquid. This is the 'batchwise' purification technique and in some cases it can be replaced by a continuous procedure. The separation of the colloid from the liquid at each step in the process always relies on the large difference in size between the colloid particles and the usual impurities.

The competitive adsorption method can be used only when the competing surface can be readily separated from the colloid, either because of its physical form (e.g. activated carbon cloth or carbon fibres) or particle size (e.g. ion exchange resins).

1.7.1 *Solute dilution procedures*

The only difficulty with these procedures is in separating the colloid from the wash liquid at each step. This is most readily done by sedimentation (either under gravity or in a centrifuge; section 1.7.1(a)) or using a membrane that is permeable to the liquid (and the impurity) but not the colloid (sections 1.7.1(b), (c)).

(a) *Sedimentation and centrifugation.* Sedimentation will be discussed in more detail in the next chapter. We need only note here that large colloidal particles (~1 μm) will sediment under gravity if their density is significantly greater than that of the liquid. They will certainly sediment quickly if the particles are aggregated together to form *floccules* or *flocs*. This rapid gravity settling of flocculated material is often used in the early stages of washing a colloid. If the wash liquid contains reagents that maintain the colloid in a flocculated condition (e.g. simple salts at fairly high concentrations (~0.1 M)) the wash liquid can be separated by decantation. More usually one must resort to centrifugation to achieve separation of solid and liquid. Although it is a rather time-consuming procedure, the repeated washing of a colloid with a pure liquid and separation by sedimentation is probably the least objectionable purification procedure. The only cause for concern is the possibility of contamination from the wash liquid and the walls of the container, both of which can be minimized. One does encounter problems, however, with very small particles of low density. The centrifugation speeds may then be so high that the sediment is subjected to very large pressures. As the deposit builds up at the base of the tube, the lower layers may become so compressed that the colloidal particles are irreversibly distorted in shape; they may also be forced into such close proximity that they can no longer be readily redispersed.

The normal centrifuge head can be replaced by a continuous flow arrangement that permits the treatment of very large volumes of dilute colloidal dispersions, the capacity then being limited only by the ability of the continuous head to store the deposited solid. This process is, however, still a batchwise one since the solid must be redispersed for each subsequent wash.

It should be obvious that effective washing by these procedures requires that the sedimented material be thoroughly redispersed by the wash liquid. It is often better to sacrifice some washing efficiency by limiting centrifuge speed and time so as to retain an easily redispersible sediment.

(b) *Dialysis.* This is the oldest procedure for separating small molecules and ions from colloidal material, since it was used by Thomas Graham in the 1860s to distinguish colloidal solutions from ordinary ones. The colloidal material is retained on one side of a membrane (often in the form of a bag), which is bathed by the washing solution. Although any of the membrane materials discussed in the next section could, in principle, be used for dialysis (see Hwang and Kammermeyer 1975, p. 161), in practice colloid scientists almost always use simple cellulose membranes (cellophane or cuprophane) produced by the dissolution and reprecipitation of cellulose. These are cheap and readily available in sheet or tubing form (for making bags) and they have the necessary flexibility and a convenient pore size. The pores are formed by a mesh of fibrils so they are irregular in shape and size but are easily able to retain the colloid whilst permitting passage of water and ions. Before use, the membrane material is boiled with several changes of distilled water to remove any readily soluble impurities and dialysis is carried out in slightly alkaline conditions to minimize hydrolysis of the polymeric cellulose.

Dialysis is a purely diffusive process in which ions and other small molecules move through the membrane as a consequence of a concentration difference. If the solutions on each side of the membrane are well mixed, the rate of dialysis for any particular species ($-\mathrm{d}C_i/\mathrm{d}t$) is proportional to the concentration difference across the membrane, and the membrane area, $\mathscr{A}$, and inversely proportional to its thickness, d:

$$\frac{\mathrm{d}C_i}{\mathrm{d}t} = \frac{-K\mathscr{A}}{d}(C_i - C_0) \tag{1.7.1}$$

where C_i is the concentration inside the dialysis bag and C_0 is the concentration outside at time t; K is a constant that measures the permeability of the membrane. Apart from using thin membranes of large area, the process may be accelerated by maintaining C_0 small by frequent changes of the external washing solution, or even by a continuous flow system.

The actual driving force for any species across the membrane is, of course, the difference in chemical potential on either side. This applies also to the solvent and it must be borne in mind if the colloid initially contains a high concentration of solute molecules or ions. The chemical potential of the water inside the bag may then be so low that rapid diffusion of water occurs into the bag (by osmosis†) with the possibility that it will rupture unless an air space is provided for expansion.

In normal laboratory dialysis it is comparatively easy to ensure good mixing of both internal and external solutions. The dialysis bags, containing a small air bubble, can for example, be mounted on a vertical disc that rotates slowly in the bathing solution. The motion of the air bubble then serves to mix the contents of the bag. In larger scale separation processes it is not so easy to ensure complete mixing and the

† Osmosis: if a solution is separated from the pure solvent by a membrane that is permeable to the solvent but not to the solute(s) (a *semi-permeable* membrane) then the solvent tends to diffuse into the solution because its chemical potential is lower there. It continues to do so until a sufficiently high hydrostatic pressure (the osmotic pressure) is built up to oppose the flow (see Appendix A6).

details of the concentration profile in the region near the membrane then become important. If osmotic flow of solvent is occurring into the bag, for example, it will create a zone of low solute concentration just inside the membrane and this will increase the diffusion rate of solute from the bulk of the sol towards the membrane. The solute will, however, now have to diffuse through the membrane pores against the (osmotic) diffusive flow of solvent and will, therefore, be retarded. Fortunately, this complication is confined to the early stages of the process, when the solute concentration is high and, hence, its dialysis rate (eqn (1.7.1)) is high. A more serious problem occurs as the solute concentration diminishes and the rate of diffusion of impurity solute from the interior of the sol becomes very slow. Adequate mixing of the material inside the bag is then very helpful in expediting the process.

Where the impurities are predominantly of an ionic type, the dialysis process can be facilitated by the application of an electric field. The typical laboratory apparatus is shown diagrammatically in Fig. 1.7.1. Such devices used to be widely used but in more recent years have fallen into disrepute, at least for the preparation of very pure sols, because of the uncertainties introduced by the possibility of contamination from the products of electrolysis at the electrodes. The phenomenon, known as *electrodialysis,* is, of course, used in the removal of salts from water (*desalination*) where the separation process is improved by the use of membranes which are selectively permeable to either anions or cations but not both (see Kesting 1971, p. 202). In Fig. 1.7.1, an anion selective (i.e. positively charged) membrane would be used on the right and a cation selective one on the left.

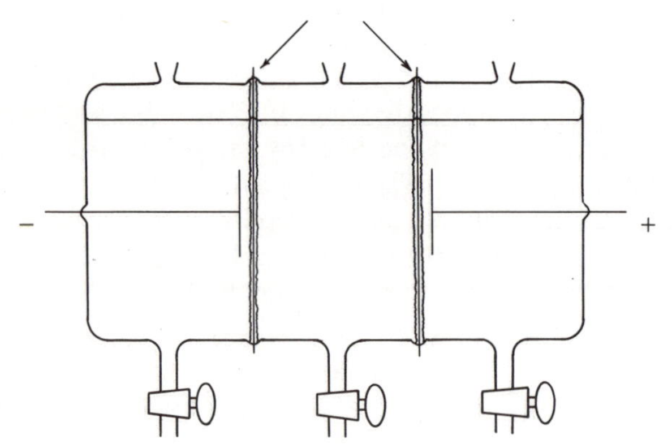

FIG. 1.7.1. An electrodialysis apparatus. The colloid is placed in the central compartment and an electrolyte solution fills the two outer compartments. The three compartments are separable and they butt together with flat ground joints (arrowed). The membranes which separate the compartments are supported on either side by a mesh of some reasonably inert material (e.g. nylon) and a d.c. electrical potential of some tens of volts is applied.

Dialysis bags can also be used to concentrate a colloidal sol, by removing them from the bathing solution and allowing the solvent to evaporate. The process can be accelerated by placing the bags in a cylinder (say a gas jar) to provide some support, and putting some masses on top to squeeze some of the solvent out. This is a crude form of the pressure filtration technique, which we will now discuss.

(c) *Filtration and ultrafiltration.* Filtration refers to any process in which suspended or dissolved material is removed from a fluid as the fluid passes through a porous system. The essential distinction between this and dialysis is that, in filtration, the fluid must flow bodily through the pores, carrying the suspended matter with it (this is called *convective transport*) until the suspended matter is caught in the pores. In some cases, especially those of interest to us, the suspended material cannot enter the pores of the filter membrane at all, but forms a cake on the surface (see Svarovsky 1977). In many important technical processes, however (such as the purification of domestic water supplies), colloidal material is filtered from a liquid by passing through a sand bed (deep-bed filtration), usually after flocculation (see Ives 1977).

Ultrafiltration is the extension of filtration to very small (i.e. colloidal) particles. This necessarily entails the use of very fine pores so the process is usually assisted by applying a pressure. It is distinguished from the very similar process of *reverse osmosis*† by the fact that the fluid is undergoing viscous flow (Chapter 9) rather than diffusion (Chapter 2); the osmotic pressures involved are, in fact, usually rather small.

There has been a considerable development in this area in recent years, with the production of membranes with very special mechanical (and electrical) properties. Improvements have also been effected by using techniques which increase filtering efficiency by limiting the deposition of the filter cake. These considerations are more important for the chemical engineer concerned with very large scale separation processes. See, for example, the discussion of cross-flow filtration and forced-flow electrophoresis in Hwang and Kammermeyer (1975). We will concern ourselves only with laboratory scale filtration operations.

The most important advance in laboratory filtration has been the development of membranes with precisely defined pore size (e.g. the ionotropic-gel membranes of the type developed by Thiele and described by Kesting (1971). These have almost perfect cylindrical pores with a radius of a few micrometres, and would be excellent for separating

† Reverse osmosis: if a solution is separated from the pure solvent by a semi-permeable membrane and the solution is subjected to a hydrostatic pressure higher than its osmotic pressure then essentially pure solvent can be forced through the membrane, leaving the solute behind. The process is used in desalination.

particles a little above the colloid size range. Of more interest for colloid separations are the nucleation-track membranes developed by Fleischer *et al.* (1963) and marketed as *Nucleopore* membranes (Fig. 1.7.2.)

Nucleopore membranes are made by irradiating a suitable material (mica or a polymer) with ionizing radiation from a radioactive source. The tracks left by the radiation as it penetrates the membrane can be etched by a suitable solvent (usually acid or base) to develop cylindrical channels of regular cross-section and well defined radius (Fig. 1.7.2). Unfortunately, the density of channels must be kept quite low to prevent the membrane becoming too brittle (and too radioactive; Kesting 1971, p. 103), but developments in this area are very rapid and the product is being continually improved. The membrane is very thin (~10 μm) and must be mounted on another supporting membrane, of larger pore size, for mechanical stability.

A rather different approach is employed in the formation of the *Diaflo* membranes, marketed by the Amicon Corporation. These are polymeric membranes having a very thin layer of microporous material (Fig. 1.7.3), again mounted on a supporting membrane made of similar material but of larger pore size and heavier wall thickness. Amicon also market a very versatile laboratory dialysis/ultrafiltration apparatus. This contains a membrane in the base, over which the pure solvent can be circulated in a spiral channel, using a peristaltic pump. The system can be subjected to

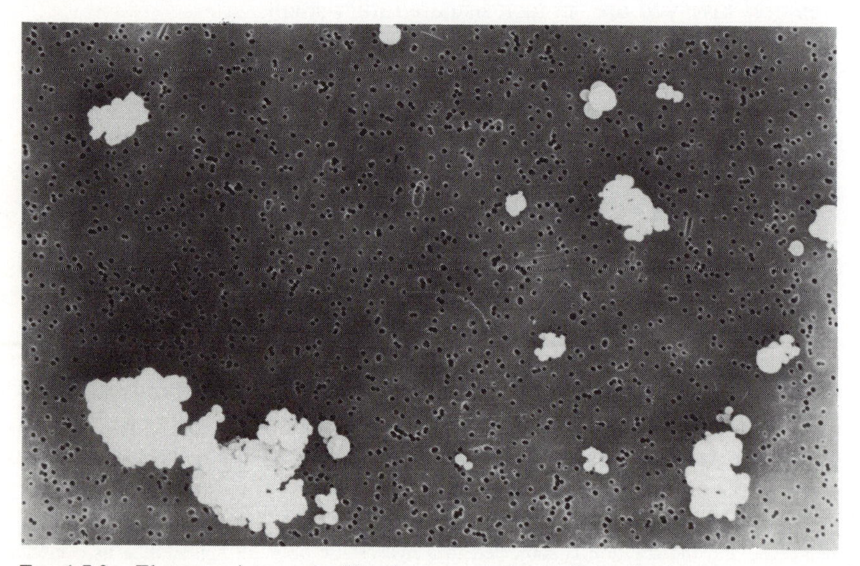

FIG. 1.7.2. Electron micrograph of Nucleopore membrane (0.4 μm) with some poly-(vinyl chloride) latex collected on it. (Courtesy of A. Rogers, University of Sydney.)

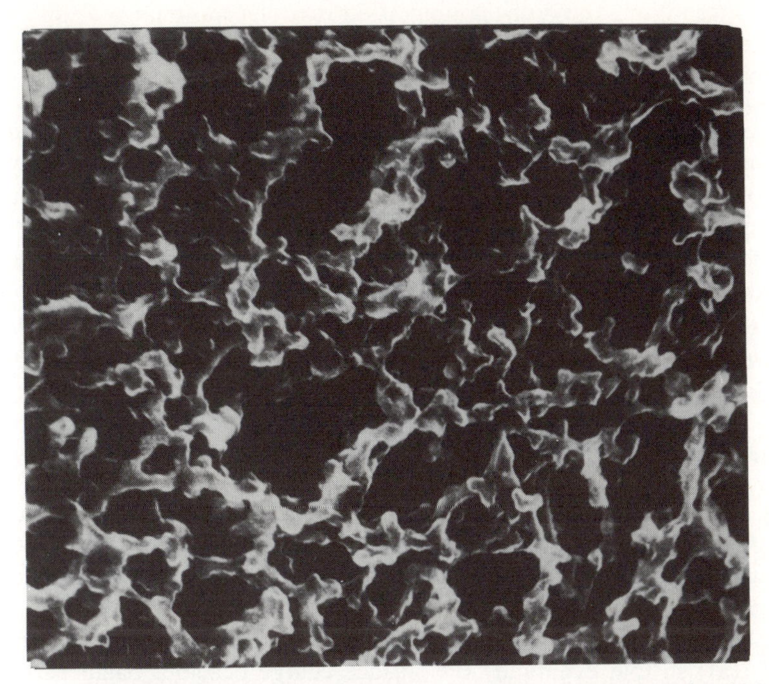

FIG. 1.7.3. Scanning electronmicrograph (SEM) of a Millipore membrane, in which the pores are produced by a cross-linked fibre matrix.

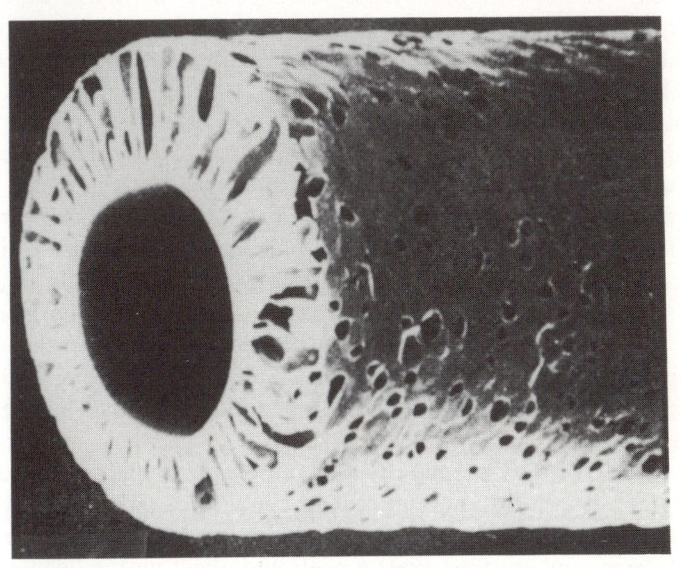

FIG. 1.7.4. SEM of a single Amicon hollow fibre. The ultrafiltration layer lines the lumen and the spaces in the spongy wall enlarge progressively outward. Exterior surface is highly porous. (From manufacturer's data sheets.)

gas pressure from a nitrogen cylinder and fresh solvent added continuously to the suspension.

The membrane can also be formed into a hollow fibre with the fine pore-size layer on the inside. The colloidal suspension is then forced down the lumen of the fibre and the clear filtrate moves radially out through the fibre wall (Fig. 1.7.4). The fibres are arranged in the form of a bundle and a laboratory dialyser designed to take such a *hollow-fibre bundle* as a filter is also available (Cross 1972). This design has the advantage of *cross-flow* filtration: the flow of the sol along the tube is at right angles to the movement through the filter and this prevents the build-up of a filter cake, which is the main limitation to flow rate in ultrafiltration.

1.7.2 *Competitive procedures*

(a) *Ion exchange.* The surface of most colloidal materials is electrically charged (by mechanisms discussed in section 6.9) and to prepare a reproducible sample of the material requires that that charge be balanced by a known *counterion* (section 1.5.3). It is for this reason that one often washes the colloid with a salt solution of moderate concentration (section 1.7.1(a)) so that eventually the surface charge becomes balanced by the cation (or anion) of that salt. This is particularly important in preparing clay–mineral suspensions because their (cation) exchange capacity is so large. All such washing procedures fall under the heading of ion exchange but we are concerned here more with the competitive ion exchange procedures.

Coleman and his co-workers used an ion exchange column saturated with hydrogen ion to prepare homo-ionic samples of clay minerals. The clay suspension was percolated down through the column in the same way as a normal solution (Coleman and Craig 1961) and emerged as H^+-clay, with all crystal charges balanced by hydrogen ions. The procedure is very fast and has the further advantage that the clay contains little associated salt. Subsequent studies can, therefore, be made at low ionic strength. The success of the procedure can be easily tested by titrating the H^+-clay against a base; it behaves as a strong acid with the equivalence point occurring when the amount of added base equals the cation exchange capacity of the clay.

The speed of the operation is very important in this case because most clay minerals are unstable under acid conditions: aluminium ions can be dissolved from the crystal edges and readily become adsorbed on the basal surfaces where they exert a profound effect on the behaviour of the clay (Eeckmann and Laudelout 1961).

More recently Vanderhoff and his collaborators have introduced a similar technique for preparing the surfaces of polymer latex particles (see Becher and Yudenfreund 1978). They use a batch rather than a column procedure, in which the thoroughly cleaned ion exchange resin beads are stirred with the sol and separated by decantation. The main problem lies in the initial cleaning of the resin and Harding and Healy (1982), as noted in section 1.6.3, have questioned the efficacy of this approach for some latex systems.

(b) *Competitive adsorption.* It is now possible to obtain a highly adsorptive surface in the form of a woven fabric. It is produced by reducing the polymer fibres in a woven cloth down to 'pure' carbon whilst retaining the integrity of the fabric. When dipped into a colloidal sol, the very large surface area and high adsorptivity of the carbon make possible the removal of large quantities of impurity materials from the sol and the surrounding liquid. Success depends upon pretreating the fabric to ensure that it does not release any contaminants to the colloid.

Large porous particles could also be used in the same way. Indeed, the ion-exchange resin beads referred to in section 1.7.2(a) are of this geometry. There does not seem to be available, however, an adsorbent with a high internal surface area, high absorptivity for neutral molecules and a reasonably hydrophilic surface (to facilitate entry of an aqueous solution into the pores).

Exercises. 1.7.1. Use eqn (1.7.1) to show that, if the external solution is continuously replaced, the time taken for the internal concentration to fall to 25 per cent of its initial value is 1.39 $d/K\mathscr{A}$

1.7.2 A membrane of thickness 0.5 mm is formed into a cylindrical bag of length 20 cm and diameter 3 cm. Estimate the permeability coefficient if the internal concentration of a solute falls to half of its initial value in 6 h.

1.7.3. Poiseuille's equation gives the volume, V, of fluid that flows through a cylindrical tube of radius r and length l in time t, under a pressure difference ΔP, as:

$$V = \frac{\pi r^4 \Delta P t}{8\eta l}$$

where η is the viscosity of the liquid. Thiele and Hallich (1959) measured the pore diameter of an ionotropic gel membrane filter by determining the volume flow rate of water through unit area of membrane under a known pressure gradient. Suppose under a pressure of 0.10 atmosphere the flow rate is 0.20 m s^{-1}. Estimate the radius, r of the pores of the membrane assuming they are parallel and cylindrical with a length: diameter ratio of five and a minimum wall thickness equal to the radius. [η for water $= 1.0 \times 10^{-3}$ N m^{-2} s].

References

Alexander, A. E. and Johnson, P. (1949). *Colloid science.* Oxford University Press, Oxford.

Becker, P. and Yudenfreund, M. N., eds (1978). *Emulsions, latices and dispersions*. Marcel Dekker, New York.
Buscall, R. (1982). *Colloids Surfaces* **5,** 269.
Coleman, N. T. and Craig, D. (1961). *Soil Sci.* **91,** 14.
Cross, R. A. (1972). *Recent advances in separation techniques* Vol. 68, p. 15 (AIChE Symposium Series 120). Am. Chem. Soc., Washington, D.C.
Davies, J. T. and Rideal, E. K. (1963). *Interfacial phenomena* (2nd edn), p. 360. Academic Press, New York.
Defay, R., Prigogine, I., Bellemans, A., and Everett, D. H. (1966). *Surface tension and adsorption,* Chapter VI. Longmans, London.
Eeckmann, J. R. and Laudelout, H. (1961). *Kolloid-Z.* **178,** 99.
Ekwall, P., Danielsson, I., and Stenius, P. (1972). Aggregation in surfactant systems. In *MTP International Review of Science* (ed. M. Kerker) Series 2, Vol. 7, pp. 97–145. Butterworths, London.
Evans, A. G. and Langdon, T. G. (1976). *Prog. Mater. Sci.* **21,** 171–441.
Fleischer, R., Price, P., and Walker, R. (1963). *Rev. Sci. Inst.* **34,** 510.
Franks, F. (1961). *Chem. Ind.* February, 204–5.
Frens, G. and Overbeek, J. Th. G. (1969). *Kolloid-Z. Z. Polym.* **233,** 922.
Freundlich, H. (1926). *Colloid and capillary chemistry*. Methuen, London. (First published in German, 1909.)
Furlong, D. N., Freeman, P. A., and Lau, A. C. M. (1981). *J. Colloid interface Sci.* **80,** 20.
Gardner, C. R. (1976). In *Membrane separation processes* (ed. P. Meares) Chapter 14, pp. 529–573. Elsevier, Amsterdam.
Goldsmith, H. and Mason, S. G. (1975). *Biorheology* **12,** 181.
Gray, G. W. and Winsor, P. A. (1974). *Liquid crystals and plastic crystals,* Vols 1 and 2. Ellis Horwood, Chichester, UK, and Wiley, New York.
Grim, R. E. (1953). *Clay mineralogy*. McGraw-Hill, New York.
Hall, D. G. and Pethica, B. A. (1967). Thermodynamics of micelle formation. In *Non-ionic surfactants* (ed. M. Schick) Chapter 16. Marcel Dekker, New York.
Harding, I. M. and Healy, T. W. (1982). *J. Colloid interface Sci.* **89,** 185.
Healy, T. W., Homola, A., James, R. O., and Hunter, R. J. (1978). *Faraday Discuss. Chem. Soc.* **65,** 156–163, and discussion pp. 182–9.
Homola, A. and James, R. O. (1977). *J. Colloid interface Sci.* **59,** 123.
Hughes, R. C., Mürau, P. C., and Gundersen, G. (1971) *Analyt. Chem.* **43,** 691–6.
Hwang, S.-T. and Kammermeyer, K. (1975). *Techniques of chemistry* (ed. Arnold Weissberger). Vol. 7, *Membranes in separations.* Wiley, New York.
Iler, R. K. (1979). *The chemistry of silica.* Wiley, New York.
Israelachvili, J. N. and Adams, G. E. (1978). *J. chem. soc. Faraday Trans. 1* **74,** 975.
Ives, K. J. (1977). Deep bed filtration. In *Solid–liquid separation* (ed. L. Svarovsky), p. 199. Butterworths, London.
Johnston, F. and McAmish, L. (1973). *J. Colloid interface Sci.* **42,** 112.
Kesting, R. E. (1971). *Synthetic polymeric membranes,* p. 104. McGraw-Hill, New York.
Kruyt, H. R., ed. (1952). *Colloid science.* Elsevier, Amsterdam. [Volume I of this two-volume work was written almost entirely by J. Th. G. Overbeek and is regarded as the definitive exposition of the subject for its time.]
Kuhn, W. E., ed. (1963). *Ultrafine particles.* Wiley, New York.
Lacey, A. R., McPhail, A. K., and Trafalski, Z. J. (1985). *J. Phys. E.* (*Sci. Inst.*) **18,** 532.

Laughlin, R. G. (1981). *J. Soc. cosmet. Chem.* **32,** 371–92.
Liu, B. Y. H. and Krieger, I. M. (1978). In *Emulsions, latices and dispersions* (ed. P. Becher and M. N. Yudenfreund) pp. 41–70. (Marcel Dekker, New York.
Matijevic, E. and Bell, A. (1973). In *Particle growth in suspensions* (ed. A. L. Smith) pp. 179–93. Academic Press, London.
McDonogh, R. W. and Hunter, R. J. (1983). *J. Rheol.*, **27,** 189.
Napper, D. H. and Hunter, R. J. (1975). Hydrosols. In *M.T.P. International Review of Science* (ed. M. Kerker) Series 2, Vol. 7, pp. 161–213. Butterworths, London and Boston.
Norrish, K. (1954). *Faraday Discuss. chem. Soc.* **18,** 120.
Ostwald, W. (1907). *Z. phys. Chem. (Leipzig)* **34,** 295.
Ottewill, R. H. and Woodbridge, R. F. (1964). *J. Colloid Sci.* **19,** 606.
Overbeek, J. Th. G. (1980). *K. Ned. Akad. Wet. Versl. genone Vergad. afd. Natuurkd.* **89,** 82.
Overbeek, J. Th. G. (1981). *Chem. Austral.* **48,** 419; (1982) *Adv. Colloid interface Sci.* **15,** 251–77.
Parfitt, G. D. and Rochester, C. H. (1976). In *Characterization of powder surfaces* (ed. G. D. Parfitt and K. S. W. Sing) pp. 57–105. Academic Press, New York.
Petrick, H.-J., Schulze, F. W., and Cammenga, H. K. (1981) *Mikrochim. acta* **II,** 277–88.
Reiss, H. and La Mer, V. K. (1950). *J. Chem. Phys.* **18,** 1; **19,** 482 (1951).
Riess, J. G. and Le Blanc, M. (1978). *Angew. Chem.* **17,** 621–34.
Schofield, R. K. and Samson, H. R. (1954). *Faraday Discuss. chem. Soc.* **18,** 135.
Sparnaay, M. J. (1972). *The electrical double layer,* Chapter 4, pp. 136–203. Pergamon Press, Oxford.
Stigter, D. (1967). *J. Colloid interface Sci.* **23,** 379.
Stöber, W., Fink, A., and Bohn, E. (1968). *J. Colloid interface Sci.* **26,** 62.
Svarovsky, L. (1977). Filtration fundamentals. In *Solid–liquid separation* (ed. L. Svarovsky) pp. 171–92. Butterworths, London.
Svedberg, T. (1928). *Colloid chemistry.* Chemical Catalog, New York.
Thiele, H. and Hallich, K. (1959). *Kolloid-Z.* **163,** 115.
van Olphen, H. (1977). *Introduction to clay colloid chemistry* (2nd edn). John Waters, New York.
Weiser, H. B. (1933). *Inorganic colloid chemistry.* John Wiley, New York. [This three-volume work provides a huge number of detailed recipes for preparing various sols.]
Weiss, A. (1963). *Angew. Chem., Int. Edn. Eng.* **2,** 134.
Zaiser, E. M. and La Mer, V. K. (1948). *J. Colloid interface Sci.* **3,** 571.

Further reading

On aerosols

Cadle, R. D. (1965). *Particle size.* Reinhold, New York. [Not restricted to size analysis but covering many aspects of aerosol science.]
Davies, C. N., ed. (1966). *Aerosol science.* Academic Press, London. [Collection of specialist articles.]
Hidy, G. M. and Brock, J. R., eds (1970–72). *Topics in Current Aerosol*

Research, vols 1–3. In *International Reviews of Aerosol Physics* and *Chemistry.* Pergamon Press, Oxford.
Liu, B. Y. H., ed. (1976). *Fine particles.* Academic Press, New York.

On the solid–gas interface

Morrison, S. Roy (1977). *The chemical physics of surfaces.* Plenum Press, New York.
Somorjai, G. A. and Kesmodel, L. L. (1975). The structure of solid surfaces. In *M.T.P. International Review of Science* (ed. M. Kerker) Vol. 7, pp. 1–45. Butterworths, London.
Kerker, M., ed. (1972). *M.T.P. International Review of Science,* Series 2. Vol. 7, *Surface chemistry and colloids.* Butterworths, London. [Contains a number of reviews of the solid/gas interface.]
Everett, D. H., ed. *Colloid science* Vol. 1 (1973), 2 (1975), 3 (1979) and 4 (1983). (Specialist Periodical Repts.) Chem. Soc. London. [Contains numerous authoritative reviews in this area.]

On the liquid–gas interface

Davies, J. T. and Rideal, E. K. (1963). *Interfacial phenomena* (2nd edn). Academic Press, New York.
Aveyard, R. and Haydon, D. A. (1973). *Introduction to the principles of surface chemistry.* Cambridge University Press, Cambridge.

On clays, sediments, and geochemistry of colloids

Yariv, S. and Cross, H. (1979). *Geochemistry of colloid systems (for earth scientists).* Springer, Berlin. [Treats some general colloid chemistry but is mainly on colloids in oceans, atmospheres and sediments, especially silica and clay minerals.]
van Olphen, H. (1977). *Introduction to clay colloid chemistry* (2nd edn). Wiley, New York.
Grim, R. E. (1953). *Clay mineralogy.* McGraw-Hill, New York.

On technological aspects of dispersing solids in liquids

Parfitt, G. D., ed. (1973). *Dispersions of powders in liquids* (with special reference to pigments) (2nd edn). Applied Science Publishers, London.
Parfitt, G. D. and Sing, K. S. W., eds (1976). *Characterization of powder surfaces.* Academic Press, London.

2

BEHAVIOUR OF COLLOIDAL DISPERSIONS

In this chapter a number of aspects of colloidal behaviour are introduced in order to provide a backdrop against which a more rigorous development can be attempted. The ideas are not elementary and the observed behaviour can be quite complex but we are interested here in giving something of an overview so that the vocabulary connected with the basic features of colloidal systems can be introduced in a semi-quantitative way.

The observation of Brownian motion and particle diffusion were among the foundation stones of the kinetic–molecular theory of matter and we will return to them again in Chapter 9.

The response of a dielectric material to an alternating electric field introduces some important concepts which can be applied immediately to the subject of light scattering by colloidal particles but that will be applied in a more definitive way in Chapter 4 to develop the theory of the attractive forces between colloidal particles (long range van der Waals forces). It is recognized that many students of colloid science have only passing acquaintance with the mathematical tools used in this section and so we have included quite a large number of practice problems to bring out the main features of the use of complex functions to describe elastic and inelastic response to an applied stress.

There follows a brief introduction to light scattering, which is a technique widely used for investigating particle size, especially in its recent manifestation (called photon correlation spectroscopy). Again we will return to a more formal treatment in Chapter 14.

The elementary functions of a complex variable are again used to describe the response of a system to a mechanical stress. An introduction to that theory is also presented at this stage to bring home the similarities between this and the dielectric response. Most of that material will not be needed again until late in Volume II and can be passed over in a first reading, although some reference is made to it in Chapter 9.

The final section on colloid charge and stability is designed to give a qualitative appreciation of what is probably the most striking feature of colloidal behaviour—the coagulation or flocculation that occurs when the attractive forces between particles dominate over the repulsive forces. The resulting aggregation process produces dramatic changes in the properties of the system, so it is essential to recognize and control this aspect of colloidal behaviour if one is to attempt any sensible study of a colloidal material.

2.1 Brownian motion and diffusion

2.1.1 *Sedimentation equilibrium under gravity*

In the absence of any external forces, the composition of a phase at equilibrium is uniform throughout and is characterized by a unique value of the chemical potential (or partial molal Gibbs free energy), μ_i:

$$\mu_i = \bar{G}_i = \left(\frac{\partial G}{\partial n_i}\right)_{p,T,n_j} = \text{a constant} \tag{2.1.1}$$

i.e.

$$\mathrm{d}\mu_i = 0$$

for each species i present. (G is the total Gibbs free energy of the system and n_i is the number of moles of component i). If external fields are important they must be incorporated into eqn (2.1.1). For example, to consider the composition of the earth's atmosphere as a function of height above the ground we would introduce the gravito-chemical potential $\tilde{\mu}_i$ defined by:

$$\tilde{\mu}_i = \mu_i + M_i\phi$$

and

$$\mathrm{d}\tilde{\mu}_i = \mathrm{d}\mu_i + M_i\,\mathrm{d}\phi = 0 \tag{2.1.2}$$

at equilibrium. Here ϕ is the gravitational potential ($=gh$) and M_i is the molar mass. Then since (cf. eqn (A6.1)):

$$\begin{aligned}\mu_i &= \mu_i^{\ominus} + RT\ln a_i \\ &\doteqdot \mu_i^{\ominus} + RT\ln c_i\end{aligned} \tag{2.1.3}$$

assuming ideal behaviour, we could solve eqn (2.1.2) to determine the concentration c_i as a function of height:

$$\mathrm{d}\ln c_i = \frac{-M_i g\,\mathrm{d}h}{RT}.$$

In colloidal dispersions, gravitational forces are much more significant than they are for molecular solutions or gases. In fact it is possible to observe the operation of gravity directly over very small distances and this was exploited by Perrin (1909) to obtain one of the first direct estimates of the Avogadro constant. Perrin and his collaborators prepared dispersions of highly monodisperse spheres of a natural colloid called gamboge. (This in itself was a very considerable feat in those days: in his account of Perrin's work, Overbeek (1982) points out that Perrin and his students would start with 1 kg of gamboge and after several months of patient fractional centrifugation might finish up with a few hundred milligrams of monodisperse particles.) When this suspension was mounted on a vertical microscope slide (Svedberg 1928) the number of

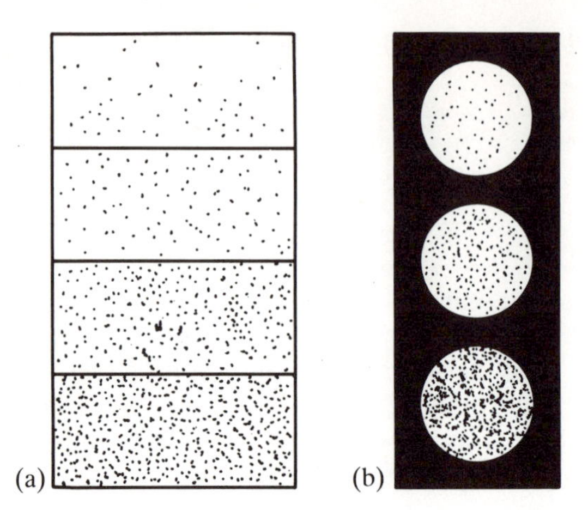

FIG. 2.1.1. Two examples of the sedimentation equilibrium observed by Perrin.
(a) Spheres of gamboge of radius 0.29 μm at levels 10 μm apart.
(b) Spheres of mastic (radius 0.52 μm) at levels 12 μm apart. (From Overbeek 1982, with permission.)

colloidal particles could be directly counted at different heights (Fig. 2.1.1) and from the resulting data a value of the Boltzmann constant, k, could be calculated (Exercise 2.1.1). Then since $k = R/\mathcal{N}_A$, where R is the universal gas constant, Avogadro's constant $\mathcal{N}_A$ could be obtained (Exercise 2.1.2). Perrin obtained a value of $6\text{–}7 \times 10^{23}\,\text{mol}^{-1}$ in his experiments and subsequent measurements by Westgren gave 6.05×10^{23}, very close to the presently accepted value of 6.022×10^{23}. (See Svedberg 1928, p. 101 for some discussion of this work.)

This procedure relies on the notion that the microscopically visible particles are being continually bombarded by the surrounding molecules and so come to have the same average translational energy ($3kT/2$). Their distribution in the gravitational field, therefore, reflects that of the surrounding molecules. Their erratic (Brownian) motion also reflects that of the surrounding molecules and Perrin also used direct measurements of this motion to estimate the Avogadro constant, with the help of some theoretical ideas suggested by Einstein and von Smoluchowski. These two sets of experiments by Perrin provided some of the very first direct evidence for the real existence of atoms, which were regarded by many scientists at the time (*ca.* 1900) as no more than convenient figments of the theoreticians' imagination. In order to properly understand this second aspect of Perrin's work we need to develop a description of the random diffusion process. For simplicity we restrict the analysis to one dimension, but it can readily be generalized to three dimensions.

2.1.2 *The one-dimensional random walk*

Suppose a particle is constrained to move in steps of equal length, l, along a line and that each step can be taken to the right or to the left with equal probability. Then after m_R steps to the right and m_L to the left the particle is a distance $x = (m_R - m_L)l$ to the right of the origin. We wish to calculate the probability that the particle is a certain distance, x, from the origin after time, t, if it takes a step of length l every τ seconds ($t \gg \tau$) and the direction of each step (whether to left or right) is entirely random.

The total number of steps taken $= m_L + m_R = t/\tau = m$, and since every step can occur in either of two directions the total number of possible positions for the particle is 2^m. Not all of these are different positions because the same outcome can result from many different sequences. In fact, the only thing that determines the final result is the total number of steps to the right and steps to the left, not the order in which they are taken. A particular outcome can be represented by a string of letters (LRRLLRRRLRLLR) and the number of possible ways of constructing a string from m *different* letters is $m!$ If m_R of them are the same they can be rearranged in $m_R!$ ways without altering the result; the same is true for the $(m - m_R)$ steps to the left, so in the present case the number of possible outcomes must be corrected by a factor $m!/m_R!\,(m - m_R)!$ The probability that the particle has reached a distance x from the origin is then:

$$\mathcal{P}(x) = \frac{m!}{m_R!\,(m - m_R)!\,2^m}. \tag{2.1.4}$$

Since $m_R = \frac{1}{2}(m + x/l)$ we can write eqn (2.1.4) as:

$$\mathcal{P}(x) = \frac{m!}{[\frac{1}{2}(m+s)]!\,[\frac{1}{2}(m-s)]!\,2^m} \tag{2.1.5}$$

where $s = x/l$.

As the number of steps, m, becomes larger it becomes increasingly more satisfactory to evaluate these factorials using the approximation formula of Stirling:

$$\ln N! \doteqdot N \ln N - N. \tag{2.1.6}$$

Then (see Exercise 2.1.3):

$$-\ln \mathcal{P} = \frac{m+s}{2} \ln\left(1 + \frac{s}{m}\right) + \frac{m-s}{2} \ln\left(1 - \frac{s}{m}\right) \tag{2.1.7}$$

and, using the approximation (valid for $s/m < 1$)

$\ln(1 \pm (s/m)) \simeq \pm(s/m) - (s^2/2m^2)$ we have (Exercise 2.1.3):

$$\ln \mathscr{P} = \frac{-x^2}{2ml^2}. \tag{2.1.8}$$

Although eqn (2.1.8) gives the correct general form for $\mathscr{P}$, the approximation eqn (2.1.6) introduces a constant error in $\ln \mathscr{P}$ (see Exercise 2.1.5 below), so we can only write

$$\mathscr{P} = k' \exp\left(\frac{-x^2}{2ml^2}\right). \tag{2.1.9}$$

Since, however, the probability of obtaining some possible outcome x lying between $-\infty$ and $+\infty$ must be unity, we have:

$$k' = \left[\int_{-\infty}^{\infty} \exp\left(\frac{-x^2}{2ml^2}\right) \mathrm{d}x\right]^{-1} \tag{2.1.10}$$

and this can be shown to be given by (Exercise 2.1.4):

$$k' = (2\pi ml^2)^{-1/2}.$$

We have then:

$$\mathscr{P}(x, m) = \frac{1}{m^{1/2} l \sqrt{(2\pi)}} \exp\left[-\frac{1}{2}\left(\frac{x^2}{ml^2}\right)\right]. \tag{2.1.11}$$

This equation is very closely related to the *normal distribution* function (Fig. 2.1.2), which occurs so widely in the theory of statistics and the treatment of errors. In fact, if we identify the mean value of x as $\bar{x}$ (=0 in this case) and the standard deviation of the distribution as $\sigma = m^{1/2} l$ we would have:

$$\mathscr{P}(x, \sigma) = \frac{1}{\sigma \sqrt{(2\pi)}} \exp\left[-\frac{1}{2}\left(\frac{x - \bar{x}}{\sigma}\right)^2\right] \tag{2.1.12}$$

which is the standard form of the normal distribution function. The connection between eqns (2.1.11) and (2.1.12) is, of course, not fortuitous. Although the variables x, m, and l are discrete in eqn (2.1.11) while x is continuous in eqn (2.1.12), we have effectively treated the distribution of x/l as continuous (i.e. the step length l as very short) by using eqn (2.1.10). A more rigorous argument connecting eqns (2.1.11) and (2.1.12) is given by Herdan (1960), p. 74.

The time-dependent expression can be obtained by substituting for m:

$$\mathscr{P}(x, t) = \left(\frac{\tau}{2\pi l^2 t}\right)^{1/2} \exp\left(-\frac{\tau x^2}{2tl^2}\right). \tag{2.1.13}$$

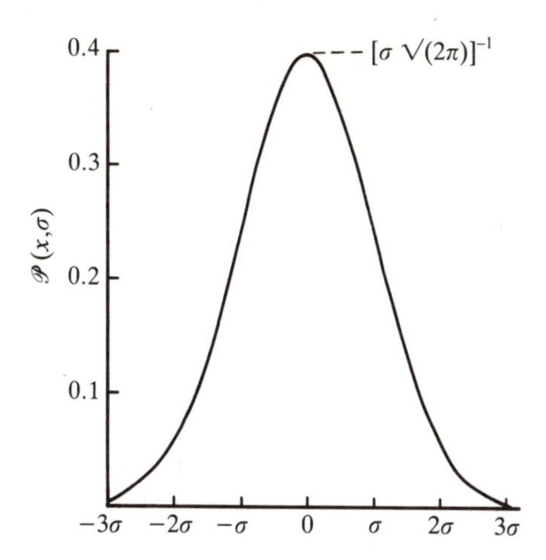

FIG. 2.1.2. The normal distribution curve. Note that although, theoretically, the distribution stretches to infinity in either direction, in practice it has become negligible after ± 3 standard deviations from the mean. The quantity $[\sigma\sqrt{(2\pi)}]^{-1}$ is called the *precision* of the distribution.

Note that in this expression $\mathscr{P}(-x, t) = \mathscr{P}(x, t)$ and because of eqn (2.1.10) this is the appropriate form for a distribution about the origin where arrival at $-x$ is treated as a different result from arrival at x (although they are equiprobable). Equation (2.1.13) then, gives the probability that the particle has arrived at a point $+x$ from the origin after 'walking' randomly for time t.

2.1.3 *The phenomenology of diffusion*

We noted above (section 2.1.1) that, in the absence of any external fields, the chemical potential, μ_i, of a substance within a phase at equilibrium is constant. If for some reason the value of μ_i varies from place to place, then substance i will tend to migrate (*diffuse*) in such a way as to equalize the values throughout the phase. The driving force for this diffusion process is the (spatial) gradient of μ_i—the larger it is the faster is the diffusion process. We write, formally, for a one-dimensional process:

$$\mathscr{F}_\mathrm{d} = -\mathrm{d}\mu_i/\mathrm{d}x$$

where $\mathscr{F}_\mathrm{d}$ is the driving force. (There is, on average, no actual force on the molecules or particles of i since they move only to maximize the entropy of the system. However, the introduction of this 'phantom force' (Atkins 1978, p. 834; 1982, p. 903) aids in the analysis.) Substituting from eqn (2.1.3) then:

$$\mathscr{F}_d = -\frac{d}{dx}(\mu_i^\ominus + RT\ln(c_i/\text{mol L}^{-1}))$$
$$= -\frac{RT}{c_i}\frac{dc_i}{dx} \tag{2.1.14}$$

since $\mu_i^\ominus$ is constant throughout the phase. The diffusion force *per molecule* can then be written:

$$f_d = -\frac{kT}{c_i}\frac{dc_i}{dx}. \tag{2.1.15}$$

This 'force' acting on the particle sets it in motion; it will be opposed by a viscous (drag) force, f_v, which for a particle of reasonably smooth shape is proportional to its velocity:

$$f_v = Bu \tag{2.1.16}$$

where B is called the friction coefficient. As the particle increases in velocity, f_v increases until it equals the diffusion force and the particle then reaches its terminal velocity, u_d. (This point is taken up in more detail in Exercise 3.4.1.) We then have $f_d = Bu_d$.

The diffusion velocity, u_d, is related to the flux (flow) of material per unit area, J_i by:

$$J_i = u_d c_i \tag{2.1.17}$$

as is evident from Fig. 2.1.3.

The flux of material is also given by *Fick's first law of diffusion*:

$$J_i = -\mathscr{D}\frac{dc}{dx} \tag{2.1.18}$$

where $\mathscr{D}$ is the diffusion coefficient. We can therefore write:

$$\mathscr{D} = -J_i/(dc/dx) = -u_d c/(dc/dx) = \frac{-f_d c}{B(dc/dx)}$$

i.e.

$$\mathscr{D} = kT/B \tag{2.1.19}$$

using eqn (2.1.15). This is one of many important equations attributed to Einstein.

The friction factor B, which measures how strongly the surrounding fluid resists the motion of the particle, is determined partly by the viscosity, η, of the fluid and partly by the shape of the particle. For a

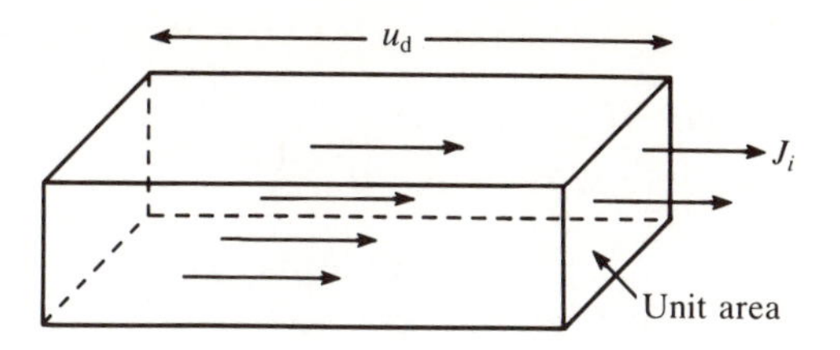

FIG. 2.1.3. The flux of material (mol s^{-1}) moving across unit area is equal to all of the material contained in the parallelepiped of length u_d and unit cross section (i.e. $u_d c_i$).

spherical particle of radius r it is given by the Stokes relation:

$$B = 6\pi r\eta \tag{2.1.20}$$

but for other shapes, the calculation of B is at best difficult and often impossible. The importance of eqn (2.1.19) lies in the fact that it enables a direct measurement to be made of B if we can determine the diffusion coefficient $\mathscr{D}$ of the particle, and that is the task to which we now turn.

2.1.4 *Time-dependent diffusion processes*

Fick's first law can be applied only when the concentration gradient is constant in time and this will be so only in rather exceptional circumstances. More usually we begin with a fixed amount of material in a limited region and as it diffuses away from its starting point, the concentration gradient is itself a function of time. To treat such a system, we proceed as follows (Fig. 2.1.4).

The amount of material which accumulates in the thin slab of material between x and $(x + \delta x)$ is $\mathscr{A}\delta x\, \partial c(x, t)/\partial t$, and is equal to the difference

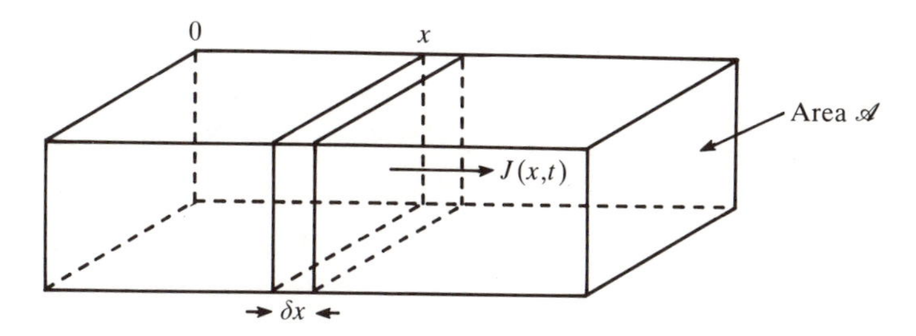

FIG. 2.1.4. Illustrating Fick's second law of diffusion as a consequence of matter conservation.

between the influx ($J_i(x, t)$) and the efflux ($J_i(x + \delta x, t)$). Hence:

$$\frac{\partial c(x,t)}{\partial t} = \frac{J(x,t)\mathscr{A}}{\mathscr{A}\delta x} - \frac{J(x+\delta x, t)\mathscr{A}}{\mathscr{A} \cdot \delta x}$$

$$= \frac{1}{\delta x}\left[-\mathscr{D}\left(\frac{\partial c(x,t)}{\partial x}\right)_x + \mathscr{D}\left(\frac{\partial c}{\partial x}\right)_{x+\delta x}\right]$$

$$= \frac{\mathscr{D}}{\delta x}\left[-\left(\frac{\partial c}{\partial x}\right)_x + \left(\frac{\partial c}{\partial x}\right)_x + \delta x\left(\frac{\partial^2 c}{\partial x^2}\right)_x\right]$$

i.e.

$$\frac{\partial c}{\partial t} = \mathscr{D}\frac{\partial^2 c}{\partial x^2} \tag{2.1.21}$$

which is referred to as *Fick's second law of diffusion.*

We do not propose to discuss the many mathematical techniques that are used to solve partial differential equations of this sort. There is, however, one solution that is of special interest. Suppose we begin with a system in which all of the solute molecules are confined to a thin slab, containing c_0 moles, at the origin and allow this material to diffuse out in both directions at right angles to the plane of the slab (Fig 2.1.5). What will be the concentration profile as a function of time and distance? It should come as no surprise to find that the solution is essentially contained in the function derived by analysis of the random walk (section

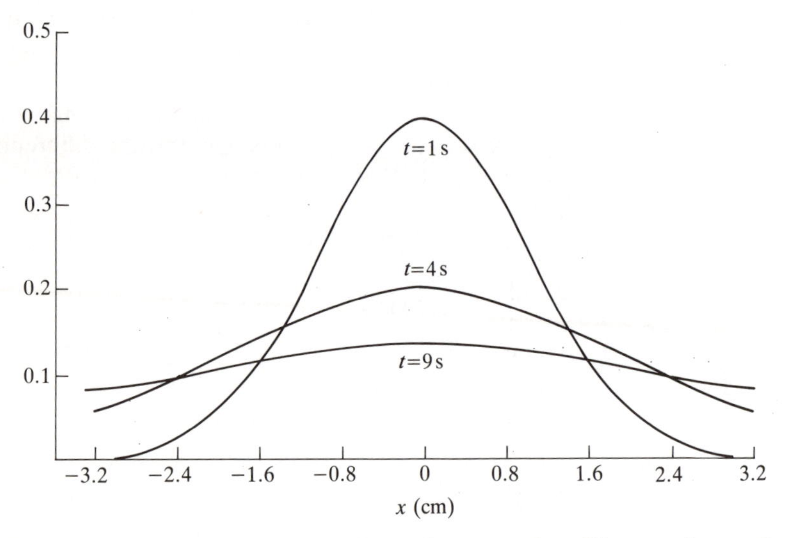

FIG. 2.1.5. Concentration profile at different times as a solute diffuses out from a plane in either direction at right angles to the plane. (From Hiemenz 1977, with permission.)

2.1.2):

$$c(x,t)=c_0\left(\frac{\tau}{2\pi l^2 t}\right)^{1/2}\exp\left(-\frac{\tau x^2}{2l^2 t}\right) \tag{2.1.22}$$

where the time for each diffusion step, τ, is given by (Exercise 2.1.6):

$$\tau = l^2/2\mathscr{D}. \tag{2.1.23}$$

We can, therefore, write eqn (2.1.22) as

$$c(x,t)=c_0\left(\frac{1}{4\pi\mathscr{D}t}\right)^{1/2}\exp-\left(\frac{x^2}{4\mathscr{D}t}\right). \tag{2.1.24}$$

Diffusion can in this case be regarded as the end result of the random motion of the molecules in one dimension.

2.1.5 *The Einstein–Smoluchowski equation*

Note that the average displacement, $\langle x\rangle$ given by eqn (2.1.24) is zero and this remains so no matter how far the material spreads. A better measure of the extent of the diffusion process is, therefore, the mean square displacement, $\langle x^2\rangle$ or the root mean square displacement, defined by:

$$\langle x^2\rangle^{1/2}=\left[\frac{\int_{-\infty}^{\infty} x^2 c(x,t)\,\mathrm{d}x}{\int_{-\infty}^{\infty} c(x,t)\,\mathrm{d}x}\right]^{1/2}$$

$$=\left[\frac{1}{c_0}\int_{-\infty}^{\infty} x^2 c(x,t)\,\mathrm{d}x\right]^{1/2}. \tag{2.1.25}$$

This quantity is exactly analogous to the standard deviation, σ, of eqn (2.1.12) (see Exercise 2.1.7). Comparing eqns (2.1.24) and (2.1.12) we can, therefore, write:

$$\langle x^2\rangle^{1/2}=(2\mathscr{D}t)^{1/2}. \tag{2.1.26}$$

This is the Einstein–Smoluchowski equation, which was used by Perrin in another procedure for estimating the Avogadro constant. He measured, directly, the root-mean-square displacement of spherical colloidal particles as a function of time by following their Brownian motion. From the resulting value of $\mathscr{D}$ he could obtain another estimate of k from the relation:

$$\mathscr{D}=\frac{kT}{B}=kT/6\pi r\eta$$

and hence obtain $\mathscr{N}_A$ from the gas constant, as in Exercise 2.1.2.

For a spherical particle of radius 100 nm, the diffusion coefficient in water at 25 °C is about $1.38 \times 10^{-23} \times 298/(6\pi \times 10^{-7} \times 9 \times 10^{-4}) \simeq 2.4 \times 10^{-12}\,\mathrm{m^2\,s^{-1}}$. Values for molecular species are normally about $10^{-9}\,\mathrm{m^2\,s^{-1}}$ and they are often determined by studying the concentration profile as a substance diffuses down a concentration gradient. One has only to solve eqn (2.1.21) with the appropriate boundary conditions to find an equation like (2.1.24) and obtain $\mathscr{D}$ from the experimental results. For colloidal particles that method is not very successful except for very small particles (like protein molecules), because $\mathscr{D}$ is so small that the diffusion experiment must be conducted for a long time, during which accurate temperature control must be maintained to avoid errors due to convection. Fortunately, there are now available more direct methods of determining $\mathscr{D}$ using photon correlation spectroscopy and these can be used for colloids of any size. We will therefore not elaborate on the older methods, which are only adaptations of the methods used for simple solutions, and will postpone further discussion of the measurement of $\mathscr{D}$ until section 2.3.2. Suffice it to say that methods are available for finding $\mathscr{D}$ accurately for any size or shape of particle, and this can be used in eqn (2.1.19) to determine the friction factor, B.

Exercises. 2.1.1 Show that for colloidal particles dispersed in a liquid, the equilibrium number of particles, N, at a height h above a reference level, h_0, is given by:

$$N = N_0 \exp[-(m - m')g(h - h_0)/kT]$$

where N_0 is the number of particles at height h_0 and m' is the mass of fluid displaced by a particle of mass m (k is the Boltzmann constant).

2.1.2 Svedberg (1928, p. 101) gives the following table of Westgren's data for the sedimentation equilibrium of a gold sol under gravity:

Height (μm)	Number of particles	Height (μm)	Number of particles
0	889	600	217
100	692	700	185
200	572	800	152
300	426	900	125
400	357	1000	108
500	253	1100	78

Assume the particles have radius 21 nm and density $19.3\,\mathrm{g\,cm^{-3}}$ and the temperature is 20 °C. Estimate k from the equation derived in Exercise 2.1.1 and then calculate $\mathscr{N}_A$ assuming $R = 8.31\,\mathrm{J\,K^{-1}\,mol^{-1}}$. Repeat the calculation with a radius of 22 nm and note how sensitive the answer is to this variable.

2.1.3 (i) Establish eqn (2.1.7) from (2.1.5) and (2.1.6).

(ii) Establish eqn (2.1.8) from (2.1.7).

2.1.4 Use the identity

$$\Gamma\left(\frac{1}{2}\right) = \int_0^\infty e^{-p} p^{-1/2}\, dp = \pi^{1/2}$$

to evaluate k' in eqn (2.1.10).

2.1.5 Use the more exact form of Stirling's approximation:

$$\ln N! = (N + \tfrac{1}{2})\ln N - N + \ln(2\pi)^{1/2}$$

to show that

$$\mathcal{P} = \left(\frac{2}{\pi m}\right)^{1/2} \exp\left(-\frac{x^2}{2ml^2}\right)$$

(Note the disappearance of l from the pre-exponential factor because this form applies to the discrete distribution, confining the particle to points distant l apart.) Why does it also contain a factor two compared to 2.1.11?

2.1.6 Verify by direct differentiation that eqn (2.1.22) is a solution of eqn (2.1.21) and, hence, establish (2.1.23).

2.1.7 Use the identity:

$$\int_0^\infty x^2 \exp(-ax^2)\, dx = \frac{1}{4a}(\pi/a)^{1/2}$$

to show that $\langle x^2 \rangle^{1/2}$ is equal to σ.

2.2 Response of a dielectric material to an electric field

If a sinusoidally varying electric field is applied to a material, the electrons in the atoms of that material will be induced to oscillate in response to the field and the analysis of their response provides valuable information about the structure of the material. The actual forces experienced by a particular electron are not simply related to the magnitude of the applied field at any time because the electron responds to the local field and that is influenced by the surrounding material. The previous history of the electric field may have led to the surrounding material being polarized in a certain way and, if the frequency of the field is high, those 'memories' may still be able to exert an influence on the electron under consideration. To discuss the response of a dielectric material to a time-varying field we will need to be able to take account of such effects. Before we develop the tools for that purpose we must briefly review the behaviour in a static electric field.

2.2.1 *Static electric fields*

Consider a dielectric material confined between two flat plates that have charges per unit area of $+\sigma_0$ and $-\sigma_0$, respectively (Atkins 1978, p. 747;

1982, p. 768). In the *absence* of a dielectric material the magnitude of the electric field $\boldsymbol{E}$, between the plates is σ_0/ε_0, where ε_0 is the dielectric permittivity of a vacuum ($\varepsilon_0 = 8.854 \times 10^{-12}\,\mathrm{F\,m^{-1}}$, i.e. $\mathrm{C\,V^{-1}\,m^{-1}}$). With the dielectric material in place, the field drops to $\sigma_0/\varepsilon = \sigma_0/\varepsilon_r\varepsilon_0$, where ε_r is a characteristic of the material (the *relative dielectric permittivity*, $\varepsilon/\varepsilon_0$, sometimes called the *dielectric constant*). Note that $\varepsilon_r > 1$; the field is always reduced because the dielectric material aligns itself as shown in Fig. 2.2.1 and this partly cancels the applied field.

The capacity of this parallel plate condenser is equal to the ratio of charge to potential difference. If the space between the plates is empty, then:

$$C_0 = \frac{\sigma_0 \mathscr{A}}{Ed} = \frac{\sigma_0 \mathscr{A} \varepsilon_0}{\sigma_0 d} = \frac{\varepsilon_0 \mathscr{A}}{d} \tag{2.2.1}$$

and when the dielectric is present it increases to:

$$C = \frac{\varepsilon_0 \varepsilon_r \mathscr{A}}{d} \tag{2.2.2}$$

so the ratio of the capacitances (C/C_0) can be used to measure ε_r.

The *polarization*, P, is in this case equal to the charge density on the surface of the dielectric adjacent to each plate (Fig. 2.2.1). The total charge on one such face is $P\mathscr{A}$ and on the other $-P\mathscr{A}$. We can regard this as a large dipole, separated by a distance d so the dipole moment is $P\mathscr{A}d$ and the *dipole moment per unit volume* is $P\mathscr{A}d/\mathscr{A}d$ or P. In more general terms we can write for the *polarization vector* at any point in a dielectric:

$$\boldsymbol{P}(\boldsymbol{r}) = \rho_N \boldsymbol{p}(\boldsymbol{r}) \tag{2.2.3}$$

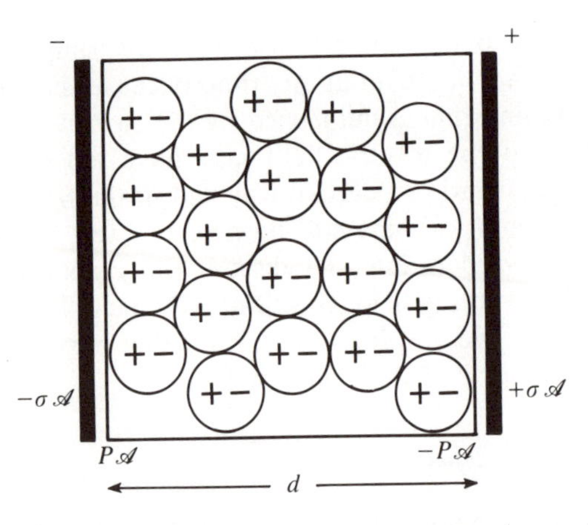

FIG. 2.2.1. Polarization of the molecules of a dielectric between two plates of area $\mathscr{A}$ on which is a charge per unit area $+\sigma$ and $-\sigma$. (After Atkins 1978.)

where $\boldsymbol{p}$ is the local dipole moment of the molecules at $\boldsymbol{r}$ and ρ_N is their local number density. $\boldsymbol{P}$ is again the dipole moment per unit volume.

The induced charge $\pm P\mathscr{A}$ near the plates reduces the effective charge to a value $(\sigma_0 - P)\mathscr{A}$ so that the new field has a magnitude (from eqn (2.2.1)):

$$E = \frac{(\sigma_0 - P)\mathscr{A}}{C_0 d} = (\sigma_0 - P)/\varepsilon_0.$$

Alternatively we can evaluate E from the effect of the permittivity: $E = \sigma_0/\varepsilon_r\varepsilon_0$. Eliminating σ_0 between these two relations gives:

$$P = \varepsilon_0 E(\varepsilon_r - 1) \tag{2.2.4}$$

$$\boldsymbol{P} = \varepsilon\boldsymbol{E} - \varepsilon_0\boldsymbol{E}. \tag{2.2.5}$$

The vector $\varepsilon\boldsymbol{E}$ is called the *dielectric displacement* $\boldsymbol{D}$, and in general

$$\boldsymbol{D} = \varepsilon_0\boldsymbol{E} + \boldsymbol{P}. \tag{2.2.6}$$

We will find in Chapter 6 that the fundamental equations of electrostatics give an expression for the distribution of the free charges in a dielectric medium in terms of $\boldsymbol{D}$ rather than $\boldsymbol{E}$ (as would be the situation *in vacuo*).

2.2.2 *Response of a bound electron to an alternating field*

In order to get an idea of the way a charge distribution responds to a time-varying electric field we consider the simplest possible situation. Suppose (Fig. 2.2.2) a negative charge $-q$, of mass m, is attached by a spring, of force constant k_0, to a positive charge (the spring represents the attractive interaction between the two and is introduced so that the motion of the electron will be simply harmonic). Newton's law gives, for

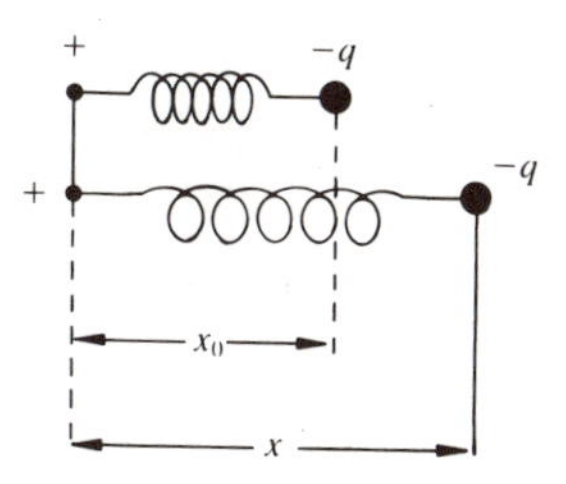

FIG. 2.2.2. Model of a charge undergoing simple harmonic motion.

the equation of motion (Richmond 1975):

$$m\frac{\mathrm{d}^2x}{\mathrm{d}t^2} = -k_0x + qE(t) \tag{2.2.7}$$

where $E(t)$ is the magnitude of the electric field which is assumed to be oscillating with a frequency ω (rad s^{-1}). This could be represented by

$$E(t) = E_0 \cos \omega t \tag{2.2.8}$$

but, for convenience, we write it as

$$E(t) = E_0\mathrm{e}^{-i\omega t} \tag{2.2.9}$$

where it is understood that only the real part of eqn (2.2.9) has any direct physical significance. ($E(t) = \mathrm{Re}[E_0\mathrm{e}^{-i\omega t}] = E_0 \cos(\omega t)$ since $\mathrm{e}^{\pm i\theta} = \cos\theta \pm i \sin\theta$ for any θ). There are two reasons for making this substitution:

(a) exponentials are rather easier to handle in integration processes than are trigonometric functions; and, more importantly,

(b) the use of the complex number procedure allows us to separate out the elastic (storage) effects during the interaction from any dissipative (frictional or viscous (lossy)) processes like absorption.

The form of eqn (2.2.7) suggests solutions of the form:

$$x = x_0\mathrm{e}^{-i\omega t} \tag{2.2.10}$$

and direct differentiation and substitution in eqn (2.2.7) gives (for $x = x_0$, $t = 0$):

$$-m\omega^2x_0 = -k_0x_0 + qE_0. \tag{2.2.11}$$

The polarization of the system is, in this simple case, equal to the dipole moment: $P(t) = qx(t)$ so that:

$$P(t) = P_0\mathrm{e}^{-i\omega t} \tag{2.2.12}$$

where:

$$P_0 = qx_0. \tag{2.2.13}$$

From eqns (2.2.11) and (2.2.13) (Exercise 2.2.1) we have:

$$(\omega_0^2 - \omega^2)P_0 = \alpha_0\omega_0^2E_0 \tag{2.2.14}$$

where

$$\omega_0^2 = k_0/m \quad \text{and} \quad \alpha_0\omega_0^2 = q^2/m.$$

Provided ω is not too close to ω_0 we can write:

$$P_0 = \left[\frac{\alpha_0\omega_0^2}{(\omega_0^2 - \omega^2)}\right]E_0. \tag{2.2.15}$$

The constant of proportionality between P_0 and E_0 is, in this case, the (frequency dependent) polarizability, α, which measures the ease with which electrons can be displaced by an applied field. This is so because the polarization is here equal to the induced dipole moment $\boldsymbol{p}$ and α is *defined* by the relation:

$$\boldsymbol{p} = \alpha \boldsymbol{E}. \tag{2.2.16}$$

It is obvious from eqn (2.2.15) that something rather dramatic occurs when $\omega = \omega_0$ since P_0 would then increase without limit. This is, of course, the natural *resonance frequency* of the system, $\omega_0 = (k_0/m)^{1/2}$, and at that frequency the dipole would be able to absorb energy from the field during every oscillation and so gradually increase its amplitude until it was infinitely large. There would be, at this frequency, an extremely sharp line in the absorption spectrum. In practice, of course, apart from the quantum effect, the spring would not remain harmonic in its action and in a real material other constraints would limit the motion of the charge so that the absorption peak would be reduced in height and broadened. The result is a Lorentzian function. (See Exercises 2.2.6 and 2.2.7.)

In order to complete the solution of eqn (2.2.7) to find $P(t)$ as a function of $E(t)$ we would have to introduce the notion of a complex polarizability ($\alpha = \alpha' + i\alpha''$) and solve for α' and α'' (Richmond 1975). Rather than do that we will now examine the more general relation between the vectors $\boldsymbol{P}$ and $\boldsymbol{E}$, which applies when a collection of interacting dipoles is involved. The vector $\boldsymbol{P}$ still measures the dipole moment per unit volume but its relation to $\boldsymbol{E}$ requires further discussion.

2.2.3 *The dielectric response function $\varepsilon(\omega)$*

The more general relation between $\boldsymbol{P}$ and $\boldsymbol{E}$ when the field varies with time has to take account of the past influence of the field on the present state of polarization and that involves the idea of a *memory function.* In place of eqn (2.2.3) we now write:

$$\boldsymbol{P}(\boldsymbol{r}, t) = \rho_{\mathrm{N}} \boldsymbol{p}(\boldsymbol{r}, t) \tag{2.2.17}$$

and the dielectric displacement becomes (compare eqn (2.2.6)):

$$\boldsymbol{D}(\boldsymbol{r}, t) = \varepsilon_0 \boldsymbol{E}(\boldsymbol{r}, t) + \boldsymbol{P}(\boldsymbol{r}, t). \tag{2.2.18}$$

Because the induced polarization takes a finite time to decay, the total polarization at time t is dependent on the electric field at all previous times $t - \tau$ $(0 < \tau < \infty)$. The most general (linear) relation for $\boldsymbol{P}(\boldsymbol{r}, t)$ is then (Landau and Lifshitz 1960):

$$\boldsymbol{P}(\boldsymbol{r}, t) = \varepsilon_0 \int_0^\infty m(\tau)\boldsymbol{E}(\boldsymbol{r}, t-\tau)\, \mathrm{d}\tau \qquad (2.2.19)$$

where $m(\tau)$ is the *memory function,* which tells us the contribution to $\boldsymbol{P}(\boldsymbol{r}, t)$ at time t because of an electric field $\boldsymbol{E}(\boldsymbol{r}, t-\tau)$ applied at a time τ before t (Parsegian 1975). We will encounter a similar memory function in section 2.4 in connection with the mechanical (rheological) properties of a colloidal material. [Such linear relations between an imposed field and the resulting response can only be expected to hold for low to moderate fields. Ultimately, at sufficiently high field strengths, the material will respond non-linearly. This will introduce further complications that are beyond the scope of the present treatment.]

The function $m(\tau)$ describes the decay of an induced polarization with time and, of physical necessity, must tend to zero for sufficiently large τ:

$$\lim_{\tau\to\infty} m(\tau) = 0. \qquad (2.2.20)$$

Parsegian (1975) gives a useful discussion on $m(\tau)$ in which he draws attention to the fact that it must also be finite for all τ (since otherwise one would have an infinite polarization produced by a finite field. Recall the problem with the undamped oscillating charge at the resonance frequency ω_0 in section 2.2.2). Note also that τ must never be negative since that would imply that the field at some future time was affecting the present polarization (i.e. the cause coming *after* the effect). (Such violations of the causality principle may be quantum mechanically possible (Wheeler and Feynman 1945) but will not concern us here.)

The function $m(\tau)$ contains all the information about local electronic, vibrational, rotational and translational *relaxations*† of the component molecules of the dielectric material and for most purposes can be taken to be independent of the position $\boldsymbol{r}$ within the material.

From eqns (2.2.18) and (2.2.19):

$$\boldsymbol{D}(\boldsymbol{r}, t) = \varepsilon_0\left[\boldsymbol{E}(\boldsymbol{r}, t) + \int_0^\infty m(\tau)\boldsymbol{E}(\boldsymbol{r}, t-\tau)\, \mathrm{d}\tau\right] \qquad (2.2.21)$$

and, once again, we represent $\boldsymbol{E}$ as a complex function (compare eqn (2.2.9)):

$$\boldsymbol{E}(\boldsymbol{r}, t) = \boldsymbol{E}_0(\boldsymbol{r})\mathrm{e}^{-i\omega t}. \qquad (2.2.22)$$

† Relaxation is the process whereby a system that has been perturbed in some way returns to its equilibrium condition.

We can then write, for the (complex) displacement vector:

$$\boldsymbol{D}(\boldsymbol{r}, t) = \varepsilon_0\left[\boldsymbol{E}(\boldsymbol{r}, t) + \int_0^\infty m(\tau)\boldsymbol{E}_0(\boldsymbol{r})\mathrm{e}^{-i\omega(t-\tau)}\,\mathrm{d}\tau\right]$$

$$= \varepsilon_0\boldsymbol{E}(\boldsymbol{r}, t)\left[1 + \int_0^\infty m(\tau)\mathrm{e}^{i\omega\tau}\,\mathrm{d}\tau\right] \tag{2.2.23}$$

since t is constant inside the integral.

Alternatively, returning to the original definition of $\boldsymbol{D}$ $(=\varepsilon\boldsymbol{E})$ in eqn (2.2.5) we can write:

$$\boldsymbol{D}(\boldsymbol{r}, t) = \varepsilon_0\varepsilon_\mathrm{r}(\omega)\boldsymbol{E}(\boldsymbol{r}, t) \tag{2.2.24}$$

where:

$$\varepsilon_\mathrm{r}(\omega) = 1 + \int_0^\infty m(\tau)\mathrm{e}^{i\omega\tau}\,\mathrm{d}\tau \tag{2.2.25}$$

and $\varepsilon(\omega) = \varepsilon_0\varepsilon_\mathrm{r}(\omega)$ as before.

Note that in this formulation $\varepsilon_\mathrm{r}(\omega)$ is a *dimensionless* generalization of the relative permittivity or dielectric constant, ε_r, appearing in eqn (2.2.4) and:

$$\varepsilon_\mathrm{r} = \varepsilon_\mathrm{r}(0) = 1 + \int_0^\infty m(\tau)\,\mathrm{d}\tau. \tag{2.2.26}$$

The dielectric response function $\varepsilon(\omega)$ is simply the extension of the familiar static relationship between $\boldsymbol{D}$ and $\boldsymbol{E}$ to take account of time-varying electric fields.

2.2.4 *The phase lag between **D** and **E***

Since the polarization at time t is the residual effect of the electric field at all previous times (and not just $\boldsymbol{E}(t)$) both $\boldsymbol{P}(t)$ and $\boldsymbol{D}(t)$ tend to be out of phase with $\boldsymbol{E}(t)$ and this is taken into account by the complex character of $\varepsilon(\omega)$. We can represent $\varepsilon(\omega)$ by real $(\varepsilon'(\omega))$ and imaginary $(\varepsilon''(\omega))$ parts so that:

$$\varepsilon(\omega) = \varepsilon'(\omega) + i\varepsilon''(\omega) \tag{2.2.27}$$

and, then, from eqn (2.2.25):

$$\varepsilon'(\omega) = \varepsilon_0\left(1 + \int_0^\infty m(\tau)\cos\omega\tau\,\mathrm{d}\tau\right) \tag{2.2.28}$$

and

$$\varepsilon''(\omega) = \varepsilon_0\int_0^\infty m(\tau)\sin\omega\tau\,\mathrm{d}\tau \tag{2.2.29}$$

$$(\text{Note that } \varepsilon(0) = \varepsilon'(0) \text{ and } \varepsilon''(0) = 0.) \tag{2.2.30}$$

To explore the phase lag, $\varepsilon(\omega)$ can be written in the alternative form (Exercise 2.2.3):

$$\varepsilon(\omega) = |\varepsilon(\omega)|e^{i\delta(\omega)} \tag{2.2.31}$$

where

$$\delta(\omega) = \tan^{-1}\left[\frac{\varepsilon''(\omega)}{\varepsilon'(\omega)}\right] \tag{2.2.32}$$

and

$$|\varepsilon(\omega)| = (\varepsilon'(\omega)^2 + \varepsilon''(\omega)^2)^{1/2}. \tag{2.2.33}$$

Substituting eqn (2.2.31) in (2.2.24) and taking the real part of the resulting equation we see that if the physical field $\boldsymbol{E}(\boldsymbol{r}, t)$ has the form of eqn (2.2.22) then the physical displacement vector has the form (Exercise 2.2.4):

$$\boldsymbol{D}(\boldsymbol{r}, t) = |\varepsilon(\omega)|\boldsymbol{E}_0(\boldsymbol{r})\cos(\omega t - \delta(\omega)). \tag{2.2.34}$$

The effect of the complex dielectric response is to cause a phase difference, δ, between the field $\boldsymbol{E}$ and the displacement vector.

The angle, δ, is a direct measure of the energy dissipation which occurs in the dielectric as the field passes through it. If $\delta = 0$, the field passes through the dielectric without loss and the medium is said to be transparent to the field at that frequency. The maximum loss angle $\delta = \pi/2$ corresponds to total absorption of the electric field energy by the medium. We will see in section 2.4 that it corresponds, for a mechanical system, to the viscous dissipation of energy when a shear wave passes through a pure liquid. From eqn (2.2.32) it is clear that it is the imaginary part, ε'', of the dielectric response that measures the extent of this loss, or absorption, of energy. If $\varepsilon'' = 0$ then $\varepsilon(\omega) = \varepsilon'(\omega)$ is purely real and there is no dissipation of energy from the field at all.

At very high (optical) frequencies, the behaviour remains the same but, for historical reasons, it is described in rather different terms. In place of ε' and ε'' we have a refractive index $n_1(\omega)$ and an absorption (or extinction) coefficient, $\kappa(\omega)$ which are related to $\varepsilon_r(\omega)$ by:

$$(n_1 + i\kappa)^2 = \varepsilon_r = \varepsilon_r' + i\varepsilon_r'' \tag{2.2.35}$$

so that

$$n_1^2 - \kappa^2 = \varepsilon_r' \quad \text{and} \quad 2n_1\kappa = \varepsilon_r''. \tag{2.2.36}$$

The optical spectrum normally records κ as a function of ω (or the wavelength, λ, of the light) and we normally measure n_1 at frequencies somewhat removed from an absorption line, where the material can behave elastically, but eqns (2.2.35) and (2.2.36) apply over the whole frequency range. Note again that ε'' is directly proportional to the absorption coefficient.

2.2.5 *The shape of* $\varepsilon(\omega)$

So far we have said nothing about the explicit mathematical form of $\varepsilon(\omega)$ except what is contained in eqn (2.2.25). It is possible to show from eqn (2.2.25), using Laplace transforms (Parsegian 1975), that any physical material can be adequately represented by a sum (or integral) of terms of the form

$$\frac{f_j + h_j(-i\omega)}{\omega_{0,j}^2 + g_j(-i\omega) + (-i\omega)^2} \tag{2.2.37}$$

where the constants f_j, h_j, g_j, and $\omega_{0,j}$ must be chosen to fit the experimental data. In practice, one can limit the types of response to just two: (a) a Debye relaxation term to characterize the response of a rotating dipole, and (b) a Sellmeier damped-oscillator form to characterize infra-red and ultraviolet absorptions.

For the Debye relaxation there is no restoring force, since the dipole simply rotates (restrained by friction) to follow the field. (This assumes that all of the dipoles can behave independently which is obviously an approximation.) Since such a rotator has no resonance frequency ($\omega_{0,j} = 0$) eqn (2.2.37) gives for such a system:

$$\frac{h_j}{g_j + (-i\omega)} = \frac{d_j}{1 - i\omega\tau_j}. \tag{2.2.38}$$

For the infra-red and ultraviolet absorptions a suitable form is (see Exercise 2.2.8):

$$\frac{f_j}{\omega_{0,j}^2 + g_j(-i\omega) + (-i\omega)^2} \tag{2.2.39}$$

and we could expect the following expression to represent a (non-conducting) polar dielectric material over the entire frequency range:

$$\varepsilon(\omega) = \varepsilon_0\left(1 + \sum_j \frac{d_j}{1 - i\omega\tau_j} + \sum_j \frac{f_j}{\omega_{0,j}^2 - \omega^2 - ig_j\omega}\right). \tag{2.2.40}$$

Exercises. 2.2.1 Establish equation (2.2.14).

2.2.2 Draw an Argand diagram to represent the functions (2.2.31)–(2.2.33).

2.2.3 Establish eqns (2.2.31)–(2.2.33).

2.2.4 Establish eqn (2.2.34).

2.2.5 Consider a capacitor (for which $C(\omega) = \varepsilon(\omega)\mathscr{A}/d$) subjected to an alternating voltage $V = V_0 \cos \omega t$. The resulting current is the real part of $I = V/Z$

where Z is the capacitive impedance or reactance and $Z = -1/i\omega C$. (This is the a.c. analogue of Ohm's Law.) Show that $I = (\omega\mathscr{A}/d)V_0[\varepsilon'' \cos \omega t - \varepsilon' \sin \omega t]$.

By considering the integral ${}_0\int^{2\pi} I(\omega t) \cdot V(\omega t)\, \mathrm{d}t$ show that the power dissipated in the capacitor is entirely determined by ε'':

$$\text{Power dissipated per cycle} = (\pi\mathscr{A}/d)\, V_0^2 \varepsilon''(\omega)$$

whereas ε' relates to the storage of electrical energy during a cycle.

2.2.6 Consider the extension of the problem treated in section 2.2.2 to include a friction (damping) term to the electron's motion. The equation of motion is then

$$m\frac{\mathrm{d}^2x}{\mathrm{d}t^2} = -k_0x - B\frac{\mathrm{d}x}{\mathrm{d}t} + qE(t).$$

Take $E = E_0 \mathrm{e}^{-i\omega t}$ and $x = A\mathrm{e}^{-i\omega t}$ where A, the amplitude of the motion, may be complex in this formulation. Verify by direct substitution in the equation of motion that

$$A = qE_0/[m(\omega_0^2 - \omega^2) - i\omega B].$$

Use the identity $1/(a - ib) = (a + ib)/(a^2 + b^2)$ to show that if $A = x' + ix''$ then

$$x'' = qE_0\frac{\omega B}{m^2(\omega_0^2 - \omega^2)^2 + \omega^2 B^2}.$$

2.2.7 Show that in the neighbourhood of the resonance frequency ($\omega \simeq \omega_0$)

$$x'' = \frac{qE_0}{2m\omega_0}\left[\frac{Q}{(\omega - \omega_0)^2 + Q^2}\right]$$

where $Q = B/2m$.

The plot of x'' as a function of ω is called a Lorentzian distribution (Fig. 2.3.3). It is commonly used to represent the shape of the lines in an absorption spectrum. Again it is the imaginary part of the displacement function which has the form of the absorption (dissipation) curve. Since the dipole moment is qx this measures the magnitude of the dipole moment out of phase with the field.

2.2.8 Show that if (eqn (2.2.39)):

$$\varepsilon(\omega) = \frac{\varepsilon_0 f_j}{\omega_{0,j}^2 + g_j(-i\omega) + (-i\omega)^2} = \varepsilon'(\omega) + i\varepsilon''(\omega)$$

then the imaginary part ε'' has exactly the same form as x'' obtained in Exercise (2.2.6). Thus eqn (2.2.39) can be expected to represent the behaviour of a bound electron moving against viscous friction in response to an applied field.

2.3 Light scattering

We noted earlier (Exercise 1.5.1) that colloidal particles when immersed in a fluid are able to *scatter* a beam of light (the Tyndall effect). The scattering pattern (i.e. the intensity of the scattered light as a function of

θ, the angle between the incident beam and the scattered beam) depends very strongly on the particle size and on the wavelength of the light. The spectral colours that are sometimes generated have fascinated investigators since before the end of the nineteenth century. It is only very recently, however, that the full potentialities of the study of light scattering have been realized, with the introduction of lasers to give coherent, monochromatic, intense and narrow incident beams, together with sensitive and stable photon-detection apparatus and rapid data analysis by computer.

2.3.1 *Basic principles*

The theory of light scattering has been extensively described by van de Hulst (1957) and by Kerker (1969) and we will examine it in some detail in Chapter 14. Here we will restrict attention to a few relationships that will be needed to address the problem of determining the particle size of colloidal dispersions (section 3.7).

When electromagnetic radiation strikes a particle it may be absorbed, transmitted, scattered, refracted or diffracted. Simple (uni-directional) refraction is a phenomenon which occurs only when the particle is very much larger than the wavelength, while diffraction occurs when the two are comparable. They can be regarded as limiting cases of the phenomenon of scattering. Scattering refers to the (elastic) interaction between the radiation and the particle in which the light is re-radiated *at the same frequency*. It is characteristic of scattering in the systems that concern us that only a small fraction of the incident beam is affected, the bulk of the light passing through the sample in the initial direction. If the size of the scattering particle is significant compared to the wavelength of the light then the (spherical) waves emanating from different regions of the particle interfere with one another to generate a complicated intensity pattern at different angles to the incident beam. Any light that is inelastically absorbed (i.e. involved in inducing quantum transitions in the particle) is usually degraded to thermal energy inside the particle although some of it may be re-radiated, at the same (or more usually a longer) wavelength as the incident light; we will not concern ourselves with that (fluorescence) behaviour here (see Exercise 2.3.1). It is the scattered light that is most intense, especially for colloidal dispersions.

A proper treatment of the interaction between an electromagnetic wave of wavelength λ and a particle consisting of many thousands of atoms would require the formalism of quantum field theory and would, in fact, be insoluble by present methods. Fortunately, however, it turns out that the problem can be tackled from a classical point of view, and was indeed solved in fairly general fashion before the quantum theory gained much credibility (Mie 1908; Debye 1909). Indeed, an approximate analysis (valid for particle diameter, $d < \sim\lambda/20$ and for refractive indices, n, very close to unity) was given by Lord Rayleigh in 1871. His equation for the scattered intensity, I, from a single particle is:

$$\frac{I}{I_{0,\mathrm{u}}} = \frac{8\pi^4}{\lambda^4 r^2}\left(\frac{\alpha}{4\pi\varepsilon_0}\right)^2 (1 + \cos^2\theta) \tag{2.3.1}$$

where $I_{0,\mathrm{u}}$ is the initial intensity of *unpolarized* light of wavelength λ in the surrounding medium and α is the particle polarizability. The angle, θ, is measured in the plane of the initial and scattered beam and the intensity I is measured at a distance r from the particle.

To see how this equation arises, it must be understood that in the classical theory of scattering one assumes that the electrons in a material in its undisturbed state are arranged in some 'equilibrium' state that is electrically neutral. When a light wave strikes such a material, the electric vector of the light wave causes the electrons to be displaced and induces a dipole moment (eqn (2.2.16)) whose magnitude is determined by the polarizability, α, of the material. It can be shown (see e.g. Atkins 1978, p. 756; 1982, p. 778) that the polarizability of a sphere, of refractive index n_1 and radius a is proportional to its volume (Exercise 2.3.3):

$$\alpha = 4\pi\varepsilon_0 \cdot a^3(n^2 - 1)/(n^2 + 2) \tag{2.3.2}$$

where n is the refractive index relative to that of the surrounding medium ($n = n_1/n_0$).

Since the particle is small compared to the wavelength of the light one can assume that at any instant in time the entire particle is being subjected to the same electric field. The electric vector of the wave is fluctuating in magnitude with some frequency, ν say, and as it does so, the dipole will fluctuate at the same frequency, though not necessarily in phase (section 2.2). Such a fluctuating dipole must, according to Maxwell's theory, continuously radiate energy in all directions in the form of electromagnetic radiation of frequency ν. This is the scattered light, the intensity of which must fall off as r^{-2} with distance from the dipole. The intensity of the radiation in any direction depends on the square of the electric field and since the field depends on the magnitude of the dipole moment the intensity would depend on a^6 (from eqns (2.2.16) and (2.3.2)). A simple dimensional argument (Kerker 1969) then suggests that the wavelength must enter as an inverse fourth power dependence. Substituting eqn (2.3.2) in (2.3.1) gives, for the total scattering at angle θ, from N_p particles per unit volume:

$$\frac{I_\theta}{I_{0,\mathrm{u}}} = \frac{9\pi^2(n_1^2 - n_0^2)^2}{2\lambda^4 r^2 (n_1^2 + 2n_0^2)^2} v^2 N_\mathrm{p}(1 + \cos^2\theta) \tag{2.3.3}$$

where v is the particle volume. In this equation, the terms 1 and $\cos^2\theta$ refer, respectively, to the vertical and horizontally polarized component of the scattered light (Fig. 2.3.1). Note that when viewed at 90°, only (vertically) polarized light is visible. In fact, if the initial beam is vertically polarized there is Tyndall scattering of this vertically polarized light at 90° in the scattering plane but none in the direction at right angles to that plane (Kruyt 1952, p. 36).

Rayleigh's eqn (2.3.3) is particularly applicable to the scattering from molecules and it is invoked to explain why the sky is blue: scattering from molecules in the air is strongest for light of short wavelength so it is the blue end of the visible spectrum that we see. At sunset it is the transmitted (red) end of the spectrum that is most obvious, and the scattering is augmented by the presence of dust particles and water droplets in the lower parts of the atmosphere.

Few colloidal particles conform to the requirements of the Rayleigh model ($d \ll \lambda$) but eqn (2.3.3) is important because it emphasizes the strong dependence of scattering on particle size, on wavelength and on angle. Since $I_\theta \propto (v^2 N_p)/\lambda^4$ at any particular angle it is apparent that for any particle size (v constant) the scattering increases directly with particle concentration and at a given mass concentration (vN_p constant) of the particles, the scattering will increase with particle size. This second property is extensively used to follow the process of particle aggregation (called *coagulation* or *flocculation* (section 2.5)).

Apart from the direct measure of intensity as a function of angle, the depolarization ratio I_H/I_V (Fig. 2.3.1) can also be used to investigate particle size. As the size increases and a more elaborate theory is

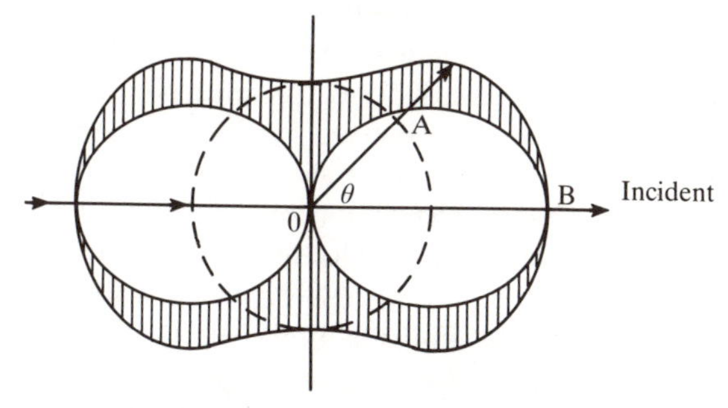

FIG. 2.3.1. Polar diagram of the light scattered in different directions from a particle at O: the inner (clear) region represents unpolarized light, while the crosshatched region represents the (vertically) polarized light. The ratio of the lengths OA/OB is the depolarization ratio ($=I_H/I_V$, the ratio of horizontally to vertically polarized light) and it varies from one for $\theta=0$ to zero for $\theta=90°$. ($I_H/I_V=\cos^2\theta$; Exercise 2.3.4). The equation of the inner boundary curves (where A lies) is $r=2\cos^2\theta$, while B is fixed at $r=2$.

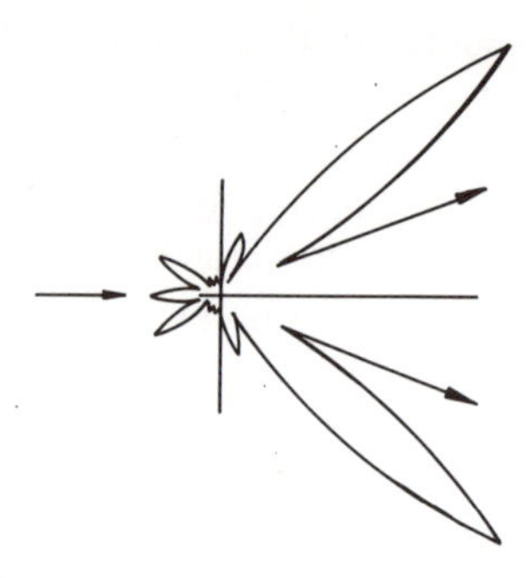

FIG. 2.3.2. Scattering pattern of light from a spherical particle for which $2\pi a/\lambda = 6$ and $n_1/n_0 = 1.44$; calculated by Johnson and La Mer (1947) from the Mie theory equations. (From Kruyt 1952, p. 100, with permission.)

required, it is only the detailed θ-dependence that changes; the dependence of I_θ on v, N_p and λ remains as indicated in eqn (2.3.3).

A second (more versatile) approximation is available if the particle diameter, $d < \lambda$. This is called the Rayleigh–Gans–Debye (RGD) region, and the scattering pattern (I_θ) in this region is more complicated than that for the Rayleigh region ($d \ll \lambda$). It is no longer symmetric about the $\theta = 90°$ line (Fig. 2.3.1) but shows a pronounced preference for forward scattering ($\theta \leqslant 90°$) compared to back scattering ($90° \leqslant \theta \leqslant 180°$) and that preference increases with increasing particle size. Measurement of the dissymmetry ratio ($I_{45°}/I_{135°}$) can therefore give a measure of particle size.

As the particle size increases, the scattering pattern becomes still more complicated as the spherical waves from each scattering centre in the particle increasingly interfere with one another so that the intensity shows pronounced maxima and minima at particular angles, θ, determined by the size parameter $2\pi a/\lambda$ and the particle refractive index (or polarizability) (Fig. 2.3.2). If white light is used for illumination of monodisperse sols, the result is the appearance of strong beams of light of particular colours (chiefly green and red) at particular angles†. These are called *higher-order Tyndall spectra* (HOTS) and their analysis can obviously lead to data on particle size. In this region ($d \sim \lambda$) the complete Mie theory must be used and the experimental scattering pattern compared to the calculated one in order to determine the particle size. The complexity of the scattering pattern clearly limits the use of this method to particles of very simple shape. A much more rewarding procedure is the use of

† This phenomenon should not be confused with the spectral colours which are produced by well dialysed monodisperse sols. In these systems it is the regular spacing *between* particles which causes the spectrum by Bragg diffraction, just as the regular atomic spacing in a crystal causes X-ray diffraction when the wavelength and spacing are comparable.

quasi-elastic light scattering (QELS) or *photon correlation spectroscopy* (PCS), which enables the diffusion coefficient to be estimated directly.

2.3.2 *Photon correlation spectroscopy (PCS or QELS)*

This procedure depends on the fact that, if the scattering particle is moving when the light photon hits it then the re-radiated light will have a slightly different frequency when viewed by a stationary observer. This is why it is called *quasi-elastic* scattering. The frequency is slightly increased or decreased depending on whether the particle is moving towards or away from the observer. This is called *Doppler broadening* (Fig. 2.3.3) and if it can be measured accurately it provides a means of determining the diffusion coefficient of the particle. The intensity I of the scattered light of frequency ω can be represented (see Fig. 2.3.3 and Exercise 2.3.2) by (Ware 1974):

$$I(\omega) = A_1 \frac{\mathcal{D}Q^2}{(\omega - \omega_0)^2 + (\mathcal{D}Q^2)^2} \tag{2.3.4}$$

where ω_0 is the centre frequency (i.e. the frequency of the incident radiation), A_1 is a constant and Q is the magnitude of the scattering vector:

$$Q = \frac{4\pi n_0}{\lambda} \sin(\theta/2) \tag{2.3.5}$$

where n_0 is the refractive index of the medium and θ is the scattering

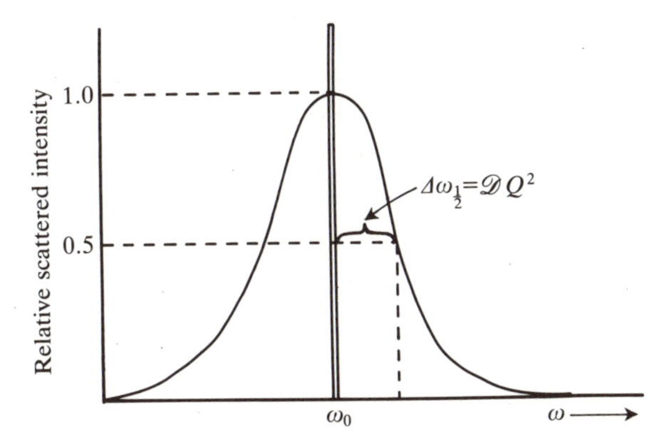

FIG. 2.3.3. Doppler line broadening. The scattered intensity at a particular angle from a *stationary* particle is given by the central sharp line. If the particle is moving the line becomes (frequency) broadened and the intensity at the central frequency (ω_0) is reduced. In colloidal systems the broadening is a direct reflection of the random motion of the particles: it amounts to about 10^3 Hz in a background $\omega_0 \simeq 10^{14}$ Hz.

angle. Q measures how strongly the light interacts with the particle. Small Q values (i.e. small θ) correspond to glancing incidence of the radiation with small momentum transfer from photon to particle and Q is a maximum for head-on collision. (The momentum transfer itself is of small consequence to the particle in this case since the photon momentum is so small, but the notion of momentum transfer in scattering processes becomes more significant when neutrons are used for the beam (Chapter 14).

The width of the peak at half height ($\Delta\omega_{1/2}$) is directly related to the diffusion coefficient (Exercise 2.3.2):

$$\Delta\omega_{1/2} = \mathscr{D}Q^2 \tag{2.3.6}$$

The function shown in Fig. 2.3.3 is called a Lorentzian (as distinct from a Gaussian; Fig. 2.1.2) distribution and is frequently encountered in radiation and absorption theory (recall Exercise 2.2.7).

It is the advent of laser sources with precisely defined frequencies that has made it possible to exploit the potentialities suggested in eqn (2.3.4). By taking the scattered signal and mixing (heterodyning) it with the incident radiation a beat frequency ($\omega - \omega_0$) is obtained and the intensity of that signal is given by eqn (2.3.4), from which $\mathscr{D}$ and consequently a mean radius can be calculated.

Exercises. 2.3.1 Describe briefly what occurs when visible light is absorbed by a particle. In what way is the energy usually stored? Why does re-radiation usually occur at a longer wavelength? When is this not the case? If re-radiation does occur at the incident wavelength is the effect on the incident beam noticeable? Why?

2.3.2 Establish eqn (2.3.6) from eqn (2.3.4).

2.3.3 Use the definitions of α and P to find a link between α and $\varepsilon(\omega)$. In a region where there is no absorption, show that

$$\alpha \simeq \varepsilon_0(n_1^2 - 1)/N$$

for a collection of gas molecules of refractive index n_1 with N molecules per unit volume. Show that this agrees closely with eqn (2.3.2) for a condensed medium (where $N = (\frac{4}{3}\pi a^3)^{-1}$) if $n_1/n_0 \approx 1$.

2.3.4 Show that the fraction of light that is unpolarized in direction θ is $2\cos^2\theta/(1+\cos^2\theta)$. Verify Fig. 2.3.1 where the length of the line from the origin to any curve is proportional to intensity. Sketch in the shape of the curve representing I_H. (The curve of I_V is shown as a broken circle.)

2.4 Response to a mechanical stress

2.4.1 *The rheology of colloidal materials*

Rheology is the study of the deformation that occurs when a material is subjected to a stress. The stress (force per unit area) can be applied in

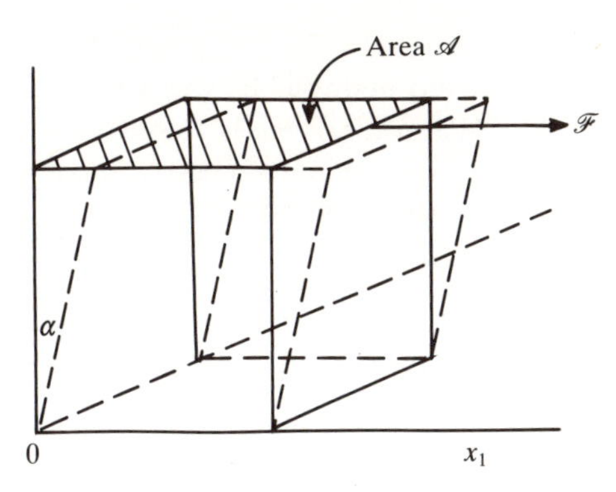

FIG. 2.4.1. Application of a shearing stress $\mathscr{S}(=\mathscr{F}/\mathscr{A})$ to a material, produces a strain $\gamma = \tan \alpha$.

various ways: as a compression, as a tension or as a shearing process (Fig. 2.4.1). In compression and tension, dilute colloidal dispersions behave very much like simple liquids, especially if the particles are rigid and/or incompressible. Only in highly concentrated dispersions does one encounter unusual behaviour under tension, while under compression most condensed materials (solid or liquid) behave rather similarly. On the other hand, even quite dilute colloidal dispersions can exhibit very unusual behaviour when subjected to a shearing stress. In particular, the simple distinction between solid (elastic) and liquid (viscous) materials becomes blurred and a whole range of intermediate behaviour patterns is exhibited. Indeed, it is often these very unusual deformation properties that are sought after in the application of a colloidal dispersion. Consider, for example, the way the 'apparent viscosity' of a paint changes during its application: it is high when the paint is held in the brush but flows freely when sheared against the surface to be painted; it must quickly increase in viscosity so that it does not run down (drip or sag) under gravity but must flow sufficiently to eliminate the brush marks. The dependence of the viscosity on time and the shearing stress to which it is subjected determine the success or otherwise of the paint. Even more stringently controlled flow characteristics are required for high speed processes like newspaper printing, paper making, electronic component dipping and encapsulation, and the preparation of photographic film and magnetic tape etc. In all of these situations (and many more) it is the rheological character of a colloidal dispersion that is important.

Consider the simple shearing regime illustrated in Fig. 2.4.1. The lower plate is held stationary and the upper plate is pulled by a force, $\mathscr{F}$ acting in the x direction over an area $\mathscr{A}$. The force per unit area or *traction*† (shearing stress) applied to the material between the plates will cause a deformation (or strain) γ. When the force is removed, we find that:

(a) the material returns to its original shape (*elastic* recovery);
(b) the material remains in the new position (*flow* has occurred); or
(c) some partial recovery occurs.

These three behaviour patterns are characteristic of solids, liquids, and plastic materials, respectively, but in practice, most materials can exhibit any or all of them, depending on the time scale involved in the application of the stress and the measurement of its effects. The time scale is measured by the Deborah number (Harris 1977, p. 21):

$$D_N = \frac{\text{relaxation time of material}}{\text{time of observation}}. \tag{2.4.1}$$

As $D_N \to 0$, materials tend to behave more like fluids and as $D_N \to \infty$ they behave like solids. Thus geologists can speak of the 'flow' of rocks over geological time, while a person falling from a great height into water encounters its solid-like characteristics when deformed over short time intervals.

2.4.2 *Ideal solids and liquids*

The ideal behaviour for an elastic solid experiencing a tensile (stretching) stress $\mathscr{S}_T$, is described by Hooke's Law:

$$\mathscr{F}/\mathscr{A} \propto \gamma \quad \text{or} \quad \mathscr{F}/\mathscr{A} = \mathscr{S}_T = Y\gamma$$

where Y is called the *Young's modulus* of the material. The strain, γ, is in that case the relative change in length. The corresponding behaviour under a shearing stress is described thus:

$$\mathscr{F}/\mathscr{A} \propto \gamma \quad \text{or} \quad \mathscr{F}/\mathscr{A} = \mathscr{S} = G\gamma \tag{2.4.2}$$

where G is the *shear modulus* of the material and γ is defined in Fig. 2.4.1. Many solids conform to eqn (2.4.2) for small stresses and provided the stress remains below some upper limit $\mathscr{S}_L$, they will recover their original shape completely when the stress is removed. If $\mathscr{S} > \mathscr{S}_L$ the material suffers permanent deformation, i.e. some *flow* or *creep* occurs

† This quantity ($\mathscr{F}/\mathscr{A}$) is usually referred to as the *shearing stress* and we will use that expression henceforth. Strictly speaking however, the traction is a vector quantity whilst the stress at a point on a surface is a *tensor* that is the aggregate of all the tractions acting on all surface elements of different orientations that contain the point (Reiner 1960, p. 5). We are assuming then that we can consider only one component of the stress on the upper surface. (The *stress tensor* will be introduced in Chapter 9.)

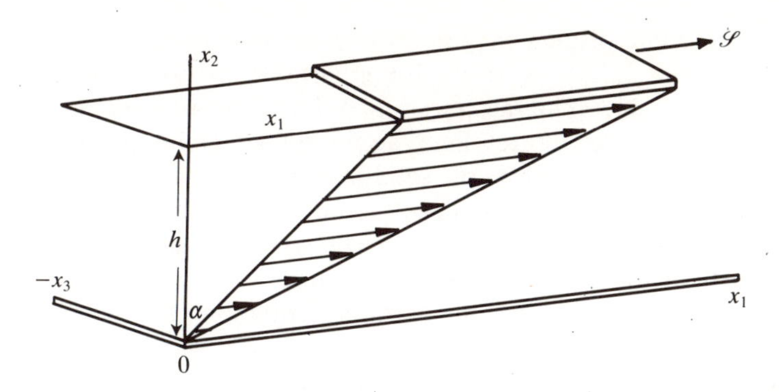

FIG. 2.4.2. Deformation (flow) of a liquid under an applied shearing stress $\mathscr{S}$. If the velocity of the upper plate is v then

$$\dot{\gamma} = \frac{d(\tan \alpha)}{dt} = \frac{1}{h}\frac{dx_1}{dt} = \frac{v}{h}.$$

(This coordinate system is used throughout Chapter 9.)

and the solid has begun to exhibit some of the characteristics of a *plastic* or a liquid.

Ideal liquid-like behaviour is described as *Newtonian behaviour* and in that case the applied shearing stress is directly proportional to the time *rate of strain* or *rate of shear* ($\dot{\gamma} = d\gamma/dt$):

$$\mathscr{S} \propto \dot{\gamma} \quad \text{or} \quad \mathscr{S} = \eta\dot{\gamma} \tag{2.4.3}$$

where the proportionality constant, η, is the (first coefficient of shear) *viscosity*. This equation was proposed by Newton to describe the flow behaviour of simple fluids (gases and liquids like water) undergoing steady shear. Consider the system shown in Fig. 2.4.2 where a simple liquid is confined between two plates. The lower plate is stationary and the upper plate is being pulled at a velocity v by the application of a force per unit area, $\mathscr{S}$. It may be assumed that the liquid in contact with the lower plate remains stationary whilst that in contact with the upper plate must move with the velocity v. Between the two a gradual variation in velocity occurs. The rate of shear, $\dot{\gamma}$, for this simple shear regime is equal to the velocity gradient v/h and is normally measured in s^{-1}. In the more general case $\dot{\gamma} = dv/dx_2$.

2.4.3 *The general response to a shearing stress*

The study of colloidal dispersions (including the more concentrated systems described as *slurries* or *pastes*) reveals that ideal behaviour (either solid-like or liquid-like) is the exception rather than the rule.

Even quite dilute dispersions can show departures from Newtonian liquid behaviour (eqn 2.4.3), especially if the particles are anisometric (like, for example, the clay minerals (section 1.5.5)). Measurements at higher concentration or on shorter time scales (higher Deborah number) may begin to reveal evidence of solid-like (elastic) behaviour in what otherwise appears to be a (viscous) liquid. Such materials are said to be *visco-elastic* and they may be intrinsically solids or liquids, depending on which of the two characteristics is dominant.

For materials of this sort, the entire deformation history may, to some extent, influence its present structure. Once again we encounter the need for a *memory function* (as in section 2.2.3) to properly describe the behaviour. An introduction to the theory is given by Ferry (1980) from which the remainder of this section is taken.

Again we restrict ourselves to the *linear theory* (compare eqn (2.2.19)) in which the effects of sequential changes in strain are assumed to be additive:

$$\mathscr{S}(t) = \int_{-\infty}^{t} G(t - t')\dot{\gamma}(t')\,\mathrm{d}t'. \tag{2.4.4}$$

Note that in this formulation of the memory process $t - t' = \tau$, so that t is the total elapsed time and the integration is carried out over all past times up to the current time t. $G(t)$ is called the (shear) *relaxation modulus* of the material.

If the function $G(t)$ approaches zero for very large t, then the substance is liquid-like and an alternative formulation of eqn (2.4.4) can be given in terms of the strain, rather than the rate of strain (Exercise 2.4.1):

$$\mathscr{S}(t) = -\int_{-\infty}^{t} m(t - t')\gamma(t, t')\,\mathrm{d}t' \tag{2.4.5}$$

in which $m(t)$, the *memory function* is $-\mathrm{d}G(t)/\mathrm{d}t$. Equation (2.4.5) is arrived at by integrating eqn (2.4.4) by parts, using the fact that $\gamma(t, t') = \int_t^{t'} \dot{\gamma}(t'')\,\mathrm{d}t''$ and taking the state of the material at time $t'' = t$ as the reference state (Bird *et al.* 1977).

If $G(t)$ remains finite for large t then the substance is solid-like and eqn (2.4.5) contains less information than eqn (2.4.4) and must be augmented by another term.

Equations like (2.4.4) and (2.4.5) are called *constitutive equations* and, as in the case of the dielectric response function, they can in principle contain all the information about the relation between an imposed simple shear stress and the resulting strain. Another form of constitutive equation links the resulting strain to the history of the time derivative of the stress:

$$\gamma(t)=\int_{-\infty}^{t} J(t-t')\dot{\mathcal{S}}(t')\,\mathrm{d}t' \tag{2.4.6}$$

where $\dot{\mathcal{S}}=\mathrm{d}\mathcal{S}/\mathrm{d}t$ and $J(t)$ is called the *creep compliance.*

A knowledge of either $G(t)$, $m(t)$ or $J(t)$ can be used to predict the effect of some imposed shear stress, provided it is not too large, nor applied too rapidly. A few typical experimental situations will now be examined (Ferry 1980):

(a) *Stress relaxation after a sudden strain.* If a sudden strain, γ, is imposed on a material over a short period of time, ξ, (Fig. 2.4.3) then, from eqn (2.4.4):

$$\mathcal{S}(t)=\int_{t_0-\xi}^{t_0} G(t-t')(\gamma/\xi)\,\mathrm{d}t' \tag{2.4.7}$$

since $\dot{\gamma}$ is zero outside of this range.

The mean value theorem tells us that the value of the integral can be written (Fig. 2.4.4 and Exercise 2.4.2):

$$\mathcal{S}(t)=\frac{\gamma}{\xi}\cdot\xi G(t-t_0+(1-e)\xi)\quad\text{for}\quad 0\leqslant e\leqslant 1.$$

For $t_0=0$ we then have, writing $e'=1-e$:

$$\mathcal{S}(t)=\gamma G(t+e'\xi)\simeq\gamma G(t) \tag{2.4.8}$$

for times that are long compared to ξ. The ratio of a stress to the corresponding strain is called a *modulus* (section 2.4.2) and for a perfectly elastic body, the equilibrium shear modulus $G=\mathcal{S}/\gamma$ from equation (2.4.2). $G(t)$ is then the time-dependent analogue of G, measured in an experiment with this sort of time pattern.

Provided the strain γ is not too large, the value of $G(t)$ should be independent of γ and it is only then that this simple procedure is of value. The concept can be applied to liquid-like and solid-like materials.

(b) *Stress relaxation after cessation of steady shear flow.* As noted earlier, liquid-like materials can be grossly deformed and still retain some structural characteristics. They can be deformed at a constant strain rate $\dot{\gamma}$, under a steady shearing stress $\mathcal{S}$ where $\mathcal{S}=\dot{\gamma}\eta_0$. ($\eta_0$ is used for the viscosity to signify that it is measured at sufficiently low shear rates for the behaviour to be linear). If the flow is suddenly stopped, the shearing stress decays with time and it can be shown (Exercise 2.4.3) that, for $t_0=0$ (Fig. 2.4.5):

$$\mathcal{S}(t)=\dot{\gamma}\int_{t}^{\infty} G(\tau)\,\mathrm{d}\tau. \tag{2.4.9}$$

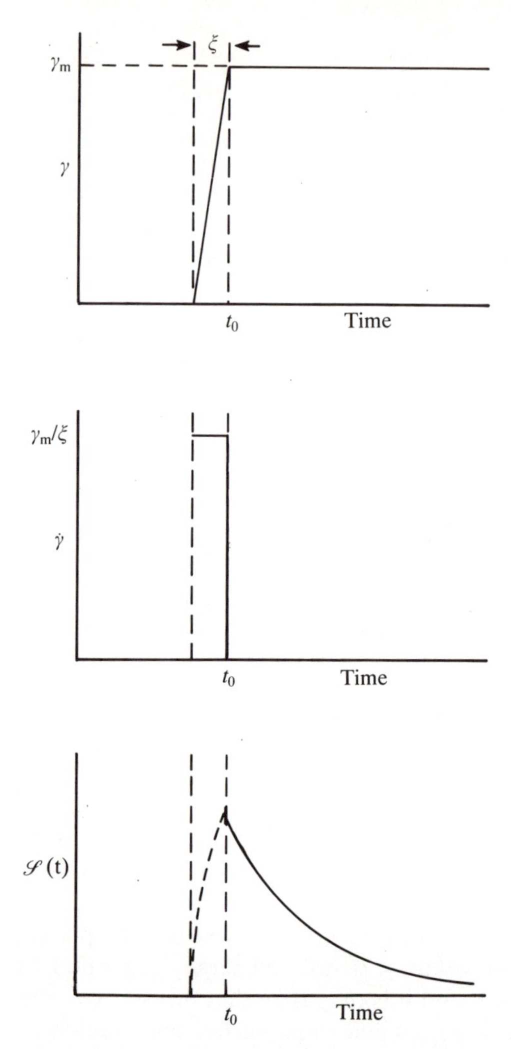

FIG. 2.4.3. Time profiles for a simple stress relaxation experiment following a sudden strain. The behaviour of $\mathscr{S}$ for $-\xi < t < t_0$ is not usually accessible but $\mathscr{S}(t)$ can be followed for $t > t_0$.

(c) *Creep after a sudden stress.* The opposite experiment to that in (a) above is to apply a sudden stress to a material and then to hold it constant and to follow the resulting strain as the material accommodates itself (Fig. 2.4.6). One can again use the mean value theorem on eqn (2.4.6) to show that (for $t_0 = 0$) (Exercise 2.4.4):

$$\gamma(t) = \mathscr{S}J(t + e'\xi) \simeq \mathscr{S}J(t) \tag{2.4.10}$$

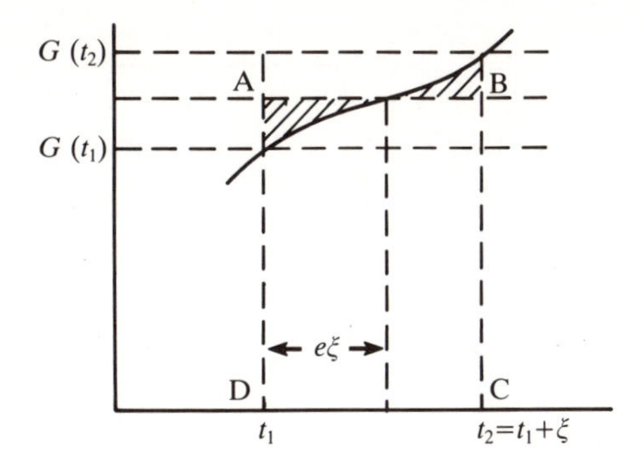

FIG. 2.4.4. The mean value theorem establishes the fact that there is a horizontal line that can be drawn in such a way as to equalize the shaded areas. Then $\int_{t_1}^{t_2} G\,\mathrm{d}t =$ area ABCD $= \xi G(t_1 + e\xi)$ for some $e(<1)$.

so that $J(t)$ is the reciprocal of a modulus. It is a monotonically non-decreasing function of time and for a perfectly elastic solid $J = 1/G$. However, $J(t) \neq 1/G(t)$ because the time course of the two experiments is different.

There are many other types of transient experiment that can be

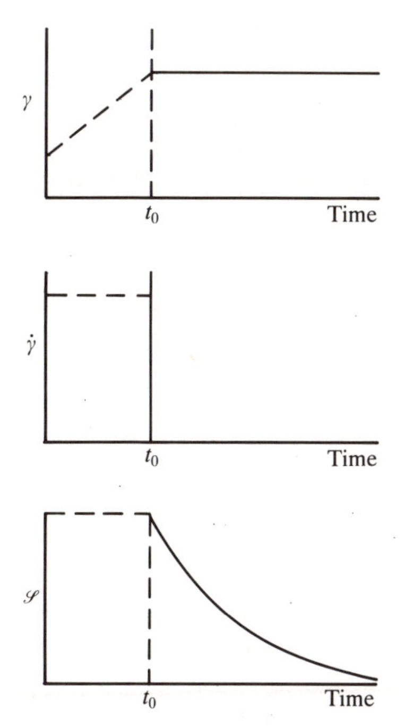

FIG. 2.4.5. A simple shear stress relaxation experiment, following cessation of flow at time $t = t_0$.

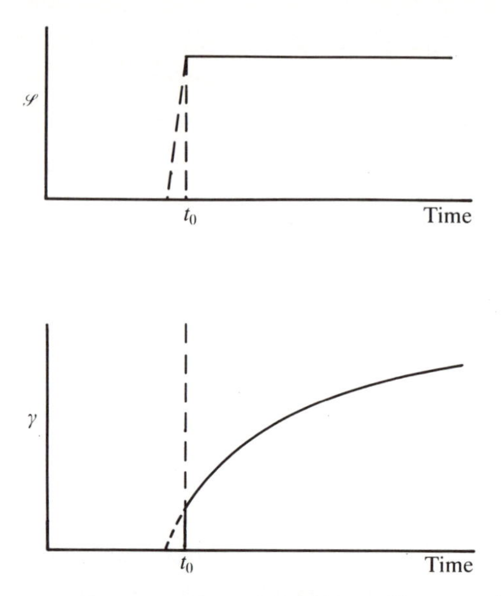

FIG. 2.4.6. A creep compliance experiment in which a sudden stress is imposed and then maintained constant while the resulting strain is measured.

performed and they all yield some information about the mechanical response of the material. Choice of the most appropriate measurement depends on what information is sought. Another type of experiment that provides useful information on these viscoelastic materials is one in which an oscillating shear regime is imposed on the system. Considering

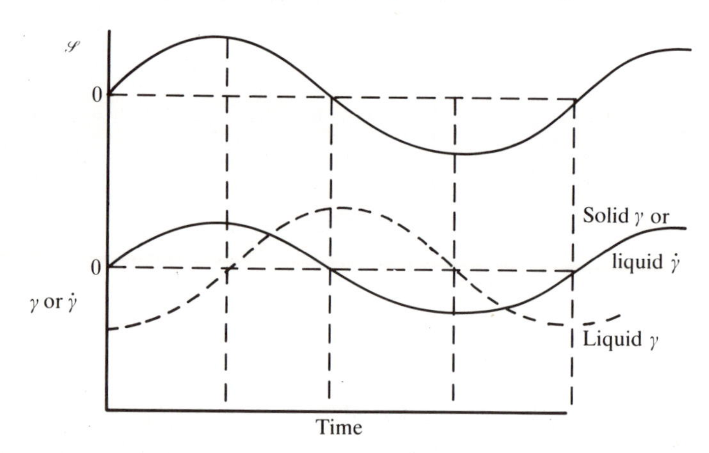

FIG. 2.4.7. Stress–strain relation for an ideal solid and an ideal liquid subjected to a sinusoidal stress. Note that for the liquid it is the *strain rate* ($\dot{\gamma}$) that is in phase with the stress and this is shown by the full curve (which also corresponds to the strain for an elastic solid). The stress in the liquid leads the strain by $\pi/2$.

the discussion in section 2.2 it should come as no surprise to find that once again the use of a complex variable enables us to keep track simultaneously of the elastic (storage) and viscous (dissipative) characteristics of the material (Fig. 2.4.7).

2.4.4 *Response to an oscillating shear field*

Suppose a material, described by eqn (2.4.4), is caused to undergo a periodic strain of frequency $\omega(\mathrm{rad\,s^{-1}})$:

$$\gamma = \gamma^0 \sin \omega t \tag{2.4.11}$$

where γ^0 is the maximum amplitude of the strain. Then

$$\dot{\gamma} = \omega\gamma^0 \cos \omega t \tag{2.4.12}$$

and substituting in eqn (2.4.4) with $t - t' = \tau$ we have

$$\begin{aligned}\mathcal{S}(t) &= \int_0^\infty G(\tau)\omega\gamma^0 \cos \omega(t-\tau)\, \mathrm{d}\tau \\ &= \gamma^0\left[\omega\int_0^\infty G(\tau)\sin \omega\tau\, \mathrm{d}\tau\right]\sin \omega t + \gamma^0\left[\omega\int_0^\infty G(\tau)\cos \omega\tau\, \mathrm{d}\tau\right]\cos \omega t\end{aligned} \tag{2.4.13}$$

(since $\cos(A-B) = \cos A \cos B + \sin A \sin B$ and t is a constant in the integration process).

The integrals will converge if $G \to 0$ as $\tau \to \infty$, and this will be so if the material is liquid-like since, by definition, a liquid cannot permanently support a shearing stress. The terms in square brackets are functions of ω but not of t and we can write

$$\mathcal{S}(t) = \gamma^0(G' \sin \omega t + G'' \cos \omega t) \tag{2.4.14}$$

where $G'(\omega)$ represents a modulus that measures the ratio of the *in-phase* stress to the strain. This is the shear *storage* modulus. The quantity $G''(\omega)$ likewise measures the ratio of the stress to the strain which is 90° out of phase; it is the shear *loss* modulus. The nomenclature here reflects the result obtained in Exercise 2.2.5: the lossy (dissipative) part of the process is represented by G''.

For a purely elastic solid the stress and strain remain in phase and so $G'' = 0$ and $G' = G$. For a purely viscous liquid it is the *rate of strain* $\dot{\gamma}$ that remains in phase with the applied stress, according to equation (2.4.3), and the material has no elastic (storage) character so $G' = 0$. Then

$$\frac{\mathcal{S}}{\dot{\gamma}} = \frac{\mathcal{S}}{\omega\gamma^0 \cos \omega t} = \frac{G''}{\omega} = \eta. \tag{2.4.15}$$

If one wants to characterize a viscoelastic material in terms of viscosity, it is necessary to use a complex viscosity function:

$$\eta = \eta' + i\eta'' \tag{2.4.16}$$

and then in eqn (2.4.15) $\eta = \eta'$ for purely viscous behaviour. For the viscoelastic material

$$\eta' = G''/\omega \quad \text{and} \quad \eta'' = G'/\omega. \tag{2.4.17}$$

An alternate method of describing a viscoelastic fluid (Fig. 2.4.8) is in terms of the phase lag, δ, between stress and strain. Writing

$$\begin{aligned}\mathcal{S} &= \mathcal{S}^0 \sin(\omega t + \delta) \\ &= \mathcal{S}^0 \cos\delta \sin\omega t + \mathcal{S}^0 \sin\delta \cos\omega t\end{aligned} \tag{2.4.18}$$

and comparing with eqn (2.4.14) we see that

$$G' = \frac{\mathcal{S}^0}{\gamma^0}\cos\delta; \qquad G'' = \frac{\mathcal{S}^0}{\gamma^0}\sin\delta \tag{2.4.19}$$

and

$$G''/G' = \tan\delta. \tag{2.4.20}$$

An oscillating measurement at frequency ω corresponds to a transient measurement over a time $t = 1/\omega$ and the result obtained gives two pieces of information: the ratio of the amplitudes of stress to strain ($\mathcal{S}^0/\gamma^0$) and the phase lag, δ, or alternatively the values of G' and G'' or of η'' and η' (from eqn (2.4.17)).

The alternative representation in terms of a complex strain $\gamma^* = \gamma^0 e^{i\omega t}$

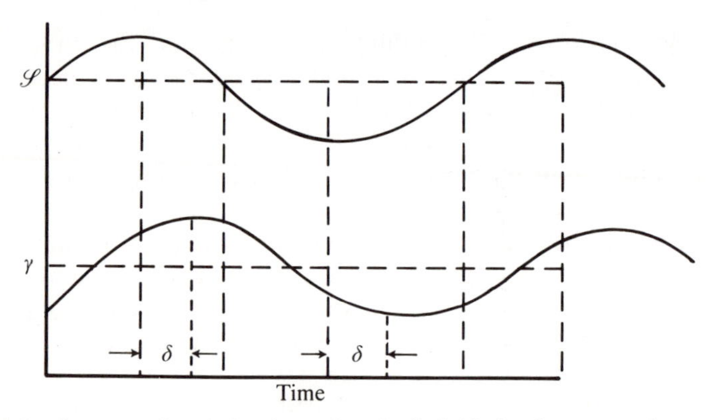

FIG. 2.4.8. Stress–strain relation for a viscoelastic fluid. $\delta = 0$ corresponds to elastic solid and $\delta = \pi/2$ corresponds to viscous liquid.

and the corresponding complex stress $\mathcal{S}^* = \mathcal{S}^0 e^{i(\omega t+\delta)}$ yields a complex modulus $G^* = \mathcal{S}^*/\gamma^*$ where $G^* = G' + iG''$ and $|G^*| = (G'^2 + G''^2)^{1/2}$.

Polymer solutions show very pronounced viscoelastic behaviour and much experimental work has been done on the phenomenon (see Bird *et al.* 1977; Ferry 1980). The elastic component is much smaller in most colloidal dispersions but coagulated colloidal sols do show some viscoelasticity (van de Ven and Hunter 1979) and we will return to that behaviour in Chapter 18.

2.4.5 *Viscous shear behaviour*

For many colloidal dispersions, the elastic effects play a rather secondary role in the behaviour, especially if the system is being sheared very strongly. Thus, in the pumping of a slurry or the high speed extrusion of magnetic iron oxide paste to make recorder tape, we are more concerned with the viscous (dissipative) aspects of the flow behaviour, even though the storage or elastic properties have to be recognized. The fundamental assumptions of linearity, small strains and small strain rates on which the memory equations (eqns (2.4.4)–(2.4.6)) are based no longer apply. In one sense this makes the analysis easier because the majority of the time effects may then have disappeared.

In many such situations it is sufficient, at least as a first step, to investigate the relationship between shear stress and shear rate, to obtain a more general form of equation (2.4.3) in which the viscosity, η, is no longer a constant. In the simplest case, when the experimental measurement is conducted on a time scale which is long compared to the relaxation time of the system, the viscosity becomes independent of time, though in general it will depend on the shear rate. We can then define an *apparent viscosity*, η_{app} as

$$\eta_{app}(\dot{\gamma}) = \mathcal{S}/\dot{\gamma} \tag{2.4.21}$$

or a *differential viscosity*:

$$\eta_{diff}(\dot{\gamma}) = d\mathcal{S}/d\dot{\gamma}. \tag{2.4.22}$$

Some of the behaviour patterns commonly exhibited by colloidal dispersions are shown in Fig. 2.4.9. If η_{app} and η_{diff} *both* decrease regularly with shear rate (curve 2) the behaviour is called *pseudoplastic* and if they both increase (curve 3) it is *dilatant*. If the material behaves like a solid until a certain value of stress is reached and then deforms like a Newtonian liquid obeying:

$$\mathcal{S} - \mathcal{S}_B = \eta_{PL}\dot{\gamma} \tag{2.4.23}$$

where η_{PL} is constant, this is called *Bingham behaviour* and it is the

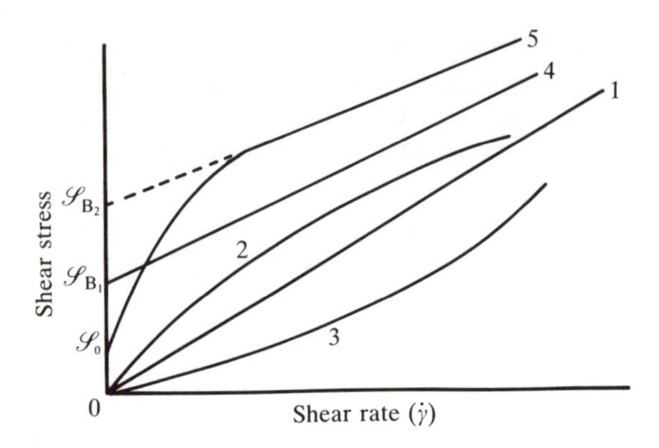

FIG. 2.4.9. Common forms of flow behaviour for colloidal dispersions: (1) Newtonian; (2) shear thinning (pseudoplastic); (3) shear thickening (dilatant); (4) (ideal) Bingham plastic, (5) non-ideal plastic. $\mathscr{S}_0$, the *primary yield value,* is usually small and may be zero. (Finite $\mathscr{S}_0$ values can arise very easily as instrument artefacts.)

'ideal' standard for plastic behaviour. η_{PL} is called the *plastic viscosity* and $\mathscr{S}_B$ is called the *Bingham yield value.* In this case η_{diff} is constant but η_{app} decreases continuously from its zero shear value ($\eta_0 \rightarrow \infty$) to some limiting (infinite shear) value η_∞. This type of behaviour is observed in, among other things, concentrated dispersions (slurries) of coal in water at fairly high volume fractions and low shear rates.

Newtonian behaviour is observed in dilute stable dispersions of spherical particles. Pseudoplastic behaviour occurs in dispersions of anisometric particles because the increasing shear rate tends to orient the particles along the fluid streamlines and this decreases the viscosity. Dilatancy is common in concentrated dispersions; in this case it is the flow of (lubricating) fluid between the particles that dominates the behaviour and the shearing process tends to drive particles together and constrict the flow channels.

Non-ideal plastic behaviour (Fig. 2.4.9, curve 5) is characteristic of coagulated colloidal dispersions in which every collision between two particles results in the formation of a (temporary or permanent) link. The extrapolated value of stress obtained from the linear high shear rate behaviour is again called a *Bingham yield value* and the differential viscosity at high shear rate can be called a *plastic viscosity*. We will discuss these forms of flow behaviour, and their interpretation in more detail in Chapter 18.

If the time scale on which the measurements are conducted is sufficiently short (that is, comparable with or shorter than the characteristic relaxation time for the structure of the flow units) then one may also observe changes in the apparent or differential viscosity *as a function of time* even at constant shear rate. The behaviour shown in Fig. 2.4.9 can then be regarded as the steady-state behaviour, arrived at after the system has had sufficient time to relax (i.e. to establish a (dynamic) structure in response to the shearing stress). Some systems show a gradual decrease and some a gradual increase in apparent or differential viscosity with time at a fixed rate of shear. These behaviour patterns are referred to as *thixotropy* and *rheopexy* respectively. Thixotropy occurs commonly in dispersions of very anisometric particles (e.g. montmorillonite (bentonite), section 1.5.5), especially at moderate and high shear rates. The increasing shear rate requires an appreciable time to break down the particle linkages to establish the steady-state structure for that shear rate. Rheopexy usually occurs under conditions of gentle stirring where the slight degree of agitation apparently allows particles to establish extensive structures, which resist the shearing action.

Exercises. 2.4.1 Establish eqn (2.4.5) from eqn (2.4.4).
2.4.2 Establish eqn (2.4.8).
2.4.3 Show that in the stress relaxation experiment of Fig. 2.4.5

$$\mathcal{S}(t) = \dot{\gamma} \int_{t-t_0}^{\infty} G(\tau)\,\mathrm{d}\tau.$$

2.4.4 Establish eqn (2.4.10).

2.5 Electrical charge and colloid stability

Up to this point we have regarded the colloidal system as consisting, for the most part, of a large number of individual particles dispersed in some continuous (usually fluid) medium. Since the particles, if small enough, will be undergoing Brownian motion, they will be continually colliding with one another and they will remain as individual particles only if those collisions do not result in permanent associations. As we noted in section 1.4 the systems of most concern to us here are what are called *lyophobic* or irreversible colloidal systems, for which the free energy is lowest when the particles are all condensed together into one large lump. Such systems can remain as individual particles for an appreciable time only if some mechanism prevents aggregation during a collision. There are two ways in which that can be done:

(a) the particles can be given an electric charge (either positive or negative) and if all have the same charge they will repel one another more or less strongly when they approach (Fig. 2.5.1);

(b) the particles can be coated with an adsorbed layer of some material (say a polymer), which itself prevents their close approach.

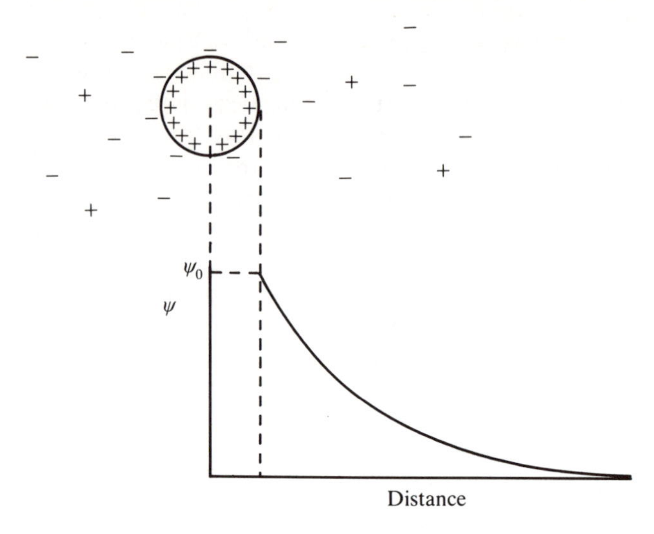

FIG. 2.5.1. The electrostatic potential near an electrically charged colloidal particle.

Mechanism (a) is referred to as *electrostatic stabilization* and mechanism (b) as *steric stabilization,* where the term stability refers to the aggregation behaviour. A system is colloidally unstable if collisions result in the formation of aggregates. Such a process is called *coagulation,* or sometimes flocculation. (We will reserve the term flocculation to refer to aggregation induced by a polymer, section 2.6.2.) The coagulation of a colloidal sol is such a striking process, and it leads to such profound changes in the behaviour of the system, that much work has gone into its study. We will review some of the modern work in that area in Chapter 7. At this point, we need introduce only a few general considerations.

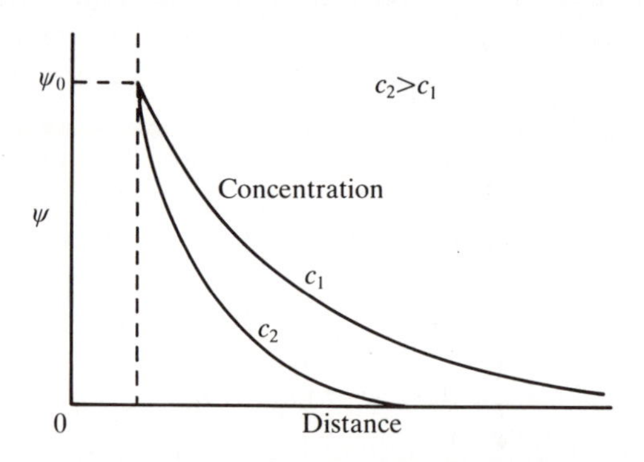

FIG. 2.5.2. Influence of electrolyte concentration on electrostatic potential.

2.5.1 *The electrical charge at a surface*

It was noted in section 1.5.5 that clay mineral particles are electrically charged as a consequence of crystal lattice defects and that those charges are balanced by the presence of ions of opposite sign (*counterions*) adsorbed on the basal cleavage planes. It has been known since the last century that most colloidal particles are electrically charged and will migrate in one or other direction when subjected to an electric field.

Indeed, it appears that, at any interface between two phases, there is a tendency for charges (electrons or ions) to accumulate. Since the relative affinities of cations and anions for the two phases are, in general, different, one phase tends to acquire a positive and the other a negative charge.

In the case of an ionic crystalline solid, like silver iodide, immersed in water, the origin of the ions is obvious and the charge on the crystal surface can be controlled by controlling the activity of the Ag^+ or I^- ions in the solution. Those two quantities are, of course, connected by the relation:

$$a_{Ag^+} \cdot a_{I^-} = K_S \qquad (2.5.1)$$

where K_S is the solubility product of silver iodide, but they can be individually varied by the addition of say, silver nitrate or potassium iodide solutions. The ions Ag^+ and I^- are referred to as *potential-determining ions* in this case since their activity can be used to control the electrostatic potential on the particle surface. This is an important concept about which we will have more to say in Chapter 6. For protein molecules, the state of charge is determined by the concentration of H^+ and OH^- (i.e. the pH) and these two ions are also the potential-determining ions (p.d.i.s) for metal oxide, polymer latex, and many other surfaces.

Figure 2.5.1 shows schematically what the electrostatic potential ψ, in the vicinity of a positively charged particle is expected to look like. (The potential profile is similar to that near an ion in an electrolyte solution.) The negative counterions are attracted towards the particle by the electric field generated by the positive charges but they are also subject to thermal motion, which tends to distribute them uniformly through the surrounding medium. The end result is usually a compromise in which a few negative ions are bound strongly to the particle surface and the concentration of the remainder falls off gradually away from the particle until it approaches the bulk concentration at distances of the order of some tens of nanometres. This charge arrangement is referred to as the *diffuse electrical double layer* around the particle and its extent depends on the electrolyte concentration: increasing electrolyte concentration causes the diffuse double layer to shrink closer in to the particle, so that the electrostatic potential falls off more quickly with distance (Fig. 2.5.2. This is sometimes called 'double layer compression'.) The reader may be familiar with these ideas from the Debye–Hückel theory of strong electrolytes, but they were developed and applied to electrode surfaces before the advent of that theory.

2.5.2 *Observation of coagulation behaviour*

There are two levels on which coagulation can be studied: the microscopic and the macroscopic. At the microscopic level we would seek to follow the individual collisions between pairs of particles, to determine the rate of formation of doublets, triplets etc. and hope to relate this to the (diffusive) motion of the particles and the forces between them. We already noted in section 2.3.1 that observation of light scattering from colloidal dispersions can be used to follow these early stages in the coagulation process and we will discuss such work in Chapter 7.

At this stage we need only note that such observations confirm the notion that in lyophobic colloidal dispersions some such coagulation is always going on, even at very low electrolyte concentrations when the sol appears, macroscopically, to be quite stable. This is, of course, what is expected if the sol is, as suggested in section 1.3, thermodynamically metastable with respect to aggregation and rendered 'stable' only in a kinetic sense, i.e. the *rate* of coagulation can be extremely slow. In this regime the presence of a *stabilizer* (either an electric charge or an adsorbed layer) sets up a repulsion barrier between approaching particles. This decreases the efficiency of collisions so that only one in, say 10^5, or 10^{10}, or even more collisions actually results in a permanent particle–particle contact.

If coagulation is allowed to continue for long enough, the particle aggregates become large enough to be macroscopically visible. They are then called *flocs* and if they differ in density from the surrounding medium they will settle quite rapidly leaving a more or less clear supernatant. If the floc density is lower than the surrounding medium, as is usual for oil in water emulsions, the aggregate accumulates at the top—a process called *creaming,* for obvious reasons. The complete description of this more complicated aggregation process is more difficult than the consideration of the initial phases of the coagulation process but some useful general remarks can be made (section 2.5.4).

2.5.3 *Coagulation by potential control*

One method of determining the iodide ion concentration in a solution is to titrate the unknown with a solution of silver nitrate of known concentration and to follow the precipitation of the AgI. The end point can be detected in a number of ways (e.g. by adding a little CrO_4^{2-}, which will precipitate as red Ag_2CrO_4 only after all of the AgI has precipitated). What concerns us here is the appearance of the precipitate in the neighbourhood of the end point.

Silver iodide is a very insoluble material, and when Ag^+ and I^- ions are mixed in solution, the precipitation occurs rapidly and the particle size is very small—frequently of colloidal dimensions. The smooth milky appearance of the initial AgI precipitate is characteristic of a stable colloidal dispersion. The sol is formed in the presence of an excess of iodide ions and so is negatively charged (section 2.5.1). As more silver ion is added the excess iodide ion concentration gradually decreases and silver ions can compete more effectively for sites on the surface. The negative charge on the surface decreases to such an extent that the repulsive force which was maintaining the sol stability disappears. The sol then beings to coagulate, and just near the end point of the titration large flocs of AgI are visible in the titration flask (Exercise 2.5.1). The sign of electric charge on the particles can be determined by examining their motion in an electric field; it changes from negative to positive as the system moves through the end point of the titration.

This is a clear case of coagulation by electrostatic *potential control.* The electric charge on the particles at some point (near the end point) is actually zero and this is called the *point of zero charge* (p.z.c.) for the colloid. It is characterized by a particular value for the activity of the Ag^+ ions (and hence for the I^- ions), and is found to occur at a silver ion concentration of 3.2×10^{-6} M, or $\text{pAg} = -\log_{10}[Ag^+]$ of 5.5. Since K_S for AgI is about 8×10^{-17}, this corresponds to an iodide ion concentration of only 2.5×10^{-11} M—indeed very near the end point of the titration.

2.5.4 *Coagulation by electrolyte addition*

The size of the repulsion barrier is determined by the nature of the material adsorbed on the particle surface. In the case of a charge-stabilized colloid it depends on the magnitude of the surface charge and on the extent of the electrical double layer and this, as noted above, depends on the total electrolyte concentration. It is necessary here to distinguish between the concentration of the potential-determining ions (p.d.i.) and that of other ions (called *indifferent ions*) that enjoy no special relationship with the surface. If the p.d.i. concentration is adjusted so that the surface has a large excess of *either* positive or negative ions then the repulsion can still be influenced by changing the extent of the diffuse double layer. If for example ions of another salt like potassium nitrate are added this leads to a less extensive diffuse double layer, a smaller repulsion between approaching particles and ultimately to the phenomenon of *rapid coagulation*. This occurs when there is no electrical barrier to the particle approach and every collision results in contact. The rate of coagulation is then determined by the rate of diffusion of particles towards one another due to Brownian motion.

The electrolyte concentration at which slow coagulation gives way to rapid coagulation can be determined experimentally and is called the critical coagulation concentration (c.c.c.). It is measured in various ways but the method recommended by Overbeek (1952, p. 289) has been found by many people to provide reproducible and reliable results. It relies on a particular sequence of mixing operations to produce a clear demarcation between slow and rapid coagulation.

One takes a series of test-tubes containing the same concentration of the sol and adds varying amounts of the coagulating electrolyte. The tubes are then shaken quickly and allowed to stand for, say, 2 h. They are then briefly reshaken in order to remix the contents and again allowed to stand. If the electrolyte range has been judiciously chosen the c.c.c. appears as the concentration above which the settling material leaves behind it a perfectly clear supernatant solution; below the c.c.c. the supernatant retains some of the uncoagulated sol. It is, incidentally, the second mixing process that is important in this procedure for it allows the flocs formed in the first stage to sweep through the suspension, collecting any uncoagulated particles to form large aggregates, which settle quickly.

The exact concentration at which the particles would experience no repulsion barrier would be expected to depend on the temperature and the nature of the sol as well as on the surface electric charge and the nature of the electrolyte and the dispersion medium. For sols in water, however, certain very striking regularities are immediately apparent.

2.5.5 *The critical coagulation concentration*

In Table 1.2, attention was drawn to the extreme sensitivity of lyophobic colloids to coagulation by electrolyte solutions. Table 2.1 gives some actual values for the concentration of various electrolytes required to cause coagulation of three typical colloidal sols, two negatively charged and one positively charged. Note that:

(a) the coagulation concentration for similar electrolyte solutions is similar but not identical;

(b) the effectiveness of an electrolyte as a coagulating agent generally increases when it contains multivalent ions; and

(c) it is the valency of the *counterion* that is of paramount importance in determining the coagulation concentration. Thus for the positively charged $Fe(OH)_3$ sol it is the chloride ion concentration in $BaCl_2$ that

Table 2.1

Coagulation concentrations of simple† *electrolytes* (*mmol* L^{-1}) (*from Overbeek* 1952, *p.* 82, *with permission*)

Valency of counterions	Sol of As_2S_3 negatively charged		Sol of Au negatively charged		Sol of Fe $(OH)_3$ positively charged	
Monovalent	LiCl	58			NaCl	9.25
	NaCl	51	NaCl	24	$\frac{1}{2}BaCl_2$	9.65
	KNO_3	50	KNO_3	23	KNO_3	12
Divalent	$MgCl_2$	0.72	$CaCl_2$	0.41	K_2SO_4	0.205
	$MgSO_4$	0.81	$BaCl_2$	0.35	$MgSO_4$	0.22
	$ZnCl_2$	0.69			$K_2Cr_2O_7$	0.195
Trivalent†	$AlCl_3$	0.093				
	$\frac{1}{2}Al_2(SO_4)_3$	0.096	$\frac{1}{2}Al_2(SO_4)_3$	0.009		
	$Ce(NO_3)_3$	0.080	$Ce(NO_3)_3$	0.003		

† Matijevic (1973) has rightly pointed out that some of these ions are 'simple' only under certain conditions. Al^{3+}, for example, is the dominant species only at low pH values ($pH \leq 3$).

determinates the coagulation behaviour, not the barium ion concentration.

The strong dependence of coagulation concentration on the valency of the counterion is referred to as the Schulze–Hardy rule and it has been recognized since the end of the nineteenth century. The theory of stability of lyophobic colloids, developed in Chapter 7, is able to provide an excellent rationale for that rule but we will find it is able to do much more than that. Like all theories, however, it has only limited applicability to real systems and works best when applied to ideal model materials. Nevertheless it provides an essential fundamental framework with which to discuss the behaviour of many more complicated systems. The theory is based on the recognition of the presence of two forces in any electrostatically stabilized colloidal system: the electrostatic repulsion, which opposes aggregation, and a universal attractive force, which acts to bind particles together if once they come into close contact (section 2.5.8). It is referred to as the DLVO theory (after the four scientists—Deryaguin, Landau, Verwey, and Overbeek—who were largely responsible for its development).

2.5.6 *Coagulation and complex formation*

The ions involved in Table 2.1 are, for the most part, simple ones which do not readily form complexes. (Aluminium is an exception to which attention has already been drawn.) The DLVO theory, referred to above, is able to treat the behaviour of such ions fairly successfully simply by calculating their effect on the extent of the diffuse double layer. Indeed, the ions are treated in that theory simply as point charges, so that their specific (chemical) properties are ignored.

It has become increasingly obvious that there are many situations in which such an approach is inadequate. Matijevic (1973), in an excellent review of the experimental data on coagulation behaviour, gives many examples of systems in which the important ions cannot be treated as simple charges but must be recognized as chemical entities with specific chemical properties. One example will serve to illustrate the approach. Figure 2.5.3 shows a plot of the critical coagulation concentration of two beryllium salts as a function of pH for a negatively charged silver bromide sol. The p.d.i.s for AgBr are Ag^+ and Br^- and one would not expect the surface to be affected by the pH. The dramatic change in effectiveness of the beryllium solution as a coagulator at different pH values does, however, bear a striking similarity to the curve showing the

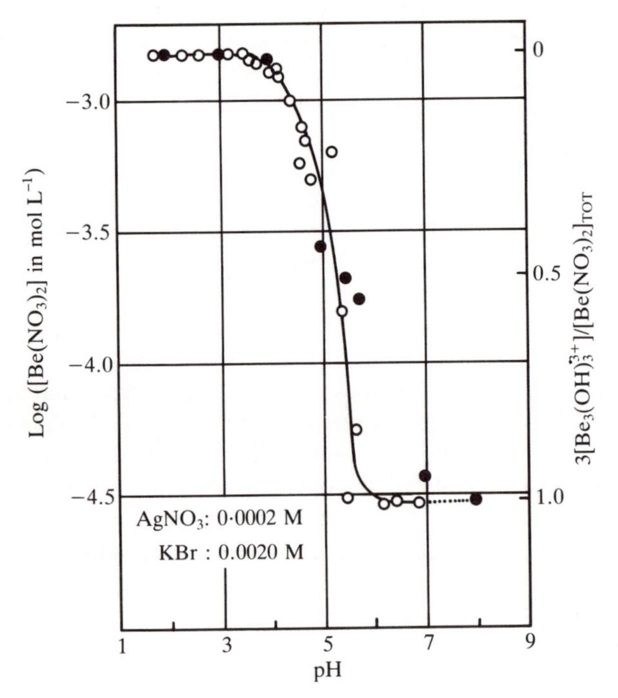

FIG. 2.5.3. Critical coagulation concentration of $Be(NO_3)_2$ (○) and $Be(ClO_4)_2$ (□) as a function of pH for negatively charged AgBr sol. Note that the experimental curve closely follows the plot of the relative concentration of the hydrolysed species $[Be_3(OH)_3^{3+}]$, which is shown thus (●). This suggests that it is the adsorption of this ion onto the particle surface that is largely responsible for the coagulation process. (From Matijevic 1973, with permission.)

concentration of the hydrolysed ion that predominates in the solution at the different pH values. Note also that the concentration of beryllium salt required to induce coagulation is very small. Obviously it is important to take account of the possibility of extreme sensitivity to the presence of a coagulating ion.

2.5.7 *Heterocoagulation*

A rather more complicated coagulation behaviour is observed with kaolinite suspensions. We noted in section 1.5.5 that at low pH values the edges of the kaolinite crystal carry a positive charge, which gradually decreases with increasing pH and is negligible above pH 7–8. Since the cleavage faces of the crystal are negatively charged at all pH values it is possible for an electrostatic interaction to occur between edges and faces at low pH but this becomes less important as the pH rises. At high pH, face to face coagulation can occur but only if the electrolyte concentration is sufficiently high to reduce the extent of the double layer.

At low pH the kaolinite can form very open flocs of a 'card house' structure (Fig. 2.5.4), which gradually break down as the pH is raised. Street (1956) has used these ideas to explain the pH dependence of the flow behaviour of kaolinite—so important in the preparation of ceramic slips and in paper making.

There may be other 'natural' colloidal systems in which the particles are so heterogeneous that they may have charges of different sign at different sites on the surface so that attraction can occur by electrostatic (Coulombic) interactions. It is much more common, however, in real systems to encounter *mixtures* of particles with different surface characteristics. It is then quite possible for one set of particles to be of opposite

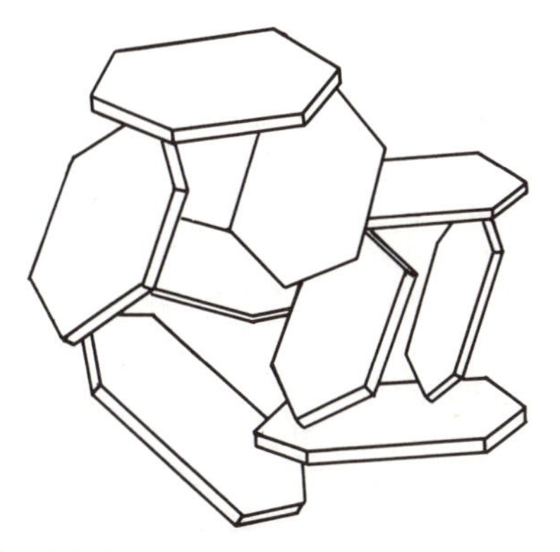

FIG. 2.5.4. Card house structure of kaolinite.

sign to another set, when immersed in a homogeneous electrolyte solution. The interaction between surfaces having significantly different charge status is called *heterocoagulation*. It is important in many technological situations, like paint and ink-making and in surface coating generally, as well as in the flotation process for mineral separation.

2.5.8 *Forces between colloidal particles*

The attractive and repulsive forces discussed in sections 2.5.7 and 2.5.3 stem from more or less simple Coulombic interactions, but so far we have not discussed the origin of the attractive forces, whose effects become obvious when the repulsion is reduced. Why do lyophobic colloidal particles tend to aggregate unless prevented from doing so by some sort of kinetic barrier?

If two neutral colloidal particles of silver iodide were to come together with two plane surfaces touching so that the interface was eliminated, the free energy of the pair would be decreased. That is what is meant by saying that silver iodide is a lyophobic colloid. The matching up of the two crystal surfaces would then be comparable to the coalescence of two bubbles or droplets to form a single larger entity. Although such a process is thermodynamically feasible (and so would provide a possible mechanism for the aggregation (coagulation) process) it almost certainly never happens, except over times that are quite long compared to the coagulation time. When aggregates have been formed by some process it is possible for some dissolution and reprecipitation to occur to form linkages between the particles. This is called *Ostwald ripening* and occurs because of the higher solubility of small particles (section 1.4) or the differential solubility of sharp asperities (i.e. of high curvature) on the particle surfaces.

The aggregates formed in the usual coagulation process are often difficult to redisperse but the particles do appear to retain their individuality (at least for times of the order days and often for much longer). The structure of the aggregates also depends strongly on the way they are formed. If the attraction between the particles is weak (or, in other words, the repulsion is significant) the particles are able to coagulate only slowly but when they do they are observed to form very compact structures with little entrapped solvent. On the other hand, when the repulsion is slight and attraction is strong (i.e. in the regime of rapid coagulation) the aggregates formed are usually voluminous with much of the dispersion medium trapped inside the loose open structure.

This behaviour can be most easily understood if the attractive forces result not from some specifically surface phenomenon but rather from an interaction between the whole of one particle and its neighbour.

Attraction can then be maximized by the particles adopting an orientation that brings the bulk of each particle close to the bulk of its neighbour. When the repulsive forces are weak this is no longer necessary and any direction of approach will lead to aggregation.

In 1932 Kallmann and Willstätter suggested that this attractive force might be of the nature of a *van der Waals or London force,* similar to that occurring between atoms and molecules, and responsible for the condensation of gases into liquids. Such a force is certainly general enough—it occurs even between neutral symmetrical atoms like those of argon—but at first sight it would appear to have the wrong characteristics. We know from considerations of gas behaviour that the van der Waals force is of quite short range. It decreases with the sixth power of the separation between the atoms and has little effect beyond a distance of about 1 nm. Since colloidal particles are often very rough and rarely approach one another to this distance, except possibly at a single point, such a force would seem inadequate to produce any significant attraction. It turns out, however, that the van der Waals forces between the atoms in two approaching colloidal particles are to some extent additive and the overall effect is an attractive force of quite long range—of the order of some tens of nanometres, quite comparable with the range of the coulombic forces. We will have much more to say about the calculation of this attractive force for different materials in Chapter 4. For the moment we need only note that the potential energy due to the London (van der Waals) effect between two colloidal particles of the same material immersed in a fluid is always negative (i.e. the force is attractive) and its magnitude decreases with about the second power of the separation (the details depend on geometry and separation distance).

The repulsive force due to coulombic interactions between colloidal particles leads to a positive potential energy, which decreases roughly exponentially with distance (Chapter 7). When both energy contributions are added, as in Fig. 2.5.5, three possible behaviour patterns emerge: curve (a) corresponds to a (meta-)stable sol in which the potential energy barrier slows down the rate of coagulation, perhaps to a very small value; (b) would be a marginally (un)stable sol, while (c) is decidedly unstable (i.e. rapidly coagulating).

Figure 2.5.5 suggests that the attraction energy becomes infinitely negative on contact so that particles once in contact could never be separated (redispersed or *repeptized*). Although redispersion is sometimes difficult, it is not normally impossible, and other forces must come into play at these very small separations. They will be discussed in Chapter 7. One other feature of Fig. 2.5.5 deserves mention: at very large distances the van der Waals force always dominates over the repulsion. The long range minimum (*called the secondary minimum*) in

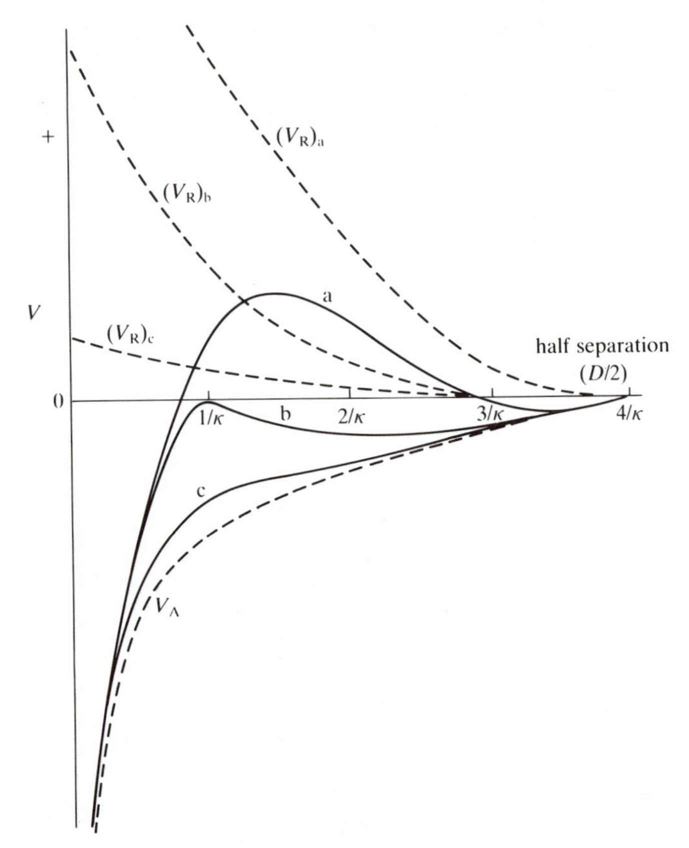

FIG. 2.5.5. The total potential energy of interaction for (a) a stable (b) a marginally stable and (c) an unstable sol. Case (b) corresponds to the critical coagulation concentration. The curves $(V_R)_{a,b,c}$ are drawn for different values of surface potential but approximately the same concentration of (indifferent) electrolyte. (From Hunter 1980.)

the energy curve can be of the order of a few kT deep for large colloidal particles and is probably responsible for some of their rather anomalous flow and aggregation properties (Chapter 18).

The force between the colloidal particles can be calculated as the derivative (with respect to separation distance) of the total potential energy curve. It is now possible to measure that force directly in a number of ways and so to explore experimentally the adequacy of the theories for double-layer forces (Chapter 7).

2.6 Effect of polymers on colloid stability

2.6.1 *Steric stabilization*

We have so far talked exclusively of electrostatic stabilization and yet the alternative procedure of steric stabilization is probably more widespread and has certainly been consciously used by mankind for a great deal longer, even though its mechanism of operation has only recently become accessible for study. The early makers of inks and paints were well aware of the value of certain natural gums in promoting the stability (in the colloidal sense) of their pigments and the early colloid scientists referred to this phenomenon as *protection*.

In effect the surface of the lyophobic colloidal particle was covered with a lyophilic material of a polymeric (or, at any rate, a long-chain) nature. When two such particles approach, the interaction between the adsorbed chains causes a repulsion that can be sufficient to induce stability. Its magnitude can be calculated by estimating the effect of the particle separation distance on the free energy of the adsorbed molecules. As they are forced together the number of chain–chain contacts is increased at the expense of chain–solvent interactions and for a lyophilic molecule that leads to an increase in free energy (or a repulsive force). Steric stabilization is important in water but it is much more important in non-aqueous systems because electrostatic stabilization is for the most part useless in such situations. We will examine it in some detail in Chapter 8.

In the meantime we merely note that the free energy of mixing of the adsorbed molecular chains as the particles come together can be broken into an enthalpic (ΔH_{mix}) and an entropic (ΔS_{mix}) part:

$$\Delta G_{mix} = \Delta H_{mix} - T\Delta S_{mix}. \qquad (2.6.1)$$

A positive value of ΔG_{mix} (repulsion) can arise from either a positive value of ΔH_{mix} (enthalpic stabilization) or a negative value of ΔS_{mix} (entropic stabilization) or both. Enthalpically stabilized systems can be made to coagulate at sufficiently high temperatures (since $\Delta S_{mix} > 0$) and the opposite is true of entropically stabilized systems. (These remarks apply particularly to systems in which the adsorbed molecules are firmly anchored to the particle surface so that they can neither desorb nor migrate from the encounter region during a collision.)

2.6.2 *Polymer flocculation*

Another very common method of inducing aggregation in a colloidal dispersion is by the use of a polymeric flocculating agent. [The term *flocculation* is used here for the formation of rather loose aggregates of particles linked together by a polymer as distinct from *coagulation* in which the particles come into close contact as a result of changes in the electrical double layer around the particles (sections 2.5.1–2.5.8).]

A polymer can adsorb on the surface of a colloidal particle either as a result of (a) a Coulombic (charge–charge) interaction, (b) dipole interactions, (c) hydrogen bonding, or (d) van der Waals interaction. A

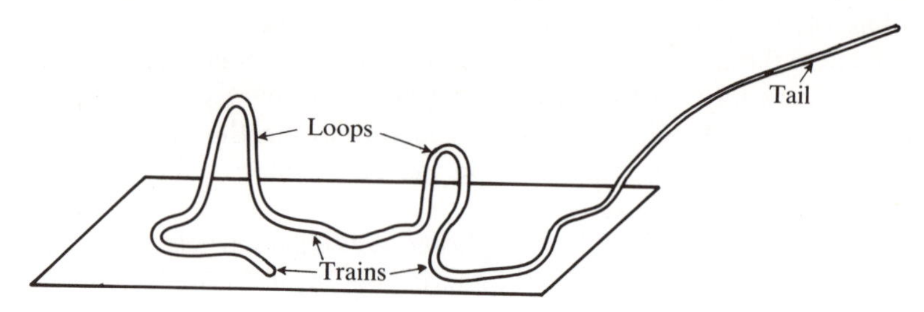

FIG. 2.6.1. Schematic diagram of polymer adsorbed on a surface.

balance must be struck between the affinity of the polymer and the particle surface for one another and for the solvent. The usual result is that the polymer is tied to the surface at a number of points but for some of its length it is able to extend into the solution (Fig. 2.6.1). Segments attached to the surface form *trains,* which are separated by *loops,* while the ends of the polymer are usually able to extend into the solution as *tails.*

When two particles are brought together it becomes possible for bridging to occur between one particle and another, especially if the adsorption density on the particle surfaces is not too high and the polymer is of very high molar mass (i.e. molecular weight). Careful mixing and control of the polymer concentration (and its nature) can result in very effective aggregation (flocculation) of a colloidal dispersion and the method is widely used in industry for the purification of water and mineral wastes. We will examine it in more detail in Chapter 8.

Exercises. 2.6.1 Set up the titration experiment suggested in section 2.5.3 and observe carefully the behaviour of the AgI precipitate in the neighbourhood of the end point.

2.6.2 What effect would you expect temperature to have on a system that was stabilized both entropically and enthalpically?

References

Atkins, P. W. (1978). *Physical chemistry*. Oxford University Press, Oxford.
—— (1982). *Physical chemistry* (2nd edn). Oxford University Press, Oxford. (3rd edition 1986.)
Bird, R. B., Armstrong, R. C., and Hassager O. (1977). *Dynamics of polymer fluids,* Vol. 1, p. 277. Wiley, New York.
Debye, P. (1909). *Ann. Physik.* **30** (4), 57.
Ferry, J. D. (1980). *Viscoelastic properties of polymers* (3rd edn). Wiley, New York.
Harris, J. (1977). *Rheology and non-Newtonian flow.* Longmans, London.
Herdan, G. (1960). *Small particle statistics.* Butterworths, London.

Hiemenz, P. C. (1977). *Principles of colloid and surface chemistry*. Marcel Dekker, New York.
Hunter, R. J. (1980). The double layer in colloidal systems. In *Comprehensive treatise of electrochemistry* (ed. J. O'M. Bockris, B. E. Conway and E. Yeager) Vol. 1, pp. 397–437. Plenum Press, New York.
Johnson, I. and La Mer, V. K. (1947). *J. Am. chem. Soc.* **69,** 1184.
Kallmann, H. and Willstätter, M. (1932). *Naturwissenschaften* **20,** 952.
Kerker, M. (1969). *The scattering of light and other electromagnetic radiation*. Academic Press, New York.
Kruyt, H. R. (1952). *Colloid science,* Vol. 1. Elsevier, Amsterdam.
Landau, L. D. and Lifshitz, E. M. (1960). *Electrodynamics of continuous media*. Pergamon, New York.
Matijevic, E. (1973). *J. Colloid interface Sci.* **43,** 217.
Mie, G. (1908). *Ann. Phys (Leipzig)* **25,** 377.
Overbeek, J. Th. G. (1952). In *Colloid science* (ed. H. R. Kruyt) Vol. 1, Elsevier, Amsterdam.
—— (1982). *Adv. Colloid interface Sci.* **15,** 251–77.
Parsegian, V. A. (1975). Long range van der Waals interactions. In *Physical chemistry: enriching topics from colloid and surface science* (ed. H. van Olphen and K. J. Mysels) Chapter 4. Theorex, La Jolla, California.
Perrin, J. (1909). *Ann. Chim. Phys.* **18** (8), 5.
Reiner M. (1960). *Deformation, strain and flow*. H. K. Lewis, London.
Richmond, P. (1975). The theory and calculation of van der Waals forces. In *Colloid science* (ed. D. H. Everett) Vol. 2, Ch. 4. Specialist Periodical Report, Chemical Society, London. [There are a number of typographical errors in the treatment of the problem given there.]
Street, N. (1956). *Austral. J. Chem.* **9,** 467.
Svedberg, T. (1928). *Colloid chemistry*. Chemical Catalog, New York.
van de Hulst, H. C. (1957). *Light scattering by small particles*. Wiley, New York.
van de Ven, T. G. and Hunter, R. J. (1979). *J. Colloid interface Sci.* **68,** 135.
Ware, B. R. (1974). *Adv. Colloid interface Sci.* **4,** 1–44.
Wheeler R. and Feynman, R. (1945). *Rev. mod. Phys.* **17,** 156; **21,** 424 (1949).

3

PARTICLE SIZE AND SHAPE

This chapter concerns itself first with the direct observation of the sizes and shapes of particles commonly found in colloidal systems. Since the size range is often very considerable we next consider how best to represent the *distribution* of particle sizes by some convenient algebraic formula. Some of the more common methods of particle size analysis are then described. The theoretical basis of each method is given in sufficient detail to appreciate its scope and limitations but to obtain practical experimental details of any particular method the reader should consult one of the many specialized manuals referred to in the text.

3.1 General considerations

The most significant characteristics of many colloidal dispersions (especially aerosols and the dispersions of solids in liquids) are the size and shape of the particles, since most other properties of the system are influenced to some extent by these factors. The idealized systems of monodisperse or highly regular particles discussed in section 1.5 are of great importance in the testing of fundamental physical models of colloid behaviour, but it must be recognized that the majority of colloidal dispersions of scientific and technological interest consist of particles that differ markedly in size (sometimes over several orders of magnitude in characteristic dimension) and may be of very irregular shape. To treat such systems at all we may have to adopt some rather drastic assumptions. We will usually try to approximate them as spheres, or more rarely as spheroids (i.e. the solid body obtained by rotating an ellipse around one of its axes (Fig. 3.1.1) or cylinders; disc-shaped particles may be regarded as cylinders of very small height, or as oblate spheroids, or sometimes as 'infinite' flat plates. Octahedral, rhomboidal or cubic crystals, especially if they are small, will often behave like spheres, and long parallelepipeds can usually be approximated as cylindrical rods.

Particles produced by dispersion methods have shapes that depend partly on the natural cleavage planes of the crystal but that also reflect the presence and disposition of imperfections, cracks and other flaws, which offer points or lines of weakness at which the imposed stress tends to concentrate. Fracture can then produce very sharp edges and asperities on the particle surface.

As the dispersion process is continued down to colloidal dimensions (say, in a colloid mill (section 1.5.1)) this effect is, to some extent, reduced because very sharp asperities have a very small radius of curvature and hence tend to dissolve preferentially, as would be expected from the analogue of the Kelvin equation (eqn (1.4.4)). Very small colloidal particles (<100 nm) therefore, often appear to be rather less jagged in outline than the larger (microscopic) particles obtained by simple grinding and crushing operations.

Particles produced by condensation have shapes that depend upon the

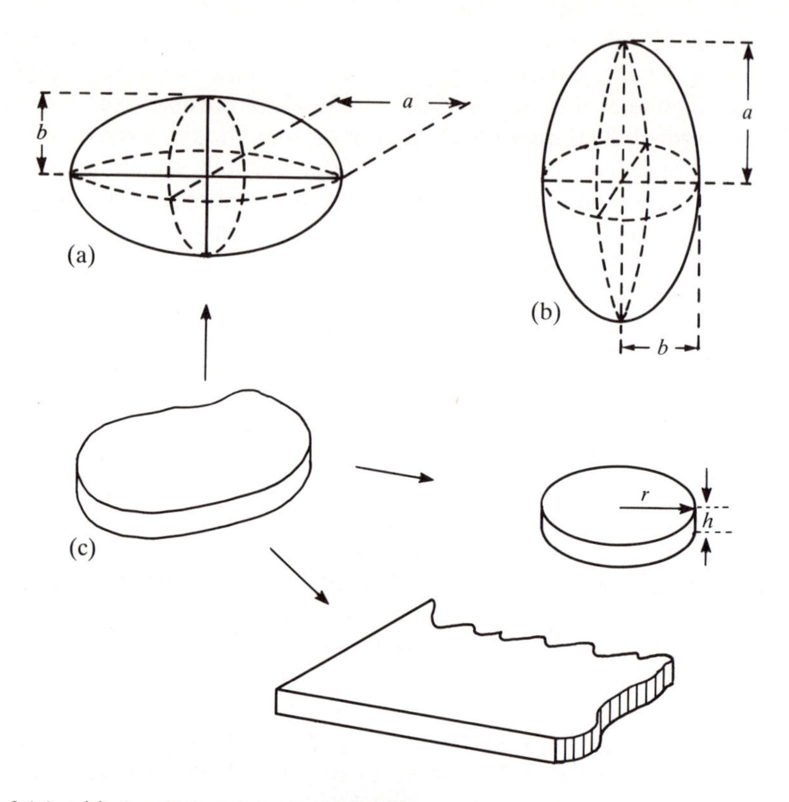

FIG. 3.1.1. (a) An oblate spheroid, obtained by rotating an ellipse around its short axis. The cross-section in the plane where a is measured is circular.
(b) A prolate spheroid, obtained by rotating an ellipse around its long axis (like a football). Sections parallel to b are circular.
(c) A disc may be approximated as an oblate spheroid, a cylinder or an 'infinite' flat plate.

rate of growth of different crystal faces. If thermodynamic equilibrium is maintained during crystal growth it can be shown (Wulff 1901) that the shape is determined by the condition that the sum $\sum \mathscr{A}_i\gamma_i$ is a minimum at constant volume of the crystal. $\mathscr{A}_i$ and γ_i are the area and surface energy of the ith face, respectively. (The surface energy of the different faces of a crystal (001, 110 etc.) is slightly different because of the differences in packing density of the atoms.) Wulff has given a geometrical construction that enables one to draw up the equilibrium crystal shape from a knowledge of the γ_is. (We will take up this question in more detail in section 5.6.) In many cases, however, the growth rate of a face is influenced by kinetic factors (e.g. rate of diffusion to the face, or rate of incorporation in it) rather than thermodynamic (equilibrium) ones. It is also observed that certain substances can be preferentially adsorbed onto

FIG. 3.1.2. (a) Electronmicrograph of calcium carbonate (calcite) crystals (normal habit). (b) Crystals of the same calcium carbonate obtained from a solution containing an organophosphorus crystal habit modifier. (Photographs obtained from the work of Leonard Dubin 1980, of Nalco Chemical Company, Illinois, which markets these modifying compounds.)

particular crystal faces, changing the surface energy and profoundly altering the shape (or *habit*) of the crystal (Fig. 3.1.2).

A very stimulating and original treatment of the concept of particle shape (including an introduction to the use of the mathematical theory of *fuzzy sets*†) is given by Beddow (1980, Chapter 6) and the interested reader is referred to that work; rather more conventional treatments of the subject are given in the standard works by Orr and Dallavalle (1959), Herdan (1960) and Allen (1975). A brief description of the more sophisticated techniques for describing shape (including the use of Fourier series) is given by Sutton (1976). Space here permits only a very brief outline of the methods of treating irregular particles.

3.2 Direct microscopic observation

The most significant development in this area has, of course, been the transmission electron microscope (TEM), and more recently the scanning electron microscope (SEM). Since colloidal particles generally fall below the limit of resolution of the optical microscope, the early colloid chemists were forced to rely on indirect evidence to obtain an idea of particle shape, and in the absence of any evidence to the contrary tended to assume that all particles were roughly spherical. There were, of course, some exceptions: the crystal structure of the common clay minerals suggested that they should be plate-like (lath-shaped) and vanadium pentoxide sols showed striking optical and viscous properties that indicated that they were composed of long rods. Thanks to the electron microscope, however, we can now determine the shape of many colloidal particles with very little residual uncertainty. Solid crystalline particles present very little problem because they are substantially unaffected by the normal methods of specimen preparation. Some of the softer polymeric materials, like poly(methyl methacrylate) do, however, have a tendency to melt in an intense electron beam. Even more difficult to deal with are systems in which the sub-microscopic structures are very sensitive to the presence of the solvent, and likely to be destroyed by the drying process which is essential in electron microscopy (since the specimen must ultimately be placed in a vacuum for viewing with the electron beam). Even here, however, the modern techniques developed for the examination of biological structures (like freeze-fracture) are increasingly being used to determine the size and shape of colloidal structures. Before dealing with these developments we will briefly examine the older techniques of optical and ultramicroscopy.

† These are sets for which the concept of membership is fuzzy (i.e. one may not be certain whether an object does or does not belong to the set). Nevertheless, they can be handled logically with the help of fuzzy set theory.

3.2.1 *Optical (light) microscopy*

The observation of colloidal particles with an optical microscope is limited by the *resolving power* of the microscope. This refers to the ability to discriminate between two closely spaced points in the field of view of the microscope. The trains of light waves emanating from two neighbouring points can interfere with one another to produce a diffraction pattern of alternate light and dark bands. The effect is particularly marked if the distance between the points is comparable with the wavelength, λ, of the light being used. If the two points are the opposite edges of a particle, then the particle will appear as an object of indeterminate shape surrounded by a halo of alternate light and dark rings, the intensity of which decreases rapidly with distance from the particle centre.

It may be shown by the methods of geometrical optics that the resolving power, d_p, of a microscope is given by:

$$d_p = \frac{0.61\lambda}{n_0 \sin\theta} \tag{3.2.1}$$

where n_0 is the refractive index of the medium and 2θ is the angle subtended by the microscope objective at the focal plane (Fig. 3.2.1). (Hiemenz 1977 gives an elementary analysis leading to an approximate relation for d_p.)

Resolving power can evidently be improved (i.e. d_p decreased) by reducing the wavelength, or increasing n_0 and θ. In practice only visible light is used ($\lambda \sim 500$ nm), but n_0 can be increased by filling the region between the lens and the sample with a transparent oil ($n_0 \simeq 1.5$) instead of air. (A drop of immersion oil is placed on top of the microscope cover

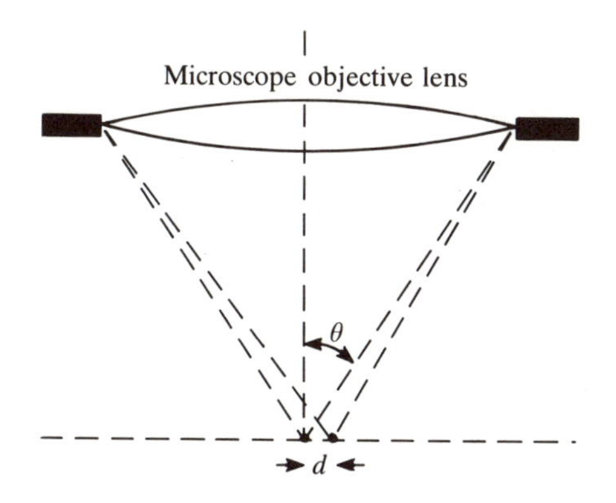

FIG. 3.2.1. Resolving power of a microscope (see text).

slip and the 'oil-immersion' lens lowered into it). Wide angle lenses are also helpful (i.e. increased θ) but the angle is limited by other optical problems like spherical and chromatic aberration. In effect the lower limit for d is about 0.2 μm, so optical microscopy is limited to the upper end of the colloidal size range.

3.2.2 *The ultramicroscope*

We noted earlier (section 1.3) the Tyndall effect, in which a beam of light shining through a colloid dispersion becomes visible when viewed at right angles to the beam, because of the scattering effect of the colloidal particles. This effect can be used to make very small colloidal particles visible, not as well defined shapes, but as pin points of light against a dark background, in the *ultramicroscope*. The principle of the method is illustrated in Fig. 3.2.2. An alternative method of obtaining dark field illumination is shown in Fig. 3.2.3.

Although the particles are not directly visible, it is possible to obtain some idea of their relative size from the amount of light scattered by each pin-point. As we have seen (section 2.3) the amount of light scattered depends not only on the particle volume but also on its relative refractive index, the wavelength of the light and the angle of observation. The minimum size of metal sols that can be seen in the ultramicroscope is about 5–10 nm but for sols of lower refractive index it is rather higher (~50 nm). This is, nevertheless, a considerable improvement on the simple light microscope, especially with the advent of laser light sources.

It is also possible to infer something of the particle's shape in the ultramicroscope. Particles that are highly anisometric are continually changing their orientation with respect to the incident light and the observer, as a result of their Brownian motion. This rapid fluctuation in

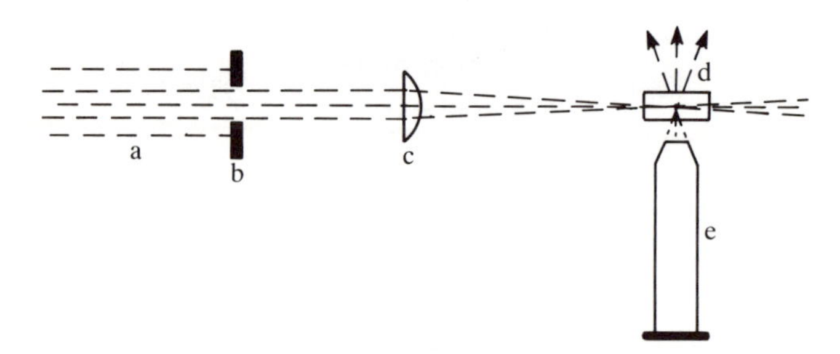

FIG. 3.2.2. The slit ultramicroscope. An intense beam of light (from an arc, xenon lamp or laser) a, emerges through a slit, b, and is focused by a lens, c, into a chamber, d, containing the colloidal sol. The light scattered from the particles can be viewed through the microscope, e, against a dark background.

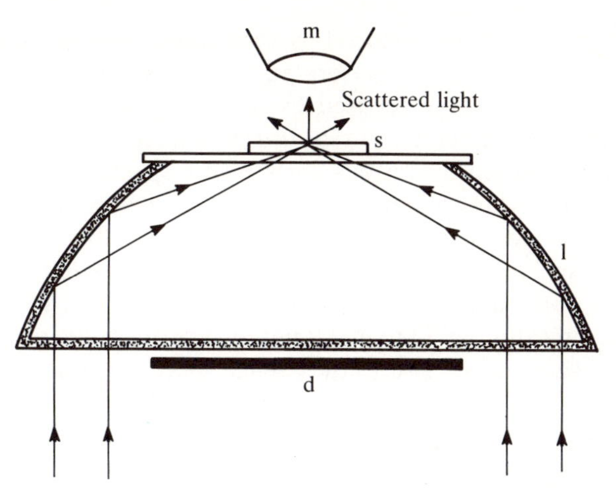

FIG. 3.2.3. The dark field (paraboloid) condenser. The diaphragm, d, prevents light from hitting the sample, s, directly, and only oblique rays reflected from the walls of the paraboloid lens, l, are able to strike the colloidal particles. The only light visible in the microscope, m, is that scattered by the particles. The cardioid condenser is a more complicated arrangement for achieving a similar result (see Kruyt 1952, p. 40).

orientation produces a twinkling appearance in the light spots as the scattered intensity varies. By contrast, spherical particles show a steady light, although their translational Brownian motion is still readily visible, especially if they are small.

It is difficult to determine the size of a colloid particle directly in the ultramicroscope (but see Davidson and Haller 1976 and Cummins *et al.* 1983). An indirect measurement can, however, be made by determining the number concentration of particles (see Exercise 3.2.1). This can be obtained either by direct counting or, more conveniently, using automatic counting devices (see sections 3.2.5 and 3.6 below).

3.2.3 *The transmission electron microscope*

This device depends for its operation on the wave nature of the electron and the fact that electric and magnetic fields of suitable geometry are able to function like lenses to refract, deflect, and focus an electron beam. The ultimate limit of resolution of an electron microscope is determined by the electron wavelength but, in practice, one can seldom resolve much below 1 nm except with very specialized techniques; the limitation in most machines is in the performance of the magnetic lenses and the maintenance of stable magnetic fields (see Exercise 3.2.2).

For a detailed description of the apparatus and techniques one should consult a standard text (e.g. Hall 1966). The following description is taken from Silverman *et al.* (1971, p. 111 *et seq.*).

The electron beam is produced by thermionic emission from a heated tungsten cathode, C, and is accelerated towards an aperture in the anode, A (Fig. 3.2.4). It is then focused by a lens L_1 and passes through the sample, which is mounted on a transparent grid, G. Electrons are absorbed or scattered by various parts of the specimen in proportion to the local electron density, and the remainder are transmitted.

An electromagnetic objective lens, L_2, collects the transmitted electrons and magnifies the image of the specimen 10 to 200 times, into the object plane of a magnetic projector lens system (L_3), which induces a further magnification of 50 to 400 times as it projects the electrons onto a fluorescent screen, S. There the image may be viewed directly or

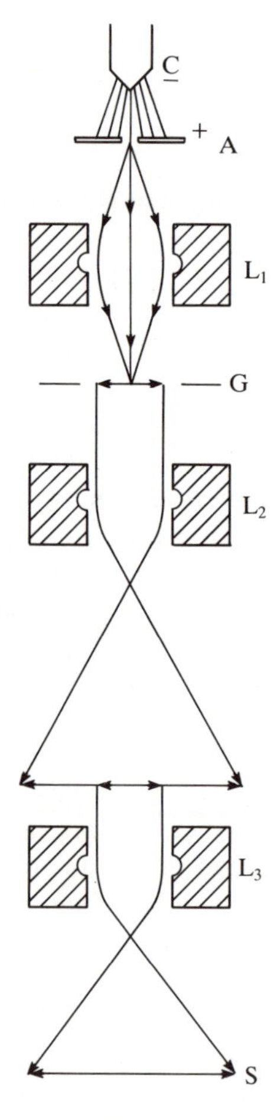

FIG. 3.2.4. Schematic diagram of the electron microscope. (For description see text.)

photographed with a fine-grained film to be enlarged a further 5 to 10 times.

The overall magnification factor ranges from about 100 to 500,000 times. Since the human eye can discriminate between points separated by about 0.2 mm, the lower limit of observation is then from 2 μm down to about 0.4 nm, though specialized procedures can actually resolve individual atoms (0.2 nm) in favourable cases.

The image formed by this procedure is a two-dimensional representation of the actual structure, and in some cases this is all that is required. In many cases, however, it is helpful to have an idea of the surface topography and this is most readily achieved by shadow-casting, as illustrated in Fig. 3.2.5. A beam of metal atoms is fired, *in vacuo,* at an angle to the sample and the deposit modifies its electron transmission characteristics. Figure 3.2.5 shows schematically how the intensity might be expected to vary in the case shown. Although the human eye will quickly interpret the resulting image in terms of a particle shape it is important to realize that the pattern of lightness and darkness on the screen or photograph is the result of a complex sequence of events and an exact shape analysis may call for more detailed consideration of the influence of shadowing angle θ and direction on the apparent shape.

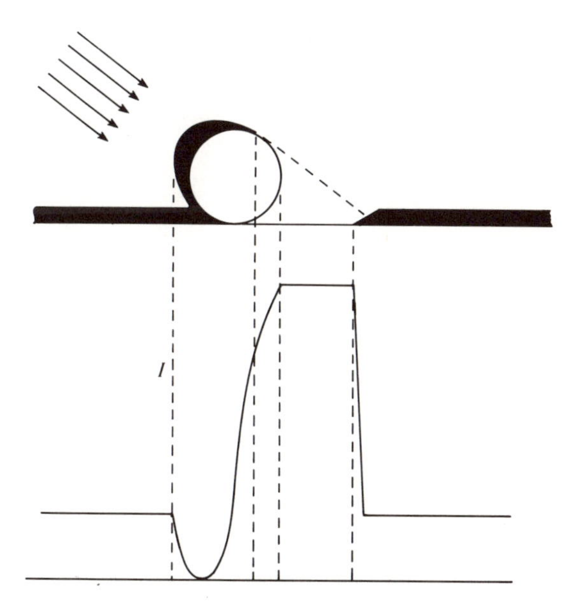

FIG. 3.2.5. Shadowing (or shadow-casting) of a spherical or cylindrical particle produces a characteristic intensity pattern in the transmitted beam, since the shadowing material (usually a metal) is a strong absorber–scatterer of electrons. I is the intensity (or number) of transmitted electrons.

Metals like gold, platinum, palladium, nickel, and chromium may be used for shadow casting.

The main problems encountered in electron microscopy concern the preparation of the sample in a way that permits it to be transferred to the evacuated (10^{-4} mmHg) chamber of the microscope and to be bombarded with electrons of high energy without undergoing changes in structure. Problems of electrostatic charging, melting, evaporation, and decomposition in the beam can be minimized by careful sample preparation and the use of techniques first introduced to study biological samples. Of these, the process of replication is the most important in colloid studies. It is illustrated in Fig. 3.2.6. By using a suitable material to form the replica a very labile surface can be converted into one of the same shape which will withstand the rigours of electron bombardment

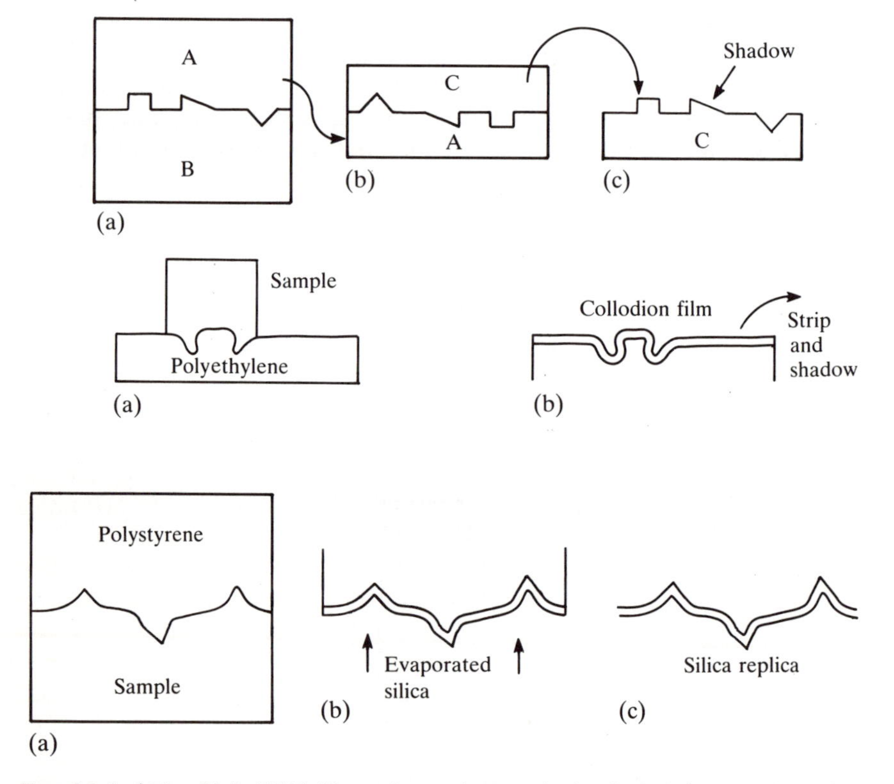

FIG. 3.2.6. (After Hall 1966.) Illustrating various methods of obtaining a replica of a surface. Top: (a) negative replication; (b) positive replication; (c) shadow casting the stripped positive replica. Middle: polyethylene method of positive replication. (a) Sample cast in polyethylene; (b) collodion positive replica. Bottom: silica replica method. (a) Negative replication with polystyrene; (b) silica evaporated onto replica surface in vacuum; (c) polystyrene dissolved to reveal positive silica replica.

(see e.g. McDonald *et al.* 1977). The technique is particularly useful for studying surface films, and is used for preparation of the final image in freeze-fracture methods.

3.2.4 *The scanning electron microscope*

Even with the use of replication and shadowing techniques, the transmission electron microscope is limited in its ability to show particle shape: it cannot, for instance, readily give information about re-entrant surfaces†. The more recently developed *scanning electron microscope* (SEM) is able to provide quite remarkable images, which are interpreted by the eye as truly three-dimensional (Fig. 3.2.7). In this instrument (see Johari 1968) an electron beam is focused to about 5–10 nm and deflected in a regular manner across the surface of the sample, which is held at an angle to the beam. The low velocity secondary electrons that are emitted as a result are drawn towards a collector grid and fall onto a sensitive detector. The output from this detector is used to modulate the intensity of an electron beam in a cathode ray tube (CRT). The beam itself is made to scan the surface of the CRT in synchronism with the scanning of the sample by the primary electron beam. The result is a reconstructed image on the CRT much like a TV picture.

The big advantage of the SEM is that the secondary electrons are emitted at low voltage and so can be easily deflected to follow curved paths to the collector. The electrons emerging from parts of the surface that are out of the line-of-sight are also collected (though at lower intensity) and it is this that contributes most to the striking realism of the three-dimensional image. The depth of field is also very large (some 300–500 times that which is available in a light microscope at the same magnification) so that the SEM is often used to examine the fine detail of quite large structures. The limit of resolution is currently some ten times larger than for the transmission electron microscope (i.e. about 5 nm).

The scanning electron beam can also be used to provide detailed information on the surface composition of the sample. The instrument is coupled to a solid-state X-ray detector, capable of determining the intensity and wavelength of the characteristic X-rays emitted by the surface atoms when bombarded with electrons. This is called *electronprobe microanalysis* and it is particularly useful for the study of composite materials. It is not very sensitive to the elements of low atomic number ($Z < 12$) but these can be detected using Auger *electron spectroscopy*. (If the primary electron beam dislodges an electron from one of the inner shells of an atom, that core electron can be replaced by

† Stereoscopic views can give a three-dimensional effect.

FIG. 3.2.7. Scanning electronmicrograph of small (~0.1 μm) polystyrene (PS) particles adsorbed on larger PS particles (~1 μm). (Photograph courtesy of Dr Brian Vincent, University of Bristol.)

an electron dropping down from one of the outer shells. When this happens, all of the excess energy is taken up by a single electron from another outer shell and this (the Auger electron) is emitted with a velocity that is characteristic of the element and independent of the energy of the primary electron. It can, therefore, be used to identify the element in the surface (see e.g. Morrison 1977, p. 87).

3.2.5 *The 'size' of irregular particles*

In the course of this chapter we will examine quite a number of ways of estimating the size of colloidal particles. The most appropriate method in any situation depends on what one requires the size for. For smooth spheres there is only one size characteristic but for most other shapes there is anything up to an infinite number of choices of 'size'. It is important, therefore, to choose a method of size measurement which is likely to reflect the aspect of the particles which is of most interest. In one case, it might be the surface area but in another case it might be more important to know the 'settling radius'.

We will begin by looking at the direct visual measurements and then introduce the elementary statistical procedures needed to handle them before going on to consider other methods of size measurement.

Whether derived from a light microscope or an electron microscope, the photographic image of the particles will represent a sample of their cross-sectional areas. For highly irregular particles there are several possible measurements (outlined by Allen 1975, p. 131). Of these the most commonly used are:

1. Martin's diameter (d_m): the length of the line that bisects the image of the particle (Fig. 3.2.8). The lines may be drawn in any direction, but the direction must be maintained constant for all the image measurements.

2. Feret's diameter (d_f): the distance between two tangents on opposite sides of the particle, parallel to some fixed direction (Fig. 3.2.8).

3. The projected area diameter (d_a): the diameter of a circle having the same area as the particle, viewed normally to a plane surface on which the particle is at rest in a stable position. (This is usually assumed to be the case for electron micrographs, since the drying process would be expected to favour a stable particle orientation.)

There are differences of opinion over the usefulness of the Martin and Feret diameters, although various relationships have been derived or

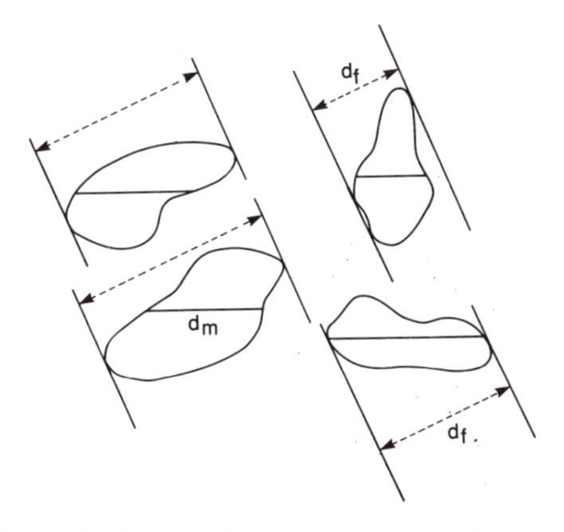

FIG. 3.2.8. Martin's diameter shown———. Feret's diameter shown ------.

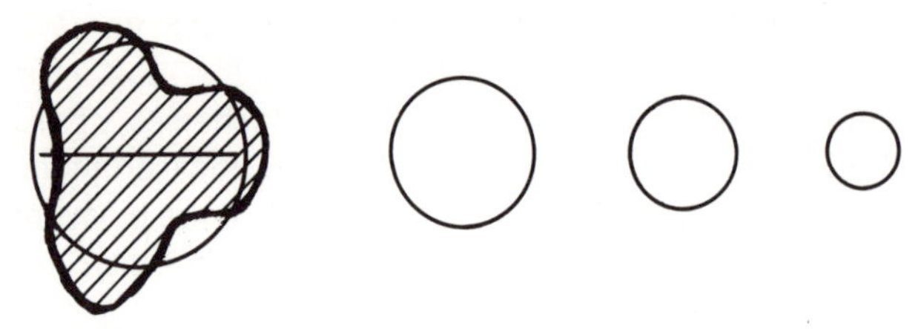

FIG. 3.2.9. Measuring the projected area diameter, d_a.

experimentally established between them and d_a, for different materials. Thus, the ratio d_m/d_f has characteristic values (a little greater than one) for different ground solids and can be used as an empirical measure of shape.

The projected area diameter, d_a, is probably the most useful parameter for general purposes and it can be fairly readily measured, either using a graticule or a more sophisticated semi-automatic procedure. In the graticule method, a glass microscope slide is engraved with circles of increasing size (usually with radii increasing by $\sqrt{2}$) and the operator is required to select the circle that best matches the projected area of the particle (Fig. 3.2.9).

The semi-automatic procedures, though still tedious, are somewhat faster. The Zeiss–Endter particle sizer (Allen 1975, p. 141), for example, is best suited to analysing photomicrographs. It projects onto the photograph a circle of light, the diameter of which is determined by an iris diaphragm controlled by the operator. When the circle is of the correct size the operator depresses a switch, which actuates one of a number of counters, each of which is associated with a pre-set size range. The machine also marks the particle with a pin-hole to avoid double counting. The entire image is projected onto a frosted screen for easier viewing.

The main advantage claimed for this instrument and the other similar ones described by Allen (1975) is that they permit the operator to exercise some judgement, both in particle selection (in heterogeneous systems) and in deciding what to do about overlapping particles. Nevertheless, there is clearly room for the introduction of microprocessor and scanning techniques at a level intermediate between the existing semi-automatic devices and the large, sophisticated and very costly devices developed for petrographic scanning. There have been some moves in this direction (Silverman *et al.* 1971 describe some of the early approaches) and rapid developments are expected.

Exercises. 3.2.1 Show that the average radius $\bar{r}$, of spherical particles is given by:

$$\bar{r} = (3CV/4\pi\rho N)^{1/3}$$

where ρ is the particle density, C is the concentration of the sol (by mass) and it is found to contain N particles in a volume V.

3.2.2 The wavelength associated with an electron is given by the de Broglie relation, $\lambda = h/p = h/(2m_e E)^{1/2}$, where h is Planck's constant, p is the electron momentum, m_e its mass and E its kinetic energy. Estimate the wavelength of an electron that has been accelerated through a voltage of 10 kV so that it has acquired an energy, E, of 10 keV. [A typical acceleration voltage would be of this order.]

3.3 Particle size distribution

Whenever we are confronted with the problem of describing the particle size of a system that is *heterodisperse* or *polydisperse* (i.e. contains many different sizes of particle) we resort to breaking the range of sizes up into convenient steps or classes, and recording the number of particles in each class. Consider, for example, the data in Table 3.1, which might represent the diameters of a sample of particles produced by a precipitation reaction. If the observed particle sizes ranged from 65 nm to 0.6 μm one might choose to break that range into 11 steps of 50 nm, as shown, and to record the number of particles in each class. The resulting data can then be plotted as a histogram or as a smoothed curve (Fig. 3.3.1), or as a curve showing the cumulative percentage equal to or smaller than a given size (Fig. 3.3.2; called the 'per cent undersize'). Rather than go to the trouble of plotting data each time, it is often sufficient to specify the main features of the distribution using a few numbers.

Table 3.1

Class range (nm)	Midpoint of class range d_i (nm)	Number of particles n_i	Fraction in this class $f_i = \frac{n_i}{\sum n_i}$	Total number with $d < d_i$	Cumulative percentage $d < d_i$
51–100	75	29	0.012	29	1.2
101–150	125	109	0.044	138	5.5
151–200	175	211	0.084	349	14.0
201–250	225	372	0.149	721	28.8
251–300	275	558	0.223	1279	51.2
301–350	325	440	0.176	1719	68.8
351–400	375	307	0.123	2026	81.0
401–450	425	223	0.089	2249	90.0
451–500	475	139	0.056	2388	95.5
501–550	525	81	0.032	2469	98.8
551–600	575	31	0.012	2500	100.0

$\sum n_i = 2500 = N$. $\sum f_i = 1.00$.

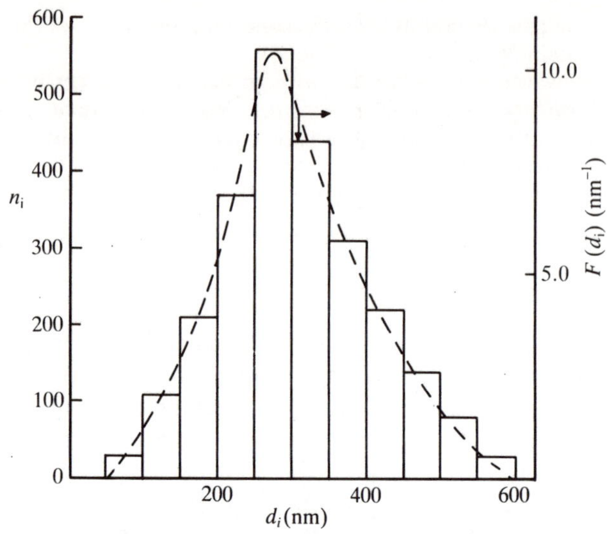

FIG. 3.3.1. The frequency histogram can be replaced by a smoothed curve. Note that the modal size is the most common one.

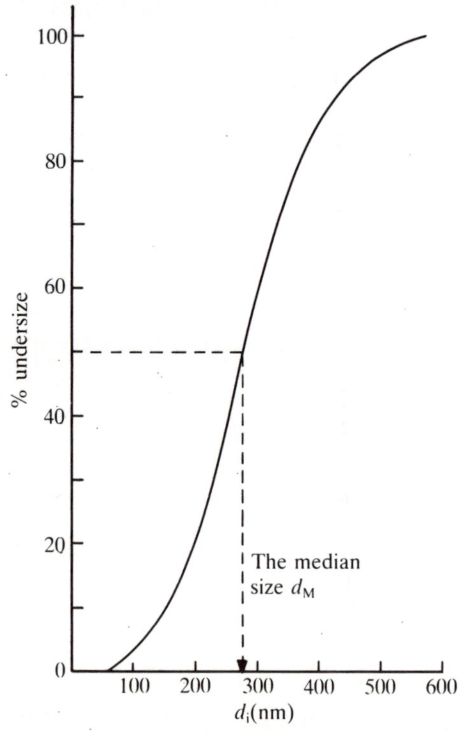

FIG. 3.3.2. The cumulative frequency curve showing the median size d_M, which divides the distribution evenly (50 per cent of the material has $d_i \leq d_M$).

3.3.1 *The mean and standard deviation*

The most important characteristics of a distribution are the *mean*, which measures the central tendency, and the standard deviation, which measures the spread of the data. The mean diameter is defined as:

$$\bar{d} = \frac{\sum n_i d_i}{\sum n_i} = \frac{\sum n_i d_i}{N} \tag{3.3.1}$$

$$= \sum f_i d_i \tag{3.3.1a}$$

where f_i is the fraction in class i (Table 3.1). (To distinguish it from other mean diameters we should strictly refer to this as the number length mean diameter.)

The standard deviation, σ, is defined as:

$$\sigma = \left\{ \frac{\sum n_i (d_i - \bar{d})^2}{\sum n_i} \right\}^{1/2} \tag{3.3.2}$$

$$= \left\{ \sum_i f_i (d_i - \bar{d})^2 \right\}^{1/2}. \tag{3.3.2a}$$

Note that again we do not use $(d_i - \bar{d})$ as a measure of the spread because it can be positive or negative†, and for a symmetrical distribution $\sum (d_i - \bar{d}) = 0$ even though the distribution might be quite broad (recall section 2.1.5 and Exercise 2.1.7).

The quantity inside the curly brackets is called the *variance* $(=\sigma^2)$ of the population. We take the square root of the sum of the squares so that σ can be more readily compared to the mean. It is easy to show (Exercise 3.3.2) that

$$\sigma = \{\overline{d^2} - (\bar{d})^2\}^{1/2} \tag{3.3.3}$$

which is often easier to compute than eqn (3.3.2). ($\overline{d^2}$ is the average value of the squares of the sample diameters.)

3.3.2 *Moments of a distribution*

It is useful at this stage to introduce the concept of a *moment of a distribution* about a point. The jth moment of the distribution of d_i about the point d_0 is defined by the relation:

$$M_j = \sum_i f_i (d_i - d_0)^j. \tag{3.3.4}$$

† One could use $|d_i - \bar{d}|$ but the modulus is more difficult to handle mathematically.

From eqn (3.3.1a) it is apparent that the first moment about the origin ($d_0 = 0$) is the mean whilst from eqn (3.3.2a) the second moment about the mean is the variance (σ^2). The third moment about the mean is a measure of the *skewness* of the distribution. Since it is an odd function ($f(x) = -f(-x)$) it will be zero for a perfectly symmetric distribution and its magnitude measures the departure from symmetry. The fourth moment about the mean measures the *kurtosis*. It very heavily weights the points far away from the mean and so is a measure of the length of the tail of the distribution.

These 'moments' may be interpreted in quite a different way in our polydisperse (heterodisperse) systems. When, for example, a measurement is made of the total surface area, $\mathscr{A}_S$, of a polydisperse system of spherical particles we can write:

$$\mathscr{A}_S = \sum_i n_i \pi d_i^2. \tag{3.3.5}$$

(This total area can be determined by measuring the capacity of the solid, S, to adsorb a gas). If the total number of particles, N, is also known then the number area mean diameter, $\bar{d}_{N\mathscr{A}}$ is defined as the diameter of the sphere for which

$$\pi \bar{d}_{N\mathscr{A}}^2 = \mathscr{A}_S / N$$

so that

$$\bar{d}_{N\mathscr{A}} = (\sum f_i d_i^2)^{1/2} \tag{3.3.6}$$

$$= (\overline{d^2})^{1/2}. \tag{3.3.6a}$$

Obviously, a system of N uniform spheres of diameter $\bar{d}_{N\mathscr{A}}$ has the same surface area as the original sample, S. Note that the number area mean diameter† is the square root of the second moment of the distribution of d_i about the origin. In a similar way one can define a series of different average sizes many of which are directly accessible to measurement (Table 3.2).

This leads us naturally to the concept of *weighting a distribution* (i.e. treating some particles as more important than others). Table 3.2 lists only a few of the possible ways of weighting a distribution. The first is weighted by number and length and the second is weighted by number and area, since

$$(\overline{d^2})^{1/2} = \bar{d}_{N\mathscr{A}} = \left(\frac{\sum n_i \mathscr{A}_i}{k' \sum n_i}\right)^{1/2}.$$

† This is often simply called the surface (or area) average diameter.

Table 3.2
Some possible average dimensions for colloidal particles

	Name	Symbol	Definition	Quantity averaged	Weighting factor
(i)	Number length mean diameter	$\bar{d}$ (or $\bar{d}_{\mathrm{NL}}$)	$\frac{\sum n_i d_i}{\sum n_i}$	Diameter	Number in each class
(ii)	Number area mean diameter	$(\overline{d^2})^{1/2}$ or $\bar{d}_{\mathrm{N}\mathscr{A}}$	$\left(\frac{\sum n_i d_i^2}{\sum n_i}\right)^{1/2}$	Particle area	Number in class
(iii)	Number volume mean diameter	$(\overline{d^3})^{1/3}$ or $\bar{d}_{\mathrm{NV}}$	$\left(\frac{\sum n_i d_i^3}{\sum n_i}\right)^{1/3}$	Particle volume	Number in class
(iv)	Mass area mean diameter	$\bar{d}_{\mathrm{m}\mathscr{A}}$	$\left(\frac{\sum m_i d_i^2}{\sum m_i}\right)^{1/2}$	Particle area	Mass in each class

Instead of weighting by volume and number as in distribution (iii) we can equally well weight the distribution by mass and area as in (iv) or by mass and volume:

$$\bar{d}_{\mathrm{mV}} = \left(\frac{\sum m_i V_i}{k'' \sum m_i}\right)^{1/3}$$

where m_i is the mass of material in class i that is characterized by a volume V_i. The constants k' and k'' in these relations are geometric factors that may be calculated for simple geometries. ($k' = \pi$ and $k'' = \pi/6$ for spheres if we retain the definition of $\bar{d}_{\mathrm{ab}}$ suggested in Table 3.2; they are often ignored.)

3.3.3 *The continuous distribution function*

Figure 3.3.1 shows that the histogram can be replaced by a smooth curve but it is important to note that *this is not a plot of n_i against d_i*. For the continuous curve, $F(d_i)$, the number of particles dn_i in the range d_i to $d_i + \mathrm{d}d_i$ is given by:

$$\mathrm{d}n_i = F(d_i)\,\mathrm{d}d_i \qquad (3.3.7)$$

where the function $F(d_i)$ is called the (number) distribution function for d_i. In Fig. 3.3.1 it will be noticed that all of the classes have the same width, so that the height of the rectangles is proportional to n_i. In the more general case, we may choose to vary the width of the classes and in that case it is preferable to draw rectangles whose *areas* reflect the numbers of particles in the class. Then we can readily relate the distribution function, F, to n_i using eqn (3.3.7). It is apparent from Fig.

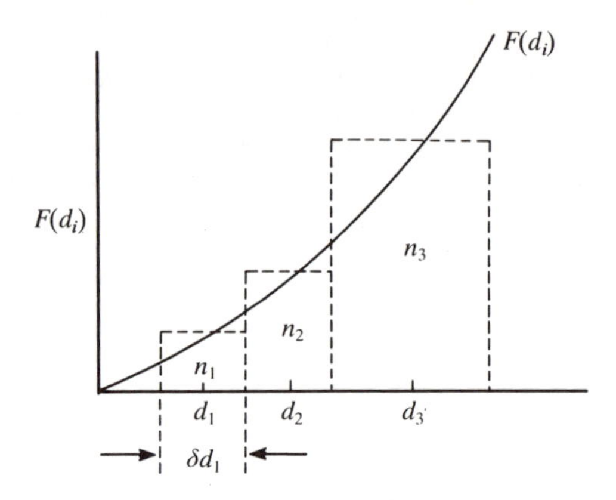

FIG. 3.3.3. The continuous distribution function, $F(d_i)$ and its relation to the frequency histogram.

3.3.3 that:

$$n_i \doteqdot F(d_i)\delta d_i \tag{3.3.8}$$

and the area under the $F(d_i)$ curve will give the total number of particles. For the data in Table 3.1 the value of $F(d_i)$ is related to that of f_i by a constant factor, since

$$f_i = \frac{n_i}{N} = \frac{F(d_i)\delta d_i}{N}.$$

One could generate the distribution function (in nm^{-1}), then, by multiplying column 4 by $(2500/50) = 50$ and it would have the same shape as the broken line in Fig. 3.3.1.

3.3.4 *Logarithmic distributions*

We noted earlier (section 3.1) that size distributions often extend over several orders of magnitude. Figure 3.3.4 shows a possible distribution that could be obtained by a direct counting operation. It extends over only two orders of magnitude but shows a very asymmetric distribution typical of material produced by grinding. In principle it is possible to describe such a curve with any desired degree of accuracy, by using a sufficient number of parameters (e.g. the mean and the second, third etc. moments). In practice, it is much better to transform the distribution into a more symmetrical shape so that two parameters still give a reasonable description.

A common procedure for curves like Fig. 3.3.4 is to convert the length to a logarithmic scale so that the spread of values is more easily

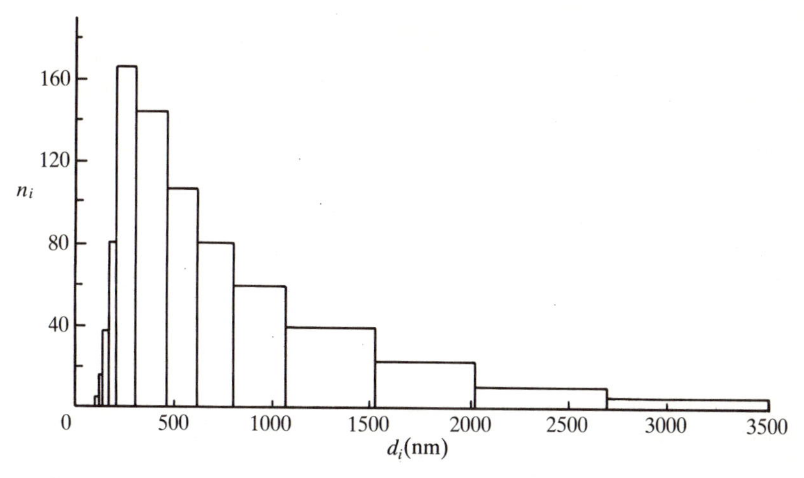

FIG. 3.3.4. A possible size distribution produced by grinding or crushing.

accommodated. Note also that the width of the rectangles in Fig. 3.3.4 is varying with d_i and the height measures the number, n_i, in each class. To convert n_i to an area basis we would simply set $\mathscr{F}(d_i) = n_i/\delta d_i$ as before. In this case, however, we wish to calculate the distribution function for $\ln d_i$ so we set

$$dn_i = \mathscr{F}(d_i)\, d \ln d_i = \frac{\mathscr{F}(d_i)\, dd_i}{d_i}$$

so that

$$n_i \doteqdot \mathscr{F}(d_i) \frac{\delta d_i}{d_i}. \tag{3.3.9}$$

In Table 3.3 the calculation is taken a stage further by dividing through by N to obtain f_i and converting this to a percentage (column 6) and this function is plotted in Fig. 3.3.5. It is called (Allen 1975) the relative percentage frequency distribution function and is useful for comparing different samples over the same size range, since the calculation takes account of differences in the sample size, N.

3.3.5 *The geometric mean*

The mean of the values of $\ln d_i$ is defined in the usual way (cf. eqn (3.3.1)):

$$\overline{\ln d} = \sum_i f_i \ln d_i \tag{3.3.10}$$

Table 3.3

Size range (nm)	Interval δd_i (nm)	Average size d_i	Number in this interval (frequency) n_i	Fraction in this interval f_i	$100\mathscr{F}/N = \frac{100 f_i d_i}{\delta d_i}$	$\ln(d_i/(\text{nm}))$
50–80	30	65	3	0.004	0.833	4.17
80–100	20	90	15	0.019	8.65	4.50
100–140	40	120	38	0.049	14.62	4.79
140–200	60	170	81	0.104	29.4	5.14
200–300	100	250	163	0.209	52.2	5.52
300–420	120	360	143	0.183	55.0	5.89
420–600	180	510	108	0.138	39.2	6.23
600–800	200	700	83	0.106	37.2	6.55
800–1100	300	950	63	0.081	25.6	6.86
1100–1500	400	1300	42	0.054	17.5	7.17
1500–2000	500	1750	23	0.030	10.32	7.47
2000–2700	700	2350	13	0.017	5.60	7.76
2700–3500	800	3100	5	0.006	2.48	8.04

$N = \sum n_i = 780.$ $\sum f_i = 1.00.$

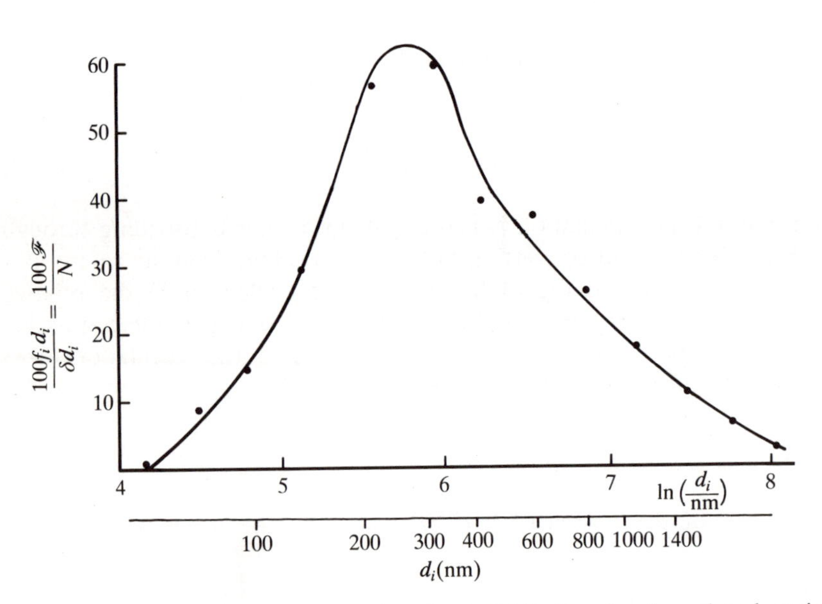

FIG. 3.3.5. The relative percentage frequency distribution function plotted against $\ln(d_i/\text{nm})$. (Data from Fig. 3.3.4.)

and this quantity is related not to the arithmetic mean of the diameters but to the *geometric mean* $\bar{d}_g$, which is the Nth root of the product of all the diameters:

$$\bar{d}_g = \left\{\prod_i (d_i^{n_i})\right\}^{1/N} \tag{3.3.11}$$

so that

$$\begin{aligned}\ln \bar{d}_g &= \frac{1}{N}\ln\left(\prod_i (d_i^{n_i})\right)\\ &= \frac{1}{N}\sum_i n_i \ln d_i\\ &= \sum_i f_i \ln d_i = \overline{\ln d}.\end{aligned} \tag{3.3.12}$$

(The symbol $\prod_i$ here requires the arithmetic product to be taken of the i terms.)

Although in these definitions it is usual to use number averaging there is no reason, in principle, why they cannot be extended to mass, area, or volume averages (section 3.3.2).

3.3.6 *The measure of polydispersity*

We noted earlier (section 3.3.2) that the spread of a distribution can be described with any desired degree of accuracy by calculating its various moments (eqn (3.3.4)). In many experimental situations however, we do not have access to the entire distribution but may only have estimates of various possible mean values. Because these are related to the moments they can be used to estimate the spread of the distribution or *degree of polydispersity,* as indicated by eqn (3.3.3) (and Exercise 3.3.2). Thus one can calculate the standard deviation from a knowledge of the area mean and length mean diameters. Alternatively, one can use the ratio of these two quantities as a measure of polydispersity, P_d (Exercise 3.3.3):

$$P_d = \bar{d}_{N\mathcal{A}}/\bar{d}_{NL} = \left\{1 + \frac{\sigma^2}{\bar{d}^2}\right\}^{1/2}. \tag{3.3.13}$$

The spread of diameters in a polymer latex system is often characterized by the coefficient of variation, defined by:

$$\text{Coefficient of variation} = \frac{\sigma}{\bar{d}} \times 100\%. \tag{3.3.14}$$

Typically, a system would be regarded as monodisperse if the coefficient of variation were less than 5 per cent (or, at most, 10 per cent).

Exercises. 3.3.1 Calculate the mean, standard deviation, and the variance of the distribution shown in Table 3.1. What is the difference between the mean and the mode in this case?

3.3.2 Establish eqn (3.3.3) from (3.3.2).

3.3.3 Calculate the number area mean diameter, $\bar{d}_{N\mathscr{A}}$ of the particles described in Table 3.1 and compare it with the number length mean diameter. Show that $\bar{d}_{N\mathscr{A}} = (\sigma^2 + \bar{d}^2)^{1/2}$ and check this with the result obtained in Exercise 3.3.1 (note the relevance to eqn (3.3.13)).

3.3.4 The accompanying figure shows a simple distribution function which can be approximated by a parabola:

$$F(d_i) = a + b\frac{d_i}{\bar{d}} + c\left(\frac{d_i}{\bar{d}}\right)^2.$$

Show that

$$F = \frac{3Nd_i}{2\bar{d}^2}\left(1 - \frac{d_i}{2\bar{d}}\right)$$

where N is the total number of particles.
[Hint: First show that $F = bu(1 - u/2)$ where $u = d_i/\bar{d}$.]

Verify that $\bar{d} = \int d_i \, dn_i / \int dn_i$. What is the maximum value of F?

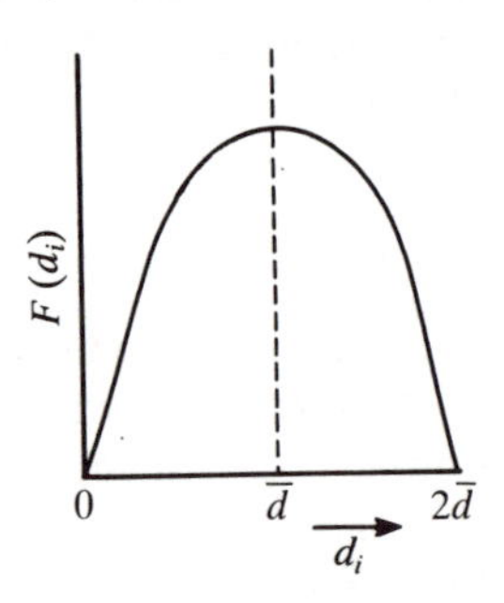

FIG. 3.3.6. Parabolic distribution (Exercise 3.3.4).

3.3.5 Calculate the number (arithmetic) mean of d_i ($\bar{d}$ or $\bar{d}_{NL}$) and its standard deviation and do the same for $\ln d_i$ for the distribution in Table 3.3. What is the (number length) geometric mean diameter of these particles?

3.3.6 Calculate the number volume mean diameter of the particles in Table 3.1 and compare it with the number area mean diameter calculated in Exercise 3.3.3. Why is it larger? Is this always true?

Show that the mass average diameter, $\bar{d}_{mL}$, is the ratio of the fourth to the third moment of the distribution.

3.4 Theoretical distribution functions

3.4.1 *The normal distribution*

Various functions have been proposed to describe the distribution function, F, obtained in particle size analysis. Some, like the Nukiyama–

Tanasawa equation (Cadle 1965, p. 36) are completely empirical, while others (e.g. the Rosin–Rammler equation; Herdan 1960, p. 86) have some theoretical basis. By far the most commonly used, and most firmly based general relationship is, however, the standard (normal or Gaussian) distribution, which we have already encountered:

$$f_G(x) = \frac{1}{\sigma\sqrt{(2\pi)}} \exp\left\{-\frac{1}{2}\left(\frac{x-\bar{x}}{\sigma}\right)^2\right\}. \qquad [2.1.12]$$

This is shown again in Fig. 3.4.1(a). Note that, in principle, the deviations from the mean extend to infinity in both directions but in practice the bulk of the distribution (68.2%) lies within $\pm\sigma$ of the mean and less than 0.3% lies outside the range $\bar{x} \pm 3\sigma$.

The normal distribution function is closely related to the error function

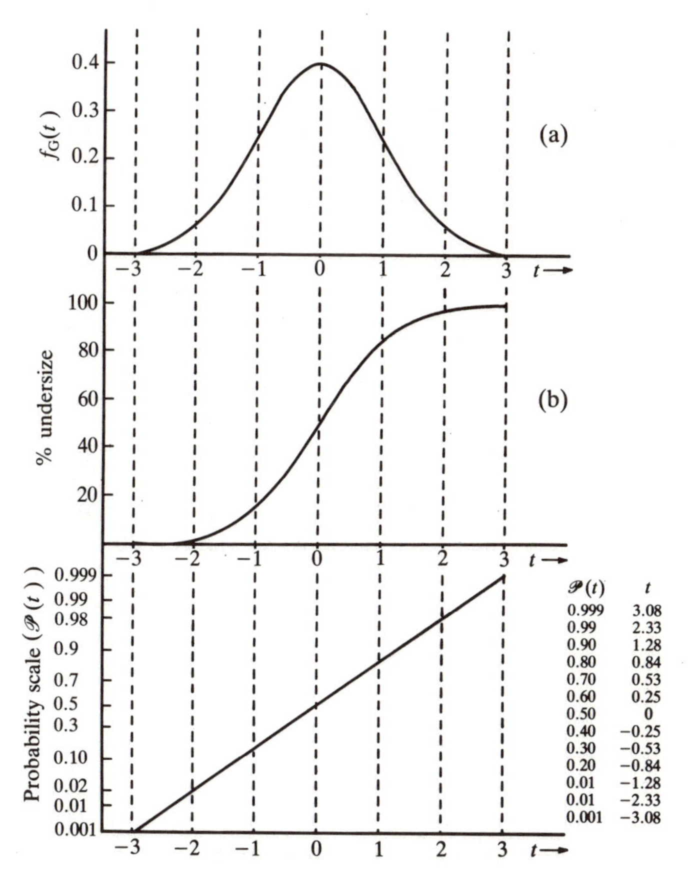

FIG. 3.4.1. The normal (Gaussian) distribution and the cumulative distribution curves.

and lies at the heart of the statistical treatment of errors. The justification for applying it to particle size distributions is that the positive or negative *differences* from the mean value that occur can be assumed to be caused by the operation of a large number of uncontrolled (and uncontrollable) influences. To the extent that those influences operate in a random fashion, eqn (2.1.12) ought to be applicable (Herdan 1960, p. 73). In fact, we know that the distribution is often very different from eqn (2.1.12) but the distribution of $\ln d$ may then be close to normal (Fig. 3.3.5). We will, therefore, confine ourselves to consideration of the normal and the log-normal distributions.

The significance of the variance (σ^2) is clear from eqn (2.1.12). It is the scaling parameter that determines how rapidly the exponential function drops to zero. The pre-exponential factor is introduced so that:

$$\int_{-\infty}^{\infty} f_{\mathrm{G}}(x)\,\mathrm{d}x = 1. \tag{3.4.1}$$

(This is the normalizing condition, and when used in this form f_{G} gives the fraction of the population in each interval. Then $\mathrm{d}n_i = Nf_{\mathrm{G}}\,\mathrm{d}x$.)

The distribution can also be represented by the *cumulative distribution curve,* which represents the fraction, P, of material that is less than a given size (Fig. 3.4.1(b)). Setting

$$t = (x - \bar{x})/\sigma \tag{3.4.2}$$

we have

$$f_{\mathrm{G}}(t)\,\mathrm{d}t = \frac{1}{\sqrt{(2\pi)}}\exp\left(\frac{-t^2}{2}\right)\mathrm{d}t$$

and

$$P(\text{all } t \leqslant t_i) = \int_{-\infty}^{t_i} f_{\mathrm{G}}(t)\,\mathrm{d}t = \frac{1}{\sqrt{(2\pi)}}\int_{-\infty}^{t_i}\exp\left(-\frac{t^2}{2}\right)\mathrm{d}t. \tag{3.4.3}$$

This function is closely related to the *error function* ($\operatorname{erf} t$), which measures the area under the error curve between the mean and some particular value of t:

$$\operatorname{erf} t_i = \frac{1}{\sqrt{(2\pi)}}\int_{0}^{t_i}\exp\left(-\frac{t^2}{2}\right)\mathrm{d}t. \tag{3.4.4}$$

In terms of the error function:

$$P(t \leqslant t_i) = \tfrac{1}{2} + \operatorname{erf} t_i. \tag{3.4.5}$$

(Note that the integral in eqn (3.4.4) is negative for $t_i < 0$ (i.e. $x < \bar{x}$) so that $P(t \leqslant t_i)$ is then less than $\frac{1}{2}$.) Values of the error function are given in standard texts (e.g. Herdan 1960, p. 77 or the CRC Handbook). The fraction of material oversize is, of course:

$$P(t \geq t_i) = 1 - P(t \leq t_i) = \tfrac{1}{2} - \operatorname{erf} t_i.$$

The relation between the normal curve and the cumulative curve is illustrated in Fig. 3.4.1. It is possible to purchase special graph paper ('probability paper') on which the ordinate scale varies in such a way as to convert the sigmoidal cumulative distribution function (Fig. 3.4.1(b)) into a straight line (3.4.1(c)). Such paper is useful for testing whether a particular set of data does or does not approximate to a normal distribution. If it does, then the standard deviation can be read from the graph by identifying the size that is bigger than 15.9 per cent (i.e. 50–(68.2/2) per cent) of the sample and subtracting this from the mean.

What, then, is the relation between the standard deviation as it appears in eqn (2.1.12) and the formula given earlier for a discrete distribution (eqn (3.3.2))? We can replace the right hand side of eqn (3.3.2) by its counterpart for a continuous variable:

$$\left[\frac{\int (d - \bar{d})^2 \, \mathrm{d}n}{\int \mathrm{d}n}\right]^{1/2} = \left\{\frac{1}{N}\int (d - \bar{d})^2 \, \mathrm{d}n\right\}^{1/2} \tag{3.4.6}$$

and substitute

$$\begin{aligned} \mathrm{d}n &= Nf \, \mathrm{d}d \\ &= \frac{N}{\sigma\sqrt{(2\pi)}} \exp\left\{-\frac{1}{2}\left(\frac{d - \bar{d}}{\sigma}\right)^2\right\} \mathrm{d}d \end{aligned}$$

to show that the expression (3.4.6) is indeed equal to σ. This is exactly analogous to the demonstration (using eqn (2.1.25)) that the r.m.s. displacement of a diffusing particle is equal to σ (Exercise 2.1.7). Any of the expressions given for discrete distributions can be converted in the same way to treat the continuous distribution function (see Exercise 3.4.2).

3.4.2 *The log-normal distribution*

We have noted already (Fig. 3.3.5) that the particle size produced by grinding often follows an approximately log-normal distribution.

This is the expected outcome if *ratios* of equal amount greater than or less than the mean are of equal likelihood rather than *differences* from the mean (Herdan 1960, p. 81). Making the transformation $z = \ln d$ we then say that d is log-normally distributed if z has the distribution function:

$$f_G(z) = \frac{1}{\sigma_z\sqrt{(2\pi)}}\exp\left\{-\frac{1}{2}\left(\frac{z-\bar{z}}{\sigma_z}\right)^2\right\}. \tag{3.4.7}$$

In that case the distribution of d is:

$$f_G(\ln d) = \frac{1}{\ln \sigma_g\sqrt{(2\pi)}}\exp\left\{-\frac{1}{2}\left(\frac{\ln d - \ln \bar{d}_g}{\ln \sigma_g}\right)^2\right\} \tag{3.4.8}$$

where $\bar{d}_g$ is the geometric mean of the values of d, and σ_g is the geometric standard deviation of the distribution of ratios around the geometric mean. Note that, by analogy with eqn (3.3.12), the logarithm of σ_g is equal to the standard deviation of $\ln d$. The number of particles between d_1 and d_2 in a log-normal distribution is (Herdan 1960, p. 81):

$$n_i = \frac{N}{p\sqrt{(2\pi)}}\int_{d_1/p}^{d_2/p}\exp\left\{-\frac{1}{2}\left(\frac{\ln d - \ln \bar{d}_g}{p}\right)^2\right\}\mathrm{d}(\ln d) \tag{3.4.9}$$

where $p = \ln \sigma_g$.

3.4.3 *Other distributions*

Although in many cases a particle size distribution can be transformed in such a way as to make it approximate a normal distribution, there are, of course, situations in which this is impossible. The most obvious case is when the sample has more than one modal size (Fig. 3.4.2). Bimodal or even polymodal distributions can occur in the preparation of a colloidal sol by a condensation method (section 1.5.1) if there are two different

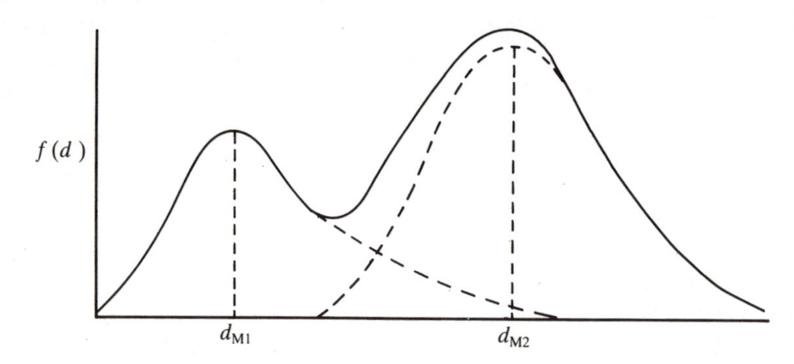

FIG. 3.4.2. A bimodal distribution resulting from the overlap of two 'normal' distributions having modes d_{M1} and d_{M2}.

nucleation periods separated by a significant time interval. Although it is, in principle, possible to represent such a shape using a polynominal:

$$f = \sum_{n=0}^{m} a_n d^n$$

it is much better to attempt to resolve the underlying distributions, as suggested in Fig. 3.4.2. This is called *deconvolution* and it is a problem often encountered in the interpretation of spectral data; computer programs can be written to perform the task if the underlying distributions have a known form (e.g. Gaussian or Lorentzian).

One final word about the normal distribution. When one first encounters the use of error concepts in the physical sciences one is tempted to regard the physical measurement (say the length of a piece of wire) as the important quantity and the associated error or standard deviation as at best merely an indication of reliability and at worst an unavoidable nuisance. It should be clear from the discussion of particle size distribution that the standard deviation is just as important as the mean in specifying the population, since in that case there is no single 'true size' that the measurement is attempting to estimate. There are in fact many situations in which the deviation is more important than the mean. The translational diffusion of a colloidal material (section 2.1) is an obvious example.

At a more fundamental level, however, the normal curve stands as one of the basic reference curves to which other natural distributions can be compared. If they correspond reasonably closely then the whole panoply of statistical methods can be applied with confidence. A few measurements then serve to define the whole population. It has been rightly said that the normal curve is to the statistician what the straight line is to the physicist.

Exercises. 3.4.1 Show that the normal distribution curve has inflexions on either side of the mean and the distance between them is 2σ.

3.4.2(i) Show that, for the distribution described in Exercise 3.3.4, the standard deviation is given by $\sigma = \bar{d}/\sqrt{5}$.

(ii) What fraction of the material lies between $\bar{d} \pm \sigma$ in this case? Compare this with the normal distribution.

3.5 Sedimentation methods of determining particle size

There are many indirect methods of estimating the particle size of colloidal dispersions, of which the oldest and most widely used depend on the principle of determining the sedimentation rate.

3.5.1 *Sedimentation under gravity*

(a) *Theory*. When a particle of mass m begins to settle through a fluid under the influence of gravity (Fig. 3.5.1) it is initially acted upon by three forces: the gravitational force, mg, the upthrust due to the displaced fluid, $m'g$, and the frictional force, f_v, due to the (viscous) drag of the surrounding fluid. From Newton's Law the net force is given by:

$$mg - m'g - f_v = m\frac{du}{dt} \tag{3.5.1}$$

where g is the acceleration due to gravity and u is the velocity at time t.

The frictional force increases with the particle velocity and for colloidal particles settling in a dense medium (like water) it very quickly balances the net downward force (see Exercise 3.5.1); the acceleration is then zero and the particles travel with their *terminal velocity* u_t. For a particle of reasonably regular shape, the drag force is, as previously stated:

$$f_v = Bu_t. \tag{2.1.16}$$

For a rigid spherical particle of radius r we can again set:

$$B = 6\pi r\eta. \tag{2.1.20}$$

Thus if ρ_s and ρ_l are the densities of the solid particles and the liquid

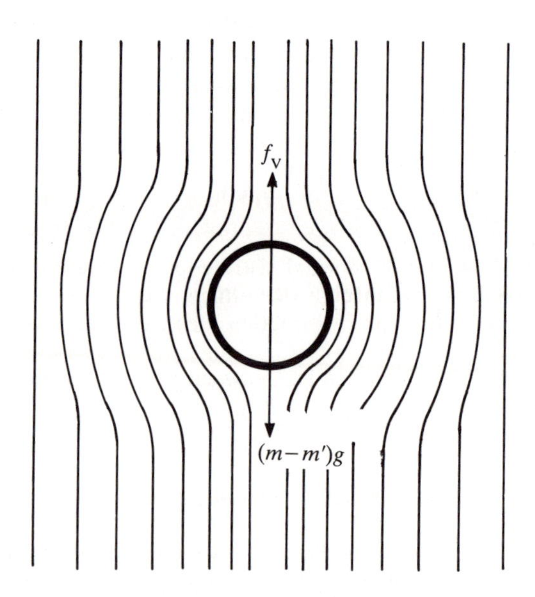

FIG. 3.5.1. Forces on a particle settling under gravity.

respectively we have, for the *Stokes settling radius, r*:

$$(\rho_s - \rho_l)\tfrac{4}{3}\pi r^3 g = 6\pi u_t r \eta \tag{3.5.2}$$

so that

$$u_t = \frac{2}{9}\frac{(\rho_s - \rho_l)gr^2}{\eta}. \tag{3.5.3}$$

Although, in principle, u_t can be measured directly with an ultramicroscope, it is more usual to follow the changes in particle concentration at a certain depth, h, below the surface of the suspension. After time t, all particles for which $u_t(r) \geq h/t$ will have settled beyond this depth and by following the concentration as a function of time, a particle size distribution can be built up.

For a collection of spheres of different radii, the 'average' settling radius $\bar{r}_s$ is, from eqn (3.5.2), given by:

$$(\bar{r}_s)^2 = (\bar{r}_{NV})^3/\bar{r}_v \doteqdot (\bar{r}_{NV})^3/\bar{r}_{N\mathcal{A}} \tag{3.5.4}$$

where r_v is the viscous (drag) radius. Since the frictional (drag) resistance acts over the surface of the particle, $\bar{r}_v$ is approximately equal to the number area mean radius, $\bar{r}_{N\mathcal{A}}$, at least for the slowly settling systems encountered in colloid science (Allen 1975, p. 171).

For non-spherical particles the value of B can sometimes be estimated theoretically, but only for simple shapes. Thus, for an oblate spheroid (Fig. 3.1.1(a)), the correction to eqn (2.1.20) is simple for small departures from sphericity:

$$B = 6\pi a\eta(1 - \varepsilon/5); \quad (\varepsilon \to 0) \tag{3.5.5}$$

where ε, the eccentricity, is $(a-b)/b$. Allen (1975, p. 169) gives a few similar relations for regular shapes, but these are of limited value for colloid settling because they require a knowledge of the initial orientation of the particles with respect to the gravitational field. Unless one or other of the principal axes of symmetry (assuming they exist) is aligned with the field, the particle's motion is very irregular and certainly not confined to a vertical direction. Fortunately, as the particle size decreases, these considerations become rather less important because the particles become subject to increasingly vigorous Brownian motion, which has two important effects: for anisometric particles the orientation becomes increasingly randomized and tends to simplify the problem, but the increasing translational Brownian motion (i.e. diffusion) interferes with the gravitational settling process.

We have already shown that when Brownian diffusion is large enough to be measured, the friction coefficient can be calculated directly from the diffusion coefficient using Einstein's equation:

$$B = kT/\mathcal{D}. \qquad [2.1.19]$$

When that occurs, however, the settling rate is significantly affected by the thermal diffusive motion of the particles. The r.m.s. displacement for a sphere is (from eqns (2.1.19), (2.1.20), and (2.1.26)):

$$(\overline{x^2})^{1/2} = (kT/3\pi r\eta)^{1/2}\sqrt{t} \qquad (3.5.6)$$

which for a 1 μm particle in water would be about 0.7 μm in 1 s. The gravitational settling by a particle of density 2×10^3 kg m^{-3} in the same time would be only about 2 μm.

For colloidal particles, then, gravitational settling is of limited use except for very dense particles. Apart from the interference of Brownian motion, there are considerable practical difficulties in ensuring the necessary degree of temperature stability over long settling periods; very small convection currents can easily vitiate the results. Settling is however, widely used in industry and agriculture for characterizing suspensions of fairly dense solids towards the upper end of the colloid size range. Even if the particles are not spherical and they are too large to have a measurable Brownian diffusion coefficient one can still obtain an estimate of the distribution of the '*equivalent settling radius*', r_e, which though difficult to interpret exactly (cf. eqn 3.5.4)), is undoubtedly a useful comparison measurement between different samples of the same material. For any particular particle, r_e is the radius of the sphere of the same density which settles at the same rate.

Before discussing the actual procedure for obtaining the distribution of r_s we should note some limitations in the above analysis. To begin with, eqn (3.5.3) is derived for a single particle settling all alone in an infinite expanse of fluid. It can only be correct when the streamlines around the separate particles do not interfere with one another (i.e. in the absence of hydrodynamic interaction). It should be applied only to the sedimenting of dilute suspensions (<1 per cent); for higher concentrations a rather more elaborate treatment (Chapter 9) is necessary. Equation (3.5.3) also requires that the particles and the fluid be characterized by their densities, ρ_s and ρ_l. For large particles (>1 μm) this presents little problem but for small colloidal particles the possibility exists that rather special effects occur at the interface between the particles and the liquid. The particle may have, associated with it, a few molecular layers of liquid that move with it as a single kinetic unit. This *solvated* particle will then have a density intermediate between ρ_s and ρ_l; even the density of this adsorbed liquid layer may differ slightly from that of ρ_l in the bulk. The extreme case of solvation occurs, of course, with lyophilic colloids, but such substances do not usually settle appreciably under gravity. A more common occurrence with particulate dispersions is that the particles are permeable to the fluid or they are aggregated into flocs, which contain trapped fluid. The sedimentation characteristics are then those of the composite. We will return to the case of flocculated material in Chapter 9.

Equation (3.5.3) is also derived for solid particles, with the condition that there is no slip between the particle surface and the fluid (i.e. the fluid velocity is the same as that of the adjoining point on the particle at each point on the surface). For emulsion droplets (which usually rise in a gravitational field) and air bubbles, the drag force can be shown to be (Frumkin and Levich 1946):

$$f_v = 6\pi r u\left(\frac{3\eta_1 + 2\eta_2}{3\eta_1 + 3\eta_2}\right)\eta_2 \tag{3.5.7}$$

where η_1 and η_2 are the viscosities of the drop (or bubble) and the fluid respectively. Equation (3.5.7) assumes that the interface is perfectly fluid so that it cannot support an applied stress. For air bubbles in water ($\eta_1 \ll \eta_2$) the drag is then reduced by one third. In actual practice, this seldom occurs because the drop or bubble surface almost always has an adsorbed layer of a surfactant (whether by accident or design) and this tends to make the interface rigid so the correction is not applicable (Levich 1962) and eqn (3.5.3) still holds.

(b) *Determining the size distribution.* The relation between particles size distribution and the concentration, C, at a depth h below the surface of a sedimenting suspension after time t is derived as follows (Allen 1975, p. 192):

Initially ($t = 0$) the concentration is uniform and given by:

$$C(h, 0) = \frac{m_s}{v_s + v_l} \tag{3.5.8}$$

where m_s and v_s are the mass and volume of the solid and v_l is the volume of the fluid. Consider a thin horizontal element at depth h below the surface. When sedimentation begins, any particles that leave this element are replaced by particles of the same size entering from above. When the largest particles, initially present throughout the suspension, have had time to fall a distance, h, there will be no more of them available to replenish the concentration and it will, therefore, fall. Hence the concentration in this thin volume element at time t is the concentration of all particles of radius less than R, where R is the size of particle with velocity h/t:

$$C(h, t) = \frac{m_s'}{v_s' + v_l} = \int_{R_{\min}}^{R} F(r)\,\mathrm{d}r. \tag{3.5.9}$$

Since

$$C(h, 0) = \frac{m_s}{v_s + v_l} = \int_{R_{\min}}^{R_{\max}} F(r)\,\mathrm{d}r \tag{3.5.10}$$

we have

$$\frac{C(h, t)}{C(h, 0)} = \left[\int_{R_{\min}}^{R} F(r)\,\mathrm{d}r\right] \Big/ \left[\int_{R_{\min}}^{R_{\max}} F(r)\,\mathrm{d}r\right]. \tag{3.5.11}$$

Note that in deriving eqn (3.5.11) we have assumed that v_s and v_s' are small compared to v_l.

A plot of $100C(h, t)/C(h, 0)$ as a function of R, calculated from the relation:

$$R = \left\{\frac{9\eta h}{2(\rho_s - \rho_l)gt}\right\}^{1/2} \tag{3.5.12}$$

will, therefore, be a plot of the cumulative percentage undersize by mass.

The concentration at depth h can be obtained by direct sampling (the pipette method) or by measuring the absorption of light or X-rays or by determining the density (using a Cartesian diver or a pressure transducer) or the hydrostatic pressure as a function of time. For details of actual experimental methods see Orr and DallaValle (1959, Chapter 3) or Allen (1975, Chapter 9). There are numerous commercial devices available for expediting such measurements.

An alternative method of determining the cumulative distribution curve is to measure the total mass of sediment accumulating on a balance pan suspended in the dispersion. The method was introduced by Oden (1926) and improved on by Bostock (1952). This description of the improved method is taken from Hiemenz (1977). Figure 3.5.2 shows a plot of the accumulated mass, M, of sediment on the pan as a function of time. This is made up of two contributions: (a) the mass, m, due to particles large enough to have sedimented through the entire depth; and (b) a fraction of the smaller particles that have only fallen a smaller distance:

$$M(t_1) = m + t_1\left(\frac{\mathrm{d}M}{\mathrm{d}t}\right)_{t_1}. \tag{3.5.13}$$

To see that the second term measures the contribution of the smaller particles, consider Fig. 3.5.2(a). The slope $\mathrm{d}M/\mathrm{d}t$ at any time, t, measures the rate of incidence of particles smaller than the cut-off size corresponding to that time, since all larger particles have already deposited. But this process has been going on at the same rate for the entire time, so it contributed a mass $t\,\mathrm{d}M/\mathrm{d}t$ to the pan. Figure 3.5.2(a) shows that the intercept of the tangent to the curve gives a direct measure

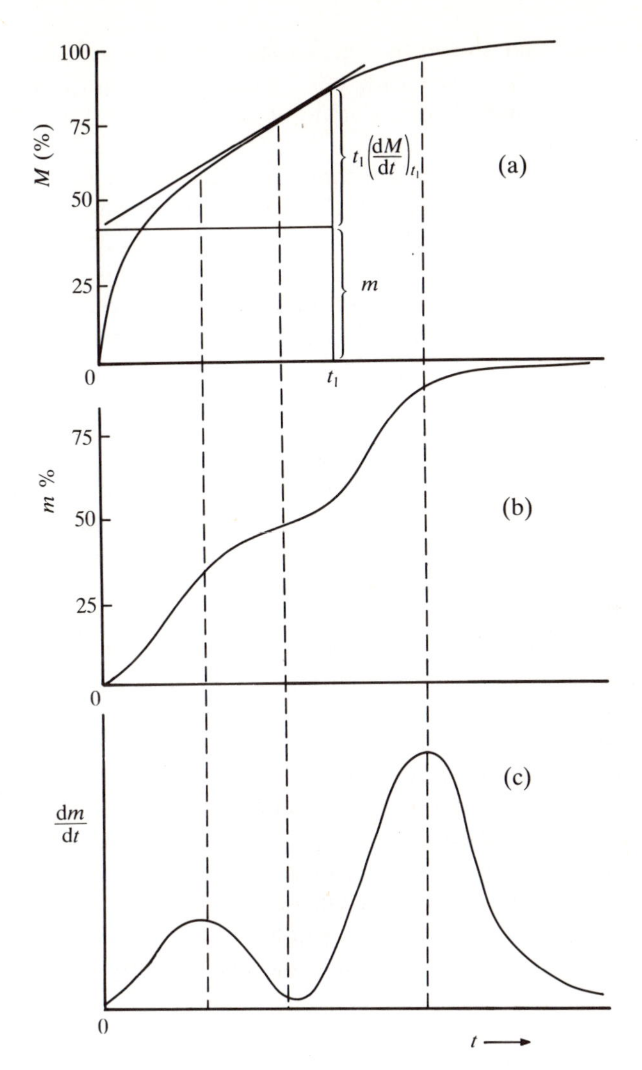

FIG. 3.5.2. Data treatment for the Oden balance method of particle size analysis (see text).

of the mass, m. A graphical differentiation of m with respect to t gives the (mass averaged) size distribution function. Note that

$$\frac{dm}{dt} = -\frac{t\,d^2M}{dt^2} \tag{3.5.14}$$

so that in preparing this function the initial data must be differentiated twice. Only data of very high accuracy could produce a reliable size distribution like that shown in Fig. 3.5.2(c), but it is usually possible to determine the modal value with fair accuracy. The time axis must, of course, be transformed into the value of R using eqn (3.5.12) as before (see Exercise 3.5.4).

One basic limitation of this technique is that the region immediately below the pan becomes depleted of particles and so has a lower density than the solution immediately above the pan. This results in a convective motion of the suspension medium, which interferes with the sedimentation process to some extent. A possible correction procedure is described by Allen (1975, p. 225) who also gives a more formal proof of eqn (3.5.13).

3.5.2 *Centrifugal sedimentation*

(a) *Time-dependent behaviour.* The time required, even for a large colloidal particle to settle through a reasonable distance under the influence of gravity alone (Exercise 3.5.2) makes that procedure rather limited. In most cases it is necessary to increase the sedimentation rate by subjecting the particles to centrifugation. Apart from the saving in time, there is then less danger of convection currents upsetting the results and the distance moved by sedimentation can be much greater than the Brownian motion.

Consider a particle immersed in a liquid at a distance x from the axis of the centrifuge head (or *rotor*), Fig. 3.5.3. If it were to stay at that distance as the head revolved, it would have to be acted on by a (*centripetal*) force directed towards the centre of the rotor and forcing the particle to travel in a circle. The magnitude of that force is equal to $(m - m')\omega^2 x$, where ω is the angular velocity (radians s^{-1}) of the rotor and $(m - m')$ is the apparent mass (corrected for buoyancy) of the particle. In the absence of that force, the particle moves away from the

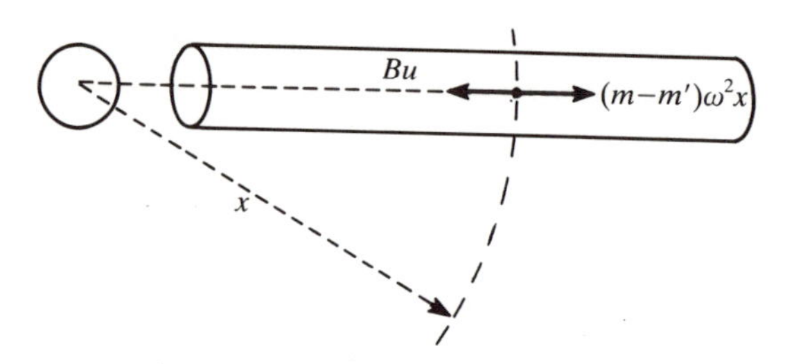

FIG. 3.5.3. Forces on a particle in a centrifuge tube. The outward force is an apparent (virtual) force invoked to explain the motion of the particle with respect to a coordinate frame attached to the rotor and moving with it.

axis of rotation as if it were acted on by a force of that magnitude acting outwards. Again it is retarded by a frictional force that is proportional to its velocity and again it takes only a very short time before these two forces are balanced (Exercise 3.5.1):

$$(m - m')\omega^2 x = Bu(x) = B\frac{\mathrm{d}x}{\mathrm{d}t}. \tag{3.5.15}$$

Note, however, that in this case the velocity is not constant but increases as the particle moves towards the outer end of the tube. The quantity:

$$S = \frac{u(x)}{\omega^2 x} = \frac{(m - m')}{B} \tag{3.5.16}$$

is called the *sedimentation coefficient* and is an important characteristic of the material. For polymeric materials (including proteins) a more appropriate form is obtained by considering the molar mass, M, of the solute, so that:

$$m - m' = \frac{1}{\mathcal{N}_\mathrm{A}}(M - \bar{V}_2\rho_1) = M(1 - \bar{v}_2\rho_1)/\mathcal{N}_\mathrm{A}$$

where $\bar{V}_2$ is the (partial) molar volume of the polymer and $\bar{v}_2$ is the volume per unit mass $(=\rho_\mathrm{s}^{-1})$; $\mathcal{N}_\mathrm{A}$ is the Avogadro constant. The sedimentation coefficient can then be written (using eqn (2.1.19)):

$$S = M(1 - \bar{v}_2\rho_1)/\mathcal{N}_\mathrm{A}B = \frac{M\mathcal{D}}{RT}(1 - \bar{v}_2\rho_1) \tag{3.5.17}$$

and, hence, a knowledge of S and the diffusion coefficient, $\mathcal{D}$, can be used to estimate the molar mass (or molecular weight) of the polymer.

S is obtained by writing eqn (3.5.15) in the form:

$$\frac{\mathrm{d}x}{x} = \frac{m}{B}\left(1 - \frac{\rho_1}{\rho_\mathrm{s}}\right)\omega^2\,\mathrm{d}t = S\omega^2\,\mathrm{d}t$$

and on integration:

$$\ln\frac{x_2}{x_1} = S\omega^2(t_2 - t_1). \tag{3.5.18}$$

Plots of $\ln x$ as a function of time at known rotation speeds can therefore be used to determine S. For colloidal particles where the notion of molar mass is inappropriate, the apparent mass of the particle can be obtained using eqn (3.5.16) and the friction factor estimated from diffusion experiments (eqn (2.1.19)). Of course, if the particles are known to be spherical one may use eqn (2.1.20) to estimate B directly and hence determine the radius:

$$r = \left(\frac{9\eta S}{2(\rho_s - \rho_l)}\right)^{1/2}. \qquad (3.5.19)$$

Values of S range from about 1 μs for large colloidal particles down to values of the order 10^{-13} s for proteins. The time unit 10^{-13} s is called 1 Svedberg, in recognition of the man who developed the ultracentrifuge and the above procedure for establishing protein molar masses.

The simplest application of eqn (3.5.18) is in the *two-layer sedimentation* technique in which the colloidal suspension is initially placed as a thin layer on top of the clear suspension medium. As centrifugation proceeds the components of that layer travel down the centrifuge tube at rates determined by eqn (3.5.18) and in favourable cases their individual progress can be followed, usually by an optical procedure. This method is particularly suited to the separation and characterization of mixtures of distinct materials with well-defined mass and density characteristics, such as mixtures of proteins. It does, however, have some problems (in particular the phenomenon of 'streaming') and more detailed treatments should be consulted (Allen 1975, pp. 266–73) for methods of minimizing their effects.

Calculation of a size distribution from the accumulated sediment is more complicated in the case of centrifugation because the particle velocity varies with distance from the rotor axis as well as with size. One way round the problem is to have the material sediment through a short distance but to start it at a point far from the axis of rotation. This procedure is used in the disc centrifuge, which enjoys a number of other advantages. For details see Allen (1975, Chapter 12).

(b) *Sedimentation equilibrium.* The Brownian motion of colloidal particles can still affect the sedimentation process, even in a centrifugation experiment. Indeed, in extreme cases the Brownian diffusion can oppose the sedimentation so effectively that the sedimentation appears to cease and an equilibrium (or steady-state) situation is reached in which the concentration profile of particles down the sedimentation tube remains constant with time. The situation is entirely analogous to the distribution of gas molecules in the atmosphere around the earth (section 2.1.1). The corresponding expression for the concentration of the colloid, as a function of distance from the axis of the centrifuge rotor, is:

$$C = C_0 \exp\left(+\frac{(m - m')\omega^2 x^2}{2kT}\right) \qquad (3.5.20)$$

where the potential energy of molecules in the Earth's gravitational field (mgh) is replaced by the particle potential energy in the centrifugal field.

(See Exercise 3.5.7 for an informal proof of this equation.) The distance x in eqn (3.5.20) is measured from some arbitrarily chosen reference line where the concentration is C_0. Note also that the exponent is positive in this case since the centrifugal field acts to concentrate the particles at larger values of x.

Equation (3.5.20) may be written:

$$\ln\frac{C_2}{C_1}=\frac{M_i\omega^2}{2RT}\left(1-\frac{\rho_1}{\rho_s}\right)(x_2^2-x_1^2) \tag{3.5.21}$$

which is the form used to calculate the molar mass (M_i) of a protein or polymer from centrifugation measurements. Note that it requires no assumptions about the friction factor B. Equation (3.5.21) does, however, assume ideal behaviour for the sedimenting material (Exercise 3.5.7). A more elaborate analysis, taking account of departures from ideality leads to:

$$\frac{\mathrm{d}\ln C}{\mathrm{d}(x^2)}=\frac{M_i\omega^2}{2RT}\left(1-\frac{\rho_1}{\rho_s}\right)\left\{1+\left(\frac{\partial\ln y_i}{\partial\ln C}\right)_{T,p}\right\}^{-1} \tag{3.5.22}$$

where y_i is the activity coefficient. This same activity coefficient correction can also be applied to equation (3.5.17) in more concentrated polymer or protein systems. The notion of activity coefficient is not really appropriate to particulate dispersions so we will not discuss this procedure further. It is more relevant to the behaviour of lyophilic than lyophobic colloids (section 1.4).

(c) *Determining size in the ultracentrifuge.* The normal centrifuge operates up to rotational speeds of a few thousand revolutions per minute (r.p.m.). High speed centrifuges take the limit up to 20 000–30 000 r.p.m. and the ultracentrifuge has its rotor spinning in a vacuum at speeds up to 60 000 r.p.m.

For particle size analysis the most important requirement is the provision of some means of measuring the concentration profile in the centrifuge tube whilst it is in motion. The most common procedures involve absorption or refraction of electromagnetic radiation (light, UV or X-ray) but other methods are possible (see Allen 1975, Chapter 12). Figure 3.5.4 illustrates diagrammatically the three main optical procedures. The Schlieren system (b) is the most useful for particulate systems at low concentration, but measurement of the absorption of X-rays is also useful, especially for very (optically) dense systems (see, for example, Ondeyka *et al.* 1978). The basic design of the ultracentrifuge (Fig. 3.5.5) has changed little since it was developed by Svedberg in the 1920s; the improvements have come in better electronic control of rotation speeds and tougher materials (especially titanium) for constructing the rotor,

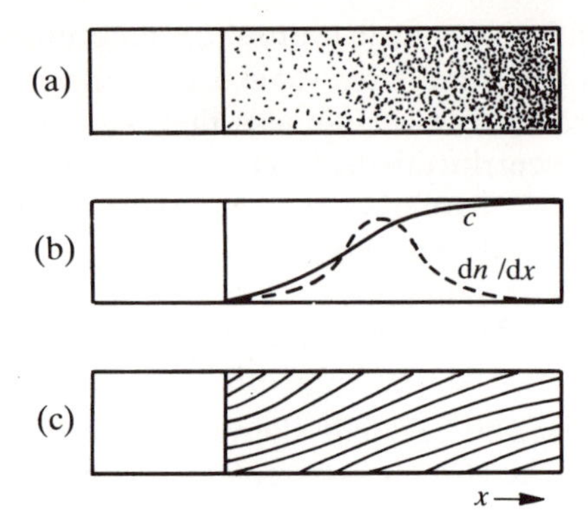

FIG. 3.5.4. Optical methods of detecting concentration profiles.
(a) The direct absorption method: Measures $c(x) \propto$ optical density.
(b) The Schlieren optical system: Detects the *gradient* of the *refractive index* (dn/dx) and is, therefore, very sensitive to the presence of a boundary between a rapidly and a more slowly settling material. Movement of the boundary with time can very easily be followed.
(c) Interference optics: the number of interference fringes crossed as one moves to a particular level down the axis of the tube is proportional to the concentration difference between that level and the meniscus.
(See Bull 1971 for a general discussion of these procedures.)

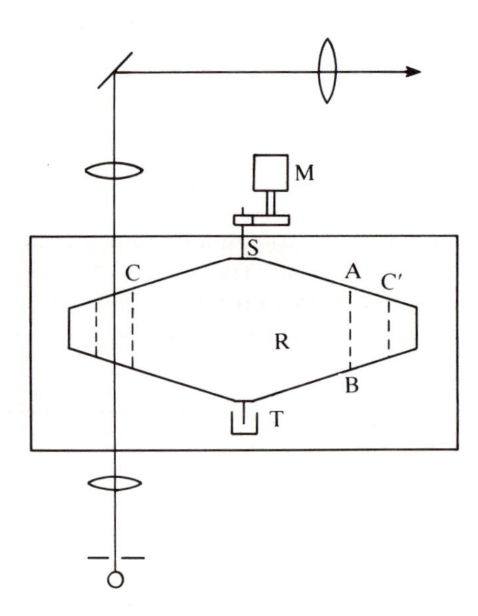

FIG. 3.5.5. Schematic diagram of an ultracentrifuge. C is the sample cell with windows, C′ is a matched cell, R is the rotor. S is the suspension wire and M is the motor. T is a thermistor connected by a mercury cup to the electrical circuit. (After Bull 1971, p. 363, with permission.)

which must withstand very high stresses at the maximum rotational speeds. The observation cell is sector-shaped rather than cylindrical so that sedimenting particles travel radially without striking the walls; this also reduces convective disturbance. The top of the cell (AB) is some distance from the axis of rotation so that the centrifugal field does not vary much down the cell. The rotor hangs by a flexible wire from the driving mechanism so as to minimize vibration and is driven by an electric motor through a train of gears. Temperature control in the evacuated space around the rotor is achieved by a flow of gas at low pressure. The optical effects are recorded photographically or photoelectrically. More detailed analytical procedures are available from a number of sources (see e.g. Bull 1971, Chapter 14).

Exercises. 3.5.1 Consider a spherical particle settling under gravity according to eqn (3.5.1). Show that during the period before it reaches terminal velocity its equation of motion is:

$$\frac{\mathrm{d}u}{\mathrm{d}t} = \rho g - Gu$$

where $\rho = (\rho_s - \rho_l)/\rho_s$ and $G = 9\eta/2\rho_s r^2$. Hence show that $u = (\rho g/G)[1 - \exp(-Gt)]$ during this period. How long does it take for a particle of radius 1 μm and density 3×10^3 kg m^{-3} to reach 99 per cent of its terminal velocity in water (density = 10^3 kg m^{-3}, viscosity = 10^{-3} N m^{-2} s)? Repeat the calculation for a radius of 0.1 μm and 0.01 μm. (Allen 1975, p. 158).

3.5.2 A suspension of silica particles ($\rho = 2.8$ g cm^{-3}) in water is allowed to settle in a cylinder at 20 °C. Calculate the time required for a particle of 2 μm radius to settle a distance of 20 cm, assuming it is spherical. (Take $\eta = 10^{-2}$ g cm^{-1} s^{-1} = 1 centipoise.) Convert the data to SI units, and repeat the calculation.

3.5.3 (Hiemenz 1977, p. 122). A suspension of Southern bean mosaic virus (SBMV) particles is placed in a thin layer on top of the clear suspension medium and then centrifuged at 12 590 r.p.m. The optical absorbance at 260 nm is measured along the settling direction as a function of time and the position, R, of the absorption maximum is obtained (Vinograd *et al.* 1963):

t(min)	R(cm)	t(min)	R(cm)
16	6.22	96	6.72
32	6.32	112	6.82
48	6.42	128	6.92
64	6.52	144	7.02
80	6.62		

Calculate the sedimentation coefficient of SBMV particles.

3.5.4 The following table gives values of the mass, M, of material deposited as a function of time on a balance pan suspended in a settling dispersion. Draw up a size distribution using the method described in connection with eqn (3.5.13). The pan has a radius of 2 cm and is exactly 15 cm below the surface of the (aqueous) suspension. The initial concentration of the suspension is 5.50 g l^{-1} and the density of the solid is 4.80 g cm^{-3}. (Take the density of water as 1.00 g cm^{-3} and assume again that $\eta = 10^{-2}$ g cm^{-1} s^{-1}.)

M(mg)	0	31	67	104	145	166	207	238
t(s)	27	40	55	74	99	122	134	148
M(mg)	290	311	363	415	487	518	549	
t(s)	181	200	270	330	403	493	545	
M(mg)	591	632	653	684	705	726	746	
t(s)	665	898	1097	1340	1810	2440	2980	

(Ignore convective effects and assume that the expected maximum deposition corresponds to the mass of the solid in the column of liquid above the pan.)

3.5.5 The time taken for a particle to reach its terminal velocity under gravity is about $5/G$ (Exercise 3.5.1). Show that the sedimentation coefficient, S, is also an approximate measure of this time.

3.5.6 In a sedimentation equilibrium experiment, Svedberg found the following concentration (C)–depth (x) profile for carboxyhaemoglobin:

x(cm)	4.61	4.56	4.51	4.46
C(%)	1.220	1.061	0.930	0.832

The rotor speed was 8710 r.p.m. and the temperature was 20.3 °C. Take the density of the water as 0.9988 g cm^{-3} and that of the dry protein as 1.338 g cm^{-3} and estimate the molar mass of the protein, assuming that the molecules sediment as individual particles.

3.5.7 An informal demonstration of the reasonableness of eqn (3.5.20) can be given by balancing the sedimentation force, $(m - m')\omega^2 x$ against the driving force for the diffusion process (eqn (2.1.15)).
Derive eqn (3.5.20) by this method. Then derive eqn (3.5.21).

3.5.8 Calculate the centrifugal acceleration at a distance of 7 cm from the axis of an ultracentrifuge rotor travelling at 20,000, 40,000, and 60,000 r.p.m. and compare this with the gravitational acceleration, g.

3.5.9 When protein A is subjected to ultracentrifugation at 20 °C, the particle boundary is found to move from a position 6.231 cm from the axis of the rotor at time $t = 0$ s to a distance of 6.823 cm after $t = 4500$ s. Calculate the sedimentation coefficient S for protein A. Given that its diffusion coefficient is 2.2×10^{-6} cm^2 s^{-1} and its partial specific volume is 0.80 g^{-1} cm^3, estimate its molar mass. (ρ_l at 20 °C for water = 0.9982 g cm^{-3}.)

3.6 Electrical pulse counters

The best known electrical pulse counter is the Coulter counter, which was originally designed simply to count the number of particles in a known volume of an electrolyte solution (~5 per cent salt) by drawing the suspension through a small orifice that had an electrode on either side of it. Passage of a particle through the orifice interferes with the current flow between the electrodes and the resistance changes can be amplified as voltage pulses and counted. A schematic diagram of the apparatus is shown in Fig. 3.6.1. With an orifice of accurately known diameter, a stable current supply and a known (variable) amplification factor, the height of each current pulse becomes proportional to the volume of the particle traversing the orifice. Using an electronic pulse height analyser then converts the apparatus into a counter and sizer. The layout of the device is described by Silverman *et al.* (1971) and a very full account of its various forms is given by Allen (1975, Chapter 13).

The mercury manometer, M, on the left is attached to a gentle suction apparatus, which creates a sufficient vacuum to draw some suspension through the sapphire orifice, O, which is set into the tube B. It also

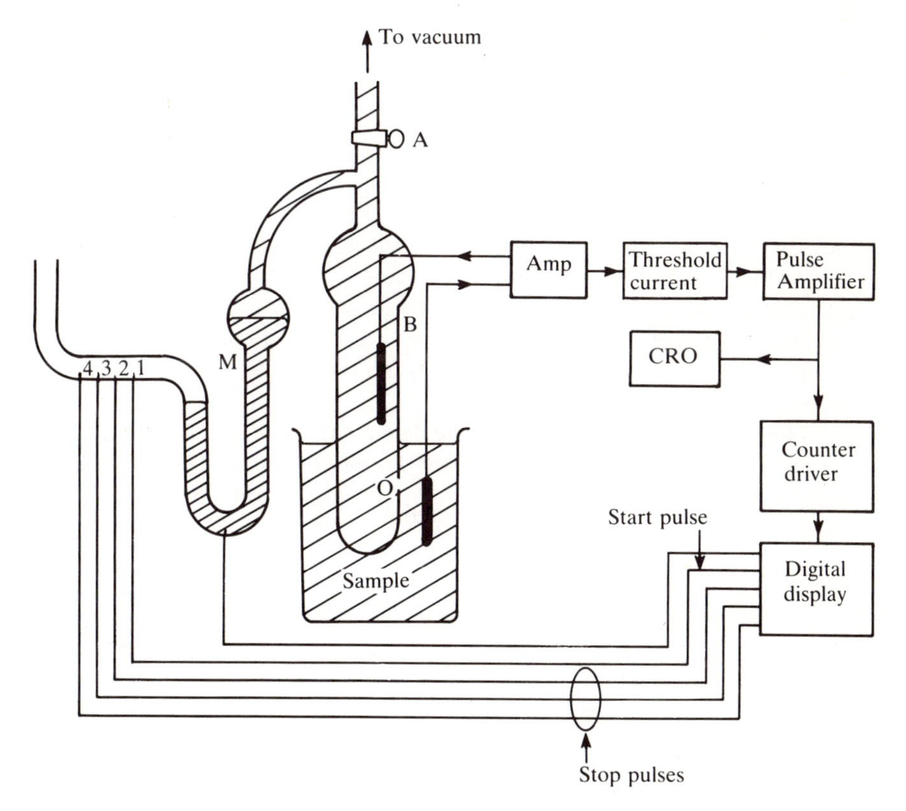

FIG. 3.6.1. Schematic arrangement of the Coulter counter. (After Allen 1975, p. 302, with permission.)

unbalances the mercury column, drawing it back to the right of the first electrical contact (1). The display sets itself to zero and the operator then closes the connection to the vacuum (tap A). The mercury siphon then takes over and draws more suspension in through the orifice. When contact (1) is reached counting starts and it can be stopped after a flow of 0.05 ml (contact 2), 0.5 ml or 2 ml (contact 4).

A setting on the instrument (the threshold) allows all pulses lower than a certain height to be ignored so that by advancing the threshold a cumulative distribution of pulse heights (proportional to particle volume) can be obtained. Some instruments also have an upper threshold, which makes possible measurement of relative frequency distribution (i.e. the number fraction of material lying between two predetermined volumes).

3.6.1 *Theory of the Coulter counter*

The instrument is calibrated using dispersions of known particle size by observing the pulse height and noting the threshold setting to which it corresponds. If the volume of the calibrating particles is v_c, then:

$$v_c = k_1 t_c \tag{3.6.1}$$

where t_c is the threshold. (This assumption, that the threshold is strictly proportional to volume, can be checked by doing calibrations at both ends of the scale. Provided one avoids the complications of coincidence (see below) and works within the recommended size range (from a diameter of about 40 per cent of the orifice diameter down to about 2 per cent, this assumption is a good one.) The calibration constant k_1 can then be used to determine the volume of an unknown particle from the corresponding threshold.

A check can also be made on whether all particles are being counted (i.e. there are no particles too small to be detected). The volume of particles, v_p, in a metered volume, v, of suspension of mass per unit volume, C is:

$$v_p = \frac{vC}{\rho_s} \tag{3.6.2}$$

where ρ_s is the particle density. A size analysis consists of decreasing numbers of counts n_i against increasing threshold settings, t_i. Now

$$v_p = \sum_i N_i v_i$$

where N_i $(=n_{i-1} - n_i)$ is the number of particles of volume v_i $(=k_1 \bar{t}_i =$

$k_1(t_{i-1}+t_i)/2$ i.e.

$$k_1 = v_p \Big/ \sum (\Delta n_i \bar{t}_i)$$

$$= \frac{vC}{\rho_s} \frac{1}{\sum \Delta n_i \bar{t}_i}. \tag{3.6.3}$$

If the value of k_1 obtained from eqn (3.6.3) agrees with that from eqn (3.6.1) then all particles are being detected. Otherwise the discrepancy is attributed to particles with a lower volume than the minimum.

The total range of the instrument is determined by the size of the orifice used: the minimum orifice size is about 16 μm diameter (so the minimum measurable size is around 0.3 μm). For routine laboratory use, however, the usual lower limit is 0.6 μm (using the 30 μm tube) because the smallest orifice is very readily blocked and larger particles (and especially flocs) need to be rigorously excluded. The largest orifice size is 1000 μm, so that the upper limit is around 400 μm—well above the colloidal size range. The concentration of the sol must also be kept low to reduce the coincidence level (i.e. the possibility of two particles going through the orifice so close together that they are interpreted as a single (larger) one. The manufacturers provide a correction for coincidence.

The pulse height is related to the volume of the particle by treating the empty orifice as a resistor (Fig. 3.6.2):

$$R_0 = \sigma_f \frac{\delta l}{A} \tag{3.6.4}$$

where σ_f is the resistivity of the fluid (usually a solution of an electrolyte at about 5 per cent concentration). When a particle is in the orifice the new resistance R can be calculated on the assumption that the particle

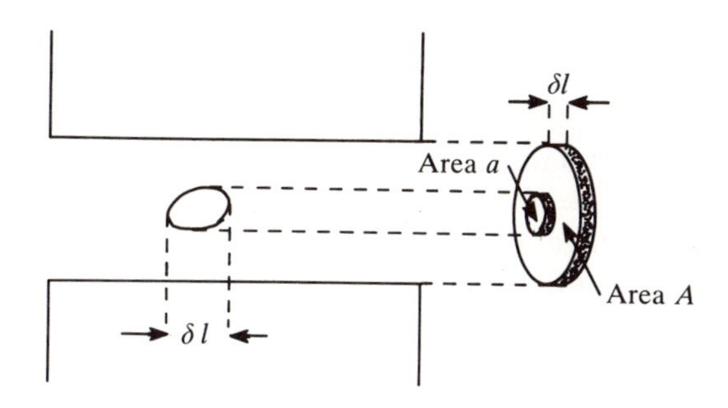

FIG. 3.6.2. Calculation of the effect of a particle on the orifice resistance.

has an infinite resistivity†, so that:

$$R = \sigma_f \frac{\delta l}{A - a}$$

and the change in resistance:

$$\delta R = R - R_0 = \sigma_f \delta l \left(\frac{1}{A - a} - \frac{1}{A} \right)$$

i.e.

$$\delta R = \frac{\sigma_f a \delta l}{A^2} \left(1 - \frac{a}{A} \right)^{-1}. \tag{3.6.5}$$

The pulse height is, therefore, not strictly proportional to particle volume ($a \cdot \delta l$) but is modified by the $(1 - a/A)$ term. The effect is obviously larger for the larger particles and leads to some distortion of the distribution (~5 per cent) at the top end (see Batch 1964).

The pulse shape also has some effect on the counting process. Particles that move near the walls of the orifice give rise to a double peaked pulse (Allen 1975, p. 309), which can cause counting errors. In the Coulter system, some models are provided with editing facilities that check the pulse shape and reject unsatisfactory ones. An alternative procedure, used in the Telefunken instrument, is called hydrodynamic focusing. By designing the delivery arrangement near the orifice in a proper way it is possible to channel all particles into the centre of the orifice. This has the additional advantage that all particles have the same transit time through the orifice so the pulse width is more reproducible (Allen 1975, p. 310).

In summary, the Coulter principle is very useful for the measurement of particle size at the upper end of the colloid range. Though it gives the number volume size distribution directly, this can readily be converted to the mass volume distribution if desired. From these the values of $\bar{d}_{NV}$ and $\bar{d}_{mV}$ can readily be calculated; for log-normal distributions, the geometric mean diameters will be more appropriate (as shown in section 3.3.5). The Coulter counter has also been used to measure floc sizes (Hunter and Frayne 1980). One assumption here is that little current flow occurs through the liquid trapped inside the floc. It must also be assumed that the floc is able to withstand the shearing forces to which it is subjected as it is drawn through the orifice—or that as it breaks up, all of the pieces traverse the orifice together. Delicate flocs, such as those produced by polymer flocculation, might not obey this assumption.

† Even for conducting particles, polarization effects at the particle/electrolyte interface prevent any current flow through the particle itself.

3.7 Light scattering methods

We noted in section 2.3 that the light scattering behaviour of a colloidal dispersion offers a number of opportunities for determining the size of the colloidal particles. The older (time average) methods using incoherent light sources are limited to particles of simple shape and fairly uniform size. For aerosol particles suspended in a gas there is also a problem of separating the scattering due to particles from the background due to the gas molecules (see Roth *et al.* 1976). Fortunately, the background scattering from a surrounding *pure* liquid is much *less* of a problem because most of the scattered light suffers destructive interference from neighbouring liquid molecules. It is, however, essential that the liquid be rigorously filtered to remove very small particulate matter (i.e. 'motes' that have nothing to do with the colloid under study). Even when coherent (laser) light sources are used there are limitations in the use that can be made of the intensity pattern as a function of angle (I_θ of section 2.3) although modern computer methods have greatly expanded the range and accuracy of the procedure (see e.g. Kerker 1969, Chapter 7). The method of *dynamic* light scattering (QELS or PCS; section 2.3.2) is now also being used much more widely.

3.7.1 *Intensity methods*

(a) *Rayleigh scattering*. We noted in section 2.3 that even in the Rayleigh scattering region ($d \ll \lambda$) the amount of light scattered at any angle was a strong function of particle size, but was symmetrical about the $\theta = \pi/2$ direction. The Rayleigh scattering region is most relevant to the study of macromolecules rather than colloids (which are usually too big to satisfy the Rayleigh criterion). It is customary, therefore to calculate a molar mass (or molecular weight) rather than a radius in such cases. To make a proper calculation of the molar mass we should undertake an analysis of the *fluctuations* that occur in the local molecule concentrations since it is these that give rise to the observed scattering. That analysis is done in Chapter 14 (Vol. II). Here we will only sketch an argument that suggests the final result (at least for ideal solutions). The scattering of initially unpolarized light from N (non-absorbing) molecules of a dilute gas per unit volume is given by (eqn (2.3.1)):

$$\frac{I_\theta}{I_{0,u}} = \frac{8\pi^4(1+\cos^2\theta)}{\lambda^4 r^2}\left(\frac{\alpha}{4\pi\varepsilon_0}\right)^2 N \tag{3.7.1}$$

and one can show (Exercise 2.3.3) that α, for isotropic particles, is given by:

$$\alpha = \varepsilon_0 \frac{(n_1^2 - n_0^2)}{N} \tag{3.7.2}$$

where n_0 is the refractive index of the surrounding medium.

We now define the *Rayleigh ratio*, R_θ:

$$R_\theta = \frac{Ir^2}{I_{0,u}}$$

$$= \frac{\pi^2}{2\lambda^4 N}(n_1^2 - n_0^2)^2(1 + \cos^2\theta) \tag{3.7.3}$$

$$= R_{90}(1 + \cos^2\theta) \tag{3.7.4}$$

where R_{90} is the value of R_θ evaluated at right angles to the incident beam.

This quantity is important in scattering measurements for a number of reasons. For one thing, a value of R_θ at different angles in conformity with eqn (3.7.4) suggests that the system is behaving as a Rayleigh scatterer and one can then use R_{90} to determine the molar mass of a dissolved (colloidal) molecule. If we substitute the following two approximations:

$$n_1^2 - n_0^2 = (n_1 - n_0)(n_1 + n_0) \simeq 2n_0(n_1 - n_0) \tag{3.7.5}$$

and

$$\frac{n_1 - n_0}{C} \simeq h_1 \text{ (a constant)} \tag{3.7.6}$$

where $C = NM_i/\mathcal{N}_A$ is the concentration of molecules (mass per unit volume), we have:

$$R_{90} = k_1 C M_i \tag{3.7.7}$$

where

$$k_1 = 2\pi^2 n_0^2 h_1^2/\lambda^4 \mathcal{N}_A. \tag{3.7.8}$$

Approximations (3.7.5) and (3.7.6) are reasonable ones in dilute solution ($n_1 \simeq n_0$) where the refractive index of the solution can be expected to increase linearly with particle concentration.

One particularly simple method of using light scattering for estimation of molar mass is to measure the *turbidity*, $\mathcal{T}$, of the sample. This is a measure of the reduction in intensity of light as it passes through the sample, due to the scattering process. The *fraction* of incident light scattered in all directions by a collection of particles is obtained by integrating the angular intensity function over the surface of a sphere (see

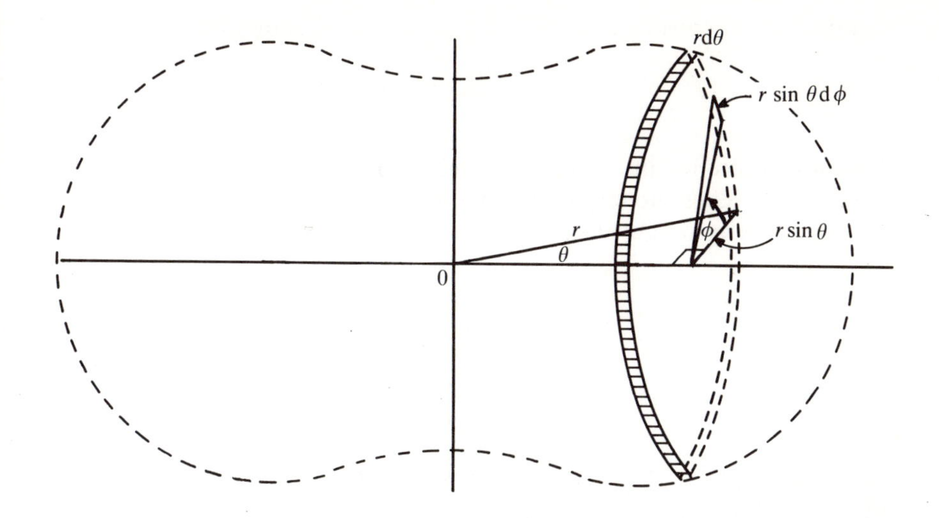

FIG. 3.7.1. Calculation of total amount of scattered light. The scattered light is normally measured at an angle to the incident direction and in the horizontal plane (Fig. 2.2.2). If, however, the initial light is unpolarised then the intensity at angle θ is the same in every plane. The total amount of light scattered is then given by the integral in eqn (3.7.9).

Fig. 3.7.1 and Exercise 3.7.2) and dividing by the initial intensity:

$$\mathscr{S} = \frac{1}{I_{0,u}} \int_{\theta=0}^{\pi} \int_{\phi=0}^{2\pi} I(\theta) r^2 \sin\theta \, \mathrm{d}\theta \, \mathrm{d}\phi$$

$$= \frac{16\pi}{3} R_{90} \qquad (3.7.9)$$

Note that this has units of reciprocal length (since $R_{90} \propto (\lambda^4 N)^{-1}$) because R_{90} measures the amount of scattering occurring from a unit path length of material.

As the beam traverses a solution, this fraction of light is (on average) being scattered at each point in the path. The attenuation of the light beam so caused is thus given by:

$$-\mathrm{d}I/\mathrm{d}x = \mathscr{S} I$$

and the fraction of light transmitted through a path length of l is (Exercise 3.7.1):

$$I_t/I_0 = \exp(-\mathscr{S} l) \qquad (3.7.10)$$

which is more usually written

$$I_t/I_0 = \exp(-\mathscr{T} l). \qquad (3.7.11)$$

Comparison of eqns (3.7.9), (3.7.10), and (3.7.11) shows that

$$\mathscr{T} = \frac{16\pi}{3} R_{90}. \tag{3.7.12}$$

Thus a measure of the turbidity (which is analogous to the extinction in Beer's Law) leads directly to R_{90} and so can be used to measure molar mass (molecular weight) using eqn (3.7.7):

$$M_i = \frac{R_{90}}{k_1 C} = \frac{\mathscr{T}}{HC} \tag{3.7.13}$$

where

$$H = \frac{32\pi^3}{3} \frac{n_0^2}{\lambda^4 \mathscr{N}_A} \left(\frac{dn}{dC}\right)^2.$$

(We have replaced h_1 (eqn (3.7.6)) by the refractive index increment, which is a little more exact.) A more elaborate treatment (Chapter 14) has been given by Debye (1944, 1947), who shows that for reasonably dilute solutions a better relation is:

$$\mathscr{T} = \frac{HC}{(1/M) + 2\mathscr{B}C} \tag{3.7.14}$$

where $\mathscr{B}$ is the second virial coefficient of the scatterer. Equation (3.7.13) is obviously the limiting form of (3.7.14) as C or $\mathscr{B}$ approaches zero. A plot of $HC/\mathscr{T}$ against C thus has an intercept of $1/M$ and a slope equal to twice the second virial coefficient.

In order to treat Rayleigh scatterers as particles it is best to return to the fundamental expression (eqn (2.3.3)) for the scattering intensity in terms of particle volume, v. The turbidity can then be expressed relative to the volume fraction ϕ ($=N_P v$) (Kerker 1969, p. 326):

$$\frac{\mathscr{T}}{\phi} = 24\pi^3 \left(\frac{n^2-1}{n^2+2}\right)^2 \frac{v}{\lambda^4} = H'v \tag{3.7.15}$$

where again, n is the relative refractive index ($=n_1/n_0$). This and a number of similar expressions are given in Exercise 3.7.7.

(b) *Rayleigh–Gans–Debye scattering*. Extension of the scattering measurements to determine I_θ as a function of θ enables one to make some assessment of shape as well as size but only when the size of the scattering molecules becomes large enough for the Rayleigh model to become invalid (say $d > 10$ nm). We then have $d < \lambda$ and the Rayleigh–Gans–Debye (RGD) theory must be invoked (section 2.3). Again we will postpone a detailed discussion until Chapter 14.

We need only note here that eqn (3.7.7) is modified to read:

$$R_\theta = k_1 C M_i P(\theta)(1+\cos^2\theta) \tag{3.7.16}$$

where the scattering factor $P(\theta)$ is a consequence of the interference which occurs between waves scattered from different parts of the molecule or particle (Debye 1915). The calculated values of $P(\theta)$ for various shapes are shown in Fig. 3.7.2. If the shape is known, a much better estimate of the molar mass can be made in the RGD regime by reference to Fig. 3.7.2. Alternatively, but less reliably, an assessment can be made of the shape of the molecule if an independent measure of size is available.

It should be noted that if the scattering solution consists of particles with many different molar masses (say a synthetic polymer solution), the measured molar mass is a mass average molar mass (usually called the weight average molecular weight; Exercise 3.7.3):

$$\bar{M} = \frac{\sum C_i M_i}{\sum C_i} = \frac{\sum N_i M_i^2}{\sum N_i M_i} \tag{3.7.17}$$

where C_i is the mass of material of molar mass M_i per unit volume, and N_i is the number of such molecules per unit volume.

In the RGD region it is also possible to make accurate estimates of the radius of monodisperse spherical particles simply by measuring the so-called dissymmetry ratio:

$$R_\mathrm{D} = I_{45°}/I_{135°}.$$

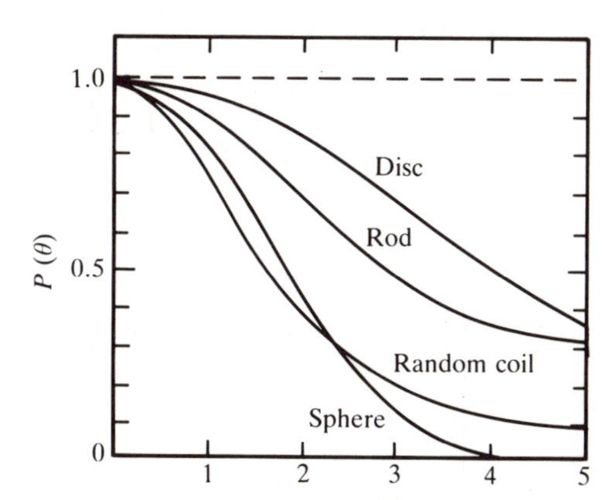

FIG. 3.7.2. Scattering function $P(\theta)$ for macromolecules of various shapes. (After Marshall 1978, p. 477.) The abscissa represents $Qd/2n_0$ where Q is the scattering vector (section 2.3) and d is the diameter for the sphere or disc, and the length for the rod. For the random coil, the abscissa represents $Qd/\sqrt{6}\,n_0$ where d is the r.m.s. end-to-end distance.

We noted in section 2.3 that as the particle size increases and the system no longer obeys the Rayleigh criterion ($d \ll \lambda$), the forward scattering becomes increasingly favoured and R_D increases (Fig. 3.7.3). Okada *et al.* (1971) used this method and in a subsequent paper (Ishizu and Okada 1978) showed that with the use of the more elaborate Mie theory one could match both particle size and refractive index, if the latter were unknown.

Alternatively, one can use the theoretical form of $P(\theta)$ to assess the particle size. For homogeneous spheres, $P(\theta)$ takes the fairly simple form (Pusey 1982):

$$P(\theta) = [3(\sin Qa - Qa \cos Qa)/(Qa)^3]^2 \quad (3.7.18)$$

where $Q = (4\pi n_0/\lambda)\sin(\theta/2)$ is again the scattering vector (cf. eqn (2.3.5)). Expansion of eqn (3.7.18) in powers of Qa gives (Exercise 3.7.4):

$$P(\theta) = 1 - (Qa)^2/5 + \ldots \quad (3.7.19)$$

and it can be shown that for particles of arbitrary shape, satisfying the RGD criterion ($a < \lambda$), Pusey (1982):

$$P(\theta) = 1 - (Qa_G)^2/3 + \ldots \quad (3.7.20)$$

where a_G is the radius of gyration of the particle. Equation (3.7.20) can be used for particles in the range $20 < a_G/(\text{nm}) < 100$: a plot of $\ln I_\theta$ as a function of Q^2 will have an initial slope related to a_G (Exercise 3.7.5). This is called a 'Guinier plot'. It has become much more useful with the advent of lasers because they permit measurements at very small angles so that the $Q \to 0$ extrapolation can be dispensed with; (see also Walstra

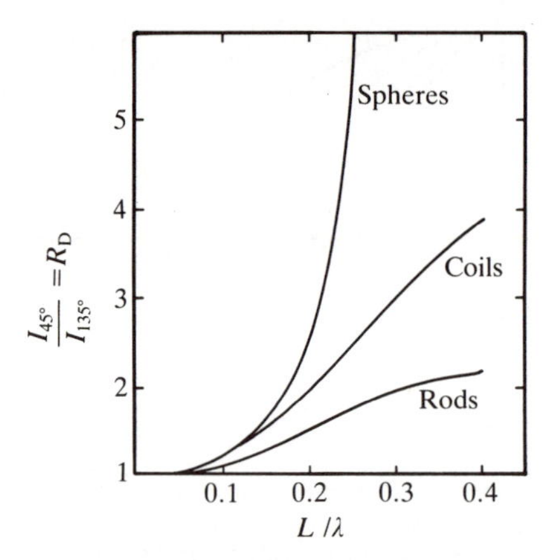

FIG. 3.7.3. The dissymmetry ratio R_D as a function of (relative) characteristic length for spheres, random coils, and rods. (From Kerker 1969, p. 432, with permission.)

1968). This procedure is called LALLS (low-angle laser light scattering).

The more complicated angular dependence of the scattering in the Rayleigh–Debye–Gans regime makes it possible to determine a particle size distribution in favourable cases. The details are given by Kerker (1969, Chapter 7) and will not be repeated here. The methods rely on measuring the scattered light (or the polarization ratio (Fig. 2.3.1)) at a large number of angles and computer matching the results to those expected from a known distribution.

The distribution must be represented in parametric form (usually a two-parameter log-normal distribution (section 3.4.2)) and measurements taken at, say, 20 different angles.

(c) *Mie scattering*. When the particles become comparable in size to the wavelength of the incident light the scattering pattern becomes extremely complex and one must then use the complete Mie theory, which is, of course, strictly applicable only to spheres. Computer matching of the anticipated scattering pattern with experimental data has again been used by Kerker and his collaborators to establish particle size distributions in this regime. The success of the method can be judged from Fig. 3.7.4. Marshall *et al.* (1976) have calculated the scattering pattern from a single spherical particle of a monodisperse polymer latex (Dow polystyrene of

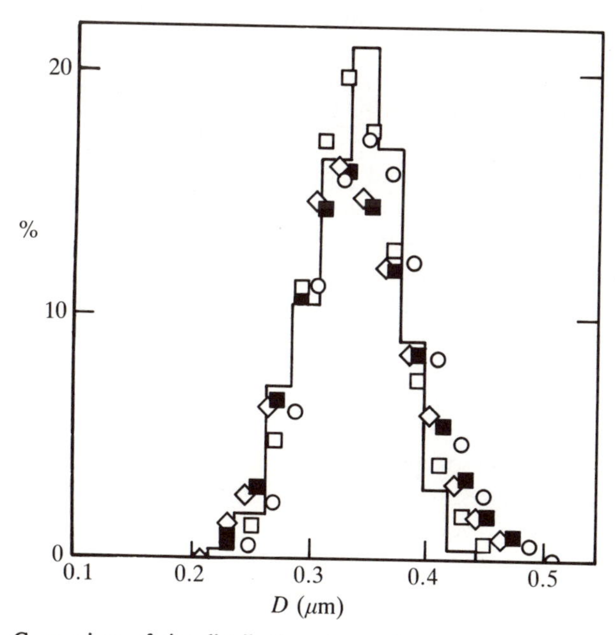

FIG. 3.7.4. Comparison of size distribution of vanadium pentoxide aerosol with electron microscope result (shown as histogram). Wavelength λ/nm: ■ 406; ○ 436; □ 546; ◇ 578. (From Kerker 1969, p. 367, with permission.)

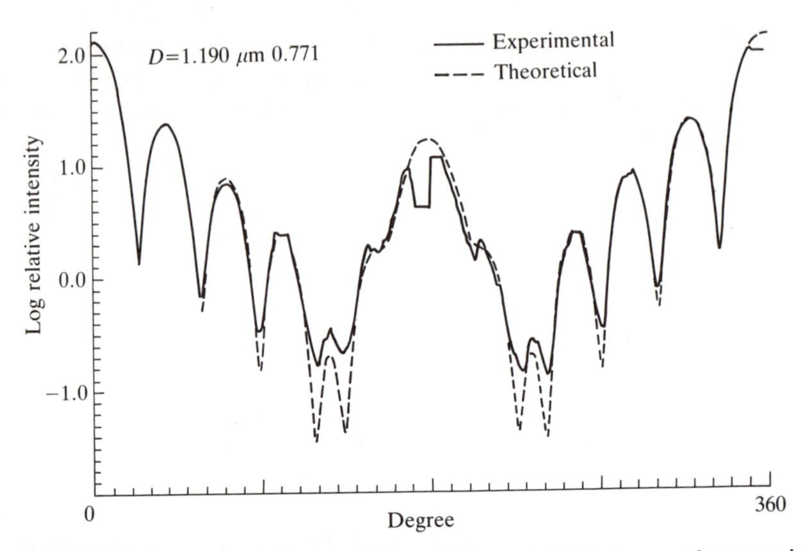

FIG. 3.7.5. Comparison of experimental data and theoretical (Mie) curves for scattering from a single polystyrene sphere. 0.771 is the least squares deviation between theory and experiment. (From Marshall *et al.* 1976, with permission.)

nominal mean diameter 1.2 μm) with the result shown in Fig. 3.7.5. The fit to the theory and the precision of the radius determination are very impressive.

These data were obtained using an ingenious device designed by Gucker *et al.* (1973), the heart of which is illustrated in Fig. 3.7.6. The light scattered by a particle at a particular angle ($\pm 2\frac{1}{2}°$) is reflected from the upper (elliptical) mirror and passes through the slot S in the rapidly rotating (3000 r.p.m.) lower disc, to reach the photomultiplier tube (PM). A complete 360° scan can be completed in about 20 ms and several scans of the same particle can be done consecutively (to check reproducibility) while it is essentially stationary in the laser beam. Only symmetric patterns (Fig. 3.7.5) are accepted and stored for analysis. (See also Davis and Ray 1980.)

A particle size distribution can, in principle be constructed by examining a sufficient number of individual particles, and this has been done for a dispersion of poly-(vinyltoluene) latex by Marshall *et al.* (1976). Excellent agreement was obtained between this and other methods. (See also Rowell *et al.* 1979.)

3.7.2 *Instrumentation*

For a description of some of the instruments available at the time of its publication see Silverman *et al.* (1971, pp. 206–224). The remarks made above, however, make it clear that developments in this area are

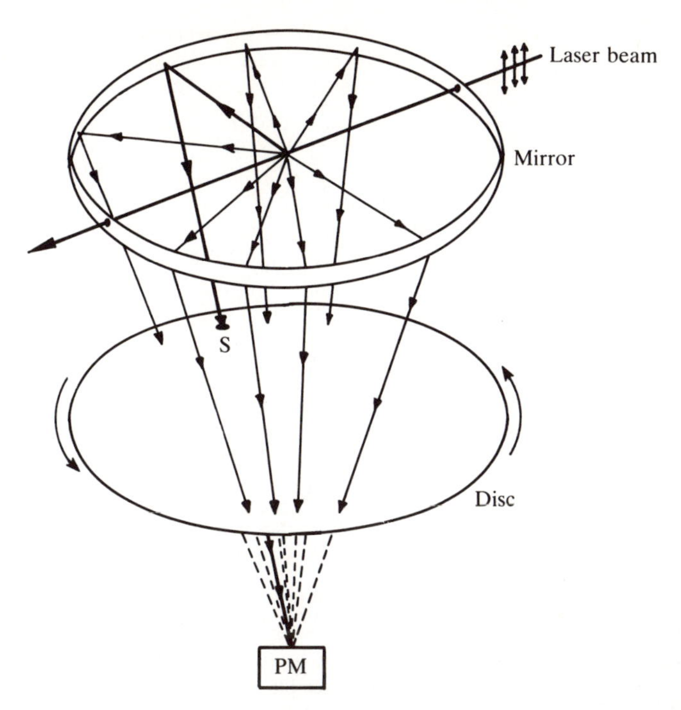

FIG. 3.7.6. Schematic diagram of the 360° scattering instrument. An aerosol stream intercepts a vertically polarized He–Ne laser beam at one of two foci of an ellipsoidal mirror. Light scattered in the horizontal plane is intercepted by a segment of the mirror and directed to a photomultiplier at the second focus, through slits in a rotating disc. (After Marshall *et al.* 1976, with permission.)

occurring very rapidly and for up-to-date information on optical methods of particle counting and sizing one needs to go directly to the manufacturers' literature. It should be obvious, also, that there are limitations in the use of simple scattering for particle size analysis of real systems. If single perfect spheres can produce the kind of complexity illustrated in Fig. 3.7.5, the analysis of a particle of arbitrary shape must be regarded as impossible, although there has been some attempt to extend the theory to long cylinders (Kerker 1969, Chapter 6; Federova 1977) and to spherical shells of different refractive index (Kerker 1969, Chapter 5; Federova and Emelyanov 1977). These latter analyses are of relevance to the study of particular problems (e.g. the structure of adsorbed layers on spherical particles) but are not relevant to our present concern.

3.7.3 *Dynamic light scattering*

We noted at the beginning of this section that a much more general method of determining particle size of colloidal dispersions was now

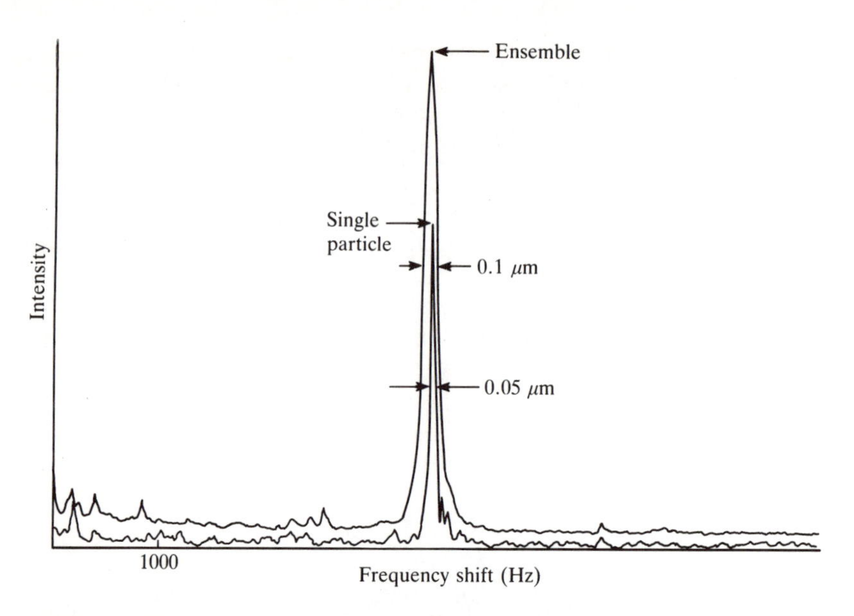

FIG. 3.7.7. Superimposed spectra of an ensemble of dioctylphthalate droplets and a single droplet from that ensemble. The precision (0.05 μm) of the measurement is indicated by the width at half height of the single particle peak. (From Chabay and Bright 1978, Fig. 3, with permission.) The breadth of the ensemble peak indicates that the precision of the method is sufficient to detect the slight variation in size of the particles.

becoming available with the advent of photon correlation spectroscopy (PCS) or quasi-elastic (or dynamic) light scattering (QELS), which permits an estimate of the diffusion coefficient from which a radius *a,* can be estimated using the Einstein equation (eqn (2.1.19)) together with Stokes' equation (eqn (2.1.20)). Then:

$$a = kT/6\pi\eta\mathscr{D} \tag{3.7.21}$$

for a spherical particle. The method has been outlined in section 2.3. It has been used by a number of people, including Thompson (1971), Fair *et al.* (1978), and Gulari *et al.* (1980); its precision has been demonstrated clearly by Chabay and Bright (1978) (Fig. 3.7.7).

A number of commercial instruments (e.g. the Brookhaven BI-90) have been marketed to take advantage of this method of analysis (see e.g. Dahneke (1983)).

3.7.4 *The fibre-optic Doppler anemometer* (*FODA*)

This is a very promising application of the phenomenon of photon correlation spectroscopy to the measurement of particle size and size distribution. The apparatus is described by Dyott (1978) and its application to particle sizing by Ross *et al.* (1978). The principal innovation is the use of a length of communication-type multimode optical fibre to carry the laser beam into the interior of a solution (Fig. 3.7.8). The light of frequency ω_0, travels down the fibre to its tip, where a small fraction is reflected back towards the detector. The remainder forms a narrow cone, which is directed into the solution; any particles near the tip can scatter this light back into the fibre so that it can be returned to the detector where it is mixed with the reflected light (frequency ω_0) and the intensity of the difference (beat) frequency determined. The advantages claimed for the instrument are that the flexible optical fibre can be of any length and can be immersed into a concentrated suspension which would otherwise be opaque to the laser beam. It samples only a small volume around the tip, and because it collects only back-scattered light it determines directly the diffusive motion in the direction perpendicular to the face of the fibre tip. The maximum anticipated error in the radius (due to the fact that a finite cone of light is sampled) is about 0.4 per cent, giving it a resolution of about 1 in 200 nm.

Ross *et al.* (1978) show that in addition to estimating the (drag) mean radius (which should be close to the number area mean radius, eqn (3.5.4)) it is possible to determine also the standard deviation from the mean, provided it is not too large ($\sigma_a < \bar{a}$). This is so because the deviations of the radius lead to a slight change in the shape of the Lorentzian curve (Fig. 2.3.3) and a computer analysis can extract the additional information.

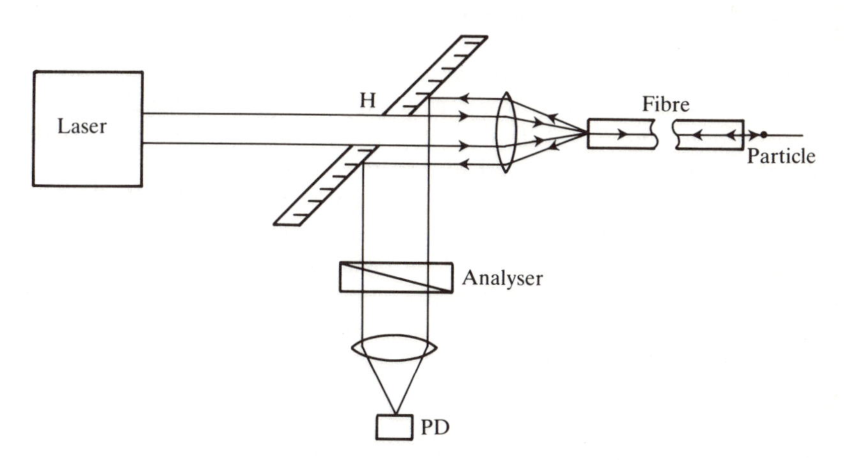

FIG. 3.7.8. The principle of the FODA (see text). H is a hole in the half-silvered mirror to permit passage of the laser beam and the analyser is interposed to remove light from spurious reflections before it falls on the photodetector, PD.

Exercises. 3.7.1 Establish eqn (3.7.10).

3.7.2 Establish eqns (3.7.7) and (3.7.9).

3.7.3 Establish eqn (3.7.17) on the assumption that

$$R_{90} = \sum_i R_{90} = \sum_i k_1 C_i M_i.$$

3.7.4 (i) Establish eqn (3.7.19).

(ii) Then show that the radius of gyration of a sphere of radius a is given by $a_G = \sqrt{(3/5)}a$ and hence reconcile (3.7.19) and (3.7.20).

$$\left[\text{Hint: } a_G = \left(\frac{\text{Moment of inertia}}{\text{Total mass}}\right)^{1/2} = \left[\frac{1}{V\rho}\int r^2\rho \, dV\right]^{1/2}.\right]$$

3.7.5 Show that the initial slope of the Guinier plot is $-a_G^2/3$.

3.7.6 Why do you think FODA (section 3.7.4) is so named?

3.7.7 Establish eqn (3.7.15) from eqn (2.3.3). Kerker (1969) gives the following alternative forms (which you can verify):

$$\frac{\mathcal{T}}{m'} = H'\left(\frac{\rho_{12}}{\rho_2}\right)v$$

where m' is the mass fraction, ρ_{12} is the density of the solution, ρ_2 is the density of the solid spheres. Since $\phi = (\rho_{12}/\rho_2)m' = C/\rho_2$, where C is the concentration (mass of solid per unit volume of solution), this can also be written

$$\frac{\mathcal{T}}{C} = H'M/\mathcal{N}_A\rho_2^2$$

where M is the molar mass.

3.8 Hydrodynamic chromatography (HDC)

Another recent development in the sizing of (particulate) colloidal dispersions is the use of hydrodynamic chromatography (Small 1974). In this procedure, the liquid dispersion is forced under pressure (~20 atm) through a long packed column of non-porous beads of radius ~10 μm. Particles of different size travel with different speeds through the bed; the effluent can be collected in fractions, as in chromatography, and the particle concentration determined in each fraction. In this way much more homogeneous size fractions are obtained and the subsequent size analysis is made more reliable than a measurement on the original unseparated example. The method is somewhat analogous to gel permeation chromatography.

It turns out that large particles travel through the column more rapidly than small ones, provided, of course, that they are all smaller than the pore size of the column. Stoisits *et al.* (1976) explain this by arguing that smaller particles diffuse further away from the axis of the pores and so sample regions of the liquid that are travelling very slowly. Large particles tend to remain nearer the line of maximum flow, down the pore axis.

The R_F value:

$$R_F = \frac{\text{Rate of transport of colloid through bed}}{\text{Rate of transport of elution fluid}} \tag{3.8.1}$$

varied from below 1.02 to slightly above 1.11 for the measurements made by Stoisits *et al.* (1976). They derived an approximate relation for R_F in the following way. They first assumed that the column, or bed, of beads could be described as a combination of many parallel cylindrical pores. (This is a common starting point for elementary descriptions of flow through porous media.) The velocity profile for laminar flow across a cylindrical pore is given by the Poiseuille equation (Chapter 9):

$$v(r) = \frac{\Delta P r_0^2}{4\eta l}\left\{1 - \left(\frac{r}{r_0}\right)^2\right\} \tag{3.8.2}$$

where ΔP is the pressure drop, r_0 is the capillary radius and l is its length. The maximum velocity $v_0 = (\Delta P)r_0^2/4\eta l$ occurs on the axis and $v = 0$ at the pore wall. The average fluid velocity $\bar{v}$ is (Exercise 3.8.1):

$$\bar{v} = \int_0^{r_0} 2\pi r v(r)\,\mathrm{d}r \Big/ \int_0^{r_0} 2\pi r\,\mathrm{d}r = \frac{v_0}{2}. \tag{3.8.3}$$

A sphere of radius a can approach the capillary wall only to within a distance a. Since these spheres cannot sample the slower fluid velocities close to the wall, they will move, on average, faster than the fluid. It is easy to show (Exercise 3.8.2) that the average velocity of a sphere is:

$$\bar{v}_p = v_0[1 - \tfrac{1}{2}(1 - \tilde{a})^2] \tag{3.8.4}$$

where $\tilde{a} = a/r_0$. We can then write:

$$R_F = \frac{\bar{v}_p}{\bar{v}} = 2 - (1 - \tilde{a})^2 = 1 + \tilde{a}(2 - \tilde{a}). \tag{3.8.5}$$

R_F can be related to the difference between the elution fluid volume of the particles, V_p, and that of the carrier fluid injected at the same time V_c:

$$\begin{aligned}\Delta V = V_c - V_p &= V_c(1 - 1/R_F)\\ &= V_c\left\{\frac{\tilde{a}(2 - \tilde{a})}{1 + \tilde{a}(2 - \tilde{a})}\right\}.\end{aligned} \tag{3.8.6}$$

This is the usual quantity used in chromatographic calibration procedures. (V_c is equal to the void volume of the column.)

Apart from the diffusion away from the axis of the pore, which leads to eqn (3.8.4), the particles will also diffuse backwards and forwards *along the axis* of the pore and this leads to a random (normally distributed) variation in the elution times about a mean value $\bar{t}_e$ for a particle of a given size. Di Marzio and Guttman (1970) have shown that the standard deviation of the elution times is given by:

$$\sigma = (2\mathcal{D}'\bar{t}_e)^{1/2} \tag{3.8.7}$$

where the effective diffusion coefficient $\mathcal{D}'$ is:

$$\mathcal{D}' = \mathcal{D} + [v_0^2 r_0^2 (1 - \tilde{a})^6 / 192\mathcal{D}]. \tag{3.8.8}$$

Note the correspondence between eqn (3.8.7) and eqn (2.1.26) for the r.m.s. displacement.

Although the theory described above can be regarded as only a first order approach to the problem, the data given by Stoisits *et al.* (1976) show that it is a reasonable description of the process, and this method of size separation and analysis can be expected to find increasing use for particles of the order of 100 nm in radius.

Exercises. 3.8.1 Verify eqn (3.8.3). Why is this average used and not simply $(1/r) \int v \, dr$?

3.8.2 Establish eqn (3.8.4) by averaging over the inner part of the capillary. Verify eqns (3.8.5) and (3.8.6).

3.9 Summary of sizing methods

We have considered particle size determination in some detail for two reasons:

(a) because the size is an important intrinsic characteristic of the colloid, important in determining its behaviour; and

(b) because the practical problem of determining size introduces us to a number of the important theoretical concepts of colloid science. We have concentrated on methods that can give some idea of the size distribution rather than just the average size, since the latter can be somewhat misleading.

There is one approach that has not yet been tackled and that is the measurement of surface area, which permits an estimate of a surface average particle size, $\bar{r}_{NA}$ if one can make some assumptions about particle shape (e.g. Exercise 1.1.3). Surface area is most readily measured by an adsorption process: in effect one determines how many molecules of a certain substance (the *adsorbate*) are required to completely cover the surface to a known depth (usually one monolayer). Then if we know the area occupied by the adsorbate molecules the area

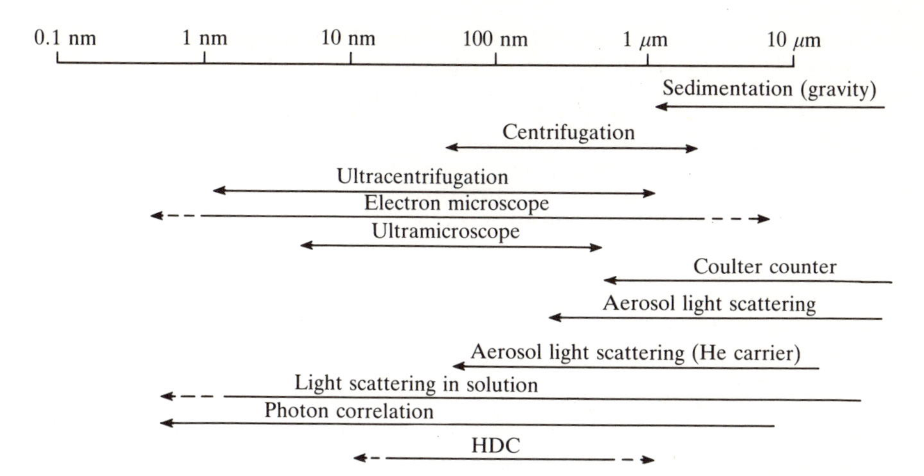

FIG. 3.9.1. Approximate range of application of the particle size methods discussed in this chapter.

of the underlying surface (the *adsorbent*) can be determined. We will be treating the phenomenon of adsorption in Chapters 5 and 6 and will then have more to say about surface area measurement (Section 6.7.4).

The range of the methods discussed above is indicated in Fig. 3.9.1 and it is apparent that there is not a great deal of choice over most of the colloid size range and even that choice may be limited by other factors. We have not, for example, considered the problem of sampling errors when dealing with, say, the tiny sample provided by an electron micrograph field. That would take us too far into the realm of statistics (see Herdan 1960, Chapter 5).

References

Allen, T. (1975). *Particle size measurement.* In the *Powder technology series* (ed. J. C. Williams). Chapman and Hall, London.

Batch, B. A. (1964). *J. Inst. Fuel* 455. (quoted by Allen 1975, p. 306).

Beddow, J. K. (1980). *Particulate science and technology*. Chemical Publishing, New York.

Bostock, W. (1952). *J. Sci. Inst.* **29,** 209–11.

Bull, H. B. (1971). *Introduction to physical biochemistry,* pp. 282–5. F. A. Davis, Philadelphia.

Cadle, R. D. (1965). *Particle size; theory and industrial applications*. Reinhold, New York.

Chabay, I. and Bright, D. S. (1978). *J. Colloid interface Sci.* **63,** 304.

CRC Handbook. *Handbook of physics and chemistry*. Chemical Rubber Publishing, New York. [Being continually revised. Not all editions contain the *t* distribution. The mathematical tables from the handbook have also been published separately by the same company.]

Cummins, P. G., Staples, E. J., Thompson, L. G., and Pope, L. (1983). *J. Colloid interface Sci.* **92,** 189–97.

Dahneke, B. E. ed. (1983). *Measurement of suspended particles by quasielastic light scattering*. Wiley, New York.

Davidson, J. A. and Haller, H. S. (1976). *J. Colloid interface Sci.* **55,** 170–80.

Davis, E. J. and Ray, A. K. (1980). *J. Colloid interface Sci.* **75,** 566.

Debye, P. (1915). *Ann. Phys.* (*Leipzig*) **46,** 809.

Debye, P. (1944). *J. appl. Phys.* **15,** 338.

Debye, P. (1947). *J. Phys. Colloid Chem.* **51,** 18.

Di Marzio, E. A. and Guttman, C. M. (1970). *Macromolecules* **3,** 131.

Dyott, R. B. (1978). *IEE J. Microwaves Opt. Acoust.* **2,** 13.

Dublin, L. (1980). *National Association of Corrosion Engineers, Annual Meeting.*

Fair, B. D., Chao, D. J., and Jamieson, A. M. (1978). *J. Colloid interface Sci.* **66,** 323.

Fedorova, I. S. (1977). *J. Colloid interface Sci.* **59,** 98.

Fedorova, I. S. and Emelyanov, V. B. (1977). *J. Colloid interface Sci.* **59,** 106.

Frumkin, A. N. and Levich, V. I. (1946). *Acta physiochim. URSS* **21,** 193.

Gucker, F. T., Tůma, J., Lin, H. M., Huang, C. M., Ems, S. C., and Marshall, T. R. (1973). *Aerosol Sci.* **4,** 389.

Gulari, E., Bedwell, B., and Alkhafaji, S. (1980). *J. Colloid interface Sci.* **77,** 202.

Hall, C. E. (1966). *Introduction to electron microscopy* (2nd edn). McGraw-Hill, New York.

Herdan, G. (1960). *Small particle statistics.* Butterworths, London.

Hiemenz, P. C. (1977). *Principles of colloid and surface chemistry.* Marcel Dekker, New York.

Hunter, R. J. and Frayne, J. (1980). *J. Colloid interface Sci.* **76,** 107.

Ishizu, Y. and Okada, T. (1978). *J. Colloid interface Sci.* **66,** 234.

Johari, O. (1968). *The scanning electron microscope—the instrument and its applications.* In Symposium on S.E.M. Illinois Institute of Technology Research Inst., Chicago Ill. (See Silverman *et al.* (1971) for further references.)

Kerker, M. (1969). *The scattering of light and other electromagnetic radiation.* Academic Press, New York.

Kruyt, H. R. (1952). *Colloid science,* Vol. I, p. 40. Elsevier, Amsterdam.

Levich, V. I. (1962). *Physicochemical hydrodynamics.* Prentice-Hall, Englewood Cliffs N.J.

Marshall, A. G. (1978). *Biophysical chemistry—principles, techniques and applications.* Wiley, New York. [This is an excellent modern treatment that gives many valuable insights into the relation between classical physical models and observed behaviour.]

Marshall, T. R., Parmenter, C. S., and Seaver, M. (1976). *J. Colloid interface Sci.* **55,** 624.

McDonald, S. A., Daniels, C. A., and Davidson, J. A. (1977). *J. Colloid interface Sci.* **59,** 342.

Morrison, S. R. (1977). *The chemical physics of surfaces* pp. 415. Plenum Press, New York.

Oden, S. (1926). In *Colloid chemistry* pp. 861–909. (ed. J. Alexander). Chemical Catalog, New York.

Okada, T., Itoh, K., and Kitani, S. (1971). *J. Colloid interface Sci.* **37,** 430.

Ondeyka, J. G., Henry, J. D. Jnr., and Verhoff, F. H. (1978). *Ind. Engng. Chem. Fundam.* **17,** 217.
Orr, Clyde Jnr., and DallaValle, J. M. (1959). *Fine particle measurement* (1st edn). Macmillan, New York. [More recent editions are available.]
Pusey, P. N. (1982). Light scattering. In *Colloidal dispersions* (ed. J. W. Goodwin) Chapter 6. Royal Society of Chemistry, London.
Ross, D. A., Dhadwal, H. S., and Dyott, R. B. (1978). *J. Colloid interface Sci.* **64,** 533.
Roth, C., Gebhart, J., and Heigwer, G. (1976). *J. Colloid interface Sci.* **54,** 265.
Rowell, R. L. *et al.* (ten authors) (1979). *J. Colloid interface Sci.* **69,** 590.
Silverman, L., Billings, C. E., and First, M. W. (1971). *Particle size analysis in industrial hygiene.* Academic Press, New York.
Small, H. (1974). *J. Colloid interface Sci.* **48,** 147.
Stoisits, R. F., Pochlein, G. W., and Vanderhoff, J. W. (1976). *J. Colloid interface Sci.* **57,** 337.
Sutton, H. M. (1976). In *Characterization of powder surfaces* (ed. G. D. Parfitt and K. S. W. Sing) Chapter 3. Academic Press, London.
Thompson, D. S. (1971). *J. phys. Chem.* **75,** 789.
Vinograd, J., Bruner, R., Kent, R. and Weigle, J. (1963). *Proc. nat. Acad. Sci. USA* **49,** 902.
Walstra, P. (1968). *J. Colloid interface Sci.* **27,** 493.
Wulff, G. (1901). *Z. Krist.* **34,** 449.

4

THE THEORY OF VAN DER WAALS FORCES

4.1 Introduction

A treatment of the theoretical calculation of the van der Waals or Hamaker function (A) will necessarily take us into mathematical areas that many will find unfamiliar. We will make every effort to explain the analysis as clearly as possible but will concentrate on those aspects that are most germane to the practical problem of calculating the value of A for a particular material. It must be admitted that not every colloid scientist will choose to undertake such a calculation for it is not a project to be entered into lightly. Nevertheless, if these quantities are to be used with confidence it is important that workers in the field are aware of the data on which they are based. If it does nothing else this should stimulate the collection of more accurate data of the kind needed to make more reliable assessments of A for different situations.

A number of reviews of the subject have appeared in recent years and reference will be made to them in what follows. We take the opportunity to develop a fairly detailed account of the theory here, partly for completeness and partly because of its intrinsic scientific interest, but particularly because it ties this aspect of colloid science in with the mainstream of physical chemistry.

Some of the more technical aspects of the theory have been relegated to appendices but the bulk of it is treated in easily digestible steps with short exercises that are intended to keep the reader involved. They require only simple algebraic manipulations or, at most, some elementary calculus, and attention is drawn to them at the appropriate stage in the text.

We begin with a discussion of the microscopic theory of London and then take up the more modern Lifshitz macroscopic treatment. The microscopic theory is based on the assumption that interactions between pairs of molecules can be added together to obtain the total interaction, i.e. that the interaction between a molecule in one colloidal particle and a molecule in another particle is unaffected by the presence of all of the other molecules. That is a gross assumption that the macroscopic approach seeks to circumvent.

The microscopic theory begins with the calculation of the potential energy between two molecules (section 4.1), then between a molecule and a large (colloidal) body (section 4.4.1) and then between two (colloidal) bodies (section 4.4.2). A number of simple geometries can be studied (flat plates, spheres, and cylinders), but more complicated shapes rapidly generate very considerable algebraic complexity. We next consider the effect of placing the particles in a dispersion medium like water (section 4.4.3).

If the particles are separated by distances larger than about 50 nm it can be shown that the interaction energy is reduced because the electric field can only be propagated from one molecule to another at the speed of light. This is the *retardation effect* and it is discussed in section 4.5. In section 4.6 we introduce the Deryaguin approximation, which is a useful device for calculating interactions between large colloidal particles (where the radius of the particle is large compared to the range of the interaction force).

The macroscopic (Lifshitz) approach is introduced (section 4.7) by treating a very simple problem that illustrates the basic ideas embodied in the theory, and, in particular, the concept of a dispersion relation. Some further remarks are then made about the dielectric response function ($\varepsilon(\omega)$), because it is the property that is of fundamental significance in the theory. No attempt is made to derive the basic equations of the theory; rather, we try to exhibit their reasonableness in the light of the ideas developed thus far. Section 4.8 outlines the procedure required to actually calculate the Hamaker function, A, while section 4.9 introduces, very briefly, some considerations relating to the effect of an intervening electrolyte.

The more peripheral sections of this chapter are rendered in smaller typeface than the remainder. That material can be ignored on a first reading. Many of the exercises are designed to show the reader how much of the mathematical manipulation should be accessible with only the tools of elementary calculus and a little algebra.

It is not the intention to expose the assumptions of the macroscopic theory to rigorous analysis but rather to indicate the nature of the experimental data that is currently used to evaluate the interaction energy. Hopefully, a wider knowledge of the limitations of that data will lead to the collection of much more useful information on the important colloid chemical materials.

The range of values of the Hamaker function (at short distances where retardation is unimportant) for substances immersed in water can be summarized as follows:

$$A = \underbrace{30 \quad 10}_{\text{metals}} \qquad \underbrace{3 \quad\quad 1}_{\text{oxides and halides}} \qquad \underbrace{0.3 \times 10^{-20}\,\text{J}}_{\text{hydrocarbons}}$$

These figures reflect the differences in the polarizabilities of the various materials, on which A largely depends. For approximate calculations, the above information may be sufficiently accurate. For more exacting analyses, the more detailed considerations of this chapter must be invoked.

4.1.1 *Interactions between molecules*

The existence of long-range attractive forces between atoms can be inferred from the observation of first order phase changes (e.g. condensation of gases to liquids) and, less spectacularly perhaps, from the deviations from the perfect gas laws of Boyle and Charles. In 1873, van der Waals proposed the famous equation of state of a gas:

$$\left(P + \frac{n^2 a}{V^2}\right)(V - nb) = nRT \tag{4.1.1}$$

in which the constant 'b' accounted for the finite volume occupied by the molecules of the gas and the constant 'a' was directly related to the strength of the intermolecular attractive forces. The success of the van der Waals equation in summarizing the properties of gaseous systems and their phase behaviour stimulated theoretical attempts to discover the origin of the forces responsible. It is interesting to note that, even in 1873 with only a vague notion of the origins of intermolecular forces, van der Waals was already separating the short-range repulsive forces that give rise to the excluded volume term b from the long range attractive forces that account for the constant a.

It has recently been pointed out (Sparnaay 1983) that as long ago as 1686, Isaac Newton in his *Principia* discussed the attraction between two bodies separated by a distance, R, in terms of a force, proportional to R^{-n}, where $n > 4$. Though he could not discuss the possible origins of such a force he could show that n must exceed four or the interaction between a small particle and a large plate would become infinitely large.

In the early years of the present century, several workers (van der Waals 1909; Reinganum 1912; Thomson 1914; Keesom (1921)) sought an explanation of the 'van der Waals' forces by postulating the existence of a permanent electric dipole in each atom or molecule of the gas. By suitably averaging the permanent dipole–dipole interaction energy over all orientations of the dipoles at a fixed separation distance R, an attractive potential energy proportional to $1/R^6$ was obtained.

This simple theoretical explanation suffered from the unfortunate fact that the fundamental postulate could be tested experimentally and the necessary permanent dipole moment was found to be absent in most of the simplest molecules. Debye (1920) proposed a less easily tested model in which atoms and molecules were assumed to possess a permanent quadrupole moment. An attractive interaction energy ($\propto 1/R^9$) was obtained by orientation averaging the permanent quadrupole-induced dipole interaction energy.

It was, however, only with the advent of quantum mechanics that a satisfactory explanation of the origin of van der Waals forces was forthcoming (London 1930).

4.2 London theory

The explanation given for the van der Waals, or dispersion, force in elementary physical chemistry texts is that it results from the interaction between a temporary dipole on one molecule and the induced dipole on a neighbouring molecule. A proper calculation of that interaction requires the use of second-order perturbation theory, and an outline of the London approach is given in Appendix A1, where it is shown that the interaction energy depends on the inverse sixth power of the separation between the molecules. The general form of the result can be obtained by the following very simplistic argument (Israelachvili 1974).

In the Bohr model of the hydrogen atom, the electron is regarded as travelling in well-defined orbits about the nucleus. The orbit of smallest radius a_0 is the ground state and Bohr calculated that

$$a_0 = e^2/8\pi\varepsilon_0 h\nu \qquad (4.2.1)$$

where e is the proton charge, ε_0 is the permittivity of free space, h is Planck's constant and ν is a characteristic frequency (in Hz) associated with the electron's motion around the nucleus ($\nu = 3.3 \times 10^{15}\ \mathrm{s}^{-1}$ for the Bohr hydrogen atom). (Note that the value of a_0 given by eqn (4.2.1) corresponds to the maximum value in the electron density distribution $|\psi_0|^2$ in the electronic ground state of hydrogen, as calculated by quantum mechanics). The energy $-h\nu$ is the energy of the electron in its ground state (relative to the separated particles) and so is equal to the ionization potential of the H atom.

Although the H atom has no permanent dipole moment it can be regarded as having an instantaneous dipole moment, p_1 of order

$$p_1 \simeq a_0 e. \tag{4.2.2}$$

The field of this instantaneous dipole, at a distance R from the atom will be of the order:

$$E \simeq \frac{p_1}{4\pi\varepsilon_0 R^3} \simeq \frac{a_0 e}{4\pi\varepsilon_0 R^3}. \tag{4.2.3}$$

If a neutral atom is nearby it will therefore be polarized by this field and acquire an induced dipole moment of strength p_2:

$$p_2 = \alpha E \simeq \frac{\alpha a_0 e}{4\pi\varepsilon_0 R^3} \tag{4.2.4}$$

where α is the atomic polarizability of the second atom. This measures the ease with which the electron distribution can be displaced (section 2.2) and is proportional to the volume of the atom:

$$\alpha \simeq 4\pi\varepsilon_0 a_0^3. \tag{4.2.5}$$

The potential energy of interaction between the dipoles p_1 and p_2 is then, using eqns (4.2.2) and (4.2.4):

$$\begin{aligned} V_{\mathrm{int}}(R) &= -p_1 p_2 / 4\pi\varepsilon_0 R^3 \\ &\simeq -\alpha a_0^2 e^2 / (4\pi\varepsilon_0)^2 R^6 \end{aligned}$$

or, using eqns (4.2.1) and (4.2.5):

$$V_{\mathrm{int}}(R) = -\frac{2\alpha^2 h\nu}{(4\pi\varepsilon_0)^2 R^6}. \tag{4.2.6}$$

The more exact expression arrived at in Appendix A1 is:

$$V_{\mathrm{int}}(R) = -\frac{C_{\mathrm{AB}}}{R^6} \tag{4.2.7}$$

where the constant C_{AB} is given by:

$$C_{AB} = \frac{3e^4\hbar}{2m_e^2(4\pi\varepsilon_0)^2} \sum_{\substack{m,n \\ (\neq 0)}} \frac{f_{0,m}^A f_{0,n}^B}{\omega_{0m}^A \omega_{0n}^B (\omega_{0m}^A + \omega_{0n}^B)} \tag{4.2.8}$$

and

$$\omega_{0m} = (E_m - E_0)/\hbar \tag{4.2.9}$$

is the frequency (rad s^{-1}) of electromagnetic radiation which would cause the transition from the ground state to the excited state ψ_m in the isolated molecule. The corresponding oscillator strength for this transition is given by:

$$f_{0,m} = \frac{2m_e\omega_{0m}}{\hbar e^2} |p_t|^2 \tag{4.2.10}$$

where $|p_t|$ is the magnitude of the transition dipole, $\hbar = h/2\pi$ and m_e is the electron mass. $f_{0,m}$ measures the probability of the transition occurring and is, therefore, the measure of the intensity of the absorption band.

The similarities between eqns (4.2.6) and (4.2.7) are obvious. Apart from the $1/R^6$ dependence we see that the magnitude of the interaction energy is determined by the ease with which the electrons in the two atoms are able to undergo displacements, since this is, in effect, what the terms in the summation calculate. The connection between V_{int} and the absorption spectrum of the material is already apparent and we will return to that point again.

The oscillator strengths, $f_{0,m}$ (which will in future be denoted f_{0m}) obey a useful sum rule (Dalgarno and Lynn 1957):

$$\sum_m f_{0m} = N \tag{4.2.11}$$

where N is the number of electrons in the molecule. Equation (4.2.11) is a quantum mechanical result whose utility we will demonstrate in the next section.

The interaction energy of a pair of neutral, non-polar molecules, eqn (4.2.7), is of the same form as in the original explanation in terms of a hypothetical permanent dipole moment on *each* molecule. This is not altogether surprising since averaging the instantaneous dipole–dipole interaction over the (perturbed) internal electronic and vibrational motions of the molecule is analogous to performing the orientational averaging of the permanent dipole–dipole interaction as originally proposed.

The London theory of the van der Waals force does not require the existence of permanent molecular multipole moments and demonstrates the universality of this long range attractive force.

Exercise. 4.2.1. Establish eqn (4.2.6). [Note that the polarizability is sometimes defined in terms of the equation $\mathbf{p}_i = \alpha\varepsilon_0\mathbf{E}$ so that it has dimensions of volume (Atkins 1978, p. 751; 1982, p. 772).]

4.3 The calculation of C_{AB}

The constant C_{AB} determines the magnitude of the van der Waals force between molecules A and B. We see from the definition of C_{AB}, eqn (4.2.8), that an exact evaluation would require a knowledge of all the absorption frequencies ω_{0m} and all the corresponding oscillator strengths f_{0m} for each molecule. While such data is, in principle, obtainable, it is obvious that in most cases of interest we must resort to approximate evaluation techniques. The simplest of these follows from the assumption that only one absorption frequency is important (i.e. has a large oscillator strength) for each molecule. For molecule A we will denote this frequency by ω_A and the corresponding oscillator strength by f_A. If f_A is dominant, then the sum rule (eqn (4.2.11)) reduces to

$$f_A = N_A \tag{4.3.1}$$

where N_A is the number of electrons in molecule A. Equation (4.2.8) simplifies to

$$C_{AB} \simeq \left[\frac{3e^4\hbar}{2(4\pi\varepsilon_0)^2 m_e^2}\right] \frac{N_A N_B}{\omega_A\omega_B(\omega_A + \omega_B)}. \tag{4.3.2}$$

This expression is useful for providing order of magnitude estimates for C_{AB} but fails badly for many-electron systems, where eqn (4.3.1) is a poor approximation. In the absence of more detailed information it is usual to estimate ω_A by

$$\omega_A = I_A/\hbar \tag{4.3.3}$$

where I_A is the lowest ionization energy of molecule A. (Compare the discussion on eqn (4.2.1) with $\omega = 2\pi\nu$.)

Before finding more exact approximations, we derive an interesting relationship between C_{AB} and the polarizability of the individual molecules. Let us consider the function $\alpha(z)$ defined by:

$$\alpha(z) = \frac{e^2}{m_e}\sum_m \frac{f_{0m}}{\omega_{0m}^2 - z^2}. \tag{4.3.4}$$

Its importance follows from the following observation. When an electric field $\boldsymbol{E}e^{-i\omega t}$ of frequency ω (real and positive) is applied to a molecule, it can be shown by first-order perturbation theory (Blokhintsev 1964) that the dipole moment induced by the field in the molecule is of the form $\boldsymbol{p}e^{-i\omega t}$ where:

$$\boldsymbol{p} = \alpha(\omega)\boldsymbol{E} \tag{4.3.5}$$

and $\alpha(\omega)$ is given by eqn (4.3.4) (compare eqn (2.2.15)).

Thus the function $\alpha(z)$ has a physical interpretation as the molecular polarizability when z is real and positive. However, eqn (4.3.4) serves to define $\alpha(z)$ for any values of z where the series converges and, in particular, for purely imaginary values of z. Thus we may write

$$\alpha(i\xi) = \frac{e^2}{m_e} \sum_m \frac{f_{0m}}{\omega_{0m}^2 + \xi^2}. \tag{4.3.6}$$

The quantity $\alpha(i\xi)$ has no direct physical significance but is clearly closely related to the physically meaningful quantity $\alpha(\omega)$ in that they are both defined in terms of absorption frequencies ω_{0m} and oscillator strengths f_{0m}. The utility of this unphysical quantity $\alpha(i\xi)$ in the present discussion follows from the substitution of the mathematical identity

$$\frac{1}{\omega_{0A}\omega_{0B}(\omega_{0A} + \omega_{0B})} = \frac{2}{\pi} \int_0^\infty \frac{d\xi}{(\omega_{0A}^2 + \xi^2)(\omega_{0B}^2 + \xi^2)} \tag{4.3.7}$$

into eqn (4.2.8). Thus we obtain, using the definition equation (4.3.4) (see Exercise 4.3.1):

$$C_{AB} = \frac{3\hbar}{\pi(4\pi\varepsilon_0)^2} \int_0^\infty d\xi\, \alpha_A(i\xi)\alpha_B(i\xi). \tag{4.3.8}$$

That the magnitude of the van der Waals force between molecules can be related to an integral over imaginary frequencies of the product of the molecular polarizabilities is the consequence of a purely mathematical rearrangement of the original eqn (4.2.8). The importance of this result lies in the useful approximation that follows from eqn (4.3.8) (see below) and in the instructive correspondence between this result and the formulae of modern dispersion force theory discussed in section 4.7.

Now making the approximation that one frequency ω_A, dominates we obtain, using eqns (4.3.1) and (4.3.6):

$$\alpha_A(i\xi) = \frac{e^2 N_A}{m_e(\omega_A^2 + \xi^2)} \tag{4.3.9}$$

and

$$\alpha_A(\omega) = \frac{e^2 N_A}{m_e(\omega_A^2 - \omega^2)}. \tag{4.3.10}$$

If α_A^0 is the (experimentally measurable) zero frequency polarizability of molecule A then

$$\alpha_A(i\xi) = \frac{\alpha_A^0 \omega_A^2}{\omega_A^2 + \xi^2} \tag{4.3.11}$$

where

$$\omega_A = \left(\frac{e^2 N_A}{m_e \alpha_A^0}\right)^{1/2}. \tag{4.3.12}$$

Inserting this form for $\alpha(i\xi)$ in eqn (4.3.8) and integrating, produces (Exercise 4.3.2):

$$C_{AB} = \left[\frac{3\hbar e}{2m_e^{1/2}(4\pi\varepsilon_0)^2}\right] \frac{\alpha_A^0 \alpha_B^0}{\left(\frac{\alpha_A^0}{N_A}\right)^{1/2} + \left(\frac{\alpha_B^0}{N_B}\right)^{1/2}}. \tag{4.3.13}$$

Since the dominant frequency is usually that of an outer orbital electronic transition, a better approximation results by taking N_A to be the *number of electrons in the outer orbital* only. Since N_A enters to the half power in eqn (4.3.13) the choice of this value is not too critical and eqn (4.3.13) (with this modification for N_A) produces satisfactory estimates of C_{AB} (Pitzer 1959; Wilson 1965). Since eqn (4.3.13) can be obtained by eliminating ω_A and ω_B from eqn (4.3.2) with the aid of eqn (4.3.12), the question as to why eqn (4.3.13) is a better approximation than eqn (4.3.2) is justified. Basically the same approximation has been made but, by introducing the zero frequency polarizabilities α_A^0 and α_B^0, which are well-defined quantities, and eliminating the imprecisely known frequencies ω_A and ω_B, we make the dependence of C_{AB} on N_A and N_B much less sensitive.

The magnitude and range of the van der Waals force between molecules follows easily from eqn (4.3.8). Let us define the frequency ω_0 by

$$\omega_0 = \frac{4}{\pi \alpha_A^0 \alpha_B^0} \int_0^\infty \mathrm{d}\xi \; \alpha_A(i\xi) \alpha_B(i\xi) \tag{4.3.14}$$

and the length R_0 by

$$R_0^6 = \frac{\alpha_A^0 \alpha_B^0}{(4\pi\varepsilon_0)^2}. \tag{4.3.15}$$

With these definitions the van der Waals interaction energy (eqn (4.2.7)) becomes

$$V_{\text{int}}(R) = -\tfrac{3}{4}\hbar\omega_0 \left(\frac{R_0}{R}\right)^6 \tag{4.3.16}$$

which is usually quoted as London's equation.

Typical values of $(\alpha_A^0/4\pi\varepsilon_0)$ range from 1 to $20\times10^{-30}\,\mathrm{m}^3$ (McClellan 1963), which yields $R_0\sim0.2$ nm. From eqn (4.3.13), ω_0 is approximately given by:

$$\omega_0=\frac{2\omega_A\omega_B}{\omega_A+\omega_B}$$

$$=\frac{2\left(\dfrac{e}{m_e^{1/2}}\right)}{\left(\dfrac{\alpha_A^0}{N_A}\right)^{1/2}+\left(\dfrac{\alpha_B^0}{N_B}\right)^{1/2}} \qquad (4.3.17)$$

Type of interaction		Interaction energy
Covalent	(HH) (HH O)	Complicated, short range
Charge–charge	Q_1 — R — Q_2	$V=+Q_1Q_2/4\pi\varepsilon_0 R$
Charge–neutral	Q — R — α	$V=-Q^2\alpha/2(4\pi\varepsilon_0)^2R^4$
Dipole–dipole (Keesom) effect	p_1 θ_1 — R — ϕ θ_2 p_2 — Fixed dipoles (low T)	$V=-\dfrac{p_1p_2}{4\pi\varepsilon_0R^3}(2\cos\theta_1\cos\theta_2-\sin\theta_1\sin\theta_2\cos\phi)$
	p_1 — R — p_2 — Freely rotating dipoles (high T)	$V=-\dfrac{p_1^2\,p_2^2}{3(4\pi\varepsilon_0)^2kTR^6}$
Charge–dipole	p θ — R — Q — Fixed	$V=-Qp\cos\theta/4\pi\varepsilon_0R^2$
	p — R — Q — Rotating	$V=-Q^2p^2/6(4\pi\varepsilon_0)^2kTR^4$
Dipole–neutral (Debye effect)	p θ — R — α — Fixed	$V=-\dfrac{p^2\alpha(1+3\cos^2\theta)}{2(4\pi\varepsilon_0)^2R^6}$
	p — R — α — Rotating	$V=-\dfrac{p^2\alpha}{(4\pi\varepsilon_0)^2R^6}$
Neutral–neutral	α — R — α	$V=-\dfrac{3h\nu\alpha^2}{4(4\pi\varepsilon_0)^2R^6}$

FIG. 4.3.1. Common types of interactions between atoms and molecules. V = interaction energy (in J), Q = electrical charge (C), p = electric dipole moment (C m), k = Boltzmann constant (1.38×10^{-23} J K^{-1}), T = absolute temperature (K), R = distance between interacting atoms or molecules (m), θ = angle, α = electric polarizability (F m^2), h = Planck constant (6.626×10^{-34} J s), ν = electron characteristic frequency (s^{-1}), ε_0 = permittivity of free space (8.854×10^{-12} F m^{-1}). The force is obtained by differentiating the energy V with respect to distance R. (Adapted from Israelachvili 1974, with permission.)

which yields $\omega_0 \sim 10^{16}$ rad s^{-1}. Thus

$$V_{\text{int}}(R) \simeq -10^{-18}\left(\frac{0.2}{R(\text{nm})}\right)^6 \text{ Joules.} \tag{4.3.18}$$

For $R \sim 0.4$ nm at $T \sim 300$ K, we see that $V_{\text{int}} \sim 2\,kT$, which emphasizes the fact that van der Waals forces are *thermodynamically* important forces under normal experimental conditions. It should be remembered, however, that other long-range interactions between molecules can be important, especially if the molecules have permanent dipole moments. Figure 4.3.1 is a summary of the main interactions which occur between atoms and molecules. The importance of the London interaction for colloid science lies in the fact that, when interactions between large collections of molecules are treated, it is the London dispersion force that dominates over the Keesom and Debye forces (if the molecules are uncharged). This is because the summation assumes that the interactions are pairwise additive and this is a rather better assumption for the London forces than for the others (Overbeek 1952, p. 265). The modern Lifshitz theory avoids this problem and calculates the change in the (Helmholtz) free energy due to *all* of the interactions at the same time.

Exercises. 4.3.1 Establish eqn (4.3.8) assuming that $\sum \int = \int \sum$ (which is true for reasonably well behaved functions).

4.3.2 Establish eqn (4.3.13).

4.3.3 Establish eqn (4.3.16) and note the similarity to eqn (4.2.6). Only the constant $\frac{3}{4}$ is missing from the simple formula.

4.4 Pairwise summation of forces (Hamaker theory)

Following London's explanation of the origins of van der Waals forces, several workers (Kallmann and Willstätter 1932; Bradley 1932; Hamaker 1937) were quick to realize that such universal long-range intermolecular forces could give rise to the long-range attractive forces between macroscopic objects that must be invoked to explain the phenomenon of colloid coagulation (section 2.5). The simplest procedure for calculating the van der Waals force between macrobodies is to use the method of pairwise summation of intermolecular forces. To elucidate the technique let us consider a set of N molecules at positions $\boldsymbol{R}_i$ ($i = 1, 2, \ldots, N$). The separation distance of molecules i and j is R_{ij} given by

$$R_{ij} = |\boldsymbol{R}_j - \boldsymbol{R}_i|. \tag{4.4.1}$$

The interaction energy of the N-body system, in the pairwise summation method, is taken to be

$$V_{\text{int}}^{1,2,\ldots,N} = \tfrac{1}{2} \sum_{i=1}^{N} \sum_{\substack{j=1 \\ \neq i}}^{N} V_{\text{int}}^{ij}(R_{ij}) \tag{4.4.2}$$

where $V_{\text{int}}^{ij}(R)$ is the interaction energy of molecules i and j separated by distance R_{ij} in the *absence* of any other molecules.

For van der Waals forces (derived from second-order quantum perturbation theory, Appendix A1) (4.4.2) is only an approximation to reality. The internal states of molecules i and j will be modified by the presence of all the other molecules of the system and they do not, therefore, interact with each other in the way that they would if the other molecules were not present. The correct treatment of these many-body effects is a difficult problem whose solution we discuss in section 4.7. Clearly the pairwise summation method will be least in error when the molecules are far from one another, so that the individual pair interactions are relatively unaffected by the presence of other molecules. The extent to which this procedure is incorrect will be examined later.

4.4.1 *Interaction of a molecule with a macrobody*

In the pairwise summation approximation let us calculate the van der Waals interaction energy of a molecule of type 1 at position $\boldsymbol{r}$ with a body of arbitrary shape comprised of molecules of type 2 at number density ρ_2. This would correspond to the energy of physical adsorption of molecule 1 on adsorbent 2. Consider a volume element $\mathrm{d}V'$ at position $\boldsymbol{r}'$ of the body containing $\rho_2\,\mathrm{d}V'$ type-2 molecules (see Fig. 4.4.1). The interaction energy of molecule 1 with the molecules of the volume element in the

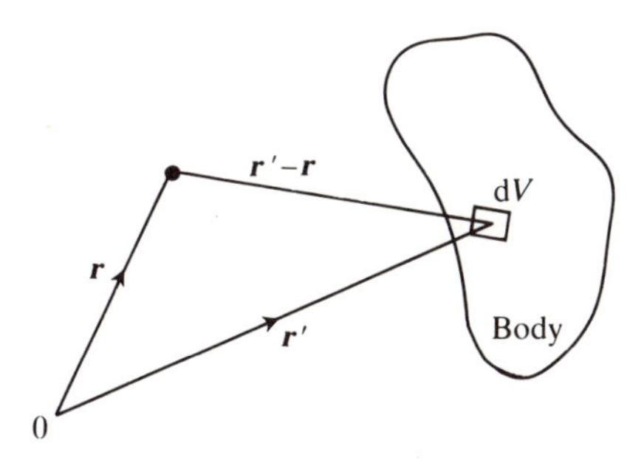

FIG. 4.4.1. The geometry of a type 1 molecule interacting with a body comprised of type 2 molecules at density ρ_2.

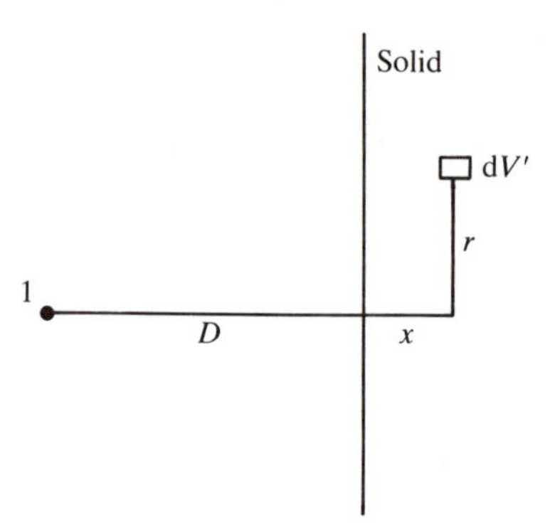

FIG. 4.4.2. Coordinate system for Hamaker summation in planar half-space.

pairwise approximation is:

$$d\phi_{12}(\boldsymbol{r}) = -\frac{C_{12}}{|\boldsymbol{r}-\boldsymbol{r}'|^6}\rho_2\,dV'. \tag{4.4.3}$$

The total van der Waals interaction energy of the molecule 1 with the body 2 is obtained by integrating eqn (4.4.3) over the volume of the body, viz.

$$\phi_{12}(\boldsymbol{r}) = -C_{12}\rho_2\int_{\text{Body 2}}\frac{dV'}{|\boldsymbol{r}-\boldsymbol{r}'|^6}. \tag{4.4.4}$$

For the case of usual interest (a molecule at a distance D from a planar half-space† we may write, using Pythagoras' theorem (see Fig. 4.4.2 and Exercise 4.4.1):

$$\begin{aligned}\phi_{12}(D) &= -C_{12}\rho_2\int_0^\infty dx\int_0^\infty\frac{2\pi r\,dr}{((D+x)^2+r^2)^3}\\ &= -\frac{\pi C_{12}\rho_2}{6D^3}.\end{aligned} \tag{4.4.5}$$

This more gradual fall-off with distance compared to the molecule–molecule interaction indicates the existence of a more long-range force or interaction energy.

4.4.2 *Interaction of two macrobodies*

If the molecule 1 is part of another body comprised of molecules of type 1 at number density ρ_1, then the interaction energy of the molecules of

† i.e. a body of infinite extent bounded by a plane surface.

type 1 in a volume element $\mathrm{d}V$ at position $\boldsymbol{r}$ is just

$$\mathrm{d}V_\mathrm{A} = \phi_{12}(\boldsymbol{r})\rho_1\,\mathrm{d}V \tag{4.4.6}$$

where $\phi_{12}(\boldsymbol{r})$ is given by eqn (4.4.4).

The total van der Waals interaction energy of body 1 with body 2 is then

$$V_\mathrm{A} = \rho_1 \int_{\text{Body 1}} \phi_{12}(\boldsymbol{r})\,\mathrm{d}V \tag{4.4.7}$$

$$= -\frac{A_{12}}{\pi^2}\int_{\text{Body 1}} \mathrm{d}V \int_{\text{Body 2}} \frac{1}{|\boldsymbol{r}-\boldsymbol{r}'|^6}\,\mathrm{d}V' \tag{4.4.8}$$

where we introduce the Hamaker constant A_{12} defined by

$$A_{12} = \pi^2\rho_1\rho_2 C_{12}. \tag{4.4.9}$$

In the limit as the bodies are separated by a distance large compared to the largest dimension of each body, the distance $|\boldsymbol{r}-\boldsymbol{r}'|$ can be replaced by R, the distance between the centres of mass of the two bodies, and eqn (4.4.8) reduces to:

$$V_\mathrm{A} \simeq -\left(\frac{A_{12}}{\pi^2}\right)\frac{V_1 V_2}{R^6} \tag{4.4.10}$$

where V_1 and V_2 are the volumes of the bodies.

Let us now consider some specific cases of interest in colloid science where analytic solutions of eqn (4.4.8) may be obtained.

For the case of two plane parallel half-spaces separated by a distance L, we may write eqn (4.4.7) as

$$V_\mathrm{A}(L) = -\frac{A_{12}}{6\pi}\int_{\substack{\text{Surface of}\\ \text{body 1}}} \mathrm{d}S_1 \int_0^\infty \frac{\mathrm{d}x}{(L+x)^3} \tag{4.4.11}$$

with the aid of eqn (4.4.5). In this case the volume element has been chosen as $\mathrm{d}x\,\mathrm{d}S_1$ where x is a coordinate measuring normal distance into body 1 from its surface and $\mathrm{d}S_1$ is an area element in that surface. Performing the x integration we have

$$V_\mathrm{A}(L) = -\frac{A_{12}}{12\pi L^2}\int_{\substack{\text{Surface of}\\ \text{body 1}}} \mathrm{d}S_1. \tag{4.4.12}$$

The remaining integral is just the surface area of half-space 1 and since the half-spaces are infinite in transverse extent, $V_\mathrm{A}(L)$ clearly diverges. However, from eqn (4.4.12), it is possible to define a finite interaction energy *per unit area of surface of half-space* 1, $E_A(L)$, as

$$E_A(L) = -\frac{A_{12}}{12\pi L^2}. \tag{4.4.13}$$

Note that the range of the interaction energy is even longer in this case, compared to that in section 4.4.1.

The Hamaker summation for more complicated geometries is, as one might expect, a mathematically involved operation. One other geometry leads to a relatively simple formula, viz. the sphere–sphere interaction.

When body 1 is a sphere of radius a_1 and body 2 is a sphere of radius a_2 with centre–centre distance R, eqn (4.4.8) reduces to (Hamaker 1937):

$$V_A(R) = -\frac{A_{12}}{6}\left\{\frac{2a_1a_2}{R^2-(a_1+a_2)^2} + \frac{2a_1a_2}{R^2-(a_1-a_2)^2} + \ln\left(\frac{R^2-(a_1+a_2)^2}{R^2-(a_1-a_2)^2}\right)\right\}. \tag{4.4.14}$$

In the limit as $R \gg a_1 + a_2$, this result yields:

$$V_A(R) \simeq -\left(\frac{16A_{12}}{9}\right)\frac{a_1^3a_2^3}{R^6}. \tag{4.4.15}$$

In terms of the distance of closest approach of the spheres $H(=R - a_1 - a_2)$, eqn (4.4.14) becomes (Exercise 4.4.2):

$$V_A(H) = -\frac{A_{12}\bar{a}}{12H}\left[\frac{1}{1+\dfrac{H}{2(a_1+a_2)}} + \frac{(H/\bar{a})}{1+\dfrac{H}{\bar{a}}+\dfrac{H^2}{4a_1a_2}} + \left(\frac{2H}{\bar{a}}\right)\ln\left\{\left(\frac{H}{\bar{a}}\right)\left(\frac{1+\dfrac{H}{2(a_1+a_2)}}{1+\dfrac{H}{\bar{a}}+\dfrac{H^2}{4a_1a_2}}\right)\right\}\right] \tag{4.4.16}$$

where

$$\bar{a} = \frac{2a_1a_2}{a_1+a_2}. \tag{4.4.17}$$

In the limit as $H \ll \bar{a}$, we have, from eqn (4.4.16) (Exercise 4.4.3):

$$V_A(H) = -\frac{A_{12}\bar{a}}{12H}\left\{1 + \frac{H}{\bar{a}}\left(1 - \frac{\bar{a}}{2(a_1+a_2)}\right) + 2\frac{H}{\bar{a}}\ln\left(\frac{H}{\bar{a}}\right) + \ldots\right\}. \tag{4.4.18}$$

The case of a sphere of radius a at distance H from a planar half-space is obtained from eqn (4.4.16) by setting $a_1 = a$ and taking the limit as $a_2 \rightarrow \infty$. We obtain (Exercise 4.4.3):

$$V_A(H) = -\frac{A_{12}a}{6H}\left\{1 + \frac{H}{2a+H} + \frac{H}{a}\ln\left(\frac{H}{2a+H}\right)\right\}. \qquad (4.4.19)$$

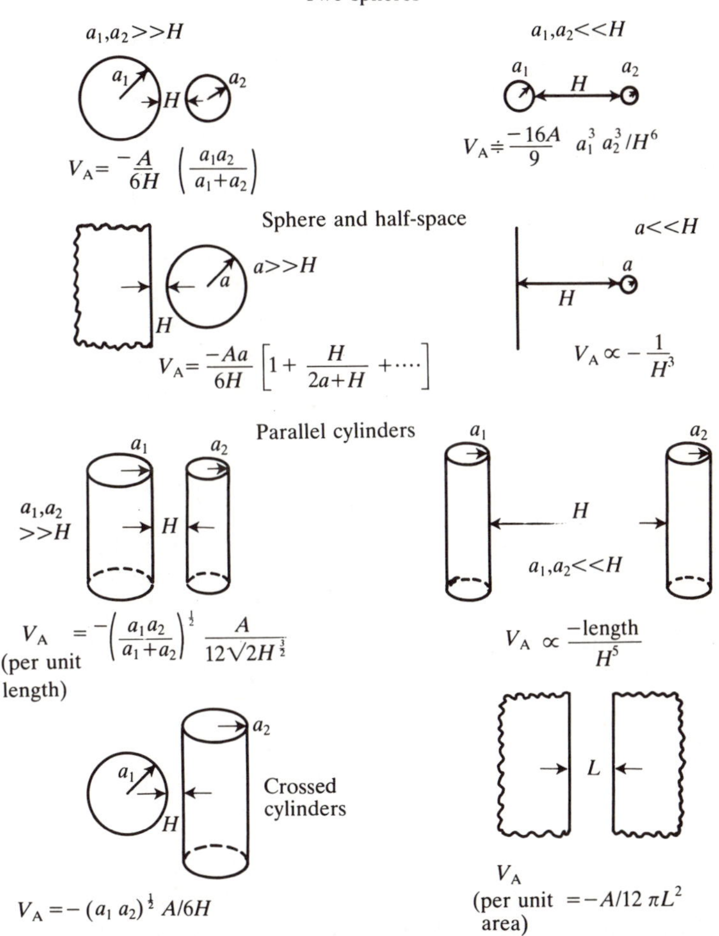

FIG. 4.4.3. Distance dependence of the *non-retarded* potential energy between pairs of solid surfaces of various geometries. The retarded interaction has a distance dependence $1/H$ times that and the proportionality constant then changes. In some cases the integration process yields quite messy results and these have not been included explicitly. Mahanty and Ninham (1976) remark that it is preferable in many cases to do the integration over the volumes of the particles by numerical methods rather than try to find analytical representations that are bound to be complicated for all but the simplest geometries.

Several other geometries yield analytic expressions for V_A. Among these are multilayers, and parallel and crossed cylinders. (See Mahanty and Ninham 1976, pp. 13–21; also Fig. 4.4.3.)

4.4.3 *Effect of the suspension medium*

The van der Waals interaction energy V_A, as calculated by the Hamaker summation method of the previous section, is of limited use in colloidal problems where the macrobodies (particles) are surrounded not by a vacuum, as assumed in eqn (4.4.2), but by a fluid medium whose molecules interact with the other molecules of the system. The presence of these extra interactions can drastically diminish the value of V_A, and even reverse its sign in some circumstances.

To see why, consider bodies 1 and 2 immersed in suspension medium 3. If they are brought from infinite separation to some distance D apart, the free energy change (the interaction energy) is not as large as in the case where medium 3 is a vacuum. When isolated, body 1 is interacting with its environment—a universe of medium 3. When brought close to body 2 it is interacting with a very similar environment, the only difference being that a number of molecules of medium 3 have been *replaced* by the molecules of body 2. The change of energy is less than that of the vacuum situation where the environment of body 1 originally contains no molecules and is then modified by the addition of the molecules of body 2.

To calculate the van der Waals interaction energy in the presence of a suspension medium, we consider the following thermodynamic path (see Fig. 4.4.4). Consider an initial state in which bodies 1 and 2 are infinitely separated in medium 3. We may regard the molecules of 3 in the volume that will finally be occupied by body 1, to be a 'body' 3 as shown. Let us remove body 1 and body 3 from the medium to vacuum. The change in free energy in this first step is just

$$\Delta F = -F_1 - (F_3 - V_{33}(D) + V_{32}(D)) \tag{4.4.20}$$

where D is the distance from 'body 3' to body 2 and F_i is the interaction energy of the isolated body i with a universe of medium 3. The energy change in removing body 3 is not $-F_3$ but $-(F_3 - V_{33}(D) + V_{32}(D))$ since its environment is not all medium 3. The energy $V_{32}(D) - V_{33}(D)$ represents the change in the interaction energy of body 3 with its environment, when the molecules of 3 that would have occupied the position of body 2 are replaced by body 2. Note in this argument we are using the notation that $V_{kj}(D)$ represents the (vacuum) interaction energy of a body the size and shape of body 1 comprised of k-type molecules with a body the size and shape of body 2 comprised of j-type molecules at separation distance D.

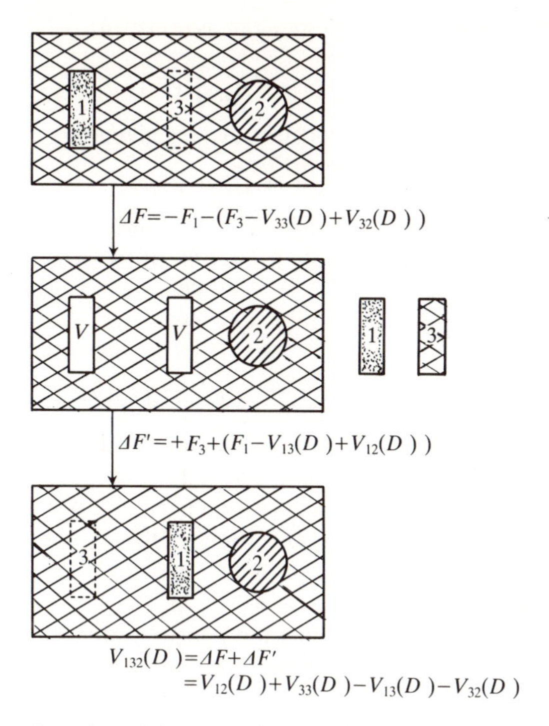

FIG. 4.4.4. Thermodynamic path for calculating the interaction energy $V_{132}(D)$ of bodies 1 and 2 in a fluid medium 3.

The second step is to return body 3 to the medium into the hole occupied originally by body 1 and to return body 1 to the hole originally occupied by body 3. The energy change in this second process is

$$\Delta F' = F_3 + (F_1 - V_{13}(D) + V_{12}(D)) \tag{4.4.21}$$

As before, the energy $V_{12}(D) - V_{13}(D)$ represents the change in the interaction energy of body 1 with its environment when the molecules of 3 that would have occupied the position of body 2 are replaced by body 2. The interaction energy $V_{132}(D)$ of bodies 1 and 2 at separation D in bathing medium 3 is given by

$$V_{132}(D) = \Delta F + \Delta F' = V_{12}(D) + V_{33}(D) - V_{13}(D) - V_{32}(D). \tag{4.4.22}$$

By the nature of the pairwise summation method we may write (see eqn (4.4.8))

$$V_{kj}(D) = -A_{kj}V(D) \tag{4.4.23}$$

where $V(D)$ is a *positive* function only of the geometry of the system and independent of the composition of the bodies 1 and 2 and A_{kj} is the vacuum Hamaker constant given by eqn (4.4.9). Thus eqn (4.4.22) becomes:

$$V_{132}(D) = -A_{132}V(D) \tag{4.4.24}$$

where the effective Hamaker constant A_{132} is given by:

$$A_{132} = A_{12} + A_{33} - A_{13} - A_{32}. \tag{4.4.25}$$

The presence of a bathing medium does not change the distance dependence of the van der Waals force but alters its magnitude through A_{132}.

Using eqns (4.4.9) and (4.3.8) we may write

$$A_{kj} = \frac{3\pi\hbar}{(4\pi\varepsilon_0)^2}\int_0^{\infty} \mathrm{d}\xi(\rho_k\alpha_k(i\xi))(\rho_j\alpha_j(i\xi)) \tag{4.4.26}$$

so that

$$A_{132} = \frac{3\pi\hbar}{(4\pi\varepsilon_0)^2}\int_0^{\infty} \mathrm{d}\xi(\rho_1\alpha_1(i\xi) - \rho_3\alpha_3(i\xi))(\rho_2\alpha_2(i\xi) - \rho_3\alpha_3(i\xi)). \tag{4.4.27}$$

When ρ_3 is set to zero, the bathing medium is a vacuum and eqn (4.4.27) reduces to (4.4.26). With an obvious notation, we may write:

$$A_{1v2} = A_{12}. \tag{4.4.28}$$

From eqn (4.4.25), it is apparent that

$$A_{132} \leqslant A_{1v2} \tag{4.4.29}$$

provided $A_{33} < A_{13} + A_{32}$ which is usually so. As Fig. 4.4.5 clearly shows, this reduction in Hamaker constant in a bathing medium can be one or even two orders of magnitude. Obviously for $A_{11} = A_{33}$, the function A_{131} is reduced to zero.

Several other generalizations can be made. In particular we note (from eqn (4.4.27))

$$A_{jkj} > 0 \tag{4.4.30}$$

i.e. like particles always attract. For unlike particles, this need not be the case. For example, if

$$\rho_1\alpha_1(i\xi) > \rho_3\alpha_3(i\xi) > \rho_2\alpha_2(i\xi) \quad \text{for} \quad 0 < \xi < \infty \tag{4.4.31}$$

then

$$A_{132} < 0 \tag{4.4.32}$$

and particles 1 and 2 repel each other in medium 3.

It is possible to derive some (historically, at least,) useful approximate

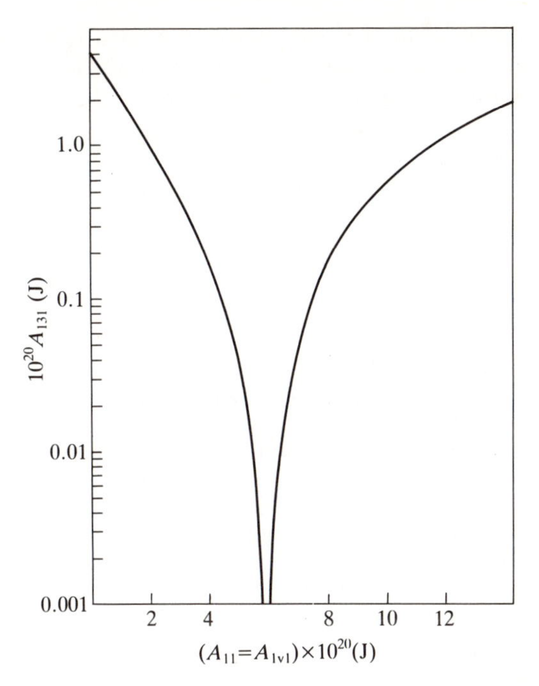

FIG. 4.4.5. The effect of an intervening medium (3) on the interaction between two semi-infinite bodies. (After Hiemenz 1977, p. 424, with permission.) Note that Hiemenz takes $A_{33} = 5.8 \times 10^{-20}$ J for water whereas the more recent figure (quoted in Table 4.2) is 3.70×10^{-20} J. (See Exercise 4.4.4.)

results. Using the definition (4.3.14) of the frequency ω_0, we may write:

$$A_{kj} = \frac{3\hbar\pi^2}{4(4\pi\varepsilon_0)^2}(\rho_k\alpha_k^0)(\rho_j\alpha_j^0)\omega_0^{kj} \tag{4.4.33}$$

so that:

$$A_{kk}A_{jj} = A_{kj}^2\frac{(\omega_0^{kj})^2}{\omega_0^{kk}\omega_0^{jj}}. \tag{4.4.34}$$

Thus, if it is assumed that

$$(\omega_0^{kj})^2 \simeq \omega_0^{kk}\omega_0^{jj} \tag{4.4.35}$$

we may write:

$$A_{kj} \simeq A_{kk}^{1/2}A_{jj}^{1/2}. \tag{4.4.36}$$

With this approximation, eqn (4.4.25) becomes (Exercise 4.4.8):

$$A_{132} \simeq (A_{11}^{1/2} - A_{33}^{1/2})(A_{22}^{1/2} - A_{33}^{1/2}). \tag{4.4.37}$$

It should be noted that the usefulness of such approximations for accurate research purposes is fast disappearing with the advent of digital computers and the rise of modern dispersion force theory.

Exercises. 4.4.1 Establish eqn (4.4.5).

4.4.2. Establish eqn (4.4.16) from (4.4.14). Also show that for two equal spheres of radius a:

$$V_A = \frac{-A_{12}}{6}\left[\frac{2}{s^2-4}+\frac{2}{s^2}+\ln\left(\frac{s^2-4}{s^2}\right)\right]$$

where $s = R/a$.

4.4.3 Establish eqn (4.4.18) and (4.4.19).

4.4.4 Calculate some points of Fig. 4.4.5 and so verify its general shape; then recalculate the figure based on $A_{33} = 3.70 \times 10^{-20}$ J for water.

4.4.5 Plot the potential energy of a molecule as a function of distance D from an infinite flat plate, using eqn (4.4.5) in the range $0 < D < 25$ nm. Take the molar volume as 20 cm^3 and estimate C_{12} from eqns (4.3.16) and (4.3.17).

4.4.6 Plot the potential energy per unit area as a function of separation L between two semi-infinite flat plates of a material 1 for which $A_{11} = 6.0 \times 10^{-20}$ J in a fluid 2 for which $A_{22} = 4.5 \times 10^{-20}$ J. (Make $1 < L < 25$ nm.)

4.4.7 Plot the function $V_A(H)/A_{12}$ for two spheres of radius 100 nm and 150 nm in the range $0 < H < 25$ nm.

4.4.8 Establish eqn (4.4.37).

4.5 Retardation effects in Hamaker theory

All the preceding calculations for the interaction energy of macrobodies are based on eqn (4.2.7)—the inverse sixth power law dependence of the long-range intermolecular attraction. This, in turn, is based on the quantum perturbation theory treatment of the interaction energy operator $\mathscr{V}_{\text{int}}$ (Appendix A1), which we represented by a set of Coulomb interactions between the electrons and nuclei of the molecules (eqn (A1.6)). This Coulombic form for $\mathscr{V}_{\text{int}}$ is an electrostatic approximation which is valid only if the molecules are sufficiently close to one another. To understand this let us consider how the attractive interaction energy arises. The molecular charge distribution A is constantly varying due to its internal electronic (and nuclear) motions and this variation propagates a complex electromagnetic field into surrounding space, the frequencies of which are those of the fundamental intramolecular motions. The field travels through space at the speed of light, c, until it reaches molecule B which is then polarized by the field. The oscillating dipole induced in B (together with the higher multipoles) re-radiates an electromagnetic field which is propagated back to A and interacts with it. If the time between A's radiating of the electromagnetic field and absorbing its reflection from B is negligible compared with the characteristic time for the internal motion, then A will be substantially in the same configuration on absorption as it was on emission, and the reflected field will, on average, be parallel to the emitting dipole moment. This will produce maximal lowering of molecule A's energy. If, on the other hand, the propagation time is comparable with the characteristic time for internal motion, the instantaneous dipole on A will have substantially altered its orientation by the time that the reflected field is received at A. The dipole moment component parallel to the reflected field will, on average, be smaller than in the nonretarded case and the energy decrease of molecule A will therefore be less. The propagation time is $\sim R/c$ and the molecular characteristic time is $\sim(2\pi/\omega_0)$. Thus, provided

$$R \ll \frac{2\pi c}{\omega_0} = \lambda_0 \tag{4.5.1}$$

the propagation can be regarded as instantaneous (non-retarded) and our present electrostatic analysis is valid. When R becomes a significant fraction of the characteristic wavelength λ_0 for the internal molecular motion, the finite propagation time (retardation) must be allowed for and the interaction energy will be smaller than that predicted by the electrostatic theory.

By modifying $\mathscr{V}_{\text{int}}$ to account correctly for the finite propagation time of the electromagnetic interaction between the molecular charge distributions, Casimir and Polder (1948) calculated the correct behaviour of $V_{\text{int}}(R)$ for all R. In particular, for $R \ll \lambda_0$ the R^{-6} form (eqn (4.2.7)) is valid and for $R \gg \lambda_0$:

$$V_{\text{int}}(R) \simeq \frac{-23\hbar c\alpha_{\text{A}}^0\alpha_{\text{B}}^0}{4\pi(4\pi\varepsilon_0)^2R^7} \tag{4.5.2}$$

The major consequence of the retardation effect is to limit the range of macroscopic (i.e. 'long-range') van der Waals forces.

Overbeek (1952, p. 266) represents the Casimir and Polder correction function in the following way (compare eqn (4.3.16)):

$$V_{\text{int}}(R) = \frac{-3\hbar\omega_0\alpha_0^2}{4(4\pi\varepsilon_0)^2R^6}f(p) \tag{4.5.3}$$

where

$$f(p) = 1.01 - 0.14p \quad \text{for} \quad 1 < p < 3$$

and

$$f(p) = \frac{2.45}{p} - \frac{2.04}{p^2} \quad \text{for} \quad 3 < p < \infty.$$

Here $p = 2\pi R/\lambda_0$ expresses the separation in terms of the characteristic wavelength (eqn (4.5.1)).

Overbeek used eqn (4.5.3) to calculate the interaction energy between two semi-infinite flat plates (Fig. 4.5.1) and Hunter (1963) (and also Vincent 1973) extended the calculation to plates of infinite area but finite thickness. Detailed calculations of the retarded interaction between two equal spheres and a sphere–plate combination have been provided by Clayfield *et al.* (1971) and that work has been summarized by Gregory (1981), who introduces some new approximation formulae and compares his results with the 'exact' Hamaker calculations. All of these approaches, however, suffer from a number of fundamental limitations. Apart from the breakdown of the additivity assumption it is obvious that the use of a single λ_0 value can at best only represent one of the characteristic modes of vibration of the electric charge in the medium. As we shall see when we deal with the Lifshitz (macroscopic) approach, every possible vibrational mode needs to be examined separately since each has its own characteristic retardation behaviour. The approximation formulae developed by Gregory can then best be used to assess which modes need to be most carefully considered in the light of the distances and geometry of the problem.

Since the retardation effect is automatically incorporated into the

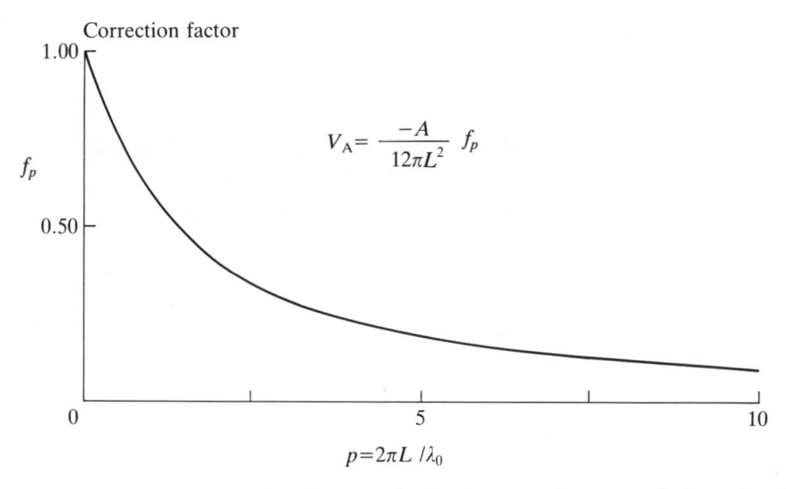

FIG. 4.5.1. Retardation correction factor to the London attraction energy between two flat plates of infinite thickness at a separation L. Note that although $f(p) = 0.59$ for $p = 3$, the total attraction energy is down by a factor of about four at that point. (From Overbeek 1952, p. 269, with permission.)

Lifshitz approach we will postpone further discussion on this point until the modern theory has been developed in section 4.7.

4.6 The Deryaguin approximation

When dealing with interactions between macrobodies, it is often the case that the range of the interaction is such that the two bodies do not interact significantly until the distance of closest approach is small compared to the radii of curvature of the bodies. Under these circumstances, a very useful approximate expression for the interaction energy of the bodies can be derived from the corresponding interaction energy per unit area of plane parallel half-spaces. The approximation is due, originally, to Deryaguin (1934).

We denote by 0 and 0′ the points on bodies 1 and 2 for which the separation distance is least and let H be this distance of closest approach. An axis system is erected on 0 with the z axis along the inward normal to body 1 and the (x, y) plane as shown in Fig. 4.6.1. The orientation of the

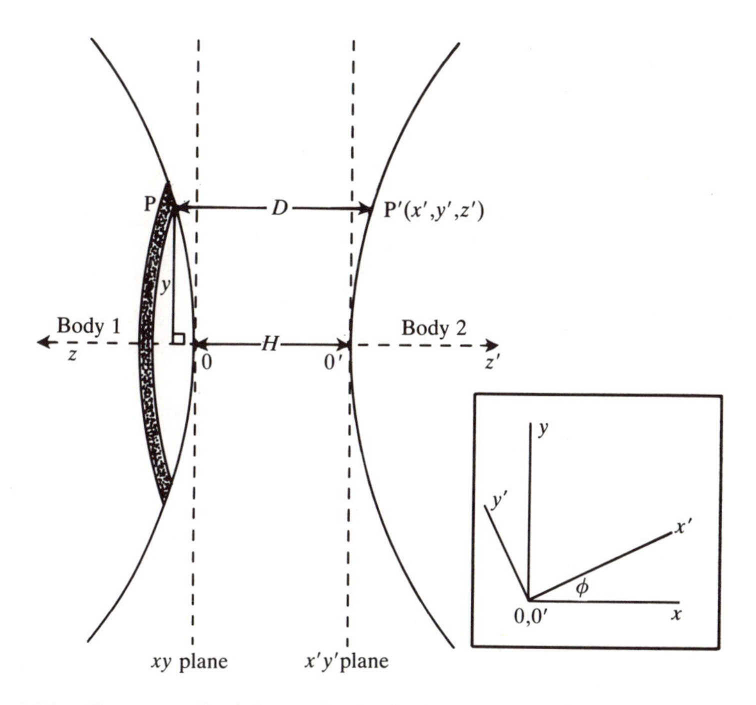

FIG. 4.6.1. Geometry of two interacting bodies illustrating appropriate principal coordinate axes for each body. The insert shows the orientation of the coordinate axes at 0 and 0′, looking along the line of centres between the two bodies.

xy axis system in this plane we will specify shortly. The bodies are assumed smooth and quadratically curved in the neighbourhood of 0 and 0′. Consider an element of area dS_1 at the point $P(x, y, z)$ on the surface of body 1. Provided the radii of curvature of bodies 1 and 2 are large compared to the separation distance H, then the area element can be approximately regarded as a surface element of a plane half-space with the properties of body 1 parallel to and separated a distance D (see Fig. 4.6.1) from another planar half-space with the properties of body 2. If $E(D)$ is the energy per unit area of half-space 1 interacting with the half-space 2, then, in this approximation,

$$dV = E(D)\, dS_1 \tag{4.6.1}$$

is the interaction energy of the area element dS_1 of body 1 with body 2. The distance D increases as the area element is taken further from the origin 0 and the associated energy dV decreases appropriately. This assignment of interaction energy to element dS_1 would be exact if the surfaces had infinite radii of curvature (i.e. were planar) and, therefore, in the case of finitely curved surfaces, is correct only to leading order in a power series in the reciprocals of the radii of curvature of the bodies.

In the Deryaguin approximation, the total interaction energy of bodies 1 and 2 is obtained by integrating dV over the surface of body 1, viz.

$$V(H) = \int_{\substack{\text{Surface}\\ \text{of body 1}}} dS_1\, E(D). \tag{4.6.2}$$

The energy $E(D)$ must decay sufficiently rapidly with distance D so that contributions to $V(H)$ are insignificant from area elements very far from the centre 0 where curvature effects are large and eqn (4.6.1) is a poor approximation. For the Deryaguin approximation to be valid, all the interaction energy should come from a small region around the point of closest approach. Deryaguin's original analysis was applied to two spheres of large radius of curvature, a, and in that case:

$$V(H) = 2\pi \int_0^\infty E(D) y\, dy. \tag{4.6.3}$$

However, from Fig. 4.6.1 $(D - H)/2 = a - \sqrt{(a^2 - y^2)}$. Hence $2y\, dy = a\sqrt{(1 - y^2/a^2)}\, dD \simeq a\, dD$ for $y \ll a$. In that case:

$$V(H) = \pi a \int_H^\infty E(D)\, dD. \tag{4.6.4}$$

White (1983) has shown that the general equation for $V(H)$ where the approaching surfaces may have arbitrary orientation and curvature (provided the latter is large) is:

$$V(H) = \frac{2\pi}{\sqrt{(\lambda_1\lambda_2)}} \int_H^\infty E(L)\,\mathrm{d}L, \tag{4.6.5}$$

where

$$\lambda_1\lambda_2 = \left(\frac{1}{R_1} + \frac{1}{R_1'}\right)\left(\frac{1}{R_2} + \frac{1}{R_2'}\right) + \sin^2\phi\left(\frac{1}{R_1} - \frac{1}{R_2}\right)\left(\frac{1}{R_1'} - \frac{1}{R_2'}\right). \tag{4.6.6}$$

The R_is and R_i's are the principal radii of curvature of the two surfaces respectively. The angle ϕ (see Fig. 4.6.1) is the angle between the principal axes on the two surfaces. Obviously, ϕ is immaterial when either of the bodies is a sphere and for two equal spheres $\lambda_1\lambda_2 = 4/a^2$, which reduces eqn (4.6.5) to (4.6.4).

From the definition of the force $F(H)$ exerted by one body on another we have that

$$F(H) = -\frac{\partial V}{\partial H} \tag{4.6.7}$$

$$= \frac{2\pi}{\sqrt{(\lambda_1\lambda_2)}} E(H). \tag{4.6.8}$$

Equation (4.6.5) is a valid approximation for *any* type of interaction energy and need not be restricted to the van der Waals interaction of this chapter. To apply eqn (4.6.5) with confidence to any particular interaction it is sufficient to require that:

$$\frac{L_0}{R_0} \ll 1 \tag{4.6.9}$$

and

$$H/R_0 \ll 1 \tag{4.6.10}$$

where L_0 is the length scale on which the interaction decays to zero and R_0 is the smallest radius of curvature of the system.

To illustrate the utility of eqn (4.6.5) let us derive the van der Waals interaction energy of two large spheres of radii a_1 and a_2 in the limit $H \ll a_i$. In this case $R_1 = R_2 = a_1$ and $R_1' = R_2' = a_2$ and eqn (4.6.6) becomes:

$$\lambda_1\lambda_2 = \left(\frac{a_1 + a_2}{a_1 a_2}\right)^2. \tag{4.6.11}$$

Thus eqn (4.6.5) can be written as

$$V_A(H) = \frac{2\pi a_1 a_2}{(a_1 + a_2)} \int_H^\infty \frac{-A_{12}}{12\pi L^2}\,\mathrm{d}L \tag{4.6.12}$$

with the aid of the planar half-space result (4.4.13). Integrating, we obtain (Exercise 4.6.1):

$$V_A(H) = -\left(\frac{2a_1a_2}{a_1+a_2}\right)\frac{A_{12}}{12H}. \tag{4.6.13}$$

Comparing this result with eqn (4.4.18) we see that eqn (4.6.13) is indeed correct to leading order in $H/\bar{a}$.

Another geometry of experimental importance is that of crossed *cylinders* of radii a_1 and a_2. In this case $R_1 = a_1$, $R_2 = \infty$, $R'_1 = a_2$, $R'_2 = \infty$ and $\phi = \pi/2$. For this geometry

$$\lambda_1\lambda_2 = \frac{1}{a_1a_2} \tag{4.6.14}$$

and the Deryaguin approximation is just

$$V(H) = 2\pi\sqrt{(a_1a_2)}\int_H^\infty \mathrm{d}L\, E(L). \tag{4.6.15}$$

Thus the force between crossed cylinders at separation distance H is

$$F(H) = 2\pi\sqrt{(a_1a_2)}E(H). \tag{4.6.16}$$

Experimentally, the measurement of $F(H)$ between crossed cylinders is used to obtain directly the interaction energy per unit area of planar halfspaces (Israelachvili and Tabor 1972). (See Chapter 7.)

Exercises. 4.6.1 Verify eqns (4.6.13) and (4.6.16).

4.6.2. Plot the correction factor $f(p)$ (eqn (4.5.3)) as a function of p for $0 < p < 10$. Note the difference between this and $F(p)$ of Fig. 4.5.1.

4.7 Modern dispersion force theory

The treatment of van der Waals forces between macrobodies suffers from two restrictions: (a) the assumption of pairwise additivity of molecular interactions of the London type; and (b) the neglect of the retardation effect. Both of these defects are remedied in the modern theory of van der Waals (dispersion) forces in macrosystems. The original theory is due to Lifshitz (1956) and was generalized by Dzyaloshinski *et al.* (1961). These treatments involve advanced statistical mechanical and quantum field theoretical arguments and cannot be repeated here. The interested reader is referred to Landau and Lifshitz (1969).

Fortunately, the results of the theory can be arrived at by a much simpler and more readily understandable approach, which was first suggested by Casimir and Polder (1948) and subsequently developed by van Kampen *et al.* (1968). It is, in effect, an extension of Planck's original treatment of a black body radiator. An excellent introductory discussion of this approach is given by Parsegian (1975) and reviews at various levels have been given by Gregory (1969), Visser (1972), Israelachvili (1973a, 1974), Richmond (1975), Mahanty and Ninham (1976), and Hough and White (1980).

In what follows we will try to indicate the main physical features of the simplified approach and skirt around the mathematical tricks that are used to solve the problem, giving only a hint as to the formalism involved. It is important to realize, from the outset, however, that this so-called 'heuristic' approach can ultimately be justified by appeal to the much more rigorous methods of quantum field theory, at least for simple geometries. The advantages of the approach are that it can be applied to many situations in which the corresponding quantum problem would be quite intractable.

The fundamental idea that lies behind the method is that the interaction we are trying to calculate is propagated as an electromagnetic wave from one body to another over distances that are large compared to atomic dimensions. Furthermore, the frequencies with which the electrons in these bodies are able to move most readily ($\omega_0 < 10^{18}\,s^{-1}$) correspond to wavelengths which are also large compared to atomic dimensions. It should be possible, therefore, to analyse the propagation of these waves by appeal to the classical equations of wave motion (i.e. Maxwell's equations). The medium could be treated as a continuum and its bulk properties introduced through the permittivity (or dielectric response, $\varepsilon(\omega)$), introduced in section 2.2. The method may break down at very close separations, where the graininess of the matter becomes evident, but that turns out to be of little importance.

Our analysis of the London dispersion interaction (sections 4.1–4.3) makes clear the importance of the fluctuating dipoles in two interacting atoms. The macroscopic approach draws heavily on the idea of *correlated* fluctuations—that is, the idea that the charge movements that occur on one atom are influenced by the movement occurring on neighbouring atoms and even those that are further away. The macroscopic body is considered to be made up of many local oscillating dipoles that are continuously radiating energy (as any vibrating dipole must do according to classical electromagnetic (e.m.) theory). These dipoles are also continuously absorbing energy from the e.m. field generated by all of their neighbours. Obviously, they are best able to absorb energy at frequencies corresponding to one of their natural resonances (recall Exercises 2.2.5–7) and these are also the frequencies at which they are most easily polarized and those at which they radiate energy. Two macroscopic bodies thus 'see' or 'feel' one another across an intervening gap as a result of the electromagnetic waves which emanate from one to the other.

Most of the energy of these correlated fluctuations remains inside the body and is part of its cohesive energy, i.e. what holds it together. The part that is of most concern to us is carried by the electromagnetic waves that escape from the surface of the two neighbouring bodies. This is the energy contained in the *surface* vibrational *modes,* and our problem is to determine how that electromagnetic field is propagated across the gap between the two bodies. We will show how that is done for the very simplest possible situation: two semi-infinite flat plates (i.e. two half-spaces) separated by a vacuum. That calculation will introduce most of the important concepts, which can then be generalized to more important situations.

It should be obvious that, in considering how one material responds to the electromagnetic field generated by a neighbouring material, we will draw heavily on the *dielectric response function* (section 2.2). That quantity contains all the important information about how the substance responds to an alternating electric field. It should also be clear that such information is implicitly contained in the electromagnetic absorption spectrum of the material because that measures exactly the same thing: what frequencies of e.m. energy can be taken up by the material and to what extent.

We will find that the function $\varepsilon(\omega)$ can, in principle, be constructed from a knowledge of the absorption spectrum, but that it is a difficult task requiring an extraordinary level of information over the whole frequency range (from microwave to far ultraviolet). Fortunately, the solution of the van der Waals problem requires much less information—merely the values of the function $\varepsilon(\omega)$ for *purely imaginary values of the frequency* (i.e. $\varepsilon(i\xi)$ where ξ is a real number). At first sight this may seem a rather strange and 'unphysical' quantity. You will recall that we introduced the corresponding quantity $\alpha(i\xi)$ in eqn (4.3.6) and remarked that it had no *direct* physical significance, but was introduced as a mathematical device to simplify an integration problem. Here $\varepsilon(i\xi)$ is used for a similar purpose, but it should be noted that if we substitute $\omega = i\xi$ in eqn (2.2.25) we have:

$$\varepsilon(i\xi) = \varepsilon_0\left(1 + \int_0^\infty m(\tau)\mathrm{e}^{-\xi\tau}\,\mathrm{d}\tau\right) \tag{4.7.1}$$

so that $\varepsilon(i\xi)$ may be regarded as the measure of the response of the system to an exponentially decaying (rather than an oscillating) field. $1/\xi$ is then the time-constant.

4.7.1 *Interaction between two flat semi-infinite bodies across a vacuum*

This problem is treated in an elementary fashion by Hunter (1975) and somewhat more formally by Richmond (1975) and Mahanty and Ninham (1976). We will adopt the elementary approach here, again because we want only to indicate how the final formalism arises. (See Fig. 4.7.1.)

We wish to find the frequencies of electromagnetic radiation that emanate from surface (1) and are propagated across the vacuum to (2) and which satisfy the Maxwell equations in the gap. For the moment we assume that the velocity of light $c = \infty$ so that we are ignoring retardation effects. The result will therefore be valid only for small values of L. The waves that can 'fit in' to this space are the analogues of the standing waves that 'fit in' to the black body radiator. If we can establish which frequencies (ω_j) produce these standing waves at a given separation, then the interaction energy can be calculated from the change in the zero point energy of the electromagnetic field:

$$V_A = U(L) - U(\infty) = \tfrac{1}{2}\sum_j \hbar\omega_j(L) - \tfrac{1}{2}\sum \hbar\omega_j(\infty). \qquad (4.7.2)$$

We assume that this is the only contribution to the energy change. This calculation of the internal energy change corresponds to London's (and Hamaker's) calculation. Equation (4.7.2) is true only at the absolute zero of temperature. The temperature must be introduced by calculating the free energy change ΔF rather than ΔU and we will introduce that correction later.

As the gap width diminishes, there is a decrease in the number and frequency of the modes which can satisfy Maxwell's equations in the gap and hence the (zero point) internal energy of the system decreases.

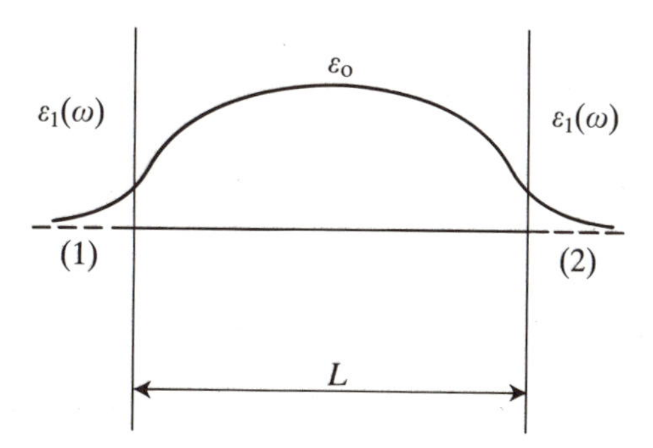

FIG. 4.7.1. Two semi-infinite materials separated by a vacuum. The approximate shape of the lowest permitted frequency wave function is shown. (The discontinuity in the slope at each surface is not obvious.)

The electrical potential can be represented as a function of the form $\phi(x, y, z)\psi(t) = \phi(x, y, z)e^{-i\omega t}$ and Maxwell's equations in a charge-free region with $c = \infty$ require that the spatial part of the wave function (ϕ) must satisfy Laplace's equation (see Appendix A3):

$$\nabla^2\phi = 0. \tag{4.7.3}$$

A suitable solution of this equation is (Exercise 4.7.1):

$$\phi(x, y, z) = f(x)e^{i(Q_2 y + Q_3 z)} \tag{4.7.4}$$

where Q_2 and Q_3 are the magnitudes of the wave vectors in the y and z directions. Q_2 and Q_3 are inversely proportional to the wavelengths in those directions). If we put $Q^2 = Q_2^2 + Q_3^2$ then $f(x)$ must satisfy:

$$\frac{d^2f(x)}{dx^2} - Q^2f(x) = 0 \tag{4.7.5}$$

so that ϕ will satisfy eqn (4.7.3). The solution to eqn (4.7.5) is:

$$\begin{aligned} f(x) &= Ae^{Qx} \qquad \text{for } x < 0 \\ &= Be^{-Qx} \qquad \text{for } x > L \end{aligned}$$

and, between the plates:

$$f(x) = Ce^{Qx} + De^{-Qx} \qquad \text{for } 0 < x < L.$$

(These functions are chosen to ensure that $f(x)$ is well behaved for large $|x|$.)

The boundary conditions at the respective surfaces require that both the potential, ϕ, and the dielectric displacement (section 2.2), $(=\varepsilon(\omega)E = -\varepsilon\, d\phi/dx)$, must be continuous. Hence (Exercise 4.7.2):

$$\begin{aligned} A - C - D &= 0 \\ e^{-QL}B - e^{+QL}C - e^{-QL}D &= 0 \\ \varepsilon_1 A - \varepsilon_0 C + \varepsilon_0 D &= 0 \\ \varepsilon_1 e^{-QL}B + \varepsilon_0 e^{QL}C - \varepsilon_0 e^{-QL}D &= 0. \end{aligned} \tag{4.7.6}$$

These equations are soluble only if the determinant of the coefficients is zero and this is so only if (Exercise 4.7.2):

$$[\varepsilon_1(\omega) + \varepsilon_0]^2 - [\varepsilon_1(\omega) - \varepsilon_0]^2e^{-2QL} = 0$$

or

$$\mathfrak{D}(\omega) = 1 - \Delta_{1v}^2 e^{-2QL} = 0 \tag{4.7.7}$$

where $\Delta_{1v} = (\varepsilon_1 - \varepsilon_0)/(\varepsilon_1 + \varepsilon_0)$.

Equation (4.7.7) is called a *dispersion* relation†. The values of $\omega(=\omega_j)$ for which eqn (4.7.7) is satisfied are the frequencies of waves which can be propagated across the gap.

For very large L values, the exponential term may be neglected and $\omega_j(\infty)$ must then satisfy:

$$[\varepsilon_1(\omega_j) + \varepsilon_0]^2 = 0. \tag{4.7.8}$$

If we take for $\varepsilon_1(\omega)$ the simplest form suggested by eqn (2.2.40), corresponding to a single undamped absorption mode at frequency ω_0

$$\varepsilon_1(\omega) = \varepsilon_0\left(1 + \frac{f_1}{\omega_0^2 - \omega^2}\right) \tag{4.7.9}$$

then eqn (4.7.8) is satisfied by (Exercise 4.7.3):

$$\omega_j(\infty) = \left(\frac{f_1}{2} + \omega_0^2\right)^{1/2}. \tag{4.7.10}$$

Thus the absorption mode gives rise to a permitted frequency that is just a little higher than the absorption frequency. Furthermore, if we add more terms in the series to represent more absorption frequencies in eqn (4.7.9), each one will give rise to an additional ω_j value like that in eqn (4.7.10), since each of the terms is important only in the neighbourhood of the absorption frequency. In this way we could construct all the ω_js required to calculate $\frac{1}{2}\Sigma\, \hbar\omega_j(\infty)$, in eqn (4.7.2), provided the absorptions were well separated.

At finite distances the solution would be a little more difficult but two things can still be said: to each absorption frequency there still corresponds an ω_j that will satisfy eqn (4.7.7) but its numerical value is lower than $\omega_j(\infty)$ and this will ensure that V_A is negative. Also, it is not too difficult to see that for finite L, eqn (4.7.7) has solutions only when $\varepsilon_1(\omega)$ is *negative*. From eqn (4.7.9) we see that this occurs at values of ω just above a resonance frequency (Fig. 4.7.2).

It might be possible, in principle, to solve some simple problems by this tortuous procedure: finding the ω_js that satisfy the dispersion equation for infinite separation and again for finite L and substituting in eqn (4.7.2). Fortunately, however, that is not necessary. By virtue of a very powerful theorem in the theory of functions of a complex variable (namely Cauchy's theorem) it is possible to evaluate the sum function in eqn (4.7.2) without knowing the separate values of ω_j. The procedure is analogous to finding the sum (4.3.4) using the integral identity (eqn

† The word 'dispersion' refers to the process whereby light of different wavelengths is dispersed on going through a triangular prism. Equation (4.7.7) gives the condition for propagation of a wave without hindrance (as distinct from absorption).

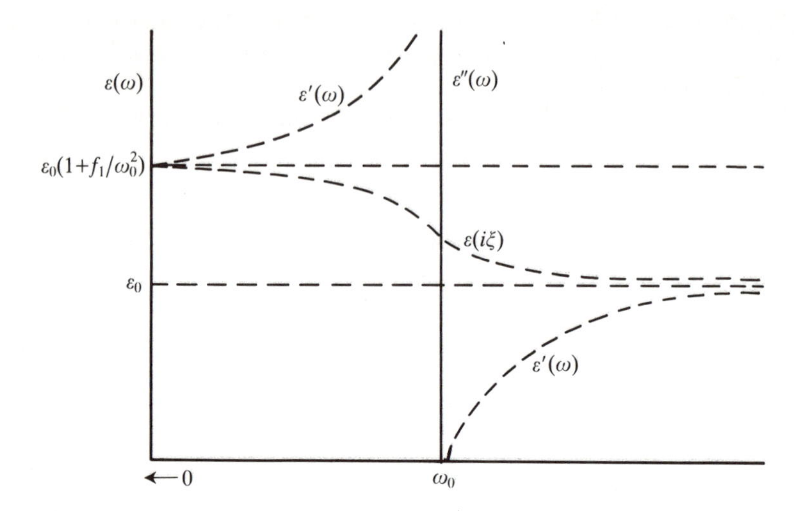

FIG. 4.7.2. The dielectric response function $\varepsilon(\omega)$ for a single undamped oscillator of resonance frequency ω_0. We will discuss the function $\varepsilon(i\xi)$ later, but note how simple it is in form compared to the behaviour of ε' in the neighbourhood of an absorption.

(4.3.7)); here again, as in that case, the sum involves only the values of ε for purely imaginary values of ω (i.e. $\varepsilon(i\xi)$). The mathematical technique introduces no new physical principles so we will not pursue it here. It is outlined in a simplified fashion in Appendix A2 in order to indicate how the integration along the imaginary axis comes about.

We will shortly examine (section 4.7.3) the more general problem of two arbitrary materials separated by a medium of different characteristics, using more rigorous procedures. Before doing so, however, we note a few more subtleties about the dielectric response function $\varepsilon(\omega)$.

4.7.2 $\varepsilon(\omega)$ *revisited*

The simple form for $\varepsilon(\omega)$ suggested by eqn (4.7.9) is fundamentally deficient. It is obviously a real function of the real variable ω. How then can it properly represent an absorption when we know that the dissipative (absorptive) part of ε is contained in ε'' (Exercise 2.2.5), which is zero in this case? The difficulty arises from the use of an *undamped* oscillator ($g_j = 0$). The form of $\varepsilon(\omega) = \varepsilon' + i\varepsilon''$ is shown in Fig. 4.7.2, where ε'' appears as a delta function (i.e. an absorption with zero line width). We will find that the calculations involve integrals of the $\varepsilon''(\omega)$ function with respect to frequency and it is a property of the delta function that such an integral is non-zero. (In a crude way we may say that the height of the absorption peak is infinitely large and this offsets, to some extent, its zero line width).

In setting up the dielectric response function we made use of the fundamental *linear* relationship between the electric field vector $\boldsymbol{E}$ and the polarizability $\boldsymbol{P}$ (eqn (2.2.19)). It is important to note that the field vector $\boldsymbol{E}$ to be used here is the local electric field strength at the molecule. This local field is *not* $\boldsymbol{E}(\boldsymbol{r}, t)$ in a dielectric material. $\boldsymbol{E}(\boldsymbol{r}, t)$ is strictly the local electric field *ensemble averaged* over a statistically significant volume element centred on position $\boldsymbol{r}$. Although linearly related, $\boldsymbol{E}(\boldsymbol{r}, t)$ and the local field are only equal when the density of the molecules is zero. We will discuss the density dependence of this relationship later. For our present purposes, it is sufficient to note that these considerations imply a linear relationship between $\boldsymbol{P}(\boldsymbol{r}, t)$ and $\boldsymbol{E}(\boldsymbol{r}, t)$. The instantaneous dipole $\boldsymbol{p}$ induced by the field will decay to zero if the field is switched off, due to the inter- and intramolecular motions, on a time scale that is characteristic of those motions. The polarization $\boldsymbol{P}(\boldsymbol{r}, t)$ will behave analogously. As we saw in section 2.2.4, this leads to a phase lag between the field $\boldsymbol{E}$ and the displacement vector $\boldsymbol{D}$. Considering the result of Exercise 2.2.5 and the discussion of section 2.4 we should not be surprised to find that such a phase lag corresponds to a dissipation of energy from the field. In terms of the real physical fields ($\boldsymbol{E}$ and $\boldsymbol{D}$), the rate at which the electromagnetic energy density changes is given by (Landau and Lifshitz 1960):

$$\frac{\partial \mathscr{E}(\boldsymbol{r}, t)}{\partial t} = -\boldsymbol{E}(\boldsymbol{r}, t)\frac{\partial \boldsymbol{D}(\boldsymbol{r}, t)}{\partial t}. \tag{4.7.11}$$

For a field with frequency ω of the form

$$\begin{aligned}\boldsymbol{E}(\boldsymbol{r}, t) &= \mathrm{Re}(\boldsymbol{E}_0(\boldsymbol{r})\mathrm{e}^{-i\omega t}) \\ &= \boldsymbol{E}_0(\boldsymbol{r})\cos \omega t \end{aligned} \tag{4.7.12}$$

we compute the rate at which the electromagnetic energy density changes, averaged over one oscillation period $(2\pi/\omega)$ and obtain:

$$\left\langle \frac{\partial \mathscr{E}}{\partial t} \right\rangle = -\frac{\omega}{2}\varepsilon''(\omega)\boldsymbol{E}_0^2(\boldsymbol{r}). \tag{4.7.13}$$

(Compare this with the result obtained in Exercise 2.2.5.) Thus the rate at which electromagnetic energy is dissipated into heat in the dielectric medium is proportional to the imaginary part of the dielectric response function. For a dielectric material a graph of $\varepsilon''(\omega)$ as a function of frequency ω would be just the absorption spectrum of that material. Schematically it has the form shown in Fig. 4.7.3(a).

The real part, $\varepsilon'(\omega)$, of the dielectric response measures the *transmission* properties of the medium through its relation to the refractive index (eqns (2.2.35) and (2.2.36)).

It should be noted that our earlier expression for $\varepsilon(\omega)$ (eqn (2.2.40)) is in the form of a real function of the complex variable $i\omega$. It can readily be converted into the form $\varepsilon = \varepsilon'(\omega) + i\varepsilon''(\omega)$ (i.e. to a complex function of

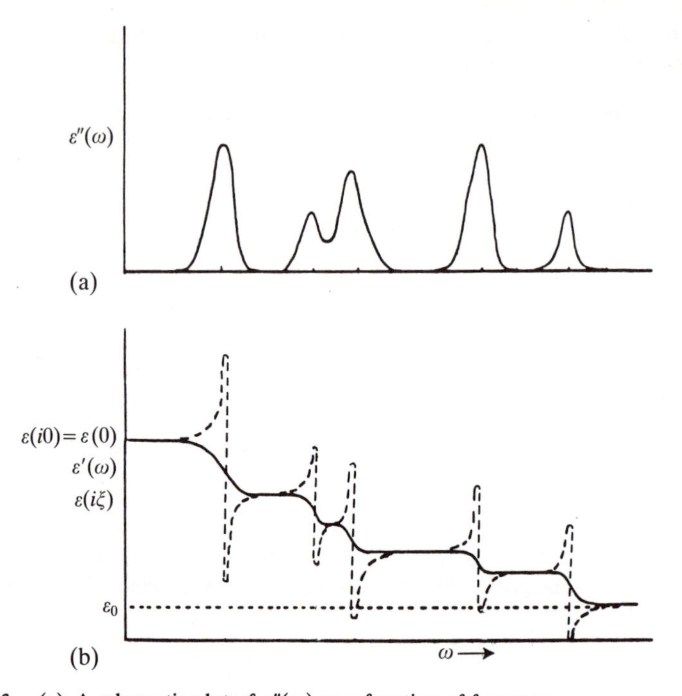

FIG. 4.7.3. (a) A schematic plot of $\varepsilon''(\omega)$ as a function of frequency.
(b) Schematic plots of $\varepsilon'(\omega)$ (dotted curve) and $\varepsilon(i\xi)$ (smooth curve) as functions of frequency. Note the coincidence of the two functions in the 'flat' regions between absorption frequencies.

the real variable ω) and that is the form we are presently discussing (see Exercise 4.7.4). That is the form that makes physical sense, since the separate components ε' and ε'' are readily measurable.

The expressions (2.2.28) and (2.2.29) for ε' and ε'' give rise to an important relation between them. This is one of the Kramers–Kronig relations and it reads (Carrier *et al.* 1966)

$$\varepsilon_r'(\omega) = 1 + \frac{2}{\pi} P \int_0^\infty \frac{x\varepsilon_r''(x)\,\mathrm{d}x}{x^2 - \omega^2}. \tag{4.7.14}$$

Since this is a purely mathematical consequence of the defining equations for ε_r' and ε_r'' its derivation will not be considered here (see Landau and Lifshitz 1960). Note, however, that the integral is a *principal value* integral, symbolized by $P\int$, since the integrand diverges when $x = \omega$:

$$P\int_0^\infty \mathrm{d}x = \lim_{\delta\to 0} \int_0^{\omega-\delta} \mathrm{d}x + \int_{\omega+\delta}^\infty \mathrm{d}x.$$

These matters need not concern us further. The Kramers–Kronig relation eqn

(4.7.14) means that if $\varepsilon''(\omega)$ (i.e. the absorption spectrum) is known for all frequencies $0 \leqslant \omega < \infty$, then $\varepsilon'(\omega)$ can, in principle, be calculated from eqn (4.7.14) (see Fig. 4.7.3(b)).

We have already assumed, in writing expressions like eqn (2.2.39), that the ε function can take complex values for its argument but the results of Exercise 2.2.8 show that this is merely a mathematical manipulation of the function $\varepsilon = \varepsilon' + i\varepsilon''$ of a *real* variable. For later use we must now formally extend the definition of ε:

$$\varepsilon(\omega) = \varepsilon_0\left(1 + \int_0^\infty m(\tau)e^{i\omega\tau}\,d\tau\right) \qquad [2.2.25]$$

to allow ω to take general complex values ($\omega = \eta + i\xi$). There are formal mathematical tests that can be applied to establish the conditions under which this can be done. In particular, we will be concerned with values of ε for purely imaginary values of ω (i.e. $\varepsilon(i\xi)$) as defined for $\xi > -\xi_0$ (where $m(\tau) \simeq e^{-\xi_0\tau}$ for large τ) when the integral certainly converges. It is possible, however, to derive a Kramers–Kronig relation (Landau and Lifshitz 1960) from the definitions (eqn (2.2.29)) for $\varepsilon''(\omega)$ and eqn (4.7.1) for $\varepsilon(i\xi)$, namely:

$$\varepsilon(i\xi) = \varepsilon_0\left(1 + \frac{2}{\pi}\int_0^\infty \frac{x\varepsilon_r''(x)\,dx}{x^2 + \xi^2}\right) \qquad (4.7.15)$$

by eliminating the unknown $m(\tau)$. This equation connects the function $\varepsilon(i\xi)$ to the imaginary part of the real (i.e. experimentally measurable) dielectric response.

Note the close similarity of eqn (4.7.14) defining $\varepsilon'(\omega)$ and eqn (4.7.15) defining $\varepsilon(i\xi)$ as functionals† of $\varepsilon_r''(\omega)$. We will make use of this similarity to construct $\varepsilon(i\xi)$ from experimental data (see section 4.8). For the present, it is sufficient to note that $\varepsilon(i\xi)$ defined by eqn (4.7.15) is an even function of ξ. The Kramers–Kronig relation has served to extend the definition of $\varepsilon(i\xi)$ onto the negative ξ axis.

Finally, we examine the relationship between molecular polarizability $\alpha(\omega)$ and the dielectric response $\varepsilon(\omega)$. For a field of frequency ω, the induced dipole on a given molecule is (eqn (2.2.16):

$$\boldsymbol{p} = \alpha(\omega)\boldsymbol{E}_{\text{loc}}(\boldsymbol{r}) \qquad (4.7.16)$$

where $\boldsymbol{E}_{\text{loc}}(\boldsymbol{r})$ is the local field at the molecule. To leading order in the density of polarizable entities in the system, we can replace $\boldsymbol{E}_{\text{loc}}$ by the average field $\boldsymbol{E}(\boldsymbol{r})$.

† A functional is a mathematical device that converts a function (in this case $\varepsilon_r''(x)$) into a single number. In this case a new number is generated for each new value of ξ and those are the (real) values of the function $\varepsilon(i\xi)$.

Thus to this order:

$$\boldsymbol{P}(\boldsymbol{r}) \simeq \rho_{\mathrm{N}}\alpha(\omega)\boldsymbol{E}(\boldsymbol{r})$$

from eqn (2.2.17). Substitution of this result in eqn (2.2.18) and comparison with eqn (2.2.24) yields

$$\varepsilon_{\mathrm{r}}(\omega) \simeq 1 + \frac{\rho_{\mathrm{N}}\alpha(\omega)}{\varepsilon_0} \tag{4.7.17}$$

to leading order in the density. A good approximation for moderate $\rho_{\mathrm{N}}\alpha(\omega)$ values is the Clausius–Mossotti formula (Jackson 1975)

$$\frac{\varepsilon(\omega) - \varepsilon_0}{\varepsilon(\omega) + 2\varepsilon_0} = \frac{\rho_{\mathrm{N}}\alpha(\omega)}{3\varepsilon_0} \tag{4.7.18}$$

which reduces to eqn (4.7.17) when $\varepsilon(\omega) \simeq \varepsilon_0$.

4.7.3 *The dispersion relation method*

The concept of the individual molecular charge distribution acting as an antenna, emitting and receiving e.m. radiation generated by the intra- (and inter-) molecular motions has been introduced already (sections 4.5 and 4.7.1). That these molecular antennae are densely packed in most materials does not change this fundamental physical picture although the e.m. fields are now propagated through a dielectric medium rather than free space. Provided we are concerned with radiation of wavelength much larger than intermolecular spacings ($\omega < 10^{18}\ \mathrm{rad\,s^{-1}}$), the effect of the medium can be described by the dielectric response function $\varepsilon(\omega)$ of the medium. That is, an e.m. field (with electric and magnetic field components $\boldsymbol{E}$ and $\boldsymbol{H}$) of frequency ω propagating in the uniform dielectric material must satisfy the macroscopic equations† of Maxwell (Landau and Lifshitz (1960)) viz:

$$\nabla \cdot (\varepsilon(\omega)\boldsymbol{E}) = 0 \tag{4.7.19}$$

$$\nabla \cdot (\mu(\omega)\boldsymbol{H}) = 0 \tag{4.7.20}$$

$$\nabla^2\boldsymbol{E} + \omega^2\mu(\omega)\varepsilon(\omega)\boldsymbol{E} = 0 \tag{4.7.21}$$

and an identical equation to (4.7.21) with $\boldsymbol{E}$ replaced by $\boldsymbol{H}$.

The role of $\mu(\omega)$ in the induction of magnetic dipole moment density in the medium by the magnetic field $\boldsymbol{H}$ is exactly analogous to the role of $\varepsilon(\omega)$ in electric dipole induction. We could discuss the properties of $\mu(\omega)$ in an identical manner to $\varepsilon(\omega)$ but, since the magnitude of the induced magnetic dipole moment

† A brief outline of the formalism of vector calculus with definitions of the operators $\nabla\cdot$ and ∇^2 is given in Appendix A3.

is small in most materials it is usually an excellent approximation to write:

$$\mu(\omega) \simeq \mu_0, \tag{4.7.22}$$

the vacuum magnetic permeability.

The error is of the order of one part in 10^6. The only exceptions to this approximation are the ferromagnetic materials, where $\mu(\omega)$ can be $\sim 10^3 \mu_0$ or more.

Note that the velocity of light *in vacuo* is given by

$$c = (\varepsilon_0 \mu_0)^{-1/2} \tag{4.7.23}$$

a result we will use later.

Thus modern dispersion force theory starts with the concept of a randomly fluctuating electromagnetic field pervading the material system. This field is driven by the motions of the individual molecular charge distributions, propagates with frequencies characteristic of these motions and is constrained to obey the macroscopic Maxwell's equations given above.

In the presence of dielectric boundaries (interfaces), where $\varepsilon(\omega)$ in particular suffers a sudden change in value over a length scale (the width of the interface) small compared to the wavelength of the e.m. radiation, the usual e.m. boundary conditions must be satisfied (compare section 4.7.1). That is, we impose continuity of: (a) the normal components of the dielectric displacement $\boldsymbol{D}(=\varepsilon \boldsymbol{E})$ and of $\mu \boldsymbol{H}$; and (b) the parallel components of $\boldsymbol{E}$ and $\boldsymbol{H}$ across the interface. These boundary conditions serve as constraints on the allowed frequencies of propagation. (A communications engineer would regard the system as a dielectric waveguide and, using conventional waveguide theory, would subsume the constraints imposed by the dielectric boundaries of the system into a single dispersion relation.) That is, for a system of given geometry and dielectric properties we can find a function $\mathfrak{D}(\omega)$ such that the allowed frequencies of e.m. propagation in the system are given by

$$\mathfrak{D}(\omega) = 0 \tag{4.7.24}$$

just as we did for the simple problem in section 4.7.1. In that case (Exercise 4.7.2) the magnetic field vector was ignored and only the continuity of potential and dielectric displacement were required. Clearly the roots of eqn (4.7.24), ω_j ($j = 1, 2, \ldots$), are functions of the geometry and dielectric properties of the system.

To make the connection with quantum mechanics, the fluctuating e.m. field is regarded as a collection of photons. If ω_j is an allowed frequency ($\mathfrak{D}(\omega_j) = 0$) then the energy associated with this *mode* is simply $(n + \frac{1}{2})\hbar\omega_j$ where n is the number of photons in the mode. As photons are emitted and absorbed by the molecules of the system, the number of photons in any given mode fluctuates along with the energy contained in that mode. The important quantity for present purposes is the Helmholtz free energy of the mode, F_j—the energy available to do work on the system's environment. From statistical thermodynamics (Landau and Lifshitz 1969) we have

$$\begin{aligned} F_j &= -kT \ln Z_j \\ &= -kT \ln \sum_{n=0}^{\infty} \exp\left(-(n+\tfrac{1}{2})\frac{\hbar\omega_j}{kT}\right) \qquad (4.7.25) \\ &= kT \ln\left(2\sinh\frac{\hbar\omega_j}{2kT}\right) \qquad (4.7.26) \end{aligned}$$

where Z_j is the partition function for the mode j (Exercise 4.7.5) and the summation is over all of the permitted quantum states of the mode j.

The total free energy of the e.m. field is obtained by summing the free energies for each mode

$$F = kT \sum_j \ln 2\sinh\left[\frac{\hbar\omega_j}{2kT}\right]. \qquad (4.7.27)$$

However, to solve eqn (4.7.24) explicitly for the ω_j is not feasible in general. It is possible to circumvent this difficulty by using the same mathematical device that was alluded to earlier (see Ninham *et al.* 1970). The free energy of the e.m. field can be written as:

$$F = kT \sum_{n=0}^{\infty}{}' \ln \mathfrak{D}(i\xi_n) \qquad (4.7.28)$$

where

$$\xi_n = \frac{2\pi kT}{\hbar} n. \qquad (4.7.29)$$

The prime on the summation sign indicates that the $n = 0$ term must be divided by two.

Equation (4.7.28) is derived from (4.7.27) using Cauchy's theorem (Appendix A2) and replaces the difficulty of explicitly obtaining the zeros of $\mathfrak{D}(\omega)$ with the difficulty of evaluating the function $\mathfrak{D}(i\xi)$ at an infinite set of discrete values (Langbein 1974).

We need to understand just what the function $\mathfrak{D}(i\xi)$ represents and we begin by extending the definition of the dispersion relation to include complex values of its argument i.e. by $\mathfrak{D}(z)$ we mean the function obtained from $\mathfrak{D}(\omega)$ by the replacement of the real positive variable ω by the complex variable z wherever it appears. Since $\mathfrak{D}(\omega)$ is derived from functions $\boldsymbol{E}$ and $\boldsymbol{H}$ that satisfy Maxwell's equations (4.7.19–22), this replacement involves ω both explicitly and implicitly through the frequency dependence of $\varepsilon(\omega)$ (and $\mu(\omega)$).

With this definition of $\mathfrak{D}(z)$, we understand $\mathfrak{D}(i\xi)$ to mean $\mathfrak{D}(z)$ evaluated on the imaginary frequency axis. In deriving eqn (4.7.28) we must invoke the following properties

$$\lim_{|z|\to\infty} \mathfrak{D}(z) = 1 \tag{4.7.30}$$

$$\mathfrak{D}(i\xi) = \mathfrak{D}(-i\xi). \tag{4.7.31}$$

The first of these is just the requirement that $\mathfrak{D}(\omega)$ be suitably *normalized* and the second follows from the observation that ω can enter $\mathfrak{D}(\omega)$ either explicitly as ω^2 or implicitly as $\varepsilon(\omega)$ (and $\mu(\omega)$). We have already seen that $\varepsilon(i\xi)$ is an even function of ξ (cf. eqn (4.7.15)).

If the system comprises two dielectric bodies immersed in a third dielectric fluid phase, the dispersion relation for the system will change as the separation of the bodies is altered. The allowed modes of the system will change correspondingly, with a consequent change in the e.m. free energy of the system. The van der Waals interaction energy $V(D)$ of the two bodies is just the work done to bring them from infinite separation to the separation distance D. Assuming that the work done manifests itself only in the change in e.m. free energy during the process, we can write:

$$\begin{aligned} V_{\mathrm{A}}(D) &= F(D) - F(\infty) \\ &= kT \sum_{n=0}^{\infty}{}' \ln \mathfrak{D}(i\xi_n \mid D) \end{aligned} \tag{4.7.32}$$

where:

$$\mathfrak{D}(i\xi \mid D) = \frac{\mathfrak{D}(i\xi)|_D}{\mathfrak{D}(i\xi)|_\infty}.$$

Clearly $\mathfrak{D}(i\xi \mid D)$ must be defined such that

$$\left.\begin{matrix} \lim\limits_{|\xi|\to\infty} \\ \lim\limits_{D\to\infty} \end{matrix}\right\} \mathfrak{D}(i\xi \mid D) = 1. \tag{4.7.33}$$

The dispersion relation method is a physically appealing approach to modern dispersion theory and provides an adequate rationale for the form of the final expression for $V_{\mathrm{A}}(D)$. In particular we see how a sum over imaginary frequencies is involved, why the theory depends on a knowledge of the $\varepsilon(i\xi)$ function for each of the dielectric materials involved, and how the temperature of the system enters, both explicitly in eqn (4.7.32) and implicitly through the definition of ξ_n.

It should be remembered that the dispersion relation $\mathfrak{D}(i\xi \mid D)$, when explicitly derived for a given system by solving Maxwell's equations, is often somewhat arbitrary, the constraints (eqns (4.7.30), (4.7.31) and (4.7.33) not being sufficient to determine $\mathfrak{D}(i\xi \mid D)$ uniquely). This normalization problem does not arise in the more sophisticated statistical mechanical method of Lifshitz (1956) which is therefore to be preferred from the point of view of rigour. The dispersion relation approach can only be verified in any given system strictly by comparison with the Lifshitz approach.

In Lifshitz theory, the temperature enters explicitly. Note that, as $T \to 0$, the spacing $2\pi kT/\hbar$ between successive values in the frequency sum (eqn (4.7.32)) tends to zero and the sum tends to a frequency integral

$$\sum_{n=0}{}' f(\xi_n) \xrightarrow[T\to 0]{} \frac{\hbar}{2\pi kT} \int_0^\infty \mathrm{d}\xi f(\xi). \tag{4.7.34}$$

At $T = 0$, we have for the interaction energy

$$V_{\mathrm{A}}(D) = \frac{\hbar}{2\pi} \int_0^\infty \mathrm{d}\xi \ln \mathfrak{D}(i\xi \mid D). \tag{4.7.35}$$

For non-zero values of T, the frequency integral will be a good approximation to the frequency sum only if the dominant contributions to the sum come from large n values. As we shall show below, this is not always the case.

4.7.4 *Modern theory for planar half-spaces*

One geometry for which the dispersion relation can be explicitly calculated is that of two plane parallel half-spaces of material 1 and 2 immersed in, and separated by, a thickness L of fluid 3. The result is algebraically rather complicated but its relationship to the expressions derived earlier will become clearer as we proceed. The interaction energy per unit area of the half-space surface is (Ninham *et al.* 1970):

$$E_{132}(L) = kT \sum_{n=0}{}' \int_0^\infty \frac{Q\,\mathrm{d}Q}{2\pi} \ln\{\mathfrak{D}_{\mathrm{E}}\mathfrak{D}_{\mathrm{M}}\} \tag{4.7.36}$$

where

$$\mathfrak{D}_{\mathrm{E(M)}} = 1 - \Delta_{13}^{\mathrm{E(M)}}\Delta_{23}^{\mathrm{E(M)}}\mathrm{e}^{-x} \tag{4.7.37}$$

$$\Delta_{j3}^{\mathrm{E}} = \frac{\varepsilon_j(i\xi_n)x - \varepsilon_3(i\xi_n)x_j}{\varepsilon_j(i\xi_n)x + \varepsilon_3(i\xi_n)x_j} \tag{4.7.38}$$

$$\Delta_{j3}^{\mathrm{M}} = \frac{\mu_j(i\xi_n)x - \mu_3(i\xi_n)x_j}{\mu_j(i\xi_n)x + \mu_3(i\xi_n)x_j} \tag{4.7.39}$$

$$x_j^2 = 4(QL)^2 + x_0^2\left[\frac{\varepsilon_j(i\xi_n)\mu_j(i\xi_n)}{\varepsilon_3(i\xi_n)\mu_3(i\xi_n)}\right](j = 1, 2, 3) \tag{4.7.40}$$

$$x = x_3$$

and

$$x_0 = 2L\xi_n[\varepsilon_3(i\xi_n)\mu_3(i\xi_n)]^{1/2}. \tag{4.7.41}$$

Recall that for most systems we may make the replacement $\mu_j(i\xi_n) = \mu_0$ with negligible error and then $\mathfrak{D}_M = 1$. The functions $\mathfrak{D}_E$ and $\mathfrak{D}_M$ are the electric and magnetic dispersion relations corresponding to e.m. radiation with transverse (i.e. parallel to the half-space surface) wave vector $\boldsymbol{Q}$. The integral in eqn (4.7.36) represents the free energy sum over all possible transverse wave vectors.

For very large Q values (i.e. for radiation of wavelengths comparable to intermolecular spacing) it is strictly necessary to regard $\varepsilon(i\xi)$ as a function of Q since the continuum treatment adopted herein will not be valid. For most purposes, $\varepsilon(i\xi)$ can be regarded as independent of Q—the dominant contributions to the Q integral coming from small Q values due to the exponential factor in the dispersion relations. A problem occurs in the limit $L \to 0$ since, in this limit, the exponential factor is replaced by unity and the Q integral diverges. This singularity in the theory is due solely to the neglect of the granularity of matter.

Let us examine the distance dependence of $E_{132}(L)$. Changing to variable x (where $x^2 = 4Q^2L^2 + x_0^2$), (4.7.36) becomes:

$$E_{132}(L) = \frac{kT}{8\pi L^2} \sum_{n=0}' \int_{x_0}^{\infty} \mathrm{d}x\, x \ln(\mathfrak{D}_E \mathfrak{D}_M). \tag{4.7.42}$$

The quantity $[\varepsilon(i\xi)\mu(i\xi)]^{1/2}$ occurring in eqn (4.7.41) for x_0 is of order $1/c$ (see eqn (4.7.23)). Thus $L[\varepsilon(i\xi)\mu(i\xi)]^{1/2}$ is a measure of the time of propagation of an e.m. field over a distance L. When this time is small compared to the period of the e.m. radiation $(\xi_n)^{-1}$ there will be negligible phase lag between the polarization fluctuation that generated the e.m. field and the reflected field when it arrives back at the fluctuation, i.e. retardation is negligible. In this limit (often called the $c \to \infty$ limit)

$$x_0 \xrightarrow[L \to 0]{} 0$$

and we have that $x_1 = x_2 = x$ and

$$\mathfrak{D}_M = 1 - \Delta_{13}^M \Delta_{23}^M e^{-x} \simeq 1$$

using the approximation $\mu_j(i\xi_n) \simeq \mu_0$. The non-retarded interaction energy ($c = \infty$) is

$$E_{132}^{\mathrm{n.r.}}(L) = -\frac{A_{132}}{12\pi L^2} \tag{4.7.43}$$

where the constant A_{132} is given by:

$$A_{132} = -\frac{3kT}{2}\sum_{n=0}{}' \int_0^\infty \mathrm{d}x x \ln\left(1 - \left(\frac{\varepsilon_1 - \varepsilon_3}{\varepsilon_1 + \varepsilon_3}\right)\left(\frac{\varepsilon_2 - \varepsilon_3}{\varepsilon_2 + \varepsilon_3}\right)\mathrm{e}^{-x}\right) \tag{4.7.44}$$

$$= \frac{3kT}{2}\sum_{n=0}{}' \sum_{s=1} \left[\frac{\varepsilon_1(i\xi_n) - \varepsilon_3(i\xi_n)}{\varepsilon_1(i\xi_n) + \varepsilon_3(i\xi_n)}\right]^s \left[\frac{\varepsilon_2(i\xi_n) - \varepsilon_3(i\xi_n)}{\varepsilon_2(i\xi_n) + \varepsilon_3(i\xi_n)}\right]^s \Big/ s^3. \tag{4.7.45}$$

(The transformation from eqn (4.7.44) to eqn (4.7.45) is considered in Exercise 4.7.6.)

Thus the non-retarded dispersion energy has exactly the same form as obtained by the Hamaker summation (eqn (4.4.13)). Lifshitz theory in the non-retarded limit still produces a Hamaker constant, but one that has a more complicated density dependence (via the $\varepsilon(i\xi)$) than the original as defined by eqn (4.4.27). Note also that, in modern theory, the bathing medium is automatically included. To complete the connection between eqn (4.4.27) and eqn (4.7.45), we note that to first order in density

$$\varepsilon_j(i\xi) = \varepsilon_0\varepsilon_{\mathrm{r}} = \varepsilon_0 + \rho_j\alpha_j(i\xi) \tag{4.7.46}$$

which follows directly from eqn (4.7.17) by replacing ω by $i\xi$. Substituting eqn (4.7.46) into eqn (4.7.45) and retaining only the leading order in density, we have that (Exercise 4.7.7):

$$A_{132} \simeq \frac{3kT}{8\varepsilon_0^2}\sum_{n=0}{}' [\rho_1\alpha_1(i\xi_n) - \rho_3\alpha_3(i\xi_n)][\rho_2\alpha_2(i\xi_n) - \rho_3\alpha_3(i\xi_n)]. \tag{4.7.47}$$

In the limit $T \to 0$, this reduces to eqn (4.4.27) (Exercise 4.7.7). In fact eqn (4.7.47) represents a finite temperature *extension* of the zero temperature Hamaker summation result. The effect of temperature is to change the magnitude of the London interaction between molecules since, at non-zero T, the molecules have a finite probability of being in excited states. Only at $T = 0$ can we be certain of finding the molecules in their ground states at all times as was explicitly assumed in the London theory of section 4.2.

Thus Lifshitz theory and Hamaker theory agree to leading order in the density and will give comparable results when the $\varepsilon(i\xi)$ values for the materials of the system are close to unity. It is precisely in this *dilute* regime that pairwise summation would be expected to be a valid approximation. Indeed, the importance of many-body effects can be gauged directly by noting the significant deviations of $\varepsilon(i\xi)$ from ε_0 that occur in liquid and solid dielectrics. In many systems of colloidal interest, Lifshitz theory and Hamaker theory can predict Hamaker constants differing by an order of magnitude or more. It is this quantitative discrepancy that has reduced Hamaker theory to a minor role in modern colloid science.

We note that the $n=0$ term in eqn (4.7.42) is always non-retarded, since x_0 is zero for $n=0$ regardless of the value of c. Thus, for finite c, we may write eqn (4.7.42) as

$$E_{132}(L) = -\frac{A_{132}^0}{12\pi L^2} + \frac{kT}{8\pi L^2}\sum_{n=1}\int_{x_0}^{\infty} \mathrm{d}x\, x \ln(\mathfrak{D}_{\mathrm{E}}\mathfrak{D}_{\mathrm{M}}) \qquad (4.7.48)$$

where

$$A_{132}^0 = \frac{3kT}{4}\sum_{s=1}^{\infty}\left(\frac{\varepsilon_1(0)-\varepsilon_3(0)}{\varepsilon_1(0)+\varepsilon_3(0)}\right)^s\left(\frac{\varepsilon_2(0)-\varepsilon_3(0)}{\varepsilon_2(0)+\varepsilon_3(0)}\right)^s\Big/ s^3 \qquad (4.7.49)$$

is the $n=0$ contribution to the Hamaker constant (eqn (4.7.45)).

To examine the distance behaviour of the $n \geqslant 1$ terms, we consider the integral

$$I_n(L) = \int_{x_0}^{\infty} \mathrm{d}x x \ln(\mathfrak{D}_{\mathrm{E}}\mathfrak{D}_{\mathrm{M}}) \qquad (4.7.50)$$

Note that $I_n(L)$ takes its non-retarded ($x_0=0$) value when $L=0$. From the definition (eqn (4.7.37)) of the dispersion relations we see that the integrand vanishes at $x=0$ and $x=\infty$ and is peaked about $x \simeq 1$. Provided the peak of the integrand is in the integration range, the dominant contribution to $I_n(L)$ will come from the neighbourhood of $x \simeq 1$. As L increases from zero, x_0 increases and the lower integration limit moves toward and through the dominant region of the integrand. As a consequence, the magnitude of $I_n(L)$ decreases—the decrease being felt first by the largest n values. For example, at $T \simeq 300$ K, for $n=100$, $x_0=1$ at $L=5$ nm whereas for $n=1$, $x_0=1$ at $L=500$ nm. With increasing L successively smaller values of n are affected by retardation and the contributions of the $n \geqslant 1$ terms to $E_{132}(L)$ decay progressively faster than L^{-2}. Ultimately this more rapid distance dependence of the $n \geqslant 1$ terms will lead to the unretarded $n=0$ term being the dominant term and $E_{132}(L)$ will appear to have a non-retarded form in the limit of large L. That $E_{132}(L)$ decays like L^{-2} asymptotically is due solely to the presence of the $n=0$ (zero frequency or *static*) term. A non-retarded asymptotic result of this type can arise only at non-zero temperatures. Although the ground state of the molecules may have no dipole moment, some of the excited states which will be occupied at temperatures above $T=0$ will possess non-zero dipole moments. It is the presence of these excited state molecules which gives rise to the non-retarded $n=0$ term in all materials. Moreover, if the molecules in the ground state do possess a dipole moment then the $n=0$ term will contain the permanent dipole–dipole and dipole-induced dipole interactions of the system as well. In these cases, the $n=0$ (static) term can be a very significant contribution to the total interaction energy, especially at larger separations when the other ($n \geqslant 1$) terms have suffered so much retardation that they have become negligible. Water is the outstanding example of this effect (Ninham and Parsegian 1970 (but see section 4.9)).

Modern dispersion force theory contains other subtleties. When the magnitudes of the $\varepsilon_j(i\xi)$ functions are close to each other, the quantity $\Delta_{13}^{E}\Delta_{23}^{E}$ can change sign as ξ_n increases. Under certain conditions, this can cause $E_{132}(L)$ to possess non-monotonic decay (local energy minima). Such intricacies are beyond the scope of the old Hamaker theory.

Various experiments have been performed to test the validity of modern dispersion force theory (Sabisky and Anderson 1974; Israelachvili and Tabor 1972) with considerable success. On the theoretical side, other geometries, e.g. spheres or cylinders, have been examined (see Mahanty and Ninham 1976 for details). Generally speaking, the explicit expressions so obtained for $V_A(H)$ are cumbersome and often difficult to compute. The planar half-space problem dealt with here in some detail is, however, a relatively straightforward calculation that all modern colloid scientists should be able to appreciate (section 4.8.2). Useful approximate results in other geometries at small separation distances can be obtained by applying the Deryaguin approximation (Fig. 4.4.3) to the planar result. A quasi-empirical procedure that yields a strikingly accurate approximation to the interaction energy of two bodies, $V_A(H)$, over the entire separation distance range, is the replacement of the Hamaker constant in the Hamaker summation expression (eqn (4.4.24)) by the quantity $(12\pi H^2 E_{132}(H))$ calculated from eqn (4.7.36) (Pailthorpe and Russel 1982).

Exercises. 4.7.1 Verify that eqn (4.7.4) is a solution of eqn (4.7.3) if $f(x)$ is given by eqn (4.7.5). $\left(\text{Note that } \nabla^2\phi = \dfrac{\partial^2\phi}{\partial x^2} + \dfrac{\partial^2\phi}{\partial y^2} + \dfrac{\partial^2\phi}{\partial z^2}.\right)$ Q_2 and Q_3 are the magnitudes of the wave vectors in the y and z direction (parallel to the face).

4.7.2 Establish the equations (4.7.6) using the stated boundary conditions and hence obtain the dispersion equation from the fact that

$$\begin{vmatrix} 1 & 0 & -1 & -1 \\ 0 & e^{-QL} & -e^{QL} & -e^{-QL} \\ \varepsilon_1 & 0 & -\varepsilon_0 & +\varepsilon_0 \\ 0 & \varepsilon_1 e^{-QL} & \varepsilon_0 e^{QL} & -\varepsilon_0 e^{-QL} \end{vmatrix} = 0.$$

4.7.3. Verify that eqn (4.7.10) is a solution of eqn (4.7.8) if $\varepsilon(\omega)$ is given by eqn (4.7.9).

4.7.4. Consider the dielectric response function:

$$\varepsilon(\omega) = \varepsilon_0\left(1 + \frac{d}{1 - i\omega\tau} + \frac{f}{\omega_0^2 - \omega^2 - ig\omega}\right)$$

and convert it into

$\varepsilon_r(\omega) = \varepsilon_r'(\omega) + i\varepsilon_r''(\omega)$ for real ω. (ε_r' and ε_r'' must both be real functions.)

4.7.5. Establish eqn (4.7.26) from eqn (4.7.25).

4.7.6 Use the series expansion for $\ln(1-p)=\Sigma_1^\infty\left(-\frac{p^s}{s}\right)$ for $0<p<1$ to show that

$$\int_0^\infty x\ln(1-ae^{-x})\,\mathrm{d}x=-\sum_{s=1}^\infty a^s/s^3.$$

(Compare eqn 4.7.45).

4.7.7 Establish eqn (4.7.47) and show that it reduces to eqn (4.4.27) as $T\to 0$.

4.8 Numerical computation of $E_{132}(L)$

Whether performing the relatively straightforward numerical task of calculating a Hamaker constant via eqn (4.7.45) or the more difficult one of evaluating $E_{132}(L)$ via eqn (4.7.36), it is obvious that the functions $\varepsilon_j(i\xi)$ must be known at the points $\xi=\xi_n$ given by eqn (4.7.29). Before discussing the calculation of $E_{132}(L)$, we must first demonstrate how $\varepsilon_j(i\xi)$ may be constructed from the available experimental data.

4.8.1 *Construction of* $\varepsilon(i\xi)$

The quantity $\varepsilon(i\xi)$ was defined for a uniform dielectric in section 4.7.2. The operational definition is the Kramers–Kronig relation, eqn (4.7.15), relating $\varepsilon(i\xi)$ to the experimentally measurable $\varepsilon_r''(\omega)$. Equation (4.7.15) serves to connect the function $\varepsilon(i\xi)$ to the physical world. Its calculation requires a knowledge of the absorption spectrum of the material over the entire real frequency range, $0\leqslant\omega<\infty$. Where such information exists, eqn (4.7.15) will give exactly the function required by Lifshitz theory. Modern electron loss spectroscopy is capable of yielding $\varepsilon_r''(\omega)$ over a very large frequency range in the ultraviolet regime and this has been used recently to construct $\varepsilon(i\xi)$ (Chan and Richmond 1977). With only a few exceptions complete curves of $\varepsilon_r''(\omega)$, however experimentally determined, are unavailable, and approximate methods of construction of $\varepsilon(i\xi)$ from available (often scanty) experimental data must be used. It is in this spirit that the method of Ninham and Parsegian (1970) was developed.

It should be understood that, unlike $\varepsilon'(\omega)$ and $\varepsilon''(\omega)$, the quantity $\varepsilon(i\xi)$ is a very *unspectacular* function of its argument. To see this we note that

$$\varepsilon_r(i0)=\varepsilon_r'(0)=\varepsilon_r(0)\tag{4.8.1}$$

and

$$\varepsilon(i\infty)=\varepsilon_0.\tag{4.8.2}$$

Furthermore we note from eqn (4.7.15) that, since $\varepsilon_r''(\omega)$ is everywhere positive, $\varepsilon(i\xi)$ is a *monotonic decreasing* function of ξ. Thus $\varepsilon(i\xi)$ decreases steadily from the static dielectric constant $\varepsilon(0)$ at $\xi = 0$ to ε_0 at $\xi = \infty$ (see Figs 4.7.2 and 4.7.3). It is this well-behaved nature of $\varepsilon(i\xi)$ that enables its construction from the minimum of experimental information as we shall show below. It should be pointed out that the behaviour of $\varepsilon(i\xi)$ in the ultraviolet frequency regime ($\xi > 10^{16}$ rad s^{-1}) is of paramount importance in the calculation of A_{132} or $E_{132}(L)$. The reason for this is that the frequencies ξ_n occur at equally spaced intervals of $2\pi kT/\hbar$($\sim 3 \times 10^{14}$ rad s^{-1} at $T \sim 300$ °K). In the microwave frequency ($\xi \sim 10^{11}$ rad s^{-1}) and infra-red frequency ($\xi \sim 10^{14}$ rad s^{-1}) regions, there are very few sampling points. For example, there are approximately 30 terms in the frequency sum (eqn 4.7.32) in the region $\xi < 10^{16}$ rad s^{-1} compared with about 300 terms in the region $10^{16} < \xi < 10^{17}$ rad s^{-1}, where $\varepsilon(i\xi)$ is still reasonably large (Fig. 4.7.3). This picture is somewhat modified in cases when the intervening medium 3 has dielectric properties and relaxation frequencies comparable to one or both of the half-spaces 1 and 2. Then the ultraviolet terms in the frequency sum do not contribute as much to the sum because either or both Δ_{13} and Δ_{23} are small. In such cases, the ultraviolet representation is not quite as critical as it would be if medium 3 were a vacuum. In all cases, however, the ultraviolet region makes a major contribution to the total frequency sum.

The relative unimportance of the infra-red contribution to $\varepsilon(i\xi)$ allows one to ignore, in all but an average sort of way, the complicated fine structure of relaxations in this region. Moreover, the ultraviolet absorption spectra of materials are usually simple if somewhat broad. If these ultraviolet and infra-red spectra are known, then one can easily extract from the data the relaxation frequencies ω_k corresponding to the important absorption peaks in the spectra. The function $\varepsilon_r''(\omega)$ can then be represented in a simple way by a discrete set of peaked functions of various heights (and widths) centred on each of the frequencies ω_k

$$\varepsilon_r''(\omega) = \sum_{k=1}^{N} F_k(\omega - \omega_k) \tag{4.8.3}$$

where $F_k(\omega - \omega_k)$ is a function peaked about ω_k that tends to zero for $|\omega - \omega_k| \gg 0$. The area under each of these peaks is a measure of the strength of the absorption. We would need to specify the functional form of $F_k(\omega - \omega_k)$ in order to use the fundamental equation (4.7.15) to construct $\varepsilon(i\xi)$. Since the function $\varepsilon(i\xi)$ is a simple function, it is possible to ignore the detailed shape of the peak and to replace the function $F_k(\omega - \omega_k)$ by the zero width, infinite height delta function without introducing too much error. Thus, we make the approximation that

$$F_k(\omega - \omega_k) = f_k \delta(\omega - \omega_k) \tag{4.8.4}$$

where

$$f_k = \int_0^\infty F_k(\omega - \omega_k)\,\mathrm{d}\omega \tag{4.8.5}$$

is the oscillator strength of the absorption at ω_k. (The oscillator strengths f_k of the dielectric are intimately related to the molecular oscillator strengths $f_{0\mathrm{m}}$ defined by eqn (4.2.10). There is not, however, a one-to-one correspondence since the absorption frequencies of the isolated molecule are not those of the molecules in the dielectric due to the perturbation of molecular energy levels by neighbouring molecules.)

We may now write, from eqns (4.8.3) and (4.8.4):

$$\varepsilon_\mathrm{r}''(\omega) = \sum_{k=1}^{N} f_k \delta(\omega - \omega_k). \tag{4.8.6}$$

Substituting eqn (4.8.6) into the fundamental relation (4.7.15) and making use of the delta function property:

$$\int_{<\omega_k}^{>\omega_k} \delta(\omega - \omega_k) G(\omega)\,\mathrm{d}\omega = G(\omega_k) \tag{4.8.7}$$

we obtain (Exercise 4.8.1):

$$\varepsilon(i\xi) = \varepsilon_0\left(1 + \sum_{k=1}^{N} \frac{C_k}{1 + (\xi/\omega_k)^2}\right) \tag{4.8.8}$$

where

$$C_k = \frac{2}{\pi}\frac{f_k}{\omega_k}. \tag{4.8.9}$$

Equation (4.8.8) is the Ninham–Parsegian representation of $\varepsilon(i\xi)$ in terms of experimentally accessible quantities, viz. relaxation frequencies ω_k and oscillator strengths f_k. Note (from eqn (4.8.1)) that C_k must satisfy the relationship

$$\varepsilon_\mathrm{r}(0) = 1 + \sum_{k=1}^{N} C_k. \tag{4.8.10}$$

The physical content of the approximation (4.8.6) is not clear. If the oscillator strengths f_k (i.e. the areas under the absorption peaks) corresponding to the relaxation frequencies ω_k are fixed, then the smaller the bandwidths of the peaks, the closer the representation eqn (4.8.3) would approximate to the delta function representation eqn (4.8.4). One is tempted to say that the approximation (4.8.4) is appropriate for materials whose spectra show sharp, non-overlapping absorption peaks. However, we only use the representation (4.8.4) inside the integral

(4.7.15) to obtain an expression for $\varepsilon(i\xi)$ which, by virtue of its simple monotonic decreasing nature, should be relatively insensitive to the precise form of $\varepsilon_r''(\omega)$. The effect of using other representations of $\varepsilon_r''(\omega)$ (perhaps Lorentzian or Gaussian line shapes) on the functional form of $\varepsilon(i\xi)$ has yet to be investigated thoroughly with a view to determining the importance of line width.

We note here the close relationship between $\varepsilon(i\xi)$ and $\varepsilon_r'(\omega)$ as evidenced by the two relations (4.7.14) and (4.7.15) connecting these functions to $\varepsilon_r''(\omega)$. If the representation (4.8.6) for $\varepsilon_r''(\omega)$ is used in eqn (4.7.14) we obtain (Exercise 4.8.2):

$$\varepsilon'(\omega) = \varepsilon_0\left(1 + \sum_{k=1}^{N} \frac{C_k}{1 - (\omega/\omega_k)^2}\right). \tag{4.8.11}$$

These two functions $\varepsilon(i\xi)$ and $\varepsilon'(\omega)$ are plotted schematically in Fig. 4.7.3(b). We see that when the relaxation frequencies are widely spaced, both $\varepsilon'(\omega)$ and $\varepsilon(i\xi)$ are characterized by large regions where those functions are essentially *constant and coincident.* This would occur in cases where there was only one significant ultraviolet relaxation and where infra-red relaxations were either unimportant (and so could all be lumped together as a single relaxation at some average frequency containing all the infra-red oscillator strength) or dominated by a single infra-red relaxation. In between the significant relaxations $\varepsilon_r''(\omega)$ is essentially zero. There $\varepsilon_r'(\omega)$ is simply the square of the refractive index $n(\omega)$ (eqn (2.2.36)), which is usually tabulated in data handbooks together with the dielectric constant. If we denote by ε_r (before ω_k) and ε_r (after ω_k), the values of the relative dielectric response (n^2) in the flat regions before and after the relaxation ω_k, then obviously,

$$\varepsilon_r(\text{after } \omega_N) = \varepsilon(\infty)/\varepsilon_0 = 1 \tag{4.8.12}$$

(since ω_N is defined as the largest relaxation frequency) and

$$\varepsilon_r(\text{before } \omega_{k+1}) = \varepsilon_r(\text{after } \omega_k). \tag{4.8.13}$$

If we evaluate eqn (4.8.11) in the flat region before the last frequency ω_N but after the frequency ω_{N-1}, we obtain

$$\varepsilon_r(\text{before } \omega_N) = \varepsilon_r(\text{after } \omega_{N-1}) \simeq 1 + C_N \tag{4.8.14}$$

since, for ω in this region and all frequencies ω_k widely spaced,

$$\left(\frac{\omega}{\omega_k}\right)^2 \gg 1 \quad (k = 1, 2, \ldots, N-1)$$

and

$$\left(\frac{\omega}{\omega_N}\right)^2 \ll 1.$$

Therefore

$$C_N = \varepsilon_r(\text{before } \omega_N) - \varepsilon_r(\text{after } \omega_N) \qquad (4.8.15)$$

using definition (4.8.12). Similarly, between ω_{N-1} and ω_{N-2} we have

$$\varepsilon_r(\text{after } \omega_{N-2}) = \varepsilon_r(\text{before } \omega_{N-1}) = 1 + C_{N-1} + C_N.$$

Using eqn (4.8.14), we obtain

$$C_{N-1} = \varepsilon_r(\text{before } \omega_{N-1}) - \varepsilon_r(\text{after } \omega_{N-1}). \qquad (4.8.16)$$

Proceeding in this manner, it is a simple matter to show (for widely spaced relaxation frequencies) that

$$C_k \simeq \varepsilon_r(\text{before } \omega_k) - \varepsilon_r(\text{after } \omega_k) \qquad (4.8.17)$$

for all k. Thus the Ninham–Parsegian representation becomes

$$\varepsilon(i\xi) = \varepsilon_0\left(1 + \sum_{k=1}^{N} \frac{\varepsilon_r(\text{before } \omega_k) - \varepsilon_r(\text{after } \omega_k)}{1 + (\xi/\omega_k)^2}\right). \qquad (4.8.18)$$

The detailed application of this approach to the calculation of $\varepsilon(i\xi)$ for fused quartz is given by Hough and White (1980) and will not be repeated here.

It should be noted that when the ω_k values are close together, no flat regions between relaxations occur. Thus, it is only possible to find ε_r (before ω_k) and ε_r (after ω_k) listed in the literature when the significant relaxation frequencies are well spaced. If one has a system where infra-red relaxations are likely to be important (due to similarities in dielectric response in the ultraviolet region) and several infra-red frequencies are significant one must use data on the relative strengths of the absorptions. The total oscillator strength C_{IR} to be assigned to the infra-red absorption can still be computed because it is given by:

$$C_{\text{IR}} \doteqdot \varepsilon_r(0) - n_1^2 \qquad (4.8.19)$$

where n_1 is the refractive index in the visible region. An inspection of the infra-red spectrum suffices to estimate the relative strengths of the different bands among which this is to be split. Again the reader is referred to Hough and White (1980) for details.

Unfortunately, experimental data in the literature for the ultraviolet region are more scanty. Where evidence exists as to a complicated ultraviolet absorption spectrum the technique of using more than one frequency with appropriate sharing of the available oscillator strength $(n^2 - 1)$, as outlined above, is clearly applicable in the absence of direct knowledge of $\varepsilon_r''(\omega)$.

At this juncture it is worth making a few remarks about the *microwave* contribution to $\varepsilon(i\xi)$. In some polar substances there is a significant permanent dipole relaxation at microwave frequencies ($\sim 10^{11}$ rad s^{-1}). For example, in water the dielectric response drops from about 80 at zero frequency to about 4 at infrared frequencies ($\sim 10^{14}$ rad s^{-1}). This relaxation has sometimes been included in the $\varepsilon(i\xi)$ construction (eqn (4.8.8)) by a Debye term (Ninham and Parsegian 1970):

$$\frac{C_{\text{MICRO}}}{1+\xi/\omega_{\text{MICRO}}}.$$

Such a term is numerically dangerous, since it becomes the dominant term in $\varepsilon_r(i\xi)$ in the far ultraviolet, where it has no right to exist.

The microwave relaxation would typically occur at 5×10^{10} rad s^{-1} and would certainly be complete by 5×10^{12} rad s^{-1}. As the function $\varepsilon(i\xi)$ is sampled first at $\xi = 0$ rad s^{-1} and next at $\xi \simeq 3\times10^{14}$ rad s^{-1}, we need only a construction that gives $\varepsilon(i\,0)$ and $\varepsilon(i\,10^{14})$ correctly (see Fig. 4.8.1). That $\varepsilon(i\,0)$ is not equal to $\varepsilon(i\,10^{14})$ is immaterial provided that these values are correct. One would need to be interested in a dipolar substance at $T\sim 2$ K before the behaviour in the microwave region was even faintly important; (only then would the $n=1$ sample point fall in the microwave region). The reader is therefore advised to omit such a term from the $\varepsilon(i\xi)$ representation.

The value of the ultraviolet relaxation frequency ω_{UV} is a critical parameter in the calculation of A_{132} and $E_{132}(L)$. In the absence of ultraviolet spectral data, a frequency $\omega = I_1/\hbar$ corresponding to the first

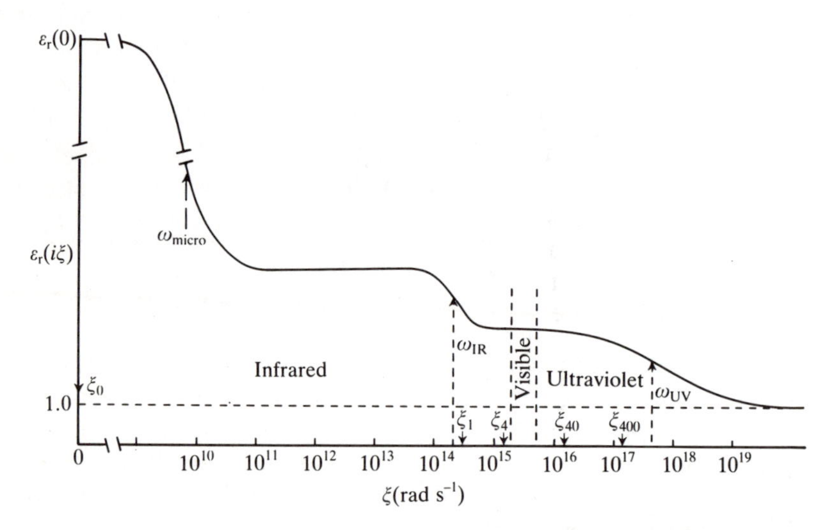

FIG. 4.8.1. A schematic representation of the function $\varepsilon(i\xi)$ for a substance with a single relaxation in the microwave region (ω_{micro}), a single infra-red absorption peak and a broad absorption in the ultraviolet. Note the positions of the sampling points ξ_n ($=(2\pi kT/\hbar)n \simeq 2.5\times10^{14}n$) s^{-1} at 300 K.

ionization potential I_1 of the material is often taken. This can be a serious error and should be avoided.

In the absence of a knowledge of $\varepsilon_r''(\omega)$ over the frequency regime $\omega > 10^{15}$ rad s^{-1}, it is still possible to obtain a reasonably accurate construction of $\varepsilon(i\xi)$ in the ultraviolet regime provided the absorption is simple; that is, only one relaxation is important.

For ξ in the visible region (provided $\xi \gg \omega_{IR}$) we can write (from eqn (4.8.8))

$$\varepsilon_r(i\xi) = 1 + \frac{C_{UV}}{1 + (\xi/\omega_{UV})^2}. \tag{4.8.20}$$

Similarly for $\varepsilon'(\omega)$ in the visible region we can write from eqn (4.8.11):

$$\varepsilon_r'(\omega) = \varepsilon_r(\omega) = n^2(\omega) = 1 + \frac{C_{UV}}{1 - (\omega/\omega_{UV})^2}. \tag{4.8.21}$$

Rearranging eqn (4.8.21), we obtain

$$n^2 - 1 = (n^2 - 1)\frac{\omega^2}{\omega_{UV}^2} + C_{UV}. \tag{4.8.22}$$

Therefore a plot of $[n^2(\omega) - 1]$ against $[n^2(\omega) - 1]\omega^2$ should yield a straight line of slope $1/\omega_{UV}^2$ and intercept C_{UV}. Experimentally it is relatively straightforward to measure the refractive index n as a function of wavelength λ (i.e. $2\pi c/\omega$) in the visible region and such information is tabulated in the literature for most common substances. Thus the plot indicated in eqn (4.8.22) can usually be made in order to obtain the vital information ω_{UV} and C_{UV}. In Fig. 4.8.2, we display such (Cauchy) plots for some common materials (from Hough and White 1980) and in Table 4.1 we list the ω_{UV}, C_{UV} data so obtained. Also listed are the appropriate infra-red and zero frequency data. Table 4.1 will serve to enable an accurate construction of $\varepsilon(i\xi)$ to be made over the frequency range of interest to us. Note, for water, the use of the more elaborate construction of Gingell and Parsegian (1972) is usually advisable in view of the significant infra-red contributions.

In the literature, use has sometimes been made of an *ultraviolet interpolation* to connect the near ultraviolet representation of $\varepsilon(i\xi)$ (eqn (4.8.8)) to the far ultraviolet (plasma) representation. For a full discussion of this procedure see Hough and White (1980). We advise strongly against the use of such procedures, which are unnecessary and often misleading.

4.8.2 *Representation of $\varepsilon(i\xi)$ for metals.*

For substances that possess conduction electrons, the dielectric response is dominated by their presence, because they are so mobile and so easily

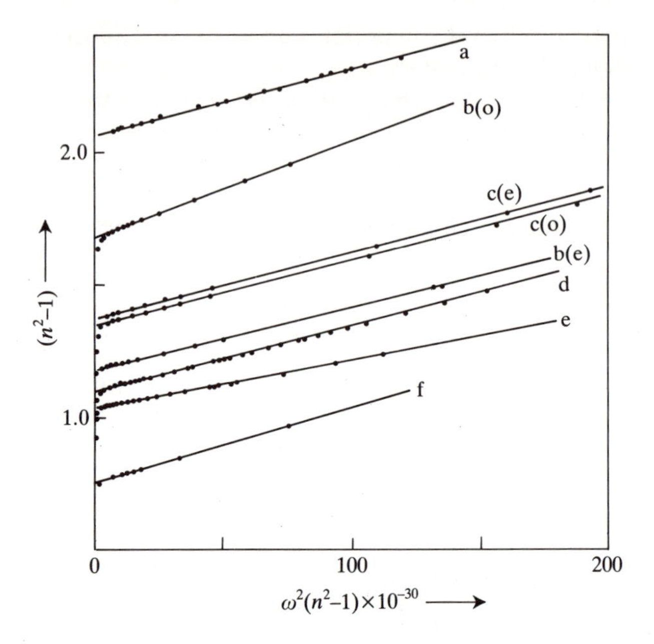

FIG. 4.8.2. Cauchy plots (from Hough and White 1980, with permission).
(a) Sapphire.
(b) Calcite (o—ordinary ray; e—extraordinary ray).
(c) Crystalline quartz.
(d) Fused quartz and fused silica.
(e) Calcium fluoride.
(f) Water.

polarizable. The behaviour of the metals is well approximated by that of a dilute electron plasma. Thus for these systems, a suitable representation of $\varepsilon(i\xi)$ is the plasma response function:

$$\varepsilon(i\xi) = \varepsilon_0(1 + \omega_p^2/\xi^2)$$

where the plasma frequency is given by (Jackson 1975):

$$\omega_p^2 = \rho_{el}e^2/m_e\varepsilon_0$$

where ρ_{e1} is the number density of conduction electrons and m_e is the electron mass. Typical ω_p values are $\sim 10^{16}$ rad s^{-1}, so that $\varepsilon(i\xi)$ for metals is extremely large until the ultraviolet region is reached. Consequently, interaction energies are much larger than for dielectric systems. The screening effect of an intervening dielectric is not as important for the interaction of metal half-spaces as for dielectric half-spaces.

Table 4.1
Dielectric data for $\varepsilon(i\xi)$ construction for a variety of common substances

Material	$\varepsilon_r(0)$ (20 °C)	n_0^2	C_{IR} $(\varepsilon_r(0)-n_0^2)$	ω_{IR} ($\times 10^{14}$ rad s^{-1})	C_{UV} (n_0^2-1)	ω_{UV} ($\times 10^{16}$ rad s^{-1})
Alkanes						
$n=5$	1.844	1.819	0.025	5.540	0.819	1.877
6	1.890	1.864	0.026	5.540	0.864	1.873
7		1.899	0.025†	5.540	0.898	1.870
8	1.948	1.925	0.023	5.540	0.925	1.863
9	1.972	1.947	0.025	5.540	0.947	1.864
10	1.991	1.965	0.026	5.540	0.965	1.873
11	2.005	1.979	0.026	5.540	0.979	1.853
12	2.014	1.991	0.023	5.540	0.991	1.877
13		2.002	0.025†	5.540	1.002	1.852
14		2.011	0.025†	5.540	1.011	1.846
15		2.019	0.025†	5.540	1.019	1.845
16		2.026	0.025†	5.540	1.026	1.848
Water	80.10	1.755	(Gingell and Parsegian 1972)		0.755	1.899
Crystalline quartz:						
ordinary ray	4.27	2.350	1.92	2.093	1.350	2.040
extraordinary ray	4.34	2.377	1.96	2.093	1.377	2.024
average	4.29	2.359	1.93	2.093	1.359	2.032
Fused quartz	3.80	2.098	1.70	1.880	1.098	2.024
Fused silica	3.81	2.098	1.71	1.880	1.098	2.033
Calcite:						
ordinary ray	8.0	2.683	5.3		1.683	1.660
extraordinary ray	8.5	2.182	6.3		1.182	2.134
average	8.2	2.516	5.7	2.691	1.516	1.897
Calcium fluoride	7.36	2.036	5.32	0.6279	1.036	2.368
Sapphire	11.6	3.071	8.5	1.880	2.071	2.017
Poly(methylmethacrylate)	3.4	2.189	1.2	5.540	1.189	1.915
Poly(vinylchloride)	3.2	2.333	0.9	5.540	1.333	1.815
Poly(styrene)	2.6	2.447	0.2	5.540	1.424	1.432
Poly(isoprene)	2.41	2.255	0.16	5.540	1.255	1.565
Poly(tetrafluoroethylene)	2.10	1.846	0.25	2.270	0.846	1.793

† Assumed value.

4.8.3 *Numerical evaluation of $E_{132}(L)$*

In the non-retarded (small L) region, it is sufficient to calculate the Hamaker constant A_{132} in order to evaluate $E_{132}(L)$. This is a simple numerical procedure involving the calculation of $\varepsilon_1(i\xi_n)$, $\varepsilon_2(i\xi_n)$ and $\varepsilon_3(i\xi_n)$ for each value of ξ_n $(n = 0, 1, 2, \ldots)$ and the subsequent evaluation of the sums in eqn (4.7.45). The s sum is very rapidly convergent and only a very few terms (less than ten) need be taken for adequate accuracy. The n sum usually requires a few thousand terms for

Table 4.2
Hamaker constants $(\times 10^{+20}/J)$ *for some common materials*

Material (M)	M |air| M	M |water| M	M |water| air	M |air| water	water |M| air
Alkanes					
$n = 5$	3.75	0.336	0.153	3.63	0.108
6	4.07	0.360	-0.368×10^{-2}	3.78	0.285
7	4.32	0.386	−0.118	3.89	0.423
8	4.50	0.410	−0.200	3.97	0.527
9	4.66	0.435	−0.275	4.05	0.624
10	4.82	0.462	−0.344	4.11	0.719
11	4.88	0.471	−0.368	4.14	0.751
12	5.04	0.502	−0.436	4.20	0.848
13	5.05	0.504	−0.442	4.21	0.855
14	5.10	0.514	−0.464	4.23	0.886
15	5.16	0.526	−0.490	4.25	0.923
16	5.23	0.540	−0.518	4.28	0.964
Fused quartz	6.50	0.833	−1.01	4.81	—
Cryst. quartz	8.83	1.70	−1.83	5.59	—
Water	3.70	0	0	3.70	0
Fused silica	6.55	0.849	−1.03	4.83	—
Calcite	10.1	2.23	−2.26	6.00	—
Calcium fluoride	7.20	1.04	−1.23	5.06	—
Sapphire	15.6	5.32	−3.78	7.40	—
Poly(methylmethacrylate)	7.11	1.05	−1.25	5.03	—
Poly(vinylchloride)	7.78	1.30	−1.50	5.25	—
Polystyrene	6.58	0.950	−1.06	4.81	—
Poly(isoprene)	5.99	0.743	−0.836	4.59	—
Poly(tetrafluoroethylene)	3.80	0.333	0.128	3.67	—
Mica (brown)	—	1.98	—	—	—
Mica (green)	—	2.14	—	—	—

a precise value of A_{132}. In Table 4.2, we list some typical A_{132} values calculated from $\varepsilon_j(i\xi)$ functions constructed from the data of Table 4.1.

A crude estimate of the non-retarded A_{132} can be obtained from the following relation (Israelachvili 1973b):

$$A_{132} = \frac{3h\nu(n_1^2 - n_3^2)(n_2^2 - n_3^2)}{8\sqrt{2}(n_1^2 + n_3^2)^{1/2}(n_2^2 + n_3^2)^{1/2}\{(n_1^2 + n_3^2)^{1/2} + (n_2^2 + n_3^2)^{1/2}\}}$$

where the ns are refractive indices and ν is a characteristic frequency. The limitations of such a procedure should, however, be obvious from the preceding analysis. One would think that it would be most useful where comparisons are to be made amongst a number of similar materials (2) interacting with a particular material (1) across an intervening medium (3). It must be noted, however, that recent calculations (Christenson 1983, private communication) reveal little correlation between the exact estimates of A_{132} and the refractive indices of materials 1, 2, and 3.

The evaluation of $E_{132}(L)$ in the general case is not a simple exercise since we must perform a numerical integration to obtain the $I_n(L)$ defined by eqn (4.7.50) and perform the sum eqn (4.7.48), viz.

$$E_{132}(L) = \frac{A^0_{132}}{12\pi L^2} + \frac{kT}{8\pi L^2} \sum_{n=1}^{\infty} I_n(L). \tag{4.8.23}$$

From the definition of the dispersion relations $\mathfrak{D}_E$ and $\mathfrak{D}_M$ we know that, asymptotically:

$$\ln(\mathfrak{D}_E \mathfrak{D}_M) \sim e^{-x}$$

for every value of n. If we define the function $F(x)$ by

$$F(x) = x \ln(\mathfrak{D}_E \mathfrak{D}_M) e^{x} \tag{4.8.24}$$

then $F(x)$ has only polynomial behaviour for large x. Equation (4.7.50) becomes

$$\begin{aligned} I_n(L) &= \int_{x_0}^{\infty} dx\, F(x) e^{-x} \\ &= e^{-x_0} \int_0^{\infty} dz\, F(x_0 + z) e^{-z} \end{aligned} \tag{4.8.25}$$

by a change of variables. Since $F(x_0 + z)$ behaves asymptotically as a polynomial,

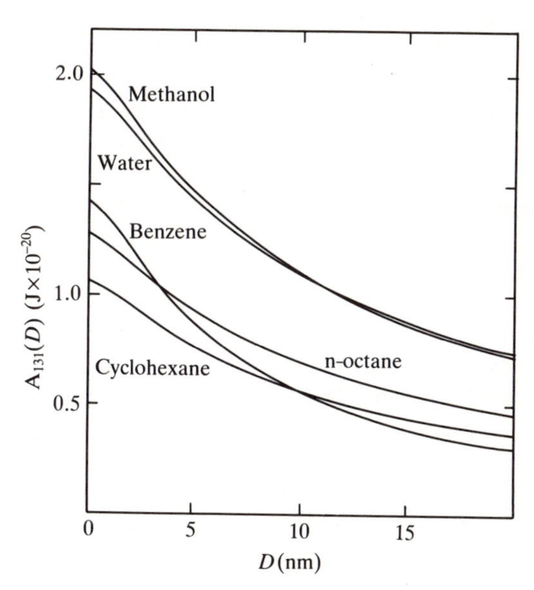

FIG. 4.8.3. Hamaker function $A_{131}(D)$ versus surface separation, D, for green mica interacting across different liquids. $T = 295$ K. (Calculations of Christenson 1983.)

we may use a Gauss–Laguerre quadrature formula

$$I_n(L) = e^{-x_0} \sum_{m=1}^{M} W_m F(x_0 + z_m) \tag{4.8.26}$$

where z_m, W_m are the sampling points and weights of the Mth order quadrature. The values of $(z_m, W_m \mid m = 1, 2, \ldots, M)$ may be obtained from tables (e.g. Rabinowitz and Weiss 1959). If $F(x_0 + z)$ is an Mth order (or smaller) polynomial in z then eqn (4.8.26) is exact. For the evaluation of $E_{132}(L)$, it is sufficient to take $M = 16$, although strictly $F(x_0 + z)$ is not really a polynomial of any degree.

As with the calculation of A_{132}, the evaluation of $E_{132}(L)$ involves the computation of a sum (over a limited number of terms) within the frequency sum. This n sum must include a few thousand terms for accuracy and testing for satisfactory convergence is advisable.

The effect of retardation with different liquids separating a pair of mica sheets can be seen from Fig. 4.8.3.

Exercises. 4.8.1 Establish eqns (4.8.8) and (4.8.9).
4.8.2 Establish eqn (4.8.11).

4.9 Influence of electrolyte concentration

One of the most important applications of the theory of van der Waals forces in colloid science is to the interaction of two materials separated by an aqueous electrolyte solution. We must now ask what effect the presence of the ions has on the propagation of the electromagnetic interactions which give rise to van der Waals forces. Although this question cannot be answered completely we can make a few observations; a more complete description of the present situation is given by Mahanty and Ninham (1976) (Chapter 7) who make it clear that there are profound problems involved in the general treatment of two charged interfaces separated by an electrolyte solution.

Richmond (1975) provides a simple analysis of the effect of an aqueous electrolyte solution between two *uncharged* surfaces. The important point to note is that ions† are so large that they are able to respond only to the lowest frequencies of the e.m. field. In fact, the main effects will be on the zero frequency (ξ_0) term. This is quite a significant term for systems containing water—especially two organic materials separated by an aqueous film (or vice versa)—because the ultraviolet contributions are similar for the two different materials and tend to cancel one another.

The zero-frequency term can then dominate and some early calculations suggested that it might be responsible for quite a large unretarded

† Mahanty and Ninham point out that this statement may not hold for the proton in water.

attraction, of importance in many biological situations. It now appears that the presence of electrolyte very considerably damps that attraction and A^0 (from eqn (4.7.49)) is not unduly large.

If the approaching surfaces are uncharged, Richmond's (linear) analysis gives, for A^0_{131}:

$$A^0_{131} = \frac{-3kT}{4} \int_0^\infty \mathrm{d}Q \cdot Q \ln\left(1 - \left(\frac{\varepsilon_1 - \varepsilon_3}{\varepsilon_1 + \varepsilon_3}\right)^2 \exp(-2(Q^2 + \kappa^2)^{1/2} L)\right) \tag{4.9.1}$$

where κ is the Debye–Hückel parameter that depends on the electrolyte concentration (section 6.3.1). Comparison with the first term of eqn (4.7.44) shows that the effect is to make the argument of the logarithm function approach unity more quickly so that A^0 is reduced in magnitude. An effect of order 50 per cent is quite possible.

The more realistic problem of two approaching *charged* surfaces has been solved only under conditions that Mahanty and Ninham (1976) regard as too restrictive to furnish unequivocal results. The theoretical point at issue is the question of whether it is permissible to treat the attraction and repulsion potential energies as separable. The success of the DLVO theory (which assumes such a separation is possible) and the lack of any viable alternative, gives us no choice. It is not the first time (nor will it be the last) where the limited insights of an approximate theory are used until a more rigorous development becomes available.

References

Atkins, P. W. (1978). *Physical chemistry*. Oxford University Press, Oxford.

Atkins, P. W. (1982). *Physical chemistry* (2nd edn). Oxford University Press, Oxford.

Blokhintsev, D. I. (1964). *Quantum mechanics*. Reidel Publishers, Dordrecht, Holland.

Bradley, R. S. (1932). *Phil. Mag.* **13,** 853.

Carrier, G. F., Crook, M. and Pearson, C. E. (1966). *Functions of a complex variable*. McGraw-Hill, New York.

Casimir, H. B. G. and Polder, D. (1948). *Phys. Rev.* **73,** 360.

Chan, D. Y. C. and Richmond, P. (1977). *Proc. R. Soc. Lond.* **A353,** 163.

Christenson, H. K. (1983). Ph.D. thesis, Australian National University.

Clayfield, E. J., Lumb, E. C., and Mackey, P. H. (1971). *J. Colloid interface Sci.* **37,** 382.

Dalgarno, A. and Lynn, N. (1957). *Proc. phys. Soc. Lond.* **A70,** 802.

Debye, P. (1920). *Phys. Z.* **21,** 178.

Deryaguin, B. V. (1934). *Kolloid-Z.* **69,** 155.

Dzyaloshinski, I. E., Lifshitz, E. M., and Pitaevski, L. P. (1961). *Adv. Phys.* **10,** 165.

Gingell, D. and Parsegian, V. A. (1972). *J. theor. Biol.* **36,** 41.
Gregory, J. (1969). *Adv. Colloid interface Sci.* **2,** 396–417.
Gregory, J. (1981). *J. Colloid interface Sci.* **83,** 138.
Hamaker, H. C. (1937). *Physics* **4,** 1058.
Hiemenz, P. C. (1977). *Principles of colloid and surface chemistry.* Marcel Dekker, New York.
Hough, D. B. and White, Lee R. (1980). *Adv. Colloid interface Sci.* **14,** 3–41.
Hunter, R. J. (1963). *Austral. J. Chem.* **16,** 774.
Hunter, R. J. (1975). Electrochemical aspects of colloid chemistry. In *Modern aspects of electrochemistry* (ed. B. E. Conway and J. O'M. Bockris) Vol. 11, Chapter 2, pp. 33–84. Plenum Press, New York.
Israelachvili, J. N. (1973a) *Q. Rev. Biophys.* **6,** 341.
Israelachvili, J. N. (1973b). *J. chem. Soc. Faraday Trans. 2* **69,** 1729.
Israelachvili, J. N. (1974). *Contemp. Phys.* **15,** 159.
Israelachvili, J. N. and Tabor, D. (1972). *Proc. R. Soc. Lond.* **A331,** 19.
Jackson, J. D. (1975). *Classical electrodynamics* (2nd edn). Wiley, New York.
Kallmann, H. and Willstätter, M. (1932). *Naturwissenschaften* **20,** 952.
Keesom, W. H. (1921). *Phys. Z.* **22,** 129 and 643.
Landau, L. D. and Lifshitz, E. M. (1960). *Electrodynamics of continuous media.* Pergamon, London.
Landau, L. D. and Lifshitz, E. M. (1969). *Statistical physics.* Pergamon, Oxford.
Langbein, D. (1974) *Theory of van der Waals attraction. Tracts in modern physics.* Springer, Berlin.
Lifshitz, E. M. (1956). *Sov. Phys. JETP.,* **2,** 73.
London, F. (1930). *Z. Phys.* **63,** 245.
Mahanty, J. and Ninham, B. W. (1976). *Dispersion forces.* Academic Press, London.
McClellan, A. L. (1963). *Tables of experimental dipole moments.* Freeman, San Francisco.
Ninham, B. W., Parsegian, V. A., and Weiss, G. H. (1970). *J. statist. Phys.* **2,** 323.
Ninham, B. W. and Parsegian, V. A. (1970). *Biophys. J.* **10,** 646.
Overbeek, J. Th. G. (1952). In *Colloid science* (ed. H. R. Kruyt) Vol 1, p. 265. Elsevier, Amsterdam.
Pailthorpe, B. A. and Russel, W. (1982). *J. Colloid interface Sci.* **89,** 563–6.
Parsegian, V. A. (1975) Long range van der Waals forces. In *Physical chemistry: enriching topics from colloid and surface science* (ed. H. van Olphen and K. J. Mysels), Chapter 4. IUPAC Commission I6, Theorex, La Jolla, California.
Pitzer, K. S. (1959). *Adv. chem. Phys.* **2,** 59.
Rabinowitz, P. and Weiss, G. (1959). *Math. Tables Aids Comput.* **68,** 285.
Reinganum, M. (1912). *Ann. Phys.* **38,** 649.
Richmond, P. (1975). In *Colloid science* (ed. D. H. Everett) Vol. 2, Chapter 4. Chemical Society, London.
Sabisky, E. S. and Anderson, C. H. (1974). Low temperature physics—LT13 Proceedings of the 13th International Conference on Low Temperature Physics (1972), Vol. 1, pp. 206–10.
Sparnaay, M. J. (1983). *J. Colloid interface Sci.* **91,** 307.
Thomson, J. J. (1914). *Phil. Mag.* **27,** 757.
van der Waals, J. H. (1873). Ph.D. Thesis, University of Leiden.
van der Waals, J. D. Jr. (1909). *Amsterdam Acad. Proc.* pp. 132, 315.

van Kampen, N. G., Nijboer, B. R. A., and Schram K. (1968). *Phys. Lett.* **26A,** 307.
Vincent, B. (1973). *J. Colloid interface Sci.* **42,** 270.
Visser, J. (1972). *Adv. Colloid interface Sci.* **3,** 331–63.
White, L. R. (1983). *J. Colloid interface Sci.* **95,** 286–8.
Wilson, J. N. (1965). *J. chem. Phys.* **43,** 2564.

5

THERMODYNAMICS OF SURFACES

5.1 Introduction

Lyophobic colloids are thermodynamically meta-stable (section 1.4). Nevertheless, the time-scale of many of the molecular exchanges occurring within a lyophobic colloidal suspension is very short compared to the lifetime of the suspension. Processes involving such molecular exchanges may, therefore, be treated by equilibrium thermodynamics. In particular, adsorption equilibrium at the particle surface is rapidly established, and so thermodynamics may be used to describe the effects of surface active materials, including ions, on the properties of the suspension.

Such thermodynamic descriptions are an essential underpinning for the more detailed molecular descriptions with which much of this book is concerned. The analysis here follows the lines laid down in the introductory text by Aveyard and Haydon (1973) and especially the more extensive analysis provided by Defay *et al.* (1966), as translated by Everett. The existence of those works and the standard texts on interfacial phenomena by Davies and Rideal (1963) and on surface chemistry by Adamson (1967, 1976) make a detailed treatment unnecessary. There is some value, however, in having a coherent account to which we can refer.

We begin with a discussion of the important concepts of surface tension and surface free energy and their origins at the molecular level. The consequent pressure difference across a curved interface is then calculated (the Young–Laplace equation). This gives rise to some important effects, which will be taken up after the basic equations of surface thermodynamics have been introduced. The thermodynamic treatment given here is intermediate between the rather oversimplified procedures, which merely demonstrate the reasonableness of the key results, and the rigorous procedures that are needed to cover all contingencies. In most cases we have chosen to simplify the notation by sacrificing a little generality rather than by sacrificing rigour on the grounds that the former is easier to recover than the latter.

The consequences of pressure differences across curved interfaces are particularly relevant to colloidal particles (sections 5.5 and 5.6) and are also important in capillary phenomena (section 5.7). A number of specific problems are then discussed: homogeneous nucleation (section 5.8), contact angle and wetting behaviour (section 5.10), and heterogeneous nucleation (section 5.10.4). We then undertake a theoretical estimation of some surface properties using the theoretical tools developed in Chapter 4 (section 5.11) and finish with a brief summary of methods of measuring surface tension and contact angle (section 5.12).

5.2 Surface energy and its consequences

5.2.1 *Surface tension and surface free energy*

The existence of surface tension can be expected from the difference in energies between molecules at the surface and molecules in the bulk phase of a material. Consider first a homogeneous liquid or solid, consisting of molecules of type A, in equilibrium with its vapour. Suppose that $v_{AA}(r)$ is the potential energy of interaction of two molecules of type A separated by distance r, when the potential energy of an isolated molecule of A is taken as zero. Assuming that nearest neighbour interactions are dominant in a condensed phase, and potential energies of interaction are pairwise additive, the energy $\mathscr{E}_{A,bulk}$ *per molecule* in the bulk phase becomes:

$$\mathscr{E}_{A,bulk} \approx \tfrac{1}{2} z_{AA,bulk} v_{AA}(r_b) \tag{5.2.1}$$

where $z_{AA,bulk}$ is the number of molecules in the shell of nearest neighbours in the bulk phase and r_b is the average distance of these neighbours from the central molecule.

The energy $\mathscr{E}_{A,S}$ per molecule at the surface is, similarly:

$$\mathscr{E}_{A,S} \approx \tfrac{1}{2} z_{AA,S} v_{AA}(r_S) \tag{5.2.2}$$

where we expect $r_S \approx r_b$ and $z_{AA,S} \approx \frac{1}{2} z_{AA,bulk}$. Remembering that $v_{AA}(r)$ is negative, it is clear that there is an increase in potential energy on taking a molecule from the bulk to the surface i.e. *work must be done in creating new surface.*

If the interface is between two condensed phases (say two liquids) consisting of molecules of types A and B respectively, then a molecule of A will lose about half of its interactions with A, but gain about an equal number of interactions with B, in moving from the bulk liquid to the interface. So

$$\mathscr{E}_{A,S} = \tfrac{1}{2} z_{AA,S} v_{AA}(r_{AA}) + \tfrac{1}{2} z_{AB,S} v_{AB}(r_{AB}) \tag{5.2.3}$$

where the meaning of the subscripts is self-evident. To a first approximation, provided the molecular species have similar sizes:

$$z_{AA,S} + z_{AB,S} = z_{AA,bulk} = z_{BB,bulk} \tag{5.2.4}$$

so that

$$\mathscr{E}_{A,S} > \mathscr{E}_{A,bulk} \quad \text{if} \quad v_{AB}(r_{AB}) > v_{AA}(r_{AA})$$

but

$$\mathscr{E}_{A,S} < \mathscr{E}_{A,\text{bulk}} \quad \text{if} \quad v_{AB}(r_{AB}) < v_{AA}(r_{AA}).$$

A similar argument applies to molecules of B. To create new surface, molecules of A and B must be brought to the interface. If the overall energy change

$$\delta\mathscr{E} = (\mathscr{E}_{A,S} - \mathscr{E}_{A,\text{bulk}}) + (\mathscr{E}_{B,S} - \mathscr{E}_{B,\text{bulk}}) \tag{5.2.5}$$

is positive, the interface will tend to shrink to its minimum possible area. However, if $\delta\mathscr{E} < 0$, the interface will tend to grow and the phases will tend to dissolve in each other. [Strictly, we should deal with the free energy that would include an entropy contribution favouring dissolution even if $\delta\mathscr{E}$ were positive, provided it was not too large.]

The number of molecules in the surface is generally a small proportion of the number in the bulk; for example, a spherical droplet of water, of volume $1\,\text{cm}^3$, has only a fraction of 2×10^{-7} of its molecules in the surface. Thus the energy of the surface molecules will make an important contribution to the total energy only for: (a) processes where there is no change in the bulk energy, or (b) systems that are so subdivided that the surface energies are, in any case, comparable to bulk energies. Case (b) does *not* apply to most lyophobic colloidal systems though it does apply to lyophilic systems. For lyophobic systems, case (a) applies; i.e. the surface energy is important because the bulk energy is substantially unaffected by most colloid processes.

From the above argument, it is clear that the work, δw, required to create new surface is proportional to the number of molecules brought from the bulk to the surface, and hence to the area, $\delta\mathscr{A}$, of the new surface:

$$\delta w \propto \delta\mathscr{A}$$

or

$$\delta w = \gamma\delta\mathscr{A} \tag{5.2.6}$$

where γ, the proportionality constant, is *defined* as the surface energy (Linford 1978) or the specific surface free energy (de Bruyn 1966). Note that it has dimensions of *force per unit length* and *for a pure liquid* it is numerically equal to the *surface tension.*

To see the relation between these two concepts, consider the following thought experiment. If an arbitrary surface is extended as in Fig. 5.2.1, the increase in area $\delta\mathscr{A}$ is given by:

$$\delta\mathscr{A} = l\delta x. \tag{5.2.7}$$

The work done, δw, in increasing the area is

$$\delta w = \gamma\delta\mathscr{A} = \gamma l\delta x. \tag{5.2.8}$$

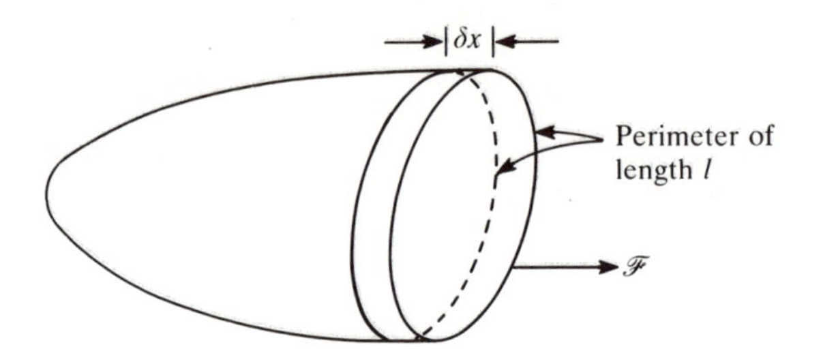

FIG. 5.2.1. Increasing the area of a surface of arbitrary shape.

If that work is done by a force $\mathcal{F}$, which is applied to the perimeter, then:

$$\delta w = \mathcal{F}\delta x \tag{5.2.9}$$

and so, comparing eqns (5.2.8) and (5.2.9):

$$\gamma = \mathcal{F}/l. \tag{5.2.10}$$

That is, γ acts like a force per unit length of the perimeter opposing any attempt to increase the area, i.e. like a restoring force or *tension* (section 2.4). More formally (Defay *et al.* 1966, p. 2), if a line AD is drawn in the surface (Fig. 5.2.2), and if the force exerted by region 1 of the surface on region 2 across the element BC, length δl, of AD is $\gamma\delta l$, then γ is called the surface (or interfacial)† tension, and is generally expressed as $mN\,m^{-1}$ (millinewton per metre, equivalent to dyne cm^{-1} in CGS units). For a liquid–liquid or liquid–vapour interface, the equilibrium value of γ is independent of the direction of BC, so that the surface is in a state of *uniform tension* in every direction if the surface is quiescent.

Typical values for the surface tensions of some pure liquids and interfacial tensions for liquid–liquid systems are given in Table 5.1.

An important element in the above argument about 'extending a surface' or 'creating fresh surface' is that we mean new surface with the *same properties* as the original surface. In other words, the surface is created by adding new molecules from the bulk, maintaining the same properties of the different molecules in a mixed system, and with the molecules in their equilibrium configuration. If the increase in surface area were to be achieved by increasing the average distances between the molecules (in the surface and bulk), then the extension of the surface would be accompanied by a change in the bulk energy of the system. This would be an elastic deformation of the body; it would, in general, require

† It is common usage to use the term 'surface tension' if one phase is a gas, but otherwise to use 'interfacial tension'.

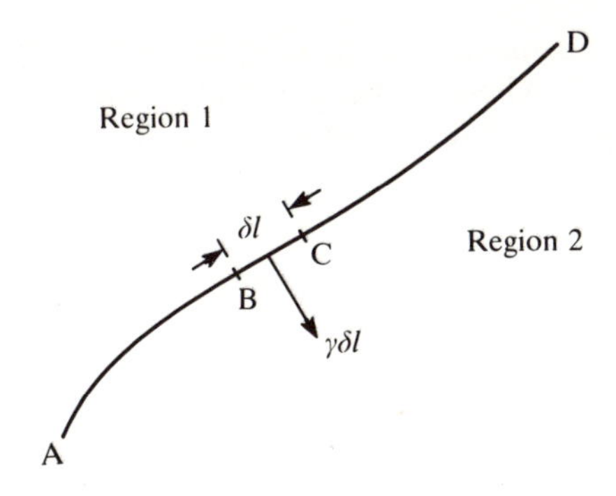

FIG. 5.2.2. The definition of surface tension (see text).

more energy input and the work done would almost certainly depend non-linearly on the area increase, $\delta\mathscr{A}$. For a pure liquid in contact with its own vapour or with another pure, immiscible liquid, the surface tension is numerically equal to the excess surface free energy, as noted above. This is *not* the case if adsorbed species are present at the interface, or if the surface is solid.

The surfaces of solids also possess a tension and confer an excess free energy on the solid. Nevertheless, it is both experimentally and conceptually difficult to treat the thermodynamics of solid surfaces in a similar way to that of liquid surfaces. The basic problem is that a freshly cleaved, or indeed very aged, solid surface is not in equilibrium, since the atoms are not, in general, free to move to positions of lower free energy. In a liquid, on the other hand, dynamic equilibrium is established very

Table 5.1

Surface and interfacial tensions of some liquids (in $mN\,m^{-1}$*) at* 293 K *(from Aveyard and Haydon* 1973, *p.* 70, *with permission)*

	Liquid–vapour	Water–liquid	$-d\gamma/dT$ (liquid–vapour) ($mN\,m^{-1}\,K^{-1}$)
Water	72.75	—	0.16
Octane	21.69	51.68	0.095
Dodecane	25.44	52.90	0.088
Hexadecane	27.46	53.77	0.085
Benzene	28.88	35.00	0.13
Carbon tetrachloride	26.77	45.0	—
Mercury	476	375	—

$d\gamma/dT$ for the hydrocarbon–water interface is 0.09 $mN\,m^{-1}\,K^{-1}$.

quickly. Thus, for example, a drop of spilt mercury will round up in a fraction of a second, whereas the pyramids have retained their shapes (except for a minimal dynamic interaction with tourists) over thousands of years.

5.2.2 *Molecular origins of surface tension*

In the previous section we deduced the existence of surface tension from the difference in energies between a molecule in the bulk of the liquid and a similar molecule at the surface of that bulk phase. That argument is a convincing proof that the properties of a surface can be represented by a uniform tension in the surface. Equivalent presentations are common in the literature and in texts concerning surface phenomena.

By contrast, there have been very few attempts to explain how an 'unbalanced' intermolecular force normal to the surface can be responsible for a stress parallel to that surface. Indeed, N. K. Adam (1941) in his celebrated text went so far as to deny the existence of any real tension in the surface of a liquid, though he acknowledged his error in the preface to a later edition (1968). Orowan (1970) was able to state: 'to this day, it seems, no textbook has given a simple and correct explanation of how the surface tension arises from cohesive forces, and the literature is full of misunderstandings caused by this deficiency'. In this section we will present a qualitative outline of Orowan's more mathematical argument for the existence of a surface tension as a consequence of the attractive (and repulsive) forces between molecules. An understanding of this material is not required for what follows in the rest of the chapter, and readers content with the energy approach may proceed directly to section 5.2.3.

The argument depends upon the balance of forces on infinitesimal cubic elements of fluid near the surface. The forces arise from the pressure within the fluid so we first need to understand the molecular basis of pressure. It will be shown how this pressure is anisotropic near the surface and that this must lead to a tension in the surface.

The pressure in a fluid in equilibrium is the time-averaged normal force per unit area exerted by all the molecules on one side of an imaginary test surface on all the molecules on the other side of the test surface. This pressure can be separated into two parts:

1. The kinetic contribution due to the transport of momentum by molecules moving across the test surface. Its value is given by:

$$p^k = \rho_N kT \tag{5.2.11}$$

where ρ_N is the number density of the molecules and k and T have their usual significance. This is the familiar pressure term in the kinetic theory of a perfect gas. It is the same for a liquid and is positive.

2. The cohesive contribution due to the time average of the net attractive and repulsive forces between molecules on opposite sides of the imaginary test surface in the body of the fluid. This second contribution, called the *static* pressure, p^s, is normally negative (i.e. the attractive force dominates) and it is particularly important for dense gases or liquids. The total pressure in a liquid is thus less than the kinetic pressure, and must be equal to the applied pressure (i.e. the vapour pressure for a one-component system) despite the fact that the kinetic pressure is so much higher in the liquid (because ρ_N is higher).

To simplify the argument we will neglect the repulsive forces when considering the static pressure; the repulsive forces have a much shorter range than the attractive forces and are important only at extremely high external pressures. The intermolecular forces can be considered to have a sphere of influence beyond which they are negligible (an idea introduced by Laplace in 1806). For molecules further from the surface than the diameter of this sphere, the pressure must be isotropic because of the symmetry of their surroundings. Near the surface between the two bulk phases, the tangential and normal contributions to the static pressure are not the same because of the asymmetric distribution of molecules within the sphere of influence.

When the sphere of influence cuts the surface (Fig. 5.2.3) there are fewer and fewer pairs of molecules attracting across the test plane as the centre of the sphere approaches the surface and hence the magnitude of the static pressure contribution decreases for both the pressure across a plane parallel to the surface (Fig. 5.2.3(a)) and for the pressure across a plane normal to the surface (Fig. 5.2.3(b)). It is apparent that when the test plane is *in* the surface (Fig. 5.2.4), the decrease in magnitude of the

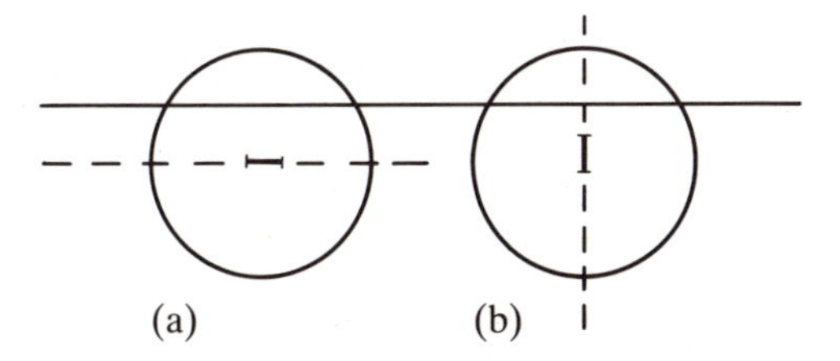

FIG. 5.2.3. Sphere of influence around an imaginary test plane near a liquid surface. (a) With test plane parallel to the surface; (b) with test plane normal to surface. In both cases the attraction between molecules on opposite sides of the test plane must be less than in the bulk of the liquid because of the deficiency of molecules in the region of vapour in the sphere of influence.

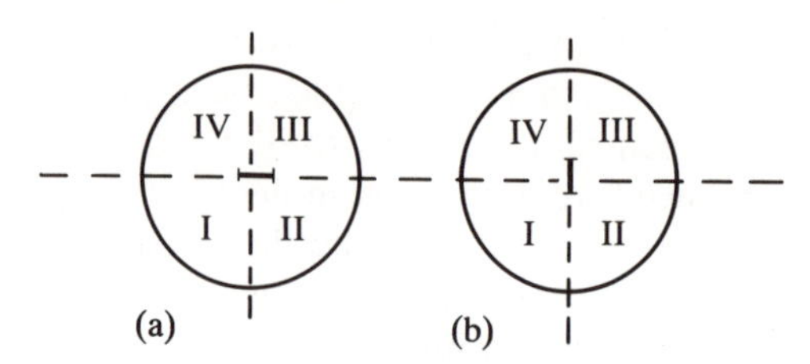

FIG. 5.2.4. Sphere of influence around imaginary planes in the liquid surface. (a) With test plane parallel to the surface; (b) with test plane normal to the surface. Note the attraction between molecules in quadrants (I + II) and (III + IV) in (a) will be very small because there are very few molecules in quadrants III and IV. Attraction between molecules in quadrants (I + IV) and (II + III) in (b) will be much greater as quadrants I and II are densely populated.

normal static pressure is greater than the decrease in magnitude of the tangential static pressure because of the smaller number of interactions remaining in the former case.

Now consider the consequences of the fact that the forces on opposite sides of an infinitesimally small cube must be equal and opposite for mechanical equilibrium. The net pressure normal to the surface must be constant right through the surface. The contributions from the kinetic and static pressures normal to the surface are shown schematically in Fig. 5.2.5(a). The net pressure is a constant equal to p_0, the pressure in the bulk phases. (We are neglecting here the hydrostatic pressure at different depths in the liquid.)

A similar graph for the pressure tangential to the surface cannot give a constant net pressure because the kinetic contribution is identical to that for the normal pressure

$$p_t^k = p_n^k = \rho_N kT$$

while the static contribution is different ($p_t^s \neq p_n^s$). The resultant pressure, p_t, is less than p_0 on passing through the surface. The resulting net stress is the surface tension and is equal to the integral of the deficit of the tangential pressure through the surface layer:

$$\gamma = \int_{-\infty}^{\infty} (p_0 - p_t(z))\, dz. \tag{5.2.12}$$

It is not, of course, an unbalanced force, but the balancing force must be applied externally, for example by elastic deformation of the walls of the container (Fig. 5.2.6). There may appear to be no balancing force available for isolated liquid droplets, but we will find that there is an increased hydrostatic pressure within the droplet, resulting from surface tension, and this provides the necessary restoring force.

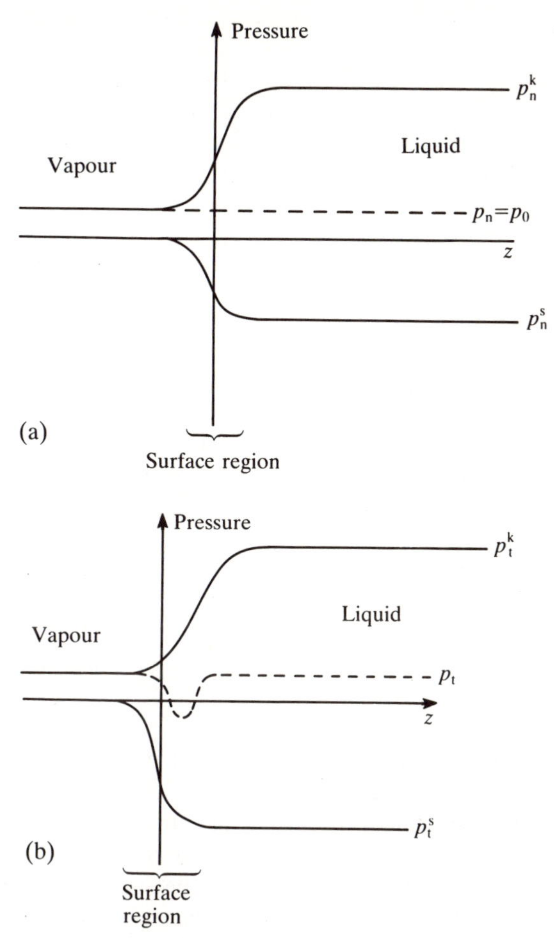

FIG. 5.2.5. (a) Variation of the kinetic (p_n^k), the static (p_n^s) and the net (p_n) pressures normal to the vapour–liquid interface.
(b) Variation of kinetic (p_t^k), static (p_t^s) and net (p_t) pressures tangential to the vapour–liquid interface as a function of position.

5.2.3 *Pressure differences across curved surfaces—the Young–Laplace equation*

The tension in a surface must be balanced by some equal and opposite force. For an isolated particle, or droplet, the balancing force must come from stresses generated within the particle or droplet by the surface tension itself. The stresses so generated† depend, for liquids, on the

† Known generically for liquids as the 'capillary pressure'.

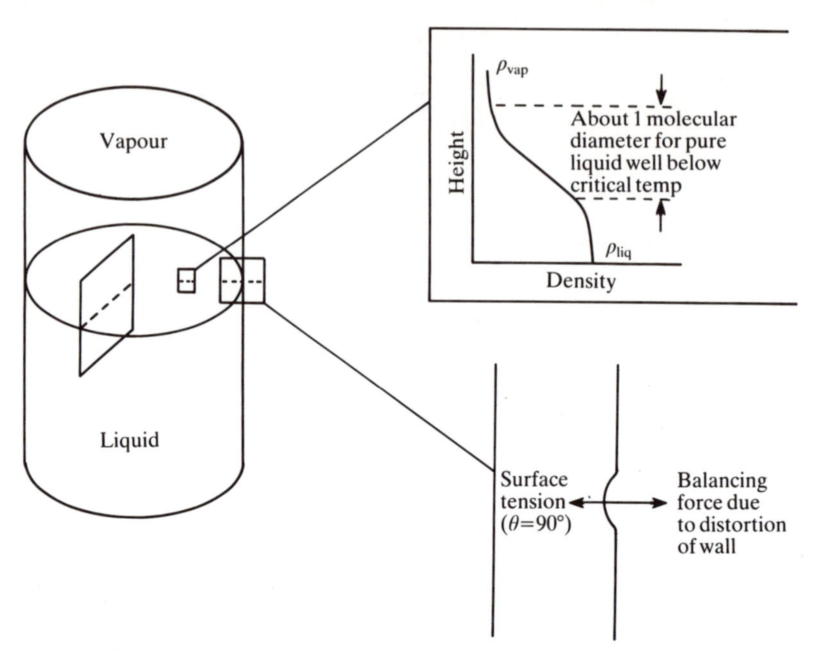

FIG. 5.2.6. Some properties of a liquid–vapour interface.

surface tension and on the curvature of the surface. (The corresponding case for solids will be discussed later (section 5.6).) This simple experimental fact was discovered by Hauksbee in 1709, but nearly a century passed before Young (1805) deduced the correct theoretical relationship between capillary pressure and the surface curvature. According to Rayleigh (1911), Young's result was 'rendered obscure by his scrupulous avoidance of mathematical symbols' and it was left to Laplace (1806), working independently, to publish the first algebraic equation (as an appendix to a ten-volume work on astronomy!) linking capillary pressure, p_c, and meniscus curvature (Klein 1974):

$$p_c = \gamma[(1/r_1) + (1/r_2)] \tag{5.2.13}$$

where γ is again the surface tension, and r_1 and r_2 are the radii of curvature of any two normal sections of the surface perpendicular to one another.

This equation is now known as the Young–Laplace equation (sometimes, unfairly, the Laplace equation (Pujado *et al.* 1972)) and is easily derived as follows (Defay *et al.* 1966).

Consider a small part ACBD of the surface of a static liquid drop (Fig. 5.2.7). The drop need not be spherical, and the circumference of the

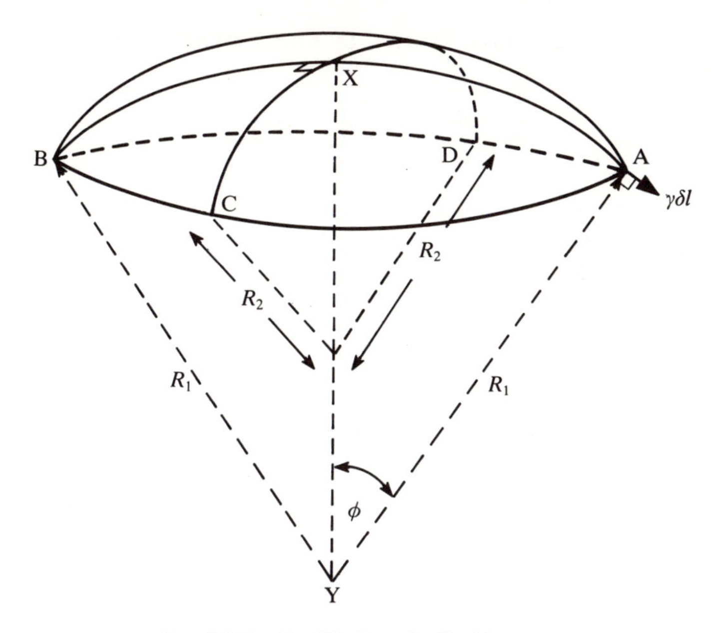

FIG. 5.2.7. Equilibrium of a liquid cap.

selected part of the surface is simply defined as a line in the surface at a constant distance, d, from a chosen point X on the surface. Through X draw any pair of orthogonal lines AB and CD in the surface. For a sufficiently small value of d, AB, and CD can be considered as parts of circles with radii r_1 and r_2 respectively.

It is known from the differential geometry of surfaces (Weatherburn 1930) that the directions of AB and CD can always be chosen so that r_1 is a maximum (R_1) and r_2 a minimum (R_2). Furthermore,

$$\frac{1}{r_1}+\frac{1}{r_2}=\frac{1}{R_1}+\frac{1}{R_2}=J \quad \text{(a quantity independent of the orientation of axes).} \tag{5.2.14}$$

(J is often referred to as the mean curvature of the surface. J^{-1} is the harmonic mean of R_1 and R_2.) Now consider the forces exerted by the cap of liquid surface ACBD. The surface has a tension γ, so that, for example, an element δl of the boundary at A exerts a resolved downward force $\mathscr{F}_{\text{vert}}$ parallel to XY and given by

$$\begin{aligned}\mathscr{F}_{\text{vert}} &= \gamma\delta l \sin\phi \\ &\rightarrow \gamma\frac{d}{R_1}\delta l \quad \text{as} \quad d\rightarrow 0.\end{aligned} \tag{5.2.15}$$

The total downward force exerted by four similar elements at A, B, C, and D is

$$\gamma\delta l\left(\frac{2d}{R_1}+\frac{2d}{R_2}\right)=2d\gamma\delta l\left(\frac{1}{R_1}+\frac{1}{R_2}\right).$$

Since this expression is independent of the choice of AB and CD, it can be integrated around the circumference (one quarter of a revolution, since there are four segments) to give the total downward force due to surface tension of

$$\pi d^2\gamma\left(\frac{1}{R_1}+\frac{1}{R_2}\right).$$

But the drop is not in motion, so there must be an upward force to balance this downward force. The upward force is provided by a pressure difference $p''-p'$ between the inside and outside of the drop. Equating the two forces:

$$(p''-p')\pi d^2=\pi d^2\gamma\left(\frac{1}{R_1}+\frac{1}{R_2}\right) \tag{5.2.16}$$

or

$$\Delta p(\equiv p''-p')=\gamma\left(\frac{1}{R_1}+\frac{1}{R_2}\right) \tag{5.2.17}$$

which is the Young–Laplace equation.

In qualitative terms, the surface tension tends to compress the droplet, increasing its internal pressure. The opposite situation arises with a concave liquid surface, as occurs with a bubble in a liquid or the meniscus of a wetting liquid in a capillary. Here the pressure in the liquid is *lower* than that outside it. The Young–Laplace equation caters for this situation since R_1 and R_2 are now *negative*, so that Δp is also negative. The radius is taken as positive if the corresponding centre of curvature lies in the phase in which p'' is measured, and negative in the converse case.

For a liquid in equilibrium Δp must be constant across all parts of its free surface, otherwise liquid flow would occur down the resulting internal pressure gradient unless an external field (e.g. gravity) were balancing the pressure change. Since γ is also a constant, it follows that, *in the absence of external fields* all liquid surfaces are surfaces of constant mean curvature $1/R_1+1/R_2$.

For an open film (e.g. a soap film on a wire frame) Δp is necessarily zero. This could mean $r_1=r_2=\infty$ (a flat film), but could also mean $r_1=-r_2$ at all points on the surface, even though both r_1 and r_2 change from point to point. This situation is illustrated in Fig. 5.2.8 which shows

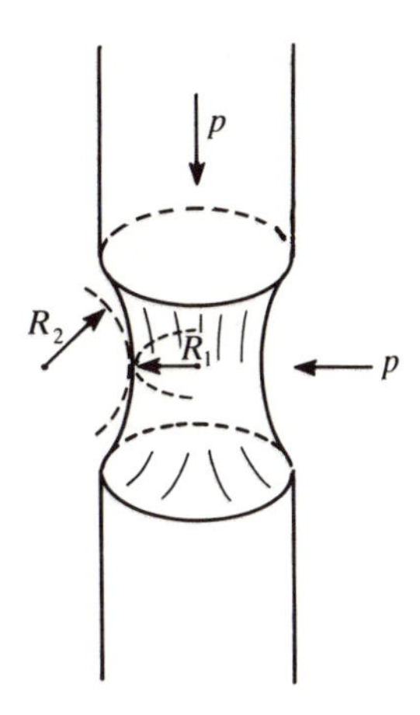

FIG. 5.2.8. A curved surface with no pressure drop across it because $1/R_1 = -(1/R_2)$ (After Adamson 1967, p. 7). (See Exercise 5.10.6.)

the shape expected for a soap film pulled between two open cylindrical pipes. Similar considerations apply to a liquid drop whose shape is unaffected by gravity, a situation that can be achieved experimentally by floating the drop in an immiscible liquid of equal density.

The Young–Laplace equation (eqn (5.2.17)) has been introduced at this stage to complete the argument of section 5.2.2 and to provide us with the important relation (see Exercise 5.2.1):

$$d\mathscr{A} = \left(\frac{1}{R_1} + \frac{1}{R_2}\right) dV \tag{5.2.18}$$

which will be used in several contexts in our discussion of surface thermodynamics. Equation (5.2.17) is actually a complicated differential equation, since R_1 and R_2 are second-order differential functions of the Cartesian coordinates and its solution under appropriate boundary conditions describes the shape of surfaces like that in Fig. 5.2.8 or the shape of liquid menisci (section 5.10.2).

Exercises. 5.2.1 Consider the increase $d\mathscr{A}$ in area of a small surface element and the concomitant volume, dV traversed by the surface (Fig. 5.2.9). Show that $d\mathscr{A} = (1/R_1 + 1/R_2)\, dV$ where R_1 and R_2 are as defined in section 5.2.3 above.

5.2.2 Use the result derived in Exercise 5.2.1 to establish the Young–Laplace equation using an energy minimization argument.

5.2.3 Calculate the excess pressure inside drops of water of radius 10^{-5} cm and 10^{-6} cm, respectively. (Take $\gamma = 70\ \mathrm{mN\ m^{-1}}$.)

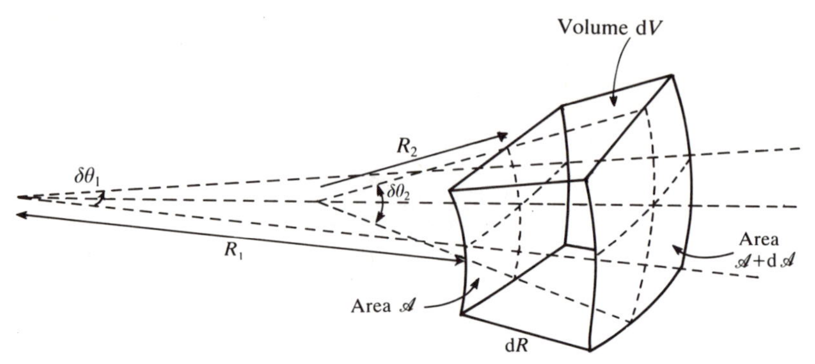

FIG. 5.2.9. Relation between increase in volume and surface area for a surface of arbitrary curvature. Both radii increase by dR at constant angle $\delta\theta_1$, $\delta\theta_2$.

5.3 Thermodynamics of surfaces

The presence of a surface introduces an additional factor to be considered in the thermodynamics of such a system, since changes in the surface area imply that there is work being done either on the system or by the system on its surroundings. The equations of bulk phase thermodynamics thus need modification if surface changes are contributing significantly to the total energy changes in the system. A summary of useful bulk phase thermodynamic relations is given in Appendix A5.

5.3.1 *Mechanical work done by a system with a surface*

If a one component, one-phase system expands by a volume dV against an external pressure p, the mechanical work done on the system is given by

$$\mathrm{d}w = -p\,\mathrm{d}V \tag{5.3.1}$$

(i.e. the work done by the system is a negative contribution to the internal energy.)

If, however, the system contains two phases, we know from the Young–Laplace equation (5.2.17) that, unless the interface between the phases has zero curvature, the pressures p' and p'' in the two phases will be different. Hence the total work done by the system will be

$$\mathrm{d}w = -p'\,\mathrm{d}V' - p''\,\mathrm{d}V'' + \gamma\,\mathrm{d}\mathscr{A}. \tag{5.3.2}$$

Consider, for example the work done during evaporation of a liquid droplet having some arbitrary shape of constant mean curvature $(1/R_1 + 1/R_2)$ (Fig. 5.3.1). If the total volume of the system is V and the external pressure is p' then $V = V' + V''$ and $\mathrm{d}V = \mathrm{d}V' + \mathrm{d}V''$. The work

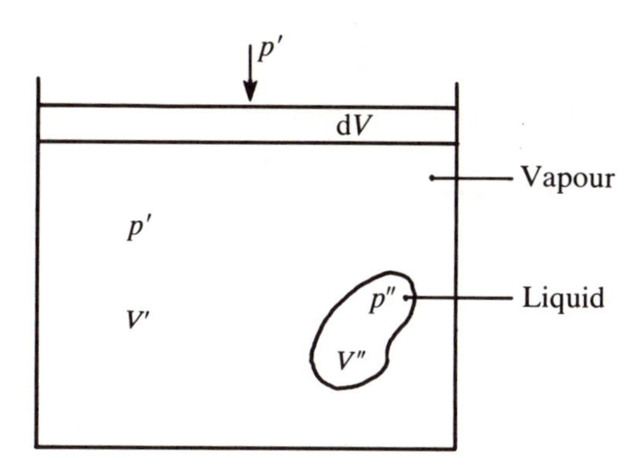

FIG. 5.3.1. Evaporation of a droplet whose surface has an arbitrary (but constant mean) curvature.

done on the system by its surrounding is:

$$\begin{aligned} dw &= -p'\,dV \\ &= -p'\,dV' - p'\,dV'' \end{aligned}$$

which may be written

$$dw = -p'\,dV' - p''\,dV'' + (p'' - p')\,dV''. \qquad (5.3.3)$$

Introducing eqn (5.2.17):

$$\begin{aligned} dw &= -p'\,dV' - p''\,dV'' + \gamma\left(\frac{1}{R_1} + \frac{1}{R_2}\right)dV'' \qquad (5.3.4) \\ &= -p'\,dV' - p''\,dV'' + \gamma\,d\mathscr{A} \end{aligned}$$

using eqn (5.2.18). One can thus identify a mechanical work term for each of the phases and a 'surface work' term for the interface, without having to introduce a hypothetical surface piston or the stretching effect depicted in Fig. 5.2.1.

It is useful at this stage to consider the droplet in Fig. 5.3.1 as a sphere of radius r. Equation (5.3.4) then becomes:

$$dw = -p'\,dV' - p''\,dV'' + \frac{2\gamma}{r}\,dV''. \qquad (5.3.5)$$

Now, according to Fig. 5.2.6 there is a gradient in density at the surface of the droplet, and for very small droplets the distance over which the gradient extends may actually be a significant fraction of the total radius of the droplet. It could reasonably be argued that the droplet 'surface' is anywhere within the region of changing density, and each arbitrary choice would give a different value of r (and of dV' and dV'').

Fortunately, however, we know that

$$V'' = 4\pi r^3/3 \quad \text{so that} \quad \mathrm{d}V'' = 4\pi r^2\,\mathrm{d}r$$

and

$$\mathscr{A} = 4\pi r^2, \quad \text{so that} \quad \mathrm{d}\mathscr{A} = 8\pi r\,\mathrm{d}r = \frac{2}{r}\,\mathrm{d}V''. \tag{5.3.6}$$

It seems, then, that eqns (5.3.2) and (5.3.5) are equivalent , no matter where we choose to place the surface. Equation (5.3.5), however, incorporates the Young–Laplace equation, which *defines* the value of r to be used, since $p'' - p'$ would be different for any other value of r. The imaginary, infinitesimally thin surface defined by this value of r is called the *surface of tension.* What we have done is to replace the real droplet, with its rather fuzzy radius, with an imaginary droplet having a sharp boundary and whose mechanical properties are equivalent to those of the real droplet.

5.3.2 *Surface excess quantities*

The presence of the surface affects virtually all of the thermodynamic parameters of a system. It is very convenient to think of a system containing a surface as being made up of three parts: two bulk phases, of volumes V' and V'', and the surface separating them. Any extensive thermodynamic property, like the energy, U, of the system, can then be apportioned between these parts as follows. If the energies *per unit volume* in the two phases are u' and u'', then the total energy of the system due to contributions from the bulk phases must be $u'V' + u''V''$. The energy to be ascribed to the surface, U^σ, must then be given by:

$$U^\sigma = U - u'V' - u''V''. \tag{5.3.7}$$

Other surface quantities such as the surface entropy, S^σ, surface Helmholtz free energy, F^σ, and surface Gibbs free energy, G^σ, may be similarly defined.

It must be clearly understood that these surface quantities can only be defined in terms of a particular model system and so their values will always depend on the model chosen. The model described above (and indicated by superscript σ) is called the *Gibbs convention.* Note that by dividing the total volume V into the precise volumes V' and V'' we have, in effect, constructed an imaginary system having the same thermodynamic properties as the real system but with the two phases separated by an infinitesimally thin dividing surface and having constant densities up to that surface. This ideal system replaces the real system, with its finite dividing region where the density is rapidly, but not discontinuously, changing. As with the droplet radius (section 5.3.1), the only reason for replacing the real system with a model is because it is easier to think of such extensive quantities as energy and volume in discrete lumps rather than as continuously varying quantities. It also means that intensive quantities like pressure and density are given definite values in each phase even when the interface is curved. The procedure was invented by Gibbs, and the imaginary dividing surface is called the *Gibbs dividing surface* (Gibbs 1874–8). Other procedures are possible (e.g. Guggenheim 1976) for planar surfaces, but generally become impossibly cumbersome if the surface is curved. Melrose (1968) presents a very full and clear discussion of the Gibbs treatment of curved surfaces and its integration with hydrostatic treatments of the stress in the surface.

The presence of the interface also affects the molecular composition. In a two-phase multicomponent system we can write c_i' and c_i'' as the concentrations of component i in each phase and if the volumes of the phases are again V' and V'' (where $V' + V'' = V$) then the numbers of moles of component i in each phase are:

$$n_i' = c_i'V' \quad \text{and} \quad n_i'' = c_i''V''. \tag{5.3.8}$$

Once again, the extra amount of i that can be accommodated in the system because of the presence of the interface is evidently:

$$n_i^\sigma = n_i - c_i'V' - c_i''V'' \tag{5.3.9}$$

where n_i is the total number of moles in the whole system. Note that all of the quantities on the right-hand side of eqn (5.3.9) are unambiguously defined. A surface concentration can be defined as:

$$\frac{n_i^\sigma}{\mathscr{A}} = \Gamma_i. \tag{5.3.10}$$

The notation Γ_i for the *surface* (excess) *concentration* of component i is used almost universally and we will use it henceforth. Note that the value of Γ_i may change dramatically not only in magnitude but even in sign, as the chosen dividing surface is varied even by a fraction of a nanometre (Fig. 5.3.2). For a curved surface the area, $\mathscr{A}$, also depends on the choice of dividing surface and this has a further effect on Γ_i.

5.3.3 *Fundamental equations of surface thermodynamics*

A general infinitesimal reversible process in a two-phase system will result in infinitesimal changes to the various thermodynamic parameters.

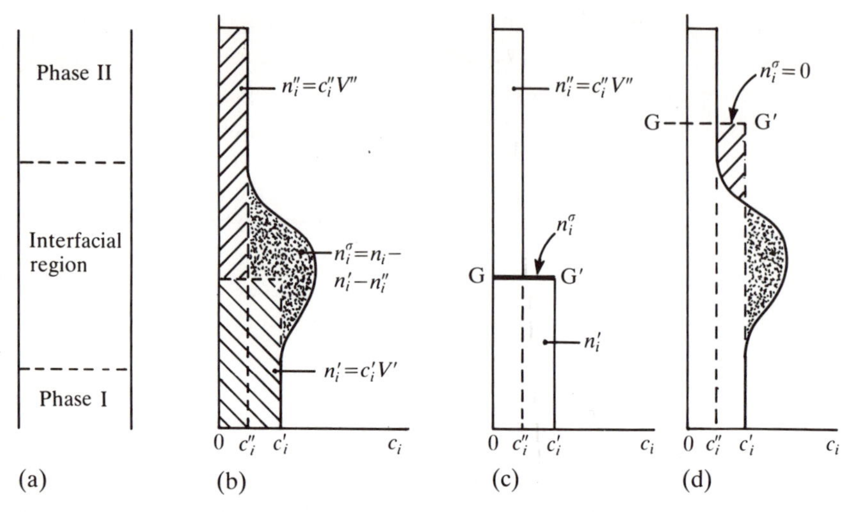

FIG. 5.3.2. The concentrations of a component i across an interface in a mixture (a) can be plotted and will generally show rapid changes through the interfacial region (b). The Gibbs model (hatched regions in (b)) ascribes constant compositions to both phases up to an arbitrarily defined interface. The excess material (dotted region in (b)) is ascribed to the infinitely thin Gibbs surface GG′ (c). A different arbitrary choice of Gibbs surface (d) can reduce the surface concentration to zero (note equality of dotted and hatched areas in (d)) or even make it negative. In practice, it is useful to choose GG′ so that n_i^σ for the *solvent* is zero. The concentrations of all other components must then be referred to this same surface. Note that for curved surfaces this choice of dividing surface is very unlikely to coincide with the surface of tension (Melrose 1968), and so the model system would not be mechanically equivalent to the real one.

Thus:

$$\begin{aligned} U &\rightarrow U + \mathrm{d}U \qquad & V' &\rightarrow V' + \mathrm{d}V' \\ S &\rightarrow S + \mathrm{d}S \qquad & V'' &\rightarrow V'' + \mathrm{d}V'' \\ n_i &\rightarrow n_i + \mathrm{d}n_i \qquad & \mathscr{A} &\rightarrow \mathscr{A} + \mathrm{d}\mathscr{A}. \end{aligned} \tag{5.3.11}$$

For the bulk phases (cf. eqn (A5.9)):

$$\mathrm{d}U' = T\,\mathrm{d}S' - p'\,\mathrm{d}V' + \sum_i \mu_i'(p', T, \{c_i'\})\,\mathrm{d}n_i' \tag{5.3.12}$$

$$\mathrm{d}U'' = T\,\mathrm{d}S'' - p''\,\mathrm{d}V'' + \sum_i \mu_i''(p'', T, \{c_i''\})\,\mathrm{d}n_i'' \tag{5.3.13}$$

while for the Gibbs surface:

$$\begin{aligned} \mathrm{d}U^\sigma &= T\,\mathrm{d}S^\sigma + \gamma\,\mathrm{d}\mathscr{A} + \sum_i \mu_i^\sigma(T, n_i^\sigma)\,\mathrm{d}n_i^\sigma \\ &= T\,\mathrm{d}S^\sigma + \gamma\,\mathrm{d}\mathscr{A} + \sum_i \mu_i^\sigma(p', p'', T, \{c_i'\}, \{c_i''\})\,\mathrm{d}n_i^\sigma \end{aligned} \tag{5.3.14}$$

Adding the above three equations, and remembering that $U^\sigma = U - U' - U''$, $S^\sigma = S - S' - S''$, we obtain a major governing equation in surface thermodynamics:

$$\mathrm{d}U = T\,\mathrm{d}S - p'\,\mathrm{d}V' - p''\,\mathrm{d}V'' + \gamma\,\mathrm{d}\mathscr{A} + \sum_i (\mu_i'\,\mathrm{d}n_i' + \mu_i''\,\mathrm{d}n_i'' + \mu_i^\sigma\,\mathrm{d}n_i^\sigma). \quad (5.3.15)$$

The first term on the right is the heat absorbed by the system, the next three terms are the mechanical work done on the system, while the last term is the chemical work done on the system.

Equation (5.3.15) leads to some interesting results. First, consider a situation in which the infinitesimal processes occur in a constant total volume, V, at fixed temperature, T. Then the change, $\mathrm{d}F$, in the Helmholtz free energy of the system is given by:

$$\begin{aligned}\mathrm{d}F &= \mathrm{d}(U - TS) = \mathrm{d}U - T\,\mathrm{d}S \\ &= -p'\,\mathrm{d}V' - p''\,\mathrm{d}V'' + \gamma\,\mathrm{d}\mathscr{A} + \sum (\mu_i'\,\mathrm{d}n_i' + \mu_i''\,\mathrm{d}n_i'' + \mu_i^\sigma\,\mathrm{d}n_i^\sigma).\end{aligned} \quad (5.3.16)$$

But the system does no mechanical work at constant volume, so the sum of the first three terms on the right must be zero. Also, since $\mathrm{d}V' = -\mathrm{d}V''$ we have:

$$-p''\,\mathrm{d}V'' - p'\,\mathrm{d}V' + \gamma\,\mathrm{d}\mathscr{A} = 0 = (p' - p'')\,\mathrm{d}V'' + \gamma\,\mathrm{d}\mathscr{A}. \quad (5.3.17)$$

Now using eqn (5.2.18):

$$(p'' - p')\,\mathrm{d}V'' = \gamma\left(\frac{1}{R_1} + \frac{1}{R_2}\right)\mathrm{d}V'' \quad (5.3.18)$$

from which the Young–Laplace equation follows immediately. Note that it is here derived on purely thermodynamic grounds.

Returning to eqn (5.3.16) we can introduce the condition

$$\mathrm{d}n_i' + \mathrm{d}n_i'' + \mathrm{d}n_i^\sigma = 0 \quad (5.3.19)$$

to rewrite it in the form:

$$\mathrm{d}F = \sum (\mu_i' - \mu_i^\sigma)\,\mathrm{d}n_i' + \sum (\mu_i'' - \mu_i^\sigma)\,\mathrm{d}n_i''. \quad (5.3.20)$$

Since F is a minimum in a closed isothermal system at fixed volume (i.e. $\mathrm{d}F = 0$) we have:

$$\mu_i' = \mu_i'' = \mu_i^\sigma \; (=\mu_i, \text{ say}) \quad \text{for all } i. \quad (5.3.21)$$

We thus reach the very reasonable conclusion that, at equilibrium, the chemical potential of any component is the same in each bulk phase and at the surface.

We can now derive a series of formulae, of varying usefulness, relating to the other thermodynamic parameters. From eqn (5.3.14):

$$\gamma = \left(\frac{\partial U^\sigma}{\partial \mathscr{A}}\right)_{S^\sigma, \{n_i^\sigma\}}. \tag{5.3.22}$$

This one is not very useful since there is no practical way to hold S^σ and n_i^σ constant whilst varying U^σ. The corresponding result from eqn (5.3.15):

$$\gamma = \left(\frac{\partial U}{\partial \mathscr{A}}\right)_{S, V', V'', \{n_i\}} \tag{5.3.23}$$

is rather more useful but a better definition flows from the Helmholtz free energy.

Recall that for a system consisting of a single phase (eqn (A5.21)):

$$U = TS - PV + \sum \mu_i n_i. \tag{5.3.24}$$

The integration procedure that leads to this equation (see Appendix A5) can be applied to each of the phases (′) and (″) and to the surface to yield:

$$U^\sigma = TS^\sigma + \gamma \mathscr{A} + \sum \mu_i n_i^\sigma. \tag{5.3.25}$$

Then from the definition of F:

$$F = U - TS = F' + F'' + F^\sigma \tag{5.3.26}$$

we have (Exercise 5.3.1):

$$F^\sigma = U^\sigma - TS^\sigma = \gamma \mathscr{A} + \sum \mu_i n_i^\sigma \tag{5.3.27}$$

and, hence (Exercise 5.3.1):

$$\begin{aligned} \mathrm{d}F^\sigma &= \mathrm{d}U^\sigma - T\,\mathrm{d}S^\sigma - S^\sigma\,\mathrm{d}T \\ &= -S^\sigma\,\mathrm{d}T + \gamma\,\mathrm{d}\mathscr{A} + \sum \mu_i\,\mathrm{d}n_i^\sigma. \end{aligned} \tag{5.3.28}$$

We can, therefore define γ thus:

$$\gamma = \left(\frac{\partial F^\sigma}{\partial \mathscr{A}}\right)_{T, \{n_i^\sigma\}}. \tag{5.3.29}$$

Alternatively, since (Exercise 5.3.1):

$$\mathrm{d}F = -S\,\mathrm{d}T - p'\,\mathrm{d}V' - p''\,\mathrm{d}V'' + \gamma\,\mathrm{d}\mathscr{A} + \sum \mu_i\,\mathrm{d}n_i \tag{5.3.30}$$

we have:

$$\gamma = \left(\frac{\partial F}{\partial \mathscr{A}}\right)_{T, V', V'', \{n_i\}} \tag{5.3.31}$$

which is probably the most useful definition since the subscripted variables can readily be held constant experimentally. Note also that (from eqn (5.3.27)):

$$f^\sigma = F^\sigma / \mathscr{A} = \gamma + \sum \mu_i \Gamma_i. \tag{5.3.32}$$

where Γ_i is defined in eqn (5.3.10) and f^σ is the Helmholtz free energy per unit area of the surface. This equation shows that f^σ is not equal to the surface tension, γ, except for a specific choice of the Gibbs dividing surface, namely that where $\sum \mu_i \Gamma_i = 0$; this is a very easy and natural choice for a one-component system, but highly unusual for multicomponent systems.

This is a convenient point to clear up a problem that often arises in the literature. It is frequently stated that surface tension and 'specific surface free energy' (or some such term) are equal only for a one-component system. Such statements can be properly interpreted only when the terms are carefully defined (Morrow 1970). The surface tension is defined by eqn (5.3.31) and if γ is independent of area (which will be so if there is an infinite reservoir of any adsorbing species) then this equation can be integrated to give:

$$(\Delta F)_{T, V', V'', \{n_i\}} = \gamma \Delta \mathscr{A} \tag{5.3.33}$$

or

$$\gamma = (\Delta F / \Delta \mathscr{A})_{T, V', V'', \{n_i\}}. \tag{5.3.34}$$

Thus, the change in Helmholtz free energy of the *system* (model independent) per unit change in surface area is numerically equal to the surface tension, γ, and this is true for a system with any number of components provided γ is independent of area.

On the other hand, for the *model system*

$$F^\sigma / \mathscr{A} = \gamma + \sum \mu_i \Gamma_i \tag{5.3.32}$$

and the specific Helmholtz free energy *ascribed* to the surface (and model *dependent*) is only equal to γ for a particular choice of dividing surface, such that $\sum \mu_i \Gamma_i = 0$. This choice of dividing surface is a natural one only for a one component system.

Exercises. 5.3.1 Establish eqns (5.3.27), (5.3.28), and (5.3.30).

5.3.2 The surface excess Gibbs free energy may be defined either as

$$\mathscr{G}^\sigma = U^\sigma - TS^\sigma + pV^\sigma$$

or

$$G^\sigma = U^\sigma - TS^\sigma + pV^\sigma - \gamma \mathscr{A}.$$

Show that the latter definition leads to:

$$G^\sigma = \sum_i \mu_i n_i^\sigma$$

which is analogous to the bulk phase equation (eqn (A5.26)). [Note that $V^\sigma = 0$ in the Gibbs model.]

5.4 Gibbs adsorption equation

The single most valuable equation in surface thermodynamics is the Gibbs adsorption isotherm. It is derived for surfaces in the same way that the Gibbs–Duhem equation (eqn (A5.23)) is derived for bulk phases. Briefly, integration of eqn (5.3.14) yields:

$$U^\sigma = TS^\sigma + \gamma\mathscr{A} + \sum \mu_i n_i^\sigma. \tag{5.4.1}$$

Differentiating eqn (5.4.1) and comparing it with eqn (5.3.14) we have the *Gibbs adsorption equation*:

$$S^\sigma\,\mathrm{d}T + \mathscr{A}\,\mathrm{d}\gamma + \sum n_i^\sigma\,\mathrm{d}\mu_i = 0. \tag{5.4.2}$$

We can then write, at constant temperature, the *Gibbs adsorption isotherm*:

$$-\mathrm{d}\gamma = \sum_i \frac{n_i^\sigma}{\mathscr{A}}\,\mathrm{d}\mu_i = \sum_i \Gamma_i\,\mathrm{d}\mu_i. \tag{5.4.3}$$

This equation is one of the most widely used expressions in surface and colloid science so we will explore its meaning at some length. It can be applied to systems where the surface tension can be measured (e.g. those containing liquid–liquid or liquid–vapour interfaces) in order to calculate the surface concentration of the adsorbed species causing the surface tension change. Equally, if the surface concentration can be measured directly but γ cannot (as occurs in many solid–gas and solid–liquid systems), the Gibbs adsorption equation can be used to calculate the lowering of γ (i.e. the spreading pressure, Π) from the measured adsorption.

Unfortunately, the absolute value of Γ_i is extremely dependent on the choice of dividing surface (see Fig. 5.3.2). We have already noted though (Fig. 5.3.2) that the Gibbs dividing surface is normally chosen so that n_i^σ and, hence, Γ_i^σ for the solvent arc equal to zero so that all other components are measured with reference to that surface, giving the *relative* surface concentrations. In the next section we will see how to define the relative surface concentration operationally so that its value is independent of the position of the dividing plane. The resulting (experimentally useful) quantity, $\Gamma_{i,1}$, (sometimes written $\Gamma_i^{(1)}$) is the surface excess of i relative to the solvent (1) and its value is numerically equal to Γ_i^σ with the convention $\Gamma_1^\sigma = 0$. The superscript σ in this case refers only to the use of a Gibbs dividing surface. We will also use the convention that $\Gamma_1^\sigma = 0$ for the solvent, so our Γ_i^σ will always be equal to $\Gamma_{i,1}$ or $\Gamma_i^{(1)}$.

5.4.1 *The relative adsorption*

The Gibbs–Duhem equation (eqn (A5.23)), when applied to the two bulk phases at constant temperature, gives (remember μ_i is the same in both phases):

$$V'\,\mathrm{d}p' = \sum_1^N n_i'\,\mathrm{d}\mu_i$$

and (5.4.4)

$$V''\,\mathrm{d}p'' = \sum_1^N n_i''\,\mathrm{d}\mu_i$$

and so, since the concentration c_i is n_i/V in each phase:

$$\mathrm{d}p' = \sum_1^N c_i'\,\mathrm{d}\mu_i$$
$$\mathrm{d}p'' = \sum_1^N c_i''\,\mathrm{d}\mu_i. \qquad (5.4.5)$$

If the interface is planar or if its curvature does not change so that $(p'' - p')$ is constant, then $\mathrm{d}p'' = \mathrm{d}p'$. Equations (5.4.5) then give†, for the variation in μ_1:

$$(c_1' - c_1'')\,\mathrm{d}\mu_1 = -\sum_2^N (c_i' - c_i'')\,\mathrm{d}\mu_i. \qquad (5.4.6)$$

We can now use this expression to eliminate the unknown μ_1 from the Gibbs adsorption equation to obtain (Exercise 5.4.3):

$$-\mathrm{d}\gamma = \sum_2^N \left\{\Gamma_i - \Gamma_1 \frac{(c_i' - c_i'')}{(c_1' - c_1'')}\right\} \mathrm{d}\mu_i. \qquad (5.4.7)$$

The quantity inside the curly brackets is defined for each component i as the relative adsorption of that component and is written:

$$\Gamma_{i,1} = \Gamma_i - \Gamma_1 \frac{\Delta c_i}{\Delta c_1} \qquad (5.4.8)$$

where $\Delta c_i = c_i' - c_i''$.

† If the curvature changes these equations become more complicated.

Although it is not essential, it is customary in the case of a liquid–vapour interface to take the solvent as component 1 and to compare the adsorption behaviour of all other components to it, since this choice leads to the definition of quantities that are directly accessible from experiments.

The definition of $\Gamma_{i,1}$ brings with it several advantages:

1. *The value of* $\Gamma_{i,1}$ *is independent of the arbitrary choice of dividing surface.* This is obvious from our derivation of eqn (5.4.7) since it depends only on general thermodynamic considerations. In a two-component system the quantity

$$\Gamma_{2,1} = \Gamma_2 - \Gamma_1 \frac{\Delta c_2}{\Delta c_1}$$

is a direct measure of $-\mathrm{d}\gamma/\mathrm{d}\mu_2$ and since this is a physically defined quantity it cannot depend on the (arbitrary) choice of dividing surface. To make this point clear we will analyse in detail the expression for $\Gamma_{2,1}$.

Using the Gibbs convention:

$$\begin{aligned} n_i^\sigma &= n_i - V'c_i' - V''c_i'' \quad \text{(by definition)} \\ &= n_i - Vc_i' + V''(c_i' - c_i'') \end{aligned} \tag{5.4.9}$$

where V' and V'' are the volumes of phases (′) and (″) respectively and $V = V' + V''$ is the total volume. n_i^σ and n_i are, as before, the number of moles ascribed to the surface and the total number of moles in the system respectively. For component 1, eqn (5.4.9) gives:

$$n_1^\sigma = n_1 - Vc_1' + V''(c_1' - c_1'') \tag{5.4.10}$$

and eliminating V'' between the two eqns (5.4.9) and (5.4.10) (Exercise 5.4.3):

$$n_i^\sigma - n_1^\sigma \frac{\Delta c_i}{\Delta c_1} = (n_i - Vc_i') - (n_1 - Vc_1')\frac{\Delta c_i}{\Delta c_1}. \tag{5.4.11}$$

None of the quantities on the right-hand side of eqn (5.4.11) depends on the choice of dividing surface, and if eqn (5.4.11) is divided by the surface area, $\mathscr{A}$, we recover the right-hand side of eqn (5.4.8), which defines $\Gamma_{i,1}$. For a curved surface it may be shown that $\mathscr{A}$ is the area of the surface of tension (Buff 1956).

2. $\Gamma_{i,1}$ *has an intuitively appealing physical meaning.* Since $\Gamma_{i,1}$ is independent of the choice of dividing surface, we are at liberty to pick any dividing surface that suits us. In particular, we can choose the surface where $\Gamma_1 = 0$ (*without* needing to be able to specify its position experimentally to a fraction of a nanometre). For that surface, $\Gamma_{i,1} = \Gamma_i$. In other words, *the relative adsorption of component i with respect to*

component 1 *is just the surface excess concentration of component i at the surface where the adsorption of component* 1 *is zero.*

As a simple example, consider adsorption at the gas–liquid interface (e.g. in a foam or aerosol). If phase (″) is the gas phase, then $c_i'' = c_1'' \doteqdot 0$, and so

$$\Gamma_{i,1} = \Gamma_i - \Gamma_1 \frac{c_i'}{c_1'}$$

$$= \Gamma_i - \Gamma_1 \frac{x_i'}{x_1'} \tag{5.4.12}$$

where the x_i's are mole fractions.

The same equation holds for adsorption from a solution onto a solid surface. In particular, if the solute is very dilute then

$$\Gamma_{i,1} \simeq \Gamma_i = n_i^\sigma / \mathscr{A} = \frac{n_i - c_i' V'}{\mathscr{A}} \tag{5.4.13}$$

i.e. one can identify the relative adsorption or the surface excess simply by determining the number of moles of solute that appear to have been removed from the bulk solution and dividing by the surface area of the solid. Such direct measurements of adsorption can be done using radio-isotopes or by measurement of concentration changes in the bulk solution when the solid adsorbent is added.

3. $\Gamma_{i,1}$ *can be introduced into the Gibbs adsorption equation, and hence related to experimentally measurable quantities.* We restrict ourselves to the simplest case, that of a two-component system, for which, from eqn (5.4.3):

$$-\mathrm{d}\gamma = \Gamma_1 \,\mathrm{d}\mu_1 + \Gamma_2 \,\mathrm{d}\mu_2. \tag{5.4.14}$$

This equation describes how the surface tension of a solution of, say, component 2 in solvent 1 is changed as the activity of substance 2 (and hence μ_2) is altered, at constant temperature.

If we choose the dividing surface so that $\Gamma_1 = 0$ then $\Gamma_2 = \Gamma_2^\sigma = \Gamma_{2,1}$ and so:

$$-\mathrm{d}\gamma = \Gamma_{2,1} \,\mathrm{d}\mu_2. \tag{5.4.15}$$

We can also write

$$\mu_2 = \mu_2^0 + RT \ln a_2 \qquad [2.1.3]$$

so that

$$\mathrm{d}\mu_2 = RT \,\mathrm{d} \ln a_2 \tag{5.4.16}$$

where a_2 is the activity of component 2.

It follows that:

$$\Gamma_{2,1} = -\frac{1}{RT}\frac{d\gamma}{d\ln a_2}. \tag{5.4.17}$$

If the activity coefficient of component 2 is y_2, so that $a_2 = y_2 x_2$ we may write eqn (5.4.17) as (Exercise 5.4.2):

$$\Gamma_{2,1} = -\frac{x_2}{RT}\frac{d\gamma}{dx_2}\left[1 + \frac{d\ln y_2}{d\ln x_2}\right]^{-1} \tag{5.4.18}$$

where x_2 is the mole fraction of component 2.

For ideal solutions:

$$\Gamma_{2,1} = -\frac{x_2}{RT}\frac{d\gamma}{dx_2} \tag{5.4.19}$$

and if the solution is fairly dilute so that $x_2 \propto c_2$ (the molar concentration) then

$$\Gamma_{2,1} = -\frac{c_2}{RT}\frac{d\gamma}{dc_2} = -\frac{1}{RT}\frac{d\gamma}{d\ln c_2} \tag{5.4.20}$$

which is more obviously an adsorption isotherm. It gives the relationship between the amount adsorbed and the solution concentration, c_2, in terms of the effect which c_2 has on the surface tension. Substances which lower the surface tension will have positive values of Γ; they are said to be *surface active*. Some typical values of γ as a function of solution concentration for various aliphatic alcohols are given in Fig. 5.4.1. Note how rapidly γ decreases with increase in solution concentration, indicating that $\Gamma_{2,1}$ is large in these cases. Note also the increase in adsorption as the chain length increases. These alcohols would be described as moderately surface active. We will find that, by contrast, electrolytes tend to raise the surface tension of water, indicating that they are negatively adsorbed at the air–water interface (i.e. they tend to be repelled towards the bulk of the liquid).

The Gibbs adsorption equation has been tested experimentally by McBain and Humphreys (1932), who used a flying microtome to scoop thin, hopefully uniform, surface layers from a series of aqueous surfactant solutions. The relative adsorptions $\Gamma_{2,1}$ of the surfactant (relative to the water) were calculated from the mole fractions of surfactant and water in the scooped-up and bulk phases, together with the volumes of scooped-up liquid and the surface areas. Comparison of these experimental values of $\Gamma_{2,1}$ with those calculated from eqn (5.4.20) showed good agreement (Defay *et al.* 1966). The remaining discrepancies are no doubt attributable to the experimental difficulties rather than the thermodynamics.

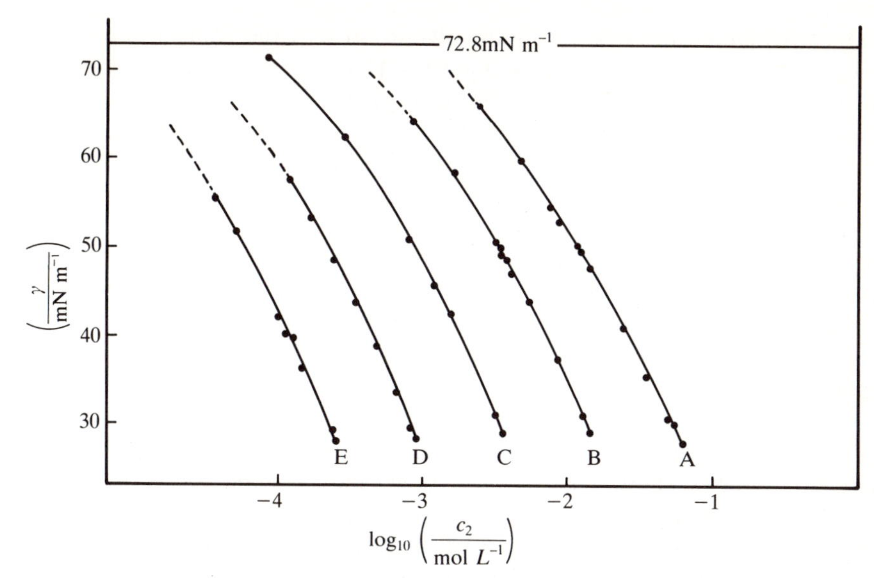

FIG. 5.4.1. Surface tension of aqueous solutions of *n*-aliphatic alcohols. A–E : C_6–C_{10} primary alkanol. (After Defay *et al.* 1966, with permission).

5.4.2 *The general form of the* γ–ln *c relation*

Figure 5.4.2 gives examples of the most commonly observed behaviour of the surface tension of a solution as a function of solute concentration.

Curve I is the behaviour that was displayed in an alternative form in Fig. 5.4.1. It is characteristic of solutions of substances that are to some

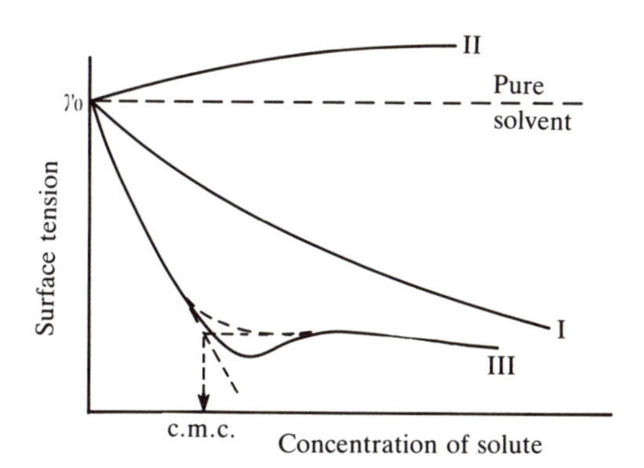

FIG. 5.4.2. The three types of curves commonly observed when the surface tension of a solution is plotted against concentration of solute. (After Davies and Rideal 1963.) The minimum in curve III is due to impurities and disappears after rigorous purification.

extent lyophobic towards the bulk solvent, so that the change in γ with increasing concentration is (from eqn (5.4.20)), negative. Such substances will tend to accumulate at the surface in preference to remaining in the bulk. If the bulk solvent is water then most polar organic molecules, if they are reasonably soluble, will behave like curve I.

By contrast curve II is the behaviour expected of a lyophilic solute. The most obvious examples in aqueous solution are the ionic salts, which are negatively adsorbed at (i.e. repelled away from) the air–water interface and so raise the surface tension. The same behaviour is also observed with hydrophilic solutes like sucrose.

Curve III shows the behaviour of a highly surface active substance—a true *surfactant*—such as a detergent (i.e. an amphiphilic substance (section 1.5)). Even at very low concentrations such solutes have a profound effect on γ and they have a great many uses in colloid science as a consequence. These we will discuss in more detail later. At quite modest concentrations (often $\sim 10^{-3}$–10^{-2} M) the decrease in surface tension ceases with the formation of micelles. The critical micellization concentration (c.m.c., section 1.5.3) is indicated in the figure. The minimum shown in the figure is caused by the presence of impurities which are usually present in commercial preparations. It disappears on rigorous purification.

5.4.3 *Particular forms of the Gibbs equation*

We have already shown in eqns (5.4.12) and (5.4.13) the appropriate forms of the Gibbs equation for adsorption of a simple solute at the liquid–gas or liquid–solid surface where there is no penetration of the solute or the solvent into the second phase. These equations become even further simplified if nearly all of the solute is adsorbed, as is the case with insoluble monolayers, for then

$$c_i' \doteqdot 0 \quad \text{and} \quad \Gamma_{i,1} \simeq \Gamma_i = n_i/\mathscr{A}. \tag{5.4.25}$$

For surface active electrolytes (e.g. an anionic detergent like $R.SO_3O^-Na^+$—an alkylsulphate) adsorbing at the air–water interface we have:

$$\mathrm{NaA} \leftrightarrows \mathrm{Na}^+ + \mathrm{A}^-$$

where $A^- = R.SO_3O^-$.

Assuming the salt is fully ionized, then

$$d\gamma = -\sum_i \Gamma_i \, d\mu_i = -\Gamma_{\mathrm{Na}^+} \, d\mu_{\mathrm{Na}^+} - \Gamma_{\mathrm{A}^-} \, d\mu_{\mathrm{A}^-} - \Gamma_{\mathrm{H_2O}} \, d\mu_{\mathrm{H_2O}}. \tag{5.4.26}$$

Using the Gibbs convention ($\Gamma_{H_2O} = 0$) and assuming that the interface as a whole is electrically neutral we have $\Gamma_{Na^+,1} = \Gamma_{A^-,1}$ and so:

$$\begin{aligned} d\gamma &= -\Gamma_{Na^+,1}\, d\mu_{Na^+} - \Gamma_{A^-,1}\, d\mu_{A^-} \\ &= -\Gamma_{A,1}\, d(\mu_{Na^+} + \mu_{A^-}) \\ &= -\Gamma_{A,1} RT\, d\ln(a_{Na^+} \cdot a_{A^-}) \end{aligned} \quad (5.4.27)$$

i.e.

$$d\gamma = -2\Gamma_{A,1} RT\, d\ln a_{\pm} \quad (5.4.28)$$

where we have introduced the mean ionic activity defined by

$$a_{\pm}^{\nu} = a_{+}^{\nu_+} a_{-}^{\nu_-} \quad (5.4.29)$$

where $\nu = \nu_+ + \nu_-$ is the number of ions produced when the molecule dissociates into ν_+ cations and ν_- anions.

For a dilute solution in which the activity coefficient y (eqn (5.4.18)) is close to unity we have:

$$\Gamma_{A,1} = -\frac{c_{NaA}}{2RT} \frac{d\gamma}{dc_{NaA}}. \quad (5.4.30)$$

Note that the reduction in γ for a given (molar) adsorption is doubled in this case: it is a colligative property like osmotic pressure and freezing point depression and depends on the number of solute particles in the interface.

If a large excess of a simple sodium salt (say NaCl) is added as well as the surfactant then the system can be held at constant ionic strength. The chloride ion can be assumed (Fig. 5.4.2, curve II) to have negligible surface activity ($\Gamma_{Cl^-} \simeq 0$) and the chemical potential of both the chloride and sodium ions can be held essentially constant as the concentration of NaA is altered. In that case the counterpart to eqn (5.4.27) is:

$$\begin{aligned} d\gamma &= -\Gamma_{A^-,1}\, d\mu_{A^-} = -RT\Gamma_{A,1}\, d\ln a_{A^-} \\ &\simeq -RT\Gamma_{A,1}\, d\ln c_{A^-} \end{aligned}$$

so that

$$\Gamma_{A^-,1} \simeq -\frac{c_{A^-}}{RT}\left(\frac{d\gamma}{dc_{A^-}}\right). \quad (5.4.31)$$

Note that the factor 2 has disappeared in this case. This difference in behaviour of a surfactant in the absence and in the presence of a swamping concentration of electrolyte is quite significant. It can be investigated with the use of radiotracers such as tritium (^3H) that undergo β-decay. The β-radiation is rapidly absorbed in water so that only radiation originating in the surface layers can escape. Using appropriately labelled surfactants one can directly measure the amount of surfactant in the surface layers (see Hiemenz 1977, p. 282).

5.4.4 *Two liquid phases in contact*

The phase rule applied to a system containing only plane interfaces reads (Defay *et al.* 1966, p. 77):

$$f = 2 + \mathscr{C} - \mathscr{R} - \mathscr{P} - \psi + \mathscr{S} \tag{5.4.32}$$

where f is the number of degree of freedom, $\mathscr{C}$ the number of components, $\mathscr{R}$ the number of chemical reactions between components, $\mathscr{P}$ the number of phases, ψ the number of surface phases and $\mathscr{S}$ the number of surfaces. Assuming $\psi = \mathscr{S}$ (the usual case) and $\mathscr{R} = 0$ we have, for two phases in contact:

$$f = \mathscr{C}. \tag{5.4.33}$$

The simple case of a two-component solution in contact with its vapour is treated extensively in most textbooks. The two independent variables in that case are usually chosen to be the temperature and the concentration of solute. For an n-component mixture, the temperature and the concentration of $(n-1)$ components are independent and one can eliminate the solvent activity (μ_1) using the Gibbs convention. The pressure in this case is a dependent variable because it is determined by the vapour pressures of the n components at the temperature in question.

For two liquids in contact the more appropriate choice of independent variables is T, p, and $(n-2)$ of the component concentrations because in this case the pressure and temperature can be arbitrarily altered. What the phase rule tells us is that we can then set the surface concentration of *two* components to zero. The most appropriate choice is the two liquid solvents that form the bulk of the contacting phases. One then has two Gibbs dividing surfaces and the surface concentration of the other $(n-2)$ components is measured relative to both components 1 and 2. It can be shown that in that case:

$$\begin{aligned}\left(\frac{\partial \gamma}{\partial_{\mu_i}}\right)_{\substack{T,P,\mu_3\ldots\mu_j\ldots\mu_n\\ j\neq i}} &= -\left\{\Gamma_{i,1} - \Gamma_2\left[\frac{c_1'c_i'' - c_1''c_i'}{c_1'c_2'' - c_1''c_2'}\right]\right\}\\ &= -\Gamma_{i,12}.\end{aligned} \tag{5.4.34}$$

In effect, one calculates the surface concentration of any component (from 3 to n) by adding the surface excess concentrations from the two phases (Exercises 5.4.4–6; Defay *et al.* 1966, p. 89) show that the difference between $\Gamma_{i,12}$ and $\Gamma_{i,1}$ is usually negligible.

Exercises. 5.4.1 Imagine a sharply defined surface layer of adsorbed material at the liquid–vapour interface. By placing the Gibbs dividing surface at the boundary between the adsorbed layer and the liquid, use eqn (5.4.8) to show that the relative adsorption $\Gamma_{i,1}$ is zero when

$$\frac{\Gamma_i}{x_i'} = \frac{\Gamma_1}{x_1'}$$

i.e. when the components i and 1 are present in the same molar ratio in the surface layer as they are in the liquid. Show also that $\Gamma_{i,1}$ is positive if the surface layer is relatively richer in i, negative if the surface layer is relatively poorer in i.

5.4.2 Establish eqn (5.4.18). Notice that the correction factor for non-ideality is the same as the one appearing in eqn (3.5.22).

5.4.3 Establish eqn (5.4.7) and (5.4.11).

5.4.4 If two phases, (′) and (″) are in contact at constant T and p, the Gibbs–Duhem equation requires that $\sum n_i \, d\mu_i = 0$ in each phase. Consider the situation where solute 3 is distributed between the bulk solvents 1 and 2. Show that the surface excess of solute 3 relative to 1 *and* 2 is given by:

$$\Gamma_{3,12} = \Gamma_3 - \left[\frac{\Gamma_1(n_2''n_3' - n_2'n_3'') - \Gamma_2(n_1''n_3' - n_1'n_3'')}{n_2''n_1' - n_2'n_1''}\right].$$

(Hint: refer back to eqn (5.4.3) and use the Gibbs–Duhem equations to eliminate $d\mu_1$ and $d\mu_2$. Note the similarity to eqn (5.4.34).)

5.4.5 Show that if the two solvents 1 and 2 in Exercise 5.4.4 are mutually insoluble then

$$\Gamma_{3,12} = \Gamma_3 - \frac{n_3'}{n_1'}\Gamma_1 - \frac{n_3''}{n_2''}\Gamma_2.$$

5.4.6 Consider the following situation (adapted from Ross and Chen 1965). A solute is distributed between a water and an oil phase (assumed to be mutually insoluble). Analysis of the oil phase shows there is 1 mol solute for every 4 mol solvent. In the water phase there is 1 mol solute for every 6 mol water. A sample of the interfacial region, of thickness 1 mm (including material from both bulk phases) and area 260 cm^2 is found to contain 1.02 mol water, 0.84 mol oil and 0.51 mol solute. Calculate the surface excess of solute (with respect to both water and oil) using the Gibbs convention. (Refer to Fig. 5.3.2(c) and calculate the excess of solute above that expected from the bulk concentrations.) Verify the result by using the expression derived in Exercise 5.4.5.

5.4.7 Consider the more realistic situation of mutual solubility indicated below, as an extension of problem 5.4.6:

	Oil	Water	Solute	
Bulk oil	0.96	0.09	0.24	mole
Surface region	0.84	1.02	0.51	mole
Bulk water	0.36	1.08	0.18	mole
	Oil	Water	Solute	

Calculate the surface excess of solute in this case, relative to both oil and water. (Again the area of the interface is 260 cm^2.)

5.5 Thermodynamic behaviour of small particles

5.5.1 *The Kelvin equation*

The existence of a pressure difference across a curved interface (governed by the Young–Laplace equation, eqn (5.2.17)) has a number of important colloid chemical consequences. For very small particles (droplets or bubbles) the pressure difference may be so great that the chemical potential of the material is affected. Taking $\gamma = 70\,\text{mN m}^{-1}$ and a (spherical) drop radius of 50 nm, eqn (5.2.17) gives for the pressure difference $(2 \times 70 \times 10^{-3}/5 \times 10^{-8})\text{Pa} = 28 \times 10^5\,\text{Pa} \doteqdot 28\,\text{atm}$. Such a pressure, applied to a liquid will raise its chemical potential (and hence its vapour pressure) by a measurable amount.

The same excess pressure inside a gas bubble can be interpreted as a *reduced* pressure in the adjoining liquid and will cause a lowering of its vapour pressure. If the curvature is caused by the fact that the liquid–vapour interface is being formed as a meniscus in a small capillary, the same lowering of vapour pressure occurs, with important consequences for adsorption in porous solids (see capillary condensation in section 5.7 below).

An idea of the magnitude of the vapour pressure changes involved can be gained from Kelvin's original, elegant physical derivation in which he considered a capillary rise experiment in a closed, isothermal vessel (Fig. 5.5.1; Thomson 1870, 1871).

For thermodynamic equilibrium to hold, the liquid and the vapour must be in equilibrium both at the flat liquid interface and the curved upper meniscus. Since the hydrostatic pressure (in the vapour phase) is lower at the upper meniscus by $\rho_v gh$ than it is at the plane liquid surface,

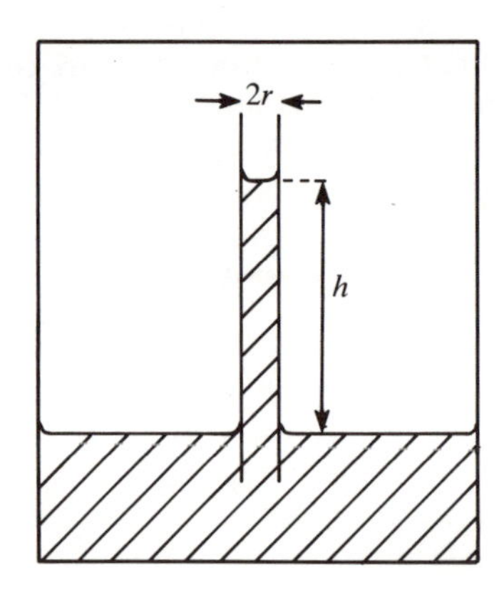

FIG. 5.5.1. Capillary rise of a wetting liquid in a cylindrical capillary tube contained in a closed isothermal chamber. The space above the liquid contains only the vapour.

it follows that the equilibrium vapour pressure at the meniscus is lower than that at the plane liquid surface by this amount (ρ_v is the density of the vapour.)

The Young–Laplace equation (eqn (5.2.17)) gives, at the meniscus:

$$\Delta p(=\rho_l gh - \rho_v gh) = \gamma\left(\frac{2}{r}\right) \tag{5.5.1}$$

so that

$$\rho_v gh = \gamma\left(\frac{2}{r}\right)\left(\frac{\rho_v}{\rho_l - \rho_v}\right) \tag{5.5.2}$$

or

$$p' - p^0 = \frac{\gamma\rho_v}{\rho_l - \rho_v}\left(\frac{2}{r}\right) \tag{5.5.3}$$

where p' is the equilibrium vapour pressure at the meniscus and p^0 is the equilibrium vapour pressure above the flat liquid surface. Equation (5.5.3) is the original (approximate) form of the Kelvin equation.

A more rigorous analysis of the effect of surface curvature on p^0 is given by Defay *et al.* (1966, Chapter 15), from which a few results are summarized below. For a single pure substance the fundamental equations are, for a sphere of radius r:

$$\Delta p = p'' - p' = 2\gamma/r \tag{5.5.4}$$

and

$$\mu'' = \mu' = \mu \tag{5.5.5}$$

where, by convention, (″) and (′) refer to the phase on the concave and convex sides of the interface, respectively. Note that eqn (5.5.4) can be generalized to any curved surface by replacing $2/r$ by $(1/R_1 + 1/R_2)$, where R_1 and R_2 are the principal radii of curvature and may have the same or different signs. The superscript convention used here allows the equations to be applied to both menisci and droplets. Note that r is positive for droplets which, therefore, have an increased vapour pressure; it is negative for a concave meniscus (Fig. 5.5.1).

For an infinitesimal process applied to a system initially at equilibrium, we have

$$\mathrm{d}p'' - \mathrm{d}p' = \mathrm{d}(2\gamma/r) \tag{5.5.6}$$

and

$$\mathrm{d}\mu'' = \mathrm{d}\mu' = \mathrm{d}\mu. \tag{5.5.7}$$

The Gibbs–Duhem equation (eqn (A5.23)) in each phase becomes:

$$\begin{aligned} \bar{S}'\,\mathrm{d}T - \bar{V}'\,\mathrm{d}p' + \mathrm{d}\mu' &= 0 \\ \bar{S}''\,\mathrm{d}T - \bar{V}''\,\mathrm{d}p'' + \mathrm{d}\mu'' &= 0 \end{aligned} \tag{5.5.8}$$

when applied to a pure substance (where $\bar{S} = S/n_i$ and $\bar{V} = V/n_i$ are the molar entropy and volume respectively). Equations (5.5.6–8) are the fundamental relations from which a large number of important results can be deduced.

For example, at constant temperature we have (from eqns (5.5.7) and (5.5.8)):

$$\bar{V}' \,\mathrm{d}p' = \bar{V}'' \,\mathrm{d}p''$$

and so, from eqn (5.5.6):

$$\mathrm{d}\left(\frac{2\gamma}{r}\right) = \left(\frac{\bar{V}' - \bar{V}''}{\bar{V}''}\right) \mathrm{d}p' \qquad (5.5.9)$$

$$= \left(\frac{\bar{V}' - \bar{V}''}{\bar{V}'}\right) \mathrm{d}p''. \qquad (5.5.10)$$

Equations (5.5.9) and (5.5.10) may be regarded as the most general forms of the Kelvin equation as it applies to spherical interfaces.

5.5.2 *Applications of the Kelvin equation*

(a) *Drops of liquid in a vapour.* In this case (″) refers to the liquid and $\bar{V}'' \ll \bar{V}'$ ($\simeq RT/p'$). Then from eqn (5.5.9):

$$\mathrm{d}\left(\frac{2\gamma}{r}\right) = \frac{RT}{\bar{V}''} \frac{\mathrm{d}p'}{p'} - \mathrm{d}p'. \qquad (5.5.11)$$

Integrating from $r = \infty$ (a flat surface where the equilibrium vapour pressure is p^0) to some finite radius where the equilibrium vapour pressure is p' we have (Exercise 5.5.2):

$$\ln p'/p^0 = \frac{\bar{V}''}{RT}\left\{\frac{2\gamma}{r} + (p' - p^0)\right\} \text{ (droplets)} \qquad (5.5.12)$$

or

$$\ln \frac{p'}{p^0} \simeq \frac{2\gamma \bar{V}''}{rRT} \quad \text{if} \quad \bar{V}'' \ll \bar{V}'.$$

Equation (5.5.12) is the exact form of Kelvin's equation, the approximate form of which was quoted in section 1.4. It shows that the equilibrium vapour pressure of the drop is higher than that of the flat liquid surface. It follows, therefore, that if a number of droplets of uniform radius are initially in equilibrium with a surrounding vapour (of infinite volume) that equilibrium must be unstable (section 1.4) because if condensation occurs on a drop its radius will increase, its equilibrium vapour pressure will decrease and it will continue to grow. Conversely, if a little evaporation occurs from a droplet, its radius decreases, its equilibrium vapour pressure increases and it continues to evaporate. If the droplets are initially of different size, then the large ones will grow at the expense of the smaller ones.

This argument applies if there is a sufficient reservoir of vapour, so that droplets do not change the partial vapour pressure as they evaporate or grow. The important point to note is that, *for a given vapour pressure* there is a critical size, equal to the size in equilibrium with the vapour pressure, above which the droplets will grow but below which they will evaporate. When a vapour is cooled so that it becomes supersaturated, condensation to the liquid cannot occur until there are formed some nuclei of liquid of a size that can continue to grow. We will return to the study of this *nucleation theory* in section 5.8.

An expression analogous to eqn (5.5.12) can be written for the solubility of a solid crystal in a surrounding liquid (as we mentioned in section 1.4):

$$\ln \frac{C(r)}{C_\infty} = \frac{2\gamma \bar{V}''}{rRT} \tag{5.5.13}$$

where $C(r)$, C_∞ are the solubilities of the particles and the bulk solid respectively and γ is the interfacial surface energy per unit area. The relation between the 'effective' radius, r, and the dimensions of the solid crystal will be examined shortly (section 5.6). Again, eqn (5.5.13) suggests that large crystals will grow at the expense of smaller ones—a phenomenon known as Ostwald ripening (section 1.4) (see Kahlweit 1975).

(b) *Bubbles in a liquid.* In this case, the phase (″) is a gas and so we have (Exercise 5.5.2):

$$\ln p''/p^0 = \frac{\bar{V}'}{RT}\left\{(p''-p^0) - \frac{2\gamma}{r}\right\} \quad \text{(bubble or concave meniscus)} \tag{5.5.14}$$

or

$$\ln p''/p^0 \simeq -\frac{2\gamma \bar{V}'}{rRT} \quad (\text{if } \bar{V}' \ll \bar{V}'').$$

Hence, the vapour pressure inside a bubble is *smaller* than the value for a flat surface. This explains the phenomenon of superheating of liquids above the normal boiling point. Remember that there has to be an *excess* pressure inside a bubble and yet the vapour pressure is actually lower than that over a flat surface at the same temperature. Boiling can only occur if either (i) the liquid can vaporize into a pre-existing bubble of reasonable size or (ii) the temperature is raised sufficiently high so that the equilibrium vapour pressure even in a small bubble is large enough to allow it to continue to grow. Case (i) is more usual, where the

Table 5.2
Influence of radius of curvature on the equilibrium vapour pressure above a spherical water surface, calculated from eqns (5.5.12) and (5.5.14), assuming that γ is independent of r

r(nm)	p'/p^0 (droplet)	p''/p^0 (bubble)
1000	1.001	0.999
100	1.011	0.989
10	1.115	0.897
1	2.968	0.337

pre-existing bubble is, say, an air bubble previously released from the liquid as its solubility was reduced. Such bubbles form on the walls of the container and can be introduced deliberately by adding porous solids ('boiling chips'). Case (ii) is called homogeneous nucleation (section 5.8) and it can only occur at temperatures well above the normal boiling point. It turns out that a temperature of almost 200 °C is required to induce boiling of water at atmospheric pressure if rigorous efforts are made to exclude any extraneous nucleation bubbles and one relies entirely on the formation of bubbles of the pure vapour.

It should be noted that the changes in vapour pressure due to surface curvature and predicted by the Kelvin equation are negligible except for very highly curved surfaces (Table 5.2).

It should also be noted that the integrations in eqns (5.5.12–14) are carried out with respect to the variable γ/r. They remain valid, therefore, even if γ changes with radius and pressure, provided that the value substituted for γ is the appropriate one for the value of r, and p' (or p'') concerned.

5.5.3 *Effect of temperature on vapour pressure—the Thomson equation*

We now want to calculate the effect of temperature change on the vapour pressure inside a bubble. We must, therefore, derive the analogue of the Clausius–Clapeyron equation for phase equilibrium across a curved interface. The resulting equation is attributed to J. J. Thomson† (the discoverer of the electron). In deriving Thomson's equation we will

† Not to be confused with William Thomson, who became Lord Kelvin and made the other important contribution in this area.

assume that the latent heat of vaporization $\Delta\bar{H}_{vap}$ for the liquid is independent of curvature. Although there is some slight dependence at very high curvature it may generally be neglected. Defay *et al.* (1966) give a full discussion of this question (pp. 230–239) and point out that even at $r = 1$ nm the decrease in $\Delta\bar{H}_{vap}$ for water is only about 6 per cent.

With this approximation we can easily determine the effect of change in radius of curvature (r) of a drop or bubble on the temperature required to establish equilibrium across the interface, if the external pressure is kept constant.

(a) *Liquid drop suspended in its vapour.* From eqn (5.5.8), with (″) referring to the liquid, we have $\mathrm{d}p' = 0$ (since the external pressure is constant) and so:

$$(\bar{S}' - \bar{S}'')\,\mathrm{d}T + \bar{V}''\,\mathrm{d}p'' = 0 \tag{5.5.15}$$

and

$$\mathrm{d}p'' = \mathrm{d}\left(2\frac{\gamma}{r}\right). \tag{5.5.16}$$

Setting $\bar{S}' - \bar{S}'' = \Delta\bar{H}_{vap}/T$ and integrating from $r = \infty$ to some finite value r, we obtain:

$$\ln\frac{T}{T_0} = -\frac{2\bar{V}''\gamma}{\Delta\bar{H}_{vap}\cdot r} \tag{5.5.17}$$

which is the Thomson equation (Thomson 1888). We have assumed here that $\bar{V}''$ and $\Delta\bar{H}_{vap}$ are unaffected by the temperature change. This is a reasonable assumption in the present case because the temperature range involved in practice is fairly small (see Table 5.3).

In this case, $T < T_0$ (since $\Delta\bar{H}_{vap}$ is always positive). Thus as r decreases the equilibrium temperature becomes lower and lower. In order to induce a vapour to condense onto a small droplet of the liquid phase it is necessary to cool it down to a temperature *below the normal condensation temperature* T_0, corresponding to the external pressure p'. This is the phenomenon called *supercooling*. Supercooled (i.e. supersaturated) vapours are used in the Wilson cloud chamber for the detection of radioactive particles. The particle ionizes the air in the chamber and this aids the formation of nuclei of sufficient size to permit further deposition (i.e. drop growth) at the temperature of the chamber.

It should be noted that the integration process leading to eqn (5.5.17) corresponds to a gradual (reversible) bending of the interface, with a corresponding reduction in temperature to maintain equilibrium between vapour and liquid at each stage. As the bending progresses the pressure inside the droplet increases and in order to maintain the chemical

Table 5.3

Influence of curvature on the equilibrium temperature of droplets and bubbles. (Adapted from Defay et al. (1966), p. 242)

	Droplet of water in water vapour at 1 atm†			Bubble of water vapour in water at 1 atm				
r(m)	T/T_0	T(K)	γ(mN m^{-1})	T/T_0	T(K)	T(°C)	γ(mN m^{-1})	$\Delta\bar{H}_{vap}$(kJ mol^{-1})
∞	1	373	55.46	1	373	100	55.46	40.40
10^{-5}	0.9999946	372.998	55.46	1.0088	376	103.3	54.75	40.23
10^{-6}	0.999946	372.980	55.46	1.0574	394.3	121.4	50.85	39.26
10^{-7}	0.99948	372.807	55.46	1.2037	449	176	39.10	36.29
10^{-8}	0.99485	371.08	55.8	1.461	545	272	18.44	31.00
5×10^{-9}	0.98906	368.92	56.34					

$\bar{V}''$ has been taken as $1.043 \times 18\ cm^3\ mol^{-1}$.

† Strictly speaking the pressure is a little less than 1 atm so the normal boiling point is 373 K and not 373.15 K as it is at 1 atm.

potential constant we make a corresponding reduction to the temperature. The actual condensation process that occurs when nucleation is induced is, of course, a highly irreversible process. The vapour deposits rapidly because as the droplet grows the equilibrium temperature rises so the driving force towards condensation tends to increase. This is offset, however, by the release of the latent heat of condensation, which tends to raise the temperature of the whole system.

(b) *Bubble immersed in a liquid.* In this case (″) refers to the gas phase and $dp' = 0$ again. Assuming ideal behaviour for the gas ($p''\bar{V}'' = RT$) and using the Young–Laplace equation (eqn (5.2.17)), we obtain from eqn (5.5.8):

$$\Delta \bar{H}_{\text{vap}}\left(\frac{1}{T_0} - \frac{1}{T}\right) = R \ln\left(\frac{2\gamma/r + p'}{p'}\right). \qquad (5.5.18)$$

(In the integration process $\Delta \bar{H}_{\text{vap}}$ is assumed constant with respect to temperature. This assumption can easily be dispensed with (Exercise 5.5.3).) This form of Thomson's equation gives the temperature at which a bubble of vapour of radius r can exist in equilibrium inside a liquid. For small values of r it follows that $T > T_0$ and, indeed, the temperature can be much higher than the normal boiling point as we noted above. Recall that there has to be an excess pressure inside the bubble so eqn (5.5.18) calculates how high the temperature must be raised to achieve that higher vapour pressure.

Table 5.3 gives values for the equilibrium temperature for drops and bubbles of various sizes. Notice that the effect on the equilibrium of bubbles is much larger than that of drops. That is not surprising when one recalls that the effect is due to the excess pressure inside the sphere: the effect of pressure on the chemical potential of a gas is much greater than its effect on a liquid. Note also that the more elaborate equation derived in Exercise 5.5.3 has been used to estimate T for bubbles because significant changes occur in $\Delta \bar{H}_{\text{vap}}$ over the temperature range involved.

5.5.4 *Application of the Kelvin and Thomson equations to solid particles*

We have already noted (section 5.5.2) that small solid particles may be treated as spheres with an equivalent radius. The resulting excess pressure experienced by the solid will increase its chemical potential and Defay *et al.* (1966) list the following consequences:

1. The vapour pressure of small crystals is greater than that of large crystals. In the presence of vapour, large crystals will grow at the expense of small crystals.
2. Small crystals will melt at a temperature lower than the normal melting

point. The m.p. of a small crystal will be given by

$$\ln\frac{T}{T_0}=-\frac{2\gamma_{SL}\cdot\bar{V}_s}{r\cdot\Delta\bar{H}_{fus}} \tag{5.5.19}$$

where T_0 is the normal melting point at the same external pressure, γ_{SL} is the interfacial energy . . . $\bar{V}_s$ is the molar volume of the solid and $\Delta\bar{H}_{fus}$ is the molar heat of fusion . . . .

3. The melting point of a substance solidified in the pores of an inert material will depend upon the size of the pores . . . .

4. Small crystals may have a heat of fusion and a heat of sublimation smaller than the value for bulk solid.

Point 1 has already been alluded to in section 1.4. Point 2 is very important in the field of ceramics where finely powdered materials are heated to high temperatures and *sintering* occurs (i.e. fusion and partial inter-diffusion occurs at localized centres where the radius of curvature is very small). Point 3 has important consequences in tertiary oil recovery and the treatment of oil sands and shales to recover high molar mass hydrocarbon fractions (i.e. heavy crude) since the melting point of such material is affected by the capillary pressure to which it is subjected in the pores. (See section 5.7.1 for further discussion of capillary pressure.)

Equation (5.5.13) for the effect of particle size on solubility can be made a little more precise by using activities rather than concentrations:

$$\ln\frac{a}{a_0}=\ln\frac{yC(r)}{y^0C_\infty}=\frac{2\gamma\bar{V}_s}{rRT} \tag{5.5.20}$$

where y and y^0 are activity coefficients.

This formulation has the advantage that it can be extended to apply to electrolyte solutions, by the introduction of the mean ionic activity:

$$a_\pm^\nu=a_+^{\nu_+}\cdot a_-^{\nu_-} \qquad [5.4.29]$$

where ν is the number of ions produced when the salt dissociates. Equation (5.4.20) then becomes (Defay *et al.* 1966, p. 272):

$$\frac{2\gamma}{r}=\nu\frac{RT}{\bar{V}_s}\ln\frac{y_\pm C(r)}{y_\pm^0 C_\infty}. \tag{5.5.21}$$

$y_\pm^0$ is the activity coefficient of the salt solution at concentration C_∞ (the solubility of large crystals) and $y_\pm$ is its value at concentration $C(r)$, which can often be calculated from extended versions of the Debye–Hückel theory.

Another important consequence of the excess pressure inside small particles is its effect on a chemical equilibrium. Defay *et al.* (1966, pp. 281–284) give an analysis of the effect of adding fine particles of Ni powder to an equilibrium mixture of Ni foil and $Ni(CO)_4$. The effect is

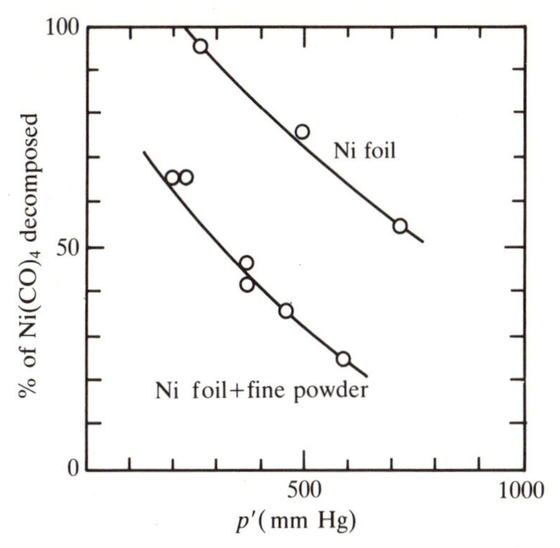

FIG. 5.5.2. Effect of particle size on the Ni–Ni(CO)$_4$ equilibrium. (From Defay *et al.* 1966, with permission.)

quite dramatic as can be seen from Fig. 5.5.2. Again it is a direct consequence of the higher chemical potential of the nickel when in the form of small particles.

Exercises. 5.5.1 Estimate the value of p'/p^0 for a droplet of water and p''/p^0 for a bubble in water with the following radii: ∞; 10^{-6} m; 10^{-7}; 10^{-8}; 10^{-9} m. (Note that the last result is unlikely to be correct. Why?) Take $t = 18\,°\mathrm{C}$; $\bar{V}_1 = 18\,\mathrm{cm^3\,mol^{-1}}$; $\gamma = 73\,\mathrm{mN\,m^{-1}}$. Calculate values for $p'' - p'$ across an interface with these radii.

5.5.2 Derive eqns (5.5.12) and (5.5.14) and show that they yield the approximate expressions under the conditions stated. How much effect does the approximation have on the calculated values for $r = 10^{-9}$ m obtained in Exercise (5.5.1) above?

5.5.3 Show that if $\Delta\bar{H}_{\mathrm{vap}} = a - bT$ where a and b are constants, the integration in section 5.5.3(b) leads to:

$$a\left(\frac{1}{T_0} - \frac{1}{T}\right) - b \ln \frac{T}{T_0} = R \ln(1 + 2\gamma/rp').$$

The form for $\Delta\bar{H}_{\mathrm{vap}}$ is suggested by Kirchhoff's law with a constant value for $\Delta\bar{C}_p = \bar{C}_{p_{\mathrm{vap}}} - \bar{C}_{p_{\mathrm{liq}}}$). This is a more precise form of eqn (5.5.18).

5.5.4 Check the values given in Table 5.3 by back-calculating the value of r from the given values of T, γ, and $\Delta\bar{H}_{\mathrm{vap}}$, using the more exact formula derived in Exercise 5.5.3.

5.6 Equilibrium shape of a crystal

It was noted in section 3.1 that crystal faces tend to grow in such a way as to minimize the total surface energy. Figure 5.6.1 shows a schematic

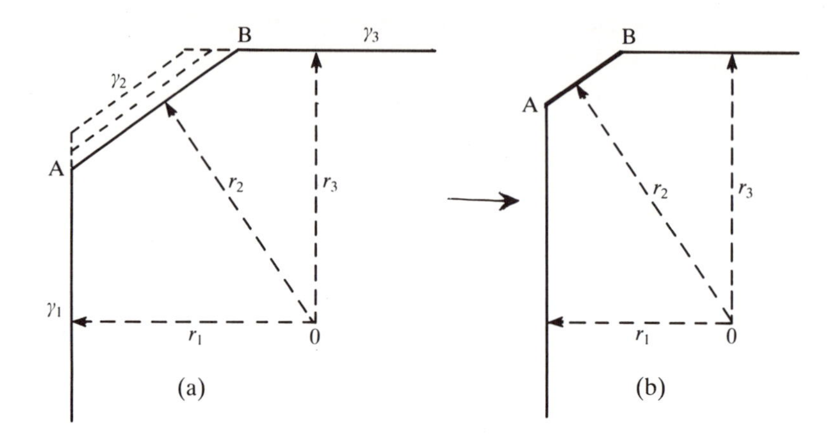

FIG. 5.6.1. Growth of a crystal to eliminate a high-energy face (AB).

diagram of the growth process. If the face AB is of high energy then growth occurs preferentially on that face so that it moves further from the origin 0 and diminishes in size. The final crystal then has faces of lowest energy. We noted in section 3.1 that Wulff established early this century (1901) that the criterion for equilibrium growth of the crystal was that:

$$\frac{\gamma_1}{r_1}=\frac{\gamma_2}{r_2}=\frac{\gamma_3}{r_3}=\text{a constant.} \tag{5.6.1}$$

A formal proof of this relation is given by Defay *et al.* (1966, pp. 292–297) but it is lengthy and rather tedious. The following analysis of a more elementary problem captures the main features of the behaviour.

Consider a prismatic crystal (β) growing by the accretion of material from a second phase (α) under conditions of constant temperature and constant total volume. If growth is to occur under equilibrium conditions then:

$$\begin{aligned} \mathrm{d}F &= \mathrm{d}F^{\alpha}+\mathrm{d}F^{\beta}+\mathrm{d}F^{\sigma}=0 \\ &= -S^{\alpha}\,\mathrm{d}T-p^{\alpha}\,\mathrm{d}V^{\alpha}+\mu^{\alpha}\,\mathrm{d}n^{\alpha}-S^{\beta}\,\mathrm{d}T-p^{\beta}\,\mathrm{d}V^{\beta} \\ &\quad +\mu^{\beta}\,\mathrm{d}n^{\beta}+\sum\gamma_i\,\mathrm{d}\mathscr{A}_i^{\beta} \end{aligned} \tag{5.6.2}$$

where the summation is carried out over each crystal face. We have:

$$\mathrm{d}T=0; \qquad \mathrm{d}V^{\alpha}=-\mathrm{d}V^{\beta}; \qquad \mathrm{d}n^{\alpha}=-\mathrm{d}n^{\beta} \quad \text{and} \quad \mu^{\alpha}=\mu^{\beta}$$

and hence, eqn (5.6.2) reduces to:

$$-(p^{\beta}-p^{\alpha})\,\mathrm{d}V^{\beta}+\sum\gamma_i\,\mathrm{d}\mathscr{A}_i^{\beta}=0. \tag{5.6.3}$$

This equation merely states the obvious fact that the work done by phase β as it expands to take up room formerly occupied by phase α is compensated for by the change in surface energy. In the present case $V^\beta = xyz$ and we assume that the opposite faces of the prism are characterized by surface energies γ_x, γ_y, and γ_z respectively (Fig. 5.6.2). (Note that γ_x refers to the surface that is perpendicular to the x-axis.) Substituting in eqn (5.6.3) we have:

$$-(p^\beta - p^\alpha)[yz\,\mathrm{d}x + xz\,\mathrm{d}y + xy\,\mathrm{d}z] + 2[\gamma_x(y\,\mathrm{d}z + z\,\mathrm{d}y) + \gamma_y(x\,\mathrm{d}z + z\,\mathrm{d}x) + \gamma_z(x\,\mathrm{d}y + y\,\mathrm{d}x)] = 0. \quad (5.6.4)$$

Collecting terms in dx, dy, and dz:

$$[-(p^\beta - p^\alpha)yz + 2\gamma_y z + 2\gamma_z y]\,\mathrm{d}x + \text{similar terms for } \mathrm{d}y \text{ and } \mathrm{d}z = 0. \quad (5.6.5)$$

Since dx, dy, and dz are arbitrary (small) changes in x, y, and z, eqn (5.6.5) can only be true if each coefficient is equal to zero, so that:

$$\begin{aligned} -(p^\beta - p^\alpha) + \frac{2\gamma_y}{y} + \frac{2\gamma_z}{z} &= 0 \\ -(p^\beta - p^\alpha) + \frac{2\gamma_x}{x} + \frac{2\gamma_y}{y} &= 0 \\ -(p^\beta - p^\alpha) + \frac{2\gamma_z}{z} + \frac{2\gamma_x}{x} &= 0\,. \end{aligned} \quad (5.6.6)$$

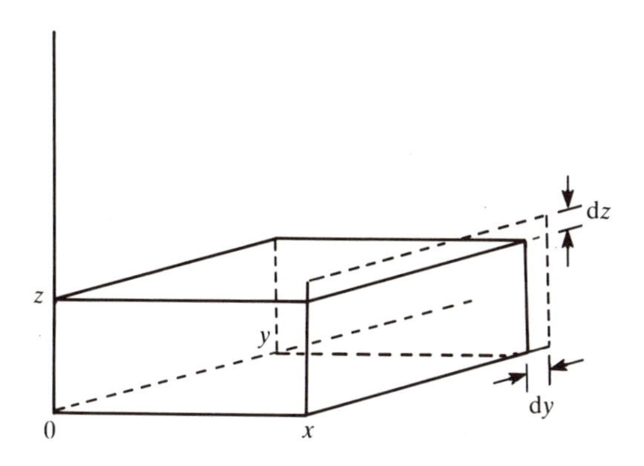

FIG. 5.6.2. A prismatic crystal growing from the vapour phase, from solution or from the melt. Increase in area of the right-hand face is $y\,\mathrm{d}z + z\,\mathrm{d}y$.

These equations are satisfied by:

$$\frac{2\gamma_x}{x} = \frac{2\gamma_y}{y} = \frac{2\gamma_z}{z} = \text{(a constant)} \tag{5.6.7}$$

which is obviously identical to eqn (5.6.1). Furthermore, we see that the excess pressure in the solid phase is given by:

$$(p^{\beta} - p^{\alpha}) = \frac{4\gamma_x}{x} \tag{5.6.8}$$

which is the analogue of the Young–Laplace eqn (5.2.17) in this case. The distance $x/2$ is evidently analogous to the radius of a spherical drop.

The equilibrium shape of a polygonal crystal can be determined from eqn (5.6.7). The method is described briefly by Dunning (1973) and (for a two-dimensional crystal) by Adamson (1967, p. 270; 1976, p. 250). It involves drawing a three-dimensional (polar) plot of the surface energy as a function of crystal plane (*hkl*) orientation and constructing a polyhedron from that plot. As we noted earlier (section 3.1) adsorption of impurities can dramatically alter the habit of crystals away from their equilibrium shape but Adamson points out that, if small enough, irregular crystals will assume an equilibrium shape on annealing (i.e. when heated to a temperature just below the bulk melting point).

5.7 Behaviour of liquids in capillaries

The difference in pressure that occurs across a curved interface has many important consequences. If the radius of curvature of the liquid–gas interface is negative (as in a bubble of gas in a liquid) then the pressure is lower in the liquid that it is in the gas. We have, so far, interpreted that as an excess pressure inside a gas bubble. When the liquid is contained in a capillary tube (Fig. 5.7.1), however, it is often more useful to think of the liquid pressure as being *negative* with respect to the gas pressure. A negative pressure (or suction pressure) of this kind is like a tension and the magnitude of the tension can be very high: it may be limited only by the tensile strength of the liquid.

5.7.1 *Capillary pressure*

Consider the liquid in Fig. 5.7.1. The height to which it will rise in the capillary tube is determined by the curvature of the meniscus. (We will return to the question of *why* there is a meniscus in section 5.10.2). If the radius of the tube, r, is sufficiently small, gravitational distortion can be neglected and we can estimate the radii of curvature, R_1 and R_2 of the

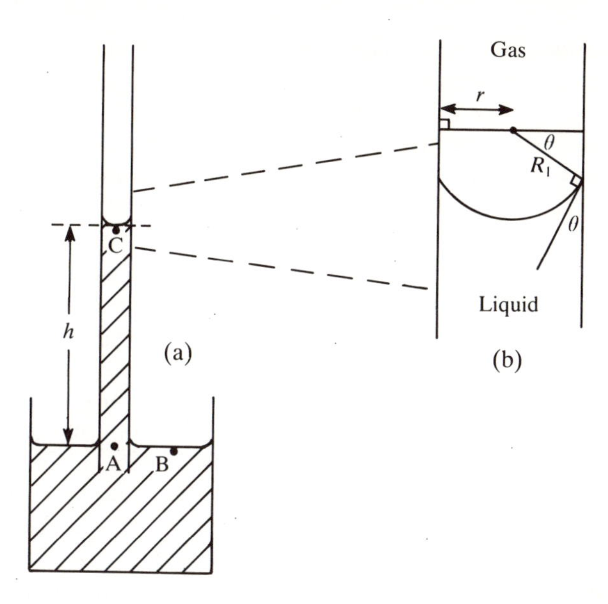

FIG. 5.7.1. Rise of liquid in a capillary tube of radius r. The pressure must be the same at points A and B.

interface as $R_1 = R_2 = r/\cos\theta$ (Fig. 5.7.1(b)) where θ is called the *contact angle*. (This is the angle between the liquid surface and the solid (measured in the liquid) and we will have much more to say about it in section 5.10. For the moment we assume only that it can be observed and measured macroscopically with sufficient accuracy.) The pressure must be the same at points A and B and is *lower* at point C by an amount $\rho_l gh$. The pressure difference across the interface is then (from eqn (5.2.17)) given by:

$$\Delta p = \rho_l gh - \rho_v gh = \gamma\left(\frac{1}{R_1} + \frac{1}{R_2}\right)$$

so that

$$h = \frac{2\gamma\cos\theta}{r(\rho_l - \rho_v)g} \approx \frac{2\gamma\cos\theta}{r\rho_l g} \tag{5.7.1}$$

where ρ_l and ρ_v are the densities of liquid and vapour respectively. This is the simple equation for the rise of a liquid in a narrow tube and it is frequently used for the measurement of surface tension, especially of pure liquids. It is really only practicable when $\theta = 0$ (that is, when the liquid completely *wets* the solid). If the contact angle is non-zero it is sometimes difficult to measure with sufficient accuracy and may be variable (see section 5.10 below). We are concerned here with the practical consequences of eqn (5.7.1) rather than its application to surface tension measurement which is treated in some detail by Adamson (1967). (See Exercise 5.7.1.)

In formal laboratory measurements the capillary radius is of the order of 1 mm and the rise, h for water would be about $(2\times70\times10^{-3}\times1/10^{-3}\times10^{3}\times9.8)$ m $\doteqdot$ 14 mm. For very tiny capillaries the rise can be very high indeed and eqn (5.7.1) has been invoked (Dixon and Joly 1894) to explain the transport of water to the top of very high trees (see e.g. the review by Pickard 1981). If capillary pressure alone were to account for the rise of water to the top of a tall tree (say 100 m) we would need to have a capillary radius of about 1.5×10^{-7} (or 150 nm) which, though small, is by no means impossible. Note that it is not necessary for the capillary to have a small bore throughout its length. Only the bore at the meniscus (i.e. in the leaf) is relevant (Fig. 5.7.2) in determining the height of the column that can be supported. It should be clear, however, that in cases (b) and (c) the liquid will not rise of its own accord to height h because it will not be able to traverse the wider parts of the tube: the pressure drop across the interface disappears as the liquid surface moves through those regions. The situation in (b) and (c) is, however, stable if the liquid is sucked up to height h and then the suction is removed.

If the suspended column is very high (as in the example of the tall tree quoted above) the limiting factor may well be the tensile strength of the water column. (Although clean water containing air has been measured to have a static breaking tension of around 46 atm, rising to 68 atm after degassing (Overton *et al.* 1982), rupture may occur earlier in columns of water in capillaries with rough walls.) The water column may break and

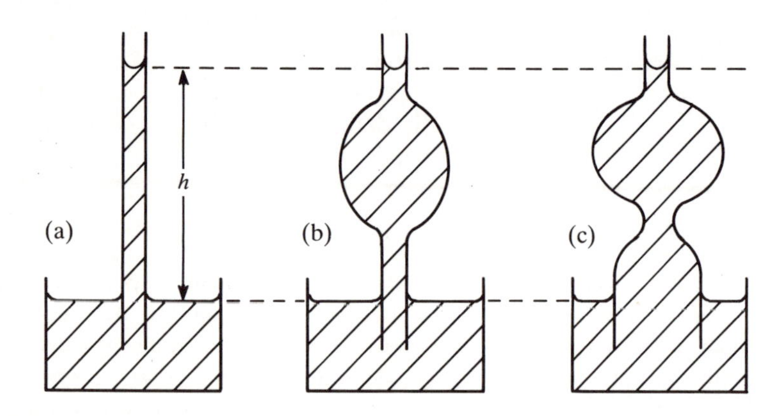

FIG. 5.7.2. Capillary equilibrium is affected only by the curvature at the interface and not by the shape of the capillary at other levels (but see text).

leave a vacuum, which will then fill with the equilibrium vapour, at a pressure determined by the Kelvin equation (eqn (5.5.14)). Even in that case, a path still remains for water to be transported through the gas phase so some connection between the roots and the uppermost leaves is maintained. Whether this kind of mechanism can account for the observed *rate* of water movement in tall trees is a matter for empirical experiment. Certainly it seems unlikely that a tree, throughout the entire course of its growth to 100 m, would be able to maintain an uninterrupted column of water and the above account provides no mechanism for removing vapour gaps.

If the local diameter of the tube where the break occurs is too large, the lower liquid level will fall until it reaches a constriction that is sufficiently narrow to generate a curvature that can support the mass of the column below it. As Adamson (1967, p. 12) shows, the exact solution of the capillary equation (taking proper account of the variation in curvature across the surface) gives, for the weight of the column of liquid that can be supported:

$$W = 2\pi r\gamma \cos\theta. \tag{5.7.2}$$

For a straight capillary this gives the same result as eqn (5.7.1) if we imagine the surface to be 'hanging' from the wall of the capillary and supported by the vertical component ($\gamma\cos\theta$) of the surface tension:

$$2\pi r\gamma\cos\theta = \pi r^2 h\rho_1 g \tag{5.7.3}$$

which leads immediately to eqn (5.7.1).

How can this be reconciled with the behaviour pictured in Fig. 5.7.2(b) and (c)? Obviously in those cases the mass of liquid apparently being supported by the interface is very much larger; the additional stress is then taken up by the walls of the container.

The movement of liquid into narrow capillaries, against the action of gravity has many important consequences. The capacity of a soil to retain water in the root zone of plants is determined by its capillary (pore) structure (Childs 1969) and this depends chiefly on the particle size of the soil constituents. Very fine (clay) particles produce very small pores that can prevent water draining away to the level of the underground water table. Too much clay, however, (i.e. too heavy a soil) can make it difficult for the plants to extract water since they must provide a still greater suction pressure to draw the water out (see Exercise 5.7.2).

It should be pointed out that these negative (suction) pressures are not limited to one atmosphere. The calculation involving the tree corresponds to a negative pressure of about 10 atm. One might set an upper limit by the consideration that r cannot be less than atomic dimensions (say 5 nm) and taking $\gamma \simeq 100\,\text{mJ m}^{-2}$ as typical we have $\Delta p_{\text{max}} \simeq 4 \times$

10^7 N m$^{-2} \simeq 400$ atm—more than enough to have a profound effect on the properties of the material in the pores.

One final point about eqn (5.7.1) concerns the contact angle θ. We have discussed, for the most part, liquids for which $\theta = 0$ i.e. perfectly wetting liquids. Whether a liquid wets (i.e. spreads freely over) another liquid or a solid surface depends upon how strong the interaction between the two materials is, compared to the cohesive forces within the liquid itself (section 5.10). Here we need only note that for a non-wetting liquid, $\theta > \pi/2$ and eqn (5.7.1) predicts a negative value for h. In that case the meniscus is convex (e.g. as in a liquid drop), the capillary pressure inside the liquid is *positive* and the surface of the liquid inside the capillary is *lower* than the free surface outside. That is the behaviour observed with mercury in a glass capillary, for which $\theta \simeq 135°$. In order to make mercury enter a porous solid it is necessary to apply an increased pressure to the liquid (since in most cases it is non-wetting). By determining the volume of mercury taken up by a solid as the pressure increases one can determine the pore size distribution, a procedure known as *mercury injection porosimetry*. Successive increases in pressure allow successively smaller pores to be entered (in accordance with the Young–Laplace equation, (5.2.17)). One must still assume a contact angle in order to relate R_1 and R_2 in that equation to the capillary radius (see Exercise 5.7.3 and Adamson 1967, p. 546; 1976, p. 529 for details).

In section 5.5.3 we referred briefly to the effect of capillary pressure on the melting point of solid materials confined inside the pores of another solid. It should be clear from the above that one cannot make sensible predictions about the *sign* let alone the *magnitude* of such effects from a knowledge of the pore size alone. The effect of the contact angle is crucial for a liquid, since it determines whether the capillary pressure is positive or negative. But how can this be applied to solids?

The problem with solids concerns not only the contact angle but the concept of surface tension or surface energy itself. Most of the arguments to date have involved the tacit assumption that the system has either reached equilibrium or that at least the surface energies involved are well defined. For the liquid–vapour interface that is usually a reasonable assumption. There may be some time effects connected with the diffusion of an adsorbate to the surface but these are minor compared to the problem with solids. As we noted in section 5.2.1 a solid surface (especially one produced by fracture) may remain in a non-equilibrium state indefinitely, with an arbitrary surface energy, since it has no means of relaxation. We should not be surprised to find, then, that problems involving the surface tension of solids give rise to considerable controversy. The solid–liquid–vapour interface is, however, so important in so many areas of colloid science that we must be prepared to make some attack on it even if that involves some sacrifice of rigour. The insights so gained, however imperfect, can help us to order our view of the behaviour of the physical world and make reasonable projections from the known to the unknown which, after all, is what we are trying to do.

5.7.2 *Capillary condensation*

One important consequence of (negative) capillary pressure is the phenomenon of *capillary condensation*. This refers to the process whereby a vapour condenses in a capillary at a vapour pressure below that at which it would condense on the free liquid surface. Capillary condensation is, of course, described by the Kelvin equation, which for present purposes may be written:

$$\ln p''/p^0 \approx \frac{-2\gamma\bar{V}'}{rRT}. \qquad [5.5.14]$$

(Recall from Table 5.2 that the overall effect of surface curvature on vapour pressure is important only at very high curvature, i.e. small radius of curvature.)

As a simple physical example of capillary condensation, consider a conical capillary tube placed in a vapour with pressure p'' (Fig. 5.7.3(a)). Condensation will immediately begin at the bottom of the tube, which will start to fill with liquid. So long as the radius of curvature of the liquid meniscus is small enough, so that the vapour pressure p above it is less than p'', the tube will continue to fill (Fig. 5.7.3(b)). When the tube has filled to the point where, from eqn (5.5.14), the equilibrium vapour pressure is p'', condensation will cease and a state of equilibrium has

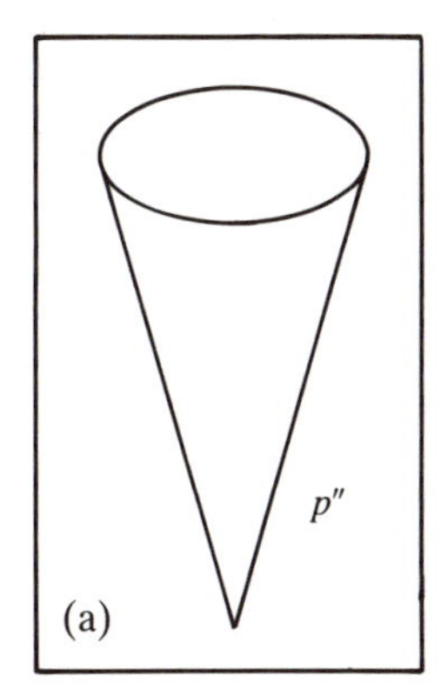

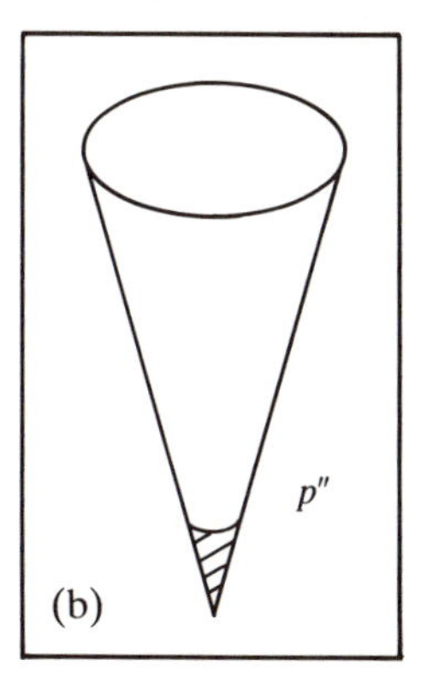

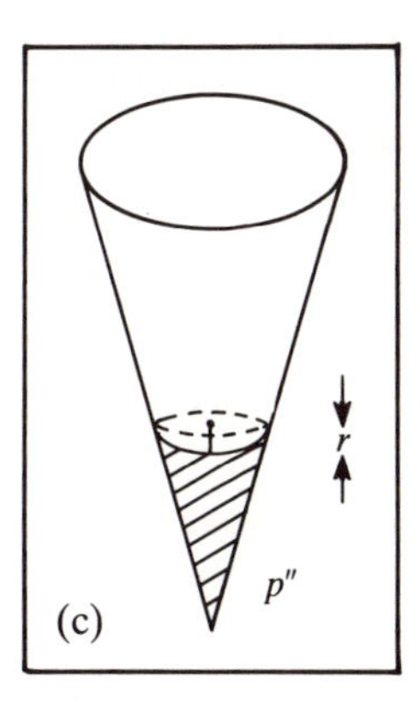

FIG. 5.7.3. Capillary condensation in a conical tube.

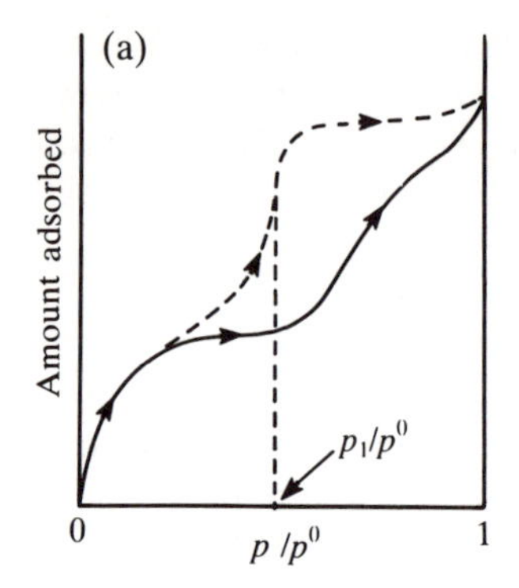

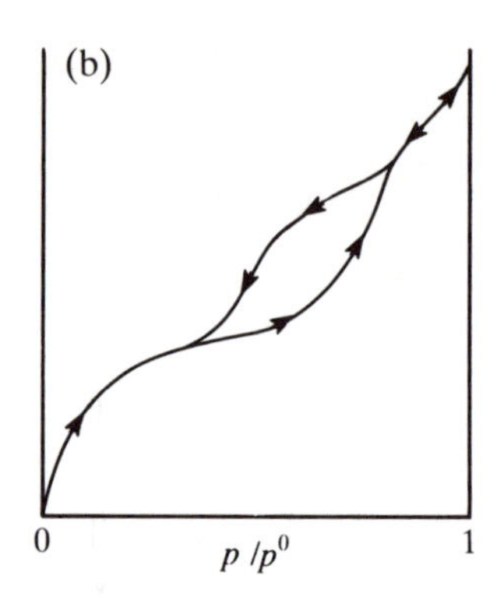

FIG. 5.7.4. Typical adsorption isotherm for vapour on a porous solid.
(a) Full line, normal behaviour; broken line, expected behaviour if all pores are uniform in radius.
(b) Hysteresis observed when adsorption is followed by desorption.
(p^0 is the equilibrium vapour pressure at the temperature of the experiment).

been reached. Any increase or decrease in p'' will result in further (reversible) filling or emptying of the tube, respectively.

Similar considerations apply to tubes of any shape, but it is important to remember (see note following eqn (5.5.5)) that r is *defined* by $2/r = 1/R_1 + 1/R_2$, where R_1 and R_2 are the principal radii of curvature of the meniscus. For a spherical meniscus, $R_1 = R_2 = r$, but even then, the value of r is only equal to the tube radius, r_c, at that point if the contact angle $\theta = 0$; otherwise $r = r_c/\cos\theta > r_c$. For a meniscus between parallel flat plates, one radius is infinitely large and so $(1/R_1 + 1/R_2) = 1/r$.

An important case, where capillary shapes are not usually known, is that of capillary condensation in porous materials. We often study the adsorption of a gas or vapour† on a porous solid (e.g. a catalyst) in order to learn more about the surface structure of the solid and its interaction with the gas. If the pores of the solid are sufficiently narrow the adsorption isotherm may look like that in Fig. 5.7.4(a) (full line) (Everett 1967). The lower part of the curve has a typical isotherm shape: adsorption occurs first on the more active sites and rapidly approaches an apparent saturation value (which may or may not correspond to a monolayer). At the top end of the scale we expect bulk liquid to condense on a flat solid surface as $p/p_0 \rightarrow 1$. If all pores have the same radius it is expected that liquid will be able to condense in those pores as soon as p reaches the equilibrium value, p_1, corresponding to the radius

† The term *vapour* is generally reserved for a gas at temperatures below the critical temperature.

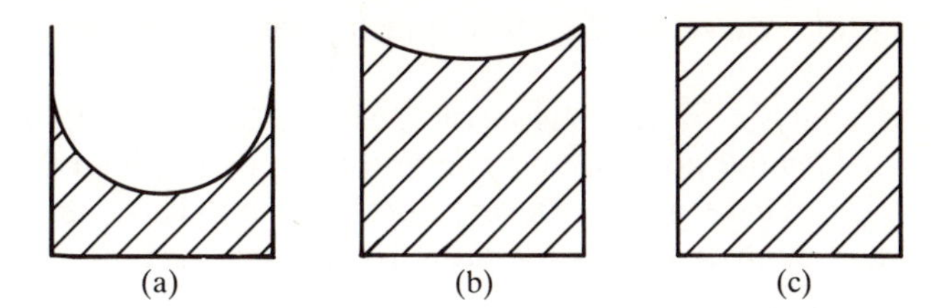

FIG. 5.7.5. Filling up the pores in a porous solid.

of curvature for that pore (broken curve in Fig. 5.7.4(a)). The curve rises very steeply because every pore can fill almost to the top at the same value of p (Fig. 5.7.5(a)). Only when the curvature starts to decrease (Fig. 5.7.5(b)) is it necessary for p to be raised further to induce more condensation. Over a very small volume change the meniscus changes from (a) to (b) to (c) and no further condensation can occur until $p = p^0$.

In practice, the pores are not of the same size and they therefore fill at different values of p, the smaller ones filling first. The hysteresis shown in Fig. 5.7.4(b) can arise in a number of ways of which the most prevalent is probably the so-called 'ink-bottle' effect (Fig. 5.7.6). As p is increased, a point is reached (p_1) at which the equilibrium curvature is that shown on the left and the pore fills as indicated in (a). If p is increased further the curvature can become flatter and the pore gradually fills up to the point indicated in (b). Now if p is reduced all of the liquid in that pore can remain in equilibrium with the vapour until the point p_1 is reached again.

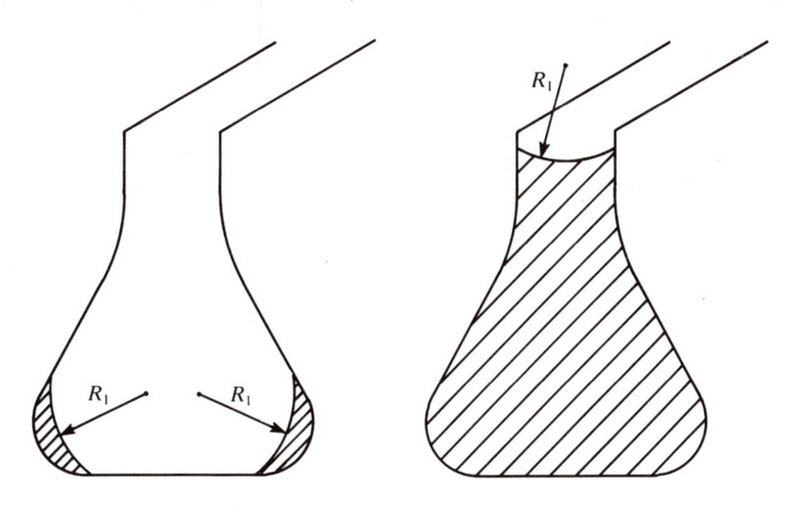

FIG. 5.7.6. The 'ink-bottle' effect (see text).

The desorption curve is therefore much higher than the adsorption branch.

Detailed analysis of the shape of these isotherms using the Kelvin equation enables the pore size distribution to be determined (Everett and Haynes 1973). There are, however, numerous pitfalls in the procedure and we will not discuss it in detail. (See Adamson 1967, pp. 637–41; 1976, pp. 618–23).

Capillary condensation can also be important in the adhesion of dust and powders, (Zimon 1969), the Laplace pressure (eqn (5.2.17)) in the capillary condensate contributing often significantly, and sometimes overwhelmingly, to the total adhesive force. It is not generally realized that the adhesive force is dependent *only on the particle radius* (for wettable, spherical particles, at least), and is independent of the mean curvature of the meniscus, and hence independent of the relative vapour pressure, within very wide limits.

Consider, for example, two rigid contacting spherical particles, each of radius R', in an atmosphere of water vapour. An annulus of capillary condensed water will form around the contact region (Fig. 5.7.7). For $r \ll x \ll R'$ (the usual case in practice) the mean curvature is

$$J = \frac{1}{r} - \frac{1}{x} \simeq \frac{1}{r}. \tag{5.7.4}$$

Since J must be constant for a given vapour pressure of water, the meniscus profile is closely approximated as an arc of a circle. Since $(R' + r)^2 = R'^2 + (r + x)^2$ we have:

$$r = \frac{x^2}{2R'} \quad (\text{for } r \ll R'). \tag{5.7.5}$$

Now the Laplace pressure Δp is given by:

$$\begin{aligned} \Delta p &\simeq \gamma J \\ &\simeq \gamma / r. \end{aligned} \tag{5.7.6}$$

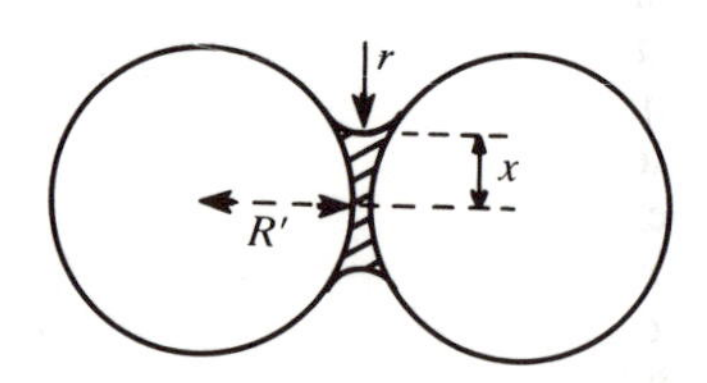

FIG. 5.7.7. Capillary condensation of a liquid between contacting spheres. (The contact angle is assumed to be zero.)

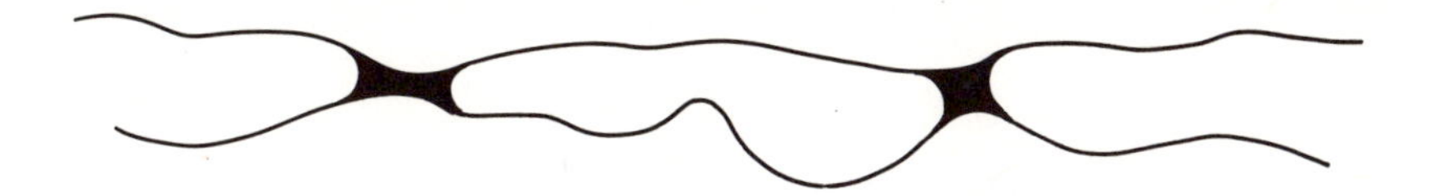

FIG. 5.7.8. Capillary condensation between rough surfaces.

This pressure is applied over an area $\mathscr{A} = \pi x^2$, and hence the force, $\mathscr{F}$, between the two spheres due to the Laplace pressure is given by:

$$\mathscr{F} = \mathscr{A}\Delta p$$
$$= 2\pi R'\gamma \tag{5.7.7}$$

which is independent of the value of r, and hence independent of the amount of capillary condensate (i.e. independent of the vapour pressure; Fisher and Israelachvili 1981a).

In practice, the adhesion between particles due to the presence of capillary condensate is often apparently much smaller than that given by eqn (5.7.7). This is generally because the particle surfaces are rough, so that capillary condensation only occurs about scattered contact points (Fig. 5.7.8).

The radii of the contacting protrusions are, of course, much smaller than the macroscopic particle radii, and hence the force (given by eqn (5.7.7)) is correspondingly smaller.

Exercises. 5.7.1 Adamson (1976, p. 10) defines the capillary constant, a, for the rise of liquid in a capillary as $a^2 = 2\gamma/(\rho_l - \rho_v)g$. For the approximate eqn (5.7.1) $a^2 = rh$. The more exact analysis taking account of the true curvature of the surface gives (Adamson 1967, p. 13):

$$a^2 = r\left(h + \frac{r}{3} - 0.1288\frac{r^2}{h} + 0.1312\frac{r^3}{h^2}\ldots\right).$$

Show that the second term in the brackets takes account of the weight of the meniscus, assuming it is hemispherical. What is the percentage error in γ using eqn (5.7.1) for $r = 1$ mm and $h = 14$ mm?

5.7.2 Consider an idealized model of a porous medium consisting of random close-packed spheres of equal radius, R'. Estimate the resulting minimum pore radius r in terms of R'. Hence estimate the order of magnitude of the capillary rise expected in such a packed bed if $R' = 100$ μm, 10 μm, and 1 μm respectively. (Note that r depends somewhat on the manner of packing but only very approximate results are required.)

5.7.3(a) The pore size distribution in a microporous solid can be determined by measuring the volume, V, of mercury which can be forced into the pores as a function of pressure. Suppose that the (volume) distribution function for particle radius is $D(r)$, so that

$$dV = D(r)\,dr$$

where dV is the volume of pores of radius between r and $r-dr$. Use the Young–Laplace equation to show that

$$D(r) = \frac{1}{r}\frac{dV}{d\ln p}$$

where p is the pressure in the mercury. (Adamson 1967, p. 546 or 1976, p. 529.)
(b) Given the following data:

Pressure (psi)	Vol. of mercury penetrated (cm^3)	
10	0.028	Sample mass = 0.493 g
20	0.032	
30	0.038	Sample density = 3.52 g cm^{-3}
50	0.046	
100	0.056	Apparent sample volume
200	0.072	(including pores) = 0.344 cm^3
300	0.096	
500	0.148	
700	0.166	Density of mercury = 13.55 g cm^{-3}
1000	0.174	
2000	0.178	

plot the pore size distribution (i.e. $D(r)$) as a function of r, assuming the pores are cylindrical. (You may assume that $\gamma_{Hg} = 473$ dyne cm^{-1} and the contact angle is 130°.) Estimate the total pore volume by integrating under an appropriate curve and compare this with the value obtained from the mass/density data.

5.7.4 Calculate the cohesive force $\mathcal{F}$ between two rigid, contacting particles of radius $R' = 1\ \mu m$, in an atmosphere of water vapour at 20 °C and a relative humidity of 90 per cent. Also estimate r (the curvature of the liquid meniscus) in this case and show that $r \ll x \ll R'$. Repeat the calculation for $R' = 0.05\ \mu m$. Imagine a string of particles dangling in moist air with the upper one fixed in position. The only cohesive force between the particles is that due to the capillary condensed water. How many particles could be supported before the string is broken? (Assume $\rho_s = 2.8$ g cm^{-3}, and $R' = 1\ \mu m$ or $0.05\ \mu m$ as before).

5.7.5 An adsorption isotherm for N_2 adsorption at 77 K on Wonkavite—a highly porous (and probably poisonous) children's confection is shown in Fig. 5.7.9. Estimate the pore diameter and the pore volume per unit surface area. The density of liquid N_2 is $\sim 10^3$ kg m^{-3} and the surface tension of the liquid N_2–air interface is $\sim 2 \times 10^{-2}$ J m^{-2}.

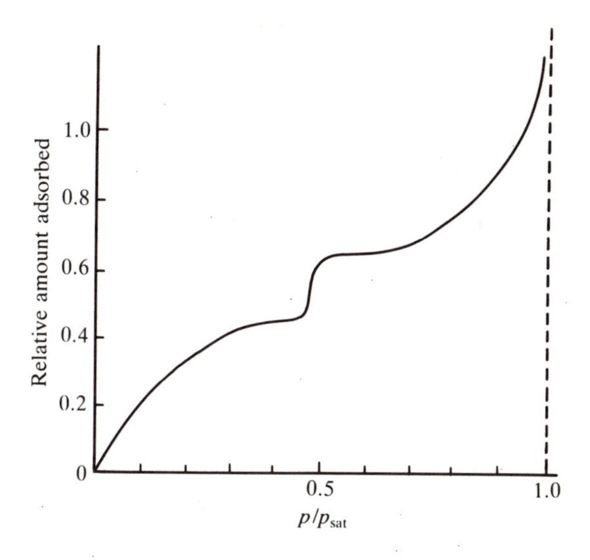

FIG. 5.7.9. Refer to Exercise 5.7.5.

5.8 Homogeneous nucleation

The Thomson equation (eqn (5.5.17)) permits the calculation of the temperature at which a drop or bubble of a given size can be in equilibrium with the surrounding phase (assuming the latter is very large). As noted earlier, that equilibrium is unstable: if the drop begins to grow by the random accumulation of a few more molecules†, its radius increases, its vapour pressure decreases and growth can continue with a further reduction in free energy. That same unstable equilibrium situation is involved in the Kelvin equation: the radius of the drop that is in equilibrium with the surrounding vapour pressure is a critical radius, r_c, because smaller drops have higher vapour pressure and will spontaneously evaporate. If the critical drop accumulates a few more molecules it can go on increasing in size because its equilibrium vapour pressure is lower than the prevailing value. It is important to note that not all such droplets *will* increase in size. A significant fluctuation in the opposite direction can reduce r below r_c before the droplet has time to grow and then the driving force is in the opposite direction. We can only say that if there are a reasonably large number of such drops, then there

† This is called a *fluctuation* and such random (positive and negative) changes are an important aspect of the kinetic molecular theory.

is a high probability that some of them will stay on the high side of r_c long enough to become so much larger than r_c that they will never evaporate.

The unstable nature of the equilibrium becomes clear if we write the equation for the free energy change involved in forming a small drop of liquid from the vapour. Adamson (1967, p. 376; 1976, p. 373) treats this as a chemical reaction of the type

$$nA \text{ (gas}, p) \rightleftarrows A_n \text{ (small liquid drop)}$$

while Davies and Rideal (1963, p. 350) treat it as a phase change. The following discussion is a simplified version of the Defay *et al.* (1966) treatment, which pays more attention to the separation of the equilibrium and non-equilibrium behaviour.

Consider the formation of a liquid phase (in the form of a droplet) from the (supersaturated) vapour in a one-component system at constant temperature and constant total overall volume, V. Before formation of the liquid we obtain the Helmholtz free energy by integrating eqn (A5.8) at constant temperature:

$$F_0 = n_i \mu_g(p_i') - p_i' V \tag{5.8.1}$$

where n_i is the initial number of moles present and p_i' is the initial pressure in the gas phase. After formation of the drops, the free energy of the entire system is obtained by integrating eqn (5.3.16) at constant temperature:

$$F = -p'V' - p''V'' + \gamma \mathscr{A} + n' \mu_g(p') + n'' \mu_l(p'') \tag{5.8.2}$$

where $V = V' + V''$, $n_i = n' + n''$, and μ_l is the chemical potential of the liquid. The dividing surface is chosen to make $n^\sigma = 0$.

The free energy of formation of the droplet phase is then

$$(\Delta F)_{T,V} = F - F_0. \tag{5.8.3}$$

Note that this is true irrespective of whether the process is an equilibrium (reversible) one or a non-equilibrium one since it only involves the initial and final state functions.

If the volume of the gas phase is sufficiently large, the amount of material needed to form the droplet phase does not appreciably alter the pressure of the gas so

$$\mu_g(p_i') = \mu_g(p') \quad \text{and} \quad p' = p_i'. \tag{5.8.4}$$

Then from eqns (5.8.1–4):

$$(\Delta F)_{T,V} = n''(\mu_l(p'') - \mu_g(p')) - V''(p'' - p') + \gamma \mathscr{A}. \tag{5.8.5}$$

This equation can be used to calculate ΔF for the formation of a droplet of any size, from our knowledge of the influence of drop size on $(p'' - p')$

(the Young–Laplace equation, eqn (5.2.17)) and on $(\mu_l - \mu_g)$ (the Kelvin equation, eqn (5.5.12)).

For example, the Young–Laplace equation gives

$$V''(p''-p') = \tfrac{4}{3}\pi r^3 \times \frac{2\gamma}{r} = \tfrac{2}{3}(4\pi r^2\gamma) = \tfrac{2}{3}\gamma\mathscr{A}. \tag{5.8.6}$$

Furthermore, at the critical radius, the liquid drop and the vapour are in equilibrium so that $\mu_l(p'') = \mu_g(p')$ and hence (from eqn (5.8.5)):

$$\begin{aligned}(\Delta F)_{\text{crit}} &= -V''(p''-p') + \gamma\mathscr{A} \\ &= \tfrac{1}{3}\gamma\mathscr{A} = \frac{4\pi r_c^2\gamma}{3}\end{aligned} \tag{5.8.7}$$

where r_c is the critical radius.

It can be shown (Exercise 5.8.1) that for larger or smaller droplets

$$(\Delta F)_{T,V} = -n''RT\ln\frac{p'}{p^0} - V''(p^0 - p') + \gamma\mathscr{A} \tag{5.8.8}$$

where p^0 is the equilibrium vapour pressure of the liquid over a plane interface at the temperature of the system. Provided the degree of supersaturation (p'/p^0) is not too large we may neglect the second term on the right and write (compare Adamson 1967, p. 376; 1976, p. 373):

$$(\Delta F)_{T,V} = -n''RT\ln\frac{p'}{p^0} + \gamma\mathscr{A}. \tag{5.8.9}$$

The first term is negative under supersaturation conditions and when it becomes large enough it dominates over the second term. In terms of the radius of the drop we have $n'' = \tfrac{4}{3}\pi r^3/\bar{V}''$ and so

$$(\Delta F)_{T,V} = -\frac{4}{3}\frac{\pi r^3}{\bar{V}''}RT\ln\frac{p'}{p^0} + 4\pi r^2\gamma. \tag{5.8.10}$$

Then

$$\begin{aligned}\left(\frac{\partial \Delta F}{\partial r}\right)_{T,V} &= -\frac{4\pi r^2}{\bar{V}''}RT\ln\frac{p'}{p^0} + 8\pi r\gamma \\ &= 4\pi r^2\left(\frac{2\gamma}{r} - \frac{RT}{\bar{V}''}\ln\frac{p'}{p^0}\right)\end{aligned} \tag{5.8.11}$$

which is obviously zero when Kelvin's equation (eqn (5.5.12)) is satisfied. Note that this is a *maximum* value for ΔF corresponding to an unstable equilibrium and that these positive values of ΔF correspond to non-spontaneous processes. The dependence of ΔF on r is plotted schematically in Fig. 5.8.1. The initial rise is a quadratic function where $\gamma\mathscr{A}$

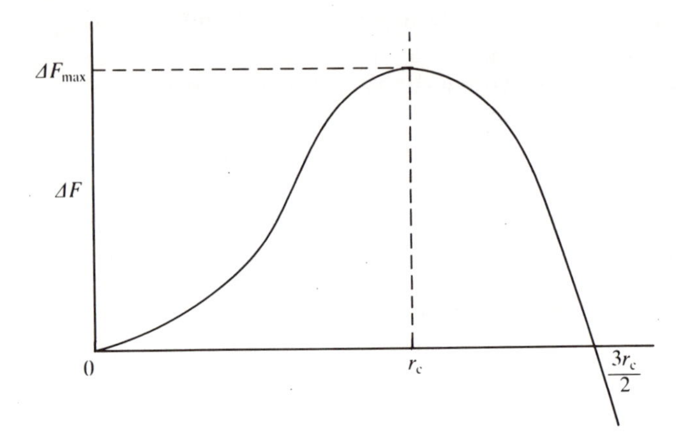

FIG. 5.8.1. Variation with radius of the free energy of formation of a nucleus in a supersaturated homogeneous phase.

dominates until r becomes large enough for the first term in eqn (5.8.10) to take over.

The value of ΔF_{max} is given by (Exercise 5.8.2).

$$\Delta F_{max} = \Delta F_{r=r_c} = \frac{16\pi\gamma^3 M^2}{3\rho_l^2(RT\ln(p'/p^0))^2} \tag{5.8.12}$$

where M is the molar mass and ρ_l the density of the liquid.

ΔF_{max} can be regarded as an activation energy for formation of a nucleus of sufficient size. The rate of nucleation, ξ, is, therefore, given by the simple Arrhenius-type relation:

$$\xi = Z\exp(-\Delta F_{max}/kT) \tag{5.8.13}$$

where Z is determined by the collision frequency of the gas molecules and so increases with $(p')^2$. A detailed calculation of the terms involved in eqn (5.8.13) is given by Adamson (1967, Table V11-3; 1976, p. 376) who shows that, as the degree of supersaturation (p'/p^0) gradually increases, a point is reached (at about $p'/p^0 \simeq 4.2$ for water at 0 °C) where the value of ξ suddenly increases extremely rapidly. This turns out to be very close to the experimentally observed degree of supersaturation which induces formation of a fog in water vapour free of dust particles or other nucleation sites.

The same effect can be achieved by lowering the temperature, which also increases the degree of supersaturation. Over the early stages of temperature reduction, the rate of condensation increases relatively slowly with decrease in temperature, but eventually a temperature is reached where the chance of forming a nucleus of critical radius, (a *critical* nucleus) by spontaneous fluctuations in molecular association increases very rapidly with a very small further decrease in temperature. The temperature at which this rapid change occurs is called the *homogeneous nucleation temperature.*

The theoretical prediction of homogeneous nucleation temperature and nucleation behaviour in general is a lively and complex area. Two major problems are the description of the properties of small clusters of molecules in other than thermodynamic terms (the macroscopic Kelvin equation may not be strictly applicable to such small clusters), and the application of statistics to random fluctuations in the size of molecular clusters. On this point it is instructive to calculate the radius of a critical nucleus. For the experiment quoted above $p'/p^0 \doteqdot 4.2$ and from the Kelvin equation (i.e. setting eqn (5.8.11) to zero):

$$r_c = \frac{2\gamma \bar{V}''}{RT \ln(p'/p^0)} \doteqdot 8 \times 10^{-10}\,\mathrm{m} = 0.8\,\mathrm{nm}.$$

This corresponds to a sphere of volume $2.1 \times 10^{-27}\,\mathrm{m}^3$, which would contain about 70 molecules of water. The next section examines in more detail the extent to which we can expect the Kelvin equation to hold under such extreme conditions.

Exercises. 5.8.1 Establish eqn (5.8.8) using the fact that $\mu_l(p'') = \mu_l^0 + \int_{p^0}^{p''} \bar{V}''\,\mathrm{d}P$ at constant temperature. μ_l^0 is the standard chemical potential of the liquid at pressure p^0 which can be taken as the equilibrium vapour pressure over a flat surface at the temperature in question. $\bar{V}'' = \bar{V}_l$ is the molar volume of the liquid.

5.8.2 Verify eqn (5.8.12).

5.8.3 Show that there is no energy barrier to growth for $r > 3r_c/2$.

5.9 Limits of applicability of the Kelvin and Young–Laplace equations

It is apparent from the previous discussion that many of the most interesting applications of the Kelvin equation and, to a lesser extent, the Young–Laplace equation, are to surfaces with radii of curvature well below 1 μm. Although it would appear reasonable to doubt the applicability of either equation to such highly curved surfaces (especially when the radii of curvature approach molecular dimensions) some recent experiments (Fisher 1982) suggest that these equations remain valid down to very small values of *r*. This is a rather surprising result because the macroscopic concepts of density, surface tension, and radius of curvature, all used in the derivation of the Kelvin equation, may not be applicable to the very small droplets or menisci encountered (Guggenheim 1940; Everett and Haynes 1973; Melrose 1966, 1972). Furthermore, three key assumptions, used in the derivation or application of both equations, may break down at very small radii. These are:

(a) that the liquid is incompressible (Melrose 1966);

(b) that higher-order terms in the Young–Laplace equation can be neglected (the Young–Laplace equation is only the first term in a series of terms containing progressively higher-order curvatures); and

(c) that, for capillary-held liquids, the presence of a solid surface does not affect the properties of the liquid–vapour interface. (Broekhoff and de Boer 1967, 1968; Deryaguin *et al.* 1976; Philip 1977). It should also be noted that although the derivations allow γ to vary with the radius of curvature, in an actual calculation we are normally forced to use the macroscopically measured value of γ (at 1 atm pressure) in the absence of other data.

Current theory predicts that, for simple liquids at least, all of the above assumptions are either valid or introduce calculable, minor changes to the predictions of the Kelvin equation for radii above a few molecular diameters so long as the macroscopic concepts of surface tension, density, radius of curvature, etc. can still be used at such radii (Fisher and Israelachvili 1981b). It has thus been up to the experimentalist to test the range of validity of the Kelvin and Young–Laplace equations in this critical region. Recent experiments have, in fact, shown that, for simple liquids, both the Kelvin and Young–Laplace equations hold down to extraordinarily small radii of curvature, sometimes as low as one or two molecular diameters.

For the Kelvin equation, direct measurement of the capillary condensation of cyclohexane between contacting crossed mica cylinders, using white light interferometry to estimate the 'mean' radius of curvature $r_m(=J^{-1})$ of the meniscus (eqn (5.2.14)), has yielded plots of $\ln(p/p^0)$ against $1/r_m$ that are linear to the lower limit of measurement ($r_m \simeq -4$ nm) i.e. a meniscus of about seven times the molecular diameter (Fig. 5.9.1); Fisher and Israelachvili, 1981b).

For the Young–Laplace equation, the force required to pull two such mica cylinders apart in the presence of an annulus of capillary-condensed liquid was measured (Fisher and Israelachvili 1981a). It was shown (eqn (5.7.7)) that the force to pull two equal spheres apart in such circumstances is just $2\pi R\gamma$, where R is the radius of each sphere—i.e. the force is independent of the amount of capillary condensate. Similarly, for a liquid condensed between contacting crossed cylinders the force is $4\pi R\gamma \cos\theta$, where θ is the contact angle. Thus the force should be independent of relative vapour pressure so long as both the Kelvin and Young–Laplace equations hold. The results for cyclohexane (Fig. 5.9.2) and other organic liquids show quite clearly that major deviations are not encountered for radii above about 0.5 nm, i.e. *one* molecular diameter.

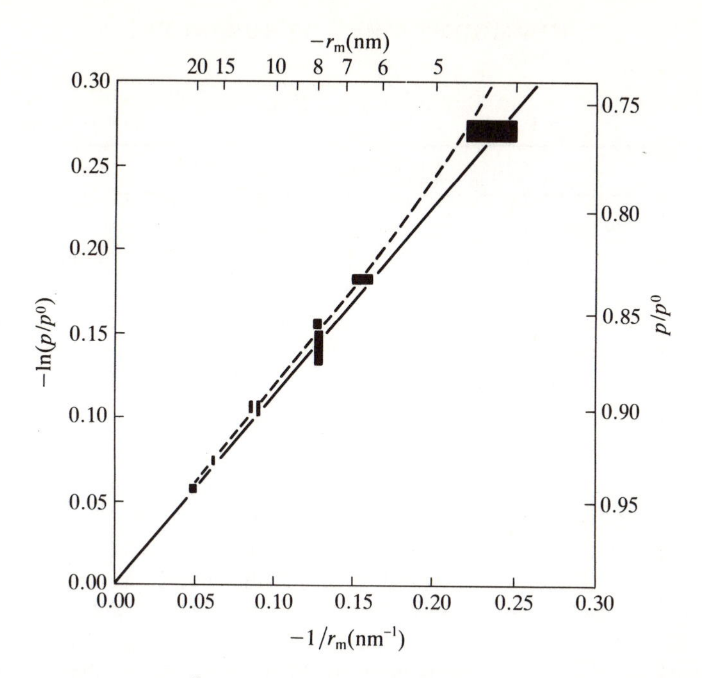

FIG. 5.9.1. Plot of $\ln(p/p^0)$ against $1/r_m$ for the capillary condensation of cyclohexane between crossed mica cylinders. (We will examine this technique in more detail in Chapter 7). The experimental points are shown to indicate estimated error. The theoretical line is drawn from the 'exact' Kelvin equation derived in Exercise 5.5.2. (From Fisher and Israelachvili 1981b, with permission.)

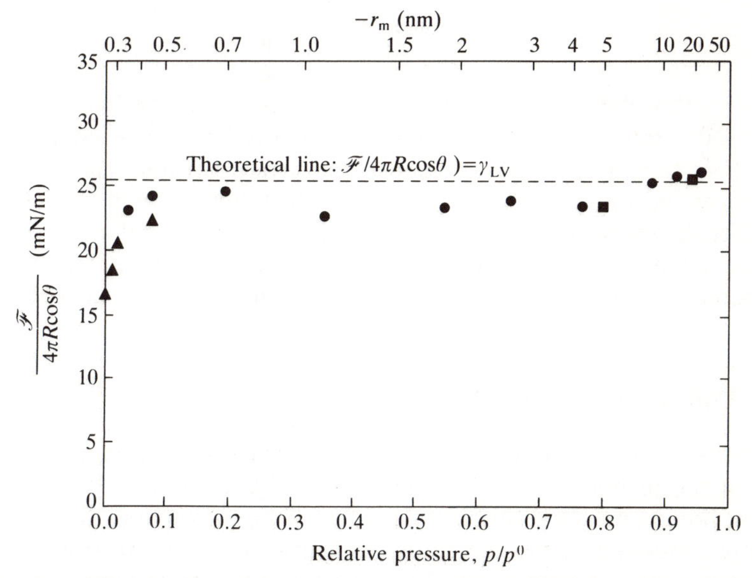

FIG. 5.9.2. Force of adhesion due to Laplace pressure in capillary-condensed cyclohexane between crossed mica cylinders. (From Fisher and Israelachvili 1981a, with permission.)

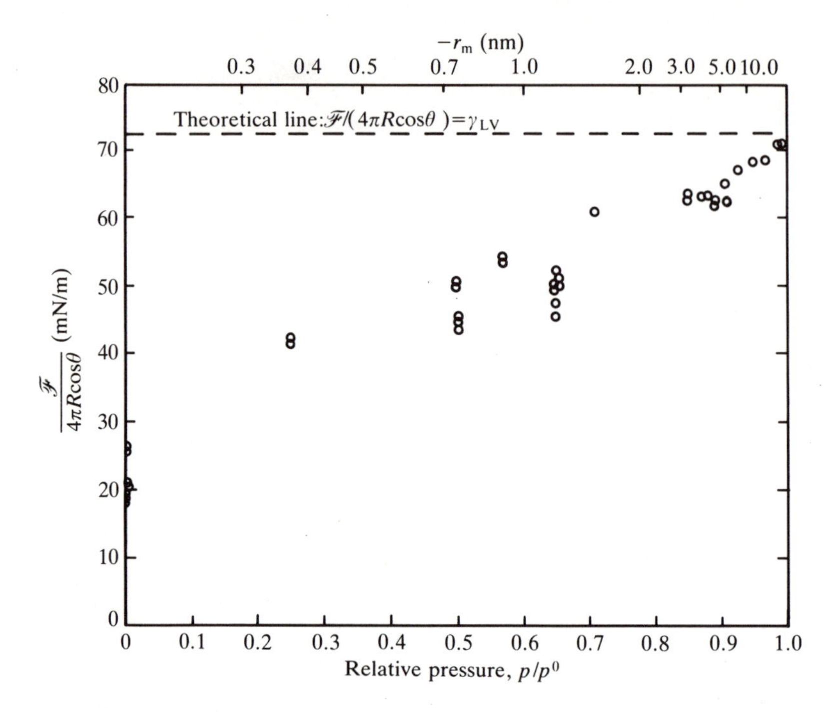

FIG. 5.9.3. Force of adhesion due to Laplace pressure in capillary-condensed water between crossed mica cylinders. (From Fisher and Israelachvili 1981a, with permission.)

Water, though, is a very different case (Fig. 5.9.3), perhaps because many water molecules are required to allow the long-range interactions that affect such macroscopic properties as surface tension. The frequent references in the literature to an empirical lower limit of relative humidity, below which the capillary condensation of water does not affect particle adhesion (Zimon 1969; Visser 1976) may be related to the breakdown of the Kelvin and Young–Laplace equations revealed by Fig. 5.9.3.

5.10 Contact angle and wetting behaviour

One of the few measurable parameters in surface science is the *contact angle* at the junction between three phases, e.g. at the contact line of a drop of liquid on a solid surface in an atmosphere of vapour. The contact angle is a measure of the competing tendencies of the drop to spread so as to cover the solid surface and to round up so as to minimize its own area. Rounding up may seem the natural process, since it also reduces the area of the liquid–solid interface but that neglects the fact that the solid–vapour interface is thereby *increased.*

The contact angle measures the *wetting* tendency of a liquid on a solid or another liquid, and is thus of particular importance in processes like fabric treatment (conditioning and dyeing), insecticide application, and mineral flotation, where wetting is important. The contact angle also determines a boundary condition for the calculation of meniscus shapes from the Young–Laplace equation. Comparison of calculated and actual shapes of sessile (sitting) drops and pendant (hanging) drops or bubbles is one good method of determining the surface tension of a liquid (section 5.12). Contact angle is also an important parameter in heterogeneous nucleation (i.e. nucleation of a new fluid phase in the presence of a solid surface; section 5.10.4).

5.10.1 *Adhesion, cohesion, and wetting*

The existence of a contact angle between a liquid and a solid (section 5.7) is a matter of considerable importance in surface chemistry and a variety of introductory and more advanced treatments of the salient ideas are available (see Aveyard and Haydon 1973; Davies and Rideal 1963; Adamson 1967; Hiemenz 1977). Here we will confine ourselves to a few concepts that prove useful in discussing adhesion and flotation behaviour and that can be linked to the calculation of the van der Waals (dispersion) forces in the previous chapter.

The reversible work, per unit area, required to break a column of liquid into two parts, creating two new *equilibrium* surfaces, and to separate the parts beyond the range of their forces of interaction is (from Fig. 5.10.1(a)):

$$W_{\mathrm{AA}} = 2\gamma_{\mathrm{AV}} \tag{5.10.1}$$

where γ_{AV} is the surface tension of liquid A in contact with its vapour. This is called the *work of cohesion* of substance A (per unit area).

By the same token, the work of adhesion per unit area between two different immiscible liquids A and B is given by the Dupré equation:

$$W_{\mathrm{AB}} = \gamma_{\mathrm{AV}} + \gamma_{\mathrm{BV}} - \gamma_{\mathrm{AB}} \tag{5.10.2}$$

since two new liquid–vapour interfaces are created and the original interface disappears. In these equations γ_{AV} represents the quantity $(\partial F/\partial \mathscr{A})_{T,V,n_i}$ (from eqn (5.3.31)). In what follows we will be concerned only with planar interfaces and for such systems many workers in the field prefer to introduce the more familiar (to chemists) Gibbs free energy function. This corresponds to using an alternative definition of γ (see e.g. Aveyard and Haydon 1973) with the pressure now constant throughout

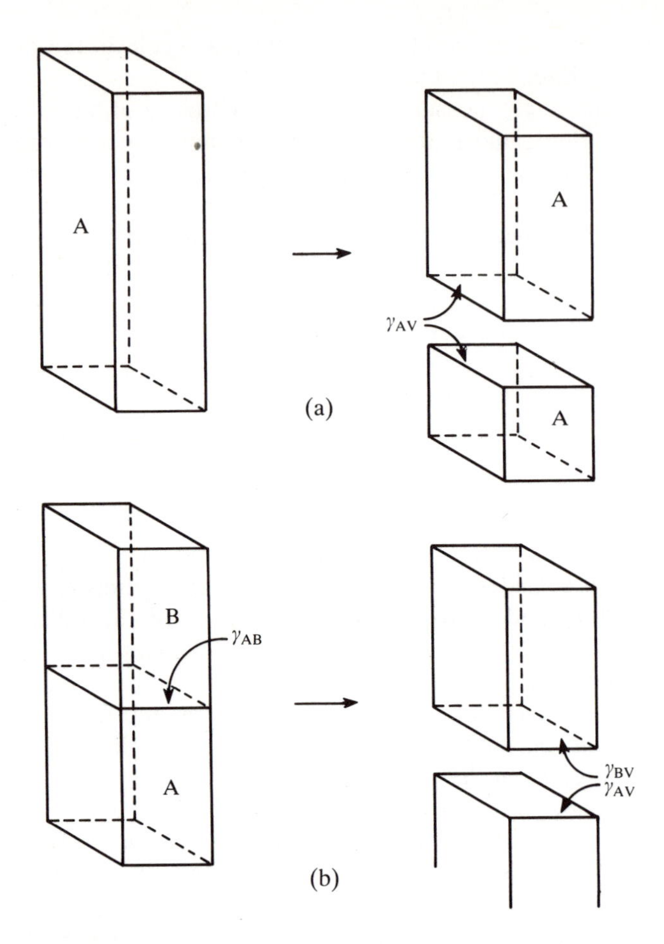

FIG. 5.10.1. Illustrating (a) the work of cohesion W_{AA} in a liquid and (b) the work of adhesion W_{AB} between two different liquids. Breaking the column creates two new interfaces in the first case. In the second case it also destroys one.

the system:

$$\gamma_{AV} = (\partial G_A / \partial \mathscr{A})_{p,T,n_i}. \tag{5.10.3}$$

In those cases where the cohesive and adhesive interactions are principally due to dispersion forces we should be able to estimate these work terms directly from the theory developed in Chapter 4. We will find in section 5.11.1 that even when other very large forces are involved (like metal bonding and hydrogen bonding) it is possible to separate out the dispersion force contribution and to use it to obtain useful correlations between the behaviour of one system and another.

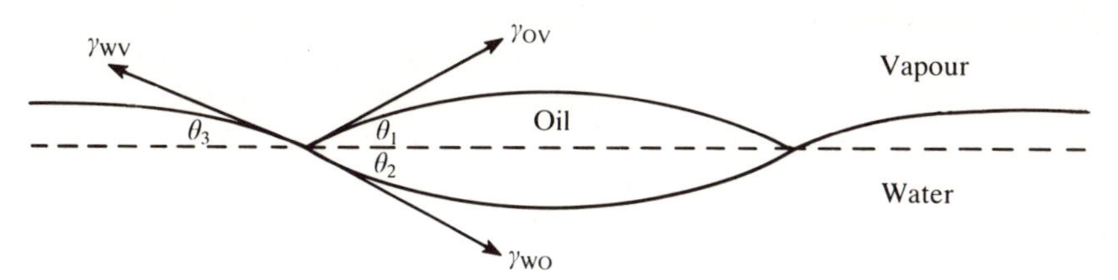

FIG. 5.10.2. Initial shape of an oil lens on a water surface. The shape may change with time if the oil is able to spread and/or if the two liquids are to some extent mutually soluble.

When an oil drop is placed on a clean water surface it may form a lens as shown in Fig. 5.10.2. The overall shape is described by the Young–Laplace equation (eqn (5.2.17)), subject to the boundary condition of a force balance at the three-phase contact line:

$$\gamma_{WV} \cos \theta_3 = \gamma_{OV} \cos \theta_1 + \gamma_{WO} \cos \theta_2. \qquad (5.10.4)$$

A more interesting version of this equation applies if the lower liquid is very dense compared with the upper liquid (like water on mercury). In that case both θ_2 and θ_3 approach zero and we have

$$\gamma_{MV} \doteqdot \gamma_{MW} + \gamma_{WV} \cos \theta_1 \qquad (5.10.5)$$

where M refers to mercury, W to water, and V to the equilibrium vapour. This corresponds to the situation one would expect if the liquid were sitting on a solid (non-deformable) substrate. Equation (5.10.5) then takes the form

$$\gamma_{SV} = \gamma_{SL} + \gamma_{LV} \cos \theta \qquad (5.10.6)$$

where S, V, and L refer to the solid, vapour, and liquid, respectively. Equation (5.10.6) was put forward by Thomas Young in 1805 and goes by his name. Its simplicity belies the considerable argument it has generated over the last couple of hundred years. The argument concerned such questions as whether it made sense to talk about a (macroscopically measured) contact angle on a real solid surface when, at the microscopic level, the surface would be rough and actual molecular contact between the phases occurs over a region, rather than along a line. There is also the problem of the influence of the vertical component of the force ($\gamma_{LV} \sin \theta$) on the underlying solid. Can the solid be considered in equilibrium when subjected to this stress? We will have more to say on such questions in section 5.10.3.

For the moment we will accept eqn (5.10.6) subject to the proviso that the γ values appearing in it are *equilibrium values* of surface free energy, obtained after the adsorption processes have ceased. In particular, at the solid–vapour interface, the surface free energy will not be that of the pure solid, γ_s^0 but will be some lower value due to adsorption of some vapour. The lowering of surface free energy (section 5.4.2) due to adsorption of a film of adsorbate is

$$\Pi_e = \gamma_s^0 - \gamma_{SV}. \tag{5.10.7}$$

It has the dimensions of an energy per unit area or a force per unit length but is often interpreted as a two-dimensional pressure called the *spreading pressure* of the adsorbate. (In simple cases on a liquid surface it can be modelled by a two-dimensional analogue of the ideal gas equation with the translational kinetic energy of the adsorbed molecules being kT (instead of $\frac{3}{2}kT$ for the three-dimensional case). (See e.g. Aveyard and Haydon 1973.)

In terms of the surface energy γ_s^0 of the pure solid (in contact with its own vapour but not with L) we have:

$$\gamma_{LV} \cos\theta = \gamma_s^0 - \Pi_e - \gamma_{SL}. \tag{5.10.8}$$

The work of adhesion W_{SL}, between a solid and a liquid can be defined, by analogy with eqn (5.10.2), as:

$$W_{SL} = \gamma_s^0 + \gamma_{LV} - \gamma_{SL} \tag{5.10.9}$$

which, using eqn (5.10.8) becomes

$$W_{SL} = \gamma_{LV}(1 + \cos\theta) + \Pi_e. \tag{5.10.10}$$

An important question relating to Fig. 5.10.2 is whether the liquid lens will spread out over the water surface or remain in the form of a lens, i.e. will the oil wet the water? In the limiting case where θ_1, θ_2, and θ_3 approach zero, the lens becomes a film, and spreading is determined by the quantity:

$$\mathrm{d}G = \left(\frac{\partial G}{\partial \mathscr{A}_{WV}}\right) \mathrm{d}\mathscr{A}_{WV} + \left(\frac{\partial G}{\partial \mathscr{A}_{OV}}\right) \mathrm{d}\mathscr{A}_{OV} + \left(\frac{\partial G}{\partial \mathscr{A}}\right)_{WO} \mathrm{d}\mathscr{A}_{WO}. \tag{5.10.11}$$

If some spreading occurs we have

$$\mathrm{d}\mathscr{A}_{WO} = \mathrm{d}\mathscr{A}_{OV} = -\mathrm{d}\mathscr{A}_{WV}$$

since the spreading film increases its contact both with the water and its own vapour but covers the WV interface. Also $(\partial G/\partial \mathscr{A}_{WV})_{P,T} = \gamma_{WV}$ etc., from eqn (5.10.3). For spreading to occur spontaneously we require $\mathrm{d}G$ to be negative at constant temperature and pressure when $\mathrm{d}\mathscr{A}_{WO}$ is positive. This will be so when the coefficient $S_c = -(\mathrm{d}G/\mathrm{d}\mathscr{A}_{WO})$ is positive.

S_c is called the *spreading coefficient* and from eqn (5.10.11):

$$S_c = \gamma_{WV} - \gamma_{OV} - \gamma_{WO}. \tag{5.10.12}$$

From the definitions of work of adhesion and cohesion given in eqns (5.10.1) and (5.10.2) it is obvious that:

$$S_c = W_{OW} - W_{OO}. \tag{5.10.13}$$

The spreading coefficient measures the difference between the adhesion of oil to water and the cohesion of the oil. It is in this sense that we use the term *wetting*. A liquid will wet another material (liquid or solid) if its own work of cohesion is less than the work of adhesion between it and the substrate. A curious example is that of benzene on water. Here S_c is positive for the benzene–water interface but negative for benzene in contact with a saturated solution of benzene in water. Thus, benzene will initially spread on water, then round up into lenses as the water becomes saturated with benzene (Aveyard and Haydon 1973).

5.10.2 *Meniscus shape and wetting*

The meniscus that forms at the junction between a water surface and the wall of a glass vessel is an indication of the spreading or wetting tendency of water on glass. The water spreads up the vessel wall and is limited in its ascent only by the gravitational force it must overcome.

The mathematics describing the shape of the meniscus is, in principle, very simple. The hydrostatic pressure at any point on the meniscus surface ($\rho g L$, where L is the height above the wide, free liquid surface) is balanced by the pressure due to the curvature at that point. That is:

$$\rho g L = \gamma\left(\frac{1}{r_1} + \frac{1}{r_2}\right). \tag{5.10.14}$$

However, when r_1, r_2, and L are expressed in terms of Cartesian coordinates, it becomes apparent that eqn (5.10.14) is deceptively simple. The full equation becomes:

$$\frac{(\rho'' - \rho')g(z - z_0)}{\gamma} = \frac{\left\{1 + \left(\frac{\partial z}{\partial x}\right)^2\right\}\frac{\partial^2 z}{\partial y^2} + \left\{1 + \left(\frac{\partial z}{\partial y}\right)^2\right\}\frac{\partial^2 z}{\partial x^2} - 2\frac{\partial z}{\partial x}\frac{\partial z}{\partial y}\frac{\partial^2 z}{\partial x\,\partial y}}{\left\{1 + \left(\frac{\partial z}{\partial x}\right)^2 + \left(\frac{\partial z}{\partial y}\right)^2\right\}^{3/2}} \tag{5.10.15}$$

where $(\rho'' - \rho')$ is the difference in densities between the two phases that form the interface, and gravity is taken as acting in the z direction. If the left-hand side is zero (no pressure across the interface as in the open soap film (Fig. 5.2.8) the solution of eqn (5.10.15) is still a classic problem (Plateau's problem) in pure mathematics (Nitsche 1975) and has only recently been analysed in full (see Almgren and Taylor (1976)). If the left-hand side is non-zero, numerical methods of integration (such as the finite element method (Orr *et al.* 1975) are the only recourse, unless the surface has some high-order symmetry.

If the surface has axial symmetry, (about z say) as is often the case for surfaces that will interest us, eqn (5.10.15) becomes:

$$\frac{(\rho''-\rho')g(z-z_0)}{\gamma}=\frac{\dfrac{\mathrm{d}^2z}{\mathrm{d}a^2}}{\left\{1+\left(\dfrac{\mathrm{d}z}{\mathrm{d}a}\right)^2\right\}^{3/2}}+\frac{\dfrac{\mathrm{d}z}{\mathrm{d}a}}{a\left\{1+\left(\dfrac{\mathrm{d}z}{\mathrm{d}a}\right)^2\right\}^{1/2}} \tag{5.10.16}$$

where a is measured axially. (See Exercises 5.10.4–6.)

Equation (5.10.16) has some approximate analytical solutions, such as those first advanced by Lord Rayleigh in 1915–1916 for menisci in very wide or very narrow capillary tubes. Generally, though, it must be solved numerically. This is a relatively easy task for modern computers, and extensive tables of solutions, together with the computer programmes that generated them, have been published by Hartland and Hartley (1976). These replace the much earlier, *hand-calculated* tables of Bashforth and Adams (1883). The reader may be interested to know that the first approximate analytical solution of eqn (5.10.16) was by Young in 1805; he chose to express his solution entirely in words without the aid of mathematical symbols.

The boundary conditions for integration of eqns (5.10.15) or (5.10.16) are established by the balance of forces at the three-phase contact line—for example, by the contact angle at the container wall. For water in a clean glass container, the equilibrium contact angle is zero and the boundary condition becomes an asymptotic one; fortunately this does not create any particular mathematical problem.

Water wets many surfaces very well and this has some important consequences. Sometimes we want to prevent wetting (as in 'waterproofing' a fabric or floating a mineral particle with an air bubble). In other cases, wetting must be promoted—as in the application of insecticides to 'greasy' (hydrophobic) animal fur or plant leaves. These effects are achieved by modifying the surface energies γ_{SV}, γ_{SL}, and γ_{LV} by the adsorption of suitable materials (*surfactants*; see Chapter 10.)

5.10.3 *Sessile and pendant drops and bubbles*

Wetting and non-wetting behaviour are most obvious when a drop of liquid is placed on a solid surface, or a gas bubble adheres to the surface of a solid in contact with a liquid (Fig. 5.10.3). The shape of such a *sessile* (sitting) drop or bubble is governed by the same principles as those governing the shape of a meniscus (section 5.10.2). That is, the profile is described by the Young–Laplace equation (eqn (5.2.17)), subject to the boundary conditions of force balance at the three-phase contact line, i.e. the contact angle. Measurements of the profile can thus be used to calculate the surface tension of the liquid if the contact angle is known or, more commonly, to calculate the contact angle for a liquid with known surface tension (Alexander and Hayter 1971; Ambwani and Fort 1979; Ramakrishnan *et al.* 1976; Fisher 1979).

The relation between the contact angle and the interfacial energies derived earlier (eqn (5.10.6)) from a force–balance argument, can be derived directly from thermodynamic arguments as follows. Consider an axially symmetric sessile drop of liquid (2)(Fig. 5.10.4), on a planar horizontal substrate (1) in an atmosphere of gas or another liquid (3) at fixed temperature and constant volume of liquid.

Suppose the droplet spreads slightly ($a \rightarrow a + \mathrm{d}a$) at constant volume. What is the change, $\mathrm{d}F$, in the Helmholtz free energy? The area changes are:

$$\mathrm{d}\mathscr{A}_{12} = 2\pi a \, \mathrm{d}a \tag{5.10.17}$$

$$\mathrm{d}\mathscr{A}_{23} = \left(\frac{1}{R_1} + \frac{1}{R_2}\right) \mathrm{d}V + 2\pi a \, \mathrm{d}a \cos \theta. \tag{5.10.18}$$

The first term in eqn (5.10.18) comes from eqn (5.2.18) and the second is

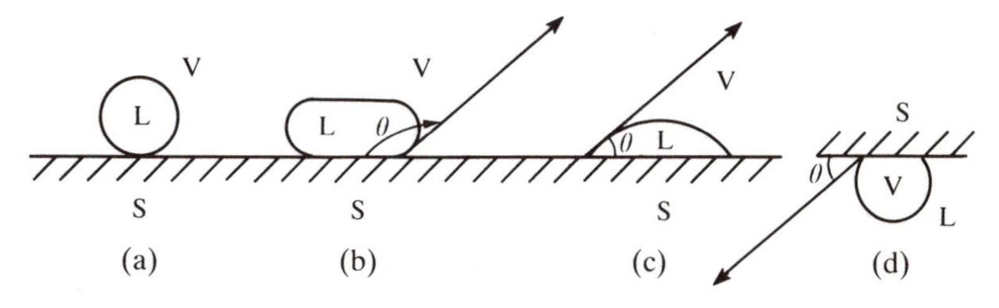

FIG. 5.10.3. Possible shapes of sessile drops or bubbles. (a) Small drops of completely non-wetting liquid of very high γ_{LV} (gravitational forces have little effect). (b) Larger drop of liquid showing partial non-wetting character ($\pi/2 < \theta < \pi$) and sagging under gravity. (c) Partially wetting drop ($0 < \theta < \pi/2$). (d) Sessile bubble with partial wetting of solid by liquid.

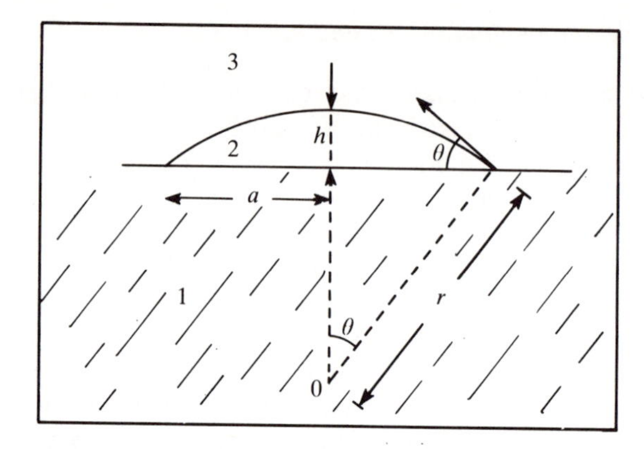

FIG. 5.10.4. Sessile drop with axial symmetry of liquid on a solid surface in a constant temperature chamber. For a sufficiently small drop, where changes in hydrostatic pressure with height are negligible, it can be shown that the solutions of the axisymmetric Young–Laplace equation (eqn (5.10.16)) yield a surface that has the shape of a spherical cap.

the area of the outer surface of the wedge illustrated in Fig. 5.10.5. Now the volume of the drop is constant ($\mathrm{d}V = 0$) so

$$\begin{aligned} \mathrm{d}\mathscr{A}_{23} &= 2\pi a \,\mathrm{d}a \cos\theta \\ &= \mathrm{d}\mathscr{A}_{12} \cos\theta. \end{aligned} \tag{5.10.19}$$

Thus

$$\begin{aligned} \mathrm{d}F &= \gamma_{13}(-\mathrm{d}\mathscr{A}_{12}) + \gamma_{12}(\mathrm{d}\mathscr{A}_{12}) + \gamma_{23}\,\mathrm{d}\mathscr{A}_{23} \\ &= (\gamma_{23}\cos\theta + \gamma_{12} - \gamma_{13})\,\mathrm{d}\mathscr{A}_{12}. \end{aligned} \tag{5.10.20}$$

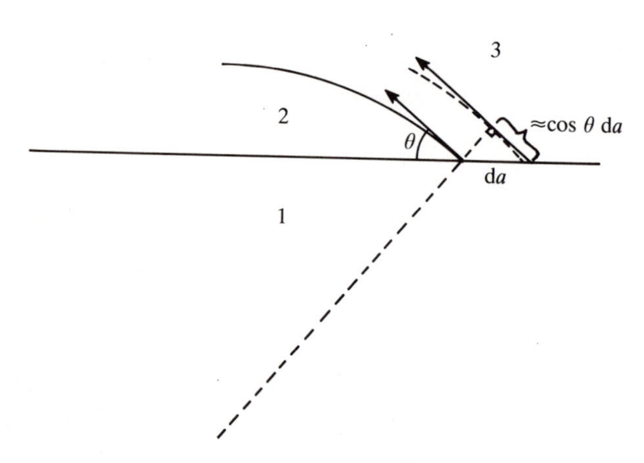

FIG. 5.10.5. Changes at the three-phase contact line for slight changes in the area of (1)–(2) contact for the sessile drop in Fig. 5.10.4.

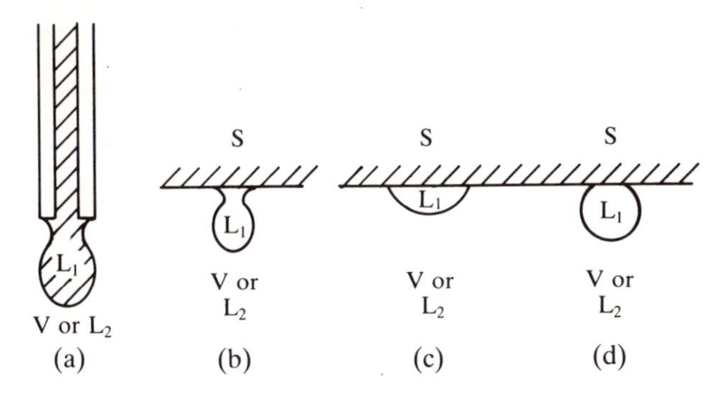

FIG. 5.10.6. Possible shapes of pendant (hanging) drops. (a) and (b) are wetting, the solid (c) is partially wetting $(0<\theta<\pi/2)$, and (d) is partially non-wetting $(\pi/2<\theta<\pi)$. Liquid L_1 is assumed to have a *higher* density than L_2 so that only the surface energies prevent the drop from falling.

Mechanical equilibrium will be established when $dF/d\mathscr{A}_{12}=0$, and this is so when

$$\cos\theta=(\gamma_{13}-\gamma_{12})/\gamma_{23} \qquad (5.10.21)$$

which is Young's equation (5.10.6).

Surface tensions are more commonly calculated from the shapes of pendant drops or bubbles (Fig. 5.10.6) rather than from sessile shapes. The detailed procedures for such measurements are discussed in many reference works and review articles. (See particularly Padday 1969; Alexander and Hayter 1971; Ambwani and Fort 1979.) Case (a) in Fig. 5.10.6 is important in connection with measurements on the dropping mercury electrode, which will be discussed in Chapter 6.

5.10.4 *Heterogeneous nucleation*

We examined in section 5.8 the criteria for homogeneous nucleation and found that it involved a significant free energy of activation for formation of the critical nucleus. This nucleus had a radius r_c and was in *unstable equilibrium,* i.e. it was at a (local) maximum in the free energy. Nucleation is much easier to induce if we provide a surface on which the new phase can begin to grow. We want now to examine this process in terms of the growth of a droplet like that shown in Fig. 5.10.4.

Equation (5.8.5) still holds for the interface between the liquid and the vapour but now, if the droplet grows there is a decrease in the area of the solid–vapour interface and an increase in the area of the solid–liquid interface. These additional contributions to the free energy must also be included. To proceed we need expressions for the volume and area of the (spherical) cap in terms of r and θ (Fig. 5.10.4). These are (Exercise 5.10.1):

$$V_L = \frac{\pi r^3}{3}(1 - \cos\theta)^2(2 + \cos\theta) \tag{5.10.22}$$

and

$$\mathscr{A}_{LV} = 2\pi r^2(1 - \cos\theta). \tag{5.10.23}$$

The free energy change involved in forming this cap is (from eqn (5.8.9)):

$$(\Delta F)_{T,V} = -\frac{\pi r^3}{3\bar{V}_L}(1 - \cos\theta)^2(2 + \cos\theta)RT \ln\frac{p'}{p^0} + \gamma_{LV}\cdot 2\pi r^2(1 - \cos\theta) + \pi r^2(1 - \cos^2\theta)(\gamma_{SL} - \gamma_{SV}) \tag{5.10.24}$$

where the last term accounts for the coverage of the solid surface.

In the neighbourhood of the critical radius ($r = r_c$) the surface free energies will obey Young's equation eqn (5.10.21) because $(\mathrm{d}F/\mathrm{d}\mathscr{A})$ is zero at that point. Equation (5.10.24) then becomes:

$$(\Delta F)_{T,V} = -\frac{\pi r^3}{3\bar{V}_L}(1 - \cos\theta)^2(2 + \cos\theta)RT \ln\frac{p'}{p^0} + \pi r^2\gamma_{LV}(1 - \cos\theta)^2(2 + \cos\theta). \tag{5.10.25}$$

Furthermore, at the critical radius p'/p^0 is given by the Kelvin equation so that eqn (5.10.25) reduces to

$$(\Delta F)_{crit} = 4\frac{\pi}{3}r_c^2\gamma_{LV}f(\theta) \tag{5.10.26}$$

where

$$f(\theta) = (1 - \cos\theta)^2(2 + \cos\theta)/4. \tag{5.10.27}$$

As in the case of homogeneous nucleation (eqn (5.8.7)), the first term in eqn (5.10.25) is the driving term and at the critical radius it is two-thirds as big as the second term. The only difference is the contact angle function $f(\theta)$, which measures the relative magnitude of $(\Delta F)_{hetero}/(\Delta F)_{homo}$ at their critical values: compare eqn (5.8.7) with eqn (5.10.26). For $\theta = \pi$, $f(\theta) = 1$, so that a completely non-wet surface has no effect on nucleation. For $\theta = 0$, $f(\theta) = 0$ and there is no activation energy for nucleation: supersaturation cannot occur if the surface is completely wet by the liquid. (This assumes, of course, that the radius of curvature of the solid surface is much larger than r_c.)

It should be emphasized that the above treatment does not assume that the droplet is in equilibrium with the vapour for any r value. We rely on eqn (5.10.20) and the fact that $(\mathrm{d}F/\mathrm{d}\mathscr{A}) = 0$ even for an unstable equilibrium in order to justify the use of Young's equation. This is important because the force balance argument for Young's equation is not very satisfactory on a solid surface where unbalanced stresses may be taken up by the solid; the derivation of eqn (5.10.21) requires only that the solid be rigid.

Exercises. 5.10.1 The volume of a spherical cap (such as that shown in Fig. 5.10.4 is given by $V = \pi h(3a^2 + h^2)/6$ and its upper surface area is $\pi(a^2 + h^2)$. Use these relations to establish eqns (5.10.22) and (5.10.23).

5.10.2 Consider a drop of liquid sitting on a solid surface as in Fig. 5.10.4. If the drop is small enough it forms a spherical cap. Use the equations obtained in Exercise 5.10.1 to show that if the drop is in (stable or unstable) equilibrium then $\gamma_{SV} - \gamma_{SL} = \gamma_{LV} \cos\theta$. (This is, of course, just a special case of the argument leading to eqn (5.10.20.)

5.10.3 Use eqn (5.10.20) to show that if $\gamma_{13} > \gamma_{12} + \gamma_{23}$ then the drop will spontaneously spread. Show also that, no matter what the initial contact angle, if $\gamma_{13} < \gamma_{12} + \gamma_{23}$, the drop will move in such a way as to establish, at equilibrium, a contact angle given by the Young equation.

5.10.4 Consider the two-dimensional curve shown in Fig. 5.10.7. $P_{(x,z)}$ is a point on the curve and θ is the tangent angle ($-\tan\theta = dz/dx$) at P as shown. Q is a neighbouring point, an infinitesimal distance ds along the curve in the direction shown. Show that the radius of curvature at P is given by

$$\frac{1}{r} = \frac{d\theta}{ds}$$

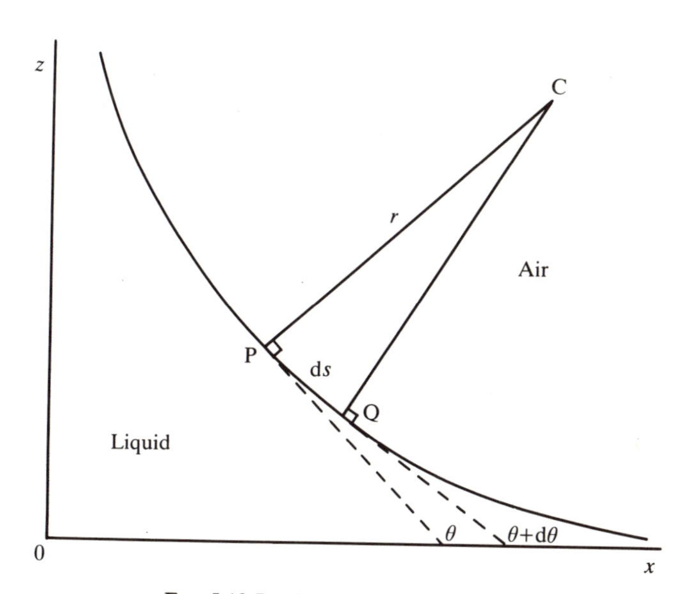

FIG. 5.10.7. Refer to Exercise 5.10.4.

and in Cartesian coordinates by

$$\frac{d^2z}{dx^2}\Big/\left[1+\left(\frac{dz}{dx}\right)^2\right]^{3/2}.$$

[Hint. Use

$$\frac{d\theta}{ds}=\frac{dx}{ds}\cdot\frac{d\theta}{dx} \quad \text{and} \quad \frac{ds}{dx}=-\sec\theta.\Big]$$

5.10.5 Suppose Fig. 5.10.7 represents the meniscus of an infinite air–liquid interface under gravity at a vertical plane wall at $x=0$. The height of the interface z at distance x from the wall is measured relative to the flat interface ($z=0$) at $x=\infty$. Show that $z(x)$ satisfies

$$\frac{\gamma}{r}=\rho_1 gz$$

where r is the radius of curvature of the interface at (x, z). Using the result of the previous exercise show that the tangent angle θ at (x, z) is given by

$$\cos\theta=-\left(1-\rho_1\frac{gz^2}{2\gamma}\right).$$

[Hint—use the result

$$\frac{d^2z}{dx^2}\Big/\left\{1+\left(\frac{dz}{dx}\right)^2\right\}^{3/2}=\frac{d}{dz}\left\{1+\left(\frac{dz}{dx}\right)^2\right\}^{-1/2}.\Big]$$

If the wall is wetting (i.e. the contact angle as measured through the liquid is zero) show that the meniscus height at the wall is

$$z=\left(\frac{2\gamma}{\rho_1 g}\right)^{1/2}.$$

5.10.6 (a) Setting the l.h.s. of eqn (5.10.16) to zero provides the solution to the zero-pressure-difference-surface illustrated in Fig. 5.2.8. Show that the solution to the resulting differential equation is $a\sin\theta=$ (a constant) where $\tan\theta=dz/da$ (Fig. 5.2.7).

(b) The zero-gravity surface (corresponding to small drops of high γ and/or low density) is obtained by setting the l.h.s. of eqn (5.10.16) equal to a constant (i.e. assuming that the variation in z is unimportant). Show that in this case the solution to the differential equation is $a/\sin\theta=$ (a constant). Prove that this equation is satisfied by a spherical surface.

[Hint:

$$\frac{1}{a}\frac{d(a\sin\theta)}{da}=\frac{d\sin\theta}{da}+\frac{\sin\theta}{a}.\Big]$$

5.11 Theoretical estimation of surface properties

5.11.1 *Surface and interfacial tension and energy*

From the above discussion it is apparent that many surface properties can be drawn into a unified conceptual framework if we can estimate, with reasonable accuracy, the interfacial energies γ_{SL}, γ_{SV}, and γ_{LV} since this will allow a calculation of the contact angle θ (from eqn (5.10.6)), the work of cohesion and adhesion and the spreading coefficient.

The most effective pioneering work in this area was done by Fowkes who extended an earlier suggestion of Girifalco and Good (1957). In his introductory review of the subject Fowkes (1965) shows the value of extracting from the surface or interfacial energy (which, incidentally he calls interfacial tension even for solids) the component due to dispersion (van der Waals) interaction, γ^{d}. He assumes that at the hydrocarbon–vapour interface it is the *only* contribution:

$$\gamma_{HV} = \gamma^{d}_{HV}. \tag{5.11.1}$$

Fowkes used the rather crude Hamaker theory to estimate surface free energies but with the analysis developed in Chapter 4 we should be able to improve on that. In order to calculate γ^{d}_{HV} from the dispersion force equations we have only to imagine the reverse process to that shown in Fig. 5.10.1(a). That is, we bring together two semi-infinite blocks of hydrocarbon (of unit area) from infinity until they are (almost) touching (separation L_c) and calculate the interaction energy per unit area, E_{HVH} where H stands for hydrocarbon and V for vacuum (or vapour), from eqn (4.7.43). Then

$$\gamma_{HV} = -\tfrac{1}{2}E_{HVH} = \frac{1}{2}\frac{A_{HVH}}{12\pi L_c^2}. \tag{5.11.2}$$

Unfortunately, to get a finite value out we must postulate a minimum (non-zero) value for L_c corresponding to the 'distance' between the bodies when they make contact. We noted in section 4.7.4 that the divergent behaviour as L approaches zero is a consequence of the failure of the heuristic method of van Kampen to take account of the 'graininess' of matter and that it is absent from the exact Lifshitz theory. Intuitively, we expect L_c in the more simplified theory to be of the order of a molecular radius. The values of A_{HVH} calculated from spectral data for the various hydrocarbons are given in Table 4.2 (in the column headed M/air/M) and using those values in eqn (5.11.2) with the experimental γ_{HC} values we find the values of L_c listed in Table 5.4. They are close to being constant and are certainly of the correct order of magnitude; the slight decrease with increasing chain length reflects the increase in density of hydrocarbons with molar mass.

Hough and White (1980) show that the critical length, L_c, calculated in this way is inversely proportional to the square root of the liquid density. This relationship can be exploited to obtain the values calculated in the last column of the table with only one adjustable parameter, the value of L_c for some reference hydrocarbon, calculated from the *measured* surface tension.

Table 5.4
Estimation of surface tension. (All but the last column come from Hough and White 1980, *with permission)*

Alkane	$10^{20}A_{HVH}$(J) (from Table 4.2)	$10^3\gamma_{HC}$ (J m^{-2})	$10^{-3}\rho_L$ (kg m^{-3}) at 20 °C	L_c(nm) (eqn 5.11.2)	$10^3\gamma_{HC}$ (theory†)
C_5H_{12}	3.75	16.05	0.6262	0.1758	15.90
C_6H_{14}	4.07	18.40	0.6603	0.1712	18.20
C_8H_{18}	4.50	21.62	0.7025	0.1660	21.40
$C_{10}H_{22}$	4.82	23.83	0.7300	0.163(8)	(23.83)
$C_{12}H_{26}$	5.03	25.35	0.7487	0.1622	25.50
$C_{14}H_{30}$	5.05	26.56	0.7628	0.1594	26.09
$C_{16}H_{34}$	5.23	27.47	0.7733	0.1587	27.39

† Calculated using $L_c = (L_c)_{n=10} \times (\rho/\rho_{n=10})^{1/2}$.

In Table 5.4 we have used the value for decane and the calculated values are then given by

$$\gamma_{HV} = \frac{A_{HVH} \times \rho}{24\pi \times (0.1638 \times 10^{-9})^2 \times 0.7300 \times 10^3)}$$
$$= 6.78 \times 10^{14} A\rho \qquad (5.11.3)$$

for A in joules and ρ in kg m^{-3}.

A much less objectionable procedure for avoiding the $L \to 0$ problem has been suggested by Hough and White (1980). They apply the cut-off to the wave vector Q (in eqn (4.7.36)), which makes more sense physically and leads to a thermodynamically self-consistent procedure for calculating interfacial free energies. It is, however, a great deal more difficult to apply than eqn (5.11.3).

In order to estimate the dispersion contribution to γ at other interfaces, Fowkes argues as follows. When two immiscible liquids (say mercury and hydrocarbon) are in contact, the dominant interaction between them is the dispersion (van der Waals) interaction and in the case of a hydrocarbon–liquid interface it may be regarded as the *only* important interaction. The surface tension of the pure hydrocarbon is reduced by the interaction that the surface hydrocarbon molecules have with the adjacent mercury atoms. Likewise, the surface tension of the mercury is reduced from its value in air (γ_M) to some lower value due to interactions between mercury atoms and the hydrocarbon molecules. Fowkes estimates the reduction in each case from the geometric mean of the dispersion component of γ (compare eqn (4.4.36)):

$$\begin{aligned}\text{For the hydrocarbon } \gamma_{HVM} &= \gamma_{HV}^0 - (\gamma_{HV}^d \gamma_M^d)^{1/2} \\ \text{and for mercury } \gamma_{MHV} &= \gamma_M^0 - (\gamma_{HV}^d \gamma_M^d)^{1/2}.\end{aligned} \tag{5.11.4}$$

Then the interfacial tension between mercury and hydrocarbon is

$$\gamma_{MHC} = \gamma_M^0 + \gamma_{HV}^0 - 2(\gamma_{HV}^d \gamma_M^d)^{1/2}. \tag{5.11.5}$$

Since $\gamma_{HV}^d = \gamma_{HV}^0$ this equation contains only one unknown (γ_M^d) and by comparing the interfacial tensions of a series of hydrocarbons against mercury it is possible to estimate the dispersion component of the surface tension of mercury. Fowkes finds a value of $\gamma_M^d = 200 \pm 7\,\text{mJ m}^{-2}$. The remainder of γ_M (284 mJ m^{-2}) would be attributed to metal bonding in this case.

The same argument can be applied to the water–hydrocarbon surface and a value for $\gamma_{H_2O}^d$ of $21.8 \pm 0.7\,\text{mJ m}^{-2}$ is obtained (Exercise 5.11.2).

The remarkable thing about these results is that if we now apply them to the mercury–water interface, assuming that again the only important interaction is the dispersion force, we find:

$$\begin{aligned}\gamma_{MW} &= \gamma_M^0 + \gamma_W^0 - 2(\gamma_M^d \gamma_W^d)^{1/2} \\ &= (484 + 72.8 - 2(200 \times 21.8)^{1/2}\,\text{mJ m}^{-2} \\ &= 424.8\,\text{mJ m}^{-2}\end{aligned}$$

which is very close to the experimental value of 426–427 mJ m^{-2}. There is, therefore, surprisingly little contribution from the permanent dipole moment of the water molecule.

This empirical finding suggests that reliable estimates of interfacial tension can be obtained directly from the dispersion energy, even for water films where hydrogen bonding might be expected to play a role. Unfortunately, a more detailed examination (Hough and White 1980) shows that one can very easily be led into false conclusions regarding the behaviour of water films if only the dispersion contribution is examined.

5.11.2 *Contact angle of liquids on low-energy solids*

According to eqn (5.10.6)

$$\begin{aligned}\cos\theta &= \frac{\gamma_{SV} - \gamma_{SL}}{\gamma_{LV}} \\ &= (\gamma_s^0 - \Pi_e - \gamma_{SL})/\gamma_{LV}\end{aligned} \tag{5.11.6}$$

from eqn (5.10.7). Normally the presence of the Π_e terms makes theoretical estimation of the contact angle from eqn (5.11.6) quite impossible. In the case of very-low-energy solids like polythene, however, it is found that $\Pi_e = 0$. That is to say, the vapour of a high surface energy liquid will not adsorb onto a low-energy solid since this will not lower the surface energy. Neither will the vapour of a hydrocarbon liquid adsorb to any appreciable extent. Low-energy solids are substances for which ΔG_{ads} is small for any adsorbate. They are usually materials with little or no dipole moment and very low polarizability so that they can interact with adsorbates only by van der Waals forces, and then only slightly. The classic example is the non-stick frying pan coated with poly-(tetrafluorethylene).

If we again assume that the interaction between the liquid and the low energy solid is predominantly due to dispersion forces we can then substitute for γ_{SL} from the analogue of eqn (5.11.5) and obtain:

$$\cos\theta = \frac{\gamma_s^0 - (\gamma_s^0 + \gamma_L - 2(\gamma_s^d\gamma_L^d)^{1/2})}{\gamma_L}$$

$$= -1 + 2(\gamma_s^d)^{1/2}\left\{\frac{(\gamma_L^d)^{1/2}}{\gamma_L}\right\}. \tag{5.11.7}$$

Fowkes (1965) shows that a plot of the cosine of the measured contact angle against $(\gamma_L^d)^{1/2}/\gamma_L$ for a number of liquids on low-energy surfaces (like polyethylene and paraffin wax; Fig. 5.11.1) does obey eqn (5.11.7), from the slope of which an estimate of the dispersion contribution to the surface energy of the solid (γ_s^d) can be obtained.

Note that again water appears to behave like other liquids in this plot.

The values of γ_s^d for the four solids in Fig. 5.11.1 should be calculable from the van der Waals constants using (cf. eqn (5.11.2)):

$$\gamma_s^d = \frac{1}{2}\frac{A_{SVS}}{12\pi L_c^2} \tag{5.11.8}$$

if a suitable value of L_c is chosen. Taking $L_c = 0.2$ nm and the extrapolated value for A_{SVS} for $C_{36}H_{74}$ from Table 4.2 ($A_{SVS} = 6.0 \times 10^{-20}$ J) we obtain $\gamma_s^d = 20\ \text{mJ m}^{-2}$, in quantitative agreement with the value obtained for paraffin wax from Fig. 5.11.1. Although this agreement is to some extent fortuitous it is very encouraging. The other values show the tendency one would expect (increasing γ_s^d with increasing molar mass) and they suggest that A_{SVS} for polyethylene should be about $37/20 \times 6 \times 10^{-20} = 1.1 \times 10^{-19}$ J: not an unreasonable figure.

The behaviour of alkanes on poly(tetrafluorethylene) (PTFE) should be amenable to more exact analysis, and this has been done by Israelachvili (1973), whose work is reviewed by Hough and White (1980). Using eqn (5.10.9) we have:

$$\gamma_{SL} = \gamma_s^0 + \gamma_{HV}^0 + E_{SVH}(0) \tag{5.11.9}$$

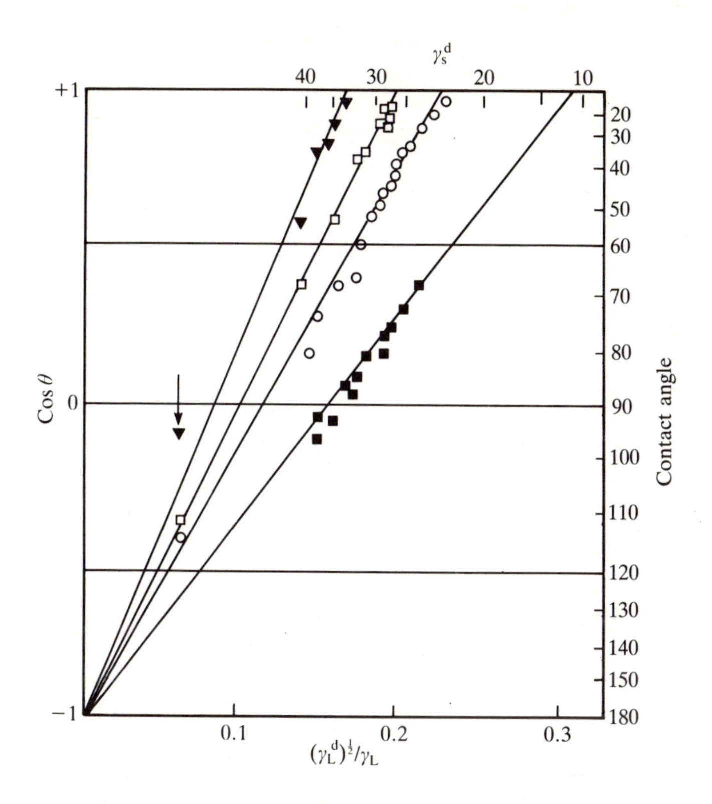

FIG. 5.11.1. Contact angle of a number of liquids on low energy surfaces. ▼: polyethylene; □: paraffin wax; ○: $C_{36}H_{74}$; ■: fluorodecanoic acid monolayer on platinum. All contact angles below the arrow are with water. (Copyright American Chemical Society. Reproduced with permission from Fowkes 1965).

where $E_{SVH}(0)$ is the van der Waals energy of interaction per unit area of the solid with the liquid hydrocarbon across a gap of essentially zero width. This term is assumed to represent the work of adhesion between the solid and the liquid. Again, we have to make a cut-off approximation and set

$$E_{SVH}(0) = -\frac{A_{SVH}}{12\pi L_c^2} \tag{5.11.10}$$

where, for simplicity, we assume that the critical separation L_c is the same as that for the liquid (eqn (5.11.2)).

Using eqns (5.11.2), (5.11.6), (5.11.9), and (5.11.10) we have

$$\cos\theta = -1 - 2\frac{A_{SVH}}{A_{HVH}}. \tag{5.11.11}$$

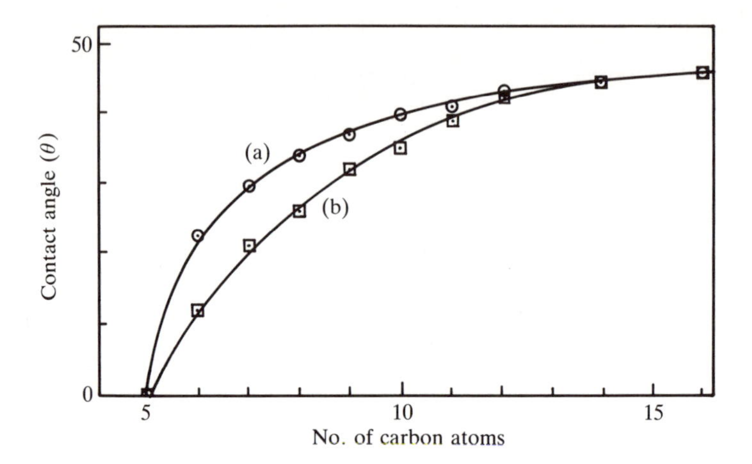

FIG. 5.11.2. A comparison of the theoretical (a) and experimental (b) curve for contact angles of alkanes on PTFE. Wetting would correspond to $\theta = 0$ and is expected to occur for $n \leqslant 5$. (From Hough and White (1980) with permission.)

Hough and White (1980) recalculate $\cos\theta$ using their more accurate assessments of A_{SVH} and A_{HVH} for a number of liquid hydrocarbons on PTFE and obtain the results shown in Fig. 5.11.2.

A cautionary note is in order at this point. The use of L_c as an adjustable parameter is not always successful; Ninham (1980) points out that it fails for liquid argon. The fact that it works reasonably well for liquid hydrocarbons must therefore be to some extent fortuitous. As the experimental spectral data on more materials becomes available, better estimates of van der Waals constants will also appear and techniques like those of Fowkes, even with their limitations, will enable a more systematic approach to the problem of wetting and contact angle. Progress will involve the combined efforts of experimentalists and theoreticians but the general procedures are now becoming much clearer and more widely understood. No doubt there will remain (at least for the forseeable future) enough observed anomalies to keep this area an active and fruitful one for further research on both the theoretical and experimental fronts.

Exercises. 5.11.1 Use the data in Table 4.2 to verify the theoretical values of $\cos\theta$ shown in Fig. 5.11.2.

5.11.2 Given the following data for the interfacial tension (γ_{12}) between water and various hydrocarbons, estimate γ_w^d.

$n =$	6	7	8	10	14
$\gamma_1(C_nH_{2n+2})(mN\,m^{-1})$	18.4	20.4	21.8	23.9	25.6
$\gamma_{12}(mN\,m^{-1})$	51.1	50.2	50.8	51.2	52.2

5.12 Measurement of surface tension and contact angle

5.12.1 *Surface tension*

We have already alluded to a number of methods of measuring surface and interfacial tension, including capillary rise (section 5.7.1) and the analysis of drop or bubble profiles (section 5.10.3). One point that must be appreciated in all techniques is that if a new surface is formed it may take some little time before the system reaches equilibrium, especially if adsorption of a long-chain surface active agent is involved.

Table 5.5 (from Padday 1969, with some additions) lists some of the common methods of measuring surface tension with comments on their suitability. The drop weight method will be referred to again in connection with the measurement of γ at the mercury–solution interface in the next chapter.

5.12.2 *Determination of contact angles*

As with techniques of surface tension measurement, the experimental determination of contact angles is well described in readily available

Table 5.5
Common methods of measuring surface tension

Method	Pure liquids	Solutions
Sessile drop profile	Very satisfactory	Very suitable when ageing occurs
Spinning drop tensiometer	Very suitable for ultra-low surface tensions	Very suitable when ageing occurs
Pendant drop profile	Very satisfactory but limited experimentally	
Capillary height	Very satisfactory	Not suitable if contact angle altered from 0°
The Wilhelmy plate	Very quick and easy to operate; susceptible to atmospheric contamination	Accuracy good; suitable when ageing occurs
The Du Nouy ring	Satisfactory	Not suitable
Drop weight	Very suitable when atmospheric contamination suspected	Poor when ageing effects suspected
Maximum bubble pressure	Somewhat difficult to operate successfully	Poor when ageing effects suspected

literature and will not be discussed in detail here. The purpose of the present section is simply to direct the reader to the literature and briefly to mention some important points to watch for.

Neumann and Good (1979) give a very detailed critical discussion of the various methods of measuring contact angles. The discussion here is confined to Table 5.6, which covers the major methods, giving brief comments on each technique.

5.12.3 *Contact angle hysteresis*

Hysteresis of contact angles is the rule rather than the exception, and it is not uncommon for the advancing angle θ_A (i.e. the contact angle

Table 5.6

Common methods of measuring contact angle

Solid surface	Technique	Comments
Flat	Direct measurement from drop profile	Either sessile drop or bubble, with measurement either via a telescope with goniometer eyepiece or photographs. Can be accurate to $\pm1°$ in the range $30° \leqslant \theta \leqslant 150°$.
Flat	Calculations from drop profile using the Young–Laplace equation	A two-parameter fit is usual (e.g. drop height and maximum diameter is usually used for $\theta > 30°$). For $\theta \leqslant 30°$, drop height and diameter of contact circle can give very accurate results ($\pm0.1°$). *Note*: Use Hartland and Hartley (1976) rather than the Bashforth–Adams tables recommended by Neumann and Good. γ can be determined simultaneously.
Flat	Wilhelmy plate	Measurement of θ is reduced to the measurement of a weight, which can be very accurate. The method is restricted to materials that can be prepared as flat plates of constant perimeter and constant surface morphology on all faces.
Flat	Capillary rise at a vertical plate	Capable of $\pm0.1°$ reproducibility i.e. an order of magnitude better than direct measurement.
Flat	Tilting plate	The apparatus is simple, but it is not easy to be sure whether a retreating or advancing angle, or something in between, is being measured.

Table 5.6 (*continued*)

Solid surface	Technique	Comments
Flat (with smooth reflecting surface)	Interference microscopy	Useful for sessile drops on polished crystal faces or lenses of one liquid on another. Very accurate for $\theta < 15°$. (Fringes become too close at higher θ.) Requena *et al.* (1975); Callaghan *et al.* (1983).
Transparent tube	Capillary rise	Restricted to small diameter tubes. Not recommended for liquid mixtures with low concentrations of an adsorbable component.
Transparent tube	Reflection from liquid surface	Particularly good for highly reflecting surfaces, e.g. liquid metals.
Cylinders, rods, fibres	Direct measurement	Rather difficult experimentally. The curvature of the solid surface can induce large experimental errors, which can be minimized by keeping the drops sufficiently small.
	Floating fibre	Good fun, but experimental accuracy is severely limited by optical resolution required to measure small vertical displacements.
	Wilhelmy method	Probably the best, provided that the perimeter of the fibre can be accurately measured (this can be a severe limitation).
Bundles and mats of fibres		There is *no* good method for measuring contact angles. Do not mistake the angle of a drop on a mat for the angle of contact with the individual fibres!
Powders (spherical particles)	Floating particle	
Powders (irregular particles)	Time taken to sink	Qualitative only.
Powders (compressed cake)	Direct measurement of macroscopic angle	At best, useless. At worst, positively misleading.
	Liquid penetration	Several methods are available (volume, pressure, rate of penetration etc.), all of which give results which at least have some meaning, although that meaning is usually model-dependent.

measured after the liquid has been caused to advance part of the way across the solid surface) to be 20°–30° greater than the corresponding receding angle θ_R. Indeed, with many systems liquid can be withdrawn from a drop placed on a solid surface, with no movement of the three-phase contact line other than a decrease in the contact angle.

A major cause of hysteresis is surface roughness (Good 1979). Old-fashioned empirical corrections (Defay *et al.* 1966) to allow for the effect of surface roughness on contact angle are not particularly useful, but more recently a good deal of experimental and theoretical work on surfaces of controlled roughness (e.g. concentric steps of known height) has been reported, (Mori *et al.* 1982) the results being related to the average contact angle on the corresponding ideally flat solid surface. This type of work is still at an early stage, but may even so be useful to the worker who has reason to suspect a particular pattern of roughness on the surface under study.

Even a molecularly smooth solid surface may display contact angle hysteresis. For example, mica may be cleaved to molecular smoothness over a large area (several square millimetres) (section 1.5.5) but the advancing and receding contact angles of cyclohexane on this surface still differ by 0.5° (5.5° vs. 5°; Fisher 1979). The cause of this residual hysteresis is almost certainly heterogeneity of the surface, in the sense that different microscopic areas have slightly different surface energies. The theory of the effect of surface heterogeneity on average contact angle has been explored by Neumann (see Neumann and Good 1979) for a number of surfaces with different geometrical arrangements of energetically different areas.

It should also be borne in mind that a freshly prepared solid surface could have quite a different surface energy, and hence a different contact angle with a particular liquid, to one that has been allowed to come to adsorption equilibrium with the vapour (Good 1979). If the liquid is volatile, time-dependence of the measured contact angle, arising from gradual adsorption of the vapour at the solid surface, is not uncommon. Time-dependent effects may also arise if the solid is slightly soluble in the liquid or if the solid swells by imbibition of the liquid.

References

Adam, N. K. (1941). *The physics and chemistry of surfaces* (3rd edn) p. 3. Oxford University Press, Oxford. Also Dover, New York (1968).

Adamson, A. W. (1967). *Physical chemistry of surfaces* (2nd edn). Wiley, New York.

Adamson, A. W. (1976). *Physical chemistry of surfaces* (3rd edn). Wiley, New York.

Alexander, A. E. and Hayter, J. B. (1971). Determination of surface and interfacial tension. In *Physical methods of chemistry* (ed. A. Weissberger and B. W. Rossiter), Vol. 1, Part A, Chapter 9. Wiley, New York.

Almgren, F. J., Jr. and Taylor, J. E. (1976). *Sci. Am.* **235** (1), 82–93.

Ambwani, D. S. and Fort, T. Jnr. (1979). Pendant drop technique for measuring liquid boundary tensions. In *Surface and colloid science* (ed. R. J. Good and R. R. Stromberg), Vol. 11, pp. 93–120. Plenum Press, New York.

Aveyard, R. and Haydon, D. A. (1973). *Introduction to the principles of surface chemistry*. Cambridge University Press, Cambridge.

Bashforth, F. and Adams, J. C. (1883). *An attempt to test the theories of capillary action.* Cambridge University Press, Cambridge.

Broekhoff, J. C. P. and de Boer, J. H. (1967). *J. Catalysis* **9,** 8–14.

Broekhoff, J. C. P. and de Boer, J. H. (1968). *J. Catalysis* **10,** 368–75.

Buff, F. P. (1956). *J. Chem. Phys.* **25,** 146–153.

Callaghan, I. C., Everett, D. H., and Fletcher, A. J. P. (1983). *J. chem. Soc. Faraday Trans. 1* **79,** 2723–28.

Childs, E. C. (1969). *An introduction to the physical basis of soil water phenomena.* Wiley, London.

Davies, J. T. and Rideal, E. K. (1963). *Interfacial phenomena.* Academic Press, New York.

de Bruyn, P. L. (1966). *Fundamental phenomena in material science—proceedings of symposium,* Vol. 3, pp. 1–36.

Defay, R., Prigogine, I., Bellemans, A., and Everett, D. H. (1966). *Surface tension and adsorption.* Longmans Green, London.

Deryaguin, B. V., Starov, V. M., and Churaev, N. V. (1976). *Colloid J. USSR* **38,** 786–789.

Dixon, H. H. and Joly, J. (1894). *Proc. R. Soc. Lond.* **57,** 3.

Dunning, W. J. (1973). Ripening and ageing processes in precipitates. In *Particle growth in suspensions* (ed. A. L. Smith), pp. 3–28. Academic Press, New York.

Everett, D. H. (1967). In *The solid–gas interface* (ed. E. A. Flood), Vol. 2, pp. 1055–113. Marcel Dekker, New York.

Everett, D. H. and Haynes, J. M. (1973). *Colloid Sci.* **1,** 123–172.

Fisher, L. R. (1979). *J. Colloid interface Sci.* **72,** 200–5.

Fisher, L. R. (1982). *Adv. Colloid interface Sci.* **16,** 117–25.

Fisher, L. R. and Israelachvili, J. N. (1981a). *Colloids Surfaces* **3,** 303–19.

Fisher, L. R. and Israelachvili, J. N. (1981b). *J. Colloid interface Sci.* **80,** 528–41.

Fowkes, F. M. (1965). Attractive forces at interfaces. In *Chemistry and physics of interfaces* (ed. D. E. Gushee), pp. 1–12. American Chemical Society, Washington.

Gibbs, J. W. (1874–1878). *Trans. Connecticut Acad. Arts Sci.* **III,** 108–248, 343–524. Reprinted (1906) in *Papers of J. Willard Gibbs* (ed. H. A. Bumstead and R. G. van Name), 2 vols, Longmans Green, London. Reprinted (1961) by Dover, New York (2 vols).

Girifalco, L. A. and Good, R. J. (1957). *J. phys. Chem.* **61,** 904.

Good, R. J. (1979). Contact angles and the surface free energy of solids. In *Surface and colloid science* (ed. R. J. Good and R. R. Stromberg), Vol. 11, pp. 1–30. Plenum, New York.

Guggenheim, E. A. (1940). *Trans. Faraday Soc.* **36,** 407–12.

Guggenheim, E. A. (1976). *Thermodynamics* (5th edn). North-Holland, Amsterdam.

Hartland, S. C. and Hartley, R. W. (1976). *Axisymmetric fluid–liquid interfaces.* Elsevier, Amsterdam.
Hauksbee, F. (the elder) (1709). *Physico-mechanical experiments,* pp. 139–169. London.
Hiemenz, P. C. (1977). *Principles of colloid and surface chemistry,* Chapter 6. Marcel Dekker, New York.
Hough, D. B. and White, L. R. (1980). *Adv. Colloid interface Sci.* **14,** 3–41.
Israelachvili, J. N. (1973). *J. chem. Soc. Faraday Trans. 2* **69,** 1729.
Kahlweit, M. (1975). *Adv. Colloid interface Sci.* **5,** 1–35.
Klein, M. J. (1974). *Physica* **73,** 28–47.
Laplace, Pierre Simon de (Marquis) (1806). Supplément au dixième livre du traité de mécanique celeste. Translated and annotated by N. Bowditch, 4 vols (1829–1839), Boston. Reprinted (1966) Chelsea, New York.
Linford, R. G. (1978). *Chem. Rev.* **78,** 81–95.
McBain, J. W. and Humphreys, C. W. (1932). *J. phys. Chem.* **36,** 300.
Melrose, J. C. (1966). *AIChE J.* **12,** 986–94.
Melrose, J. C. (1968). *Indust. Engng Chem.* **60,** 53–70.
Melrose, J. C. (1972). *J. Colloid interface Sci.* **38,** 312–22.
Mori, Y. H., van de Ven, T. G. M., and Mason, S. G. (1982). *Colloids Surfaces* **4,** 1–15.
Morrow, N. R. (1970). *Indust. Engng Chem.* **62,** 32–57.
Neumann, A. W. and Good, R. J. (1979). Techniques of measuring contact angles. In *Surface and colloid science* (ed. R. J. Good and R. R. Stromberg), Vol. 11, pp. 31–92. Plenum, New York.
Ninham, B. W. (1980). *J. phys. Chem.* **84,** 1423.
Nitsche, J. C. C. (1975). *Vorlesungen über Minimalflächen.* Springer, Berlin.
Orowan, E. (1970). *Proc. R. Soc. Lond.* **A316,** 473–91.
Orr, F. M., Jr., Scriven, L. E., and Rivas, A. P. (1975). *J. Colloid interface Sci.* **52,** 602–10.
Overton, G. D. N., Edwards, M. J., and Trevens, D. H. (1982). *J. Phys.* **D15,** 129–131.
Padday, J. F. (1969). Surface tension II. The measurement of surface tension. In *Surface and colloid science* (ed. E. Matijevic) Vol. I, pp. 100–49. Wiley, New York.
Philip, J. R. (1977). *J. Chem. Phys.* **66,** 5069–75.
Pickard, W. F. (1981). *Prog. Biophys. mol. Biol.* **37,** 181–229.
Pujado, P. R., Huh, C., and Scriven, L. E. (1972). *J. Colloid interface Sci.* **38,** 662–3.
Ramakrishnan, S., Scholten, P., and Hartland, S. (1976). *Ind. J. Pure appl. Phys.* **14,** 633–8.
Rayleigh, Lord (1911). *Encyclopaedia Britannica* (11th edn), Vol. 5, pp. 256–275.
Rayleigh, Lord (1915–1916). *Proc. R. Soc. Lond.* **A92,** 184–95.
Requena, J., Billett, D. F., and Haydon, D. A. (1975). *Proc. R. Soc. Lond.* **A347,** 141–59.
Ross, S. and Chen, E. S. (1965). Adsorption and thermodynamics at the liquid–liquid interface. In *Chemistry and physics of interfaces* (ed. D. E. Gushee) pp. 44–56. American Chemical Society, Washington, D.C.
Rusanov, A. I. and Krotov, V. V. (1976). *Colloid J. USSR* **38,** 176–9.
Thomson, J. J. (1888). *Applications of dynamics to physics and chemistry.* Macmillan, London.
Thomson, W. (Lord Kelvin) (1870). *Proc. R. Soc. Edinburgh* **7,** 63–8.

Thomson, W. (1871). *Phil. Mag.* **42,** 448–52.
Visser, J. (1976). *Surface Colloid Sci.* **8,** 3–79.
Weatherburn, C. D. (1930). *Differential geometry of three dimensions.* Cambridge University Press, Cambridge.
Wulff, G. (1901). *Z. Krist.* **34,** 449.
Young, T. (1805). *Phil. Trans. R. Soc. London* **95,** 65–87.
Zimon, A. D. (1969). *Adhesion of dust and powder,* pp. 81–90. Plenum, New York.

6

ELECTRIFIED INTERFACES: THE ELECTRICAL DOUBLE LAYER

6.1 The mercury–solution interface

We noted in section 2.5 that many important properties of colloidal systems are influenced by the electric charges on the particle surface. When immersed in an electrolyte solution a charged colloidal particle will be surrounded by ions of opposite sign so that, from a distance, it appears to be electrically neutral. The surrounding ions are, however, able to move under the influence of thermal diffusion so that the region of charge imbalance, due to the presence of the particle, can be quite significant, relative to the size of the particle itself. Indeed, for very small particles (~50 nm) the disturbance it creates can stretch out to several particle diameters. The arrangement of electric charge on the particle, together with the balancing charge in the solution, is called an electrical *double layer,* and it has been studied on various surfaces for well over a century.

There are many recent reviews of the structure of the double layer, since it is the basis of the entire field of electrochemistry: the same double layer forms at the surface of an electrode and determines its behaviour in an electrochemical cell, an electrolysis apparatus or an electroanalytical chemical device. Again we undertake a review here to provide a coherent description on which to base the remainder of our work in colloid chemistry. More detailed treatments are given in the texts by Bockris and Reddy (1970), Delahay (1966), Sparnaay (1972), Hunter (1981) and the recent treatise edited by Bockris *et al.* (1980).

The most reliable information on the components of charge at an electrified interface comes from the study of the mercury–aqueous electrolyte solution interface. A small drop of mercury issuing from a glass capillary under the surface of an electrolyte solution is probably as close as we can get to an ideal system for study (Fig. 6.1.1). The mercury is contained in a reservoir, M, and as it drops from the lower end of capillary $\mathscr{C}$ a new (and clean) surface is continually created. The surface is also molecularly smooth, which makes interpretation of results easier.

If the solution contains no ions that can readily undergo oxidation or reduction then there is no mechanism for transport of electric charge across the mercury interface. The mercury is then said to be *perfectly polarized*†. It is then possible to adjust the electrical potential difference across the mercury interface by altering the setting on the potentiometer, P. The potential drop between the interior of the reference electrode, R, and the solution is determined by the activity of H^+ ions in solution (and the pressure of the H_2 gas). Any change dE in the setting of the

† The word *polarization* is used by electrochemists to mean any process that leads to a limitation in current flow.

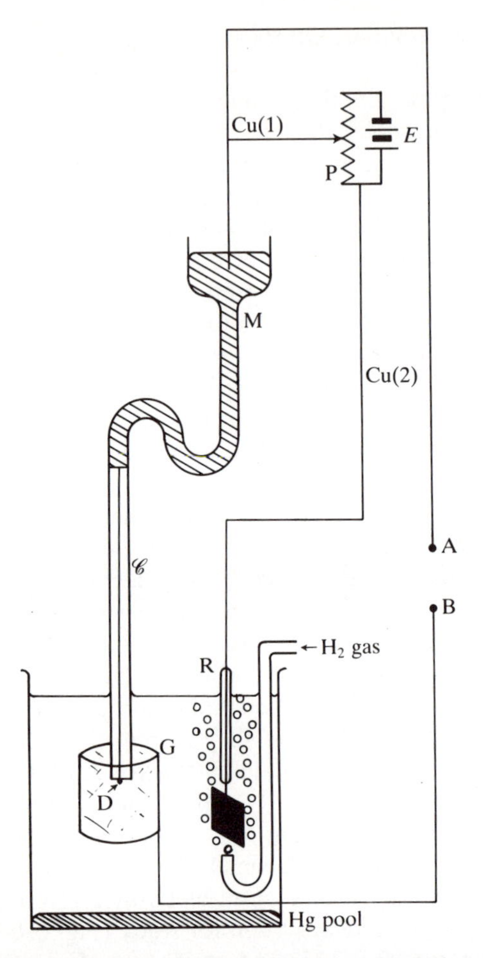

FIG. 6.1.1. Schematic arrangement for determining the electrical capacitance across the surface of a mercury drop D (see text.) G is an electrode made of, say Pt gauze and R is a H_2/Pt reference electrode.

potentiometer P is, therefore, immediately transmitted to the surface of the drop. Since no current can flow through the drop interface, the only effect is a gradual build up of charge on the drop, with a counterbalancing charge in the surrounding solution. This double layer of charge (Fig. 6.1.2) behaves like an electrical capacitor and the magnitude of the capacitance can be measured with a suitable instrument at terminals AB. The system shown in Fig. 6.1.1 could be used to study the behaviour of, say, $HClO_4$ or HNO_3 solution, since it turns out that the H^+ ion is not very easily reduced on the mercury–solution interface (i.e. negligible current flows through the interface so long as the mercury is not made

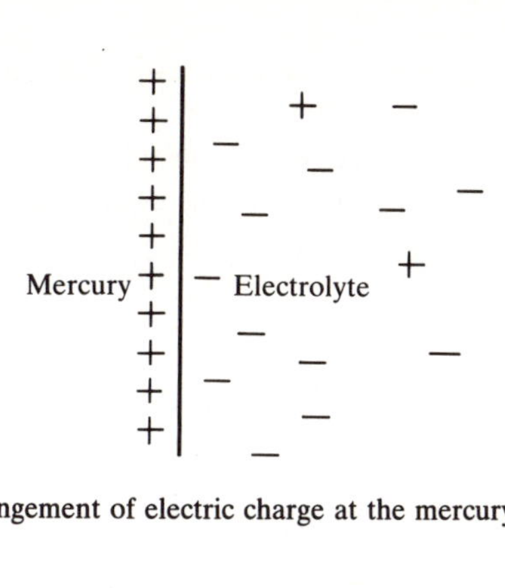

FIG. 6.1.2. Arrangement of electric charge at the mercury–solution interface.

too negative with respect to the reference electrode). Typical experiments of this simple kind take two forms. Either the bulk activity of $HClO_4$ is kept constant and the setting, E, on the potentiometer is varied, or E is kept fixed and the activity (or concentration) of $HClO_4$ in the solution is changed. In both cases we want to know how the surface tension of the mercury–solution interface is affected. We will find that such studies provide a wealth of information about adsorption in charged systems and they can, in some important cases, be extended to the solid–solution interfaces of interest in colloid chemistry.

The thermodynamics of such charged interface can be analysed using the form of the Gibbs adsorption equation appropriate to charged species (Exercise 6.1.1):

$$-\mathrm{d}\gamma = \sum_i \Gamma_i \, \mathrm{d}\tilde{\mu}_i + \sum_j \Gamma_j \, \mathrm{d}\mu_j. \tag{6.1.1}$$

The electrochemical potential $\tilde{\mu}_i$ is defined in eqn (A5.20) where it is pointed out (Appendix 5) that in systems involving electric charges, the equilibrium condition between two phases at constant temperature requires equality of the electrochemical potential for any species (including electrons) that has access to both phases. In particular it should be noted that the potentiometer P measures the difference in the electrochemical potential of the electrons in the wires Cu(1) and Cu(2):

$$E = -\frac{1}{\mathcal{F}}(\tilde{\mu}_e\mathrm{Cu}(1) - \tilde{\mu}_e\mathrm{Cu}(2)) \tag{6.1.2}$$

$$= -\frac{1}{\mathcal{F}}(\tilde{\mu}_e(\mathrm{M}) - \tilde{\mu}_e(\mathrm{Pt})). \tag{6.1.3}$$

Equation (6.1.3) involves the electron equilibrium with the metal electrodes. We now seek to apply eqn (6.1.1) to the simple system of

$HClO_4$ in aqueous solution in contact with the mercury surface. The procedure is that of Parsons (1975a).

The components in the mercury may be taken to be Hg^{2+}, e^- and Hg. They are not independent, of course, and equilibrium requires that

$$\tilde{\mu}_{Hg^{2+}} + 2\tilde{\mu}_e(M) = \mu_{Hg}. \qquad (6.1.4)$$

Likewise in the water:

$$\tilde{\mu}_+ + \tilde{\mu}_- = \mu_{salt}. \qquad (6.1.5)$$

Also at the reference electrode

$$2H^+ + 2e \rightleftarrows H_2(g)$$

and so

$$2\tilde{\mu}_+ + 2\tilde{\mu}_e(Pt) = \mu(H_2). \qquad (6.1.6)$$

From eqn (6.1.1) we have

$$-d\gamma = \Gamma_{Hg^{2+}}\, d\tilde{\mu}_{Hg^{2+}} + \Gamma_e\, d\tilde{\mu}_e(M) + \Gamma_+\, d\tilde{\mu}_+ + \Gamma_-\, d\tilde{\mu}_- + \Gamma_{H_2O}\, d\mu_{H_2O} \qquad (6.1.7)$$

and now we need to introduce the equilibrium conditions in both phases to eliminate quantities which are not accessible to independent measurement. We note that:

$$d\tilde{\mu}_{Hg^{2+}} = d\mu_{Hg} - 2d\tilde{\mu}_e(M) = -2d\tilde{\mu}_e(M) \qquad (6.1.8)$$

$$d\tilde{\mu}_+ = \tfrac{1}{2} d\mu_{H_2} - d\tilde{\mu}_e(Pt) = -d\tilde{\mu}_e(Pt) \qquad (6.1.9)$$

and

$$d\tilde{\mu}_- = d\mu_{salt} - d\tilde{\mu}_+. \qquad (6.1.10)$$

(Equations (6.1.8) and (6.1.9) assume that in the experiments referred to above the activity of the mercury and the H_2 gas are kept constant. The subscript salt in this case refers to $HClO_4$.)

Substituting eqns (6.1.8–10) in eqn (6.1.7) and rearranging terms (Exercise 6.1.2) gives:

$$-d\gamma = (-2\Gamma_{Hg^{2+}} + \Gamma_e)\, d\tilde{\mu}_e(M) - (\Gamma_+ - \Gamma_-)\, d\tilde{\mu}_e(Pt) + \Gamma_-\, d\mu_{salt} + \Gamma_{H_2O}\, d\mu_{H_2O}. \qquad (6.1.11)$$

The coefficient of $d\tilde{\mu}_e(M)$ measures the excess of electrons in the interface above that required to neutralize the Hg^{2+} in the interface. It corresponds, therefore, to $\frac{-\sigma_0}{\mathscr{F}}$, where σ_0 is the charge per unit area on the mercury surface. Likewise $(\Gamma_+ - \Gamma_-)\mathscr{F}$ is the balancing charge in the aqueous solution (where $\mathscr{F}$ is Faraday's constant $= 96\,485\ \mathrm{C\,mol^{-1}}$).

Substituting these quantities in eqn (6.1.11) and using eqn (6.1.3) we have:

$$\begin{aligned} -\,\mathrm{d}\gamma &= \sigma_0\,\mathrm{d}E_+ + \Gamma_-\,\mathrm{d}\mu_{\text{salt}} + \Gamma_{\text{H}_2\text{O}}\,\mathrm{d}\mu_{\text{H}_2\text{O}} \\ &= \sigma_0\,\mathrm{d}E_+ + \Gamma_-^{\sigma}\,\mathrm{d}\mu_{\text{salt}} \end{aligned} \qquad (6.1.12)$$

where the subscript + on E refers to the fact that E is measured in a system with a reference electrode *reversible to the cation*. If the reference electrode is made reversible to the anion, a similar argument gives:

$$-\mathrm{d}\gamma = \sigma_0\,\mathrm{d}E_- + \Gamma_+^{\sigma}\,\mathrm{d}\mu_{\text{salt}}. \qquad (6.1.13)$$

(The superscript σ on Γ_+ or Γ_- refers to the use of the Gibbs convention with $\Gamma_{\text{H}_2\text{O}}$ set equal to zero to get relative surface excesses of cations and anions (section 5.4.1). We will drop it in future to avoid confusion with σ_0.)

Equation (6.1.12) leads directly to the Lippmann equation:

$$\left(\frac{\partial\gamma}{\partial E_+}\right)_{\mu_{\text{salt}},P,T} = -\sigma_0 \qquad (6.1.14)$$

which was put forward over a century ago and forms the basis of the study of *electrocapillarity*. The interfacial tension can be measured in a variety of ways (section 5.12) of which the most common for the dropping mercury electrode (DME) is the drop weight method. The high surface tension of mercury makes the drop almost exactly spherical in shape (Perram *et al.* 1973a) and by weighing a known number of drops the (relative) value of γ can be estimated. To obtain exact data, however, a numerical solution of the curvature equations is required and that is beyond the scope of the present discussion (see Perram *et al.* 1973a). In the capillary electrometer (Fig. 6.1.3), the mercury–electrolyte interface

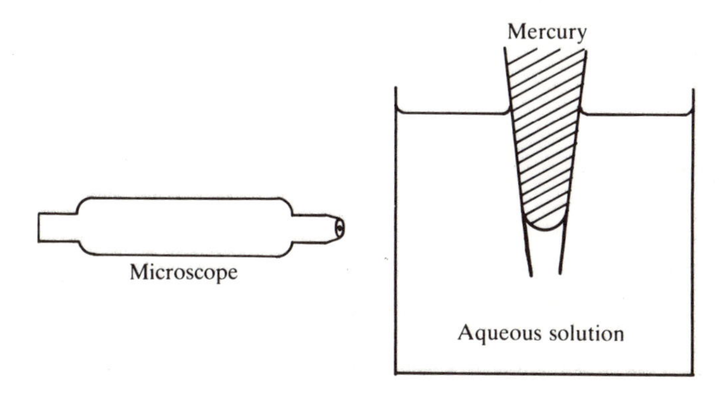

FIG. 6.1.3. Principle of the Lippmann capillary electrometer.

is formed in a tube of varying radius. As γ varies with E the hydrostatic pressure required to return the interface to the same position in the capillary is measured. Since that is determined only by the diameter at that point (and γ) one can calculate γ (or a relative γ) from the Young–Laplace equation (section 5.2.3). For details of the modern apparatus see for example Mohilner and Kokiuchi (1981), which references earlier work on computer-controlled devices.

Note that when there is no charge on the mercury surface $(\partial\gamma/\partial E)_\mu = 0$ and at this value of E the interfacial tension is a maximum. Figure 6.1.4 shows some typical plots of γ against the applied potential difference. When E is made negative (*cathodic polarization*) the mercury surface is negative with respect to the solution. The predominant counterions on the solution side are then the cations and Fig. 6.1.4 shows that Na^+ and K^+ behave essentially identically with respect to γ. On the other hand, under *anodic polarization*, the counterions are anions and it is clear from the figure that even halide ions behave very differently from one another. It is perhaps not surprising that anions show some of their

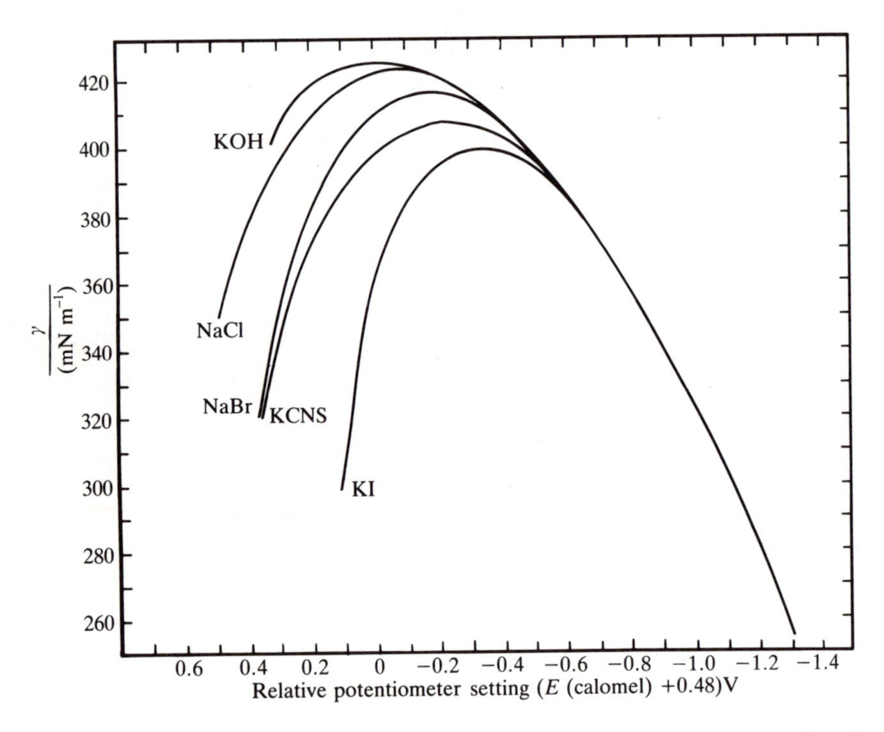

FIG. 6.1.4. Electrocapillary curves (after Grahame 1947). The potentiometer reading E has been adjusted so that $E = 0$ at the electrocapillary maximum for sodium fluoride (see text). (Copyright American Chemical Society.)

chemical character against a metal surface whereas cations tend to behave more like simple positive charges with no special affinity for the surface. We will discuss this behaviour in more detail below (section 6.3).

The most striking feature of the curves in Fig. 6.1.4 is the maximum in γ, called the *electrocapillary maximum* (e.c.m.). The maximum value identifies the point at which $\sigma_0 = 0$ (from eqn (6.1.14)) and so is also referred to as the *point of zero charge* (p.z.c.). Putting a charge on the mercury surface, whether positive or negative, has the same effect as does a surfactant at other liquid surfaces (section 5.4.2).

The charge at any other value of E can be obtained by differentiation of the γ versus E curve. At the dropping mercury electrode, however, it is more usual to measure the *electrical capacitance* of the drop surface by surrounding it with a counter-electrode (G in Fig. 6.1.1), usually made of platinum gauze. The total impedance between the mercury and the electrode G is made up of capacitive effects at each electrode plus a resistive component, $\mathcal{R}$, through the solution. Since the electrode capacitances are in series the large area of G makes its contribution negligible and provided the electrolyte concentration is not too low (so that $\mathcal{R}$ is not too high), an accurate determination of the capacitance, C, of the dropping mercury electrode can be made. C is measured by imposing a small a.c. signal across the electrode system and the measured value, therefore, corresponds to a *differential* capacitance:

$$C = \left(\frac{\partial \sigma_0}{\partial E}\right)_\mu \tag{6.1.15}$$

which, from eqn (6.1.14) is equal to $-\mathrm{d}^2\gamma/\mathrm{d}E^2$. The importance of the capacitance lies in the fact that, being a differential quantity, it contains a great deal more detailed information than does the surface tension γ.

To see this more clearly we note that the curves in Fig. 6.1.4 are very nearly parabolic. If they are represented by an equation of the form

$$\gamma = \gamma_{\mathrm{ecm}} - b'(E - E_{\mathrm{ecm}})^2 \tag{6.1.16}$$

(where the subscript ecm refers to the electrocapillary maximum) we would have (from eqn (6.1.14)):

$$-\left(\frac{\mathrm{d}\gamma}{\mathrm{d}E}\right) = 2b'(E - E_{\mathrm{ecm}}) = \sigma_0$$

and $C = \mathrm{d}\sigma_0/\mathrm{d}E = 2b$ (i.e. a constant). The actual experimental values of C are far from constant as is shown in Fig. 6.1.5. This data is for sodium fluoride, which is now recognized to be about the simplest possible electrolyte behaviour to interpret. We will attempt some interpretation in section 6.3.

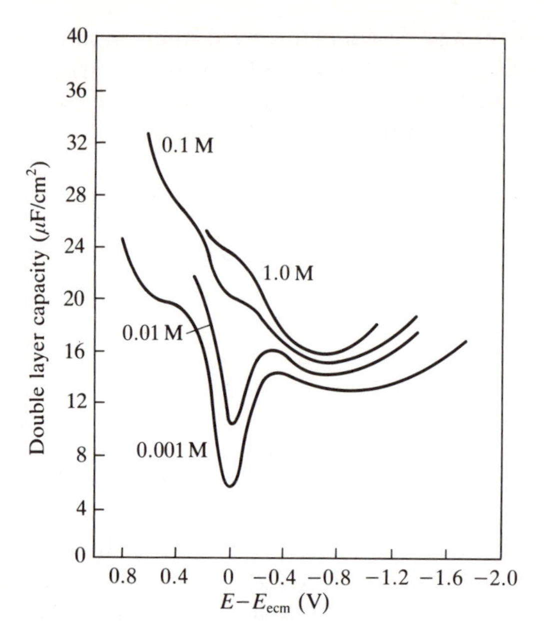

FIG. 6.1.5. Differential capacitance at the DME in contact with NaF solutions at 25 °C. (After Grahame 1947.) (Copyright American Chemical Society.)

The data in Fig. 6.1.5 can be used to determine the charge on the electrode at any value of E since (from eqn (6.1.15)):

$$\sigma_0 = \int_{E=E_{ecm}}^{E} C \, \mathrm{d}E. \tag{6.1.17}$$

A second integration can be used to evaluate γ from the Lippmann equation:

$$\gamma - \gamma_{ecm} = -\int \sigma_0 \, \mathrm{d}E = -\int_{E_{ecm}} \int^{E} C(\mathrm{d}E)^2 \tag{6.1.18}$$

and this value can be compared with the direct measurements as a check on the reliability of the data.

If measurements are done at a variety of salt concentrations (as shown in Fig. 6.1.5) one can determine the values of Γ_+ and Γ_- from eqn (6.1.12) or (6.1.13):

$$\Gamma_+ = -\left(\frac{\partial \gamma}{\partial \mu}\right)_{E_-}; \qquad \Gamma_- = -\left(\frac{\partial \gamma}{\partial \mu}\right)_{E_+}. \tag{6.1.19}$$

Only one of these is needed since the electroneutrality condition requires that $\mathscr{F}(\Gamma_- - \Gamma_+) = \sigma_0$. In this way a huge body of data has been built up on the relation between applied e.m.f. and interfacial adsorption for a large number of systems.

The (perfectly polarized) mercury–solution interface has been examined in some detail because it is important to establish the sound thermodynamic basis of the experimental work in that area: it is the underpinning for the models of the electrical double layer to be discussed below (section 6.3). In colloid systems the presence of the solid–solution interface introduces all the problems concerning the quantity γ_{SL} (section 5.5.4). They can be circumvented almost entirely, however, by appeal to basic thermodynamic relationships like eqns (6.1.12–19) with the interpretation of γ_{SL} as a surface free energy (section 5.2.1). The surface charge density, σ_0, is a relatively easy quantity to measure in a solid–liquid system.

This does not mean that our descriptions of the solid–liquid interface rest solely on the certainties of classical thermodynamics. An understanding of the microscopic structure at the interface can only come from the introduction of extra-thermodynamic assumptions. The thermodynamics assists in manipulation of experimental data without introducing new and possibly erroneous assumptions. In that respect, modern electrochemistry may be regarded, in the words of one of its most respected modern exponents as 'the most remarkable of the applications of classical thermodynamics' (Parsons 1980).

In the derivation of the important expressions (6.1.12) and (6.1.13) it was tacitly assumed that each of the phases in contact could be characterized by an electrostatic potential ϕ so that the electrochemical potential of any species in that phase was fixed by eqn (A5.20). This potential is called the *inner* or *Galvani potential* of the phase and although it is a perfectly well-defined thermodynamic concept it is only directly measurable under certain prescribed conditions. In particular, the difference $\Delta\phi$ in Galvani potentials between two phases can *only* be directly measured if the phases have the same chemical composition. We will examine this point in more detail in section 6.2. For the present we note only that the system shown in Fig. 6.1.1 can be represented by:

$$\begin{array}{ccccc} \mathrm{Cu(2)}\,| & \mathrm{H_2(g), Pt}\,| & \mathrm{HClO_4(aq)}\,| & \mathrm{Hg}\,| & \mathrm{Cu(1)} \\ \phi_5 & \phi_4 & \phi_3 & \phi_2 & \phi_1 \end{array}$$

and

$$E = \phi_1 - \phi_5. \tag{6.1.20}$$

(The vertical lines here represent phase boundaries.) The electrochemical potential of an electron in a phase whose Galvani potential is ϕ is given by (from eqn A5.20):

$$\tilde{\mu}_e = \mu_e(\mathrm{chem}) - \mathscr{F}\phi. \tag{6.1.21}$$

If two phases have an identical chemical composition (like the copper wires connected to the potentiometer), then μ_e(chem), the 'chemical' part of the electron's energy, will be the same in both and hence:

$$\tilde{\mu}_e(\mathrm{Cu}(1)) - \tilde{\mu}_e(\mathrm{Cu}(2)) = -\mathcal{F}(\phi_1 - \phi_5) \tag{6.1.22}$$

which leads directly to eqn (6.1.2). Now writing:

$$E = (\phi_1 - \phi_2) + (\phi_2 - \phi_3) + (\phi_3 - \phi_4) + (\phi_4 - \phi_5) \tag{6.1.23}$$

we see that the first and last terms are in the nature of metal–metal contact potentials that are characteristic of the metals and are, therefore, constant. Likewise $(\phi_3 - \phi_4)$ is determined by the activity of H^+ in the solution and if it is kept constant while the potentiometer is adjusted we have:

$$\mathrm{d}E = \mathrm{d}(\phi_2 - \phi_3) = \mathrm{d}(\Delta\phi) \tag{6.1.24}$$

where $\Delta\phi$ is the potential difference between the interior of the mercury (or its surface) and the interior of the aqueous solution. It is this potential $\Delta\phi$ which we will wish to investigate at the microscopic level (section 6.3).

Exercises. 6.1.1 The Helmholtz free energy function for an electrified surface phase can be written (compare eqn (5.3.28):

$$\mathrm{d}F^\sigma = -S^\sigma\,\mathrm{d}T + \gamma\,\mathrm{d}\mathcal{A} + \sum \mu_i\,\mathrm{d}n_i^\sigma + \phi \sum \mathrm{d}q$$

where $\phi \sum \mathrm{d}q = \phi z_i \mathcal{F} \sum \mathrm{d}n_i^\sigma$. Use this together with the usual ξ integration procedure (Appendix 5) to obtain eqn (6.1.1). [Note: ϕ is the electrostatic potential characterizing the surface and the $\sum \phi\,\mathrm{d}q$ term represents the work done in placing charged species in the interface.]

6.1.2 Establish eqn (6.1.11).

6.2 Potential differences between phases

The galvani potential difference $\Delta\phi$ between two phases, can be measured *only* when the two phases are of identical composition. It is then obtained from the difference in electrochemical potential of the electron in the two phases (eqn 6.1.22) because that is experimentally measurable. When the two phases are not the same we cannot assume that the 'chemical' part of $\tilde{\mu}_e$ is the same in both phases because μ_e (chem) would measure the work done in bringing an electron (reversibly) from infinity into the interior of the (uncharged) phase. When the electron is in the phase it produces an effect on surrounding dipoles and other electrons and the extent of that effect depends upon the properties of the material concerned. The problem has been examined by many people including Guggenheim (1929) and Grahame (1947) and a very clear exposition given by Parsons (1954).

As Guggenheim points out, the problem stems from the very notion of an electrostatic potential at a point. It measures the work done in bringing an *infinitesimal* charge from infinity up to the point in question. The charge must be so small that it has no effect on surrounding charges or dipoles. Clearly, the smallest charge we use *experimentally* for a measurement of potential is an electron and it is certainly not able to satisfy the fundamental requirement.

There is no theoretical difficulty about measuring the potential *difference* between two points in the same (uniform) phase because the 'chemical' effects can be assumed to be fixed but as soon as the test charge has to move across an interface problems arise. In the expression

$$\tilde{\mu}_i^\alpha = \mu_i^\alpha + z_i e \phi^\alpha \qquad \text{[A5.20]}$$

the quantity μ_i^α can be regarded (Parsons 1954) as the work done in transferring the charged particle i from infinity (*in vacuo*) into the interior of the *uncharged* phase α whilst the additional term accounts for the work done in crossing the interface, where all the charge and dipolar effects are assumed to be located. The Galvani potential ϕ^α is made up of two parts:

$$\phi^\alpha = \Psi^\alpha + \chi^\alpha \qquad (6.2.1)$$

The term Ψ^α is due to any excess of charge on the phase (and that is called the *outer* or Volta potential). The second term is called the 'chi' or 'jump' potential of the phase and it measures the work done in taking the infinitesimal test charge through the layer of dipoles at the interface; these may be oriented in a special and quite different way from the bulk material.

The Volta potential, Ψ, is a measurable quantity because the excess charge can be sensed without going into the phase. One merely has to bring a probe up near (but not too near) to the (clean) surface. The potential at a distance r from the surface of a charged sphere of radius a, *in vacuo*, is given by Coulomb's law:

$$\Psi = \frac{Q}{4\pi\varepsilon_0(a+r)} \qquad (6.2.2)$$

where Q is the charge; Ψ is essentially constant for distances $r \ll a$ and if the sphere is macroscopic in size (say $a > 1$ mm) then Ψ is fairly constant for values of r between about 0.01 and 0.001 mm ($= 1$ μm). This is still outside the region of the double layer and makes it (theoretically) possible to bring a probe into the region of constant potential without interfering with the arrangement of charges and dipoles on the particle surface. The constant (plateau) potential so measured is the outer (or Volta) potential of the phase. For the dropping mercury electrode, where we consider a certain charge to reside on the (macroscopic) drop and a balancing charge to be present in the solution, the outer potential Ψ of the mercury is actually zero because the charge in the aqueous solution reduces the potential to zero over the double layer region. Equation (6.2.1) implies that the galvani potential difference between phases 2 (the mercury) and 3 (the aqueous solution) can be written:

$$\phi_2 - \phi_3 = \Delta\phi = \Delta\Psi + \Delta\chi. \tag{6.2.3}$$

The quantity $\Delta\Psi$ is the Volta potential difference between the two phases and it is, in principle, measurable. It arises because of differences in the electron affinity between the two phases and it is tempting to think of it as being equal to the potential difference due to the total transfer of a certain number of charges (ions or electrons) from one phase to the other. Unfortunately, that identification cannot be made rigorously.

It is, however, useful to break the term $\Delta\phi$ into two parts: a contribution due to the separation of the free charges ($\Delta\psi$) and the contribution ($\Delta\chi_{\text{dipole}}$) due to the orientation of dipolar molecules at the interface. (Note, however, that in general $\Delta\psi \neq \Delta\Psi$ and $\Delta\chi_{\text{dipole}} \neq \Delta\chi$ in eqn (6.2.3).) The potential due to the ion distribution in the aqueous phase can then be assumed to change from zero, in the bulk, to some value ψ_0 on the mercury surface with $\psi_0 \approx \Delta\psi$ and

$$\Delta\phi = \psi_0 + \Delta\chi_{\text{dipole}}. \tag{6.2.4}$$

ψ_0 is often referred to as the *surface potential* by colloid chemists.

The surface potential, as that term is used in *surface* chemistry, is measured by bringing a probe up near a water surface before and after spreading some surfactant material at the interface. Its relationship to the potentials discussed here is examined by Davies and Rideal (1963, pp. 64–79). Essentially it is a measure of the change in Volta potential caused by a controlled degree of contamination (the surface film).

The significance of the Volta potential, Ψ, is much smaller in the case of a disperse phase since it is no longer possible to apply the argument of eqn (6.2.2). The particles are so small that it makes no sense to speak of a macroscopic charge separation that is measurable from outside the solid phase. In that case it is again preferable to treat the entire double layer region as an arrangement of charges and dipoles that is *uncharged* overall. It is, however, still profitable to separate out the contribution due to the disposition of free charges from that due to dipole orientation (eqn (6.2.4)).

Exercise. 6.2.1. Sketch the curve of Ψ against $\log_{10} r$ from eqn (6.2.2) for $10^{-4} \leq r\ (\text{cm}) \leq 100$ and $a = 1$ cm, $Q = 1$ pC.

6.3 Potential distribution at a flat surface—the Gouy–Chapman model

The earliest theoretical studies of the behaviour of an electrified interface were made by Helmholtz well over a century ago. His equations were interpreted by Perrin as implying a simple charge distribution in the solution, opposite to that on the solid (Fig. 6.3.1). The equations for the electrical potential as a function of distance into the solution can readily be solved for this simple model of the double layer and they were able to explain some features of the behaviour of double-layer systems.

The success of the kinetic theory of molecular behaviour made it clear, however, that the Helmholtz model was unrealistic—especially in the treatment of the electric charge in the solution. Since the metal is an electronic conductor it is reasonable to assume that the charge on it is confined to the surface and that that surface can be regarded as a surface of constant potential. In the solution, on the other hand, ions of opposite sign predominate over ions of the same sign but the latter are not completely excluded from the surface region (Fig. 6.1.2). That surface region is also of significant thickness. We will anticipate the final result and state (as we did in section 2.5) that this region of varying charge density stretches over distances of order 100 nm in dilute electrolyte solution and rather less at higher concentrations. The curvature of the mercury drop is of the order of 100 μm so it can be reasonably approximated as flat so far as the electrical double layer is concerned. We must now determine the electrical charge and potential distribution in this diffuse charge region by solving the relevant electrical and statistical thermodynamic equations. The problem was first tackled by Gouy (in 1910) and, independently, by Chapman (1913) and the result is referred to as the *Gouy–Chapman model.* Solutions of that model are available in the standard texts (e.g. Adamson 1967; Overbeek 1952) and reviews (Grahame 1947). They depend on the solution of what is called the

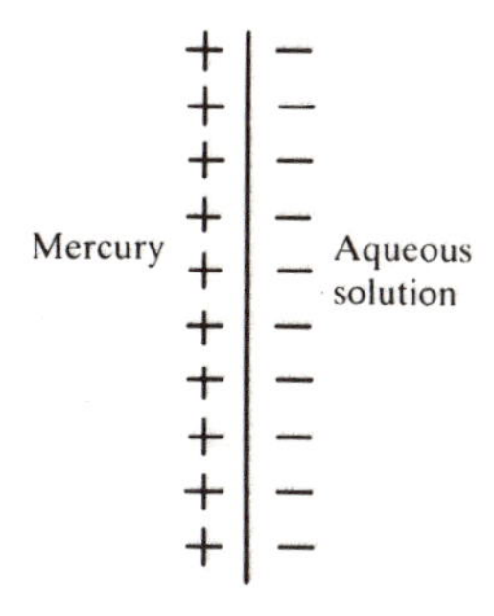

FIG. 6.3.1. The Helmholtz model of the double layer.

Poisson–Boltzmann equation, one of the most important equations of statistical physics. Although some criticism can be levelled at this equation on strictly statistical mechanical grounds it has been shown to be remarkably accurate in its representation of the diffuse double layer so we will save the criticism for later (section 6.6). Rather than repeat the treatment so readily available elsewhere we will discuss a rather better model than the simple one proposed by Gouy and Chapman. We will recognize, from the outset, that the ions in the solution (whether bare or hydrated) have a finite size and so are not able to get closer than a certain distance from the metal surface. There is, therefore, a charge-free region near the surface, which must be treated differently from the rest of the double layer. The thickness of the charge-free region varies from about one bare ion radius (say 0.1 nm) up to or a little beyond one hydrated ion radius (~0.5 nm).

The transition from ϕ_2 to ϕ_3 across the interface occurs over a finite distance and it may be assumed that ϕ_2 remains constant almost right up to the interface (Fig. 6.3.2). This is a reasonable assumption for a metal. The potential at the plane $x=0$ is $\phi^{\mathrm{M}}(=\phi_2)$ and for $x>0$ it is determined by the Poisson equation (see appendix 3 for a discussion of the meaning of this equation):

$$\operatorname{div} \boldsymbol{D} = \operatorname{div} \varepsilon \boldsymbol{E} = \rho \tag{6.3.1}$$

where $\boldsymbol{D}$ is the dielectric displacement vector (eqn (2.2.6)) and ρ is the local volume density of charge (i.e. the number of charges per unit volume). Now

$$\boldsymbol{E} = -\operatorname{grad} \phi \tag{6.3.2}$$

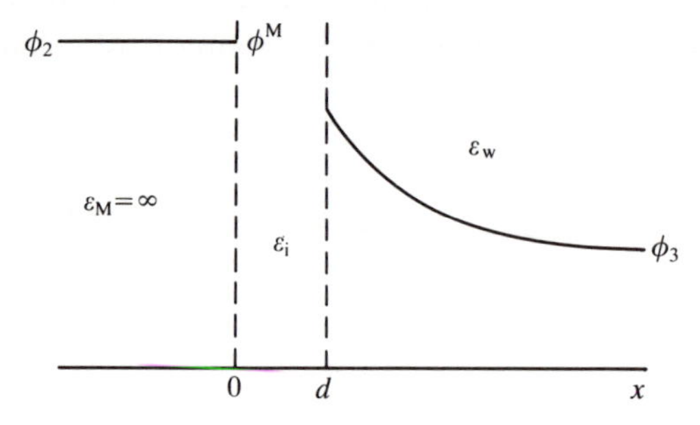

FIG. 6.3.2. A possible potential distribution across the metal–electrolyte interface. The region between $x=0$ and $x=d$ is assumed to be free of any charges and to have a permittivity ε_{i} which differs from the bulk value ε_{w} and may be a function of position. Outside that layer, ε takes its bulk values in each phase. (The potential distribution in the inner region $(0 \leqslant x \leqslant d)$ will be considered later.)

and so from eqns (6.3.1) and (6.3.2):

$$\text{div}(\varepsilon \text{ grad } \phi) = -\rho. \tag{6.3.3}$$

For the region $x > d$, where $\varepsilon(=\varepsilon_w)$ is constant:

$$\text{div grad } \phi = \nabla^2\phi = -\rho/\varepsilon_w = \frac{-\rho}{\varepsilon_0\varepsilon_r} \tag{6.3.4}$$

where ∇^2 is the Laplace operator $(= \partial^2/\partial x^2 + \partial^2/\partial y^2 + \partial^2/\partial z^2)$ and $\varepsilon_r(=\varepsilon_w/\varepsilon_0)$ is the (dimensionless) relative permittivity (section 2.2).

In the region $x > d$, the ions are influenced by the local electrostatic potential. If the metal surface is charged there will be an accumulation of oppositely charged ions given by the Boltzmann equation:

$$n_i = n_i^0 \exp(-w_i/kT) \tag{6.3.5}$$

where w_i represents the work done in bringing an ion i up from the bulk solution (where $\phi = \phi_3$) to a point in the double layer where the potential is, say, ϕ and n_i^0 is the bulk concentration of ions of type i.

As a first approximation, we assume that

$$w_i = z_i e(\phi - \phi_3) = z_i e\psi \tag{6.3.6}$$

where

$$\psi = \phi - \phi_3. \tag{6.3.7}$$

In other words, the only work done in bringing the ion near the surface is the electrical work done on or by the ion as it moves in response to the field. This ignores the energies involved in moving aside other ions or creating a hole in the solvent, or any effect which the ion might have on the local structure of the solvent or the distribution of other ions. The ion is simply treated as a *point charge*.

The volume density of charge, ρ, is given by

$$\rho = \sum_i n_i z_i e \tag{6.3.8}$$

where the summation is over all the species of ion present and the valency, z_i, may take positive or negative values. From eqns (6.3.4–8), assuming ϕ_3 is constant:

$$\nabla^2\psi = \frac{-1}{\varepsilon_0\varepsilon_r}\sum_i n_i^0 z_i e \exp(-z_i e\psi/kT). \tag{6.3.9}$$

This equation is the Poisson–Boltzmann equation referred to earlier; it is one of the most important equations we will encounter since it is the basis of our understanding of electrolyte solutions, electrode processes, colloid interaction, membrane transport, nerve conduction, transistor behaviour, and even plasma physics.

6.3.1 *The Debye–Hückel approximation*

If the electrical energy is small compared to thermal energy ($|z_i e\psi| < kT$) it is possible to expand the exponential in eqn (6.3.9) neglecting all but the first two terms, to give:

$$\nabla^2\psi = -\frac{1}{\varepsilon_0\varepsilon_r}\left(\sum_i z_i e n_i^0 - \sum_i z_i^2 e^2 n_i^0 \psi / kT\right). \tag{6.3.10}$$

The first summation term must be zero to preserve electroneutrality in the bulk solution, so:

$$\nabla^2\psi = \left(\frac{\sum z_i^2 e^2 n_i^0}{\varepsilon_0\varepsilon_r kT}\right)\psi$$
$$= \kappa^2\psi \tag{6.3.11}$$

where

$$\kappa = \left(\frac{e^2 \sum n_i^0 z_i^2}{\varepsilon kT}\right)^{1/2}. \tag{6.3.12}$$

This simplification of assuming ψ to be very small is called the Debye–Hückel approximation because it was used by those two workers in their theory of strong electrolytes. The solution of eqn (6.3.11) is of the form $\psi = \text{const} \cdot \exp(-\kappa x)$ (Exercise 6.3.3). The quantity κ (which has the dimensions of $(\text{length})^{-1}$) is called the *Debye–Hückel parameter* and it plays a prominent part in the theory of the double layer. The extent of the double layer is measured by the size of $1/\kappa$: the region of variable potential shown in Fig. 6.3.2 (from $x = d$ out to the bulk solution) is of the order of $3/\kappa$ to $4/\kappa$. Note that, apart from some fundamental constants, κ depends only on the temperature and the bulk electrolyte concentration. At 25 °C in water the value of κ is given by (Exercise 6.3.1):

$$\kappa = \left(\frac{2000\mathscr{F}^2}{\varepsilon_0\varepsilon_r RT}\right)^{1/2} \sqrt{I} \text{ in units of m}^{-1} \tag{6.3.13}$$

$$= 3.288\sqrt{I}\,(\text{nm}^{-1}) \tag{6.3.14}$$

where $\mathscr{F}$ is Faraday's constant and I is the ionic strength ($= \frac{1}{2}\sum c_i z_i^2$ where c_i is the ionic concentration in mol L^{-1}). In 10^{-3}M 1:1 aqueous electrolyte solution $1/\kappa = 9.6$ nm and for the systems of interest in colloid science $1/\kappa$ ranges from a fraction of a nanometre to about 100 nm.

6.3.2 *Solution of the complete Poisson–Boltzmann equation*

Unfortunately, in most situations of interest in colloid science and electrochemistry it is not possible to assume that $|ze\psi| < kT$. The range of values of E shown in Figs 6.1.4 and 6.1.5 suggest that $(\phi_2 - \phi_3)$ will be of order one volt so that $e\psi \simeq 1.6 \times 10^{-19}$ J which is about $40kT$ at room temperature. Under these conditions the complete Poisson–Boltzmann (eqn (6.3.9)) must be solved. Fortunately, for the case of a flat surface that is relatively straightforward. To simplify the algebra we set $z_i = z_+ = -z_- = z$ so that the analysis is limited to a symmetric $z:z$ electrolyte. It turns out that this is not a very serious restriction because in most situations of interest in colloid science, the behaviour of the system is governed almost entirely by the ions of sign opposite to that of the surface (see e.g. section 2.5). Equation (6.3.9) can then be written (Exercise 6.3.4):

$$\frac{d^2\psi}{dx^2} = \frac{2n^0 ze}{\varepsilon} \sinh \frac{ze\psi}{kT} \tag{6.3.15}$$

using the identity $\sinh p = (\exp p - \exp(-p))/2$.

This can be integrated by multiplying both sides by $2(d\psi/dx)$:

$$\frac{2d\psi}{dx}\frac{d^2\psi}{dx^2} = \frac{4n^0 ze}{\varepsilon} \sinh \frac{ze\psi}{kT}\frac{d\psi}{dx}. \tag{6.3.16}$$

The left-hand side is the differential (with respect to x) of $(d\psi/dx)^2$. Integrating with respect to x then gives:

$$\int \frac{d}{dx}\left(\frac{d\psi}{dx}\right)^2 dx = \int \frac{4n^0 ze}{\varepsilon} \sinh \frac{ze\psi}{kT} d\psi. \tag{6.3.17}$$

Integrating from some point out in the bulk solution where $\psi = 0$ and $d\psi/dx = 0$ (see Fig. 6.3.2) up to some point in the double layer $(x > d)$, we have (Exercise 6.3.4):

$$\left(\frac{d\psi}{dx}\right)^2 = \frac{4n^0 kT}{\varepsilon}\left[\cosh \frac{ze\psi}{kT} - 1\right] \tag{6.3.18}$$

or (Exercise 6.3.5):

$$\frac{d\psi}{dx} = -\left[\frac{8n^0 kT}{\varepsilon}\right]^{1/2} \sinh \frac{ze\psi}{2kT} \tag{6.3.19}$$

$$= -\frac{2\kappa kT}{ze} \sinh ze\psi/2kT \tag{6.3.20}$$

from eqn (6.3.12). Note that the negative sign is chosen so that $d\psi/dx$ is always negative for $\psi > 0$ and positive for $\psi < 0$. This ensures that $|\psi|$ always decreases going towards the bulk solution and becomes zero far from the surface.

Equation (6.3.20) can be integrated from the bulk solution up to the plane $x = d$ to give (Exercise 6.3.6).

$$\tanh(ze\psi/4kT) = \tanh(ze\psi_d/4kT)\exp(-\kappa(x-d)). \qquad (6.3.21)$$

For very low potentials the substitution $\tanh p \simeq p$ can be made and eqn (6.3.21) reduces to

$$\psi = \psi_d \exp(-\kappa(x-d)) \qquad (6.3.22)$$

which is the solution to the linear equation (6.3.11) (see Exercise 6.3.3). A comparison between eqns (6.3.21) and (6.3.22) is shown in Fig. 6.3.3. Note that for $z\bar{\psi}_d = \dfrac{ze\psi_d}{kT} < 2$ (i.e. $|z\psi_d| < 51.4$ mV at room temperature) the approximate expression is reasonably accurate. Note also that far out in the double layer where the potential is low we can substitute $\tanh p \simeq p$ and then:

$$\psi = \frac{4kT}{ze} Z \exp(-\kappa(x-d)) \qquad (6.3.23)$$

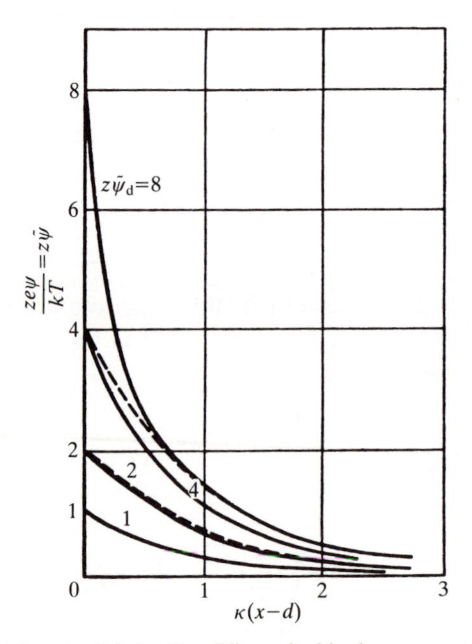

FIG. 6.3.3. Electrical potential in the diffuse double layer according to the Gouy–Chapman model. Full curves are from eqn (6.3.21) and broken lines from eqn (6.3.22) for $z\bar{\psi}_d = 2$ and 4. ($\bar{\psi}$ is a dimensionless quantity called the *reduced potential*.) (After Overbeek 1952.)

where

$$Z = \tanh ze\psi_d/4kT$$
$$= \frac{\exp(ze\psi_d/2kT) - 1}{\exp(ze\psi_d/2kT) + 1}. \tag{6.3.24}$$

Since Z approaches unity for high values of $z\psi_d$ one can expect the potential far from the wall to resemble that for a wall of potential $\psi_d = 4kT/ze$ *irrespective* of the actual potential provided it is sufficiently high (compare eqn (6.3.23) with (6.3.22). In colloid situations, any measurement of a highly charged system at ordinary temperatures will suggest that $\psi_d \simeq (100/z)$ mV if the measuring method only samples the outer region of the double layer. Likewise, if one wants to predict the behaviour of a highly charged system in a situation in which only the outer part of the diffuse layer is important, the approximation $\psi_d \simeq (100/z)$ mV should be a good one.

6.3.3 *The diffuse layer charge*

The total charge, per unit area of surface, in the diffuse layer is given by:

$$\sigma_d = \int_d^\infty \rho \, dx \tag{6.3.25}$$

and substituting for ρ from eqn (6.3.4) (since $d^2\phi/dx^2 = d^2\psi/dx^2$):

$$\sigma_d = \int_\infty^d \varepsilon_0\varepsilon_r \frac{d^2\psi}{dx^2} dx = \varepsilon\left[\frac{d\psi}{dx}\right]_\infty^d.$$

Now $(d\psi/dx)_{x\to\infty} = 0$, so that:

$$\sigma_d = \varepsilon\left(\frac{d\psi}{dx}\right)_{x=d} \tag{6.3.26}$$

and from eqn (6.3.20):

$$\sigma_d = \frac{-2\kappa kT\varepsilon}{ze} \sinh ze\psi_d/2kT$$
$$= \frac{-4n^0 ze}{\kappa} \sinh ze\psi_d/2kT. \tag{6.3.27}$$

Note that the sign of σ_d is opposite to that of ψ_d (since $z > 0$).

For a symmetric electrolyte in water at 25 °C, eqn (6.3.27) gives (Exercise 6.3.8):

$$\sigma_d = -11.74c^{1/2}\sinh(19.46z\psi_d) \tag{6.3.28}$$

in $\mu C\,cm^{-2}$ when ψ_d is in volts and c is in $mol\,L^{-1}$.

For very small potentials, where the linear eqn (6.3.11) can be used, a similar analysis leads to (Exercise 6.3.10):

$$\sigma_d = -\varepsilon\kappa\psi_d \tag{6.3.29}$$

so that

$$\frac{-\sigma_d}{\psi_d} = K_d = \varepsilon\kappa. \tag{6.3.30}$$

The quantity K_d is called the *integral* capacitance of the (diffuse) double layer and eqn (6.3.30) shows that for low potentials the diffuse layer behaves like a parallel plate condenser with a spacing of $1/\kappa$ between the plates, a charge of $+\sigma_d$ and $-\sigma_d$ on them, and a potential difference of ψ_d. For low potentials then, the Helmholtz model (Fig. 6.3.1) is quite satisfactory for many purposes. [It may be noted in passing that in this

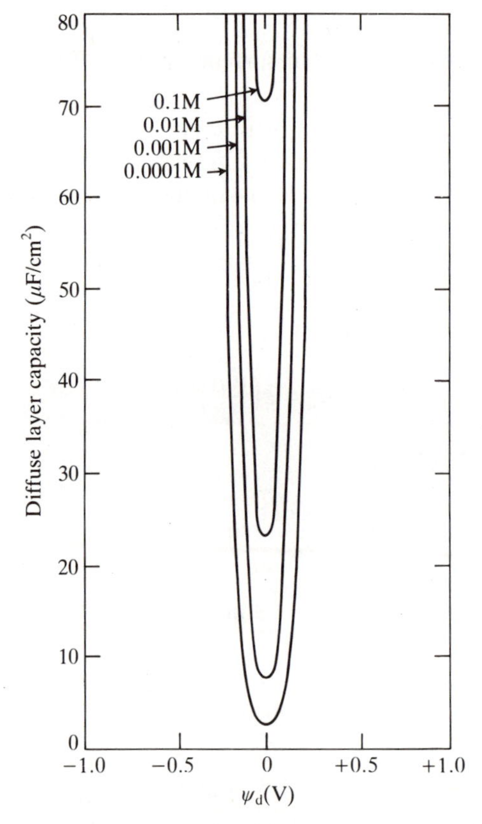

FIG. 6.3.4. Differential capacity of the diffuse double layer (from eqn (6.3.31)).

case also the integral capacitance is equal to the differential capacitance $(-\mathrm{d}\sigma_\mathrm{d}/\mathrm{d}\psi_\mathrm{d})$.]

One further quantity of importance is the differential capacitance of the diffuse double layer, C_d, defined by (from eqn (6.3.27)):

$$C_\mathrm{d} = \frac{-\mathrm{d}\sigma_\mathrm{d}}{\mathrm{d}\psi_\mathrm{d}} = \frac{2n^0 z^2 e^2}{\kappa kT} \cosh ze\psi_\mathrm{d}/2kT$$

$$= \varepsilon\kappa \cosh ze\psi_\mathrm{d}/2kT \qquad (6.3.31)$$

$$= 228.5zc^{1/2}\cosh(19.46z\psi_\mathrm{d})\,\mu\mathrm{F\,cm^{-2}} \qquad (6.3.32)$$

in water at 25 °C, for c in mole $\mathrm{L^{-1}}$ and ψ_d in volts. Values of C_d as a function of ψ_d at different electrolyte concentrations are given in Fig. 6.3.4. Note the similarity in the shape of C_d about the point $\psi_\mathrm{d} = 0$ and the experimental curve for C at $E \doteqdot E_\mathrm{ecm}$ and low electrolyte concentration in Fig. 6.1.5.

6.3.4 *The inner (compact) double layer*

We must now consider the potential distribution in the region $0 < x < d$. In the mercury–aqueous solution system we will find that d is of the order of 0.5 nm so this region can accommodate only a few layers of solvent molecules. Nevertheless, we assume, at least for the moment, that the concept of a dielectric permittivity remains valid and that Poisson's equation (6.3.3) is satisfied in this region. There are several models of varying complexity that can now be investigated, and their predictions compared with experiment.

(a) *Charge-free inner region.* From eqn (6.3.3) we have:

$$\mathrm{div}(-\varepsilon_\mathrm{i}\,\mathrm{grad}\,\psi) = 0$$

or

$$\frac{\mathrm{d}}{\mathrm{d}x}(\varepsilon_\mathrm{i}\,\mathrm{grad}\,\psi) = 0 \quad \text{so} \quad \varepsilon_\mathrm{i}(x)\frac{\mathrm{d}\psi}{\mathrm{d}x} = Q_1 \qquad (6.3.33)$$

where Q_1 is a constant. To evaluate Q_1 we note that, at $x = d$ using eqn (6.3.26):

$$\varepsilon_\mathrm{i}(d)\left(\frac{\mathrm{d}\psi}{\mathrm{d}x}\right)_{d_-} = \varepsilon_\mathrm{w}\left(\frac{\mathrm{d}\psi}{\mathrm{d}x}\right)_{d_+} = \sigma_\mathrm{d} \qquad (6.3.34)$$

(The symbol d_- means that $\dfrac{\mathrm{d}\psi}{\mathrm{d}x}$ is evaluated on the left-hand side of the line $x = d$ whilst d_+ is evaluated on the right-hand side. At points where the permittivity changes value there must obviously be a discontinuity in the derivative of ψ (cf. section 4.7)). The quantity on the left of eqn

(6.3.34) is fixed throughout the region $0<x<d$ and so:

$$\int_{\psi_d}^{\psi_0} \mathrm{d}\psi = \int_d^0 \frac{\sigma_d}{\varepsilon_i(x)} \mathrm{d}x$$

$$\psi_0 - \psi_d = \frac{-\sigma_d \cdot d}{\bar{\varepsilon}_i} \tag{6.3.35}$$

where $\bar{\varepsilon}_i$ is an average permittivity over the inner layer region defined by

$$\frac{1}{\bar{\varepsilon}_i} = \frac{1}{d}\int_0^d \frac{\mathrm{d}x}{\varepsilon_i(x)}. \tag{6.3.36}$$

The capacity of this inner layer region is then given by:

$$K_i = C_i = \frac{-\sigma_d}{\psi_0 - \psi_d} = \frac{\sigma_0}{\psi_0 - \psi_d} = \frac{\bar{\varepsilon}_i}{d} \tag{6.3.37}$$

where σ_0 is the charge on the metal (which balances the charge in the diffuse layer). Note that here again the integral and differential capacities are the same and the inner region behaves as a simple parallel plate capacitor of plate separation d, and some characteristic average permittivity.

(b) *Adsorbed charge in the inner region.* The simplest model to account for the possibility of additional charge adsorbed in the inner region is that proposed initially by Stern (1924) and refined by Grahame (1947). All of the ions are assumed to be confined to a layer at a distance $x=b$ from the metal surface and again they are treated as point charges, in a first approximation (Fig. 6.3.5). Since the regions $0<x<b$ and $b<x<d$ are again free of any charge, the same procedure can be used to arrive at expressions for the potential drop across each region:

$$\psi_0 - \psi_i = \sigma_0 \frac{b}{\bar{\varepsilon}_1} \tag{6.3.38}$$

and

$$\psi_i - \psi_d = -\sigma_d \frac{(d-b)}{\bar{\varepsilon}_2} \tag{6.3.39}$$

with average values of permittivity defined in each region. Charge balance also requires that

$$\sigma_0 + \sigma_i + \sigma_d = 0. \tag{6.3.40}$$

It is common practice to assume that ε_1 and ε_2 are constant so that eqns (6.3.38) and (6.3.39) predict a linear change in potential in each region

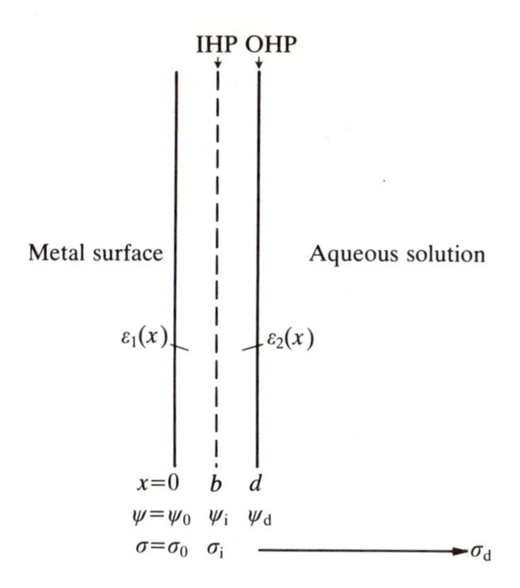

FIG. 6.3.5. The Gouy–Chapman–Grahame model of the electrical double layer. The plane where the diffuse layer begins (at $x = d$) is called the outer Helmholtz plane (OHP) and additional adsorbed charge is assumed to be confined to another plane called the inner Helmholtz plane (IHP). At the mercury–aqueous solution interface the IHP is assumed to be the locus of the centres of adsorbed (dehydrated) anions whilst the OHP is the plane of closest approach of (hydrated) cations.

but the above derivation shows that this is not necessary. If $\varepsilon_i(x)$ is allowed to vary smoothly from $x = 0$ to $x = d$, then ψ will also vary smoothly, rather than discontinuously. Such models have been examined by Buff and Goel (1969), Levine (1971) and Robinson and Levine (1973).

A complete solution of the potential and charge distribution in the double layer would give values for ψ_0, ψ_i, ψ_d, σ_0, σ_i, and σ_d. So far we have four equations—(6.3.27), (6.3.38), (6.3.39), and (6.3.40)—and six unknowns. In the case where there is no charge in the plane $x = b$, there are four unknowns (ψ_0, ψ_d, σ_0, and σ_d) and three equations (eqns (6.3.27), (6.3.35 or 37), and (6.3.40) with $\sigma_i = 0$). One further equation comes from the imposed external e.m.f. E, via eqn (6.1.24). Using eqn (6.2.4) we have

$$d(\Delta\phi) = d\psi_0 + d(\Delta\chi_{\text{dipole}}) \tag{6.3.41}$$

and for the moment we will ignore the last term. Then, using eqn (6.1.24):

$$dE = d(\phi_2 - \phi_3) = d\psi_0. \tag{6.3.42}$$

The assumption that $(\Delta\chi_{\text{dipole}})$ is constant would appear at first sight to be a difficult one to justify. We will use eqn (6.3.42) to develop eqn (6.4.9), relating the measured capacitance of the mercury drop to the capacitances C_i and C_d of the model. Fortunately, the error introduced by this procedure is small because the contribution of the dipoles to the capacity of the interface is so large that, in a series arrangement, as we assume, the effect on the total capacitance is negligible. (We will return to this point in section 6.4.)

The four equations would appear to be sufficient to completely describe the system if $\sigma_i = 0$. But what if $\sigma_i \neq 0$? That means that some ions are sitting very close to the mercury surface. It is assumed that the plane $x = d$ is the plane of closest approach of hydrated ions so any ion in the plane $x = b$ must be dehydrated (at least on the side next to the mercury surface). It can be in that position only if the free energy of adsorption more than compensates for the work done in dehydrating it.

To describe such adsorbed ions requires an isotherm of the form:

$$\sigma_i = z_i e n_i^s = z_i e f(N_s, a_i, \psi_i, \theta_i) \tag{6.3.43}$$

where n_i^s is the number of ions adsorbed per unit area, which is expected to be a function of (a) the number of adsorption sites on the surface (N_s), (b) the activity of ion i in the solution (a_i), (c) the local electrostatic potential, and (d) θ_i, an extra free energy term that takes into account all other special effects that the ion experiences when it is in the plane $x = b$. This will involve its special interaction with the metal, which will include purely physical interactions (like the image force†) but may also involve 'chemical' effects (i.e. effects that are only described in terms of molecular orbital overlap and bond formation).

Exercises. 6.3.1 Establish eqns (6.3.13) and (6.3.14) using the fact that $\mathcal{F} = 96\,485\ \mathrm{C\,mol^{-1}}$, $R = 8.31\ \mathrm{J\,K^{-1}\,mol^{-1}}$, $\varepsilon_0 = 8.85_4 \times 10^{-12}\ \mathrm{F\,m^{-1}}$ (i.e. $\mathrm{CV^{-1}\,m^{-1}}$) and $\varepsilon_r = 78.5$.

6.3.2 Calculate the value of $1/\kappa$ in each of the following solutions: (i) 10^{-2} M KCl; (ii) 10^{-4} M KCl; (iii); 10^{-6} M KCl; (iv) 10^{-3} M NaCl + 10^{-4} M Na_2SO_4; (v) 10^{-3} M K_2SO_4; (vi) 5×10^{-3} M $MgSO_4$.

6.3.3 Show that the solution to eqn (6.3.11) for a flat surface can be written:

$$\psi = \psi_{x=d} \exp(-\kappa(x-d)).$$

6.3.4 Establish eqns (6.3.15) and (6.3.18).

6.3.5 Establish eqn (6.3.19) using the identity $\cosh p = 2\sinh^2(p/2) + 1$. Also derive eqn (6.3.20).

6.3.6 Establish eqn (6.3.21) using the fact that $\sinh p = 2\sinh p/2 \cdot \cosh p/2$; $\operatorname{sech} p = 1/\cosh p$ and $\mathrm{d}\tanh p/\mathrm{d}p = \operatorname{sech}^2 p$.

† The image force of a charge or dipole near a conductor is an attractive force generated by the interaction between the charge and its (oppositely charged) image in the conductor.

6.3.7 Show that for an asymmetric electrolyte:

$$\left(\frac{d\psi}{dx}\right)^2 = \frac{2kT}{\varepsilon}\sum n_i^0[\exp(-z_i e\psi/kT) - 1]$$

and so

$$\sigma_d = (-\text{sgn}\ \psi_d)\{2\varepsilon kT \sum n_i^0[\exp(-z_i e\psi_d/kT) - 1]\}^{1/2}$$

where (sgn ψ_d) is the algebraic sign of the potential (i.e. sgn $\psi_d = \psi_d/|\psi_d|$).

6.3.8 Establish the value of the conversion factors in eqns (6.3.28) and (6.3.32).

6.3.9 The diffuse layer charge σ_d can be broken up into a contribution σ_d^+ from the cations and σ_d^- from the anions. Use the fact that

$$\sigma_d^+ = \int_d^\infty (\rho_+ - zen_+^0)\, dx$$

and $\rho_+ = zen_+^0 \exp(-ze\psi/kT)$ to show that

$$\sigma_d^+ = \frac{2zen^0}{\kappa}[\exp(-ze\psi_d/2kT) - 1].$$

Similarly

$$\sigma_d^- = \frac{2zen^0}{\kappa}[1 - \exp(ze\psi_d/2kT)].$$

6.3.10 Establish eqn (6.3.29) using the results obtained in Exercise 6.3.3.

6.4 Comparison with experiment

6.4.1 *Presence or absence of specific adsorption*

The similarity in the electrocapillary curves on the cathodic side (Fig. 6.1.4) for sodium and potassium salts suggests that these two ions behave in the same way at the interface. They are said to be *indifferent* ions and they are not specifically adsorbed at the interface. This can be explained by assuming that they remain hydrated as they approach the mercury surface and never get closer than the plane $x = d$. Their adsorption is described purely by their response to the electric field in the diffuse part of the double layer.

By contrast the monovalent anions all behave as individuals. The model assumes that this is because such ions are able to penetrate into the inner region, presumably because their hydration energy is lower than that of cations, but also, possibly, because anions can interact more effectively with the metal. They are then said to be *specifically adsorbed*. Not all anions are able to do this, however; the evidence given below suggests that the fluoride ion, at least at modest concentrations, is not able to penetrate into the inner region.

The question is: can we identify when specific adsorption is occurring? It would be preferable if we could do this in a completely model-independent way. Although that is not quite possible it is certainly possible to offer an experimental criterion that can distinguish those situations in which there is *no need* to invoke specific adsorption. If this criterion is obeyed it means that the adsorption behaviour can be quite adequately accounted for by appealing to diffuse double layer theory alone.

The test for the absence of specific adsorption relies on an examination of the dependence of the electrocapillary maximum (e.c.m.) or point of zero charge (p.z.c.) on the concentration (or activity) of the background electrolyte solution.

The influence of electrolyte on the e.c.m. was studied by Esin and Markov (1939) and the coefficient, β, defined by:

$$\beta = \mathcal{F}\left(\frac{\partial E}{\partial \mu}\right)_{\sigma_0} \tag{6.4.1}$$

is called the Esin and Markov coefficient. When β is evaluated at the e.c.m. (i.e. where $\sigma_0 = 0$), it is found to have a value of zero in some cases, and for such electrolytes we can assert that there is no *need* to invoke specific adsorption to account for their behaviour in the neighbourhood of the e.c.m. The behaviour can be quite adequately described in terms of the ion concentrations in the *diffuse* part of the double layer.

Of course, we are interested in the behaviour for all values of σ_0 and the complete analysis of the problem is not restricted to the immediate neighbourhood of the e.c.m. Delahay (1966) sets out the development, as it was applied by Parsons (1957), and we will follow that approach here.

The e.m.f., E, appearing in the expression (6.4.1) for β is normally measured in a cell *with a liquid junction.* (That means that the test solution is separated from the reference electrode solution by a membrane or porous plug which permits ion transport but prevents physical mixing. The galvani potential of the reference electrolyte is then different from that of the test solution and the difference is called the liquid junction potential. Considerable effort goes into minimizing (or at least maintaining constant) this potential difference. The exact thermodynamic analysis of the Esin and Markov effect is best done using a slightly modified coefficient, β_0, involving a reference electrode *without a liquid junction* (like the one treated in section 6.1).

It is not difficult to show from eqn (6.1.12) that (Exercise 6.4.1):

$$\beta_0 = \mathcal{F}\left(\frac{\partial E_+}{\partial \mu}\right)_{\sigma_0} = -\mathcal{F}\left(\frac{\partial \Gamma_-}{\partial \sigma_0}\right)_{\mu}. \tag{6.4.2}$$

The coefficient thus measures how the surface charge is balanced by an excess of anions or a deficit of cations. In the absence of specific adsorption the balance is determined *solely* by the Poisson–Boltzmann equation and it can be shown (Parsons 1957) that, in that case (Exercise 6.4.3):

$$\beta_0 = \left(\frac{\partial \sigma_d^-}{\partial \sigma_0}\right)_\mu = -\left(\frac{\partial \sigma_d^-}{\partial \sigma_d}\right)_{a_\pm} = \frac{|z|\mathscr{F}}{RT}\left(\frac{\partial E_+}{\partial \ln a_\pm^2}\right)_{\sigma_0}$$
$$= -\tfrac{1}{2}\exp\left[\sinh^{-1}\left(-\frac{\sigma_d}{2B}\right)\right]\left[1+\left(\frac{\sigma_0}{2B}\right)^2\right]^{-1/2}. \qquad (6.4.3)$$

(Note that $\sinh^{-1}$ is the inverse function (sometimes written arcsinh) and *not* the reciprocal.)

Here σ_d^- is the amount of negative charge per unit area in the diffuse layer and $a_\pm$ is the mean ionic activity of the electrolyte. B is a constant ($=2zen^0/\kappa$ from Exercise 6.3.9). This equation can be applied to data over the entire range of μ and E_+ values for a particular salt to determine whether its behaviour can be explained without invoking specific adsorption. (See Delahay 1966, p. 55). In particular, it can be seen from eqn (6.4.3) that as the surface is polarized more positively or more negatively, the limiting values of β_0 are (Exercise 6.4.2):

$$\left(\frac{\partial \sigma_d^-}{\partial \sigma_0}\right)_\mu = -z\mathscr{F}\left(\frac{\partial \Gamma_-}{\partial \sigma_0}\right)_\mu \to 0 \quad \text{for} \quad \begin{cases}\sigma_d \to \infty \\ \sigma_0 \to -\infty\end{cases}$$

and

$$\left(\frac{\partial \sigma_d^-}{\partial \sigma_0}\right)_\mu \to -1 \quad \text{for} \quad \begin{cases}\sigma_d \to -\infty \\ \sigma_0 \to \infty\end{cases}. \qquad (6.4.4)$$

Parsons (1957) shows that for sodium fluoride on mercury eqn (6.4.3) is valid for all of the data obtained by Grahame.

In colloid systems it is rare to apply such a strong test. More usually we examine the behaviour near the point of zero charge (or the e.c.m.). From eqn (6.4.3), the β_0 coefficient is then $-\tfrac{1}{2}$ *and it is independent of electrolyte concentration.* In simple physical terms this means that the diffuse layer charge is made up of a certain quantity of anions and an *equal* deficit of cations (or *vice versa*). This is a direct consequence of the linear form of the Poisson–Boltzmann equation for low potentials.

Now suppose that instead of using a reference electrode which is reversible to the cation, as we have done so far, we use a reference electrode of fixed electrolyte activity (like a calomel electrode) with a liquid junction leading to the aqueous solution. The e.m.f. imposed is then say E_r, where (Exercise 6.4.7):

$$E_r = E_+ + \frac{RT}{z\mathscr{F}}\ln a_\pm + \text{a constant} \qquad (6.4.5)$$

and the constant includes the liquid junction potential difference between the reference electrode and the aqueous solution. Equation (6.4.2) now becomes

$$\left(\frac{\partial E_{\mathrm{r}}}{\partial \mu}\right)_{\sigma_0} = \frac{1}{RT}\frac{\partial E_{\mathrm{r}}}{\partial \ln a_{\pm}^2} = \left(\frac{\partial E_+}{\partial \mu}\right)_{\sigma_0} + \frac{1}{2z\mathscr{F}}$$

$$= \frac{1}{z\mathscr{F}}\left\{\frac{\partial \sigma_{\mathrm{d}}^-}{\partial \sigma_0} + \frac{1}{2}\right\} \tag{6.4.6}$$

i.e.

$$\left(\frac{\partial E_{\mathrm{r}}}{\partial \mu}\right)_{\sigma_0} = \frac{1}{z\mathscr{F}}\{\beta_0 + \tfrac{1}{2}\}. \tag{6.4.7}$$

Thus, at the e.c.m. where $\beta_0 = -\frac{1}{2}$, the quantity

$$\mathscr{F}\left(\frac{\partial E_{\mathrm{r}}}{\partial \mu}\right)_{\sigma_0=0} = \beta' = 0.$$

That is to say, the point of zero charge is not affected by the electrolyte concentration *if it is measured with respect to a reference electrode with a liquid junction.*

This is the usual test that is used in colloid chemistry. If the point of zero charge can be determined at a number of electrolyte concentrations and it turns out to be independent of the electrolyte concentration, this serves to establish that the electrolyte being used is an indifferent one (i.e. is not specifically adsorbed).

How is this definition related to our ideas about the nature of specific adsorption? If an ion is specifically adsorbed it has a special relation with the surface and we expect some adsorption to occur even when the charge on the metal is zero (at the e.c.m. or p.z.c.). Suppose it is the anion that is specifically adsorbed (as is often the case on mercury). Then the presence of anions near the surface tends to drive electrons back into the bulk mercury and in order to restore the condition $\sigma_0 = 0$ it is necessary to give the mercury a more negative polarization. In other words, the position of the e.c.m. shifts to more negative values. As the activity of that ion in the aqueous solution is increased, the tendency to adsorb is increased and the e.c.m. moves further to the cathodic side. If there is no specific adsorption this effect is absent and so:

$$\left(\frac{\partial E}{\partial \mu}\right)_{\sigma_0=0} = 0 \quad \text{if} \quad \sigma_{\mathrm{i}} = 0. \tag{6.4.8}$$

The difference in the surface tension behaviour is illustrated schematically in Fig. 6.4.1. Note that if there is no specific adsorption, γ_{ecm} is

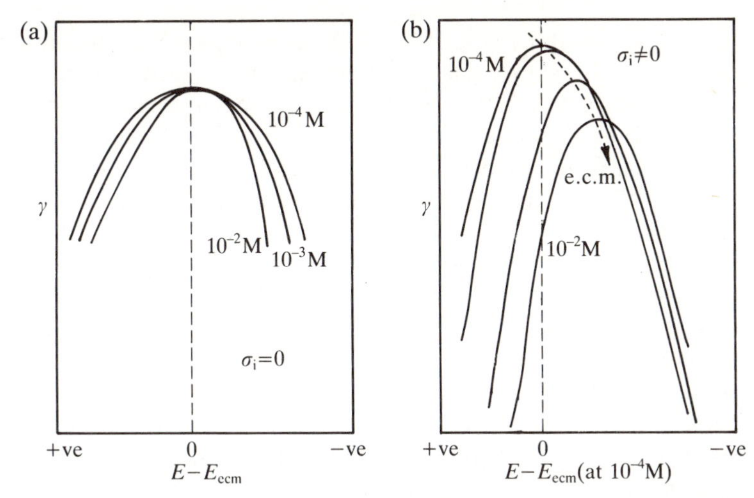

FIG. 6.4.1. Illustrating the effect of specific adsorption of an anion on the electrocapillary curves at different salt concentrations. (a) No specific adsorption. (b) With specific adsorption.

unaffected by the salt concentration whereas when $\sigma_i \neq 0$ the maximum is *lowered* and *shifted* as μ increases and so specific adsorption is readily identified.

6.4.2 *No specific adsorption*

The behaviour of sodium fluoride solutions can be analysed on the assumption that $\sigma_i = 0$ (i.e. neither the sodium nor the fluoride ion is specifically adsorbed at the mercury–solution interface). In that case $\sigma_0 = -\sigma_d$ so that (from eqns 6.3.31, 37, and 42):

$$\frac{1}{C_T} = \frac{dE}{d\sigma_0} = \frac{d\psi_0}{d\sigma_0} = \frac{d(\psi_0 - \psi_d)}{d\sigma_0} + \frac{d\psi_d}{d\sigma_0}$$
$$= \frac{1}{C_i} + \frac{1}{C_d} \tag{6.4.9}$$

where C_T is the total differential capacity of the double layer. The system behaves as a pair of capacitors in series. (This is obvious when $\sigma_i = 0$ but is not strictly true for $\sigma_i \neq 0$ (see Exercise 6.4.4).)

It is apparent from eqn (6.4.9) that the smaller of the two capacitances on the right is the one that determines the observed value C_T. Incorporating into the model an inner double layer region of limited capacitance has the effect of limiting C_T so that the very large values calculated for C_d (Fig. 6.3.4) are not observed. In fact, C_d only influences the observed C_T in the neighbourhood of E_{ecm} and then only at low electrolyte concentrations (Fig. 6.4.2). A more detailed examination of the data in Fig. 6.1.5 shows that C_i is not constant, as is suggested in Fig. 6.4.2, but depends upon the charge on the metal surface. Grahame (1947) made the important suggestion that C_i could be estimated from the total capacitance at high electrolyte concentration (~1 M) when C_d is so large that $C_T \approx C_i$. He then calculated the expected capacitance at lower concentrations, from eqn (6.4.9), assuming only that C_i was the same *when the charge on the metal was the same.*

Figures 6.4.3 and 6.4.4 show that these calculations go a long way towards accounting for the capacitance data for the mercury/sodium fluoride solution interface. The value of C_i can be calculated as a function of σ_0 at various temperatures (Fig. 6.4.5). To interpret these in terms of eqn (6.3.37) we need some idea of the distance d which we have earlier suggested is of the order of the radius of a hydrated cation. If $d = 0.5$ nm then a C_i value of 32 μF cm^{-2} corresponds to $\varepsilon_i = \varepsilon_0\varepsilon_r = 32 \times 10^{-6} \times 10^4$ F m$^{-2} \times 5 \times 10^{-10}$ m $= 1.6 \times 10^{-10}$ F m^{-1} so that $\varepsilon_r = 18$ compared to the normal value for bulk water of about 80. Is it reasonable to expect such a low value for the relative permittivity in this region? The high value of ε_r in bulk water is due to the ability of the water molecules to orient themselves in an applied field (section 2.2) and to reorient themselves to follow the field if it is changing. The measurement of C_i is done with an alternating applied field (section 6.1) but the water

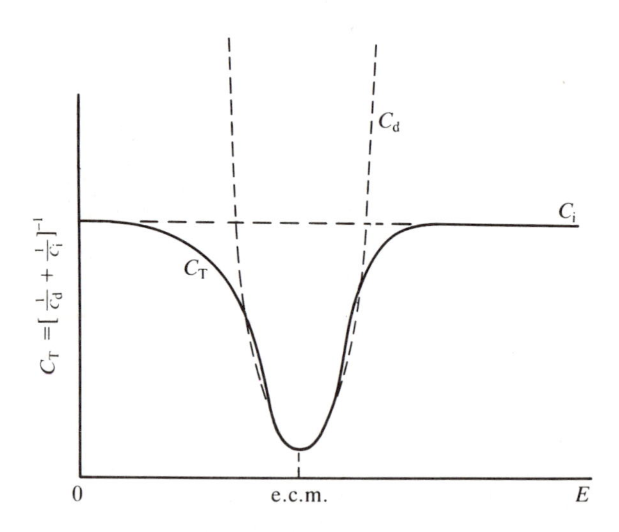

FIG. 6.4.2. Schematic diagram of the effect of adding capacitances in series: at low concentrations and near the e.c.m. the differential capacitance determines the behaviour but in other situations C_i is more important.

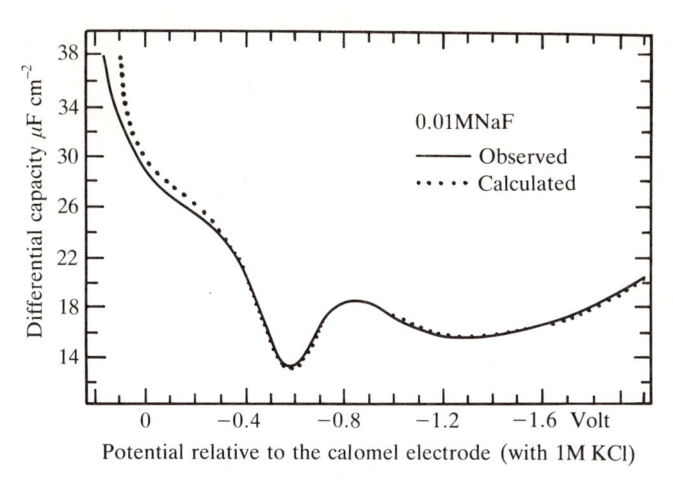

FIG. 6.4.3. Comparison of calculated and experimental differential capacitance at 25 °C in water using Grahame's method. (After Payne 1972, with permission.)

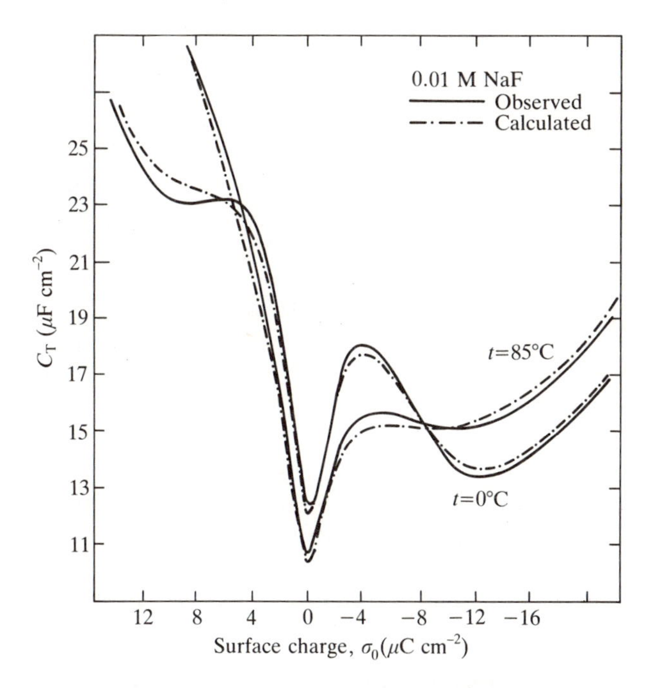

FIG. 6.4.4. Differential capacitance as a function of surface charge calculated by Grahame's method at two different temperatures. (After Payne 1972, with permission.)

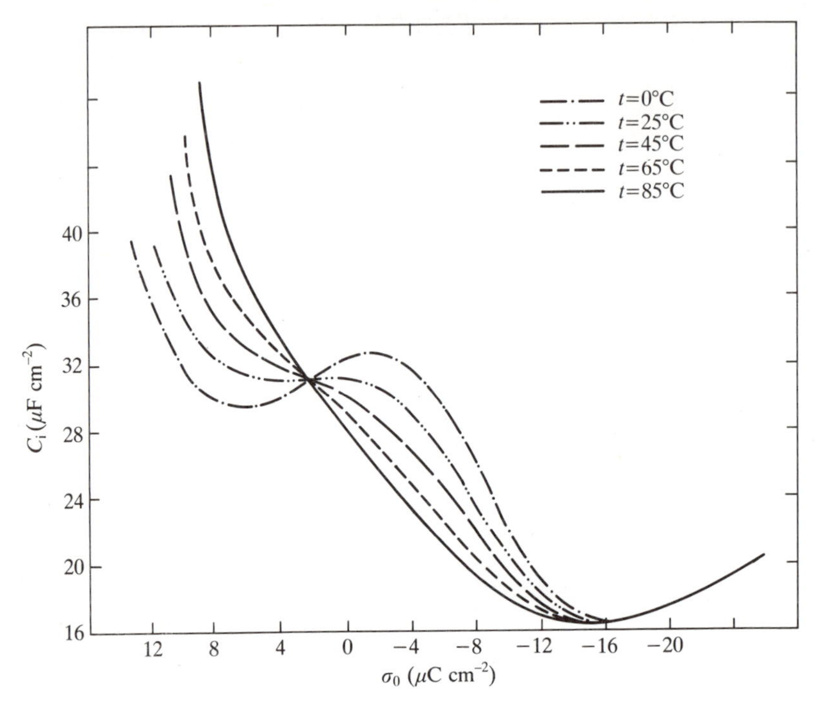

FIG. 6.4.5. Capacity of the inner region of the double layer on mercury in the presence of NaF.

molecules near the mercury surface are not able to follow that field as easily as those in bulk water because they are already oriented to a considerable extent by the very high electric field near the surface. The potential drop across the inner layer is, from eqn (6.3.37) equal, to σ_0/C_i; for $C_i = 32\ \mu F\ cm^{-2}$ and $\sigma_0 = 16\ \mu C\ cm^{-2}$ this has a value of 0.5 v. If that potential drop occurs across a distance of 0.5 nm the field strength is $10^9\ V\ m^{-1}$. The energy of a dipole in an electric field is given by $\boldsymbol{p} \cdot \boldsymbol{E}$ where $\boldsymbol{p}$ is the dipole moment which, for water, is 6×10^{-30} C m. Its potential energy in the field is, therefore, of order 6×10^{-21} J which is about 1.5 kT. Although this effect alone might not be expected to lower ε_r from 80 to 18 there are other effects (including the local field of the ions in the OHP and the image force in the mercury) which further restrict the orientational motion of the water molecules in response to the field. In the extreme case of a completely oriented layer, the anticipated value of ε_r is about 6 for water, so a mean value of 18 is not unreasonable. At least in this respect the model is self-consistent.

It should also be noted that since $\sigma_i = 0$, and $\sigma_0 = -\sigma_d$ the potential in the double layer can be calculated from eqn (6.3.27). It turns out that, even at extreme polarizations, $|\psi_d|$ is never more than about 0.2 V, and then only at very low electrolyte concentrations (Exercise 6.4.6). As the electrolyte concentration increases $|\psi_d|$ falls to less than 50 mV in 1 M solution. This is why the diffuse layer capacitance is well described by the simple Poisson–Boltzmann equation, at least at modest electrolyte concentrations.

Returning now to the neglect of the $\Delta\chi_{dipole}$ term in eqn (6.3.42) we can see how Grahame's procedure (Figs 6.4.3–5) largely circumvents the problem. Using the more exact expression for dE in eqn (6.4.9) we have:

$$\frac{1}{C_T}=\frac{dE}{d\sigma_0}=\frac{d[\psi_0+\Delta\chi_{dipole}]}{d\sigma_0}$$

$$=\frac{d(\phi_2-\psi_d)}{d\sigma_0}+\frac{d\psi_d}{d\sigma_0} \qquad (6.4.10)$$

and the first term can still be identified with $1/C_i$. Grahame's procedure amounts to assuming that the charge on the metal is much more important than, say, the electrolyte concentration, in determining the detailed structure (including dipole orientation) of the inner layer. It does not assume that $\Delta\chi_{dipole}$ is constant under all conditions but only that at any particular value of σ_0, $\Delta\chi_{dipole}$ is unaffected by the electrolyte concentration. Figure 6.4.5 shows clearly that the inner layer capacitance varies significantly with the charge on the metal and if the variation is attributed largely to dipole orientation (so that the thickness parameters, b and d, are constant) then $\Delta\chi_{dipole}$ will also vary significantly with charge.

Even at the e.c.m. in the absence of specific adsorption when $\psi_0=0$ (from eqns (6.3.27) and (6.3.37)), the value of $(\phi_2-\phi_3)_{ecm}=\Delta\chi_{dipole}$ may be quite large (of the order of tens or hundreds of millivolts). Further discussion of this point can be found in Bockris and Reddy (1970) and Sparnaay (1972). It is of particular importance in the study of the adsorption of uncharged (organic) molecules because they usually act as dipoles, which compete with water molecules for sites at the surface and hence profoundly affect the χ-('chi'-) potential.

A complete molecular model of the interfacial region would require a description, in molecular terms, of the permittivity ε_i and distance d as functions of the polarization (σ_0) and temperature. Quite a lot of work has been done in this area. The early models are reviewed by MacDonald and Barlow (1964), by Levine *et al.* (1967), and by Bockris and Reddy (1970), while Sparnaay (1972, pp. 92–104) gives a very good description of the models current up to that time. More recently, Parsons (1975b) and Oldham and Parsons (1977) have developed a four-state model for the water molecules, based on an earlier suggestion of Damaskin and

Frumkin (1974). Salem (1976) and Damaskin (1977) also offer simple models of the same system. All are attempting, with varying degrees of success, to describe the dependence of ε_i on σ_0 and temperature. A more general description would need to take account of the metal surface (Trasatti 1971; Gardiner 1975). The subtlety of the behaviour even when $\sigma_i = 0$ makes it obvious that a complete description of the inner region when $\sigma_i \neq 0$ will be a very difficult task indeed. A good model description should also allow one to transfer the calculation to other solvents (on which there is also a great deal of data) with a realistic adjustment of the distance parameter d and the use of independently determinable properties like dipole moment.

In colloid chemical systems it is rare to find solid surfaces that are atomically smooth. There is, therefore, seldom much point in attempting a detailed description of ε_i and d in the inner region. It often suffices to postulate a value for the capacitance C_i and to use that as a fundamental parameter of the system. The main point we will take from this analysis is that a good deal of information about the nature of a charged interface can be obtained from the study of the adsorption of non-specifically adsorbed ions. These are called *indifferent* electrolyte ions and a systematic study of any colloidal system always begins with a study of its behaviour towards such electrolytes (commonly the alkali nitrates on silver iodide or alkali halides on other systems). Indifferent ions are assumed to interact with the surface only in response to the 'long-range' forces that operate beyond the Outer Helmholtz Plane (OHP). Any ion that can penetrate into the inner region becomes subject to other short range and much more highly specific interactions. Because they are more strongly hydrated, cations tend to be indifferent on the mercury surface whereas anions are often specifically adsorbed.

6.4.3 *Interpretation of specific adsorption*

Figure 6.1.4 suggests that the e.c.m., when referred to the same reference electrode, is different for the different halides of sodium. Measurements on the chloride, bromide, and iodide all give curves like those shown in Fig. 6.4.1(b), indicating that for all of these systems, in contrast to the fluoride, the anion is specifically adsorbed. The amount of specific adsorption of the anion can be determined if it can be assumed that the cation is not specifically adsorbed. Measurements of Γ_+ $(= -(\partial\gamma/\partial\mu)_{E_-})$ can then be set equal to $\sigma_d^+/z\mathscr{F}$ where σ_d^+ is the contribution of cations to the diffuse layer charge. This allows calculation of ψ_d (from Exercise 6.3.9) and hence σ_d^-. Then since $\sigma_0 = -(\sigma_i + \sigma_d)$ we can obtain σ_i (which will be negative in this case). Values of σ_i for a variety of anions on the mercury surface are shown in Fig. 6.4.6.

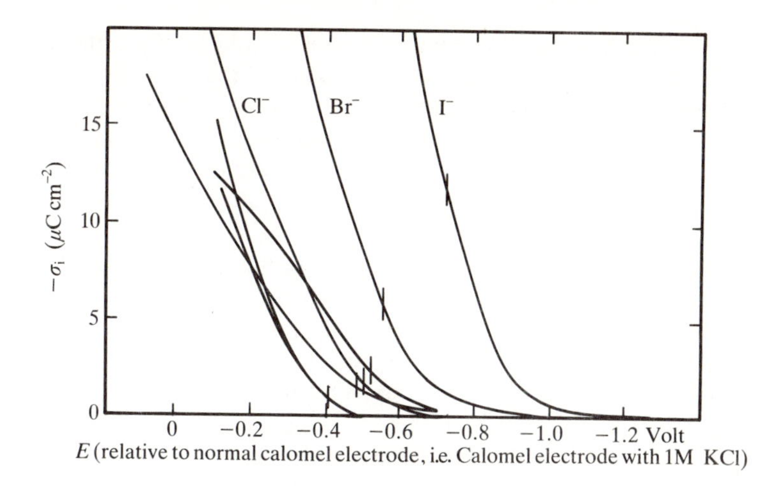

FIG. 6.4.6. Amount of specifically adsorbed anions in various electrolytes (0.1 M) in contact with mercury at 25 °C. Curves computed by Parsons from Grahame's data. Vertical lines indicate p.z.c. (From Mott and Watts-Tobin 1961, with permission.)

A great deal of work has been done on the development of adsorption isotherms to describe curves of this sort in terms of equations of the form of eqn (6.3.43). Since our primary concern is with colloidal systems we will not examine this material in detail because it is adequately treated elsewhere. (See e.g. Delahay 1966.) There are, however, some general ideas that come out of this work and that have a bearing on the treatment of colloidal systems.

Most, if not all, treatments assume that the adsorbed ion is, in effect, in a separate phase and that its electrochemical potential is equal to that in the bulk. They differ only in the degree of sophistication that is brought to bear in calculating $\tilde{\mu}_i$ for the specifically adsorbed ions. The original Stern isotherm can be derived as an extension of the Langmuir isotherm by incorporating an expression for the adsorption equilibrium constant, $\mathscr{K}$. Langmuir's isotherm (derived in any physical chemistry text, e.g. Atkins 1978, p. 944) is:

$$\eta = \frac{n_i^s}{N_s} = \frac{\mathscr{K}x_i}{1+\mathscr{K}x_i} \tag{6.4.11}$$

where η is the coverage and x_i is the mole fraction of i in the bulk solution. Then putting $\Delta G^{\ominus}_{\text{ads}}$ (per mole) $= -RT \ln \mathscr{K}$ we have (from eqn (6.3.43)):

$$\sigma_i = \frac{z_i e N_s x_i \exp(-\Delta G^{\ominus}_{\text{ads}}/kT)}{1 + x_i \exp(-\Delta G^{\ominus}_{\text{ads}}/kT)} \tag{6.4.12}$$

where the $\Delta G^{\ominus}_{ads}$ values now refer to one ion rather than a mole. The term in the denominator accounts for the effect of ions already present in the layer and is important when the sites are almost fully occupied. This is seldom the case for adsorbed charges (because of lateral repulsion) so Grahame (1947) neglects that term and uses a simpler expression:

$$\sigma_i = 2z_i ern_i^0 \exp(-\Delta G^{\ominus}_{ads}/kT) \tag{6.4.13}$$

where r is the radius of the adsorbed ion. This amounts to estimating N_s, the number of adsorption sites per unit area on the surface, as equal to $2rN_w$ where N_w is the number of water molecules per unit volume.

Both of these isotherms when expressed in terms of the coverage, $\eta(=n_i^s/N_s)$, can be put in the form (Exercise 6.4.8):

$$\ln \eta + h(\eta) = \ln a_i - \frac{\Delta G^{\ominus}_{ads}}{kT} \tag{6.4.14}$$

where $h(\eta) = -\ln(1-\eta)$ for the Stern (Langmuir) isotherm and a_i (the activity of ion i in the bulk) is usually set equal to x_i. Improvements in the description can then involve either the function $h(\eta)$, which is an entropic correction for ion size (put equal to zero in eqn (6.4.13)) or in the calculation of $\Delta G^{\ominus}_{ads}$. The simplest analysis would begin with:

$$\Delta G^{\ominus}_{ads} = z_i e\psi_i + \theta_i \tag{6.4.15}$$

where ψ_i is the electrostatic potential in the Inner Helmholtz Plane (IHP) (called the *macro*potential) and θ_i incorporates all interactions other than the 'macroscopic' electrical one.

When this is done it is found that θ_i depends upon the state of charge or polarization of the surface (but see section 6.4.4 below). When testing a particular isotherm (i.e. the relation between η and a_i) it is, therefore, essential to hold either σ_0 or E constant. A review of the various isotherms is given by Delahay (1966, Chapter 5). The relative merits of maintaining constant E (the Frumkin (1964) approach) or constant σ_0 (the Parsons (1964) approach) still provoke some argument but we will adopt the latter procedure, since it is more consistent with the way the boundary conditions were applied to solve the Poisson equation (6.3.33).

Delahay (1966, p. 89–91) shows how data like that in Fig. 6.4.6 can be used to determine the value of $\Delta G^{\ominus}_{ads}$ for an adsorbed anion as a function of charge on the electrode. For the iodide ion, Parsons (1959) has shown that Grahame's data can be described if $\Delta G^{\ominus}_{ads}$ varies linearly with adsorbed charge from about $-27.5kT$ at $\sigma_0 = 20\,\mu C\ cm^{-2}$ to $-4.5kT$ at $\sigma_0 = -10\,\mu C\,cm^{-2}$. Some of that variation is due to the $z_i e\psi_i$ term but there remains a significant variation in θ_i. According to Levine *et al.* (1962), that remaining variation can be accounted for quantitatively by the *ion self-atmosphere* effect, which is analogous to the 'self-atmosphere' effect appearing in the Debye–Hückel theory of strong electrolytes. This correction then, is in the nature of an activity coefficient correction for the adsorbed ions.

6.4.4 *The discreteness of charge (or adsorbed ion self-atmosphere) effect*

We noted above (eqn (6.4.4)) that in the absence of specific adsorption the modified Esin–Markov coefficient, β_0 varies from -1 for very positive metal charge to zero for very negative metal charge and is $-\frac{1}{2}$ at the e.c.m. Viewed in terms of equation (6.4.2) that means that when the metal is highly positively charged essentially all of the balancing charge is due to a surface excess of anions (and very little is caused by cation repulsion). When the metal surface is highly negatively charged the anions are of little importance ($\beta_0 \rightarrow 0$). At the e.c.m. both cations and anions have equal significance:

$$\beta_0' = -\tfrac{1}{2}(= -\mathcal{F}\,|\partial\Gamma_-/\partial\sigma_0| = -\mathcal{F}\,|\partial\Gamma_+/\partial\sigma_0|).$$

The behaviour is then described by the Debye–Hückel theory (Exercise 6.4.9), and β $(=\beta')$ is then equal to zero. (The ($'$) refers to quantities measured at the e.c.m. or p.z.c..)

What happens to the Esin–Markov coefficient, β, when there *is* specific adsorption? Experimentally it is found that β is often fairly constant over the whole range of electrode charge. The data calculated by Parsons (1957) for the halides (other than fluoride) on mercury, suggest that β_0 remains at about -1 for all values of the metal charge. Substituting this in eqn (6.4.7) would give:

$$\mathcal{F}\left(\frac{\partial E_\mathrm{r}}{\partial \mu}\right)_{\sigma_0} = \beta \doteqdot -\frac{1}{2z} \tag{6.4.16}$$

so that

$$\left(\frac{\partial E_\mathrm{r}}{\partial \ln a_\pm}\right)_{\sigma_0} \simeq -\frac{RT}{z\mathcal{F}} \tag{6.4.17}$$

or

$$\left(\frac{\partial E_\mathrm{r}}{\partial \log a_\pm}\right)_{\sigma_0} \simeq -2.303RT/z\mathcal{F}. \tag{6.4.18}$$

Esin and Markov (1939) were the first to study this effect, in the particular case of $\sigma_0 = 0$. They showed that the e.c.m. shifted in a regular (linear) fashion with the logarithm of the electrolyte concentration. Parson's analysis shows that the same result holds at other values of charge. The importance of Esin and Markov's work was that they were able to show that one would not expect such a large shift if the Stern theory were adequate to describe the specific adsorption. The observed rate of change of e.c.m. with $\log a_{\pm}$ is about twice as great as can be accounted for on Stern's theory. The theoretical explanation of the discrepancy was provided initially by Esin and Shikov (1943), and the analysis has been developed by Ershler (1946), Grahame (1958), and many others; an account of that work is given by MacDonald and Barlow (1964) and by Sparnaay (1972, pp. 112–119). The discrepancy is due to the initial assumption that the charge in the IHP is smeared out in a uniform sheet.

If instead of attributing a uniform charge density, σ_i, to the IHP, it is recognized that the ions are present as discrete charges with spaces between them, then the resulting potential profile through the interfacial region is altered: it becomes smoother (and more physically sensible). Ershler introduced the term *micropotential* to distinguish the calculated potential on the assumption that the ions were discrete, as distinct from the *macropotential* calculated on the basis of a smeared-out charge.

The potential ψ_i that appears in eqn (6.4.15) is, on the Stern theory, the macropotential. If instead we replace it with the *micropotential,* ψ_i', defined by

$$\psi_i' = \psi_i + \phi_a \tag{6.4.19}$$

where ϕ_a is the correction to the potential due to discreteness effects, then the new expression for $\Delta G^{\ominus}_{ads}$ is:

$$\Delta G^{\ominus}_{ads} = z_i e \psi_i' + \theta_i'. \tag{6.4.20}$$

If the correction, ϕ_a, has been correctly calculated θ_i' should become essentially independent of the state of charge of the electrode (Levine 1971); it then measures the purely chemical effects at the interface and is a function of temperature and pressure only.

The micropotential, ψ_i', is defined by MacDonald and Barlow (1964) as 'The potential, relative to the solution, at the position of the centre of an adsorbed ion with that ion absent but all other charges, including image charges, as they would be in the presence of the ion.' The earliest estimates by Esin and Shikov (1943) of the magnitude of the correction ϕ_a were based on two static hexagonal arrays of ions of opposite sign (representing adsorbed and diffuse layer charge) and this gave rise to too big an effect; it was more than enough to account for the anomalous values of Esin and Markov. Later calculations by Ershler (1946) used a more realistic treatment of the diffuse layer and this gives very much better agreement with the experimental results.

The application of discreteness-of-charge corrections to colloidal systems was pioneered by Levine and his collaborators (Levine *et al.* 1962, 1967) so we will outline his results here.

Levine's calculation takes some account of the properties of the diffuse layer and assumes the surface to be a conductor. He obtains for ϕ_a:

$$\phi_a = -\frac{\sigma_i K_i}{K_1 K_2} \frac{(1 + K_2/C_d)}{(1 + K_i/C_d)} \tag{6.4.21}$$

where K_i is the integral capacity of the whole inner layer, K_1 and K_2 refer to the inner and outer parts of that layer (Fig. 6.3.5), and C_d is the differential capacity of the diffuse layer (eqn 6.3.32). This calculation slightly overestimates ϕ_a but that will be less serious for a colloidal material because the value of ϕ_a is expected to increase if the wall is non-conducting, as is usually the case in colloidal systems.

At high concentrations or away from the p.z.c. the value of C_d will dominate over K_i and K_2 so that eqn (6.4.21) becomes:

$$\phi_a = -\sigma_i K_i / K_1 K_2 \tag{6.4.22}$$

Equations (6.4.19–22) imply that (Exercise 6.4.10):

$$\theta_i = \theta_i' - \frac{z_i e \sigma_i}{K_1 + K_2} \tag{6.4.23}$$

and since z_i and σ_i are of the same sign this implies that the apparent 'chemical' adsorption potential θ_i, is more negative than the true value, θ_i'. The discreteness of charge effect, therefore, leads to higher adsorption densities in the IHP than would otherwise be expected.

In colloidal systems, the surface charge is also, in many cases, discrete and this may be of considerable significance. Some of the models used to describe the behaviour of solid–electrolyte interfaces have been criticized because they require large (unrealistic) adsorption potentials for simple ions. It may be that much of the explanation lies in the neglect of the discreteness of charge effect.

Exercises. 6.4.1 Derive eqn (6.4.2) from eqn (6.1.12).

6.4.2 Establish the limiting behaviour quoted in eqns (6.4.4).

6.4.3 Use the results of Exercise 6.3.9 to establish eqn (6.4.3).

6.4.4 Show that if there is a charge in the IHP and we define integral capacitances of regions 1 and 2 (Fig. 6.3.5) as follows:

$$K_1 = \frac{\sigma_0}{\psi_0 - \psi_i} \quad \text{and} \quad K_2 = -\frac{\sigma_d}{\psi_i - \psi_d}$$

then the total differential capacity, C_T, is given by:

$$\frac{1}{C_T} = \frac{1}{K_1} - \left(\frac{1}{K_2} + \frac{1}{C_d}\right) \frac{d\sigma_d}{d\sigma_0}$$

if K_2 is independent of σ_0. Hence show that if σ_i is small and can be assumed to be constant

$$\frac{1}{C_T} \doteq \frac{1}{K_1} + \frac{1}{K_2} + \frac{1}{C_d}.$$

6.4.5 Show that the integral and differential capacities of the double layer are linked by the expression

$$C = K + E\left(\frac{\partial K}{\partial E}\right)_\mu$$

and hence, that they are equal at the e.c.m.

(Hint: $K \overset{\text{def}}{=} \sigma_0/(E - E_{ecm})$)

6.4.6 Estimate ψ_d as a function of σ_0 for the data given below. Plot ψ_d against σ_0 and also against $(E - E_{ecm}) = \psi_r$ (what Grahame calls the rational potential scale). (Assume that NaF is not specifically adsorbed.)

	NaF in 1 M solution at 25 °C†						
σ_0 (μC cm^{-2})	+7	+4	0	−4	−8	−12	−16
$E - E_{ecm}$ (V)	0.270	0.195	0	−0.130	−0.295	−0.530	−0.765

6.4.7 Justify eqn (6.4.5) by examining the cell Cu(2) | Hg | Hg_2Cl_2, KCl ‖ $HClO_4$ | Hg | Cu(1), where ‖ is a liquid junction. The left-hand electrode is a *calomel* (Hg_2Cl_2) *electrode* and the phase |Hg_2Cl_2, KCl| consists of an aqueous solution of the two salts. It is usually saturated with respect to both of them and maintained in that condition by being in contact with an intimate mixture of the two solids.

6.4.8 Show that eqn (6.4.14) can represent eqns (6.4.12) and (6.4.13).

6.4.9 Show that for the linear (Debye–Hückel) description of the double layer $\sigma_d^- = \sigma_d^+ = -\kappa\varepsilon\psi_d/2$, and so $\sigma_0 = -2\sigma_d^+$ where σ_d^+ is the contribution of cations to the diffuse layer charge.

6.4.10 Establish eqn (6.4.23).

6.5 Adsorption of (uncharged) molecules at the mercury–solution interface

A great deal of work has been done on the analysis of the adsorption of various organic materials at the mercury–solution interface which serves as a standard against which to compare adsorption at other (metallic)

† Caution: these figures have been read from published graphical data and are only approximate.

electrodes used in electrolysis and electrochemical cell manufacture. We will not examine that material here at all since it is covered in considerable detail elsewhere. (See e.g. Delahay 1966, pp. 91–119; Payne 1972, pp. 117–25; Bockris and Reddy 1970, pp. 791–801; Sparnaay 1972, pp. 119–32.) Suffice it to say that the adsorption can be clearly detected from the shape of the differential capacity curve (Fig. 6.5.1) and the

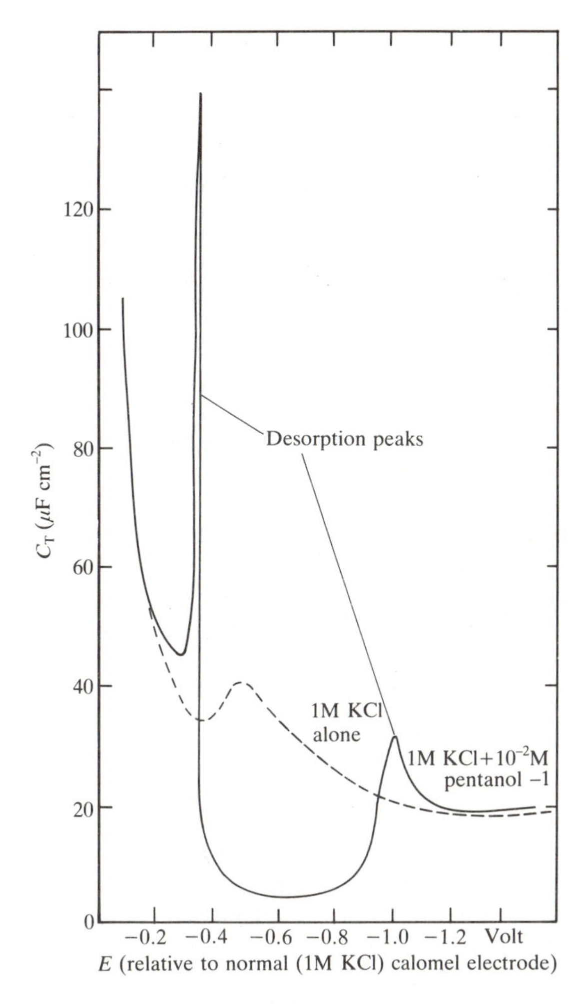

FIG. 6.5.1. Differential capacity in the presence of a (surface active) organic adsorbate. Note the pronounced adsorption and desorption peaks as the potential is swept from anodic to cathodic values. Note also the low flat capacitance in the region on either side of the e.c.m. Near the e.c.m. $C = K$ (Exercise 6.4.5) and the low capacitance can be immediately interpreted as a lower value of ε_i and/or a larger value of d for the inner layer capacitance. (C_d is very large and hence unimportant in 1 M KCl solution.)

interpretation of the adsorption process, as remarked above, is done in terms of the effect of the adsorbing molecules in displacing water dipoles from the interface. For most dipolar organic materials, adsorption is strongest at or near the e.c.m. and the prominent peaks shown in Fig. 6.5.1 correspond to displacement of the organic molecules by the water dipoles. As the potential is made sufficiently negative or positive the large water dipole can compete effectively with the organic molecule, even though the latter has a high surface activity (since the hydrophobic portion will want to avoid the aqueous environment).

6.6 Limitations of the Gouy–Chapman equation

The theory of the double layer, as presented in sections 6.3 and 6.4, has frequently been criticized on fundamental statistical mechanical grounds. The main problems concern

(a) the arbitrary separation of the double layer into a compact and a diffuse region; and

(b) the identification of $w_i = z_i e\psi$ in eqn (6.3.5) to describe the diffuse layer.

A proper statistical mechanical approach would lead to a unified description across the whole double-layer region with the finite size of the ions and molecules entering by way of the distribution functions for those species in the neighbourhood of the interface (Cooper and Harrison 1977). We will have more to say about such procedures in Chapter 11 (Vol. II) but qualitatively it should be obvious that very close to a plane interface the distribution of ions and molecules is dramatically affected by their size. They cannot approach closer than a distance x_i and the number of molecular centres at that distance is likely to be very high as the liquid presses up against the wall. The distribution function $g(x)$ takes the form suggested in Fig. 6.6.1, where $g(x)$ measures the probability of finding a molecule in a small volume element at a distance x from the wall. In the more intuitive approach adopted in sections 6.3 and 6.4 we attempted to overcome this problem by separating off the region where $g(x)$ fluctuates wildly and treating it in a special way.

The other approximation ($w_i = z_i e\psi$) was noted at the time (section 6.3) to be a rather crude one. The w_i value which appears in Boltzmann's equation (6.3.5) should measure the total work done in bringing the ion from the bulk solution up to a point in the double layer, in short (reversible) steps, where the effect of the ion on the rearrangement of all of the other ions (and dipoles) is taking into account after each step. To put it equal to $z_i e\psi$ is to ignore all of the ancillary effects of the ion and to assume that only the electrical work, estimated by bringing up an *infinitesimal* charge, is important. Fortunately, it turns out that many of

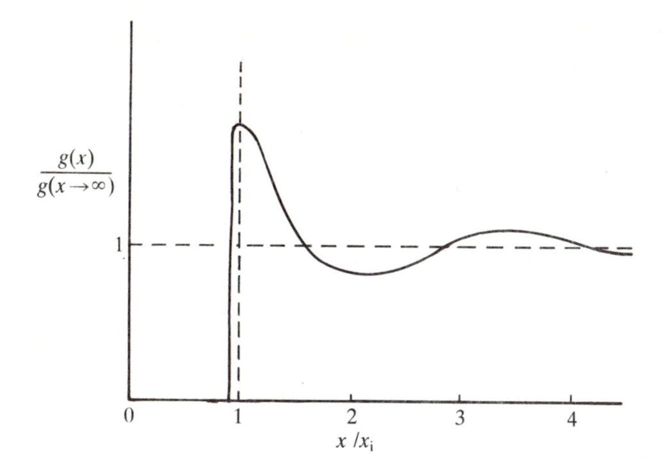

FIG. 6.6.1. Distribution function for centres of liquid molecules near a wall.

the other effects tend to cancel one another out provided that neither $|\psi|$ nor the electrolyte concentration is too large (Levine and Bell 1966). Under those conditions the Gouy–Chapman theory remains quite satisfactory. Its success in accounting for the behaviour at the mercury–solution interface stems from the fact that we restricted its application to the diffuse layer (where $|\psi|$ is low; Exercise 6.4.6). At high concentrations, when it is likely to break down, the diffuse layer has a negligible effect on the double layer capacitance, so deficiencies in the diffuse layer theory are not apparent. In colloidal systems the same situation applies: at very high electrolyte concentrations, double-layer effects become much less important and the limitations of the Poisson–Boltzmann equation are not too serious.

A critique of the simple Poisson–Boltzmann equation was given by Blum (1977) and a critique of the breakdown into inner and outer regions was made by Cooper and Harrison (1977). Levine and Outhwaite (1978) have replied to the latter criticism (for the case $\sigma_i = 0$) and the Monte Carlo calculations of Torrie and Valleau (1979) suggest that the kind of model we have presented, at least for the case $\sigma_i = 0$, is quite satisfactory up to 0.1 M salt concentrations. It can also be justified from the theory of continuum mechanics (Cade 1978).

Finally, we note that it is frequently asserted that the Poisson–Boltzmann equation must be erroneous because it fails to satisfy the principle of linear superposition of fields. It is true that the potential due to a static array of point charges can be calculated by linear superposition of the potentials due to the individual charges. The situation here is, however, quite different. We begin with an electrically charged surface and ask how the ions in the neighbourhood will *respond* to that charge. Nordholm (1983 private communication) argues that the linear relation of electric field to charge is satisfied in the Poisson equation (6.3.3) and that the resulting potential/charge relation need not be linear. Indeed, all of the evidence suggests that it is not.

6.7 The silver iodide–solution interface

The most fundamental studies of the structure of the electrical double layer on a *colloidal* system have been carried out on silver iodide, over a long period of time, notably by the Dutch school of colloid scientists. Although we will find that some of the ideas that came out of that work cannot readily be applied to other colloidal systems, it is essential that they be understood. We must also examine the relationship of this system to the mercury–solution interface. Before doing so, however, we note a most important characteristic of the silver iodide system and one on which much of colloid chemical theory rests. This is the notion of *potential-determining ions.*

6.7.1 *Potential-determining ions*

When silver iodide crystals are placed in water, a certain amount of dissolution occurs to establish the equilibrium $AgI \rightleftarrows Ag^+ + I^-$, and then $\mu_{AgI} = \mu_{Ag^+} + \mu_{I^-}$. The concentrations of Ag^+ and I^- in solution are very small because the solubility product is in this case very small ($K_{sp} = a_{Ag^+} \cdot a_{I^-} \simeq 10^{-16}$). Nevertheless, these very small concentrations are very important because slight shifts in the balance between Ag^+ and I^- ions cause profound changes in the behaviour of the silver iodide crystals. The surface of the crystal can be regarded as a more or less regular array of Ag^+ and I^- ions in cubic close pack and if there are exactly equal numbers of each ion, the crystal is at its *point of zero charge.* It turns out that this does *not* correspond to the point where there are equal numbers of Ag^+ and I^- ions in solution (i.e. $[Ag^+] = [I^-] = 10^{-8}$). Rather, it seems that the iodide ions have a higher affinity for the surface and tend to be preferentially adsorbed. In order to reduce the charge to zero it is found that the Ag^+ concentration must be increased to about $10^{-5.5}$ (so that $[I^-] \simeq 10^{-10.5}$—a ratio of 100 000 to 1). This can be done by adding a very small amount of silver nitrate. The charge on the crystal surface can thus be altered from highly positive through zero to highly negative values by the addition of very small amounts ($\sim 10^{-6}$ mol L^{-1}) of these two ions. They are called the *potential-determining ions* for this surface, for the following reason.

The condition for equilibrium between the solution and the surface of the crystal is (from Appendix A5, eqn (A5.20)):

$$\mu_{Ag^+}(w) + \mathscr{F}\phi(w) = \mu_{Ag^+}(c) + \mathscr{F}\phi(c) \qquad (6.7.1)$$

where (w) refers to the water solution and (c) to the AgI crystal, i.e.

$$\mu^{\ominus}(w) + RT \ln a(Ag^+)_w + \mathscr{F}\phi(w) = \mu^{\ominus}(c) + RT \ln a(Ag^+)_c + \mathscr{F}\phi(c). \qquad (6.7.2)$$

As the bulk activity of Ag^+ is changed, the potential $\phi(c)$ on the crystal will change and at the point of zero charge we have

$$\mu^{\ominus}(w) + RT \ln a'(Ag^+)_w = \mu^{\ominus}(c) + RT \ln a'(Ag^+)_c + \mathscr{F}\Delta\chi'_{dipole} \qquad (6.7.3)$$

where (′) refers to quantities measured at the point of zero charge and it is assumed that at that point only the chi-(dipole)-potential difference (section 6.2) remains. Subtracting eqn (6.7.3) from eqn (6.7.2) we have:

$$\mathscr{F}\{(\phi(c) - \phi(w)) - \Delta\chi'_{dipole}\} = RT \ln \frac{a(Ag^+)_w}{a'(Ag^+)_w} \qquad (6.7.4)$$

where we have assumed that $a(Ag^+)_c = a'(Ag^+)_c$. The quantity inside the curly brackets may be defined as the surface potential of the AgI crystal (ψ_0) and so:

$$\psi_0 = \frac{RT}{\mathscr{F}} \ln a(Ag^+)_w / a'(Ag^+)_w$$

$$\simeq \frac{RT}{\mathscr{F}} \ln [Ag^+]/[Ag^+]_{pzc} \qquad (6.7.5)$$

$$\psi_0(\text{volt}) = 0.0598 \log_{10}[Ag^+]/[Ag^+]_{pzc}$$
$$= 0.0598(\text{pAg(pzc)} - \text{pAg})$$

at 25 °C in water, where $\text{pAg} = -\log_{10}[Ag^+]$.

The important assumption made in deriving eqn (6.7.4) is that $a(Ag^+)_c = a'(Ag^+)_c$. That is to say, the activity of the silver ions on the crystal surface does not change as the surface is charged up. The justification for this is that at the p.z.c., there are a large number of silver ions and an equal number of iodide ions. Only a relatively small (percentage) increase in the number of Ag^+ ions is required to establish the surface potentials that are normally encountered (Exercise 6.7.1). The chemical environment of these extra ions is, therefore, fairly constant over a wide range of surface potentials. (We will find that this is not so for most oxide systems. On the SiO_2 surface, for example, there are very few charges on the surface at the p.z.c. so the chemical environment of the potential-determining ions varies as the surface is charged up. The Nernst equation for ψ_0 then breaks down.)

Equation (6.7.5) shows that ψ_0 changes by about 60 mV per ten-fold change in the concentration of the potential-determining ion. This is referred to as ideal Nernstian behaviour and eqn (6.7.5) is referred to as the Nernst equation for this colloidal sol (compare eqn (A5.16)). In order to study in more detail the relation between the surface charge on silver iodide crystals and their surface potential, we turn now to an examination of the electrochemistry of the Ag/AgI electrode system in the presence of another electrolyte.

6.7.2 *The completely reversible electrode*

It was pointed out in section 6.1 that, in the presence of many electrolytes, the mercury solution interface behaves as a perfectly polarizable electrode. That is, it can be subjected to a polarizing e.m.f. and no current will flow through the interface.

The silver–silver iodide electrode is an example of the opposite extreme of behaviour. It is not possible to impose an arbitrary polarizing potential across the interface between the AgI and an aqueous solution because that potential difference is determined by the activity of Ag^+ and I^- ions in the solution. Any attempt to alter the potential difference, by applying an external e.m.f., results in a current flow.

If the potential is not too different from the equilibrium one, a new equilibrium will be established when the silver or iodide ion concentration in the solution has been adjusted by the electrode reaction (e.g. $AgI + e^- \rightarrow Ag + I^-$) to restore equality of the electrochemical potentials.

The behaviour of a silver/silver iodide electrode immersed in an electrolyte solution (say HNO_3) would involve the cell (compare eqn (6.1.20)):

$$\begin{array}{cccccc} \text{Cu(2)} \mid & \text{H}_2\text{(g), Pt} \mid & \text{HNO}_3, \; \text{AgNO}_3 \mid & \text{AgI} \mid & \text{Ag} \mid & \text{Cu(1)} \\ & & a_1 \qquad a_2 & & & \\ \phi_6 & \phi_5 & \phi_4 & \phi_3 & \phi_2 & \phi_1 \end{array} \qquad (6.7.6)$$

We will almost always be able to assume that $a_2 \ll a_1$, since only minute amounts of the potential-determining ions (usually less than 10^{-5} M) are required to establish the usual surface potentials. Increasing concentrations of $AgNO_3$ would permit the exploration of increasingly positive values of ψ_0.

Using the same procedure as previously (section 6.1) it is not difficult to show (Exercise (6.7.2) that:

$$-\mathrm{d}\gamma = \mathscr{F}(\Gamma_{Ag^+} - \Gamma_{I^-})\,\mathrm{d}E_+ + \Gamma_{NO_3^-}\,\mathrm{d}\mu_{HNO_3} \qquad (6.7.7)$$

where, again, E_+ refers to the fact that the e.m.f. is measured with respect to an electrode reversible to the cation. The first term on the right-hand side can again be identified with $\sigma_0\,\mathrm{d}E_+$ if we acknowledge the special role of Ag^+ and I^- ions. Since they are the normal components of the crystal lattice it is reasonable to assume that the net excess of those ions in the interface is actually *on* the solid surface and is balanced by the excess of (indifferent) nitrate ions, $\Gamma_{NO_3^-}$.

How can we now establish the relation between surface charge and surface potential for the system? Although from eqn (6.7.7) we can write, formally

$$\sigma_0 = -(\partial\gamma/\partial E_+)_\mu \qquad (6.7.8)$$

this relation is of little help because one cannot readily determine the surface energy γ as a function of E_+ for the solid–solution interface. Nor can we proceed from a directly measured differential capacitance by way of eqn (6.1.15) to obtain σ_0 because the a.c. measuring signal causes a current flow and a significant amount of electrode reaction, which interferes with the determination of the capacitance. [Actually, this problem has been overcome very recently (see Pieper *et al.* 1975) but much of the work of the Dutch school was completed before that.]

Fortunately, the surface charge can be measured directly in this case because the effective surface area of the electrode can be made very large. Instead of just a silver wire covered with AgI, the cell can contain also a significant quantity of silver iodide in the form of a dispersion of large surface area. The measurements are then done by adding small volumes of $AgNO_3$ solution of known concentration to the cell, stirring until equilibrium is established and then measuring E_+:

$$E_+ = E^{\ominus} - \frac{RT}{\mathscr{F}}\ln a_{H^+}a_{I^-} \qquad (6.7.9)$$

$$= E^{\ominus} - \frac{RT}{\mathscr{F}}\ln(K_{sp}a_{H^+}/a_{Ag^+}).$$

Since the amount of Ag^+ added is known, and the amount remaining in solution can be calculated from the measured E_+ by using eqn (6.7.9) (assuming $a_{Ag^+} \simeq [Ag^+]$) the amount that has disappeared is assumed to be on the surface of the crystal.

The resulting values of total charge can be converted into σ_0 values only after the area of the solid has been measured. We will return to that shortly. It must also be noted that these are *relative* charge values, since the charge status of the solid is not known at the beginning of the titration unless we know the point of zero charge. For silver iodide we have already noted that it occurs at $pAg = -\log_{10}[Ag^+] \approx 5.5$ but that was difficult to determine. Overbeek (1952, p. 160) surveys the early work in the area and it is clear from that that a reliable general method of identifying the p.z.c. on any colloid would be of great value. That can now be done using the Esin and Markov coefficient, β, (section 6.4; Lyklema 1972).

6.7.3 *Determining the point of zero charge*

If the titration is done *in a cell with liquid junction* (*l.j.*) such as:

$$\begin{array}{cccccccc} \mathrm{Cu(2)} \,|\, \mathrm{Hg} \,|\, & \mathrm{Hg_2Cl_2} \,|\, & \mathrm{KNO_3} \,|\, & \mathrm{HNO_3, AgNO_3} \,|\, & \mathrm{AgI} \,|\, & \mathrm{Ag} \,|\, & \mathrm{Cu(1)} \\ & \mathrm{KCl} & \mathrm{l.j.} & & & & \end{array}$$

$$\phi_8 \quad \phi_7 \quad \phi_6 \quad \phi_5 \quad \phi_4 \quad \phi_3 \quad \phi_2 \quad \phi_1$$

then the measured e.m.f. E_r is now (c.f. eqns (6.1.23–24)):

$$E_r = (\phi_3 - \phi_4) + \text{const. and } dE_r = d\psi_0$$

whereas in the previous cell (eqns (6.7.6) and (6.7.9)) it was

$$E_+ = (\phi_3 - \phi_4) + \left(\text{const} - \frac{RT}{\mathscr{F}} \ln a_{H^+}\right) + \text{const.}$$

Again we have (compare eqn (6.4.5)):

$$E_r = E_+ + \frac{RT}{\mathscr{F}} \ln a_{H^+} + \text{a constant.} \tag{6.7.10}$$

Then

$$\begin{aligned} \left(\frac{\partial E_r}{\partial \mu}\right)_{\sigma_0} &= \frac{1}{RT} \frac{\partial E_r}{\partial \ln a_\pm^2} = \frac{1}{RT} \frac{\partial E_+}{\partial \ln a_\pm^2} + \frac{1}{2\mathscr{F}} \\ &= \left(\frac{\partial E_+}{\partial \mu}\right)_{\sigma_0} + \frac{1}{2\mathscr{F}}. \end{aligned} \tag{6.7.11}$$

At the point of zero charge, the quantity $\mathscr{F}(\partial E_+/\partial \mu)_{\sigma_0=0} = \beta_0'$ is again equal to $-\frac{1}{2}$ if there is no specific adsorption (section 6.4) and so $\mathscr{F}(\partial E_r/\partial \mu)_{\sigma_0=0} = \beta' = 0$. The importance of doing the measurement in a cell with liquid junction is not always realized. Note that β' would not be zero at the p.z.c. if the titration were done in a cell without liquid junction, but the activity of H^+ was kept constant; we would then have, from eqn (6.7.10), $dE_r = dE_+$ and, from eqn (6.4.3), $\beta' = -\frac{1}{2}$.

It can be shown (Lyklema 1972) that when the Esin and Markov coefficient, β, is zero then so too is the quantity (Exercise 6.7.4):

$$\left(\frac{\partial \sigma_0}{\partial \mu}\right)_{\psi_0} = \left(\frac{\partial \sigma_0}{\partial \mu}\right)_{E_r} = \left(\frac{\partial \sigma_0}{\partial \mu}\right)_{\mathrm{pAg}} \tag{6.7.12}$$

and this is so only at the point of zero charge.

Thus, if there is no specific adsorption, the point of zero charge can be identified as the only point at which $\left(\frac{\partial \sigma_0}{\partial \mu}\right)_{\mathrm{pAg}}$ is zero. A plot of the relative surface charge σ_0'' looks like Fig. 6.7.1(a) and it can be converted into a plot of actual surface charge σ_0 (Fig. 6.7.1(b)) by recognizing that the common point of intersection must be at the p.z.c. because that is the *only* point at which a change in the activity of the background electrolyte has no effect on the surface charge (Exercise 6.7.4). The behaviour in Fig. 6.7.1 therefore serves to identify *both* the p.z.c. and the fact that this electrolyte is not specifically adsorbed at the AgI–solution interface. If no common intersection point occurs it is assumed that the electrolyte is not indifferent and one must find another electrolyte that does give a common intersection point.

6.7.4 *Determination of surface area*

A surface area can be estimated from the particle size distribution determined by any of the methods discussed in sections 3.1–3.8. There are advantages, however, in doing area measurements directly on the suspension without having to dry it out. One can then be confident that the measured area relates to the same system as the other electrochemical measurements. There are two possible procedures: positive adsorption and negative adsorption.

(a) *Positive adsorption.* In this method a highly surface active material is added in small amounts and its concentration in the solution is monitored. The difference between the amounts added and the amount remaining in solution is equal to that adsorbed and a suitable adsorbate will show an adsorption isotherm of the Langmuir type (Fig. 6.7.2). To determine the surface area it is necessary to assume that the plateau corresponds to a monolayer of adsorbate and to have a cross-sectional area for the adsorbing molecule. It is rarely possible to find a suitable adsorbate of known cross-section and adsorption properties, but fatty acids and dyestuffs can be used (Adamson 1967, pp. 412–4) and the concentration of the latter is obviously easily monitored in the solution. The usual form of the Langmuir isotherm (eqn (6.4.11)) can be recast in terms of solution concentration (c_i) rather than mole fraction and N_s then becomes the number of adsorption sites (per gram of adsorbent)

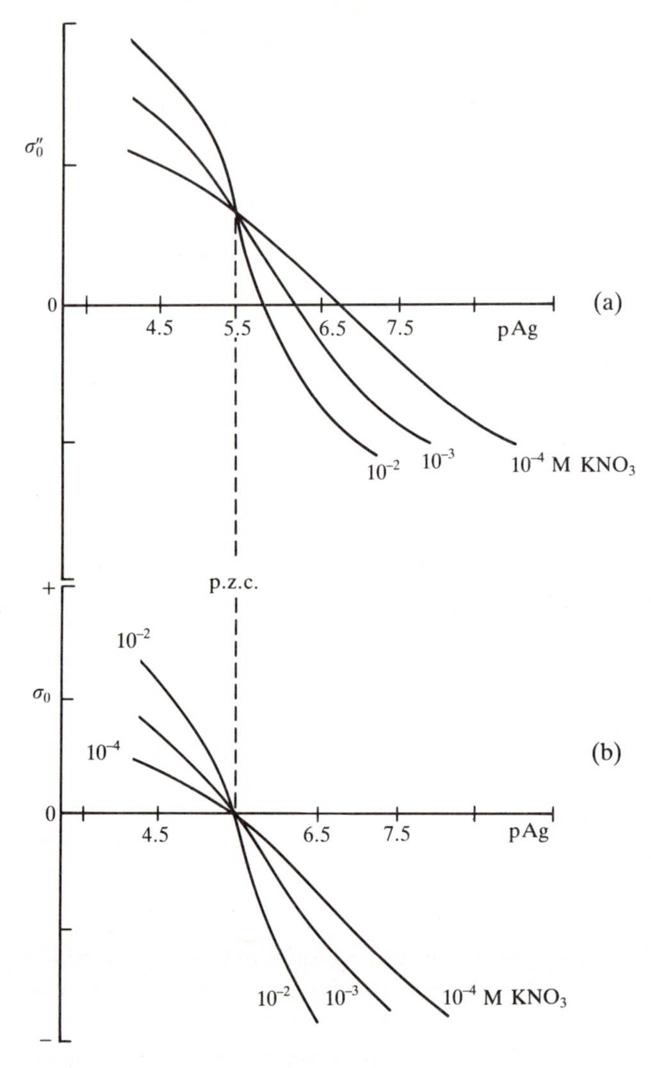

FIG. 6.7.1. Relative surface charge σ_0'' (a) can be converted into absolute charge σ_0 (b) by recognizing that the common point of intersection of the curves at different ionic strengths must be the p.z.c. because only there is $(\partial\sigma_0/\partial\mu)_{\mathrm{pAg}}$ equal to zero.

available for a single monolayer of adsorbate. Equation (6.4.11) then gives (Exercise 6.7.3):

$$\frac{c_i}{n_i^s} = \frac{1}{\mathcal{K}' N_s} + \frac{c_i}{N_s} \tag{6.7.13}$$

where n_i^s is the amount adsorbed *per gram of adsorbent* and $\mathcal{K}'$ is the new

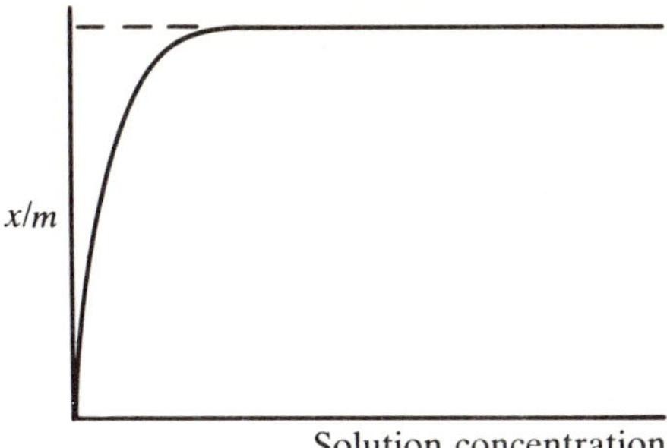

FIG. 6.7.2. Adsorption of a strongly adsorbed (organic or inorganic) charged or neutral species on a surface. (x is the number of moles of adsorbate taken up on m grams of the solid.)

'equilibrium' constant. A plot of c_i/n_i^s against c_i should be linear if the isotherm is obeyed, and the value of N_s obtained from the slope can be used to estimate the area (per gram) of the solid (Exercise 6.7.3).

Herz (1974) has made a special study of this method on silver halide surfaces. The method is also used with cetyl pyridinium bromide as adsorbate for measuring the surface area of clay minerals (Greenland and Quirk 1964).

(b) *Negative adsorption*. This procedure was pioneered by Schofield (1947) for use on clay minerals for which it is particularly suitable, but has been used more recently by van den Hul and Lyklema (1967, 1968) for the silver iodide system. It depends on the fact that a highly charged surface repels ions of the same sign from the region around it and so increases their concentration in the bulk. There is, in effect a region to which the ions do not have access, as is illustrated in Fig. 6.7.3.

The extent of *negative* adsorption is defined by van den Hul and Lyklema (1967) as:

$$\Gamma_i = \int_0^\infty (c_i^0 - c_i(x))\,dx. \tag{6.7.14}$$

Then if $\mathscr{A}$ is the area of the solid phase and V_t is the total volume of liquid:

$$\mathscr{A}\Gamma_i = V_t \Delta c_i \tag{6.7.15}$$

since the total number of moles of the salt i remains unchanged. For a symmetrical electrolyte we know from the result of Exercise (6.3.9) that

$$\Gamma_- = \frac{\sigma_d^-}{ze} = \frac{2n^0}{\kappa}[1 - \exp(ze\psi_d/2kT)]$$

$$\simeq \frac{2n^0}{\kappa} \text{ if } \psi_d \text{ is large and negative.}$$

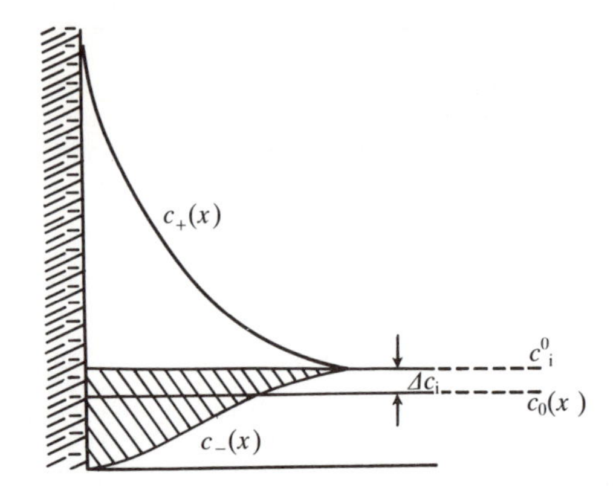

FIG. 6.7.3. Ion distribution in the neighbourhood of a negative surface. $c_0(x)$ is the concentration if the surface is uncharged whereas c_i^0 is the bulk concentration in the presence of the charged surface. (Note the very pronounced asymmetry in the behaviour of the (positive) counterions and the (negative) co-ions. This is a consequence of the non-linearity of the Poisson–Boltzmann equation: it disappears at low potentials when the Debye–Hückel approximation is valid.)

According to Lyklema (1977) the more detailed analysis of van den Hul, for a mixture of a 1–1 and 1–2 electrolyte gives for $\mathscr{A}(=V_t\Delta c_i/\Gamma_i)$, from diffuse double layer theory:

$$\mathscr{A} = V_t \frac{\Delta c_1}{c_1} \kappa / f(c_2/c_1) \tag{6.7.16}$$

where c_1 is the concentration of the 1–2 electrolyte whose negative adsorption is to be measured and c_2 is the concentration of the potential-determining ion solution used to charge up the surface. Using a mixture of KI and K_2SO_4 it is possible to arrange things so that the surface is highly charged (negatively) but the amount of KI remaining in solution is very low. The function f then takes a simple limiting form ($\sqrt{3}$ in this case) and direct chemical analysis of the increment in sulphate concentration Δc_1 allows one to calculate the area from

$$\mathscr{A} = \frac{V_t \Delta c_1}{g c_1^{1/2}} \tag{6.7.17}$$

where g is a constant calculated from double layer theory. (Recall that the Debye–Hückel parameter $\kappa \propto c^{1/2}$ and only the (1–2) electrolyte remains in the solution at a significant concentration.) The volume from which salt is excluded $\Delta V = \Delta c_1 V_t / c_1$ corresponds to an area $\mathscr{A}$ and thickness $g/c_1^{1/2}$ and this latter quantity turns out to be several times the double layer thickness κ^{-1}. Apart from its use in determining surface areas, negative adsorption is important in salt-sieving (reverse osmosis), membrane permeability, filtration of solutions through charged porous media, and in light scattering from colloidal suspensions.

The plot of ΔV against g/κ is a straight line if eqn (6.7.16) is obeyed and its slope gives $\mathscr{A}$ directly. A typical plot is shown in Fig. 6.7.4 for sulphate and phosphate on AgI. It should be noted that it is not necessary to know the surface potential for these calculations but it must be known to be high because the derivation of the constant g requires that assumption.

The chief advantage of this method is that it measures the electrochemical surface area, on which the ions are adsorbed. Other measurements, such as gas adsorption, may sample parts of the surface that are not accessible to ions, or on which an electrical double layer cannot properly develop (such as the interior of surface cracks). The main difficulty, in the case of AgI, is that the areas are very small so the change in salt concentration is very small and it must be measured with great accuracy to get reliable results. Nevertheless, the best results by this method agree well with those obtained by other procedures.

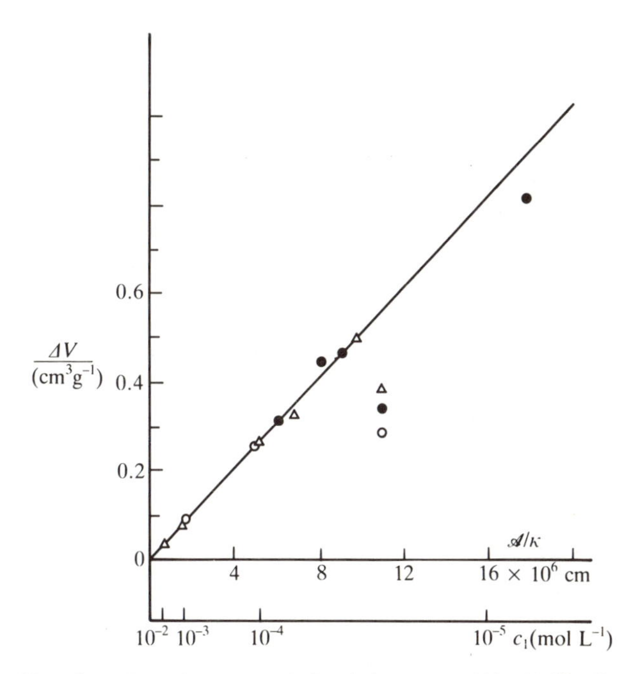

FIG. 6.7.4. Negative adsorption on an AgI sol for $\psi_0 \sim -300$ mV (●, ○: sulphate △: phosphate). (From van den Hul and Lyklema 1967, with permission.)

6.7.5 *The capacitance of the AgI interface*

We can now draw up plots of the surface charge density, σ_0, as a function of silver ion concentration, pAg or ψ_0 (using eqn (6.7.5)). The resulting curves (Fig. 6.7.5) are rather similar to those found on mercury. They show that as the indifferent electrolyte concentration is raised, more potential-determining ions are driven onto the surface in order to maintain the surface potential constant at any particular pAg value.

Differential capacitances can be estimated from these curves by graphical differentiation and the resulting curves are very similar to those on the mercury–solution interface (Fig. 6.1.5). They show the same behaviour at low electrolyte concentration in the neighbourhood of the p.z.c. This is hardly surprising since the diffuse layer capacitance is then dominant. In the absence of specific adsorption we can again set $\sigma_0 = -\sigma_d$ and so calculate ψ_d and hence C_d and so obtain C_i from $1/C_T = 1/C_i + 1/C_d$ (eqn (6.4.9)). Values of the inner layer capacitance, C_i, can then be plotted as a function of the surface charge (Fig. 6.7.6). The observed behaviour is very similar to that found on mercury at high temperatures (Fig. 6.4.5). It should be noted, however, that the data in

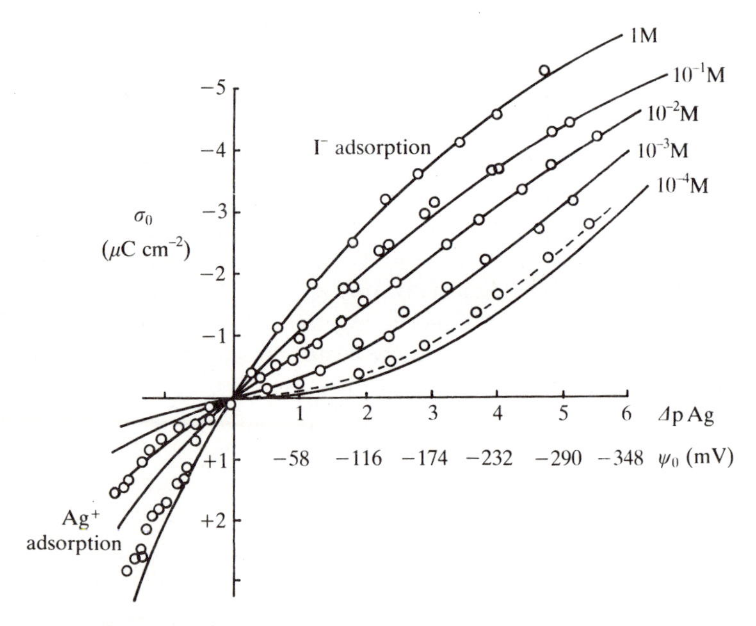

FIG. 6.7.5. Adsorption of iodide and silver ions onto AgI. $\Delta pAg = pAg - pAg(p.z.c.) = pAg - 5.52$. The concentrations shown on the graph are those of the *indifferent electrolyte*. Drawn curves are for a 7 : 1 KNO_3, $NaNO_3$ mixture (van Laar), $NaClO_4$ (Mackor), and $NaNO_3$ (Mackor). (From Overbeek 1952, p. 162, with permission.)

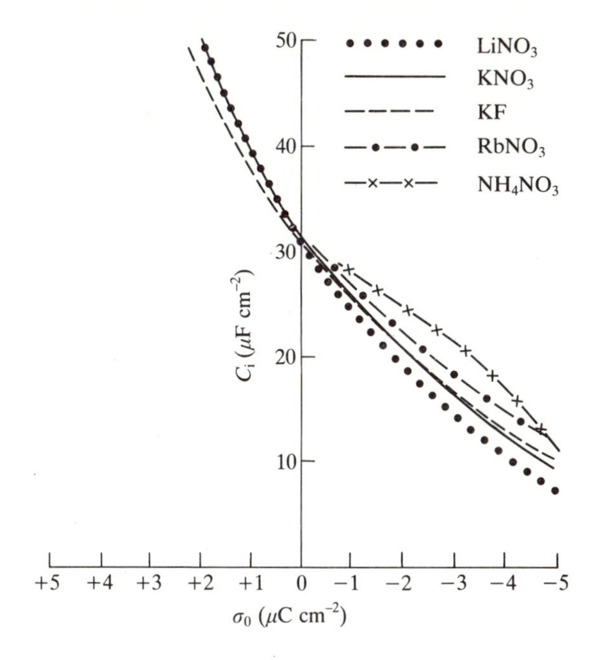

FIG. 6.7.6. Inner layer capacity at the AgI–solution interface (from Lyklema and Overbeek 1961, with permission.)

Fig. 6.7.6 are necessarily less precise than those on mercury because they are determined by graphical differentiation.

We will discuss the structure of the AgI–solution interface in more detail when we come to the study of adsorption generally (Chapter 12). At this stage we need only note that it differs from the mercury–solution interface in two respects (Sparnaay 1972):

1. If the surface is made more negative, the capacity of the AgI–solution interface decreases much faster than the capacity of the Hg–solution interface and

2. There is no 'hump' in the inner layer capacity as is observed on mercury at ordinary temperatures (Fig. 6.4.5).

Both of these effects suggest that there is rather more ordering of the local water structure on the AgI surface than on mercury. That is hardly surprising when it is recalled that AgI crystals are used as seeds for inducing rain formation in supersaturated clouds. They are chosen for the purpose because the ice crystal lattice can fit very precisely onto the surface AgI crystal structure.

One other important difference with the AgI surface is evident from Fig. 6.7.6. There is a slight, but significant, difference in the C_i values for different alkali cations. The difference is of the order of some tens of per cent on AgI compared to only a few per cent on mercury (Lyklema 1977). Since the capacitance C_i remains at values comparable with that of mercury it seems reasonable to assume that the cations remain hydrated and are not adsorbed into the plane $x = b$ (Fig. 6.3.5). Rather the differences are to be ascribed to small changes in the inner layer thickness, d, and slight differences in the degree of ordering of water in the inner region and, hence, differences in ε_i in the presence of different alkali metal cations.

6.7.6 *The suspension effect*

Before leaving the silver iodide–solution interface (for the present) we should note one problem that arises in the titration experiments used to determine σ_0. It was noted that measurement of the potential E_r must be done using a reference electrode with a liquid junction if one wishes to determine the p.z.c. by the application of the Esin and Markov coefficient method ($\beta' = 0$). This gives rise to an effect (called the *Pallman effect*) that is important in many electrometric measurements on colloidal systems, including the measurement of pH. It is discussed by Overbeek (1952, p. 185) and very little need be added to that discussion.

The effect is generated by the presence of the liquid junction and is most obvious if we consider the cell:

$$\mathrm{Hg}\,|\,\mathrm{Hg_2Cl_2, KCl}\left\|\begin{matrix}\text{colloidal}\\ \text{suspension}\end{matrix}\right.\ \vdots\ \text{supernatant}\,\Big\|\,\mathrm{KCl, Hg_2Cl_2}\,|\,\mathrm{Hg}$$

l.j.1 l.j.2

The vertical broken line is used to indicate the boundary between the settled colloidal dispersion and its supernatant since the Galvani potential, ϕ, (section 6.2) can be different in these two regions.

Even when the colloidal suspension is in equilibrium with its supernatant, the cell above exhibits an e.m.f. (called the Donnan e.m.f.), which can be considered to arise from the difference in liquid junction potentials at l.j.1 and l.j.2. The irreversibility introduced by the presence of the junctions enables (a small amount of) work to be extracted from the system because of the difference in the diffusion potentials of the ions in the two junctions, brought about by the differences in their environment.

The effect is not necessarily small; the e.m.f. generated can be over 100 mV. If now, we wished to determine the pH of this suspension, we would use one or other of the calomel half-cells in conjunction with a glass electrode. Whether the glass electrode is placed in the suspension or the supernatant makes no difference to the resulting e.m.f. since the glass electrode functions reversibly with respect to hydrogen ions. There is, however, certainly a difference (of up to two pH units) depending on whether the calomel electrode liquid junction is placed in the suspension or the supernatant.

The thermodynamic analysis given above for the Esin and Markov coefficient is best interpreted on the assumption that the liquid junction is in the equilibrium supernatant above the suspension. In cases where little or no settling occurs, the suspension effect can be minimized by using a reference electrode with a double liquid junction. The inner jacket contains the usual saturated KCl and the outer jacket is filled with a solution of the same composition as the indifferent electrolyte in the colloidal dispersion. The liquid junction potential should then be small and due solely to the effect of the colloid particles.

Exercises. 6.7.1 Calculate the surface charge density of a silver iodide crystal in 10^{-3} M salt solution, from diffuse double layer theory for $\psi_d = 100$ mV. Compare this with the density of Ag^+ ions on the crystal at the p.z.c. assuming that Ag^+ and I^- ions each occupy 0.2 nm^2. [This calculation assumes there is no specific adsorption of other ions.]

6.7.2 Establish eqn (6.7.7) using the procedure of section 6.1 and noting that, for this cell, the reaction is $AgI + \frac{1}{2}H_2(g) \rightarrow Ag + I^- + H^+$, so $E_+ = E^\ominus - (RT/\mathscr{F}) \ln a_{H^+}a_{I^-}$. (Also $\mu_{Ag^+} + \mu_{I^-} = \mu_{AgI}$ is constant in the solution and $\sum_i \Gamma_i = 0$.)

6.7.3 Use the Langmuir isotherm (eqn (6.4.11)) in the form $\eta = \mathscr{K}'c_i/(1 + \mathscr{K}'c_i)$ to establish eqn (6.7.13). Check the following data for conformity to the Langmuir isotherm and estimate the area of the solid assuming that the cross-sectional area of the adsorbate molecules is 0.20 nm^2.

c_i(mmol L^{-1})	0.20	0.81	1.20	1.70	2.00
n_i(mmol g^{-1})	0.035	0.081	0.105	0.102	0.103

(Note that c_i is the equilibrium concentration *after* adsorption is complete.)

6.7.4 The analogue of the Esin and Markov coefficient for the AgI system is $\mathscr{F}(\partial\psi_0/\partial\mu)_{\sigma_0} = 2.303RT(\partial \text{pAg}/\partial\mu)_{\sigma_0}$ since $dE_r = d\psi_0$. Show that if this is zero then so, too is $(\partial\sigma_0/\partial\mu)_{\text{pAg}}$.

6.8 Other Nernstian surfaces

The crucial assumption in the derivation of the Nernst equation (eqn (6.7.5)) for the silver iodide surface is that the activity of the potential-determining ions on the surface of the crystal is unaffected by the charging process that establishes the surface potential, ψ_0. This is a reasonable assumption for any solid surface for which the potential determining ions are components of the crystal lattice. It would certainly be expected to hold for the other insoluble silver halide surfaces (AgBr and AgCl).

Freyberger and de Bruyn (1957) have shown that the same electrochemical procedures can be carried over to the study of the silver sulphide surface. By examining the behaviour at two different pH values they were able to establish that only the Ag^+ and S^{2-} are important potential-determining ions (p.d.i.) and the p.z.c. occurs at $pAg = 10$. HS^- and H_2S are evidently not strongly adsorbed at this surface. The behaviour was qualitatively similar to that observed on silver iodide but with some significant differences—notably a very steep rise in the capacitance when the surface was positively charged.

The calcium oxalate monohydrate surface has also been studied by Curreri *et al.* (1979) using the Nernst equation to calculate ψ_0. Calcium oxalate is an important constituent of kidney stones and this work showed that its electrochemical behaviour could be adequately described using the Nernst equation and the Stern equation (eqn (6.4.12)) but only if it was also assumed that the hydrated oxalate and calcium ions could function as specifically adsorbable ions (as well as being potential determining in their dehydrated state). This is a rather unusual situation due, presumably, to the higher solubility of calcium oxalate. Normally, the concentration of the p.d.i. opposite in sign to that of the surface can be ignored, because it is so low, but that is apparently not so for this system.

Another system for which the Nernst equation is probably obeyed is that of barium sulphate. Although it has not been subjected to a complete electrochemical examination it has been studied by electrokinetic techniques (see Chapter 9). Hall and Rendall (1980) present a general technique for determining in simple situations ($\sigma_i = 0$) whether the Nernst equation is obeyed and show that this appears to be so for calcite ($CaCO_3$) and $Ca_3(PO_4)_2$.

6.9 Mechanisms of surface charge generation

The Nernstian systems treated in sections 6.7 and 6.8 owe their surface charge to the presence of a slight imbalance in the number of crystal lattice cations or anions on the surface. Such a mechanism of charge generation is possible only for ionic crystals. The great majority of colloidal materials are charged by other mechanisms, of which the most important are:

(a) surface dissociation;
(b) ion adsorption from solution; and
(c) crystal lattice defects.

We will examine ion adsorption from solution in more detail in Chapter 12. The particular case of an ionic crystal (like AgI) supplying a small number of ions in solution and then readsorbing one species in preference to the other is a highly specific process. There is, however, a wide range of colloid materials that respond strongly to the pH of the solution in which they are immersed. Many of these latter have surface chemical groups that can undergo dissociation or can interact with H^+ and OH^- ions from solution:

$$MO^- + H^+ \leftarrow M\text{–}OH \xrightarrow{OH^-} M\text{–}O^- + H_2O$$

so that there is little to distinguish between (a) and (b) above. This is particularly true of the metal oxides and many polymer latex systems. We will have much to say about them shortly.

There is also now available a wide range of ionic surface active agents that can be relied upon to adsorb strongly on the surface of colloidal particles and so to confer a charge on them or to modify the charge already present. Such substances are widely used to control the charge status and, hence, the stability (in the colloidal sense (section 2.5)) of the colloidal system. It has also been noted already (section 2.5) that some hydrolysed metal ions adsorb very strongly onto colloidal surfaces and may be largely responsible for determining their charge status. Polymeric charged materials (especially proteins) have also been used for many centuries as adsorbates whose presence on the surface of colloidal particles can modify the electrical and other, properties of the system. A review of much of the work in this area is provided elsewhere (Hunter 1981).

These more complex adsorbates are best regarded as specifically adsorbed species rather than as contributors to the *surface charge*. They are assumed to be located in the inner (compact) part of the double layer and are not potential-determining ions, even though they dramatically effect the potential in the diffuse part of the double layer.

Crystal lattice defects are responsible for the very large charge densities observed on many clay mineral systems (section 1.5.5). In those cases the defect is an isomorphous replacement of one ionic species by another of lower charge. In some other systems, a more subtle form of crystal lattice defect has been invoked to explain some features of the electrochemical behaviour. Sparnaay (1972, Chapter 4) for example, reviews the work on silver halides, which suggests that the surface charge is probably not confined to the actual solid surface but is distributed as a (diffuse) space charge in the solid. This does not affect the validity of the Nernst equation but does have a significant effect on kinetic processes, especially the rate of adjustment of the charge (or potential) when the potential-determining ion concentration is changed.

The most important recent development in the understanding of the charge–potential behaviour of colloidal systems concerns the dissociation of ionizable surface groups. Extensive reviews of this work have appeared recently (Healy and White 1978; Hunter 1981; and James and Parks 1982) so only a very brief resumé will be attempted in the next section. The most important dissociable groups are of the strong acid (sulphate, $-O.SO_2.OH$; sulphonate, $-SO_2.OH$), weak acid (sulphite, $-O.SO.OH$; carboxyl, $-COOH$), weak base (amine, $-NH_2$) and strong base (quaternary ammonium, $-N^+R_3$) type and they may occur alone at the surface or in various combinations. The particular case of a weak acid/weak base combination, such as occurs in proteins, is of special importance: the corresponding zwitterionic surface has some interesting properties.

Apart from describing the behaviour of adsorbed proteins at interfaces, models of the surface dissociation process are very useful for describing polymer latex systems, which often have carboxyl, sulphate, and sulphonate groups on their surfaces and which can also be made with zwitterionic surfaces (Homola and James 1977). More importantly, they are also widely used to interpret the behaviour of oxide surfaces; a proper understanding of such systems is, of course, vital in many areas of mineral preparation, agriculture, ceramics, and surface coatings as well as other technologically and scientifically important systems.

6.9.1 *Dissociation of a single site*

Where the surface contains only dissociable groups of one type (say an acid), the concept of a point of zero charge with positive and negative charge branches (Fig. 6.7.1) is no longer applicable. The charge simply decreases to zero at some sufficiently low pH and becomes increasingly negative with increase in pH. The dissociation of the surface site may be represented:

$$AH \rightleftarrows A^- + H^+$$

with a dissociation constant given by:

$$K_a = \frac{[A^-]a(H^+)_s}{[AH]}. \tag{6.9.1}$$

The activity of the protons on the surface, $a(H^+)_s$, is given by a Boltzmann equation:

$$a(H^+)_s = a(H^+)_b \exp(-e\psi_0/kT) \tag{6.9.2}$$

where the subscript b refers to the bulk solution phase. The surface charge, σ_0, is given by:

$$\sigma_0 = -e[A^-] \tag{6.9.3}$$

and if $N_s = [A^-] + [AH]$ is the total number of surface sites per unit area, then it is not difficult to show (Exercise 6.9.1) that:

$$-\sigma_0 = \frac{eN_s}{1 + (a_b/K_a)\exp(-e\psi_0/kT)}. \tag{6.9.4}$$

Then in the absence of specific adsorption (from eqn 6.3.27):

$$\sigma_d = -\sigma_0 = -\frac{eN_s}{\xi}\sinh(e\psi_d/2kT) \tag{6.9.5}$$

where

$$\xi = 10^3 N_s\kappa/4\mathcal{N}_A c. \tag{6.9.6}$$

The parameter ξ is characteristic of the system and includes both site density (N_s) and indifferent electrolyte concentration (c and κ) effects. In order to reduce the degree of arbitrariness of the model, Healy and White (1978) set $\psi_0 \simeq \psi_d$ and solved eqns (6.9.4) and (6.9.5) simultaneously to find ψ_0 as a function of pH. The result can be put in the form (for ψ_0 in mV and $T = 298$K):

$$\psi_0 = 59.8(pK_a - pH) - 59.8\log_{10}\left\{\frac{-\xi}{\sinh(0.0195\psi_0)} - 1\right\} \tag{6.9.7}$$

which clearly shows the inadequacy of the Nernst equation for this type of surface. ($d\psi_0/dpH$), instead of being constant at 59.8 mV (compare eqn (6.7.5)) depends strongly on the indifferent electrolyte concentration, through the parameter ξ. Healy and White (1978) showed that this model was able to account qualitatively for the experimental data of Ottewill and Shaw (1966) on polymer latices, but there were some quantitative discrepancies, which is hardly surprising considering the assumption that $\psi_0 \simeq \psi_d$.

6.9.2 *Two-site dissociation models*

There are three common situations involving two-site models:

1. The surface may have two distinct (acid) dissociable groups each characterized by a particular dissociation constant. This is often the case with polymer latices where both sulphate and carboxylate groups can be generated in the polymerization process.

2. A single (amphoteric) surface group may be able to function as both acid and base:

$$-AH_2^+ \underset{K_+}{\overset{H^+}{\leftrightarrows}} -AH \underset{K_-}{\rightleftarrows} -A^- + H^+ \tag{6.9.8}$$

where K_+, K_- are the equilibrium constants.

3. Zwitterionic surfaces in which two groups are present: one functions as an acid and the other as a base.

The first type of surface cannot show a p.z.c. but when titrated with base clearly shows a two-step dissociation process (Fig. 6.9.1). The actual shape of the curve depends upon the relative number of strong (N_A) and weak (N_B) acid sites per unit area.

The amphoteric-type surface (2) is of particular interest in describing the behaviour of oxide systems, which are discussed below (section 6.10). The zwitterionic model was examined in some detail by Healy *et al.* (1977) and applied by Rendall and Smith (1978) to describe the electrokinetic behaviour of nylon. That work will be outlined in Chapter 12. A more detailed analysis of the same system was also given by Healy and White (1978). Again the Nernst equation is not obeyed and the variation of ψ_0 with pH is given by:

$$\frac{d\psi_0}{dpH} = -2.303\frac{kT}{e} + \frac{kT}{2e}\left\{\frac{d\alpha_-}{dpH}\left(\frac{1}{\alpha_-} + \frac{1}{1-\alpha_-}\right) - \frac{d\alpha_+}{dpH}\left(\frac{1}{\alpha_+} + \frac{1}{1-\alpha_+}\right)\right\} \tag{6.9.9}$$

where α_+, α_- refer to the fraction of groups ionized at each pH. The first term is the Nernst slope and it is often very significantly altered by the second term. It should be noted, however, that the zwitterionic model

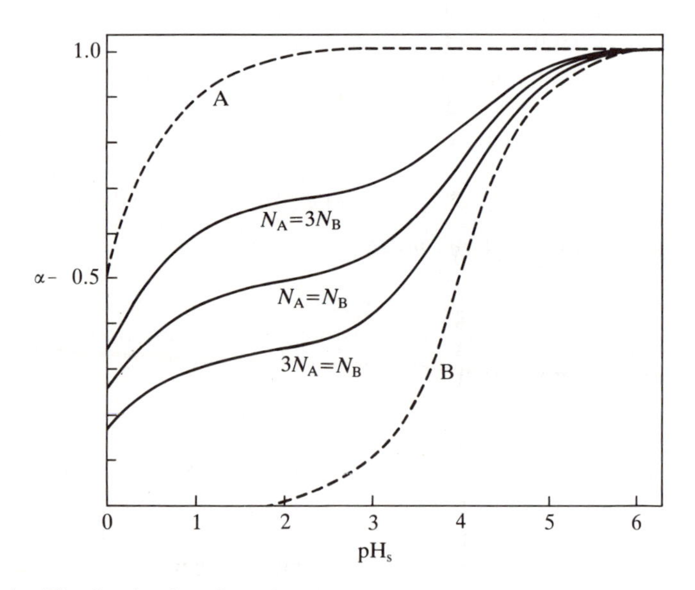

FIG. 6.9.1. The fractional surface charge α_- $(=(n_A^- + n_B^-)/(N_A + N_B))$ as a function of surface pH, pH_s $(= -\log_{10} a(H^+)_s)$, for a two-site model.

can be used to describe the silver iodide surface with reactions like (Healy and White 1978):

$$\begin{aligned} S_X + X^- &\rightleftarrows S_X X^- \\ S_{Ag} + Ag^+ &\rightleftarrows S_{Ag} Ag^+. \end{aligned} \tag{6.9.10}$$

In that case the second term in eqn (6.9.9) turns out to be zero and the Nernst equation is obeyed.

In order to apply these models to a real system it is necessary to develop techniques for adjusting the parameters of the model to obtain the best fit to the experimental data. The more data that are available, the more exacting is this parametrization and techniques have now been developed to enable this to be done by computer. The most elaborate procedures have been applied to the oxide systems, which we will now discuss.

Exercise. 6.9.1 Establish eqn (6.9.4).

6.10 The double layer on oxide surfaces

Some of the most important colloidal systems from the technological point of view are the metal oxides. Some reference has already been made to their uses (section 1.2) in industry and they are, of course, commonly encountered in soils and in mineral ores.

The surface charge and potential of an insoluble metal oxide is determined in part by the pH of the solution in which it is immersed. For such systems the H^+ and OH^- ions are *potential-determining ions,* presumably as a result of reactions like (compare eqn (6.9.8)):

$$-M^+{-}OH_2 \overset{H^+}{\underset{K_+}{\longleftarrow}} M{-}OH \overset{OH^-}{\underset{K_-}{\longrightarrow}} M{-}O^- + H_2O. \tag{6.10.1}$$

(The presence of hydroxyl groups on metal oxide surfaces has been amply demonstrated by infrared spectroscopy. They can often be removed by heating to high temperatures but gradually return when the surface is exposed to water, either as liquid or vapour.)

Although some early attempts were made to describe oxide systems using the Nernst eqn (6.7.5) it is now generally recognized (Hunter and Wright 1971) that that approach is doomed to failure. The problem lies in the assumption that the activity of the potential-determining ions on the surface is independent of the surface potential (Wright and Hunter 1973). For silver iodide that is a reasonable assumption because there are always large numbers of Ag^+ and I^- ions on the surface and only slight imbalances are necessary to generate the observed surface potentials. On oxide surfaces that is not so. There must be equal numbers of positive and negative charge sites present at the p.z.c. but that number may be very small so that as the number of, say, positive sites is increased, the environment of those sites changes significantly. In that case $a(H^+)$ (surface) is no longer constant and one cannot derive the Nernst equation.

It is still possible to determine the surface charge unequivocally, as a function of pH by the same sort of titration procedure as is used for silver iodide (section 6.7.2) but using a glass electrode to determine the pH after successive additions of an acid or base. Again the amount of OH^- or H^+ that disappears is assumed to be present as a charge on the surface and the p.z.c. is identified by the point at which the Esin and Markov coefficient, $(\partial\sigma_0/\partial\mu)_{pH}$, is zero (Fig. 6.7.1). The procedure is described in some detail by Hunter (1981, p. 225). The thermodynamic argument for finding the p.z.c. by this method is developed in detail by Lyklema (1972). Fortunately, it does not require the existence of a reversible oxide electrode, nor the validity of the Nernst equation for these systems.

The resulting plots of surface charge ($\sigma_0 = e(\Gamma_{H^+} - \Gamma_{OH^-})$) against pH are as shown in Fig. 6.10.1, together with a typical curve for silver iodide, for comparison. The difference between the oxides and the AgI is rather striking.

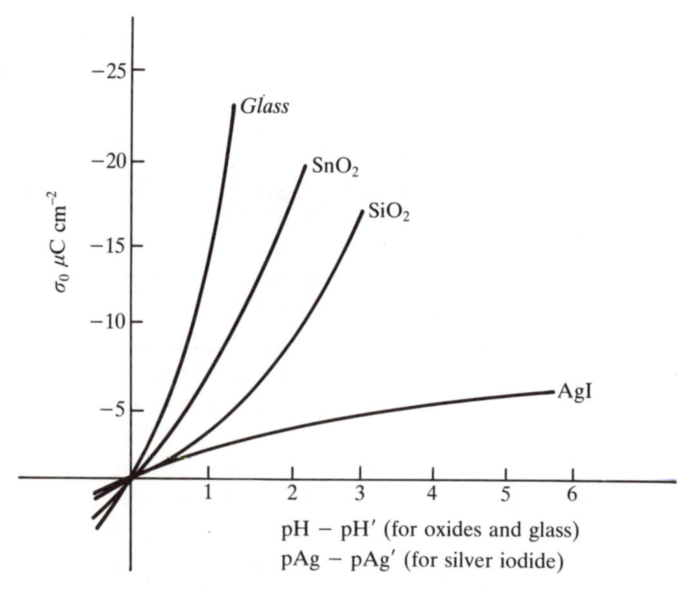

FIG. 6.10.1. Surface charge as a function of ΔpH on the negative side of the p.z.c. (From Lyklema 1981, with permission.)

The much higher charges on the oxides and the steep rise with increase in ΔpH require some explanation, especially in view of the fact that other measurements on oxide systems suggest that the diffuse double layer potentials are quite low–comparable to and even less than those of silver iodide. Essentially this means that a large part of the surface charge must be balanced by the adsorption of counterions into the region between the 'surface' and the outer Helmholtz plane.

Two different approaches have been proposed to account for this effect: (a) the porous surface model, and (b) the surface complex models. The first was suggested by van Lier *et al.* (1960) and an approximate theory was developed by Lyklema (1968). A more elaborate examination was undertaken by Perram *et al.* (1973b), (1974) who showed that a model of this type could account for the two major features of oxide behaviour: high surface charge and low diffuse layer potentials. The problem amounts to solving the Poisson equation throughout a porous (gel) layer of finite thickness in which all of the surface charge and a significant fraction of the counterion charge is located (Fig. 6.10.2).

Even for the simple case of a 1:1 electrolyte, the solution is messy and involves the simultaneous solution of a number of transcendental equations involving Jacobi elliptic functions. It is described by Hunter (1975). Suffice it to say here that it is possible to account for the experimental data on the titanium dioxide surface with reasonable values for the gel thickness ($L = 2$–6 nm) and small values for the chemical adsorption potentials of indifferent ions ($-2kT < \theta_i < 0$).

There is, however, little independent evidence for the existence of porous layers on oxide surfaces, with the possible exception of silica. There has, therefore, been much more interest in the alternative descriptions, which we have described as 'surface complex' models. There are at least five of these with numerous variations and they have

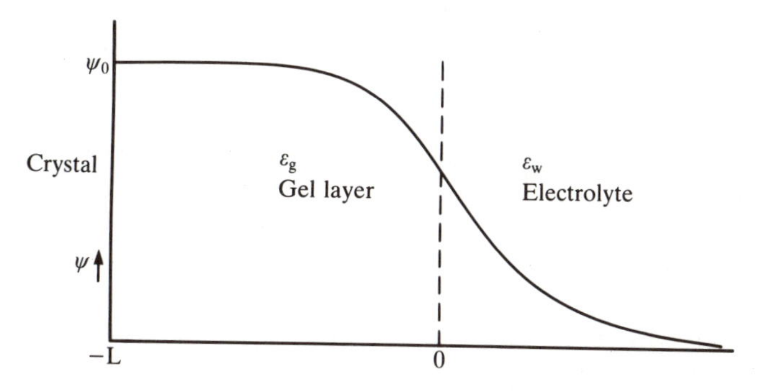

FIG. 6.10.2. Anticipated potential distribution in a porous gel model of the oxide–solution interface.

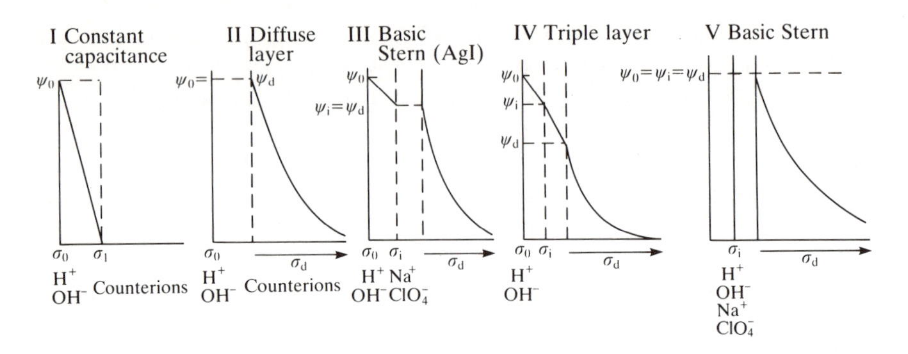

FIG. 6.10.3. Schematic representation of various models of the oxide–solution interface showing the potential as a function of distance from the surface and the planes of adsorption. At each plane, or layer, is shown the charge, the potential and the ions assigned to that layer. (After Westall and Hohl 1980, with permission.) The models are described in the following references: I: Hohl and Stumm (1976); II: Huang and Stumm (1973); III: Bowden *et al.* (1977); IV: Yates *et al.* (1974) and Davis *et al.* (1978); V: Stern (1924).

been compared by Westall and Hohl (1980). Figure 6.10.3, from their review, shows a comparison of the assumed potential distributions for each of the models. Obviously, only model IV, the triple layer model, attempts a full description in terms of the concepts developed on the mercury and silver iodide surfaces. The other models were developed to 'explain' the charge (i.e. titration) data on oxide surfaces and their authors did not wish to introduce too many arbitrary parameters, since they did not have sufficient data to characterize the surface completely. Indeed, Westall and Hohl show that if one wishes *only* to describe the titration data then any one of the five models will suffice, with appropriate choice of its parameters. It was made clear by the work of Wright and Hunter (1973) and the subsequent much more wide-ranging studies of Healy and his collaborators that a reliable description of the oxide surface could best be gained by a simultaneous matching of both the charge data and the electrokinetic data on a single oxide system as a function of pH at different concentrations of indifferent electrolyte. It has now become accepted (Hunter 1981) that the electrokinetic or zeta potential (ζ) measures the electrostatic potential at, or very near to, the beginning of the diffuse double layer (i.e. $\zeta \simeq \psi_d$). A satisfactory model of the interface should therefore, reproduce both the σ_0/pH and the ζ/pH curves with reasonable values for the parameters. It should be obvious from Fig. 6.10.3 that *only* the triple layer model makes any attempt to do this. We will, therefore, concentrate attention on it.

As noted in section 6.9 the surface charge in oxide systems is caused by the dissociation of amphoteric surface hydroxyl groups (eqn (6.10.1)) and this is determined by the pH. The surface potential, ψ_0, is then a function of both pH and the indifferent electrolyte concentration (compare eqn (6.9.9)). The main feature of model IV (Fig. 6.10.3) is that it incorporates both the amphoteric dissociation process, with constants K_+ and K_- (eqn (6.10.1)) together with a series of binding constants for the other ions (like Na^+ and NO_3^-) in the Stern layer. These ions, although normally treated as 'indifferent' are, on the oxide surface, regarded as forming complexes with the surface, the dissociation constants required to account for the amount of binding in the Stern layer are quite small, indicating quite strong binding. It is only in this way that the high surface charge on the oxides can be reconciled with the low measured zeta potentials.

6.10.1 *The site-dissociation–site-binding model*

The dissociation of surface amphoteric sites can be regarded as a simplification of the zwitterionic system (Levine and Smith 1971) with:

$$N_{s_-} = N_{s_+} = N_s \tag{6.10.2}$$

At the p.z.c., where the number of negative and positive sites is equal, the fraction of negative or positive sites ϕ_0 is:

$$\phi_0 = \frac{\alpha'_- N_{s_-}}{N_s} = \alpha'_-(=\alpha'_+). \tag{6.10.3}$$

The most important parameter in the model is the quantity (Exercise 6.10.2):

$$\Delta pK = pK_- - pK_+ = 2\log_{10}\left(\frac{1-2\phi_0}{\phi_0}\right) \tag{6.10.4}$$

which measures, in effect, how much charge is present at the interface *at the p.z.c.* Negative values of ΔpK ($<\sim -3$) correspond to surfaces like AgI with a large (but equal) number of positive and negative charges present at the p.z.c. As ΔpK rises to positive values, the number of charges present at the p.z.c. becomes smaller. For oxide surfaces it turns out that ΔpK is about three in most cases, but values as high as ten are required to describe the charge behaviour of silica (Healy and White 1978; Fig. 6.10.4).

It follows from eqn (6.10.4) that in oxide systems the majority of amphoteric sites are uncharged at the p.z.c. ($10^{-1} > \phi_0 > 10^{-5}$). It should be noted that the dissociation constants are not independent but are

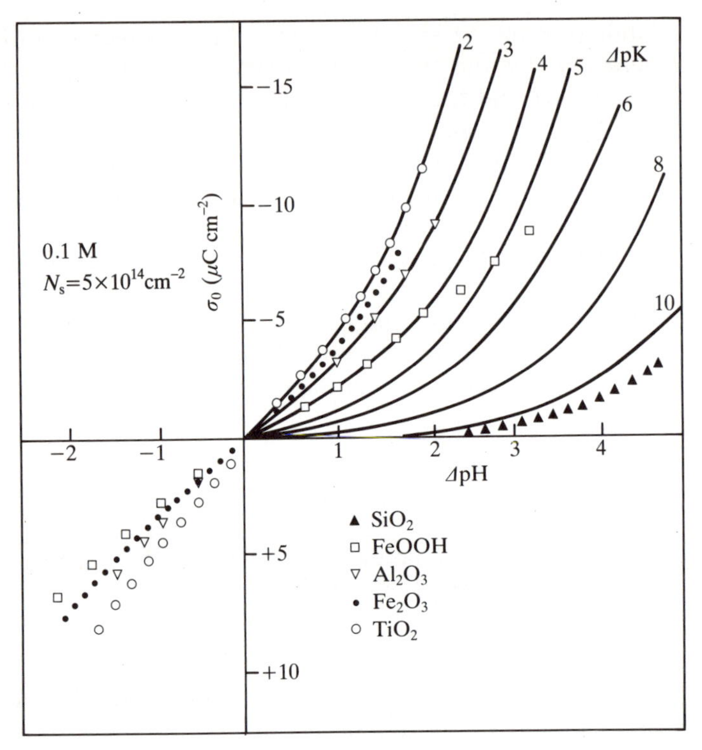

FIG. 6.10.4. Surface charge as a function of pH for a simple amphoteric site dissociation model with $N_s = 5 \times 10^{14}$ sites cm^{-2} and $c = 0.1$ M in all cases. (After Healy and White 1978.)

related by the expression (Exercise 6.10.1):

$$\text{pH(p.z.c.)} = \frac{(\text{p}K_+ + \text{p}K_-)}{2}. \tag{6.10.5}$$

It can also be shown (Smith 1976) that (for $\phi_0 \ll 1$):

$$\left(\frac{d\psi_0}{\text{dpH}}\right)_{\sigma_0 \to 0} = N - \frac{kT}{2N_s e^2}\frac{1}{\phi_0}\frac{\text{d}\sigma_0}{\text{dpH}} \tag{6.10.6}$$

where N is the Nernst slope (=59.8 mV at 25 °C). The second term is only important for small values of ϕ_0 and as it approaches its maximum value (0.5) the system approaches Nernstian behaviour.

Although this simple site dissociation model can go some way towards describing both the charge and electrokinetic potential data on oxides (see Healy and White 1978), it cannot describe the data quantitatively. The site-binding model, first introduced by Yates *et al.* (1974) and extended and improved by Davis *et al.* (1978) provides a more quantitative description at the expense of introducing a set of exchange reactions of the type:

$$AH + Na_s^+ \rightleftarrows A^-Na^+ + H_s^+. \tag{6.10.7}$$

Dissociation (or complexing) constants for these reactions must be postulated:

$$^*K_{Na^+}^{int} = \frac{[A^-Na^+][H^+]_s}{[AH][Na^+]_s} \tag{6.10.8}$$

$$= \frac{[A^-Na^+][H^+]}{[AH][Na^+]} \exp\left(\frac{e(\psi_i - \psi_0)}{kT}\right) \tag{6.10.9}$$

where the subscript s again refers to surface concentrations. Equation (6.10.9) is obtained from eqn (6.10.8) by assuming that the surface charge is contained in the plane of ψ_0 while the sodium ions are at the IHP. A very detailed description of this model is provided by James and Parks (1982), which should be consulted for details of the method of determining $^*K^{int}$ values. The values of the intrinsic surface binding constants, p^*K^{int}, required to account for the experimental data using this model are of the order of 3–10 (Fig. 6.10.5). When interpreted in terms of the normal Stern equation (6.4.12) this would correspond to values of the chemical part of the adsorption free energy (θ_i in eqn (6.4.15)) of the order of $-10kT$ and this is the major point of criticism of the site-binding model. It seems unreasonable to postulate a 'chemical' binding energy of such strength for non-specifically adsorbed ions. As indicated earlier, the solution to this problem may lie in a better description of the electrostatic interactions in the surface layers. The discreteness of charge effect (section 6.4.4), if it were properly calculated for these systems, might go a considerable way towards accounting for these surprising effects. That would amount to an estimation of the surface activity coefficients for the adsorbed ions in the Stern layer. Although some progress is now being made in the study of surface activity coefficients for uncharged species (see Lane 1982) there is still much to be done in this area for charged species. (See Sposito (1983)).

One final comment about the triple layer model (IV of Fig. 6.10.3). Apart from a knowledge of the pH at the p.z.c. it is necessary to select a suitable value for either K_+ or K_- (and get the other from eqn (6.10.5)) and also to select values for $^*K^{int}$ for each of the indifferent ions. One must also postulate a value for the two capacitances in the compact region. In oxides it turns out that a very large value is required for C_1 ($\sim$150 μF cm^{-2}) and a much smaller value for C_2 ($\sim$20 μF cm^{-2}). In order to keep the arbitrariness of the model within reasonable bounds these

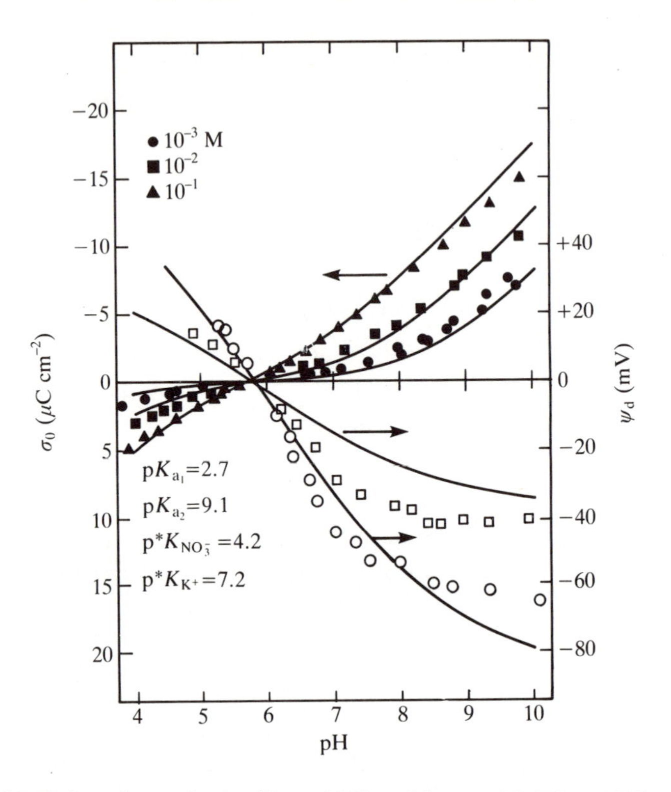

FIG. 6.10.5. Surface charge density (Yates 1975) and ζ-potential (Wiese 1973) of a TiO_2 dispersion as a function of pH at various concentrations of KNO_3 and at 25 °C. Solid lines are calculated from the site binding model with constants indicated on the figure. (From James and Parks 1982.)

values are taken to be constant and independent of the charge on the surface. Considering Fig. 6.7.6 for the AgI surface this would seem to be a rather unlikely situation. It should also be noted that the extrapolation procedure advocated by Davis *et al.* (1978) to obtain values of $^*K^{int}$ produces quantities which are independent of the solution concentration. Again this is justifiable on the grounds that it reduces the arbitrariness of the model but the connection between $^*K^{int}$ and the chemical adsorption potential, θ_i, makes it likely that $^*K^{int}$ will, in reality, depend upon the state of charge of the interface until the discreteness of charge effect is properly taken into account. (See the discussion concerning eqns (6.4.19) and (6.4.20).)

Really definitive experimental evidence on these points is difficult to obtain. Foissy *et al.* (1982), for example, show that, by separately following the adsorption of Na^+ and Cl^- on a titanium dioxide surface, using radioisotopes, it is possible to build up a complete picture of the

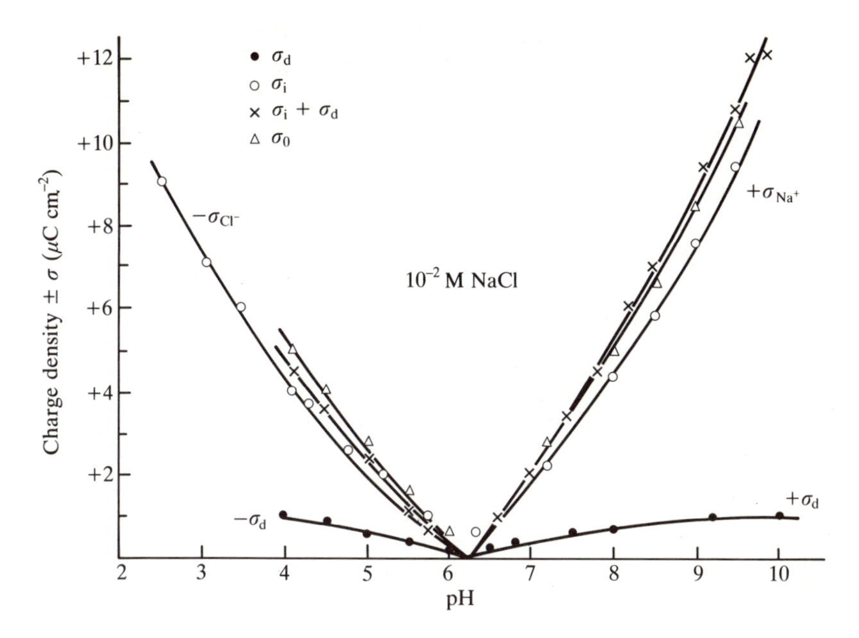

FIG. 6.10.6. Sum of diffuse layer charge (σ_d) and Stern layer charge (σ_{Cl^-}, σ_{Na^+}) with the measured surface charge (σ_0). (From Foissy *et al.* 1982, with permission.)

disposition of the counter charge. They find (Fig. 6.10.6) that less than 10 per cent of the charge is in the diffuse layer and that the adsorption of the counter-ion (Cl^- below the p.z.c. and Na^+ above the p.z.c.) can be adequately described by the Stern equation (6.4.12) with θ_i values that are independent of pH (and, hence, of the state of charge of the surface). Their estimate of ψ_i made no attempt to introduce the discreteness of charge correction so it is possible that their result is somewhat fortuitous, and it is not clear that their washing procedure allows them to separately estimate the amount of charge in the inner (non-diffuse) part of the double layer. Despite these reservations, the agreement between theory and experiment is encouraging. Nevertheless, similar measurements of the components of charge in the double layer will have to be made on many systems before we can be confident that the models discussed above give an adequate account of the structure of the oxide–solution interface.

6.10.2 *Clay mineral systems*

The structure of clay minerals (section 1.5.5) suggests that they may be described by a modification of the model for oxide surfaces. The fixed charge inside the crystal lattice can be accommodated by assuming the presence of a very strong acid group that remains dissociated at the lowest accessible pH. Certainly the fact that highly charged clay minerals often show quite small values for $|\zeta|$ would suggest some real similarities with the oxide solution interface and James and Parks (1982) have used the site-dissociation–site-binding model (Model IV of Fig. 6.10.3) to treat some literature data on clay mineral systems. Again it turns out that quite large values of the 'chemical' binding energy would be required to account for the amounts of K^+ and Cl^- that are apparently adsorbed inside the outer Helmholtz plane. Perhaps, again, the discreteness of charge effect has some rôle to play here, but it seems much less likely because the lattice charge is often buried below the surface so that the outer layer of the solid must be very like a constant-potential surface (see Hunter 1981, p. 21).

The fact that a significant fraction of the surface charge in these systems is due to crystal lattice defects (1.5.5) makes the kind of analysis developed in 6.10.1 somewhat suspect. A more general approach has recently been developed by Sposito (1981) and we will take that up in more detail in Chapter 12 (Vol. II).

Exercises. 6.10.1 Establish eqn (6.10.5).

6.10.2 Establish eqn (6.10.4). (Note that K_+ and K_- should both be defined as acid dissociation constants). Use eqn (6.9.4) to obtain K_- and derive a corresponding expression for K_+.

6.10.3 How would you interpret the very large values (~150 μF cm^{-2}) that are postulated for the capacitance C_1 between the plane of the surface charge and the IHP in oxide systems?

6.11. Double layer around a sphere

Equation (6.3.9) remains valid for the diffuse part of the double layer around a spherical particle except that the Laplace operator† for the case of spherical symmetry has a different form:

$$\frac{1}{r^2}\frac{\mathrm{d}}{\mathrm{d}r}\left(r^2\frac{\mathrm{d}\psi}{\mathrm{d}r}\right) = \boldsymbol{\Delta}\psi = -\frac{1}{\varepsilon_0\varepsilon_r}\sum_i n_i^0 z_i e \exp\left(-\frac{z_i e\psi}{kT}\right). \quad (6.11.1)$$

Unfortunately, this equation cannot be solved analytically and recourse must be had to the Debye–Hückel approximation (section 6.3.1), valid for small values of the potential. Expanding the exponential ($e^{-x} \simeq 1 - x$) we again obtain (compare eqn (6.3.11)):

$$\boldsymbol{\Delta}\psi = \kappa^2\psi. \quad (6.11.2)$$

† Note that the form ∇^2 for the Laplace operator is best confined to the Cartesian coordinate system. (See Appendix A3.)

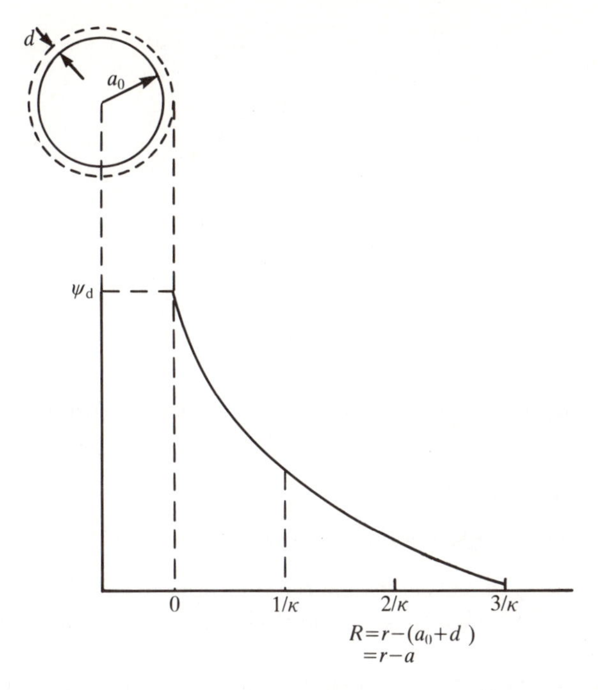

FIG. 6.11.1. The diffuse double layer around a sphere of low potential. (Note the change of coordinate system.)

The solution of this equation, for a particle of radius a_0, with a compact layer of thickness d (Fig. 6.11.1) is (Exercise 6.11.1):

$$\psi = \psi_d \frac{a}{r} \exp[-\kappa(r-a)] \tag{6.11.3}$$

where ψ_d is the potential characterizing the beginning of the diffuse double layer and $a = a_0 + d$.

6.11.1 *Charge in the diffuse layer*

The total charge in the diffuse layer around the particle is given by

$$Q_d = \int_a^{\infty} 4\pi r^2 \rho \, dr \tag{6.11.4}$$

and assuming that ρ is given by (compare eqn 6.3.4):

$$\rho = -\varepsilon \boldsymbol{\Delta}\psi = -\varepsilon\kappa^2\psi \tag{6.11.5}$$

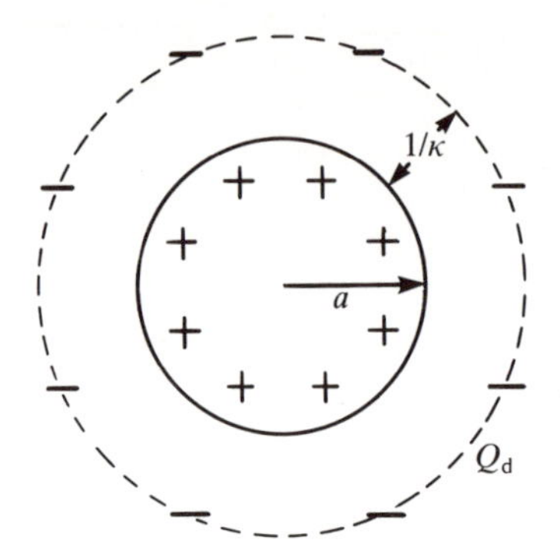

FIG. 6.11.2. Apparent distribution of charge around a spherical particle at low potential.

we obtain (Exercise 6.11.2):

$$Q_d = -Q_0(1 + \kappa(a_0 + d)) = -4\pi\varepsilon_0\varepsilon_r\psi_d a[1 + \kappa a] \qquad (6.11.6)$$

where $Q_0 = 4\pi\varepsilon\psi_d a$. ($Q_0$ is the charge on a sphere of radius a and surface potential ψ_d in a dielectric of permittivity ε ($=\varepsilon_0\varepsilon_r$) but with no diffuse space charge around it.)

If there is no charge in the compact layer ($a_0 < r < a$) then the charge on the particle $Q_p = -Q_d$ and substituting for ψ_d in eqn (6.11.3) we have:

$$\psi = \frac{1}{4\pi\varepsilon}\frac{Q_p}{1+\kappa a}\frac{\exp(-\kappa(r-a))}{r} \qquad (6.11.7)$$

ψ_d, from eqn (6.11.6) can be written in the form (exercise 6.11.3):

$$\psi_d = \frac{Q_p}{4\pi\varepsilon a} + \frac{Q_d}{4\pi\varepsilon(a + 1/\kappa)} \qquad (6.11.8)$$

$$= \psi_d^p + \psi_d^a. \qquad (6.11.9)$$

The first term in eqn (6.11.8) represents the potential on the surface at $r = a$ due to the particle charge, while the second term (which is opposite in sign) represents the potential due to the space charge in the diffuse layer. It corresponds to the potential due to a spherical shell of charge $Q_d = -Q_p$ at a radius $(a + 1/\kappa)$ so that, once again, the function $1/\kappa$ can be interpreted as the 'thickness' of the double layer (Fig. 6.11.2). Remember, though, that this analysis holds *only* for low potentials.

6.11.2 *Behaviour at higher potentials*

Although it is impossible to provide an analytical solution for the complete Poisson–Boltzmann equation (6.11.1) in the spherical case there are a number of analytical approximations available. These are briefly reviewed by Hunter (1981) who concludes that the semi-empirical equation proposed by Loeb *et al.* (1961) is the most useful:

$$Q_{\mathrm{d}} = -4\pi\varepsilon\frac{kT}{ze}\kappa a^2\left\{2\sinh\frac{ze\psi_{\mathrm{d}}}{2kT}+\frac{4}{\kappa a}\tanh\frac{ze\psi_{\mathrm{d}}}{4kT}\right\}. \qquad (6.11.10)$$

A theoretical justification of this equation is given by Dukhin *et al.* (1970) and by Stokes (1976). Extensive tables of the potential and charge relation have been computed by Loeb *et al.* (1961) and a compact series representation of those numerical results is provided by Stigter (1972) whose results are also presented in Hunter (1981, pp. 46–9). White (1977) has developed a good approximate analytical expression using a novel perturbation scheme and, more recently, Ohshima *et al.* (1982) have produced an even better approximation, in the course of which they also provide an analytical justification for eqn (6.11.10).

Exercises. 6.11.1 Substitute $\psi = u/r$ in eqn (6.11.1) and hence show that it can be written $\mathrm{d}^2u/\mathrm{d}r^2 = \kappa^2 u$. Verify that the general solution of this equation is $u = A\mathrm{e}^{\kappa r} + B\mathrm{e}^{-\kappa r}$. Hence establish eqn (6.11.3) using Fig. 6.11.1.

6.11.2 Establish eqn (6.11.6) using eqns (6.11.3) and (6.11.5).

6.11.3 Establish eqns (6.11.7) and (6.11.8).

6.12 Double layer around a cylinder

The complete form of the Poisson–Boltzmann equation, for a symmetrical electrolyte, in cylindrical coordinates is:

$$\Delta\psi = \frac{1}{R}\frac{\mathrm{d}}{\mathrm{d}R}\left\{R\frac{\mathrm{d}(ze\psi/kT)}{\mathrm{d}R}\right\} = \sinh\frac{ze\psi}{kT} \qquad (6.12.1)$$

where $R = \kappa r$ (if end effects are neglected) and r is the radial distance from the surface. There is no general analytical solution of eqn (6.12.1) and even the corresponding linear equation (valid for $ze\psi < kT$) has a solution that involves zero and first-order modified Bessel functions of the second kind (Dube 1943). Some aspects of this solution are described by Hunter (1981) who also discusses the more recent calculations by Philip and Wooding (1970) and by Stigter (1975). The papers by White (1977) and Ohshima *et al.* (1982) referred to in 6.11.2 also address this problem.

References

Adamson, A. W. (1967). *Physical chemistry of surfaces,* Chapter 4. Wiley-Interscience, New York.

Atkins, P. W. (1978). *Physical chemistry.* Oxford University Press, Oxford.

Blum, L. (1977). *J. phys. Chem.* **81,** 136.

Bockris, J. O'M., Conway, B. E., and Yeager, E. (1980). *Comprehensive treatise of electrochemistry,* Vol. 1. *The double layer.* Plenum, New York.
Bockris, J. O'M. and Reddy, A. K. (1970). *Modern electrochemistry* Vol 2. Plenum, New York.
Bowden, J. W., Posner, A. M., and Quirk, J. P. (1977). *Austral. J. Soil Res.* **15,** 121.
Buff, F. P. and Goel, N. S. (1969). *J. chem. Phys.* **51,** 4983, 5363.
Cade, R. (1978). *J. Colloid interface Sci.* **66,** 358.
Chapman, D. L. (1913). *Phil. Mag.* **25,** 475.
Cooper, I. L. and Harrison, J. A. (1977). *Electrochim. acta* **22,** 519.
Curreri, P., Onoda, G. Y., Jr., and Finlayson, B. (1979). *J. Colloid interface Sci.* **69,** 170.
Damaskin, B. B. (1977). *J. electroanal. Chem.* **75,** 359.
Damaskin, B. B. and Frumkin, A. N. (1974). *Electrochim. acta* **19,** 173.
Davies, J. T. and Rideal, E. K. (1963). *Interfacial phenomena,* Chapter 2. Academic Press, London.
Davis, J. A., James, R. O., and Leckie, J. O. (1978). *J. Colloid interface Sci.* **63,** 480–99.
Delahay, P. (1966). *The double layer and electrode kinetics.* Wiley-Interscience, New York.
Dube, G. P. (1943). *Ind. J. Phys.* **17,** 189.
Dukhin, S. S., Semenikhin, N. M., and Shapinskai, L. M. (1970). *Dokl. Akad. nauk SSR* **193,** 385.
Ershler, B. V. (1946). *Zh. fiz. Khim.* **20,** 679.
Esin, O. A. and Markov, B. F. (1939). *Acta physicochim. URSS* **10,** 353.
Esin, O. A. and Shikov, V. (1943). *Zh. fiz. Khim.* **17,** 236.
Foissy, A., M'Pandou, A., LaMarche, J. M., and Jaffrezic-Renault, N. (1982). *Colloids Surfaces* **5,** 363–8.
Freyberger, W. L. and de Bruyn, P. L. (1957). *J. phys. Chem.* **61,** 586–92.
Frumkin, A. N. (1964). *J. electroanal. Chem.* **7,** 152.
Gardiner, C. L. (1975). *J. electroanal. Chem.* **61,** 113.
Gouy, G. (1910). *J. Phys. Radium* **9,** 457.
Grahame, D. C. (1947). *Chem. Rev.* **41,** 441.
Grahame, D. C. (1958). *Z. Elektroch.* **62,** 264.
Greenland, D. J. and Quirk, J. P. (1964). *J. Soil Sci.* **15,** 178.
Guggenheim, E. A. (1929) *J. phys. Chem.* **33,** 842.
Hall, D. G. and Rendall, H. M. (1980). *J. chem. Soc. Faraday Trans. 1* **76,** 2575–86.
Healy, T. W. and White, L. R. (1978). *Adv. Colloid interface Sci.* **9,** 303–45.
Healy, T. W., Yates, D. E., White, L. R., and Chan, D. (1977). *J. electroanal. Chem.* **80,** 57.
Herz, A. H. (1974). *Photogr. Sci. Engng* **18,** 323.
Hohl, H. and Stumm, W. (1976). *J. Colloid interface Sci.* **55,** 281.
Homola, A. and James, R. O. (1977). *J. Colloid interface Sci.* **59,** 123.
Huang, C. P. and Stumm, W. (1973). *J. Colloid interface Sci.* **43,** 409.
Hunter, R. J. (1975). Electrochemical aspects of colloid chemistry. In *Modern aspects of electrochemistry* (ed. B. E. Conway and J. O'M. Bockris) Vol. 11, Chapter 2, pp. 33–84. Plenum, New York.
Hunter, R. J. (1981). *Zeta potential in colloid science.* Academic Press, London.
Hunter, R. J. and Wright, H. J. L. (1971). *J. Colloid interface Sci.* **37,** 564–80.
James, R. O. and Parks, G. A. (1982). Characterization of aqueous colloids by

their double layer and intrinsic surface. In *Surface and colloid science* (ed. E. Matijevic) Vol. 12. Plenum, New York.
Lane, J. E. (1982). In *Adsorption from solution.* Royal Society of Chemistry and Society of Chemical Industry, London.
Levine, S. (1971). *J. Colloid interface Sci.* **37,** 619.
Levine, S. and Bell, G. M. (1966) *Discuss. Faraday Soc.* **42,** 69.
Levine, S., Bell, G. M., and Calvert, D. (1962) *Can. J. Chem.* **40,** 518.
Levine, S., Mingins, J., and Bell, G. M. (1967). *J. electroanal. Chem.* **13,** 280.
Levine, S. and Outhwaite, C. W. (1978). *J. chem. Soc. Faraday Trans. 2* **74,** 1670.
Levine, S. and Smith, A. L. (1971). *Discuss. Faraday Soc.* **52,** 290.
Loeb, A. L., Wiersema, P. H., and Overbeek, J. Th.G. (1961). *The electrical double layer around a spherical colloidal particle.* M.I.T. Press, Cambridge Mass.
Lyklema, J. (1968). *J. electroanal. Chem.* **18,** 341.
Lyklema, J. (1972). *J. electroanal. Chem* **37,** 53.
Lyklema, J. (1977). The electrical double layer on silver iodide. In *Trends in electrochemistry* (ed. J. O'M. Bockris, D. A. J. Rand, and B. J. Welch) pp. 159–76. Plenum Press, New York.
Lyklema, J. (1981). *Pure appl. Chem.* **53,** 2199–209.
Lyklema, J. and Overbeek, J. Th. G. (1961). *J. Colloid Sci.* **16,** 595.
MacDonald, J. R. and Barlow, C. A. Jr (1964). In *Electrochemistry,* pp. 199–247. Proceedings of the First Australian Conference in Electrochemistry, 1963. Pergamon, London.
Mohilner, D. M. and Kakiuchi, T. (1981). *J. electrochem. Soc.* **128** (2), 350–1.
Mott, N. F. and Watts-Tobin, R. J. (1961). *Electrochim. acta* **4,** 79.
Ohshima, H., Healy, T. W., and White, L. R. (1982). *J. Colloid interface Sci.* **90,** 17–26.
Oldham, K. B. and Parsons, R. (1977). *Soviet Electrochem.* **13,** 732.
Ottewill, R. H. and Shaw, J. N., (1966). *Discuss. Faraday Soc.* **42,** 154.
Overbeek, J. Th. G. (1952). In *Colloid science,* Vol. 1, pp. 128–32. Elsevier, Amsterdam.
Parsons, R. (1954). Equilibrium properties of electrified interphases. In *Modern aspects of electrochemistry* (ed. J. O'M. Bockris and B. E. Conway) Vol. 1, pp. 103–179. Butterworths, London.
Parsons, R. (1957). *Proceedings of the Second International Congress on Surface Activity,* Vol. 3, p. 38. Butterworths, London.
Parsons, R. (1959). *Trans. Faraday Soc.* **55,** 999.
Parsons, R. (1964). *J. electroanal. Chem.* **8,** 93.
Parsons, R. (1975a). Thermodynamics of electrified interphases. In *Physical chemistry: enriching topics from colloid and surface science,* Chapter 17, pp. 251–269. Theorex, La Jolla, Cal.
Parsons, R. (1975b). *J. electroanal. Chem.* **59,** 229.
Parsons, R. (1980). Thermodynamic methods for the study of interfacial regions in electrochemical systems. In *Comprehensive treatise of electrochemistry* Vol. 1 *The double layer* (ed. J. O'M. Bockris, B. E. Conway, and E. Yeager) Chapter 1. Plenum Press, New York.
Payne, R. (1972). In *Techniques of electrochemistry* (ed. E. Yeager and A. J. Salkind) Vol. 1, Chapter 2. Wiley-Interscience, New York.
Perram, J. W., Hayter, J. B., and Hunter, R. J. (1973a). *J. electroanal. Chem. interfacial electrochem.* **42,** 291–298.

Perram, J. W., Hunter, R. J., and Wright, H. J. L. (1973b). *Chem. Phys. Lett.* **23,** 265.
Perram, J. W., Hunter, R. J., and Wright, H. J. L. (1974). *Austral. J. Chem.* **27,** 461–75.
Philip, J. R. and Wooding, R. A. (1970). *J. chem. Phys* **52,** 593.
Pieper, J. H. A., de Vooys, D. A., and Overbeek, J. Th. G. (1975). *J. electroanal. Chem.* **65,** 429.
Rendall, H. M. and Smith, A. L. (1978). *J. chem. Soc. Faraday Trans. 1* **74,** 1179.
Robinson, K. and Levine, S. (1973). *J. electroanal. Chem.* **47,** 395.
Salem, R. R. (1976). *Russ. J. phys. Chem.* **50,** 656.
Schofield, R. K. (1947). *Nature* **160,** 408.
Smith, A. L. (1976). *J. Colloid interface Sci.* **55,** 525.
Sparnaay, M. J. (1972). *International Encyclopedia of Physical Chemistry and Chemical Physics* (ed. D. H. Everett) Topic 14, Vol. 4. *The electrical double layer.* Pergamon, Oxford.
Sposito, G. (1981). *Soil Sci. Soc. Am. J.* **45,** 292–7.
Sposito, G. (1983). *J. Colloid interface Sci.* **91,** 329–40.
Stern, O. (1924). *Z. Elektrochem.* **30,** 508.
Stigter, D. (1972). *J. electroanal. Chem.* **37,** 61.
Stigter, D. (1975). *J. Colloid interface Sci.* **53,** 296.
Stokes, A. N. (1976). *J. chem. Phys.* **65,** 261.
Torrie, G. M. and Valleau, J. P. (1979). *Chem. Phys. Lett.* **65,** 343.
Trasatti, S. (1971). *J. electroanal. Chem.* **33,** 351.
van den Hul, H. J. and Lyklema, J. (1967). *J. Colloid Sci.* **23,** 500.
van den Hul, H. J. and Lyklema, J. (1968). *J. Am. chem. Soc.* **90,** 3010.
van Lier, J. A., de Bruyn, P. L., and Overbeek, J. Th. G. (1960). *J. phys. Chem.* **64,** 1675.
Westall, J. and Hohl, H. (1980). *Adv. Colloid interface Sci.* **12,** 265–94.
White, L. R. (1977). *J. chem. Soc. Faraday Trans. 2* **73,** 577–96.
Wiese, G. R. (1973). Ph.D. Thesis, University of Melbourne.
Wright, H. J. L. and Hunter, R. J. (1973). *Austral. J. Chem.* **26,** 1183–90, 1191–206.
Yates, D. E. (1975). Ph.D. Thesis, University of Melbourne.
Yates, D. E., Levine, S., and Healy, T. W. (1974). *Trans. Faraday Soc.* **70,** 1807.

7

DOUBLE LAYER INTERACTION AND PARTICLE COAGULATION

In this chapter we consider the various forces that come into play when two colloidal particles or two double-layer systems approach one another. The behaviour is determined to some extent by how rapidly the approach occurs, since the double layers may take a significant time to adjust to the new situation. Equilibrium behaviour may then be somewhat different from transient behaviour.

The repulsive interaction that occurs between double layers of like sign can be analysed by two alternative procedures by considering:

(a) the free energy change involved when overlap occurs; or

(b) the osmotic pressure (section A6) generated by the accumulation of ions between the particles. The case of large flat plates can be treated fairly rigorously but some useful approximation formulae are also derived for the case of low potentials or for small degrees of overlap. We then look at situations where the potentials on the two approaching surfaces are different. This is relevant to such processes as flotation, mixed colloidal suspensions, and the drainage of wetting films on solids. The theory for spherical particles is more complicated algebraically and will not be discussed in great detail.

The experimental evidence in support of the DLVO Theory (section 2.5) comes from a variety of sources but the most definitive results have been obtained since about 1976, using atomically smooth sheets of mica, interacting in an aqueous electrolyte solution. That work clearly establishes the general validity of the theory presented below, at the same time suggesting that when particles approach one another very closely there are quite dramatic new effects that come into play.

The chapter closes with a brief discussion of the kinetics of the coagulation process and how it is affected by the double layer interaction.

7.1 Surface conditions during interaction

When two colloidal particles (or two charged interfaces, in general) approach one another so that their electrical double layers begin to overlap the result is usually a repulsive force, which tends to oppose further approach (section 2.5). For flat plates, this effect can be understood in terms of the osmotic pressure created by the difference in ion concentration in the region between the two approaching surfaces compared to the bulk (or reservoir) concentration (Fig. 7.1.1). To

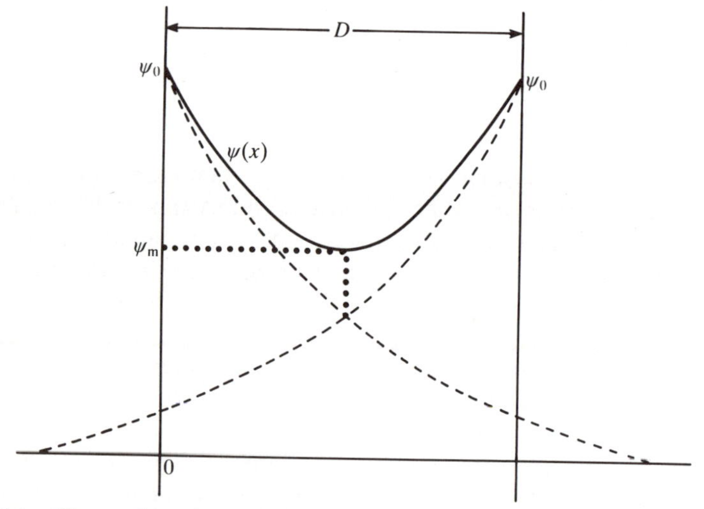

FIG. 7.1.1. The overlap of two diffuse double layers. The potential distribution in the neighbourhood of a single double layer is shown ----. The full line is the anticipated potential distribution for the pair of particles. ψ_m is the potential at the minimum (which in this case is also in the median plane). The high potentials between the plates ($\psi_m > 0$) lead to high counterion concentrations and this in turn leads to a large osmotic pressure, tending to push the particles apart.

simplify the exposition we will initially assume that the surfaces are planar and that the Poisson–Boltzmann equation holds over the entire region between them. In most interaction situations it is only the diffuse layers that interact so that the boundary electrostatic potential (section 6.3) is ψ_d rather than ψ_0, but we will leave that aside for the moment.

When two surfaces approach one another there are several possible situations that might arise. The approach may be slow, so that equilibrium can be established between the ions on the surface and in the bulk. For silver iodide particles under those conditions one would expect the surface potential to remain constant during the approach. On the other hand, if the particle charge is caused by built-in crystal defects, as in some clay minerals, it might be more sensible to assume that the surface charge is constant during approach. In the case of oxide surfaces, the interaction may itself influence the degree of dissociation of surface groups (section 6.10) so that neither ψ_0 nor σ_0 is constant. The condition known as charge regulation (Ninham and Parsegian 1971) may then be more appropriate. There are other possibilities if specific adsorption in a Stern layer is involved. In general, however, these three possibilities can adequately cover most experimental situations.

The constant potential case was extensively studied by the early workers in the field and is the basis of the DLVO theory of colloid stability (section 2.5) described in detail in the monograph by Verwey and Overbeek (1948). The other possibilities are best understood as fairly straightforward generalizations of the constant potential case.

We will examine first the case where the approaching surfaces are identical and subsequently deal with the case of unlike systems (heterocoagulation). The repulsion can be calculated either from the osmotic pressure, as noted above, or from the increase in Helmholtz free energy, ΔF, that occurs as two double layers overlap. The osmotic pressure method cannot be applied to the approach of spherical particles because another factor (the Maxwell stress, section 7.3) enters in that case. The free energy method is rather more general in its application. We will begin, therefore, with an examination of the free energy of a single (i.e. isolated) double layer.

7.2 Free energy of formation of a double layer

When a silver iodide crystal is immersed in an aqueous solution, the double layer forms spontaneously on its surface (section 6.7). The free energy of formation (ΔG or ΔF) is, therefore, negative. It can be calculated by determining the effect of the double layer on the capacity of the system to do work and that in turn can be calculated by determining the work that would have to be done to establish the double-layer structure from an initially uncharged system. The procedure is similar to that employed by Debye and Hückel in their theory of strong electrolytes but since that is seldom described clearly in elementary texts and the original treatise by Verwey and Overbeek (1948) is now out of print we will reproduce the argument here.

The double layer arises because of a preferential adsorption of potential-determining ions of one sign over those of the other. The adsorption occurs against an increasingly unfavourable electrostatic potential because the chemical potential of the ion (Ag^+ or I^-) is lower on the surface than it is in the bulk solution. If we begin with an initially uncharged system and imagine the ions being adsorbed one at a time, then for each ion adsorbed there is a reduction, $\Delta\mu$, in the chemical potential. This 'chemical' part of the free energy change can be evaluated by noting that, when the last ion goes on, the chemical free energy decrease, $\Delta\mu$, is exactly compensated by the electrical free energy increase: $e\psi_0$ per ion or $\sigma_0\psi_0$ per unit area of surface, i.e.

$$\Delta\mu_{\text{chem}} = -\sigma_0\psi_0 = \Delta F_{\text{chem}}. \tag{7.2.1}$$
(per unit area of surface)

(This inherently assumes that ΔF_{chem} is independent of the state of charge of the surface. It is strictly true only at low charge density and corresponds to the condition introduced in section 6.7.1 to derive the Nernst equation. Chan and Mitchell (1983) show how to remove this restriction to generate a more general expression for ΔF but we will not need that in the present treatment.)

During the charging process, as the double layer builds up, the *electrical* work done can be calculated by imagining the transfer of infinitesimal amounts of charge $d\sigma$ (per unit area) through a potential ψ_0', where ψ_0' increases from zero to its final value ψ_0 when the last piece of charge is installed. The electrical work done is then:

$$\Delta F_{\text{elec}} = \int_0^{\sigma_0} \psi_0' \, d\sigma. \tag{7.2.2}$$

(This is analogous to the electrical work done in charging a condenser).

There is another 'work' term involved in the arrangement of the charges in the solution, after each step in the charging process, so that the balancing charge in the solution can be established. It is not difficult to show, however, (Exercise 7.2.1) that this does not involve any further change in free energy if the equilibrium ion distribution obeys the Boltzmann equation. The total free energy change involved in establishing the double layer is then

$$\Delta F = \Delta F_{\text{chem}} + \Delta F_{\text{elect}} = -\sigma_0\psi_0 + \int_0^{\sigma_0} \psi_0' \, d\sigma$$

$$= -\int_0^{\psi_0} \sigma_0' \, d\psi_0'. \tag{7.2.3}$$

(Note that the chemical term is always larger in magnitude than the electrical term so that ΔF is negative since σ_0 and ψ_0 always have the same sign.)

Equation (7.2.3) may also be derived directly from the Lippmann equation (6.1.14), expressed in the form:

$$\left(\frac{\partial\gamma}{\partial\psi_0'}\right)_\mu = -\sigma_0'.$$

Then

$$\Delta F = \gamma_0 - \gamma = \int \mathrm{d}\gamma = -\int_0^{\psi_0} \sigma_0' \, d\psi_0'.$$

The free energy of a single (isolated) diffuse double layer is readily calculated from eqn (7.2.3) because σ_0 can be expressed as a function of ψ_0 using the appropriate form of eqn (6.3.27). Then (Exercise 7.2.2):

$$\Delta F = -\frac{8n^0kT}{\kappa}\left(\cosh\frac{ze\psi_0}{2kT} - 1\right). \tag{7.2.4}$$

(Note that no distinction is made between ΔF and ΔG in these systems: the $p\Delta V$ contribution to the work done in the charging process is negligible.) Some simple expressions for ΔF can be established when the potential on the surface is low (Exercise 7.2.3).

We can now readily see why the interaction is repulsive. As the particles approach under constant potential conditions the potential profile (Fig. 7.2.1) becomes increasingly shallow. The absolute value of $\mathrm{d}\psi/\mathrm{d}x$ at the surface, therefore, decreases. This can be interpreted as a decrease in the surface charge density since (compare eqn (6.3.26)):

$$\sigma_0 = -\varepsilon(\mathrm{d}\psi/\mathrm{d}x)_{x=0}. \tag{7.2.5}$$

As the particles approach, the surface potential can only remain constant if the potential determining ions are gradually driven off the surface until the double layer finally disappears on contact. Since the double layer was initially formed spontaneously, this forcible discharge of the plates results in an increase in free energy. The repulsive potential energy V_R, due to

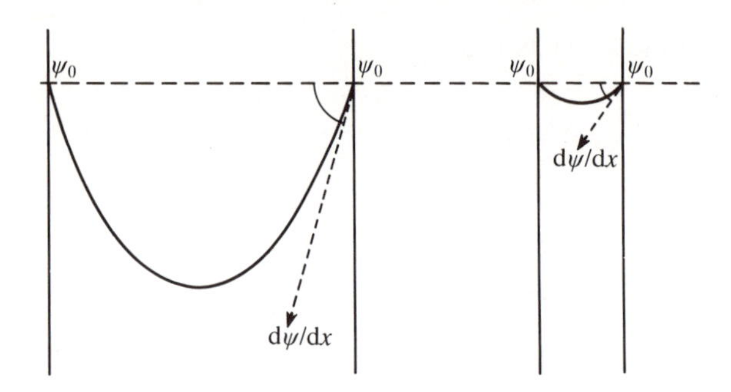

FIG. 7.2.1. As two plates approach at constant potential the slope $d\psi/dx$ at $x=0$ decreases in absolute value, indicating a progressive discharge of the surface potential-determining ions.

approach of the two plates is thus given by:

$$V_R = (\Delta F(D) - \Delta F(\infty)) \tag{7.2.6}$$

where $\Delta F(\infty)$ is the free energy (per unit area) associated with the isolated double layer and $\Delta F(D)$ is the corresponding free energy when the plates are at a distance D apart.

The corresponding expression for the repulsive potential energy, evaluated from the osmotic pressure, $\bar{p}$, between the plates is:

$$V_R = -\int_{\infty}^{D} \bar{p}\, dD \tag{7.2.7}$$

where $\bar{p}$ is the *difference* in the osmotic pressure midway between the plates and in the bulk solution.

Both the free energy and the osmotic pressure method can be used to calculate V_R for flat plates and each has advantages in different situations. Given V_R, the total potential energy of interaction V_T can then be obtained by adding V_A (Chapter 4) and V_T can then be used to describe the stability behaviour (section 2.5).

Exercises. 7.2.1 Show that the free energy change involved in arranging the diffuse counter charge is zero if the ions obey the Boltzmann equation. (Hint: note that $\Delta S = k \ln n_i^0/n_i$ where n_i is the counterion concentration).

7.2.2 Establish eqn (7.2.4) assuming that

$$\sigma_0 = \frac{4n^0 ze}{\kappa} \sinh ze\psi_0/2kT.$$

7.2.3 Show that at low potentials, where the Debye–Hückel approximation holds, the free energy of a flat double layer is:

$$\Delta F = -\tfrac{1}{2}\sigma_0\psi_0 = -\tfrac{1}{2}\varepsilon\kappa\psi_0^2$$

and of a sphere is $-2\pi\varepsilon a(1+\kappa a)\psi_0^2$.

7.3 Overlap of two flat double layers

It was noted in section 2.5 that the stability behaviour of colloidal sols was determined almost entirely by the concentration of the counterion. It is possible, therefore, to restrict attention to symmetric electrolyte systems because only the valency of the counterion is of any great importance. A salt like $MgCl_2$ will then be expected to behave more like a 1 : 1 electrolyte with respect to a positive surface and more like a 2 : 2 electrolyte with respect to a negative surface.

The potential profiles drawn in Figs 7.1.1 and 7.2.1 must still satisfy the Poisson–Boltzmann equation (6.3.15) which, for a symmetric electrolyte, can be written:

$$2n^0 ze \sinh\frac{ze\psi}{kT} - \varepsilon\frac{d^2\psi}{dx^2} = 0. \qquad (7.3.1)$$

This equation can be integrated (for a fixed value of D) as before to obtain (compare eqn (6.3.18)):

$$2n^0 kT \cosh\frac{ze\psi}{kT} - \frac{\varepsilon}{2}\left(\frac{d\psi}{dx}\right)^2 = p_m \quad \text{(a constant).} \qquad (7.3.2)$$

The first term on the left is related to the osmotic pressure between the plates. That pressure varies from point to point because of the variation in ψ (and, hence, of the local ion concentration). It can be most readily evaluated at the midplane between the particles, because at that plane $(d\psi/dx)_{\psi=\psi_m} = 0$ and so the osmotic pressure is equal to p_m:

$$\begin{aligned} p_m &= 2n^0 kT \cosh\frac{ze\psi_m}{kT} \\ &= kT\left[n^0\exp\left(\frac{ze\psi_m}{kT}\right) + n^0\exp\left(-\frac{ze\psi_m}{kT}\right)\right] \\ &= kT(n_+ + n_-)_m. \end{aligned} \qquad (7.3.3)$$

The higher osmotic pressure at other points between the plates is partly offset by the other term $\left(\frac{\varepsilon}{2}\left(\frac{d\psi}{dx}\right)^2 = \frac{\varepsilon}{2}|\mathbf{E}|^2\right)$, which is called the *Maxwell stress*. The actual force per unit area exerted on the plates is given by the difference in osmotic pressure between the solution in the midplane between the plates and that outside (in the reservoir of electrolyte; see Appendix A6):

$$\bar{p} = kT(n_+ + n_- - 2n^0)$$

$$\bar{p} = 2n^0kT\left(\cosh\frac{ze\psi_m}{kT} - 1\right) \tag{7.3.4}$$

$$= 2n^0kT(\cosh y_m - 1) \tag{7.3.5}$$

where $y_m = ze\psi_m/kT$ is a dimensionless potential.

To calculate V_R as a function of D from eqns (7.2.7) and (7.3.4) we would need to know ψ_m as an explicit function of D; unfortunately that is only possible if certain assumptions are made (see section 7.3.1). For spheres, there is no plane where the Maxwell stress is zero; the osmotic pressure method is then inapplicable.

The alternative method for flat plates, using eqn (7.2.6), also involves a similar problem.

In that case we have an expression for $\Delta F(\infty)$ in eqn (7.2.4) but $\Delta F(D)$ depends on the surface charge σ_0 at each value of D and, again, it is not possible to obtain an analytical expression for that. It is, however, possible to solve eqn (7.3.2) and to express the potential profile in terms of elliptic integrals of the first kind. It is then possible to calculate $\Delta F(D)$ and, hence V_R in terms of elliptic integrals of the second kind. The resulting expressions are rather lengthy and will not be repeated here. The complete analysis is given by Verwey and Overbeek (1948), and the final result is quoted by Overbeek (1952). An outline of the procedure for finding the potential profile is given by Hunter (1981) (appendix 5), while a detailed discussion of the procedures is given in the compilation of tables by Devereux and de Bruyn (1963). For our purposes, we need only record the final result, in the form of Table 7.1 (from Verwey and Overbeek 1948) which gives values of $f(y_m, \psi_0) = (z^2/\kappa)V_R$ for various values of ψ_0, ψ_m, and κD. From this table it is possible to calculate V_R as a function of D for any given ψ_0 value and electrolyte concentration (Exercise 7.3.1).

The ready availability of computing facilities makes it worthwhile to explore the use of direct numerical procedures to evaluate V_R accurately as a function of D. The following method was described recently by Chan *et al.* (1980). Writing eqn (6.3.15) in terms of the dimensionless variables $y = ze\psi/kT$ and $X = \kappa x$ we have (Exercise 7.3.2):

$$\frac{d^2y}{dX^2} = \sinh y. \tag{7.3.6}$$

Table 7.1

$f(\psi_m, \psi_0) = (z^2/\kappa)\, V_R$ in units of $10^{-12}\, J\, m^{-1}$ for κ in m^{-1}. Corresponding values of κD are also given for different values of y_0 $(= ze\psi_0/kT)$. The midplane potential $\psi_m = kTy_m/ze$ can also be read off for any y_0 and κD. (Note that D is the total separation between the plates). (Adapted from Overbeek 1952, p. 254, with permission.) The temperature is taken as 25 °C and the relative dielectric permittivity $\varepsilon_r = 78.55$. (κD values obtained by doubling Overbeek's values and so may contain some round-off error.)

y_0		$y_m = y_0$	$y_0 - 0.1$	$y_0 - 0.3$	$y_0 - 0.6$	9	8	7	6	5	4	3	2	1	0.5	0.25	0.1
10	f	268.3	228.2	192.6	160.0	127.1	75.4	44.1	25.4	14.1	7.36	3.42	1.26	0.26	0.06	0.015	0.0023
	κD	0.0000	0.00868	0.01672	0.0268	0.0408	0.0874	0.1626	0.286	0.488	0.824	1.380	2.296	3.924	5.442	6.880	8.732
9	f	161.5	135.2	115.2	95.6		76.3	44.3	25.4	14.1	7.36	3.42	1.26	0.26	0.06	0.015	0.0023
	κD	0.0000	0.0146	0.0276	0.0442		0.0674	0.1442	0.268	0.472	0.806	1.358	2.278	3.906	5.424	6.862	8.714
8	f	96.52	80.56	68.56	56.60			44.8	25.4	14.1	7.36	3.42	1.26	0.26	0.06	0.015	0.0023
	κD	0.0000	0.0242	0.0454	0.0728			0.111	0.238	0.442	0.776	1.330	2.248	3.878	5.396	6.834	8.686
7	f	57.13	47.46	40.18	32.89				25.8	14.17	7.36	3.42	1.26	0.26	0.06	0.015	0.0023
	κD	0.0000	0.0398	0.075	0.120				0.183	0.392	0.728	1.282	2.202	3.830	5.348	6.786	8.636
6	f	33.27	27.47	23.04	18.66					14.38	7.39	3.42	1.26	0.26	0.06	0.015	0.0023
	κD	0.0000	0.0654	0.1236	0.198					0.3018	0.646	1.202	2.122	3.752	5.270	6.708	8.560
5	f	18.83	15.32	12.69	10.07						7.52	3.43	1.26	0.26	0.06	0.015	0.0023
	κD	0.0000	0.1082	0.2036	0.3264						0.4976	1.066	1.990	3.622	5.140	6.580	8.430
4	f	10.13	8.07	6.51	4.97							3.50	1.26	0.26	0.06	0.015	0.0023
	κD	0.0000	0.1782	0.336	0.5384							0.821	1.768	3.404	4.924	6.362	8.214
3	f	4.962	3.793	2.913	2.061								1.291	0.26	0.06	0.015	0.0023
	κD	0.0000	0.2942	0.5548	0.891								1.362	3.036	4.560	5.996	7.848
2	f	1.993	1.413	0.966	0.584									0.265	0.06	0.015	0.0023
	κD	0.0000	0.487	0.9286	1.502									2.356	3.916	5.360	7.216
1	f	0.4682	0.271	0.135	0.0348										0.063	0.015	0.023
	κD	0.0000	0.8712	1.710	3.064										2.558	4.070	5.942

This can be integrated once to give (Exercise 7.3.2):

$$\frac{\mathrm{d}y}{\mathrm{d}X} = \mathrm{sgn}(y_\mathrm{m})Q \tag{7.3.7}$$

$$= \mathrm{sgn}(y_\mathrm{m})\{2(\cosh y - \cosh y_\mathrm{m})\}^{1/2} \tag{7.3.8}$$

where the coordinate system is placed with the origin for X in the midplane. The function $\mathrm{sgn}(y_\mathrm{m}) = y_\mathrm{m}/|y_\mathrm{m}|(=\pm 1)$ gives the algebraic sign to be attached to $\frac{\mathrm{d}y}{\mathrm{d}X}$ so that there is a minimum in $|\psi|$ at the midplane.

Now note that (Exercise 7.3.3):

$$\frac{\mathrm{d}Q}{\mathrm{d}y} = \frac{\sinh y}{Q}$$

$$= \frac{\mathrm{sgn}(y_\mathrm{m})}{Q}\left\{\left(\frac{Q^2}{2} + \cosh y_\mathrm{m}\right)^2 - 1\right\}^{1/2} \tag{7.3.9}$$

and hence (Exercise 7.3.3):

$$\frac{\mathrm{d}X}{\mathrm{d}Q} = \left\{\left(\frac{Q^2}{2} + \cosh y_\mathrm{m}\right)^2 - 1\right\}^{-1/2}. \tag{7.3.10}$$

This equation can be used to calculate the relation between X and y_m provided the value of Q on the surface Q_s, is known. The variable Q is defined by eqns (7.3.7) and (7.3.8) and if the surface potential, ψ_0, is known then

$$Q_\mathrm{s} = \{2(\cosh y_0 - \cosh y_\mathrm{m})\}^{1/2} \tag{7.3.11}$$

where $y_0 = ze\psi_0/kT$. Given any arbitrary y_m (with $|y_\mathrm{m}| < |y_0|$) one can integrate (numerically) eqn (7.3.10) from the midplane where $X = 0$ and $Q = 0$ to obtain corresponding values of Q and X. When Q reaches the value Q_s, the corresponding X value corresponds to half the separation between the plates (i.e. $X(Q = Q_\mathrm{s}) = \kappa D/2$) for the given initial value of y_m. By choosing different y_m values from near zero (corresponding to large separations) up to near y_0 (small separations) one can obtain y_m as a function of separation D for the given surface condition (Q_s). Since the force per unit area between the plates is then given by eqn (7.3.5), the potential energy of repulsion can be calculated by numerical integration using eqn (7.2.7).

If the interaction occurs under conditions of constant surface charge, σ_0, then Q_s is given by:

$$Q_\mathrm{s} = \mathrm{sgn}(y_\mathrm{m})\left(\frac{\mathrm{d}y}{\mathrm{d}X}\right)_{y=y_0} = \mathrm{sgn}(y_\mathrm{m})\frac{ze}{kT\kappa}\left(\frac{\mathrm{d}\psi}{\mathrm{d}x}\right)_{\psi=\psi_0}$$

$$= \left|\frac{ze\sigma_0}{\varepsilon kT\kappa}\right| \tag{7.3.12}$$

(from eqn (7.2.5)). Again, the integration of eqn (7.3.10) is pursued until this value of Q_s is reached and the corresponding value of X is recorded for subsequent evaluation of V_R from eqn (7.2.7).

The same procedure can be used for the condition of *charge regulation*. In that case the surface groups undergo some form of dissociation and the degree of dissociation is influenced by the interaction. Using the single-site model of section 6.9.1 it is necessary to ensure that the surface condition satisfies eqn (6.9.4). For a positive surface that releases a negative ion (A^-) into solution the corresponding relation is:

$$\sigma_0 = \frac{eN_s}{1 + \frac{[A^-]}{K_a}\exp(y_0)}. \qquad [6.9.4]$$

Hence, when two such identical surfaces interact this condition must be fulfilled at all separations. In order to apply the numerical algorithm described in the previous section the finishing condition Q must always comply with eqn (6.9.4). Calculation of the repulsive interaction with this surface condition gives the regulating case—where both surface potential and charge will vary with separation in order to satisfy the surface equilibria. The precise regulation of the interacting surface will depend on the concentration of adsorbing ions, the dissociation constant, and the site density. An example is shown in Fig. 7.3.1. It is found that the

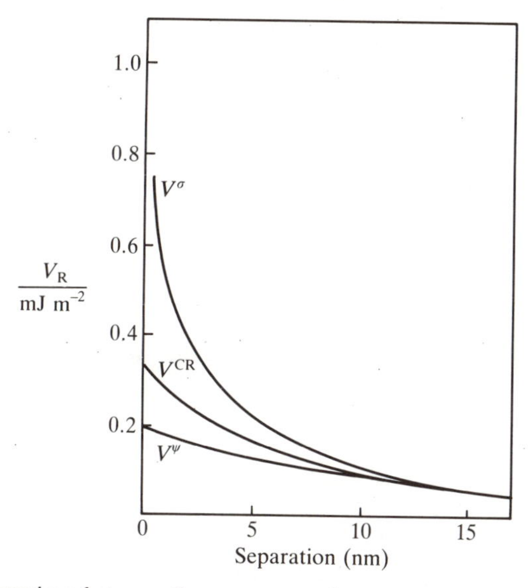

FIG. 7.3.1. Comparison between the constant surface potential (lowest curve), constant surface charge (uppermost curve) and charge regulation (CR) calculation of V_R. The calculation is for a surface of $\Delta pK = 6$ (see section 6.10) at pH 7 and 10^{-3} M in 1 : 1 electrolyte. (From Healy *et al.* 1980, with permission.)

regulation case must always lie between the limits of constant charge and potential (Chan and Mitchell 1983). For some systems, the difference in interaction energy between the range of surface conditions is slight. However, for spherical colloids near the point of coagulation the effect of regulation can be of the utmost importance. Under these conditions it is necessary to obtain values for the surface dissociation parameters, e.g. site density and dissociation or binding constants.

This same numerical procedure can also be used for calculating the interaction between two dissimilar plates. Two separate integrations are done up to Q_{s1} and Q_{s2} and the resulting values of D_1 and D_2 are added together to obtain the total separation between the plates.

7.3.1 *Approximate relations for overlap of flat double layers*

If the potential in the midplane is sufficiently small, the cosh term in equation (7.3.5) can be expanded to obtain (Exercise 7.3.4):

$$\bar{p} = \frac{\kappa^2 \varepsilon}{2} \psi_{\mathrm{m}}^2. \tag{7.3.13}$$

For small degrees of double-layer overlap (i.e. $D > 1/\kappa$) the midplane potential can be approximated by adding the potentials due to each plate. If $x = 0$ on one plate, then:

$$\psi_{\mathrm{m}} = 2\psi(x = D/2).$$

Under these conditions, eqn (6.3.23) is a very good approximation for ψ and so:

$$\psi_{\mathrm{m}} = \frac{8kT}{ze} Z \exp(-\kappa D/2) \tag{7.3.14}$$

where $Z = \tanh ze\psi_0/4kT$.

The repulsion potential energy is then given by (Exercise 7.3.4):

$$V_{\mathrm{R}}^{\psi} = -\int_{\infty}^{D} \bar{p}\, \mathrm{d}D = -\int_{\infty}^{D} 64n^0 kTZ^2 \exp(-\kappa D)\, \mathrm{d}D$$

$$= \frac{64n^0 kTZ^2}{\kappa} \exp(-\kappa D). \tag{7.3.15}$$

This is one of the most widely used approximate expressions for V_{R}. It is also valid for conditions of constant charge, since little discharge occurs if the degree of double layer overlap is small.

If the potential in the region between the plates is small everywhere (i.e. $|y_0| < 1$) so that the Debye-Hückel equation (6.3.11) holds then it can be shown (Exercise 7.3.5) that the midplane potential is given by:

$$\psi_m = \psi_0/\cosh(\kappa D/2). \tag{7.3.16}$$

Then using the appropriate form for the osmotic pressure when $|\psi_m|$ is small (eqn (7.3.13)) it can be shown that (Exercise 7.3.6):

$$V_R^\psi = \frac{2n^0kT}{\kappa} y_0^2\{1 - \tanh(\kappa D/2)\} \tag{7.3.17}$$

$$= \frac{4n^0kT}{\kappa} y_0^2 \frac{\exp(-\kappa D)}{1+\exp(-\kappa D)}. \tag{7.3.18}$$

Comparing this with equation (7.3.15) it would seem that a very good approximation, valid for all potentials provided the interaction is not too strong, would be (Verwey and Overbeek 1948, p. 97) (Exercise 7.3.7):

$$V_R^\psi = \frac{32n^0kT}{\kappa} Z^2\{1 - \tanh(\kappa D/2)\}. \tag{7.3.19}$$

Figure 7.3.2 shows a comparison between the 'exact' value of V_R calculated from Table 7.1 (or by numerical integration) with that obtained using eqn (7.3.15). The log-linear relation between V_R^ψ and κD is evident over a considerable range of κD values, though the approximate equation tends to overestimate the value of V_R.

The corresponding expression for V_R when the plates interact under conditions of constant charge V_R^σ rather than constant potential is given

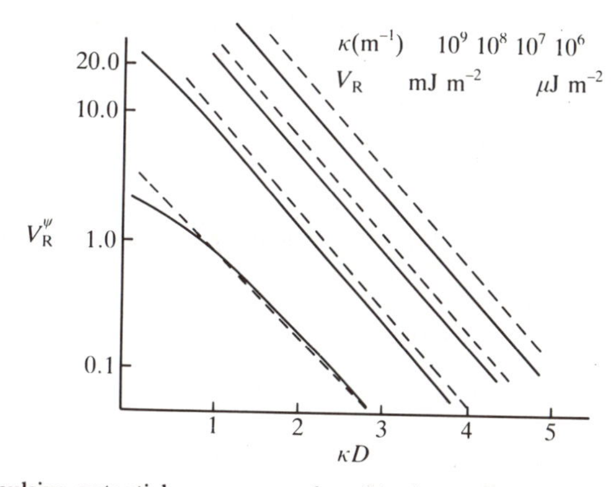

FIG. 7.3.2. Repulsive potential energy on a logarithmic scale against distance separating the plates. Full curves: Table 7.1. Dotted curves: eqn (7.3.15). (After Verwey and Overbeek 1948, p. 85, with permission.)

by Parsegian and Gingell (1972):

$$V_{\mathrm{R}}^{\sigma}=\frac{\sigma_0^2}{\varepsilon\kappa}\left[\frac{1+\exp(-\kappa D)}{\sinh \kappa D}\right]. \tag{7.3.20}$$

Note that under low potential conditions where $\sigma_0=\varepsilon\kappa\psi_0$ and small degrees of double layer overlap ($\exp(-\kappa D)\ll 1$) this expression reduces to

$$V_{\mathrm{R}}^{\sigma}=2\varepsilon\kappa\psi_0^2\exp(-\kappa D) \tag{7.3.21}$$

which is the same as the result obtained under constant (low) potential conditions (Exercise 7.3.9). This equality of V_{R}^{ψ} and V_{R}^{σ} for low potentials holds only at large D. At small D values the repulsion *energy* is different in the two cases (though the force is the same).

A more elaborate expression, valid for higher surface potentials, has been derived by Gregory (1973). He calculates the volume density of charge between the plates by adding together the effects of (i) the compression process limiting the available volume and (ii) the presence of the other plate with its counterions. The potential in the midplane can then be written:

$$\sinh y_{\mathrm{m}}=y_0 \operatorname{cosech} \kappa D/2 \tag{7.3.22}$$

where $y_0=ze\psi_0/kT$ and ψ_0 is the potential when the plates are infinitely far apart. The repulsion potential energy under constant charge conditions is then:

$$V_{\mathrm{R}}^{\sigma}=\frac{2n^0kT}{\kappa}\left[2y_0\ln\left\{\frac{B+y_0\coth(\kappa D/2)}{1+y_0}\right\}\right.$$
$$\left.-\ln(y_0^2+\cosh \kappa D+B\sinh \kappa D)+\kappa D\right] \tag{7.3.23}$$

where $B=\{1+y_0^2\operatorname{cosech}^2(\kappa D/2)\}^{1/2}$. This gives values in close agreement with the exact calculation.

Exercises. 7.3.1 Calculate the potential energy of repulsion V_{R} as a function of separation ($0\leqslant D\leqslant 20$ nm) for two flat plates of surface potential 51.4 mV in a 2 : 2 electrolyte at a concentration of 10^{-4} M in water at 298 K.

7.3.2 Establish eqn (7.3.6) and integrate it to obtain eqn (7.3.8) using the boundary condition $\mathrm{d}y/\mathrm{d}X=0$ at $X=0$ and $y=y_{\mathrm{m}}$.

7.3.3 Establish eqns (7.3.9) and (7.3.10).

7.3.4 Expand the cosh term in eqn (7.3.5) to establish eqn (7.3.13). Hence establish eqn (7.3.15).

7.3.5 Integrate the Debye–Hückel equation $\mathrm{d}^2\psi/\mathrm{d}x^2=\kappa^2\psi$ with the appropriate boundary conditions to obtain $\mathrm{d}\psi/\mathrm{d}x=\kappa(\psi^2-\psi_{\mathrm{m}}^2)^{1/2}$. Use the substitution $\psi=\psi_{\mathrm{m}}\cosh w$ to perform the second integration and hence show that $\psi_0=\psi_{\mathrm{m}}\cosh \kappa D/2$.

7.3.6 Use eqn (7.3.13) along with the result of Exercise (7.3.5) and eqn (7.2.7) to establish eqns (7.3.17) and (7.3.18).

7.3.7 Calculate the value of V_R from eqn (7.3.19) for the same conditions as in Exercise 7.3.1 and plot the results on the same graph.

7.3.8 Show that for small ψ_0 (in the Debye–Hückel approximation) the potential profile between two flat plates may be written (for large D):

$$\psi(x) = \psi_0 \frac{\cosh(\kappa(D/2 - x))}{\cosh(\kappa D/2)}$$

(where $x = 0$ on one plate).

7.3.9 Show that if the Debye–Hückel approximation holds throughout the region between two approaching flat plates then

$$V_R = 2\varepsilon\kappa\psi_0^2 \exp(-\kappa D)$$

if the surface potential is constant during the interaction.

7.3.10 The potential profile between two flat plates is given by the implicit relation (Verwey and Overbeek 1948):

$$\kappa x = 2\exp(-y_m/2)\{\mathscr{F}(\phi_2, k) - \mathscr{F}(\phi_1, k)\}.$$

y_m can be interpolated from Table 7.1 for any given ψ_0 and D value and the value of x can then be found for any ψ using a table of elliptic functions of the first kind ($\mathscr{F}(\phi, k)$) with

$$\mathscr{F}(\phi, k) = \int_0^{\phi} (1 - k^2 \sin^2\theta)^{-1/2}\, d\theta$$

$$\phi_1 = \arcsin[\exp\{-(y_0 - y_m)/2\}]$$

$$\phi_2 = \arcsin[\exp\{-(y - y_m)/2\}]$$

and $k = \exp(-y_m)$.

Draw up the potential profile for the case where $\psi_0 = 4kT/ze (= 102.8/z \text{ mV})$ and $D = 1.768/\kappa$ and compare it with the result obtained by assuming simple additivity of the potentials from each plate: $\psi(x) = 4(kT/ze)\{\operatorname{arc\,tanh}(Ze^{-\kappa x}) + \operatorname{arc\,tanh}(Ze^{-\kappa(D-x)})\}$ where $Z = \tanh ze\psi_0/4kT$ (eqn (6.3.24)).

7.4 Interaction between dissimilar flat plates

It was noted in connection with eqn (7.3.12) that the numerical procedure for evaluating V_R can also be used in the case where the plates have different potentials. The potential profile then is as shown in Fig. 7.4.1 and the two parts $(\psi_{01} - \psi_m)$ and $(\psi_m - \psi_{02})$ can be treated separately.

It is also possible to calculate the 'exact' value of V_R from the information in Table 7.1 by using the method of *isodynamic curves* as suggested by Deryaguin (1954). The method is described in some detail by Hunter (1975). It depends on the fact that the quantity

$$\bar{p} = (p_m - 2n^0kT) = p - \frac{\varepsilon}{2}\left(\frac{d\psi}{dx}\right)^2 \qquad (7.4.1)$$

is constant everywhere between the plates (eqn (7.3.2)). Here $p = 2n^0kT(\cosh ze\psi/kT - 1)$ and $\bar{p}$ is the value of p for $\psi = \psi_m$ (as defined by eqn (7.3.5)).

The potential profile between the plates A and B is simply a part of the profile between two plates of *equal potential,* if the second plate is placed at the appropriate position (C in Fig. 7.4.1). A plate B of surface potential ψ_{02} placed as shown in Fig. 7.4.1 is repelled by the same force as would be exerted on a plate, C, of potential ψ_{01} placed further away, at a distance D from the first plate. The total potential energy of repulsion between the plates A and B is then given by:

$$V_R(AB) = \tfrac{1}{2}V_R(AC) + \tfrac{1}{2}V_R(EB) \tag{7.4.2}$$

and the quantities on the right depend only on interactions between identical plates.

If the second plate were placed at the position E in Fig. 7.4.1 it would still experience the same force if its potential were ψ_{02} but this would correspond to a plate with a charge of opposite sign to the first plate (A). (Note that $(d\psi/dx)$ at the plate E has the opposite slope to that at plate B.) Thus it is possible for plates of opposite sign of charge to repel one another, if they are interacting under constant potential conditions.

The force between the two plates A and B at separation h is determined by adjusting the position of the plate C (called the control plate) until the potential profile between A and B passes through ψ_{02} at a distance h from plate A. Then, from eqn (7.4.2):

$$V_R = \tfrac{1}{2}[V_R(\psi_{01}, \kappa, D) + V_R(\psi_{02}, \kappa, 2h - D)]. \tag{7.4.3}$$

The force between the two plates, for any particular values of ψ_{01} and

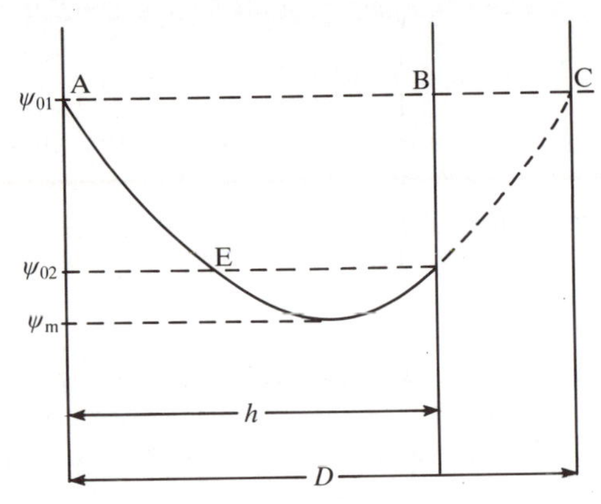

FIG. 7.4.1. The potential profile between two dissimilar plates A and B of surface potential ψ_{01} and ψ_{02}. Note that the minimum no longer occurs in the midplane.

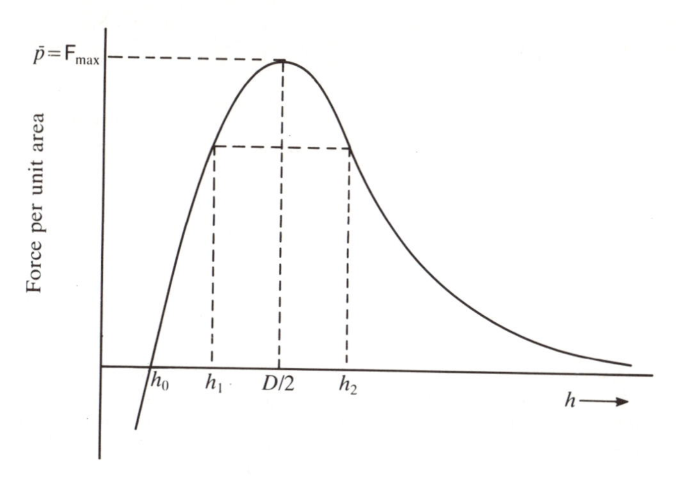

FIG. 7.4.2. Force–distance relation for the interaction between two plates of unequal (but constant) potential; both potentials have the same sign.

ψ_{02} has the same magnitude at two different values of h (corresponding to positions B and E in Fig. 7.4.1). The plot of force as a function of h is shown schematically in Fig. 7.4.2. Notice that it rises to a maximum, which must correspond to the situation when plate 2 is a distance $D/2$ from the first plate and $\psi_{02} = \psi_m$ (where ψ_m is the potential midway between two plates of potential ψ_{01} and separation D). Note also that when plate B is in that position (Fig. 7.4.1) the function $\frac{d\psi}{dx} = 0$ at the surface of B and so B must be uncharged (from eqn (6.3.26)). The maximum pressure is then given by eqn (7.3.5).

This at first sight rather surprising result (that the force is a maximum when the plate B is uncharged) can be readily understood when one realizes that constant potential plates, as they move through the position of the maximum, change their surface charge from positive to negative values (assuming ψ_{01} and ψ_{02} are both positive to begin with). Indeed, if the charge on the second plate is made sufficiently negative then the repulsion force diminishes to zero. This corresponds to a situation where the potential on the second plate at distance h_0 is exactly equal to the potential at a distance h_0 from an isolated plate of potential ψ_{01} (see Exercise 7.4.5):

$$\ln \tanh[ze\psi_{02}/4kT] = \ln \tanh[ze\psi_{01}/4kT] - \kappa h_0 \qquad (7.4.4)$$

The potential profile between the plates is then identical to that of an isolated plate and there is no force (Fig. 7.4.3). The charge on the second

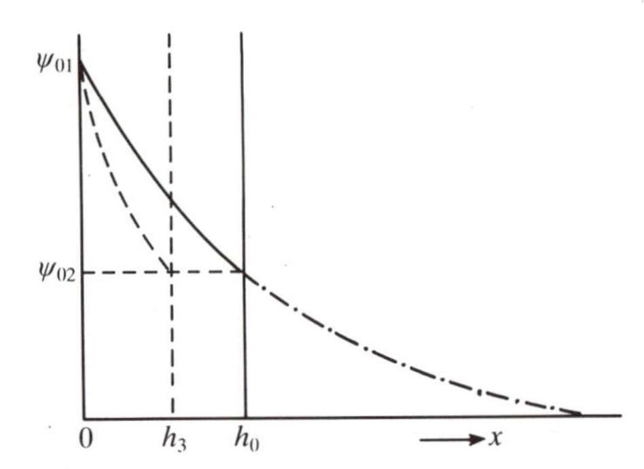

FIG. 7.4.3. The condition for zero force between two plates of different potential at separation h_0. The force between the plates becomes attractive for $D<h_0$ if the potentials remain the same. (The curve shown thus –·–·–· is the potential profile for a single plate.)

plate compensates for all the charge which would have been associated with plate 1 in the region $h_0<x<\infty$.

Finally, it should be noted that if the second plate is moved even closer to the first plate, then its potential can only be kept constant by piling even more negative charge onto it. (Note the higher slope of $(d\psi/dx)$ for $x=h_3$.)

This gives rise to an *attractive* force between the plates, even though they have the same sign of potential. The converse behaviour is not possible—if both surfaces have the same sign of charge then the interaction is *always* repulsive.

The above analysis assumes that the surface potential is constant on each plate. For a discussion of the constant charge behaviour see Jones and Levine (1969). The charge regulation behaviour is treated in Chan *et al.* (1976).

7.4.1 *Approximation formulae for the interaction between dissimilar plates*

So far we have used the osmotic pressure method to evaluate the repulsive potential energy in most cases. Returning to the free energy method it should be noted that eqn (7.2.3) remains valid for the free energy associated with a pair of dissimilar interacting double layers, provided the integration is done for both plates. The problem is that this is rarely possible because σ_0 is not known as an explicit function of D for constant ψ_0. There is, however, one situation in which it is possible and that is when the potential is sufficiently small for the Debye–Hückel approximation to hold between the plates. We can then write (compare

$$\psi_0' = \sigma_0'/\varepsilon\kappa \tag{7.4.5}$$

and substituting in eqn (7.2.3) and integrating gives (Exercise 7.2.3):

$$\Delta F = -\frac{\psi_0\sigma_0}{2}. \tag{7.4.6}$$

Hogg *et al.* (1966) used this procedure with two dissimilar double layers of (constant) potential ψ_{01} and ψ_{02} to obtain:

$$\Delta F(D) = -\tfrac{1}{2}(\sigma_1\psi_{01} + \sigma_2\psi_{02}) \tag{7.4.7}$$

where the surface charges σ_1 and σ_2 are, of course, functions of the separation distance D. Then if we assume that the Debye–Hückel equation (6.3.11) applies:

$$\frac{\mathrm{d}^2\psi}{\mathrm{d}x^2} = \kappa^2\psi, \tag{7.4.8}$$

the potential between the plates can be written (Exercise 7.4.2):

$$\psi(x) = A_1 \cosh \kappa x + A_2 \sinh \kappa x \tag{7.4.9}$$

where A_1 and A_2 are constants to be obtained using the boundary conditions given by the two potentials ψ_{01} and ψ_{02}. Thus at $x = 0$ and $x = D$ we obtain the results:

$$\psi_{01} = A_1$$

and

$$\psi_{02} = A_1 \cosh \kappa D + A_2 \sinh \kappa D. \tag{7.4.10}$$

Hence combining (7.4.10) with eqn (7.4.9) (Exercise 7.4.2):

$$\psi = \psi_{01} \cosh \kappa x + \left(\frac{\psi_{02} - \psi_{01}\cosh \kappa D}{\sinh \kappa D}\right)\sinh \kappa x, \tag{7.4.11}$$

which describes the potential distribution as a function of distance x from the ψ_{01} potential surface. Using eqn (7.4.11) we can now obtain by differentiation the potential gradient at each surface (Exercise 7.4.2):

$$\left.\frac{\mathrm{d}\psi}{\mathrm{d}x}\right|_{x=0} = \kappa(\psi_{02}\,\mathrm{cosech}\,\kappa D - \psi_{01}\coth \kappa D) \tag{7.4.12}$$

and

$$\left.\frac{\mathrm{d}\psi}{\mathrm{d}x}\right|_{x=D} = -\kappa(\psi_{01}\,\mathrm{cosech}\,\kappa D - \psi_{02}\coth \kappa D).$$

Hence the double-layer charge on each surface is given by†:

$$\left.\begin{aligned}\sigma_1 &= -\varepsilon\kappa(\psi_{02}\,\mathrm{cosech}\,\kappa D - \psi_{01}\coth\kappa D)\\ \text{and}\quad \sigma_2 &= +\varepsilon\kappa(\psi_{02}\coth\kappa D - \psi_{01}\,\mathrm{cosech}\,\kappa D).\end{aligned}\right\} \tag{7.4.13}$$

Using eqn (7.4.7) we obtain:

$$\Delta F(D) = \frac{\varepsilon\kappa}{2}[2\psi_{01}\psi_{02}\,\mathrm{cosech}\,\kappa D - (\psi_{01}^2 + \psi_{02}^2)\coth\kappa D] \tag{7.4.14}$$

and if D is very large (no interaction):

$$\Delta F(\infty) = -\frac{\varepsilon\kappa}{2}(\psi_{01}^2 + \psi_{02}^2). \tag{7.4.15}$$

Hence using the interaction energy definition (eqn 7.2.6):

$$V_R^\psi = \frac{\varepsilon\kappa}{2}[(\psi_{01}^2 + \psi_{02}^2)(1 - \coth\kappa D) + 2\psi_{01}\psi_{02}\,\mathrm{cosech}\,\kappa D]. \tag{7.4.16}$$

Thus we obtain an explicit expression for the potential energy of repulsion between two planar surfaces of constant unequal (low) potential. For the case of $\psi_{01} = \psi_{02}$ and weak interaction ($\kappa D \gg 1$) this equation reduces to the relation derived in Exercise 7.3.9. (But see Exercise 7.4.6.)

Equation (7.4.16) allows computation of the effect of unequal potentials on the interaction, which corresponds to the case of heterocoagulation. Under conditions of constant dissimilar potential we noted in section 7.4 above that it is possible to obtain attractive double-layer forces below some particular separation. The separation (in terms of κ^{-1}) at which the interaction energy V_R becomes zero can easily be obtained for particular values of ψ_{01} and ψ_{02} using eqn (7.4.16) (Exercise 7.4.4). For example, for $\psi_{01} = 25$ mV and $\psi_{02} = 50$ mV, $V_R = 0$ at $0.2\,\kappa^{-1}$. (The separation at which an energy maximum occurs can similarly be found from eqn (7.4.16)).

The separation at which the *force* between the plates changes from being positive (repulsive) to negative (attractive) is a rather more significant quantity and that is examined in Exercise 7.4.5. As noted above (section 7.4) there is no force between the plates when the potential of the second one satisfies:

$$\psi_{02} = \psi_{01}\exp(-\kappa D)$$

so that the profile between the plates is the same as for an isolated double layer at low potential. For an extension of this treatment to higher potentials, see Ohshima *et al.* (1982, 1983).

† Note that on plate 1 the charge is $-\varepsilon(\mathrm{d}\psi/\mathrm{d}x)$ whereas on plate 2 it is $+\varepsilon(\mathrm{d}\psi/\mathrm{d}x)$.

In summary, if the charge on the plates is fixed during an interaction then the interaction free energy is always positive (repulsive) if the signs are like and negative (attractive) if they are unlike. If the potential is constant during an interaction and is different on each plate then an attraction can occur at small separations, even when the potentials have the same sign.

Exercises. 7.4.1 If the approaching plates are of opposite sign of potential, the control plate must have potential $-\psi_{01}$ and the potential in the midplane is then zero. Discuss the resulting forces in terms of the Maxwell stress in the region between the plates (refer to Hunter 1975).

7.4.2 Verify that eqn (7.4.9) is a general solution for eqn (7.4.8). (Compare the result of Exercise 7.3.5). Derive eqns (7.4.10–12).

7.4.3 Establish eqns (7.4.14–16).

7.4.4 Verify that $V_R^\psi = 0$ for $\psi_{01} = 25$ mV and $\psi_{02} = 50$ mV for $\kappa D \simeq 0.2$.

7.4.5 The force between two flat plates is given by $\mathsf{F} = -\mathrm{d}V_R/\mathrm{d}D$. Use eqn (7.4.16) to show that

$$\mathsf{F} = \frac{-\varepsilon\kappa^2}{2\sinh\kappa D}[(\psi_{01}^2 + \psi_{02}^2)\operatorname{cosech}\kappa D - 2\psi_{01}\psi_{02}\coth\kappa D]$$

and hence prove that $\mathsf{F} = 0$ when $\psi_{02} = \psi_{01}\exp(\pm\kappa D)$. (i.e. plate 1 cannot 'see' plate 2 if the potential there is the same as it would be for an isolated plate).

7.4.6 Show that for any value of κD, eqn (7.4.16) reduces to:

$$V_R^\psi = \varepsilon\kappa\psi_0^2[1 - \tanh(\kappa D/2)]$$

when $\psi_{01} = \psi_{02} = \psi_0$.

7.5 Interaction between two spherical particles

7.5.1 *For large values of κa*

When two identical spherical particles of radius a approach under conditions of constant potential, the repulsive potential energy can be calculated using the Deryaguin procedure (section 4.6) if κa is sufficiently large. We then write (Verwey and Overbeek 1948, p. 138; compare eqn (4.6.4)):

$$V_R^\psi = \pi a \int_H^\infty V_R(\text{flat plate})\,\mathrm{d}D. \tag{7.5.1}$$

Introducing eqn (7.3.17) as an approximation for V_R^ψ (valid for low potentials) then (Exercise 7.5.1):

$$V_R^\psi = \pi a\varepsilon\kappa\psi_0^2 \int_H^\infty (1 - \tanh(\kappa D/2))\,\mathrm{d}D \tag{7.5.2}$$

$$= 2\pi\varepsilon a\psi_0^2 \ln(1 + \exp(-\kappa H)) \tag{7.5.3}$$

where H is the distance of closest approach (Fig. 4.6.1). If the centre-to-centre distance is r and we set $s = r/a$ and $\tau = \kappa a$, equation (7.5.3) can be expressed in the alternative form:

$$V_R = 2\pi\varepsilon a\psi_0^2 \ln\{1 + \exp[-\tau(s-2)]\}. \tag{7.5.4}$$

A more accurate calculation of V_R can be made from eqn (7.5.1) by a numerical integration procedure in which the 'exact' value for V_R is used at each separation. In actual practice, Verwey and Overbeek (1948) used the approximate formula (eqn (7.3.15)) for V_R (flat plate) and integrated this directly to get an approximate expression for V_R (spheres) like eqn (7.5.4). They could then introduce a correction function and integrate the *correction* numerically. In this way only a small part of the final result is involved in the rather tedious numerical integration process. With

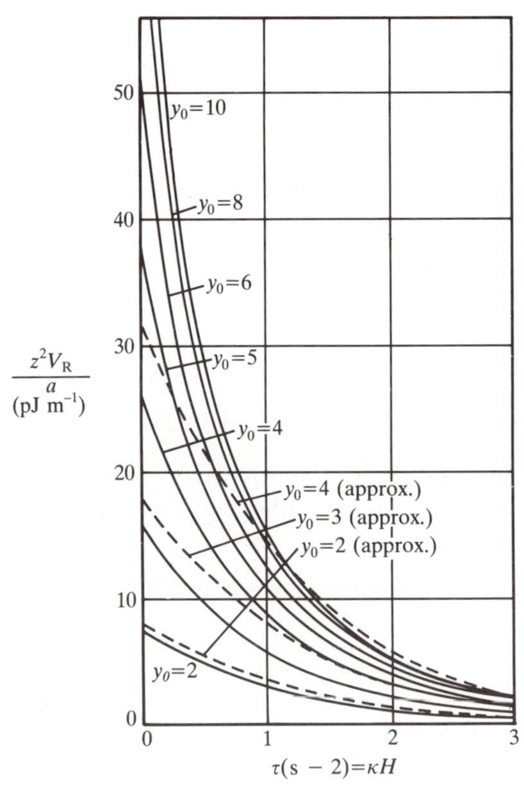

FIG. 7.5.1. The repulsive potential energy, V_R, between two large spherical particles when the exact expression for V_R (flat plates) is used in eqn (7.5.1). Broken lines refer to the approximate expression (eqn (7.5.4)). (Adapted from Verwey and Overbeek 1948, p. 141, with permission.)

modern computer facilities the whole process can be done quite quickly and accurately.

Wiese and Healy (1970) have shown that, for low potentials, the repulsion energy under conditions of constant charge is given by (Exercise 7.5.1):

$$V_R^\sigma = V_R^\psi - 2\pi\varepsilon a\psi_0^2 \ln[1-\exp(-2\kappa H)] \tag{7.5.5}$$

$$= -2\pi\varepsilon a\psi_0^2 \ln[1-\exp(-\kappa H)]. \tag{7.5.6}$$

The Deryaguin method may be used provided κa is reasonably large (say $\kappa a > 10$) and the results of 'exact' and approximate calculations of V_R are shown in Fig. 7.5.1.

7.5.2 *For small values of κa*

When the double layer around the particles is very extensive ($\kappa a < 5$) the Deryaguin procedure begins to break down and an alternative approach

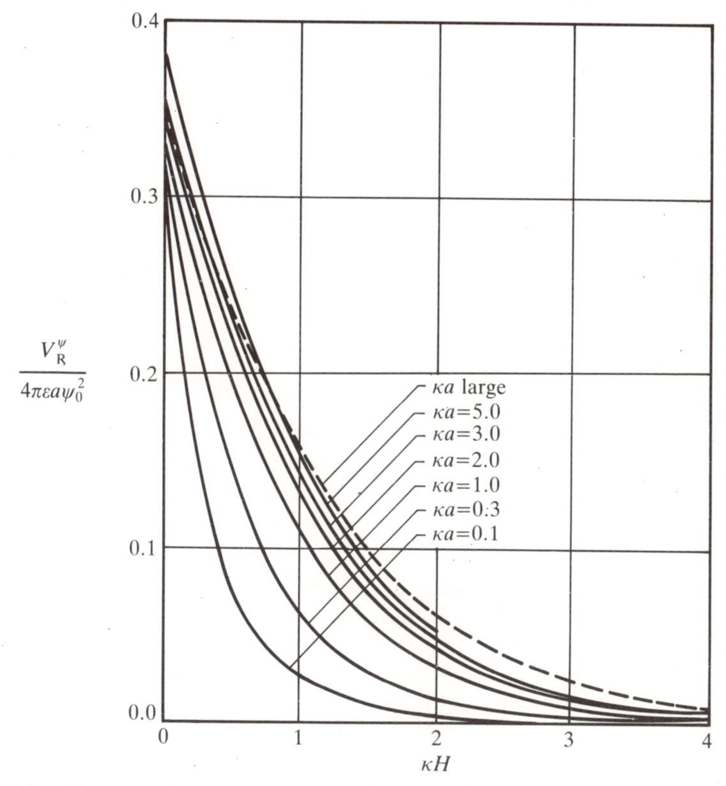

FIG. 7.5.2. The potential energy of repulsion for approach at constant surface potential when κa is small. (Adapted from Overbeek 1952, p. 260, with permission.)

is necessary. Verwey and Overbeek (1948) developed an approximate procedure for low surface potentials but the resulting expressions are not simple. They did, however, show that an approximate value for V_R can be calculated from:

$$V_R^\sigma = V_R^\psi = \frac{4\pi\varepsilon a\psi_0^2}{s}\exp(-\tau(s-2))$$

$$= 4\pi\varepsilon\frac{a}{s}\psi_0^2\exp(-\kappa H) \quad (7.5.7)$$

$$\simeq 2\pi\varepsilon a\psi_0^2\exp(-\kappa H) \quad (7.5.8)$$

if an error of up to about 40 per cent can be tolerated. The results of the more complete calculation for V_R^ψ are shown in Fig. 7.5.2.

This expression is valid only for low surface potentials but a number of other approximate expressions have been developed by Honig and Mul (1971) for small and large separations, and moderate potentials or charges, on the basis of both the constant potential and constant charge assumption.

Exercises. 7.5.1 Establish eqns (7.5.2–4). Also show that eqn (7.5.5) leads to eqn (7.5.6).

7.6 Total potential energy of interaction

We have already discussed, in a qualitative way, the total potential energy of interaction between surfaces given by eqn (2.5.8):

$$V_T = V_A + V_R \quad (7.6.1)$$

where V_A is obtained using the appropriate expression for the non-retarded van der Waals interaction (Chapter 4) between two infinite flat plates:

$$V_A = -\frac{A}{12\pi D^2}. \quad [4.4.13] \; [4.7.43]$$

The repulsive term due to the double layer interaction between identical surfaces is given, approximately, for the case of weak interaction ($\kappa D > 1$), by the relation:

$$V_R = \frac{64n^0kTZ^2}{\kappa}\exp(-\kappa D) \quad [7.3.15]$$

where Z is equal to tanh $(ze\psi_0/4kT)$. The combination of these two functions is the essence of the DLVO theory. The particular (approximate) expressions used here give rise to the curves shown in Fig. 2.5.5. It should be noted that the van der Waals attraction always dominates at both large and small separations; in the former case however, it may be too weak to be of significance. At small separations V_R must approach a finite magnitude, whereas $|V_A|$ increases very markedly and hence is expected to pull the surfaces into a deep attractive well, called the *primary minimum.* This well is not infinitely deep, as expected from the equation for V_A, because of a very steep, short-range repulsion between the atoms on each surface (see Fig. 7.6.1). The secondary minimum that occurs at larger distances ($\sim 7\kappa^{-1}$) is responsible for a number of important effects in colloidal suspensions and these will be alluded to later.

Experimental investigations of the coagulation properties of a wide range of colloidal solutions suggest that not all systems can be explained using the DLVO theory. Many experiments now indicate that there is an extra, so-called structural term that must be included in eqn (7.6.1). This

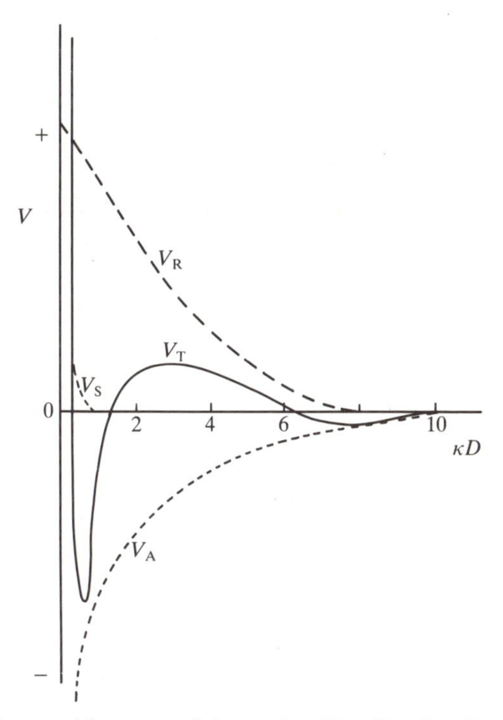

FIG. 7.6.1. Total potential energy of interaction $V_T = V_S + V_R + V_A$, where V_S is the potential energy of repulsion due to the solvent layers. V_S is assumed to be negligible until $D < \sim 10$ nm.

term arises because of the influence of a surface on adjacent solvent layers. Depending on the type of surface this can give rise to either repulsive or attractive forces. We may therefore generally define the total interaction as:

$$V_T = V_A + V_R + V_S \tag{7.6.2}$$

where V_S is the solvent-structural term.

The early evidence for the existence of a V_S term came from the observation that some colloids (e.g. silica) could not be coagulated even at very high electrolyte concentrations, where the double layer should be completely compressed. Also, the phenomenon of repeptization (i.e. redispersal of coagulated particles by dilution of the electrolyte) cannot be explained by the simple DLVO theory (eqn (7.6.1)), since an increase in V_R by diluting the coagulating electrolyte cannot significantly affect the depth of the primary minimum. Evidence has recently been obtained for both repulsive and attractive solvent mediated forces which are significant at separations of up to about 5 nm (see section 7.7). These forces may well be responsible for repeptization and for the stability of some surfaces at high electrolyte concentrations. At present, the lack of an adequate theory for the structure of water prevents theoretical evaluation of the important V_S term. Its magnitude can be estimated, at least for the mica surface (section 7.7.4) on the assumption that the DLVO theory, which holds well for moderate and large $D(>10\,\text{nm})$, can also describe the V_A and V_R terms for small values of D. The discrepancy, for $0<(D/\text{nm})<5$ is then attributed to V_S. In aqueous media, positive values of V_S indicate the presence of 'hydration forces', whereas negative, attractive values are believed to be due to the 'hydrophobic interaction' (Chapter 10). Both types of interaction apparently decay exponentially with decay lengths typically of the order of one nanometre.

7.6.1 *The Schultz–Hardy rule*

In section 2.5.5 we discussed the concept of a critical coagulation concentration (c.c.c): that is the concentration of indifferent electrolyte that induces rapid coagulation. The data in Table 2.1 indicate that the c.c.c. is a very strong function of the counterion valency, an observation known as the Schultz–Hardy rule. It was one of the early triumphs of the DLVO theory that it was able to account for that strong valence dependence on the basis of equations like (4.4.13) and (7.3.15). It is not difficult to show (Exercise 7.6.1) that the potential energy barrier that prevents rapid coagulation is reduced to zero when $\kappa D = 2$ (curve b of Fig. 2.5.5). Substituting this value into the expression for V_T (eqn

(7.6.1)):

$$V_T = -\frac{\kappa^2 A}{48\pi} + \frac{64n^0 kT}{\kappa} Z^2 \exp(-2) \tag{7.6.2}$$

allows an estimate of the c.c.c. (Exercise 7.6.1);

$$\text{c.c.c. (mol L}^{-1}) = \frac{(4\pi\varepsilon_0)^3 \times 0.107\varepsilon_r^3 (kT)^5 Z^4}{\mathcal{N}_A A^2 (ze)^6} \tag{7.6.3}$$

where $\mathcal{N}_A$ is the Avogadro number and ε_r is the relative dielectric permittivity. The quantities on the right are in SI units. (For c.g.s. units the constant is 1.07×10^5 and $4\pi\varepsilon_0 = 1$ statfarad cm^{-1}.) At 25 °C in water (taking $\varepsilon_r = 80$) if the potential is high ($Z \simeq 1$):

$$\text{c.c.c. (mol L}^{-1}) = 87 \times 10^{-40}/z^6 A^2 \tag{7.6.4}$$

where A is in joules.

The variation of c.c.c. with valency does depend approximately on the inverse sixth power of Z. The data for As_2S_3 in Table 2.1 for example, give the ratio (for $z = 1$, 2, and 3):

$$50 : 0.7 : 0.09 \simeq 1 : 0.014 : 0.0018$$

compared to the 'theoretical' 1 : 0.016 : 0.0014. Although this agreement is impressive there is some doubt as to its significance. For one thing the values of van der Waals or Hamaker constant estimated from eqn (7.6.4) are rather high ($A \simeq (87 \times 10^{-40}/0.05)^{1/2} = 4 \times 10^{-19} J$).

More seriously, it must be noted that eqn (7.6.4) is derived on the assumption that the surface potential is high ($Z \approx 1$). This latter condition is very difficult to reconcile with the general experimental observation that coagulation usually occurs between low potential surfaces. If the more realistic assumption of low potentials is used, for example, the result obtained is (Exercise 7.6.2):

$$\text{c.c.c.} \propto \frac{\psi_0^4}{z^2} \tag{7.6.5}$$

where now we have a less dramatic dependence on the valency and a strong dependence on the surface potential. The experimental data could easily be explained by this result if $\psi_0 \propto 1/z$. That is, if the surface potential were reduced by, for example, adsorption of oppositely charged ions, where the adsorption energy increases with valency of the ion. At present there is some interest in applying ion adsorption models to quantitative coagulation studies in a similar manner to the model used in the surface regulation section (section 7.3).

This raises the question of which potential is to be used in the equations for the repulsive potential energy. Since the interaction occurs between the diffuse layers it would seem to be more reasonable to assess the magnitude of V_R from the value of the diffuse layer potential, ψ_d (section 6.3). That potential can be estimated from the electrokinetic (or zeta-) potential (ζ), which is discussed in Chapter 9. A large body of experimental evidence shows that when rapid coagulation occurs in a colloidal sol, the ζ-potential is commonly around ±25–50 mV, which is indeed too small to assume $Z \simeq 1$ if we define Z in terms of ψ_d (eqn (7.3.14)). Such a result is, however, quite consistent with eqn (7.6.5) above, since ζ would measure the potential, ψ_d, after any counterion adsorption in the Stern plane (section 6.4).

Exercises. 7.6.1 The condition for the potential energy barrier to just disappear is that $V_A + V_R = 0$ and $\mathrm{d}(V_A + V_R)/\mathrm{d}D = 0$ simultaneously. Show that this occurs when $\kappa D = 2$. Hence establish the relations (7.6.3) and (7.6.4).

7.6.2 Use the approximate expression

$$V_R = 2\pi\varepsilon a \psi_0^2 \exp(-\kappa H) \qquad [7.5.8]$$

together with the appropriate attractive energy for spheres of radius a (Chapter 4) to establish the relation (7.6.5) for the c.c.c. under conditions of low surface potential.

7.6.3 Calculate the repulsive potential energy (under constant potential conditions) between two spherical particles of radius 0.5 μm, of surface potential 35 mV, when the electrolyte concentration is (a) 10^{-4} M NaCl and (b) 10^{-2} M NaCl. (Use H values from 0 to 20 nm.)

7.6.4 Calculate the attraction potential energy between the particles in Exercise 7.6.3 (assuming that $A_{121} = 5 \times 10^{-20}$ J) as a function of H for $0 < H <$ 20 nm. Combine this with the curves for V_R found in Exercise 7.6.3 to produce curves for V_T and comment on the result. (Ignore V_S.)

7.7 Experimental studies of the equilibrium interaction between diffuse double layers

The theory developed in sections 7.1–7.6 has been applied to a large variety of problems over the past 45 years. Only in the last few years, however, has it been possible to set up systems that are sufficiently well-defined and controllable to provide quantitative tests of the interaction force or energy as a function of distance between the approaching surfaces. In this section we will examine some of the experiments conducted on double layers that are approaching sufficiently slowly to enable equilibrium to be maintained. We will also restrict attention here for the most part to the interaction of macroscopic surfaces. Studies of the equilibrium interaction between microscopic (colloidal) particles can be conducted by a variety of methods (osmotic pressure measurement;

Barclay and Ottewill 1970; Barclay *et al.* 1972), light scattering and neutron scattering (Ottewill 1982) and centrifugation (El-Aaser and Robertson 1971, 1973) and general agreement is obtained between the experimental results and the DLVO theory. The proper discussion of those experiments requires, however, a knowledge of the particle–particle distances in concentrated suspensions and will be deferred (Volume II) until we have more fully examined (Chapter 11) the notion of distribution functions (Fig. 6.6.1) and their measurement using scattering theory (Chapter 14). In section 7.8 the more rapid interaction which occurs during coagulation is examined by the study of the kinetics of the coagulation process.

The systems investigated in this chapter are (a) adsorbed liquid films, (b) soap films, (c) swelling of clay minerals, and (d) interaction between immersed solid bodies. Clay minerals are, of course, microscopic particles but the distance between them can be estimated from macroscopic observations (assuming that they are aligned parallel to one another) and confirmed by low angle X-ray diffraction without recourse to radial distribution functions.

7.7.1 *Adsorbed liquid films*

In 1938 Langmuir applied double-layer theory to explain the apparent decrease in surface tension of water on addition of salt, when measured by the capillary rise method. This initial decrease in the tension at low salt was known as the Jones–Ray effect (see Fig. 7.7.1). Langmuir explained the effect by consideration of the capillary rise system, where the water rising to height H above the reservoir must also wet the inside of the glass capillary to some thickness h (since we assume that the glass is clean and hydrophilic; see Fig. 7.7.2). The effective radius of the

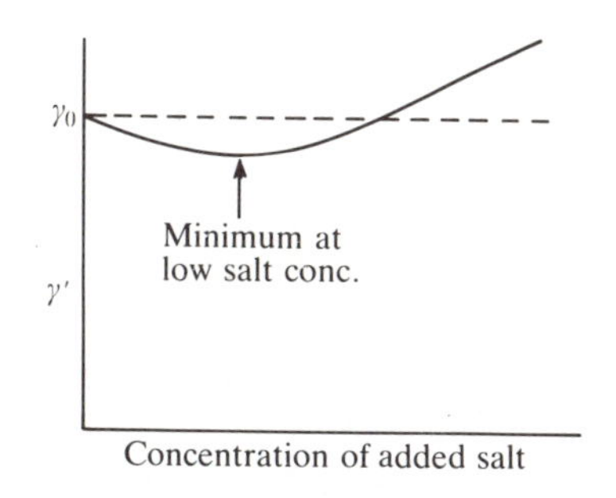

FIG. 7.7.1. The Jones–Ray effect. Addition of small amounts of electrolyte reduces the *apparent* surface tension γ' when it is measured by the capillary rise method.

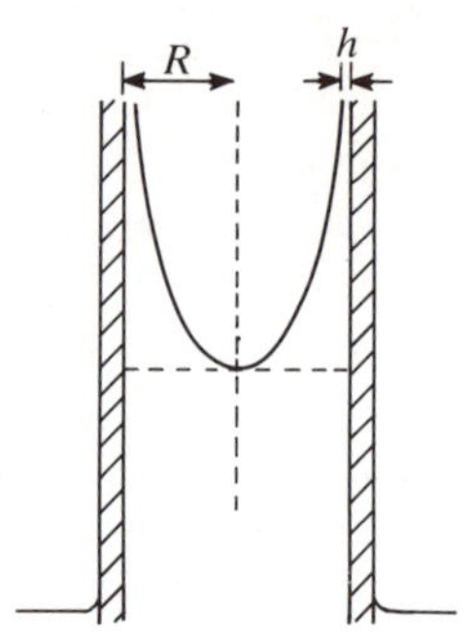

FIG. 7.7.2. Langmuir's model to explain the Jones–Ray effect (Fig. 7.7.1) postulates a thin film of water on the capillary surface reducing the effective capillary radius.

capillary is now $(R-h)$ and if the aqueous film is stabilized by double-layer forces (between the air–water and glass–water interfaces) the equilibrium film thickness will be reduced by the addition of salt until at high concentration the thickness h becomes negligible. Since in the original experiments the capillary radius was considered constant at R we can now explain the initial apparent decrease in surface tension. At low levels of added salt the effective capillary radius will be increased and the water level will fall, whereas at higher concentrations h will be negligible and the level will rise with the real increase in surface tension of the water.

More recent studies (see e.g. Adamson 1967, p. 78) cast doubt on this explanation of the Jones–Ray effect, but the existence of thin aqueous films on glass, stabilized by double-layer forces, is not in question.

These thin film experiments were extended by Deryaguin and Kussakov (1939) to the study of the film of liquid between an air bubble and a flat glass surface (Fig. 7.7.3) immersed in an electrolyte. The excess pressure in the film is given by the Laplace pressure in the bubble $2\gamma/r$ (section 5.2.3) and the corresponding equilibrium film thickness can be measured by optical interference techniques. Analysis of the measurement for the case of *water* on glass showed rough agreement with the Langmuir equation (7.3.5). Addition of electrolyte (NaCl) produced, as in the Jones–Ray effect, a marked reduction in film thickness, but this was not as great as expected for reasonable values of the surface potential (i.e. $\psi_0 < 100$ mV). At high concentrations (10^{-1} M NaCl where $\kappa^{-1} =$ 0.95 nm) the observed film thickness was 48 nm at a pressure of $1\ \mathrm{N\ m^{-2}}$. For this electrolyte concentration and pressure the corresponding mid-plane (or air–water surface) potential, ψ_m, given by eqn (7.3.5) is very low (<1 mV). The corresponding value of the film thickness ($D/2$) is only

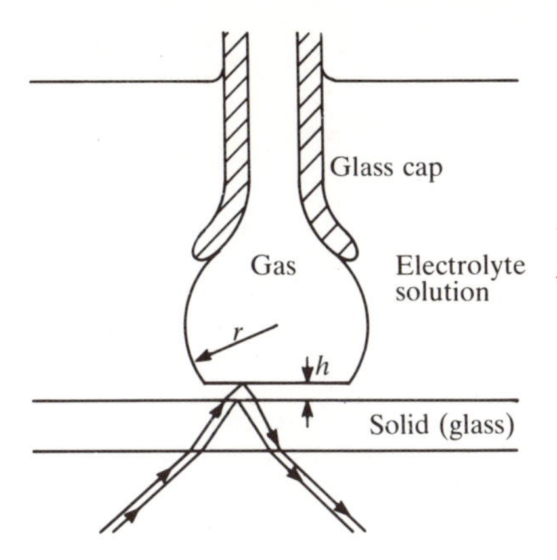

FIG. 7.7.3. Study of thin liquid film by the captive bubble method.

about 11 nm, even for the case of very high silica surface potentials. This value is much less than that observed.

Experiments involving water and silica (especially glass) have long been criticized for the possibility of the presence of swollen silica gel layers on the surface, which would be expected to strongly influence surface forces. It is quite likely that in these early experiments using lead glass such layers were the cause of much thicker films, especially at high electrolyte concentrations.

The early experiments with water films on silica were repeated by Read and Kitchener (1969), who concluded that wetting films were stabilized by double-layer forces but that there was only broad agreement between theory and experiment for films of about 30–130 nm thickness. For both monovalent and divalent electrolyte the films were significantly thicker (~10–20 nm) than expected from Gouy–Chapman theory using the Langmuir model (i.e. assuming that the potential gradient at the air–water interface is zero). This discrepancy can be explained by the assumption that there is a low double-layer charge at the air interface. Although the experimental evidence for this suggestion is somewhat variable, the charge at the air–water interface is believed to be of the same sign as the silica surface due to excess adsorption of OH^- ions. Such a charge would cause a significantly thicker film to be stabilized on the surface (see Fig. 7.7.4).

7.7.2 *Soap films*

Soap films are very easily formed by bubbling gas through surfactant solution and are stabilized by the repulsive forces between layers of

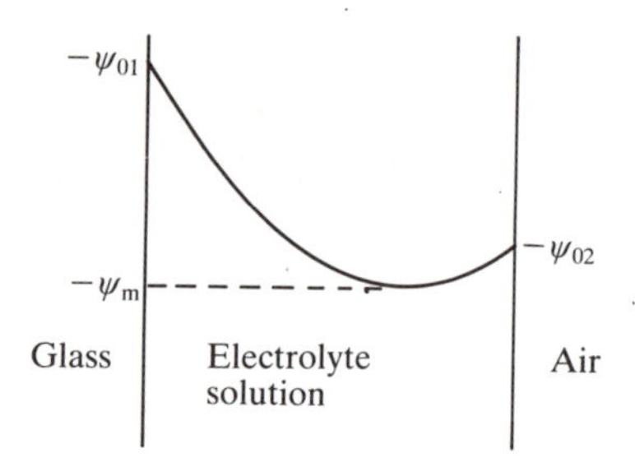

FIG. 7.7.4. Presence of a (negative) charge at the air–solution interface would be expected to cause an increase in the thickness of the equilibrium wetting film. Note that the glass surface is also negatively charged.

surfactant molecules adsorbed at the air–solution interface. These repulsive forces are often sufficiently strong to prevent drainage of the water layer by combined action of attractive van der Waals forces and gravitational forces. For films with a water-layer thickness greater than about 10 nm the dominant force is, for the case of ionic surfactants, due to double-layer repulsion; it is usually balanced by hydrostatic pressure (see Fig. 7.7.5(a)). A soap film can also be simply formed by drawing a wire frame vertically upwards through the surface of a surfactant solution. At each height H above the solution the double-layer pressure must be balanced by the hydrostatic pressure ($H\rho g$) tending to drain the film (Fig. 7.7.5(b)). Hence, in a region of film 10 cm above the solution there must be a repulsive double-layer pressure of about 10^3 N m^{-2}. If the water-layer thickness D at this height can be measured we are then in a position to study the double-layer interaction. The film thickness can be measured by the observation of reflected light which produces colours by interference between the front and rear surfaces. To obtain equilibrium measurements it is necessary to carefully control the environment in which the film is drawn with regard to temperature, humidity, and vibration.

Deryaguin and Titievskaya (1957) were the first to suggest that soap films could be used to investigate the forces that stabilize hydrophilic colloids, and they obtained reasonable film thicknesses for soap films in the pressure range 30–200 N m^{-2}. They showed that soap films of greater than about 20 nm thickness were stabilized by the overlap of diffuse double layers, with a surface potential of about 30 mV. Although the potential could not be obtained independently, the results gave clear evidence for the validity of the double layer model. Scheludko, Lyklema, and Mysels continued this work to investigate more comprehensively the effect of salt concentration on the equilibrium film thickness. That work

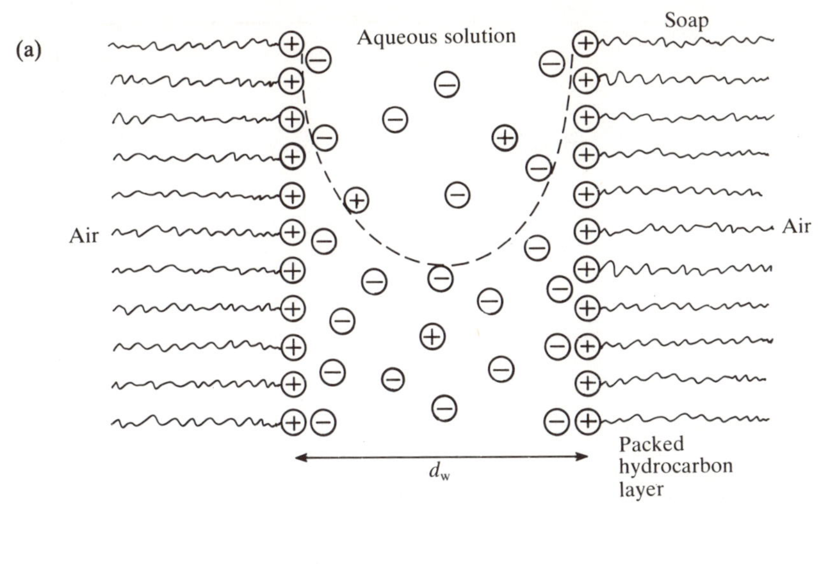

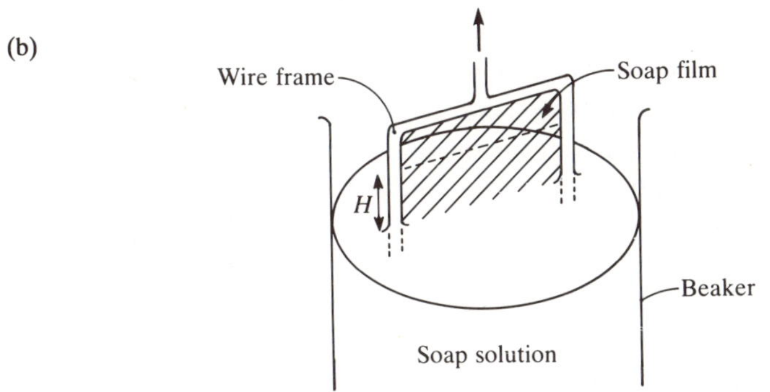

FIG. 7.7.5. (a) Soap film stabilized by cationic surfactant. (b) Soap film produced by drawing a wire frame out of a solution.

is reviewed by Lyklema (1967). The agreement between calculated and experimental film thickness (Fig. 7.7.6) is very reasonable.

Donners *et al.* (1977) recently used light scattering from soap films in order to investigate surface forces. By measuring the power spectrum of light scattered from ripples or fluctuations at the soap-film interface, information can be obtained both from single interfaces (i.e. surface tension and viscosity data) and from interacting charged soap layers. The latter case gives direct information about the double-layer interaction. Initial results obtained for the case of CTAB (cationic surfactant cetyl trimethylammonium bromide) show that, in 10 mM monovalent electrolyte, the double layer interaction is accurately described by an

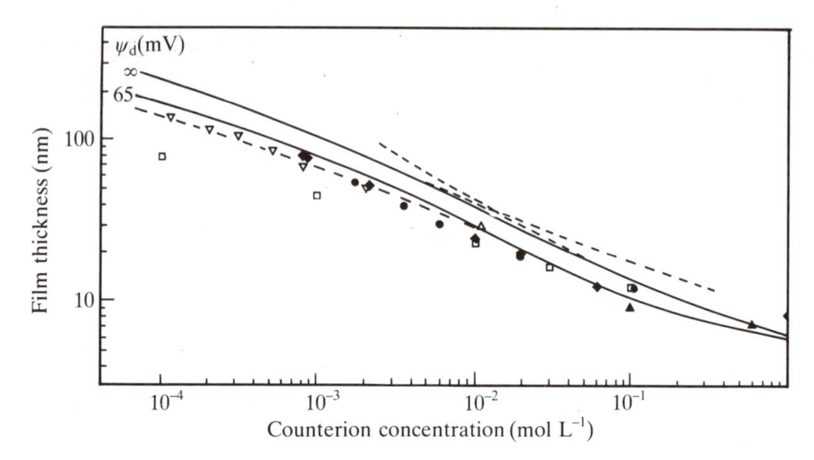

FIG. 7.7.6. Calculated and experimental thicknesses of soap films as a function of ionic strength (copyright American Chemical Society). (From Lyklema and Mysels 1965, with permission.)

exponentially decaying pressure, at separations greater than 60 nm. In all experiments the apparent double-layer potential was found to be in the range 60–90 mV. From the Gouy–Chapman model (eqn (6.3.27)) these potentials correspond to charge in the range of 1 per 14–18 nm^2 which indicates that the great majority of the head group charge (~98 per cent) is balanced by counterions in a compact layer, and the appropriate potential determining the magnitude of V_R is the diffuse layer potential, ψ_d.

7.7.3 *Swelling of clays*

The structure of clay minerals was described in section 1.5.5 where we noted their significance as model systems for the study of double-layer interaction. The actual swelling behaviour of clays is, of course, very important in agriculture, and in civil engineering (dam, road, and building construction) and in the making of ceramics and other clay products.

In montmorillonite and vermiculite the alumino-silicate sheets are separated by water layers whose thickness varies with the concentration and type of electrolyte. In the crystalline state these extremely thin, negatively charged sheets are held together by the electrostatic forces between alternate layers of bridging cations (typically Na^+, K^+ or Ca^{2+}). It is the dissociation of some of these ions in aqueous solution that gives rise to a double-layer repulsion between adjacent sheets.

Norrish (1954) made one of the first careful studies on the 'free' swelling of montmorillonite and found that for a water interlayer spacing of greater than about 2 nm the equilibrium spacing was dependent on electrolyte concentration. He obtained an empirical relation to describe the water-layer thickness:

$$d_w \simeq 1.14(1 + C^{-1/2}) \tag{7.7.1}$$

where d_w is in nm and C is the concentration of monovalent electrolyte $[0.01 < C(\text{mol L}^{-1}) < 0.25]$. As indicated by this equation it was found that even at high levels of salt the montmorillonite did not collapse completely but retained about 1 nm of water. This very short range region was referred to as the crystalline swelling stage, where the hydration energy of the specific cation dominates the interaction. Beyond this initial region, dilution of electrolyte causes swelling of the plates up to at least 12 nm.

We can analyse these results using the approximate equations already derived (section 7.3.1). Since the equilibrium spacing of the plates is large compared with κ^{-1} over the concentration range studied, we can safely apply the weak overlap approximation, that is:

$$V_R = \frac{64n^0kTZ^2}{\kappa}\exp(-\kappa d_w) \qquad [7.3.15]$$

which for the case of high potentials ($Z^2 \simeq 1$) and substituting for κ becomes

$$V_R = \frac{64}{ze}(kT)^{3/2}(\varepsilon n^0/2)^{1/2}\exp(-\kappa d_w). \tag{7.7.2}$$

In the swelling region discussed, $\kappa d_w \gg 1$, so the exponential term must dominate over the square root term with respect to the influence of electrolyte concentration, n^0, on the interaction energy V_R. Thus we can reduce the equation to:

$$V_R \simeq \text{const. } \exp(-\kappa d_w). \tag{7.7.3}$$

If we now make the rather arbitrary assumption that facing platelets will always come to equilibrium at a separation distance corresponding to some fixed attractive energy $V_A = -V_R$ (e.g. $10kT$ per interacting pair of plates of known size) then

$$\begin{aligned} d_w &\propto \kappa^{-1} \\ &\propto C^{-1/2} \end{aligned} \tag{7.7.4}$$

which was what was observed in Norrish's 'free' swelling experiments. It should be noted here that this extra attraction is not due to van der Waals forces, which are in fact negligible at these separations. The

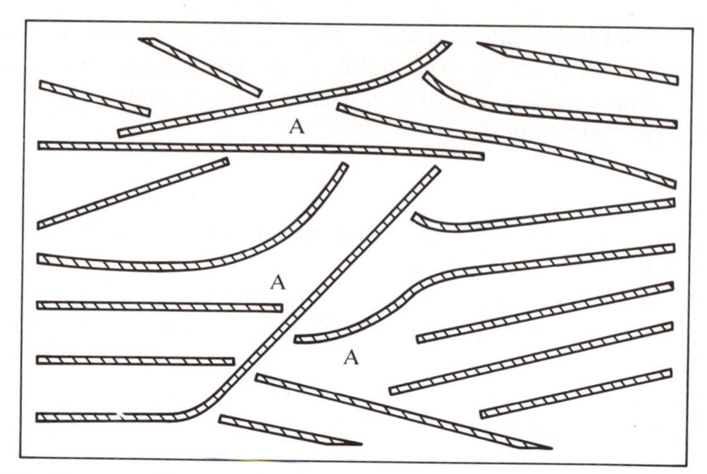

FIG. 7.7.7. Presence of improperly aligned clay particles carrying an edge positive charge can reduce the total swelling of a stack of clay particles. Edge–face interactions occur at A and they can resist both expansion and compression. (After Norrish and Rausell-Colom 1963).

attractive energy is probably due to some misalignment of the plates (Fig. 7.7.7), which introduces a mechanical resistance into the swelling gels. (This is why we put 'free' in inverted commas.) The mechanical resistance also produces quite large hysteresis in pressure–separation curves measured on these gels, because the presence of such structures inhibits both expansion and compression.

It is interesting to analyse how we expect this resistance to vary with the plate–plate separation. Since the attractive component apparently corresponds to a fixed interaction energy, the corresponding attractive pressure P_A is given by

$$P_A = \frac{dV_R}{d(d_w)} = \text{const.}(\kappa). \tag{7.7.5}$$

Hence the attractive pressure due to this resistance is, from eqn (7.7.4), proportional to d_w^{-1} and, therefore, falls off rapidly with plate separation as we might have expected.

In subsequent experiments Garrett and Walker (1962) and Norrish and Rausell-Colom (1963) used the mineral vermiculite, which suffers less from edge effects since it is better aligned. The pressure–separation curve measured by Norrish and Rausell-Colom for the swelling of Li-vermiculite in 0.03 M LiCl solution is given in Fig. 7.7.8. As can be seen from this figure there is little sign of hysteresis and the pressure decays exponentially with separation with a Debye length of about 1.7 nm, which agrees closely with the value calculated from the known salt

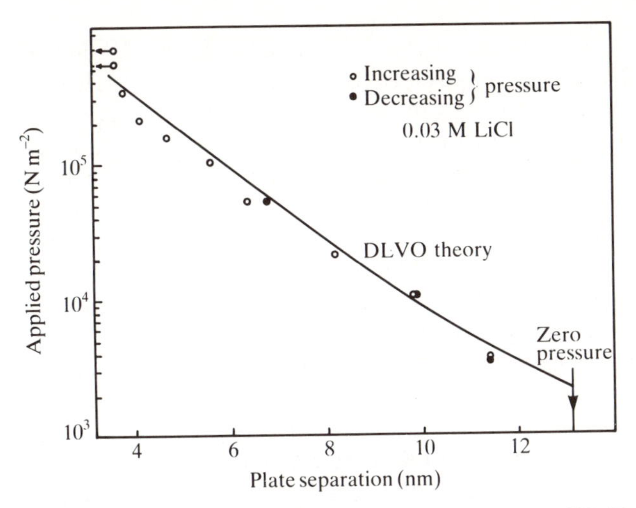

FIG. 7.7.8. Swelling of lithium vermiculite under pressure in 0.03 M LiCl. The theoretical line is calculated from the total crystal charge, which is almost certainly much higher than the diffuse layer charge in this system. (From Norrish and Rausell-Colom 1963, with permission.)

concentration (using eqn (6.3.12)). However, the magnitude of these pressures presents us with a problem since it can only be explained assuming that (a) the double-layer charge is given by the (high) lattice charge of the mineral and (b) there is a 0.5–0.6 nm layer of structured water on each surface and that for this reason the diffuse double layer originates at this distance from the surface. The lattice charge corresponds to a surface potential of about −150 mV at this electrolyte concentration, whereas zeta-potential measurements suggest a much more reasonable value of about −60 mV (see Friend and Hunter 1970). This zeta-potential corresponds to a much lower degree of dissociation at the vermiculite surface, which is consistent with observations on other clay minerals, e.g. mica and montmorillonite. The potential of −60 mV would give a much lower repulsion than that observed in the experiments reported by Norrish. Since the hysteresis was negligible in these experiments it is difficult to reconcile the magnitude of the pressures observed with our expectations from double-layer theory. In the following section we will discuss a much less ambiguous method for measurement of the double-layer interaction between clay mineral surfaces.

7.7.4 *Direct force measurements between mica surfaces*

In the last few years a new method has been developed for accurately measuring the total interaction force between molecularly smooth sheets of cleaved mica. The original apparatus built by Tabor and Winterton (1969) has been greatly improved by Israelachvili, such that both force and separation can be measured accurately in aqueous electrolyte solution over a wide range (from 10^{-7} to 10^{-4} N force and from 0.2 to greater than 200 nm separation). The technique and results obtained are described in some detail here because they afford the most thorough test of double-layer theory so far completed. The use of a crystalline layered silicate material allows the measurement of forces down to very small separations—at the level of the smoothness of the surface (~0.2 nm), whereas even well-polished surfaces (e.g. quartz) have been found to be rough on the microscale (having a random distribution of hills of the order of at least 5 nm) which of course affects the measured, short-range interaction.

A schematic diagram of the force measuring apparatus is shown in Fig. 7.7.9. Thin, parallel-cleaved mica sheets (1–3 μm thick) are silvered on

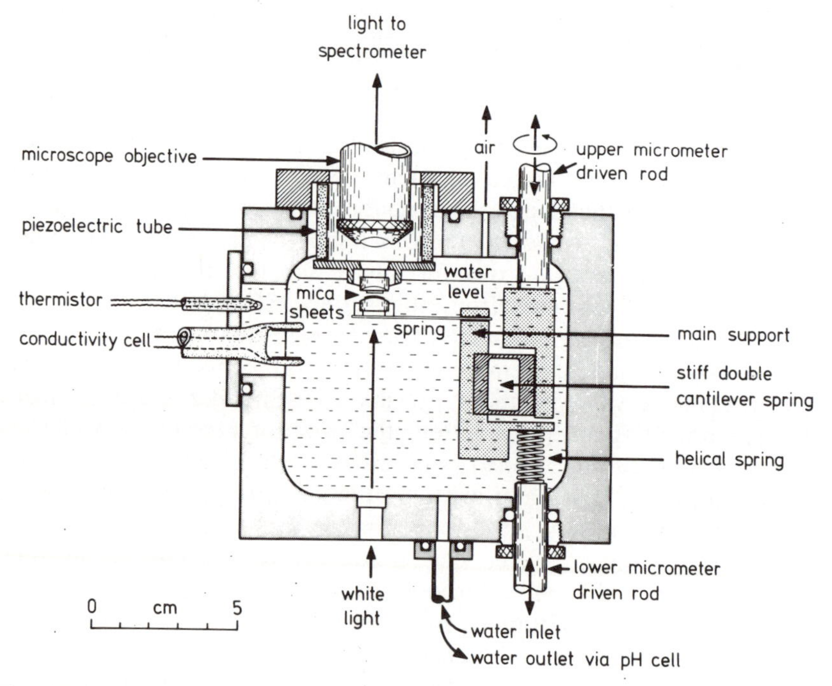

FIG. 7.7.9. Schematic drawing of apparatus to measure long-range forces between two crossed cylindrical sheets of mica (of thickness ~1 μm and radius of curvature ~1 cm) immersed in liquid. By use of white light and multiple beam interferometry the shapes of, and separation between, the two mica surfaces may be independently measured. The separation between the two mica surfaces may be controlled by use of two micrometer-driven rods and a piezoelectric crystal to better than 0.1 nm. (From Israelachvili and Adams 1978, with permission.)

the back face with a 50 nm layer and are then glued down onto curved transparent silica discs which are polished to give a single radius and positioned to orientate the surfaces in the geometry of crossed cylinders, the thin sheets following the curvature of the discs. White light passing through the thin sheets is multiply reflected between the silvered back surfaces such that only certain wavelengths (FECO fringes, i.e. fringes of equal chromatic order) are transmitted and these can be measured in a spectrometer. Analysis of the shift in wavelength from when the mica surfaces are in molecular contact to when they are moved apart gives an accurate measure of the separation (± 0.2 nm). Using the apparatus illustrated in Fig. 7.7.9 the lower mica surface can be moved towards the upper one in a well-controlled motion via the lower rod, which compresses a fairly weak helical spring. This spring in turn acts on the left-hand side of the much more rigid double-cantilever spring and hence the lower disc. Since the latter spring is about 1000 times stiffer than the helical spring a movement of, say, 10 μm in the lower rod moves the lower disc only about 10 nm. This movement can be accurately monitored from the corresponding shift in fringe wavelength. For this same regular movement of the rod the lower surface will continue to move the same distance (10 nm).

When the mica surfaces are pushed so close together that their diffuse double-layers overlap, a repulsion arises. This reduces the actual distance moved by the lower disc by bending the weak leaf spring to which it is attached. Clearly at this point the leaf spring must have been bent by exactly the difference in distance between what the surfaces should have moved (i.e. 10 nm) and that actually observed, say, 8 nm. The spring must then have bent 2 nm. On further pushing the surfaces together the corresponding deformation at each separation is similarly obtained, with the spring deformation increasing during approach. Since the spring constant can be directly measured, the total force $\mathcal{F}$ between the mica surfaces as a function of separation distance D can be obtained.

Muscovite mica has a similar layer-silicate structure to the swelling clays already discussed (section 1.5.5). Because of its high mechanical strength, it is possible to mine large sheets that can then be cleaved to give reasonably large areas ($\sim 10\ cm^2$) of thin, exactly parallel sheets. These cleaved sheets are cut up to give roughly $1\ cm^2$ pieces, which are silvered and glued down so that two molecularly smooth basal plane surfaces face each other. Like most of the layer silicates the aluminosilicate lattice has a large negative charge ($\sim 2e$ per nm^2) which, in the case of muscovite mica, is balanced by K^+ ions (Fig. 1.5.11(b)). When cleaved sheets are immersed in aqueous solution some fraction of these exposed ions dissociates to form a diffuse double layer and it is the interaction between these layers that can be accurately measured using the force apparatus. In addition, the exposed basal plane acts as an ion-exchanger with other cations in aqueous solution, where the degree of exchange depends on the specific adsorption preference and relative bulk concentration. Hence, the K^+ ion initially held on the mica surface can be almost completely exchanged for a wide range of other cations (e.g. H^+, Na^+, Ca^{2+} and La^{3+}). Variation in bulk concentration of electrolyte therefore affects adsorption both at the surface (and hence ψ_d) and in the diffuse layer (κ^{-1}). Measurement of the interaction forces under a wide range of solution conditions can therefore give a very good test of the double-layer theory. Since forces can be measured down to small separations, both the attractive van der Waals force and any solvent structural forces can also be measured.

The total force results obtained for the case of mica immersed in conductivity water are shown in Fig. 7.7.10. In this figure the force $\mathscr{F}$ is divided by the mean radius R' of the curved surfaces and this is plotted against the separation distance D. Using the Deryaguin approximation (eqn (4.6.16)) $\mathscr{F}/R'$ is equivalent to $2\pi V_R$, where V_R is the corresponding interaction energy between flat surfaces per unit area, which is also given.

This is the function usually computed theoretically (as in the earlier part of this chapter). From Fig. 7.7.10 it is clear that the agreement between theory and experiment is excellent over separations from one-fiftieth of κ^{-1} to greater than κ^{-1}. The interaction energy decays roughly exponentially as predicted from the results of the approximate theories.

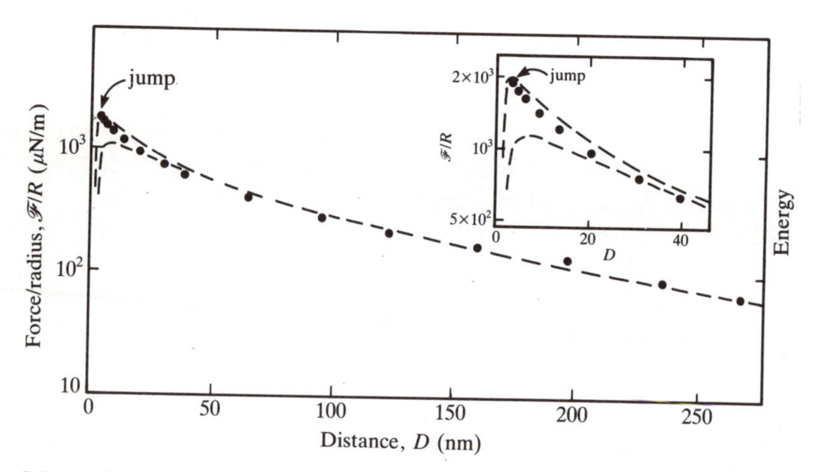

FIG. 7.7.10. The force–distance (and energy–distance) curve for mica plates in water. κ^{-1} is in this case about 150 nm. The upper and lower curves (inset) are calculated for constant charge and constant potential respectively. In this figure and Figs 7.7.11–14 the value of $\mathscr{F}/R$ is equal to $2\pi E$, where E is the interaction energy per unit area of two flat plates at the same separation, D. (From Pashley 1981a, with permission.)

The exact numerical procedure already described has been used to obtain the best fit at large separations ($>0.5\kappa^{-1}$). At shorter distances the experimental results clearly follow the constant charge curve, which falls above the curve for constant potential. The Debye length obtained agrees within experimental error with that estimated from conductivity measurements. The double-layer potential is obtained (or measured) by fitting the experimental results with the 'best fit' theoretical curves at large separations ($>\sim\kappa^{-1}$). At short distances (<3 nm) the van der Waals attraction pulls the surfaces into primary minimum contact, as expected. This attraction creates a force maximum (see Fig. 7.7.10) and at slightly closer distances the surfaces 'jump' into contact at $D = 0$ because of the leaf spring to which the lower surface is attached. This jump prevents measurement of the rapidly varying forces in this region.

For water the Debye length is very large (~150 nm) but this can be reduced by addition of electrolyte. If KCl is added to the dilute solution both the double-layer potential and the Debye length are altered. The forces measured under these conditions are shown in Fig. 7.7.11. The solid lines are calculated using the numerical procedure and are fitted to the experimental measurements. The Debye lengths agree within experimental error (±10 per cent) with those calculated from the concentration of the KCl solutions. The double-layer potentials obtained from the theoretical curves show a gradual decrease with increase in KCl concentration. These potentials can also be measured from streaming potential experiments (Chapter 9). The results obtained on large, cleaved sheets of mica are in good agreement (±10 mV) with the potentials

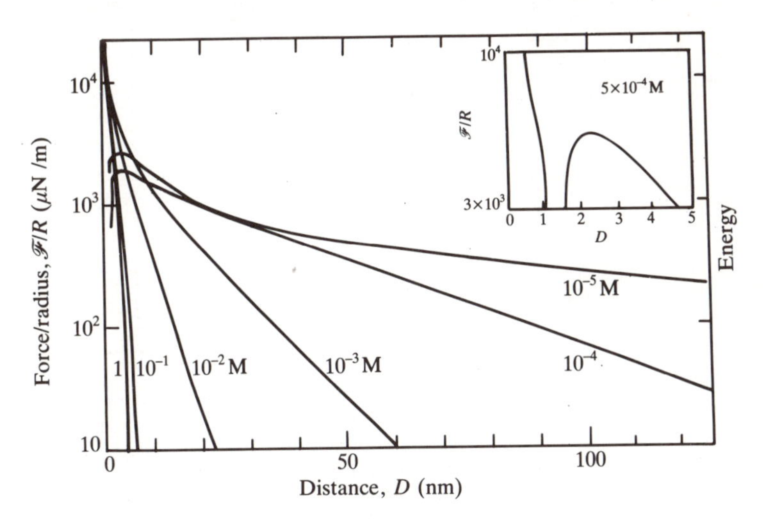

FIG. 7.7.11. Double-layer repulsion in the presence of potassium chloride.

obtained from the measured force. It seems clear that not only do the force measurements follow the precise form expected by exact, non-linear double-layer theory but also agree closely with the expected diffuse layer potentials and Debye lengths.

At fairly high electrolyte concentrations ($\gtrsim 10^{-2}$ M) and for surfaces with a significant double-layer potential ($\gtrsim 50$ mV) a secondary minimum may be observed (see Fig. 7.6.1). Measurement of the total force in the region of this minimum allows determination of the van der Waals attraction, since V_R is small and can be estimated accurately from DLVO theory. Israelachvili and Adams (1978) report measurements of the

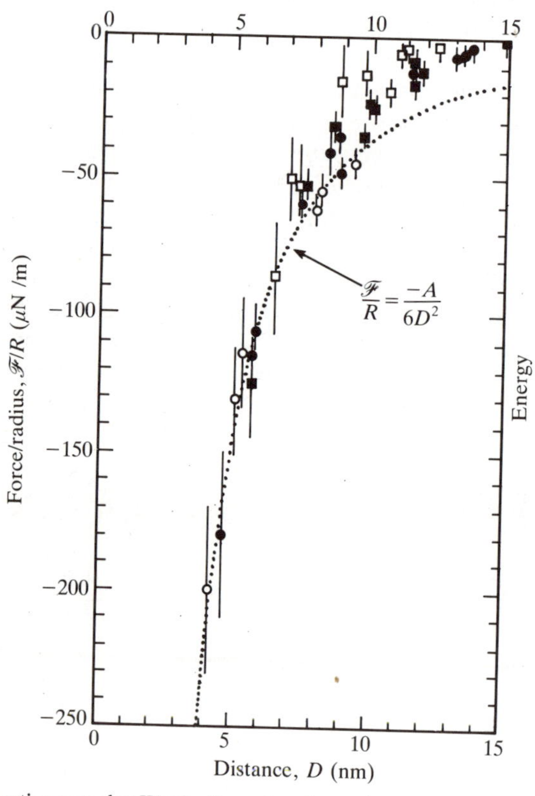

FIG. 7.7.12. Attractive van der Waals dispersion forces between mica surfaces measured in the region of secondary minima in various aqueous solutions. The dotted line represents a purely non-retarded inverse square van der Waals force law, inserted for comparison. Below ~6.5 nm the forces are effectively non-retarded; above ~6.5 nm they decay more rapidly with increasing separation. $A = 2.2 \times 10^{-20}$ J. ●, 10^{-1} mol dm^{-3} KNO_3; ○, 1 mol dm^{-3} KNO_3; ■ 10^{-2} mol dm^{-3} $Ca(NO_3)_2$; □, 10^{-1} mol dm^{-3} $Ca(NO_3)_2$. (From Israelachvili and Adams 1978, with permission.)

attraction between mica surfaces in aqueous solution obtained using this method (Fig. 7.7.12). For separations between 4 and 9 nm the results are in very good agreement with the non-retarded equation (4.4.13). At greater separations, however, retardation effects reduce the attraction below the non-retarded values, as expected (see Chapter 4).

At close distances the van der Waals attraction should pull the mica sheets into adhesive contact in the primary minimum. However, in KCl solutions more concentrated than about 10^{-4} M (at pH ~5.6) there arises an additional short-range repulsion that is stronger than the attractive forces (compare Fig. 7.7.13). This repulsion has been shown to be due to the hydration of adsorbed K^+ ions at the mica surface and corresponds to the crystalline swelling of clays (Pashley (1981a, b)). Because of the short-range nature of this force, which decays exponentially with a decay length of about 1 nm, it does not affect the diffuse double-layer interaction, except at short distances (<5 nm). Clearly in very high salt concentrations the diffuse layer decay length (also exponential) is less than 1.0 nm and the hydration force must dominate the total interaction; this is the case for 1 M KCl.

Since the double-layer values κ^{-1} and ψ_d are obtained from the theoretical fit to the experimental results beyond a separation of about

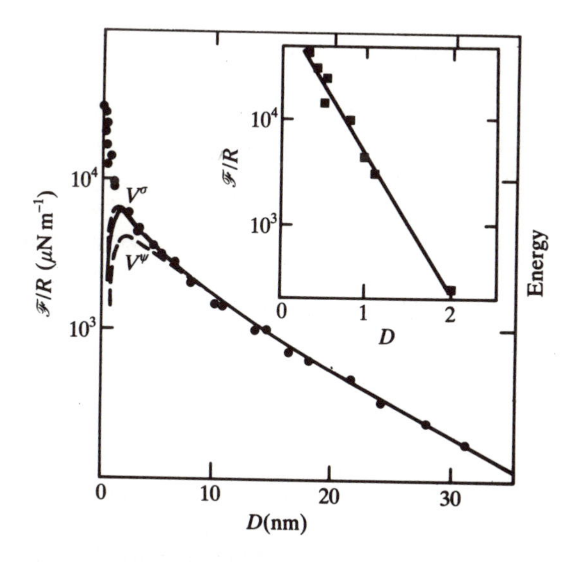

FIG. 7.7.13. The influence of hydration forces on the interaction at very small separations in 10^{-3} M KCl. This figure also shows a comparison of the constant charge V^σ, constant potential, V^ψ, and surface regulation (middle curve) models for double-layer interaction (cf. Fig. 7.3.1). (From Pashley 1981b, with permission.)

one Debye length, the hydration of the surface can be ignored in dilute electrolyte at separations greater than about 5 nm.

Repulsive hydration forces have been observed for a wide range of cations (e.g. Na^+, Ca^{2+}, and La^{3+}) adsorbed at the mica surface (Pashley and Quirk 1984). Similar forces, but of shorter range, were also measured between adsorbed bilayers of CTAB, cetyltrimethylammonium bromide (Pashley and Israelachvili 1981)†. These results for hydrated lipid-like surfaces agree well with those reported earlier on soap films (Clunie *et al.* 1967) and multi-bilayers (Le Neveu *et al.* 1977). Silica surfaces also appear to give rise to hydration forces (Peschel *et al.* 1982), presumably because of the favourable interaction between surface silanol groups and adjacent water layers.

The hydration forces observed between mica surfaces are apparently caused by adsorption of hydrated species (ions) rather than by the inherent nature of the surface, which is the case for lipid bilayer and silica surfaces. The former case has been referred to as 'secondary hydration'. In comparison, the adsorption of hydrophobic molecules would be expected to give rise to quite different surface properties. In a recent study Israelachvili and Pashley (1982) were able to measure an additional attractive force between mica surfaces made hydrophobic by adsorption of a packed *monolayer* of CTAB. This 'hydrophobic attraction' between hydrocarbon surfaces across aqueous solutions was found to be substantially stronger than the expected van der Waals attraction at separations less than about 8 nm. The attractive force decayed exponentially with a decay length of about 1 nm. The results reported indicate that the hydrophobic interaction is not sufficient by itself to cause coagulation of a highly charged surface but that as the Debye length approaches 1 nm the attraction will dominate and cause coagulation, even for quite high surface potentials. Addition of monovalent electrolyte at concentrations greater than about 0.07 M should, therefore, cause coagulation of most hydrophobic colloids, though coagulation might occur at lower concentrations due to a decrease in double-layer potential.

It is clear from these recent results that although both the double-layer and van der Waals (Lifshitz) theories are correct they do not completely describe the interaction between all types of surfaces. The effect of a particular surface on adjacent layers of solvent must also be considered. Precise theories for this solvent interaction depend critically on adequate models for the liquid state which at present, are not available.

The importance of solvent structure in the interactions between surfaces has been highlighted by a recent (1983) investigation of very

† These bilayers produce a hydrophilic surface because one layer of head groups lies against the mica surface and the other surfactant layer points head groups towards the water.

short-range (<~1 nm) forces between mica sheets in aqueous solution by Israelachvili and Pashley. In this work it was found that the solvent imposed an oscillatory perturbation on the monotonic repulsive hydration force. The periodicity of the oscillation is related to the finite size of the solvent molecules, which for water is about 0.25 nm; this is presumably a truly 'structural' contribution to the force, corresponding to the 'crystal hydrates' observed in the X-ray analysis of clay swelling. It can be seen as a macroscopic repulsion force only when the approaching surfaces are extremely smooth and solid (Fig. 7.7.14).

One final comment is called for. The differences that are observed in

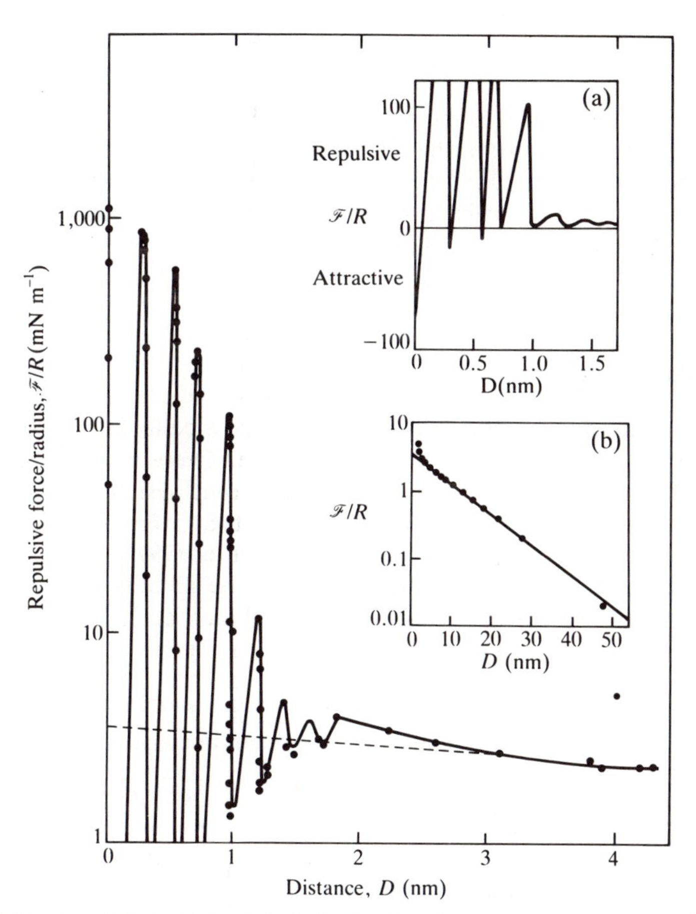

FIG. 7.7.14. Detailed structure of the hydration force between mica plates in 10^{-3} M KCl at very short distances. Against a solid surface the packing arrangement of the solvent molecules is revealed in an oscillating force law. (From Israelachvili and Pashley 1983, with permission.)

the short-range repulsion forces caused by different counterions have been attributed to differences in their degrees of hydration. There are many processes in which this more subtle difference between ions of the same valency is made manifest. It can be seen in the data for c.c.c. values of monovalent ions (Table 2.1) and in the adsorption of anions at the mercury–water interface (Fig. 6.4.6).

There are many other situations in colloid science where the ions of a particular valency follow this same (or a very similar) order. For anions the order of adsorbability from water is:

$$CNS^- > I^- > Br^- > Cl^- > F^- > NO_3^- > ClO_4^-$$

whilst for cations it is

$$Cs^+ > Rb^+ > K^+ > Na^+ > Li^+.$$

These are called the Hofmeister or *lyotropic series* of the ions. They evidently reflect the size of the ions and their polarizabilities. Such factors are important in determining both the mobility and adsorbability of an ion. It is, therefore, hardly surprising that they should determine aspects of both the kinetic and equilibrium behaviour of colloidal systems. We will have more to say about these series in Chapter 10.

7.8 Kinetics of coagulation

We noted in section 2.5 that the rate at which a colloidal sol coagulated was one of its most important characteristics. There exists a very large body of experimental work, on many different systems, showing the extent to which DLVO theory can account for the observed kinetic behaviour. See, for example, Napper and Hunter (1972, 1974). Here we want only to introduce the central concept of the stability ratio, W, which measures the effectiveness of the potential barrier in preventing the particles from coagulating:

$$W = \frac{\text{Number of collisions between particles}}{\text{Number of collisions that result in coagulation}}. \tag{7.8.1}$$

We will find that values of W in excess of 10^5 are easily obtained with quite modest potential barriers [$(V_T)_{max} \simeq 15\,kT$) and values in excess of 10^{10} are not impossible. The rate of coagulation in the absence of a potential barrier, R_f, is limited only by the rate of diffusion of the particles towards one another; that is the domain of fast or rapid coagulation. The rate of slow coagulation R_s is given by:

$$R_s = R_f/W \tag{7.8.2}$$

and we seek to establish the relation between W and the height and extent of the potential energy barrier (Fig. 7.6.1).

7.8.1 *Rate of rapid coagulation*

The problem of calculating the rate of coagulation in the absence of any repulsive barrier was first examined by von Smoluchowski (1916, 1917), and his treatment is discussed very fully by Overbeek (1952, pp. 278–86), from which this summary is derived.

Von Smoluchowski considered a central particle and calculated the number of particles that diffuse towards that particle as a result of Brownian motion. The flux, J_f, of particles whose centres pass through every sphere of radius r surrounding the central particle, and eventually come into contact with it, is given by Fick's Law (section 2.1.3):

$$J_f = \mathscr{D} 4\pi r^2 \frac{\partial \nu}{\partial r} \qquad (7.8.3)$$

where ν is the particle concentration and $\mathscr{D}$ is the diffusion coefficient. The solution to this equation for the appropriate boundary conditions is (Exercise 7.8.1):

$$\nu = \nu_0 - \frac{J}{\mathscr{D} 4\pi r} \qquad (7.8.4)$$

where ν_0 is the (bulk) particle concentration far from the central particle. The number of collisions with the central particle is, therefore (Exercise 7.8.1):

$$J_f = 8\pi \mathscr{D} a \nu_0 \qquad (7.8.5)$$

where a is the particle radius. (This analysis ignores the very beginning of the coagulation process, and concentrates attention on the steady-state situation, which is very quickly established.)

If the central particle is also undergoing Brownian motion, the appropriate value for $\mathscr{D}$ is (Exercise 7.8.2):

$$\mathscr{D} = \mathscr{D}_1 + \mathscr{D}_2 \qquad (7.8.6)$$

and so, when all the particles are of the same size, the rate of rapid coagulation is (Exercise 7.8.2):

$$-\frac{\mathrm{d}\nu}{\mathrm{d}t} = \tfrac{1}{2} \times J_f \times \nu_0 = 8\pi \mathscr{D} a \nu_0^2 = R_f. \qquad (7.8.7)$$

This relation can apply, of course, only in the early stages of coagulation before the formation of triplets etc. must be taken into account. The rapid coagulation time is often characterized by the time, $t_{1/2}$, required for the number of particles to be reduced to half of its initial value (Exercise 7.8.3):

$$t_{1/2} = \frac{1}{8\pi \mathscr{D} a \nu_0}. \tag{7.8.8}$$

Substituting for $\mathscr{D}$ from the Einstein relation (eqn (2.1.19)) and using eqn (2.1.20) gives:

$$t_{1/2} = 3\eta / 4kT\nu_0 \tag{7.8.9}$$

so that

$$t_{1/2} \simeq (2 \times 10^{11}/\nu_0) \text{ seconds}$$

in water at room temperature (with ν_0 in particles per cm^3). Very concentrated sols (~5 per cent) would have $\nu_0 \simeq 10^{14}\, cm^{-3}$ and so have coagulation times in the millisecond range. The more usual systems investigated in coagulation studies have rapid coagulation times in the range of seconds to minutes.

7.8.2 *Rate of slow coagulation*

The presence of an energy barrier to coagulation is somewhat analogous to the existence of an activation energy barrier for an ordinary chemical reaction. There are, however, very important differences between the two phenomena. The collision of two molecules, and the resulting bond rearrangements that result in reaction, take place on a very short time scale (of the order of picoseconds). Although the surrounding molecules may to some extent moderate the interaction between the reacting molecules, the activation energy is determined almost entirely by the distortion that must occur to the reacting molecules before reaction can occur. If they possess sufficient energy and it can be channelled into the right places, then the activation barrier can be surmounted and reaction becomes possible.

It might be argued then, that coagulation should occur between two colloidal particles if they have sufficient mutual kinetic energy to surmount the potential energy barrier (V_T) that separates them (Fig. 7.6.1). Such a procedure can however, easily be shown to be inadequate because any particle having such a large amount of energy would lose it by frictional interactions with the dispersion medium long before it was able to penetrate through the barrier (see e.g. Verwey and Overbeek 1948, p. 164) (Exercise 7.8.4). How, then, does coagulation occur?

The approaching particles experience the barrier as their double-layers overlap, and this occurs over distances of the order of 1–100 nm. The diffusion over this distance is accomplished as a result of many individual

Brownian events, some of which bring the particles closer together and some of which take them further apart. The presence of the barrier merely makes the probability of movement in one direction higher than that in the other. One must, therefore, solve the diffusion equation (7.8.3) in a slightly more general form that takes account of the presence of this force field.

This problem was first examined by Fuchs (1934), and again the analysis here follows Overbeek's (1952) treatment. The problem can be greatly simplified with negligible error by again neglecting the very earliest stage and concentrating on the steady-state situation, which is quickly established. The flux J then becomes constant and eqn (7.8.3) can be modified to take account of the force field as follows:

$$J = 4\pi r^2\left(2\mathscr{D}\frac{\mathrm{d}\nu}{\mathrm{d}r} + \frac{\nu}{B}\frac{\mathrm{d}V_\mathrm{T}}{\mathrm{d}r}\right) \tag{7.8.10}$$

where B is, as before (section 2.1.3), the friction factor. The second term is simply the product of particle concentration and the mutual particle velocity (Fig. 2.1.3) induced by the field of force. Because of the mutual motion of the two particles, however, we must replace B by $kT/2\mathscr{D}$ (rather than $kT/\mathscr{D}$). The solution of eqn (7.8.10) for the appropriate boundary conditions is (Exercise 7.8.5):

$$\nu = \nu_0 \exp(-V_\mathrm{T}/kT) + \frac{J\exp(-V_\mathrm{T}/kT)}{8\pi\mathscr{D}}\int_\infty^r \exp(V_\mathrm{T}/kT)\frac{\mathrm{d}r}{r^2}. \tag{7.8.11}$$

To satisfy the condition that $\nu = 0$ when $r \simeq 2a$ we must have, for the flux:

$$J_\mathrm{s} = \frac{8\pi\mathscr{D}\nu_0}{\displaystyle\int_{2a}^\infty \exp(V_\mathrm{T}/kT)\frac{\mathrm{d}r}{r^2}} \tag{7.8.12}$$

and the limiting value of J when there is no potential between the particles ($V_\mathrm{T} = 0$), except an infinitely strong attraction when they actually make contact is given again by (Exercise 7.8.6):

$$J_\mathrm{f} = 8\pi\mathscr{D}a\nu_0. \tag{7.8.13}$$

Since the rates of rapid and slow coagulation are directly proportional to the fluxes J_f and J_s it follows from eqn (7.8.2) that the stability ratio is given by:

$$W = R_\mathrm{f}/R_\mathrm{s} = J_\mathrm{f}/J_\mathrm{s} = 2a\int_{2a}^\infty \exp(V_\mathrm{T}/kT)\frac{\mathrm{d}r}{r^2}$$

$$= 2\int_2^\infty \exp(V_\mathrm{T}/kT)\frac{\mathrm{d}s}{s^2} \tag{7.8.14}$$

where $s = r/a$. This integral must be evaluated graphically or numerically but Verwey and Overbeek (1948) showed that W was determined almost entirely by the value of V_T at the maximum (see Fig. 7.6.1) (V_{max}). This is hardly surprising when V_T enters the integrand through an exponential function (Fig. 7.8.1). The detailed calculations of Verwey and Overbeek (1948, p. 169) suggest that for $V_{max} = 15\,kT$, W is of the order of 10^5 and for $V_{max} = 25\,kT$, W is $\sim 10^9$. A dilute sol for which the rapid coagulation time is of the order of 1 s can, therefore, be converted into a very stable sol (with a coagulation time of the order of several months) if its double-layer potential is raised sufficiently to produce a barrier height of the order of $20\,kT$.

When W is calculated for typical values of the other parameters (ψ_0 or ψ_d, and the van der Waals constant A) it is found that $\log_{10} W$ decreases linearly with the logarithm of the electrolyte concentration, C, over quite a wide range ($9 \geqslant \log_{10} W \geqslant 0.5$) and this is the relationship that has been most extensively tested experimentally. Reerink and Overbeek (1954)

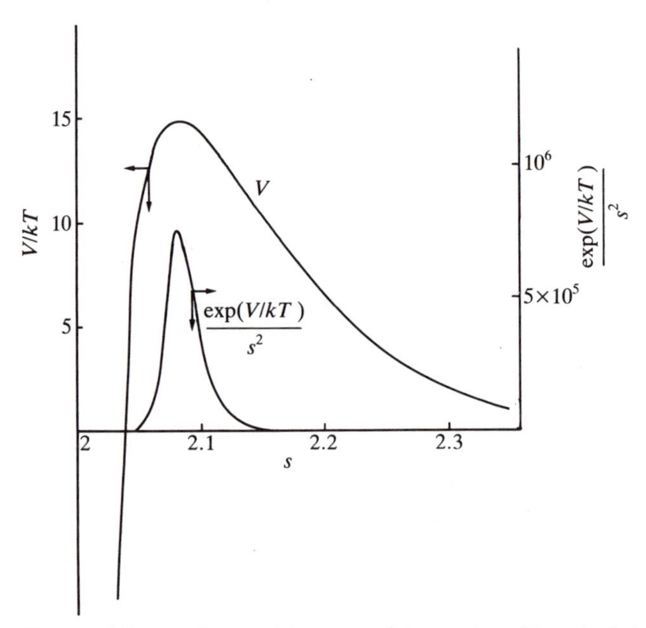

FIG. 7.8.1. Curves of the total potential energy of interaction, V, and of the function $s^{-2}\exp(V/kT)$ for a sol with $a = 10^{-5}$ cm, $\kappa = 10^6\,\text{cm}^{-1}$, $A = 10^{-19}$ J, $\psi_0 = 28.2$ mV. In this case $W \simeq 5.4 \times 10^4$ (obtained by graphical integration). (After Verwey and Overbeek 1948, with permission.)

provided a theoretical justification for the approximate relation:

$$\log_{10} W = -k_1 \log_{10} C + k_2 \tag{7.8.15}$$

by expanding the potential function V_T as a Taylor's series about the maximum. (This amounts to replacing the actual potential energy by a Gaussian curve (Fig. 2.1.2).) Using a light-scattering technique (sections 2.3, 3.7) they followed the slow coagulation process at various electrolyte concentrations and showed that the value of the stability ratio, W, did indeed follow the relation (7.8.15) (Fig. 7.8.2) and this has since been amply confirmed. Their approximate theoretical analysis gave for the constant k_1, a value of

$$k_1 = 2.15 \times 10^7 aZ^2/y_d^2 \tag{7.8.16}$$

where $y_d = ze\psi_d/kT$ and a is the particle radius (in cm). The dependence of k_1 on the particle radius was, however, *not* confirmed and this remains one of the outstanding puzzles in this area. Some more exacting work using monodisperse latex particles by Ottewill and Shaw (1966) also failed to demonstrate any relation between $d\log_{10} W/d\log_{10} C$ and the particle radius. Experimentally W seems to be quite independent of particle size.

One important modification to the simple theory described above is the introduction of the *hydrodynamic correction,* first suggested by Deryaguin (1966). This attempts to take account of the fact that in the final stages of approach of two particles, they are slowed down because it is difficult for the remaining film of liquid to escape. Various authors have examined the problem (see Overbeek 1977) but the most useful approximate relation is that derived by Honig *et al.* (1971) for the

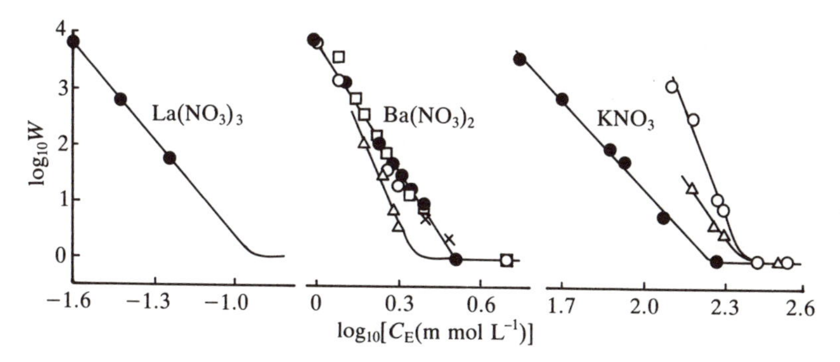

FIG. 7.8.2. Stability curves for various (negative) silver iodide sols at pI = 4. The different symbols refer to slight variations in the method of preparation. C_E is the electrolyte concentration. (From Reerink and Overbeek 1954, with permission.)

modified diffusion coefficient:

$$\frac{\mathscr{D}(H)}{\mathscr{D}(H\to\infty)}=\frac{2H}{a}\left(\frac{1+3H/2a}{1+13H/2a+3H^2/a^2}\right)$$

$$=\frac{1+2a/3H}{1+13a/6H+a^2/3H^2} \tag{7.8.17}$$

where H $(=r-2a)$ is the distance of closest approach.

Note that $\mathscr{D}$ approaches zero as $2H/a$ so that Brownian encounters could not occur unless they were assisted by an attraction which was at least proportional to a/H, which is, of course, true of the van der Waals attraction.

This effect was tested experimentally by Lichtenbelt *et al.* (1974a, b) who found that the rate of rapid coagulation was reduced to a little less than half of the von Smoluchowski value (eqn (7.8.13)), and in agreement with eqn (7.8.17). It should also be noted that the introduction of the hydrodynamic correction *does not* explain the fact that W is essentially independent of particle size. That remains something of a mystery.

The Deryaguin–Landau–Verwey–Overbeek (DLVO) theory of colloid stability occupies a central position in colloid science because it allows the Gouy–Chapman theory of the equilibrium double layer to be extended to situations involving double layer overlap. This is essential not only for the study of particle–particle interactions and coagulation but also for such situations as sedimentation and filtration, the behaviour of electrolyte solutions in the pores of a solid (where the pore walls are normally charged) and the movement of ions through membranes.

An increasing body of evidence is developing to show that the DLVO theory is certainly valid in broad outline and that it is able to account almost quantitatively for the main features of the equilibrium repulsion and attraction between smooth macroscopic surfaces and colloidal particles. Although there remain some uncertainties concerning the kinetics of the coagulation process, the main features are again well described by DLVO theory. Most analyses of the more complex areas of filtration, sedimentation, and electrical conduction in porous media also begin from this same fundamental standpoint. DLVO theory is to colloid science what the van der Waals equation is to the theory of real gases. With all its faults that latter equation has provided many useful insights, stretching from its initial domain into the theory of liquids and of electrolyte solutions.

Exercises. 7.8.1 Show that eqn (7.8.4) is a solution of eqn (7.8.3) with the boundary conditions $v=v_0$, $r=\infty$. Establish eqn (7.8.5), assuming $v=0$ for $r=2a$ (i.e. the approaching particle is assumed to be swallowed up on contact).

7.8.2 Use the Einstein–Smoluchowski eqn (2.1.26) to establish eqn (7.8.6). (Hint: consider the average value of $(x_1 - x_2)^2$ where $(x_1 - x_2)$ is the relative displacement and note that $\langle x_1 x_2 \rangle = 0$.) Establish eqn (7.8.7) assuming that in the early stages of coagulation the rate is the same as the collision frequency. (Why is the $\frac{1}{2}$ included?)

7.8.3 Establish eqn (7.8.8) by integrating eqn (7.8.7) with ν_0 treated as a variable. A more rigorous approach, which makes an approximate allowance for collisions of particles of all sizes, gives for the total number of particles of all sizes after time t (Overbeek 1952, p. 281):

$$\sum_{k=1}^{\infty} \nu_k = \frac{\nu_0}{1 + t/t_{1/2}}$$

where ν_k is the number of k-fold particles (i.e. particles produced by the aggregation of k particles of equal radius).

7.8.4 Consider a particle of mass m travelling with velocity v in a fluid. Newton's law gives $-m\,\mathrm{d}v/\mathrm{d}t = Bv$ so that $v = v_0 \exp(-Bt/m)$ where B is the coefficient of frictional resistance (cf. Exercise 3.5.1). Show that if the particle starts with a kinetic energy of $p_0 kT$ then it will come to rest after travelling a distance of the order of $(2p_0 kTm)^{1/2}/B$. Use reasonable values for these quantities to show that a colloidal particle of initial energy $10kT$ will come to rest in water after travelling at most only a fraction of a nanometre.

7.8.5 Consider the differential equation (7.8.10) under steady-state conditions:

$$\frac{J}{8\pi\mathscr{D}} = r^2 \frac{\mathrm{d}\nu}{\mathrm{d}r} + \frac{\nu r^2}{kT}\frac{\mathrm{d}V}{\mathrm{d}r} = \text{constant}.$$

Show, using the differential d $(\mathrm{e}^{V/kT}\nu)$ that the solution is given by eqn (7.8.11) for the boundary conditions $V = 0$, $\nu = \nu_0$ at $r = \infty$.

7.8.6 Establish eqn (7.8.13). (Remember to avoid double counting.)

7.8.7 Overbeek (1952) gives as an approximation for W the expression:

$$W \simeq \frac{1}{2\kappa a} \exp[V_{\max}/kT]$$

Estimate W from this relation for the data in Fig. 7.8.1 and compare it with the exact result.

References

Adamson, A. W. (1967). *Physical chemistry of surfaces* (2nd ed). Interscience, New York.

Barclay, L. and Ottewill, R. H. (1970). *Special Discuss. Faraday Soc.* No. 1, p. 164.

Barclay, L., Harrington, A., and Ottewill, R. H. (1972). *Kolloid-Z. Z. Polym.* **250,** 655.

Chan, D. Y. C., and Mitchell, D. J. (1983). *J. Colloid interface Sci.* **95,** 193.

Chan, D. Y. C., Healy, T. W., and White, L. R. (1976). *J. chem. Soc. Faraday Trans. 1* **72,** 2844–65.

Chan, D. Y. C., Pashley, R. M., and White, L. R. (1980). *J. Colloid interface Sci.* **77,** 283.

Clunie, J. S., Goodman, J. F., and Symons, P. C. (1967). *Nature (Lond.)* **216,** 1203.
Deryaguin, B. V. (1954). *Discuss. Faraday Soc.* **18,** 85.
Deryaguin, B. V. (1966). *Discuss. Faraday Soc.* **42,** 317.
Deryaguin, B. V. and Kussakov, M. (1939). *Acta physicochim.* **10,** 25, and 113.
Deryaguin, B. V. and Titievskaya, A. S. (1957). *Proceedings of the International Congress on Surface Activity* Vol. 1, p. 211. Butterworths, London.
Devereux, O. F. and de Bruyn, P. L. (1963). *Interaction of plane parallel double layers.* MIT Press, Cambridge, Mass.
Donners, W. A. B., Rijnbout, J. B., and Vrij, A. (1977). *J. Colloid interface Sci.* **61,** 249.
El-Aasser, M. S. and Robertson, A. A. (1971). *J. Colloid interface Sci.* **36,** 86.
El-Aasser, M. S. and Robertson, A. A. (1973). *Kolloid-Z. Z. Polym.* **251,** 241.
Friend, J. P. and Hunter, R. J. (1970). *Clays Clay Minerals* **18,** 275.
Fuchs, N. (1934). *Z. Physik,* **89,** 736.
Garrett, W. G. and Walker, G. F. (1962). *Clays Clay Minerals* **9,** 557.
Gregory, J. (1973). *J. chem. Soc. Faraday Trans. 2* **69,** 1723.
Healy, T. W., Chan, D. Y. C., and White, L. R. (1980). *Pure appl. Chem.* **52,** 1207–19.
Hogg, R., Healy, T. W., and Fuerstenau, D. W. (1966). *Trans. Faraday Soc.* **62,** 1638.
Honig, E. P. and Mul, P. M. (1971). *J. Colloid interface Sci.* **36,** 258.
Honig, E. P., Roebersen, G. J., and Wiersema, P. H. (1971). *J. Colloid interface Sci.* **36,** 97.
Hunter, R. J. (1975). *Modern aspects of electrochemistry,* Vol. 11, Chapter 2. Plenum Press, New York.
Hunter, R. J. (1981). *Zeta potential in colloid science.* Academic Press, London.
Israelachvili, J. N. and Adams, G. E. (1978). *J. chem Soc. Faraday Trans. 1,* **78,** 975.
Israelachvili, J. N. and Pashley, R. M. (1982). *Nature* **300,** 341.
Israelachvili, J. N. and Pashley, R. M. (1983). *Nature* **306,** 249.
Jones, J. and Levine, S. (1969). *J. Colloid interface Sci.* **30,** 241.
Langmuir, I. V. (1938). *J. Chem. Phys.* **6,** 893.
Le Neveu, D. M., Rand, R. P., Parsegian, V. A., and Gingell, D. (1977). *Biophys. J.* **18,** 209.
Lichtenbelt, J. W. Th., Ras, H. J. M. C., and Wiersema, P. H. (1974a). *J. Colloid interface Sci.* **46,** 522.
Lichtenbelt, J. W. Th., Pathmamanoharan, C., and Wiersema P. H. (1974b). *J. Colloid interface Sci.* **49,** 281.
Lyklema, J. (1967). *Pontif. acad. sci. scripta varia* **31,** 221.
Lykelma, J. and Mysels, K. J. (1965). *J. Am. chem. Soc.* **87,** 2539.
Napper, D. H. and Hunter, R. J. (1972). Hydrosols in *M.T.P. Int. Review of Science. Phys. Chem. Series 1.* (ed. M. Kerker) **7,** 241. Butterworths, London. (1974) *Series 2,* **7,** 161–213.
Ninham, B. W. and Parsegian, V. A. (1971). *J. theor. Biol.* **31,** 405–428.
Norrish, K. (1954). *Discuss. Faraday Soc.* **18,** 120.
Norrish, K. and Rausell-Colom, J. A. (1963). *Clays Clay Minerals* **10,** 123.
Ohshima, H., Chan, D. Y. C., Healy, T. W., and White, L. R. (1983). *J. Colloid interface Sci.* **92,** 232–42.
Ohshima, H., Healy, T. W., and White, L. R. (1982). *J. Colloid interface Sci.* **89,** 484–93.

Ottewill, R. H. (1982). Concentrated dispersions in *Colloidal dispersions* (ed. J. W. Goodwin) Chapter 9. Royal Society of Chemistry, London.
Ottewill, R. H. and Shaw, J. N. (1966). *Discuss. Faraday Soc.* **42,** 154–63.
Overbeek, J. Th. G. (1952). In *Colloid Science* (ed. H. R. Kruyt) Vol. 1. Elsevier, Amsterdam.
Overbeek, J. Th. G. (1977). *J. Colloid interface Sci.* **58,** 408–22.
Pashley, R. M. (1981a). *J. Colloid interface Sci.* **80,** 153.
Pashley, R. M. (1981b). *J. Colloid interface Sci.* **83,** 531.
Pashley, R. M. and Israelachvili, J. N. (1981). *Colloids Surfaces* **2,** 169.
Pashley, R. M. and Quirk, J. P. (1984). *Colloids Surfaces* **9,** 1–17.
Parsegian, V. A. and Gingell, D. (1972). *Biophys. J.* **12,** 1192.
Peschel, G., Belouschek, P., Müller, M. M., Müller, M. R., and Konig, R. (1982). *Colloid Polym. Sci.* **260,** 444.
Read, A. D. and Kitchener, J. A. (1969). *J. Colloid interface Sci.* **30,** 391–8.
Reerink, H. and Overbeek, J. Th. G. (1954). *Discuss. Faraday Soc.* **18,** 74–84.
Tabor, D. and Winterton, R. H. S. (1969). *Proc. R. Soc. Lond.* **A312,** 435–50.
Verwey, E. J. W. and Overbeek, J. Th. G. (1948). *Theory of stability of lyophobic colloids*. Elsevier, Amsterdam.
von Smoluchowski, M. (1916). *Physik. Z.* **17,** 557 & 585.
von Smoluchowski, M. (1917). *Z. phys. Chem.* **92,** 129.
Wiese, G. and Healy, T. W. (1970). *Trans. Faraday Soc.* **66,** 490.

8

POLYMERIC STABILIZATION AND FLOCCULATION

8.1 Historical prologue

The stabilization of colloidal dispersions by naturally occurring polymers has been exploited continuously by man for almost five millennia, since methods for the preparation of ink were first devised in ancient Egypt and China around 2500 BC. This primitive but highly ingenious ink was prepared by mixing lamp black, obtained by combustion, with a solution of a natural polymeric stabilizer (e.g. casein from milk, egg albumin or gum arabic). The carbon black was then moulded into a pencil-like shape and dried. When the ink was required, the mould was simply dipped into water, whereupon the carbon black particles coated by the biopolymer redispersed spontaneously ('instant' ink). Such spontaneous redispersion is a characteristic feature of colloidal particles that are stabilized by polymeric species.

Nowadays stabilization by either natural or synthetic polymers is exploited in a diverse range of industrial products: paints, glues, inks, pharmaceutical and food emulsions, detergents, lubricants, etc. Furthermore, polymeric stabilization is operative in many biological systems, such as blood and milk. It is also important in a wide range of industrial and agricultural processes (e.g. water treatment, soil stability, coal washing, oil recovery, etc.).

The honour of conducting the first properly designed scientific study of the effects of biopolymers on colloid stability probably belongs to Michael Faraday. As early as 1857 he reported that in the absence of gelatin, a gold sol could be induced to aggregate by the addition of sodium chloride, as evidenced by a characteristic red-to-blue colour change. This aggregation was not observed, however, when gelatin was present. Faraday even speculated that this lack of aggregation in the presence of high concentrations of salt (which is now known to eliminate electrostatic stabilization) is a consequence of the gold particles becoming associated with envelopes of 'that animal substance' (Faraday 1857).

Zsigmondy (1901) later quantified the relative efficacy of different biopolymers in preventing the coagulation of a gold sol by electrolyte in his now classic gold number experiments. The gold number of a macromolecule was defined as the amount (expressed in mg) of polymer which, when added to a sample of a gold sol ($10.0\,cm^3$), just prevents its coagulation on the addition of $1.0\,cm^3$ of a 10 per cent sodium chloride solution. Some gold numbers for different biopolymers are presented in Table 8.1. Note that the smaller gold numbers correspond to the more effective polymeric stabilizers. Zsigmondy coined the term *protective colloid* (*Schutzkolloide*), which was said to function by *protective action* or *protection*. These terms are still widely used, although 'protection' is slowly being supplanted by the term *steric stabilization*, especially for the

Table 8.1
The gold numbers of some natural polymers

Biopolymer	Gold number ($mg/10\ cm^3$)
Gelatin	0.005–0.01
Sodium caseinate	0.01
Albumin	0.1–0.2
Gum arabic	0.15–0.25
Dextrin	6–20
Potato starch	~25

effects that adsorbed non-ionic synthetic polymers exert on colloid stability. It must be stressed that this more recent terminology, which was introduced by Heller and Pugh (1954), is something of a misnomer in that steric stabilization has little, if anything, in common with the steric effects that abound in organic chemistry. The latter have their origins in the electron repulsion that arises from the operation of the Pauli exclusion principle. In contrast, steric stabilization, as will become apparent, has a quite general thermodynamic basis. Note that in what follows the term *coagulation* will be applied to aggregation that is induced by van der Waals attraction between the colloidal particles, whereas *flocculation* will be reserved for polymer-induced aggregation (section 2.6). Coagulation usually gives rise to compact aggregates whereas flocculation produces more open structures but this is not always the case.

8.2 Methods for imparting colloid stability

It will be apparent from the concepts that have been elaborated in the preceding chapters that, in general, dispersions of naked uncharged colloidal particles undergo rapid coagulation (sections 2.5, 7.8). The number of particles in such colloidal dispersions is reduced by a factor of two in less than a few seconds at normal particle concentrations. The reasons for this are straightforward: the particles are constantly undergoing Brownian collisions and these collisions are 'sticky'. This aggregation is a consequence of the attractive van der Waals forces that are operative between the particles. As explained in Chapter 4, the van der Waals attraction between colloidal particles is relatively long range in character, extending over a distance of a few nanometres. It follows that to prepare stable colloidal dispersions it is necessary to provide a repulsive interaction that is at least comparable, both in its range and magnitude, to that of the van der Waals attraction.

Table 8.2
The spatial extensions of electrical double layers at different ionic strengths and of polymers of different molecular weight

1 : 1 Electrolyte concentration ($mol\ dm^{-3}$)	Double layer thickness (nm) κ^{-1}	Polymer molecular weight	Spatial extension (nm)
10^{-5}	100	1,000,000	60
10^{-4}	30	100,000	20
10^{-3}	10	10,000	6
10^{-2}	3	1000	2
10^{-1}	1		

At the present time, there exist remarkably few methods by which colloid stability can be imparted in a controlled fashion. The primary reason for this is the difficulty of ensuring that the repulsion extends over the required distance of several nanometres. Two general methods are currently exploited to impart colloid stability: electrostatic stabilization (section 2.5) and polymeric stabilization (section 2.6). In electrostatic stabilization, which was discussed in Chapter 7, the long-range repulsion between the particles is provided by the Coulombic interaction between the electrical double layers. At low ionic strengths, these layers extend over distances of several nanometres. Some values for the double layer thickness ($1/\kappa$) calculated for different concentrations of a 1 : 1 electrolyte are presented in Table 8.2 to illustrate representative orders of magnitude. Also shown in Table 8.2 for comparative purposes are typical spatial extensions (as expressed by the r.m.s. end-to-end distance of the chain) for non-ionic polymer molecules of different molecular weights. These clearly demonstrate that, like electrical double layers, macromolecules of at least a few thousand molecular weight also extend in space over distances comparable to, or greater than, the van der Waals attraction. It follows that, provided non-ionic macromolecules become mutually repulsive under suitable conditions (which indeed they do in so-called 'good' solvents), they can be utilized to impart colloid stability.

Exercises. 8.2.1 Use the formula for the double layer thickness in a 1 : 1 electrolyte (eqn (6.3.14)):

$$(1/\kappa)/\text{nm} = 0.304/(I/\text{mol L}^{-1})^{1/2}$$

to check the values shown in Table 8.2.

8.2.2 The 'diameter' of a linear polymer molecule is crudely given by its r.m.s.

end-to-end length, which for many carbon backbone polymers is given approximately by

$$\langle r^2 \rangle^{1/2}/\mathrm{nm} \simeq 0.06\, M^{1/2}$$

where M = molecular weight (Flory 1969). Use this crude formula to check the values for the spatial extension given in Table 8.2.

8.2.3 Calculate the van der Waals attraction between two naked polystyrene latex particles of diameter 100 nm dispersed in water at room temperature when their minimum surface-to-surface distance of separation is given by twice the polymer spatial extensions listed in Table 8.2. Express your results in units of kT at room temperature. Any comments? (Hint: see Exercise 4.4.2.)

8.3 Effects of polymers on colloid stability

The most widely exploited effect of nonionic polymers on colloid stability is called steric stabilization, in which stability is imparted by polymer molecules that are adsorbed onto, or attached to, the surface of the colloid particles. It is of course, possible to impart stability by combinations of different stabilization mechanisms: e.g. an attached polyelectrolyte can impart stability by a combination of electrostatic and steric mechanisms. This has been termed *electrosteric stabilization.* It is commonly encountered in biological systems and is being increasingly utilized in industrial dispersions, such as paints, where environmental considerations have caused a dramatic shift from organic-based dispersion media to water-based systems. Polymer that is in free solution can also influence colloid stability. Stabilization by free polymer is well documented (Napper 1983) and is called *depletion stabilization*.

8.4 Effects of attached polymer: steric stabilization

8.4.1 *Advantages*

Steric stabilization is widely exploited industrially because it offers several distinct advantages over electrostatic stabilization. These are summarized in Table 8.3.

The first advantage is that aqueous sterically stabilized dispersions are comparatively insensitive to the presence of electrolytes because the dimensions of non-ionic chains vary relatively little with the electrolyte concentration. This contrasts sharply with the spatial extensions of electrical double layers, which are strongly dependent upon the ionic strength. It is apparent from Table 8.2 that at ionic strengths greater than *ca.* 10^{-2} mol dm^{-3}, electrical double layer thicknesses have shrunk to such an extent that the electrostatic repulsion may no longer outweigh the van der Waals attraction. This accounts for the coagulation of

Table 8.3
Comparison of the properties of electrostatically and sterically stabilized dispersions

Steric stabilization	Electrostatic stabilization
1. Insensitive to electrolyte	1. Coagulates on addition of electrolyte
2. Equally effective in aqueous and non-aqueous dispersion media	2. Mainly effective in aqueous dispersion media
3. Equally effective at high and low volume fractions of particles	3. More effective at low volume fractions
4. Reversible flocculation common	4. Coagulation often irreversible
5. Good freeze–thaw stability	5. Freezing often induces irreversible coagulation

electrostatically stabilized dispersions on the addition of electrolyte (see section 2.5).

A second advantage of steric stabilization is that it is equally effective in both aqueous and non-aqueous dispersion media. This contrasts with electrostatic stabilization, which is relatively ineffective in non-polar dispersion media.

Third, steric stabilization is equally effective at both high and low volume fractions of the dispersed phase, the high volume fraction dispersions displaying relatively low viscosities.

A fourth advantage is that sterically stabilized dispersions can often be flocculated reversibly whereas this is less common with electrostatically stabilized dispersions.

Finally, sterically stabilized dispersions frequently display good freeze–thaw stability, which can be a desirable attribute in some practical applications (e.g. paint systems).

8.4.2 *The best steric stabilizers*

It has been shown experimentally that the best steric stabilizers are amphipathic block or graft copolymers (Barrett 1975). These consist of at least two chemically bound homopolymer components, one of which is soluble in the dispersion medium, while the other is nominally insoluble. The latter, termed the *anchor polymer,* serves to attach the soluble species to the colloidal particles (see Fig. 8.4.1). The soluble chains, which project away from the particle surface into the dispersion medium, are responsible for the observed colloid stability and thus are referred to as the *stabilizing moieties.* The precise details (e.g. sequence and

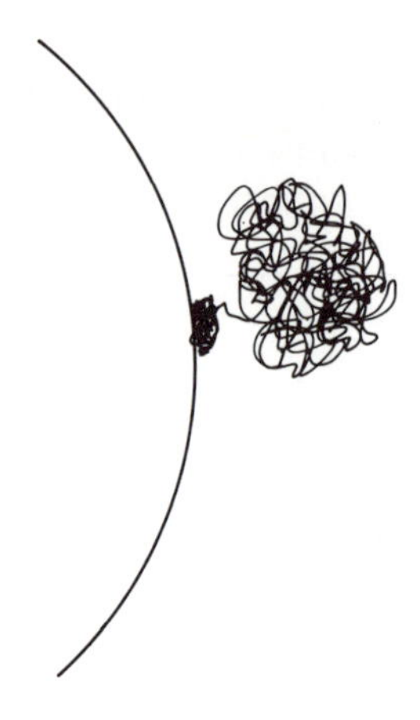

FIG. 8.4.1. Schematic representation of steric stabilization by an amphipathic copolymer (not to scale).

sequence length by which the anchor polymer segments and the stabilizing moieties are connected) may be quite complex. The manufacturers of non-ionic surfactants have long recognized the necessity for polymer surfactants to possess amphipathic character. Indeed, this concept was embodied by Griffin (1954) in his development of the so-called 'HLB scale'. This scale endeavours to express in a semi-quantitative fashion the balance between the hydrophilic and lipophilic ('oil loving') character of the surfactant (section 1.5.3). The foregoing discussion is not meant to imply that homopolymers themselves cannot impart steric stabilization. This they certainly can do, especially if the homopolymer interacts specifically with the surface of the colloidal particles. Nevertheless, the stability that is imparted by homopolymers is usually only a pale shadow of that generated by tailor-made amphipathic copolymers, suggesting the important part played by the attachment of the polymer chain to the particle surface. Reasons for this will be presented below.

The stability imparted by amphipathic copolymers is optimized when the surfaces of the colloidal particles are completely coated by polymer chains. When two colloidal particles approach one another in a Brownian encounter, the stabilizing moieties attached to one particle must repel those attached to the other if stability is to be observed. The stress thus generated on the stabilizing moieties can be relieved by movement of those moieties away from the interaction zone. One way in which this could be accomplished is by desorption of the stabilizing chains from the surfaces of the particles. It is to prevent this desorption that the anchor polymer is incorporated into the amphipathic stabilizer. A second way in which the stress could be relaxed is by lateral movement of the stabilizing chains over the surface of the particles. Such surface migration is precluded by ensuring that the surface is completely coated with polymer. Complete surface coverage also prevents the polymer chains attached to one particle from inducing bridging flocculation by adsorption onto another particle. When desorption, lateral surface diffusion, and inter-particle bridging are all precluded, optimum stability is imparted by the polymer. The stress generated in the interaction zone is then relieved by diffusion of the particles, together with their associated steric layers, away from one another.

8.4.3 *Phenomenology of flocculation of sterically stabilized dispersions*

The preceding discussion can be summarized by saying that thermodynamic factors limit the stability of sterically stabilized dispersions and emulsions provided that the following three requirements are fulfilled: first, the anchoring must be adequate; second, the surface coverage must be complete; and, third, the thickness of the steric layers must be sufficiently large. The thermodynamic limit to stability under these conditions has been shown experimentally to be largely independent of the chemical nature of the anchor polymer and, commonly, the nature and size of the colloidal particles. It is also independent of the number concentration of colloidal particles. As will be discussed later, exceptions to this general pattern of behaviour do occur, especially with large particles coated by thin steric layers.

The thermodynamic limit to the stability of many sterically stabilized systems is determined primarily by the chemical nature of both the stabilizing moieties and the dispersion medium. Instability can be induced by decreasing the solvency of the dispersion medium for the stabilizing moieties. This can be achieved in at least three different ways: by changing the temperature, by increasing the pressure, and by adding a miscible non-solvent for the stabilizing moieties to the dispersion medium. The transformation from long-term stability to catastrophic flocculation for sterically stabilized dispersions occurs abruptly at the critical flocculation point (CFPT). This may be a critical flocculation temperature (CFT), a critical flocculation pressure (CFP) of a critical flocculation volume (CFV) (for a non-solvent that induces flocculation).

An example of the sharp flocculation point that occurs on heating poly-(oxyethylene) stabilized latices in 0.48 M $MgSO_4$ is shown in Fig. 8.4.2. Below 290 K, the dispersion displayed long-term (indeed true thermodynamic) stability; above 290 K, catastrophic flocculation was apparent. The lowest temperature at which instability was observed

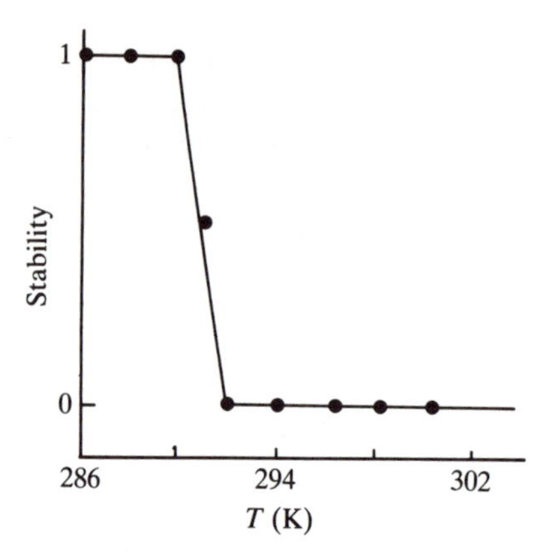

FIG. 8.4.2. The effect of temperature on the stability of latices stabilized by poly(oxyethylene) in 0.48 M $MgSO_4$. The stability scale is arbitrary and would be reflected by a measurement of e.g. optical density. (Modified from Napper 1970.)

corresponds to the critical flocculation temperature. Note that as flocculation is induced, in this instance by heating, the CFT corresponds to an upper critical flocculation temperature (UCFT). When flocculation is induced by cooling, it is said to occur at a lower critical flocculation temperature (LCFT). In principle, all sterically stabilized dispersions display both a UCFT and a LCFT, although both are not always experimentally accessible (see Fig. 8.5.1 below).

The sharp transition from stability to catastrophic flocculation observed with sterically stabilized dispersions on changing the temperature contrasts with the relatively sluggish temperature response of electrostatically stabilized dispersions. The latter, as Faraday was one of the first to note, usually become less stable on heating.

8.4.4 *Identification of the critical flocculation point*

Evidence for the existence of sharp critical flocculation points for sterically stabilized dispersions was presented in the preceding section. The question then arises as to whether such critical flocculation points can be identified in any general unified fashion. Fortunately, this identification can be accomplished in many systems, irrespective of whether flocculation is induced by heating or by cooling, by an increase in pressure, or by the addition of a miscible non-solvent for the stabilizing moieties. For many sterically stabilized dispersions, the practical thermodynamic limit to their stability is the theta (or Flory)-point for the stabilizing moieties, free in a solution of the dispersion medium. As a result, there is a strong correlation between the critical flocculation point and the theta-point (θ-point) of the stabilizing moieties. Some results supporting this correlation are presented in Table 8.4.

These all relate to critical flocculation temperatures. It is stressed that similar correlations have been reported for both critical flocculation volumes and pressures and their corresponding θ-points.

8.4.5 *Theta-point of polymer molecules*

The θ-point of polymer molecules is of special importance in polymer solution thermodynamics. It is closely analogous to the Boyle point for real gases (Flory 1953). An ideal gas, which notionally consists of non-interacting point molecules, would obey the equation of state

$$P\bar{V} = RT \tag{8.4.1}$$

where the bar denotes the molar quantity. Real gas molecules deviate from ideal behaviour in at least two respects: first, they occupy a finite volume in space; second, van der Waals attractive forces are operative between the molecules. The van der Waals equation of state

$$(P + a/\bar{V}^2)(\bar{V} - b) = RT \tag{8.4.2}$$

endeavours to account for these deviations from ideality, by introducing correction terms which have opposite signs. The pressure correction term arises from the van der Waals attraction between the gas molecules and may be considered to be enthalpic in origin. The volume correction term arises from the fact that real gas molecules occupy a finite volume which influences their configurational entropy (Exercise 8.4.3). At lower temperatures and pressures, the van der Waals attractive term is dominant and negative deviations from ideality are observed (see Fig. 8.4.3). At higher temperatures, the excluded volume term manifests itself and positive deviations from ideality are observed. Since the non-ideality correction terms are opposite in sign, it follows that a temperature can be found at which the non-ideality terms cancel. This is termed the *Boyle temperature.* The real gas molecules then obey the ideal equation of state over a wide, but limited, range of pressures (up to, say, a few hundred atmospheres; see Fig. 8.4.3).

The analogue of the ideal gas equation for a polymer solution is the van't Hoff osmotic pressure equation (see Appendix A6)

$$\Pi/c_2 = RT/\langle M_n\rangle \tag{8.4.3}$$

where Π = osmotic pressure at polymer concentration c_2 and $\langle M_n\rangle$ = number average molecular weight. This equation would be obeyed by an ideal solution of small molecules. It is scarcely surprising that, except in the limit of infinite dilution, large polymer molecules do not obey this

Table 8.4
Comparison of critical flocculation temperatures with theta-temperatures for sterically stabilized dispersions

Stabilizing moieties	Molecular weight	Dispersion medium	CFT(K)	*U*(L)†	θ(K)	References
Poly(oxyethylene)	10,000	0.39 M $MgSO_4$	318 ± 2	UCFT	315 ± 3	Napper (1970)
	96,000		316 ± 2			
	1,000,000		317 ± 2			
Poly(acrylic acid)	9,800	0.2 M HCl	287 ± 2	LCFT	287 ± 5	Evans *et al.* (1972)
	51,900		283 ± 2			
	89,700		281 ± 1			
Poly(*iso*-butylene)	23,000	2-methylbutane	325 ± 1	UCFT	325 ± 2	Evans and Napper (1975)
	150,000		325 ± 1			Croucher and Hair (1981)
	760,000		327 ± 2			
Poly(dimethylsiloxane)	3,200	*n*-heptane/ethanol	340 ± 2	LCFT	340 ± 2	Dawkins and
	11,200	(51 : 49 v/v)	340 ± 2			Taylor (1980)
	23,000		341 ± 2			
	48,000		338 ± 2			

† Nature of the critical flocculation temperature whether upper (U) or lower (L).

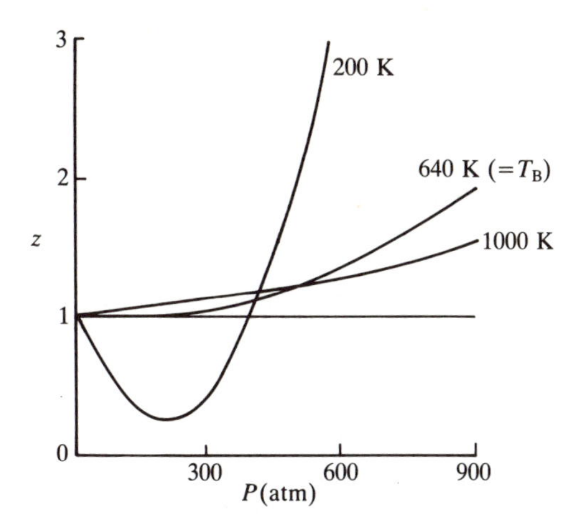

FIG. 8.4.3. The experimentally determined pressure dependence of the compressibility factor ($z = P\bar{V}/RT$) of methane at temperatures above and below the Boyle temperature, T_B. (Note that all temperatures are above the critical temperature of the gas.)

equation. In good solvents (i.e. solvents where polymer–solvent contacts are energetically favoured over polymer–polymer and solvent–solvent contacts), the polymer chains repel one another. Since there is a finite excluded volume, the effective concentration of the chains is greater than that nominally expected with zero excluded volume and so positive deviations from ideality result (see Fig. 8.4.4).

In contrast, a poor solvent is such that segment–segment contacts are energetically favoured, leading to association of the polymer molecules. Since osmotic pressure is a colligative property, intermolecular association leads to fewer osmotically active species in solution and this leads to

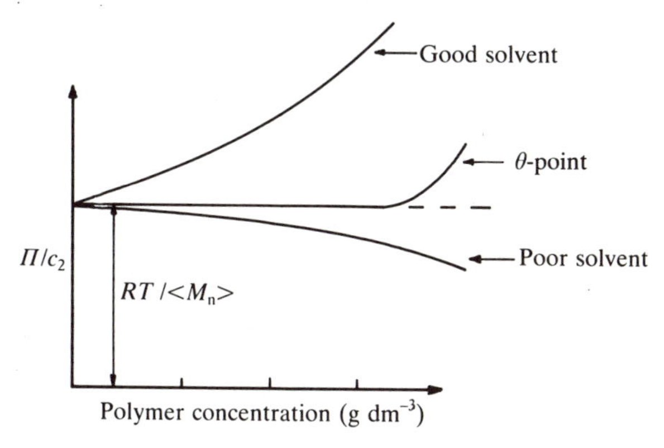

FIG. 8.4.4. The reduced osmotic pressure of polymer solutions as a function of the concentration in good, poor, and θ-solvents.

negative deviations from ideality. As with gas molecules, polymer chains can show either positive or negative deviations from ideal behaviour. It follows that a temperature can be found at which the solvency is such as to lead to ideal osmotic pressure behaviour, at least up to a polymer concentration of a few per cent. This is the θ-temperature. As Flory (1953) has pointed out, under θ-solvency conditions, polymer molecules of, say, a million molecular weight behave as if they were ideal small molecules (see Fig. 8.4.4). At this temperature, the effects of attractive and repulsive interactions balance one another out. This implies that the polymer chains can telescope one another without any change in Gibbs free energy. Thus the θ-point represents the transitional point with respect to segment–solvent interactions: at this point, the polymer segments change from exhibiting a net mutual repulsion to a net mutual attraction. A fuller description of that change would involve a consideration of segment–segment, segment–solvent and solvent–solvent interactions.

8.4.6 *Implications of the θ-point for steric stabilization*

It was noted earlier that in order to impart steric stabilization, the stabilizing moieties must, in effect, be repulsive. According to the previous discussion, this occurs in dispersion media for better solvency than θ-solvents. As the solvency of the dispersion medium is reduced, however, so the steric repulsion would decrease. Ultimately, in worse than θ-solvents, the 'stabilizing' moieties actually become mutually attractive. It is this mutual attraction of the steric layers in slightly worse than θ-solvents that induces the flocculation of many sterically stabilized dispersions. Only with thin steric layers and/or large particles is the van der Waals attraction sufficiently strong to cause coagulation. The unimportance of the van der Waals attraction between the core particles in inducing aggregation of many sterically stabilized dispersions contrasts with its dominant role in causing the coagulation of electrostatically stabilized dispersions.

The discussion set forth above as to the nature of the θ-point implies that it depends solely upon a property of the polymer, being determined by the chemical nature of the polymer segments and solvent in question. The θ-point should thus be independent of the molecular weight of the polymer, an inference that has been borne out by experiment (Flory 1953). The critical flocculation point is likewise predicted to be independent of the molecular weight of the stabilizing moieties. Data relating to this prediction have already been presented in Table 8.4. Perhaps the most convincing results in this respect are those for poly(dimethylsiloxane), because the stabilizing moieties in this instance were monodisperse in molecular weight. The other results in the table, however, which pertain to polydisperse stabilizing moieties, provide additional confirmatory evidence that the critical flocculation point is indeed independent of the molecular weight of the stabilizing moieties. This does not, of course, mean that the magnitude of the steric repulsion between particles is independent of molecular weight; rather, it implies that the thermodynamic factors that control the onset of instability are insensitive to the polymer molecular weight.

Finally, we note that one reason why homopolymers are usually poor steric stabilizers is that they place conflicting demands on the dispersion medium: the dispersion medium must be a good solvent for the homopolymer in order to provide strong steric repulsion yet it must concomitantly be a poor solvent to ensure the strong adsorption of the homopolymer onto the surface of the colloidal particles. These conflicting demands are difficult to satisfy concurrently. Of course, if the homopolymer were attached terminally to the surface by covalent bonding, these conflicting demands would no longer arise.

Exercises. 8.4.1 Use eqn (8.4.2) to show that the Boyle temperature T_B is given to good approximation by $T_B = a/bR$.

8.4.2 Discuss the factors that would influence your choice of colloid stabilizer in dispersing

(a) an oil slick in sea water;
(b) particulates in a dry cleaning system;
(c) droplets of grease in a dish washing system;
(d) particulate carbon in an automotive oil.

8.4.3 Show that for a gas obeying the van der Waals equation (8.4.2):

$$\left(\frac{\partial S}{\partial V}\right)_T = \frac{1}{T}\left\{\left(\frac{\partial U}{\partial V}\right)_T + p\right\} = R/(\bar{V} - b).$$

Hence compare the entropy change involved in an isothermal expansion with that occurring for an ideal gas. [This is the change in configurational entropy.]

8.5 Thermodynamic basis of steric stabilization

The temperature dependence of the stability of sterically stabilized dispersions provides a clue to the thermodynamic factors that control stability near to the CFT. In principle, every sterically stabilized dispersion should undergo flocculation both on heating and on cooling: the reasons for this will be presented below. In practice, however, one or even both of these CFTs is often not experimentally accessible.

As two sterically stabilized colloidal particles undergo a Brownian collision, there is a change in Gibbs free energy ΔG_F of the pair of particles. (Note that the subscript F denotes flocculation, so that ΔG_F is the free energy of flocculation of a pair of particles, although any change in configurational entropy is ignored in this approach.) This free energy

Table 8.5
Possible combinations of signs for the overall thermodynamic components of steric stabilization

ΔG_F	ΔH_F	ΔS_F	$\|\Delta H_F\|/\|T\Delta S_F\|$	Stabilization type	Flocculation
$\geqslant 0$	+	+	$\geqslant 1$	enthalpic	heating to UCFT
$\geqslant 0$	−	−	$\leqslant 1$	entropic	cooling to LCFT
$\geqslant 0$	+	−	$\lesseqgtr 1$	combined enthalpic–entropic	not accessible

change can be resolved into its respective enthalpy (ΔH_F) and entropy (ΔS_F) components such that

$$\Delta G_F = \Delta H_F - T\Delta S_F. \tag{8.5.1}$$

In order for the polymer-coated particles to be stable, ΔG_F must be positive (or at least non-negative). The various possible combinations of signs and magnitudes for ΔH_F and ΔS_F that give rise to steric stabilization are summarized in Table 8.5.

Three different types of steric stabilization can be usefully identified according to this classification: enthalpic, entropic, and combined enthalpic–entropic stabilization. In enthalpic stabilization, the enthalpy change on close approach of the particles promotes stabilization, whereas the corresponding entropy change promotes flocculation. Their relative contributions to the overall free energy change are such that the enthalpy contribution predominates. In entropic stabilization, the roles of the enthalpy and entropy terms are reversed: the entropy change promotes stability whereas it is disfavoured by the enthalpy change. In this case, however, their relative contributions to the overall free energy change are such that the entropy term prevails. Combined enthalpic–entropic stabilization is different from the two previously mentioned categories in that both the enthalpy and the entropy changes contribute to stability. Enthalpic, entropic, and combined enthalpic–entropic stabilization are differentiated by their different responses to temperature changes. At the CFT, ΔG_F must change sign from positive to negative as the dispersion passes from a stability regime to one exhibiting incipient flocculation. Since $\partial(\Delta G_F)/\partial T = -\Delta S_F$, this temperature dependence of stability allows the sign of ΔS_F near to the CFT to be assigned unambiguously.

For a dispersion that flocculates on heating, $\partial(\Delta G_F)/\partial T$ must be negative, which implies that ΔS_F must be positive close to the UCFT.

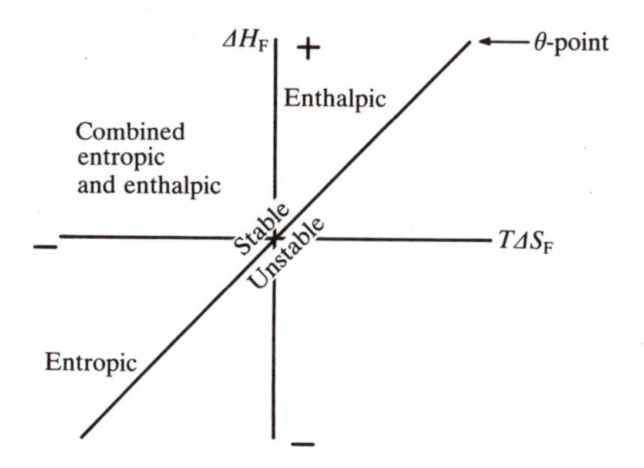

FIG. 8.5.1. Schematic representation of the different types of steric stabilization on an enthalpy–entropy diagram.

This corresponds to enthalpic stabilization (see Table 8.5). Conversely, ΔS_F for a sterically stabilized dispersion that flocculates on cooling must be negative, which corresponds to entropic stabilization just above the LCFT. A dispersion that is stabilized by combined enthalpic–entropic stabilization cannot in principle be flocculated at any accessible temperature. Variations in temperature may, however, convert it to an enthalpically or entropically stabilized dispersion whereupon it may become susceptible to flocculation by temperature changes.

The three different types of steric stabilization can be represented schematically by an enthalpy–entropy diagram of the type shown in Fig. 8.5.1. In this diagram, ΔH_F is plotted against $T\Delta S_F$ so that the line of unit positive slope corresponds to $\Delta G_F = 0$. It thus represents the locus of all θ-points for the stabilizing moieties. Dispersions located to the left of this line are stable; those located to the right are unstable. Stable dispersions located in the upper octant are enthalpically stabilized whereas those located in the lower octant are entropically stabilized. Those located in the intermediate quadrant exhibit combined enthalpic–entropic stabilization. Note that Fig. 8.5.1 shows clearly that it is not possible for a dispersion to be moved from a domain of enthalpic stabilization to a region of entropic stabilization by a change in temperature without passing through a region of combined enthalpic–entropic stabilization. This implies that any dispersion that flocculates both on heating and on cooling necessarily displays combined enthalpic–entropic stabilization over some intermediate temperature range. All sterically stabilized dispersions in principle exhibit this behaviour.

Table 8.6 presents the classification of some typical aqueous and non-aqueous dispersions stabilized by a range of stabilizing moieties at room temperature and pressure (and also near to their CFT). Ottewill

Table 8.6
Classification of some sterically stabilized dispersions at room temperature and pressure

Stabilizing moieties	Dispersion medium			Type of stabilization
	Type	Example	Flocculation	
Poly(oxyethylene)	aqueous	0.39 M $MgSO_4$	heating to UCFT	enthalpic
Poly(vinyl alcohol)	aqueous	2 M NaCl	heating to UCFT	enthalpic
Poly(acrylic acid)	aqueous	0.2 M HCl	cooling to LCFT	entropic
Poly(acrylamide)	aqueous	2.1 M $(NH_4)_2SO_4$	cooling to LCFT	entropic
Polystyrene	non-aqueous	cyclopentane	cooling to LCFT	entropic
Poly(*iso*-butylene)	non-aqueous	2-methylbutane	heating to UCFT	enthalpic

(1973) was the first to point out that at room temperature and pressure entropic stabilization is more common than enthalpic stabilization in non-aqueous dispersion media.

Indeed, the high incidence of entropic stabilization under these conditions has sometimes led to the mistaken belief that all sterically stabilized dispersions are entropically stabilized. The mere observation of flocculation on heating disproves this misapprehension. Enthalpic stabilization is probably more common than entropic stabilization in aqueous dispersions. It should be stressed, however, that both enthalpic and entropic stabilization have been observed in both aqueous and non-aqueous dispersion media. Indeed, most, if not all, dispersions should in principle display both enthalpic and entropic stabilization under different conditions of temperature and pressure in the same dispersion medium. By way of specific example, we cite polyacrylonitrile latex particles stabilized by polystyrene stabilizing moieties and dispersed in cyclopentane; these particles flocculate both on heating to 410 K and on cooling to 280 K (Croucher and Hair 1980). These observations show that the dispersion is enthalpically stabilized at higher temperatures and entropically stabilized at lower temperatures. An intermediate domain of combined enthalpic–entropic stabilization must separate these upper and lower temperature regions.

8.6 Thermodynamics of polymer solutions

The preceding discussion of the thermodynamic basis of steric stabilization is independent of any models that might be devised to explain the signs of the changes involved. This is both a strength and a weakness: a strength in that the thermodynamic discussion has very general validity but a weakness in that such a macroscopic treatment provides no insight into the microscopic processes that generate steric repulsion. This shortcoming can be remedied by resort to the methods of statistical mechanics, although these will not be elaborated in full detail here.

Any discussion of the microscopic origins of steric repulsion must of necessity be couched in the statistical mechanical theories describing the thermodynamics of polymer solutions. The current status of these theories will now be reviewed.

8.6.1 *Flory–Huggins theory*

The most celebrated theory of polymer solution thermodynamics is that developed independently by Flory and Huggins in the early 1940s (Flory 1941; Huggins 1941). This theory endeavours to calculate the free energy of mixing of pure amorphous polymer molecules with pure solvent. This is accomplished by the separate calculation of the entropy of mixing (which is a combinatorial term) and the enthalpy of mixing (which is a contact dissimilarity term). These two terms are then combined in the usual manner:

$$\Delta G^{\mathrm{M}} = \Delta H^{\mathrm{M}} - T\Delta S^{\mathrm{M}} \tag{8.6.1}$$

where the superscript M denotes mixing.

The combinatorial entropy change was originally calculated by Flory using a lattice approach. It can, however, be more simply derived by a crude free volume approach. This approach, which will be followed here, renders the assumptions inherent in the Flory–Huggins theory more transparently obvious. Further, it allows the scene to be set for a subsequent discussion of free volume theories.

The free volume of a substance can be considered crudely to represent that fraction of the total (external) volume not occupied by the geometrical volumes of the constituent molecules. It is therefore space that is accessible to the centres of mass of the molecules of the substance. In what follows, the solvent and the polymer will be denoted by the subscripts 1 and 2 respectively. Moreover, it will be assumed, somewhat unphysically, that the amorphous polymer and the solvent possess the same free volume fraction, f_{V}, and further, that this free volume is conserved on mixing (i.e., no overall volume change occurs on mixing). (Note that f_{V} is a (dimensionless) relative volume; the actual volume of empty space associated with a polymer molecule is, of course, much larger than that associated with a solvent molecule.)

The free volume accessible to n_1 solvent molecules and n_2 polymer molecules before mixing is simply $n_1 V_1 f_{\mathrm{V}}$ and $n_2 V_2 f_{\mathrm{V}}$, respectively. Here V_i $(i = 1, 2)$ is the measured external volume per molecule ($V_i = \bar{V}_i/\mathcal{N}_{\mathrm{A}}$, where $\bar{V}_i$ = partial molar volume and $\mathcal{N}_{\mathrm{A}}$ = Avogadro's constant). After mixing, the total accessible free volume becomes $(n_1 V_1 + n_2 V_2) f_{\mathrm{V}}$.

By analogy with gaseous systems, the thermodynamic probability W in the Boltzmann equation $S = k \ln W$ may be taken to be proportional to the free volume accessible to the centres of mass of the species constituting the system. This implies that the entropy change of the solvent on mixing with the polymer is simply $n_1 k \ln\{(n_1 V_1 + n_2 V_2)/n_1 V_1\} = -n_1 k \ln v_1$, where v_1 = volume fraction of the solvent in the polymer solution. By analogy, the entropy change of the amorphous polymer molecules on mixing is simply $-n_2 k \ln v_2$, where v_2 = volume fraction of polymer. It follows that the combinatorial entropy of mixing polymer and solvent is

$$\Delta S^{M} = -k\{n_1 \ln v_1 + n_2 \ln v_2\} \tag{8.6.2}$$

which is positive since $v_1, v_2 < 1$.

The enthalpy of mixing is evaluated in the Flory–Huggins theory in the spirit of the classical van Laar approach. Mixing is considered to be a quasi-chemical reaction between the dissimilar solvent contacts and segment contacts:

$$1-1+2-2 \rightarrow 2(1-2). \tag{8.6.3}$$

For these contacts, an interaction parameter χ_1 is defined such that $\chi_1 kT$ is the difference in energy of a solvent molecule (hence the subscript 1) immersed in pure polymer compared with that in pure solvent. For n_1 solvent molecules, each immersed in pure polymer, the energy change would be $n_1 \chi_1 kT$. In a polymer solution, the probability of a solvent molecule being in contact with a polymer segment is simply v_2. It follows that the contact dissimilarity energy change, which in condensed phases can be equated to the corresponding enthalpy change, is given by

$$\Delta H^{M} = n_1 v_2 \chi_1 kT. \tag{8.6.4}$$

The total free energy of mixing according to the Flory–Huggins theory is obtained by combining eqns (8.6.1), (8.6.2) and (8.6.4), whereupon

$$\Delta G^{M} = kT\{n_1 \ln v_1 + n_2 \ln v_2 + n_1 v_2 \chi_1\}. \tag{8.6.5}$$

Note that as χ_1 is often positive for non-aqueous solvents (i.e. mixing is endothermic), the contact dissimilarity term often opposes mixing. In contrast, both of the combinatorial entropy terms are negative and promote mixing. Since the polymer is of high molecular weight, n_2 is comparatively small and the dominant term promoting mixing is $n_1 k \ln v_1$. The primary reason why polymer and solvent molecules mix to form polymer solutions is now apparent: the entropy of the solvent molecules is increased as a result of the additional space available when the domains of the polymer molecules becomes accessible to the solvent. The detailed structure of the polymer is irrelevant according to the precepts of the Flory–Huggins theory, rods being just as effective in providing space for the solvent molecules as polymer coils (Flory 1970). This point is stressed because it is sometimes erroneously asserted that polymer molecules dissolve primarily because of an increase in their configurational entropy.

Note that although $\chi_1 kT$ was originally introduced as a change in internal energy, the arguments presented above would be essentially unchanged if $\chi_1 kT$ were a free energy change. In this way, χ_1 as determined experimentally may incorporate both enthalpy and non-combinatorial entropy contributions.

Finally, we point out that eqn (8.6.5) is closely analogous to the Bragg–Williams equation derived for the mixing of small molecules:

$$\Delta G^{\mathrm{M}} = kT\{n_1 \ln X_1 + n_2 \ln X_2 + n_1 v_2 \chi_1\} \qquad (8.6.6)$$

where X_i = mole fraction of species i. The primary difference between the Flory–Huggins equation and that of Bragg–Williams resides in the use of volume fraction statistics for polymer solutions as against mole fraction statistics for small molecules. The latter is inappropriate for polymer molecules because the high molecular weight (molar mass) of the polymer results in the mole fraction of the solvent in a polymer solution always being close to unity, except at extremely high polymer volume fractions.

8.6.2 *Free volume theories*

Since its publication, the Flory–Huggins theory has been widely, and in many respects successfully, used to account for the behaviour of polymer solutions. Indeed the theory has had such a profound impact on polymer science that it has been referred to as a ‘paradigm of polymer science’ (Derham *et al.* 1974). This is not to say, however, that it is without serious shortcomings. Whilst it embodies concepts that remain central to the theory of polymer solution thermodynamics, it nevertheless ignores certain features that are vital to the explanation of the entire range of observed phenomena.

There are three experimental observations that point to serious inadequacies in the Flory–Huggins theory, all of which are crucial to the understanding of the thermodynamic basis of steric stabilization.

First, experimental studies of the temperature-dependence of χ_1 allow it to be resolved into its enthalpy and entropy components. Although χ_1 was originally introduced into the theory as an enthalpy term, experiments show that for many non-aqueous polymer–solvent systems, the positive values for χ_1 are determined primarily by entropic considerations. The corresponding enthalpy terms are relatively small and of variable sign. The experimental results imply that there is a non-combinatorial entropy change that opposes the mixing of polymer and solvent and is not accounted for by the Flory–Huggins theory.

Second, χ_1 is found experimentally to depend upon the polymer concentration. Usually χ_1, whose value lies in the range 0.1–0.5, becomes more positive as the polymer concentration increases (e.g. poly(*iso*-butylene) in benzene) but some exceptions to this generalization are known (e.g. polystyrene in toluene).

Third, the Flory–Huggins theory predicts that as mixing of polymer and solvent is an entropically driven process, mixing should be favoured as the temperature is increased. Yet experimentally it is found that most if not all polymer solutions can be induced to undergo phase separation as the temperature is raised to near the critical point of the solvent. In this regard, polymer solutions differ significantly from solutions of small molecules.

The observation of phase separation on heating polymer solutions provides a clue to one major inadequacy of the Flory–Huggins theory. This is its failure to account properly for the difference in free volume between the solvent and the polymer. This difference is dramatically magnified near the critical point of the solvent when the solvent molecules become relatively gas-like, with a large free volume. The polymer molecules, in contrast, undergo only a comparatively small increase in free volume on heating because the segments are constrained by their connectivity. Crudely speaking, the polymer segments, when placed in contact with the gas-like solvent molecules cause the solvent to undergo 'condensation' (Patterson 1969). This results in a significant decrease in the entropy of the solvent molecules on contact with the polymer, one that can outweigh the combinatorial entropy of mixing. It is stressed that this difference in free volume can be shown to persist down to room temperature.

The foregoing concepts provide a plausible explanation for most of the observed phenomena that lie outside the scope of the Flory–Huggins theory. First, the entropic contribution to χ_1 that opposes mixing at room temperature is clearly associated with the difference in free volume between the solvent and the polymer; mixing (with the help of a little imagination) is akin to the 'condensation' of a gas (the solvent) in a dense medium (the polymer). Second, the increase in χ_1 on increasing the volume fraction of polymer merely reflects the greater average 'condensational' ordering of the solvent molecules on increasing the segment concentration. Finally, the observed phase separation on heating polymer solutions to near the critical temperature of the solvent is also a consequence of the decrease in entropy on mixing the solvent and polymer under these conditions.

Note that aqueous polymer solutions often show phase separation on heating but at temperatures well below the critical temperature of water. The explanation for this phenomenon must be sought elsewhere. It appears to be associated with the directionality of the H-bonds formed between the water molecules and the polymer but this is far from certain.

Flory (1970) has developed an elaborate equation-of-state theory for non-aqueous polymer solutions that incorporates the change in free volume on mixing, as well as the combinatorial and contact dissimilarity contributions. This provides, for example, a semi-quantitative description of the concentration dependence of χ_1 with but one adjustable parameter. The innate complexity of this theory precludes a detailed discussion of it here (see Napper 1983). However, the concepts that underpin the theory can be readily adapted to provide a qualitative discussion of the microscopic processes that generate steric stabilization in both non-aqueous and aqueous dispersions.

Exercises. 8.6.1 Calculate the relative contributions of each of the terms in eqn (8.6.5) for the mixing of poly(methyl methacrylate) (5.0 g) of molecular weight 500 000 with toluene (200 g) at 20 °C, assuming that the interaction parameter is 0.45 under these conditions. Take the densities of the polymer and solvent as 1.190 and 0.866 g cm^{-3}, respectively, and assume ideal mixing. Specify which terms favour and which terms disfavour mixing.

8.7 The microscopic origins of steric stabilization

8.7.1 *Non-aqueous dispersions*

The signs of the Gibbs free energy changes (and their component enthalpy and entropy contributions) associated with the combinatorial, contact dissimilarity, and free volume dissimilarity terms for non-aqueous sterically stabilized dispersions are summarized in Table 8.7. It is immediately apparent that the microscopic origins of steric stabilization are quite complex. Nonetheless, it is clear that it is the combinatorial entropy that is responsible for the stability of non-aqueous dispersions at normal temperature, while both the free volume dissimilarity and (often) the contact dissimilarity promote flocculation. The origins of entropic stabilization are represented schematically in Fig. 8.7.1. As the two colloidal particles undergo a Brownian encounter, the polymer sheaths interpenetrate. This leads to an increase in the segment concentration in the interaction zone and to the exclusion of solvent molecules into the bulk dispersion medium. Clearly, there is a decrease in the entropy of mixing of the polymer segments with solvent molecules on close approach of the particles. This loss of entropy generates the observed entropic stabilization. Note that the increase in entropy associated with the

Table 8.7
Microscopic contributions to the steric interactional free energy for non-aqueous dispersions

	Combinatorial		Contact dissimilarity		Free volume dissimilarity	
	ΔH	ΔS	ΔH	ΔS	ΔH	ΔS
	0	−ve	−ve†	~0	+ve (small)	+ve
ΔG_F	+ve		−ve†		−ve	
Promotes	stabilization		flocculation		flocculation	
$\partial(\Delta G)/\partial T$	+ve		+ve (small)		−ve	

† May be of opposite sign for some non-aqueous systems.

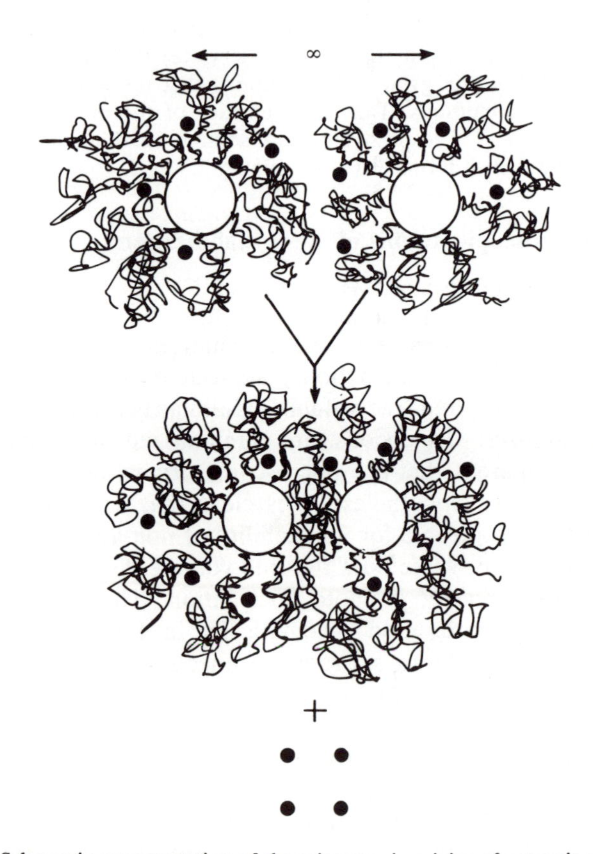

FIG. 8.7.1. Schematic representation of the microscopic origins of entropic stabilization.

difference in free volumes promotes flocculation. The net entropy change is nonetheless negative and opposes flocculation.

On heating non-aqueous dispersions, the free volume dissimilarity contribution changes faster than the combinatorial term, resulting in flocculation at higher temperatures near to the critical temperature of the dispersion medium. Just below the UCFT, the enthalpy change associated with the free volume dissimilarity term gives rise to enthalpic stabilization. On the other hand, the contact dissimilarity term becomes relatively more important on cooling. Both the free volume and the contact dissimilarity contributions may lead to the occurrence of flocculation on cooling.

8.7.2 *Aqueous dispersions*

It should first be acknowledged that at present there is no satisfactory equation-of-state theory for the thermodynamics of aqueous polymer solutions. Indeed, there is no satisfactory theory for aqueous polymer solutions. The principal reason for this is the strong interactions (e.g. hydrogen bonding) that can occur between the polar water molecules and many water-soluble polymers. As a result, some of the discussion to follow is necessarily speculative.

The signs of the various contributions to the overall free energy of interpenetration for some aqueous sterically stabilized dispersions are presented in Table 8.8. Comparison with the entries in Table 8.7 for non-aqueous dispersion media shows that the combinatorial term (which promotes stabilization) and the free volume dissimilarity term (which favours flocculation) emerge relatively unscathed when the dispersion medium is changed from being non-aqueous to being aqueous. The contact dissimilarity term, however, is radically transformed. Instead of promoting flocculation, it now favours stabilization. This is a consequence of the specific interactions (such as hydrogen-bonding) that water

Table 8.8
Microscopic contributions to the steric interactional free energy for some aqueous dispersions

	Combinatorial		Contact dissimilarity		Free volume dissimilarity	
	ΔH	ΔS	ΔH	ΔS	ΔH	ΔS
	0	−ve	+ve	+ve	+ve	+ve
ΔG_F	+ve		+ve		−ve	
Promotes	stabilization		stabilization		flocculation	
$\partial(\Delta G)/\partial T$	+ve		−ve		−ve	

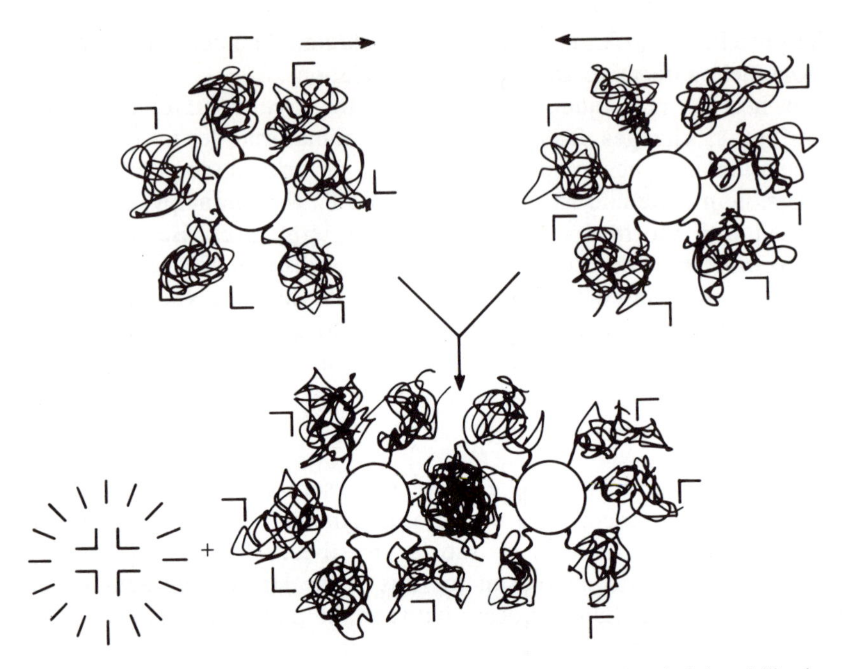

FIG. 8.7.2. Schematic representation of the microscopic origins of enthalpic stabilization.

molecules can undergo with suitably polar stabilizing moieties (e.g. the ether oxygens in poly(oxyethylene) or the hydroxyl groups in poly(vinyl alcohol)). Such specific interactions imply that when interpenetration of the steric sheaths occur, bound water molecules are released from the interaction zone into the bulk dispersion medium. The energy required to populate the new degrees of freedom (principally translational and rotational) in the bulk dispersion medium generates a contact dissimilarity enthalpy that opposes flocculation (see Fig. 8.7.2). The associated increase in entropy promotes flocculation but is outweighed by the enthalpy term. As a result, the overall contact dissimilarity free energy favours stabilization.

At temperatures just below the UCFT, the enthalpy change associated with the contact dissimilarity gives rise to enthalpic stabilization, the net entropy change being positive under these conditions and promoting flocculation. On heating, the free volume contribution to the free energy becomes more negative and the contact term less positive, resulting in flocculation.

Note that the foregoing discussion of aqueous dispersions is only applicable to certain common stabilizing moieties, such as poly(oxyethylene) and poly(vinyl alcohol). Aqueous dispersions stabilized by poly(acrylamide) or poly(acrylic acid) seem more likely to follow the entries set forth for non-aqueous dispersions in Table 8.7.

8.8 Quantitative theories of steric stabilization

It should be recognized immediately that no *ab initio* theory of steric stabilization yet exists that is the counterpart to the DLVO theory of electrostatic stabilization. The reasons for this are twofold. First, any quantitative theory for steric stabilization must be couched in terms of one or other of the theories of polymer solution thermodynamics, the most refined of which could at best be described as semi-quantitative. Second, in order to calculate the distance dependence of the steric interaction, it is necessary to be able to predict the conformation of close-packed polymer molecules that are sandwiched between two colloidal particles. In a good solvent, this is a many-bodied problem that has so far defied exact solution. Indeed, it is not yet possible to predict quantitatively the conformation of a polymer molecule free in solution, let alone one attached to a planar interface, except in certain limiting cases, such as that where the bathing liquid is a θ-solvent (Flory 1969). Despite these difficulties, some progress has been made in developing pragmatic models capable of predicting the distance dependence of the steric repulsion. These are more plausible for high molecular weight polymers with simple mechanisms of attachment to the particle surface.

8.8.1 *The three domains of close approach*

It is useful to distinguish three domains of close approach for sterically stabilized particles. For simplicity, the particles will be considered to be parallel flat plates, the impenetrable surfaces of which are separated by a distance d and coated by steric layers of thickness L (see Fig. 8.8.1). The three domains are specified by the relative magnitudes of d and L.

(i) $d > 2L$. If the separation between the plates exceeds twice the barrier layer thickness, the stabilizing moieties are unable to interact. Closer approach of the plates in this domain involves no free energy change. This is the *non-interpenetrational domain*.

(ii) $L \leqslant d \leqslant 2L$. Once the distance between the plates is less than twice the steric layer thickness, the stabilizing moieties attached to the opposing surfaces undergo interpenetration. This is the *interpenetrational domain*. The increase in the polymer segment density in the interpenetrational zone on closer approach forces molecules of the dispersion medium into the bulk phase, so reducing the mixing of segments and solvent. In good solvents, this demixing of segments and solvent raises

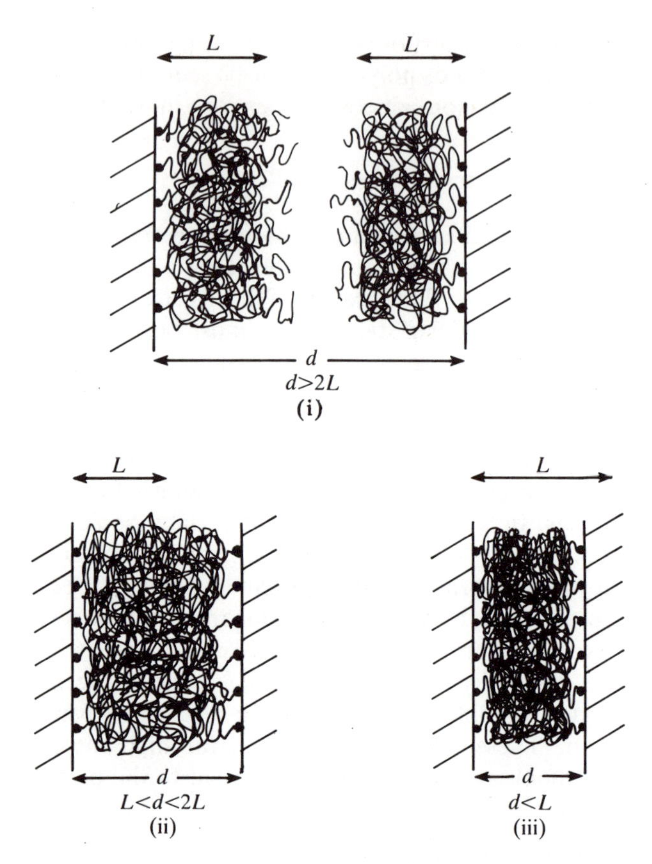

FIG. 8.8.1. The three domains of close approach of sterically stabilized flat plates: (i) Non-interpenetration. (ii) Interpenetration. (iii) Interpenetration plus compression.

the free energy of the system, leading to repulsion; in worse than θ-solvents demixing lowers the free energy, resulting in attraction.

(iii) $d < L$. When the separational distance is less than the barrier layer thickness, not only do the polymer chains undergo interpenetration but the stabilizing moieties attached to one plate are compressed by the opposing impenetrable surface. This is the *interpenetrational-plus-compressional* domain. The free energy of interaction in this domain is derived from two components: a solvent–segment mixing term together with an elastic component. The latter is associated with the compression of the stabilizing moieties by the opposing surfaces. This reduces their configurational entropy and, irrespective of the quality of the solvent, invariably results in repulsion.

Considerable experimental evidence has now been amassed to show that the flocculation properties of many sterically stabilized dispersions are determined primarily by the sign of the free energy of interaction in the interpenetrational domain. For this reason the following discussion will focus on the quantitative prediction of this quantity. Note, however, that entry into the interpenetrational-plus-compressional domain may occur if a concentrated dispersion is compressed. The magnitude of the change in free energy in this domain is also required to construct the total potential energy curve so that the extensions required to comprehend this domain will also be briefly indicated.

The free energy of mixing of amorphous polymer with pure solvent according to the Flory–Huggins theory is given by eqn (8.6.5):

$$\Delta G^{\mathrm{M}} = kT\{n_1 \ln v_1 + n_2 \ln v_2 + n_1 v_2 \chi_1\}.$$

This equation can be readily adapted to provide the free energy of mixing of segments of the stabilizing moieties with solvent in a small volume element δV in the stabilizer sheath (see Fig. 8.8.2):

$$\delta(\Delta G^{\mathrm{M}}) = kT\{\delta n_1 \ln v_1 + \delta n_1 v_2 \chi_1\} \qquad (8.8.1)$$

where δn_1 = number of solvent molecules in the volume element. Note that in eqn (8.8.1), n_2 has been set equal to zero because not one complete polymer molecule is contained within such small volume elements. Note further that $v_1 = 1 - v_2 = 1 - \rho_2 V_s$, where ρ_2 = segment density distribution function of the polymer and V_s = volume of a polymer segment. Further, $\delta n_1 = v_1 \delta V / V_1 = ((1 - \rho_2 V_s)/V_1)\delta V$, where V_1 = volume of a solvent molecule. It follows that

$$\delta(\Delta G^{\mathrm{M}}) = (kT/V_1)\{\ln(1 - \rho_2 V_s) + \rho_2 V_s \chi_1\}(1 - \rho_2 V_s)\delta V. \qquad (8.8.2)$$

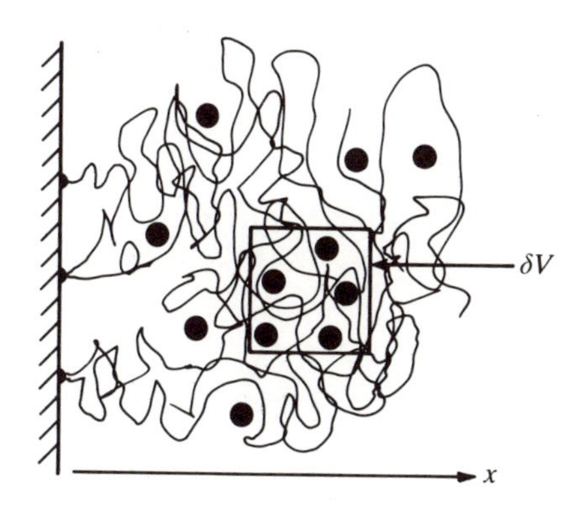

FIG. 8.8.2. The mixing of solvent and polymer segments at an interface.

On expanding the logarithmic function and ignoring terms of order higher than the second, it is found that (Exercise 8.8.1):

$$\delta(\Delta G^{\mathrm{M}}) = (kT/V_1)\{-(1-\chi_1)\rho_2 V_s + (\tfrac{1}{2}-\chi_1)\rho_2^2 V_s^2\}\delta V. \quad (8.8.3)$$

This equation is applicable to all domains of close approach.

8.8.2 *The interpenetrational domain*

The steric free energy of interaction for two parallel flat plates in the interpenetrational domain can be calculated by considering the change in free energy that occurs when two equal volume elements located in the steric sheaths are caused to overlap, having initially been at infinite separation. To facilitate this calculation, it is convenient to assume that the polymer segment density parallel to the surface is uniform. This is a reasonable assumption at high surface coverage. All the distance dependence of the segment density is then manifest in the x-direction, normal to the interfaces (see Fig. 8.8.3). The total free energy of mixing per steric layer at infinite separation is readily obtained from eqn (8.8.3) by integration over all volume elements:

$$\Delta G_\infty^{\mathrm{M}} = kT(V_s^2/V_1)(\tfrac{1}{2}-\chi_1)\int_0^\infty \rho_\infty^2\,\mathrm{d}V - kT(V_s/V_1)(1-\chi_1)\int_0^\infty \rho_\infty\,\mathrm{d}V \quad (8.8.4)$$

where the subscript ∞ denotes the value at infinite separation. For unit area of plate, $\mathrm{d}V = A\,\mathrm{d}x = \mathrm{d}x$. Further, if the stabilizing moieties are assumed to be attached terminally ('tails') to the plates and there are assumed to be ν chains per unit area, each composed of i segments, then it follows that for unit area $\int_0^\infty \rho_\infty\,\mathrm{d}V = \nu i = \nu i \int_0^\infty \hat{\rho}_\infty\,\mathrm{d}x$. Here the hat denotes the normalized distribution function (i.e., $\int_0^\infty \hat{\rho}_\infty\,\mathrm{d}x = 1$). Thus for

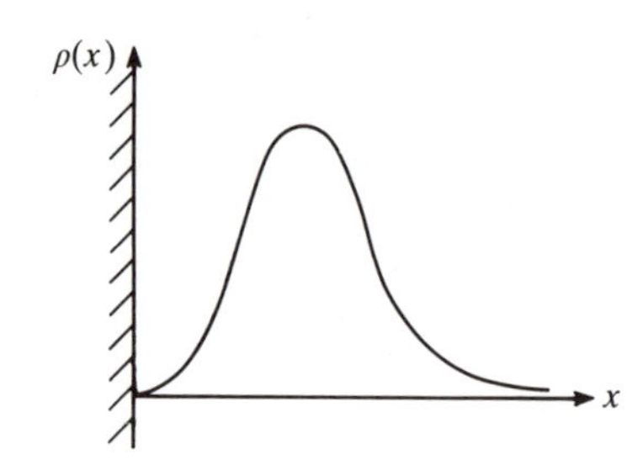

FIG. 8.8.3. The segment density distribution function for tails.

unit area of a *pair* of plates, eqn (8.8.4) may be recast as

$$\Delta G_{\infty}^{\mathrm{M}} = 2kT(V_{\mathrm{s}}^2/V_1)v^2i^2(\tfrac{1}{2} - \chi_1)\int_0^{\infty} \hat{\rho}_{\infty}^2\, \mathrm{d}x$$
$$- 2kT(V_{\mathrm{s}}/V_1)(1 - \chi_1)vi. \tag{8.8.5}$$

The free energy of interaction per unit area of the two plates at a separation d is given by

$$\Delta G_d^{\mathrm{M}} = kT(V_{\mathrm{s}}^2/V_1)v^2i^2(\tfrac{1}{2} - \chi_1)\int_0^{d} (\hat{\rho}_d + \hat{\rho}_d')^2\, \mathrm{d}x$$
$$- 2kT(V_{\mathrm{s}}/V_1)(1 - \chi_1)vi \tag{8.8.6}$$

where the prime denotes the segment density of the second steric layer such that $\hat{\rho}_d'(x) = \hat{\rho}_d(d - x)$. Note that it is assumed that the chains are irreversibly attached to the plates.

The steric free energy change on moving the two parallel flat plates from infinite separation to a separation d is thus (Exercise 8.8.2)

$$\Delta G_{\mathrm{FP}}^{\mathrm{M}} = \Delta G_d^{\mathrm{M}} - \Delta G_{\infty}^{\mathrm{M}}$$
$$= 2kT(V_{\mathrm{s}}^2/V_1)v^2i^2(\tfrac{1}{2} - \chi_1)\left[\int_0^{\infty} (\hat{\rho}_d^2 - \hat{\rho}_{\infty}^2 + \hat{\rho}_d\hat{\rho}_d')\, \mathrm{d}x\right]. \tag{8.8.7}$$

Note that in this equation (first derived by Meier 1967), the first two terms in the integral arise from *intra*molecular segmental penetration (i.e. interpenetration of segments from the *same* molecule) whereas the third term results from *inter*molecular segmental interpenetration. In the interpenetrational domain, $\hat{\rho}_d = \hat{\rho}_{\infty}$ and thus

$$\Delta G_{\mathrm{FP}}^{\mathrm{M}} = 2kT(V_{\mathrm{s}}^2/V_1)v^2i^2(\tfrac{1}{2} - \chi_1)\left[\int_0^{d} \hat{\rho}_d\hat{\rho}_d'\, \mathrm{d}x\right]. \tag{8.8.8}$$

This equation predicts that if $\chi_1 < \frac{1}{2}$ (i.e. the dispersion medium is a better solvent for the stabilizing moieties than a θ-solvent), $\Delta G_{\mathrm{FP}}^{\mathrm{M}}$ is positive and steric repulsion is operative between the plates. If $\chi_1 > \frac{1}{2}$, $\Delta G_{\mathrm{FP}}^{\mathrm{M}}$ is negative. Attraction will thus occur between the steric layers in worse than θ-solvents. In a θ-solvent, for which $\chi_1 = \frac{1}{2}$, $\Delta G_{\mathrm{FP}}^{\mathrm{M}}$ is zero in the interpenetrational domain. Broadly speaking, these predictions are in conformity with the experimental results described above.

By way of a specific yet simple example of the calculation of $\Delta G_{\mathrm{FP}}^{\mathrm{M}}$, we consider the case where the segment density is constant in the steric layer. This is a reasonable approximation for a highly branched polymer. Further, it should be realized that the distance dependence of the steric free energy is relatively insensitive to the precise details of the segment density distribution function. For a constant segment density, distribution function, $\hat{\rho}_d = \hat{\rho}'_d = 1/L$, as demanded by the normalization requirement $\int_0^L \hat{\rho}_d \,\mathrm{d}x = 1$. It follows that (Exercise 8.8.3):

$$\int_0^d \hat{\rho}_d \hat{\rho}'_d \,\mathrm{d}x = \int_{d-L}^{L} (1/L^2)\,\mathrm{d}x = 2(1/L - d/2L^2) \tag{8.8.9}$$

so that

$$\Delta G_{\mathrm{FP}}^{\mathrm{M}} = 4kT(V_{\mathrm{s}}^2/V_1)v^2 i^2(\tfrac{1}{2} - \chi_1)(1/L - d/2L^2). \tag{8.8.10}$$

The constant segment density distribution function is admittedly an oversimplified model for most steric barriers, a more realistic segment density distribution function for tails being that shown schematically in Fig. 8.8.3. This form of the distribution function, which has been confirmed by recent neutron scattering measurements, can be treated analytically as well but the results are too complex to present here (Napper 1983).

It should be pointed out that eqn (8.8.10) given above for flat plates can be readily extended to describe spherical colloidal particles by using the Deryaguin approximation (section 4.6) (Deryaguin 1934):

$$\Delta G_{\mathrm{s}}^{\mathrm{M}} = \pi a \int_{d_0}^{d_{\mathrm{m}}} \Delta G_{\mathrm{FP}}^{\mathrm{M}} \,\mathrm{d}d \tag{8.8.11}$$

where a is the radius of the spheres and the subscripts 0 and m refer respectively to the minimum and the maximum distances between the surfaces of the spheres at which interactions between the layers are possible. For the constant segment density steric layers treated previously, it can readily be shown (Exercise 8.8.4) that

$$\begin{aligned}\Delta G_{\mathrm{s}}^{\mathrm{M}} &= \pi a \int_{d_0}^{2L} 4kT(V_{\mathrm{s}}^2/V_1)v^2 i^2(\tfrac{1}{2} - \chi_1)(1/L - d/2L^2)\,\mathrm{d}d \\ &= 4\pi kTa\omega^2 \mathcal{N}_{\mathrm{A}}(\bar{v}_2^2/\bar{V}_1)(\tfrac{1}{2} - \chi_1)(1 - d_0/2L)^2 \end{aligned} \tag{8.8.12}$$

where $\omega = viV_{\mathrm{S}}/\bar{v}_2$ = mass of stabilizing moieties per unit area, $\bar{v}_2$ = partial specific volume of these moieties and $\bar{V}_1$ = molar volume of the dispersion medium. Note that eqn (8.8.12) implies that the magnitude of the steric repulsion in the interpenetrational domain is directly proportional to both the particle radius and the thermodynamic factor $(\frac{1}{2} - \chi_1)$. It also depends upon the square of both the number density of the stabilizing moieties and the geometrical factor $(1 - d_0/2L)$.

8.8.3 *The interpenetrational-plus-compressional domain*

Equation (8.8.7) derived above can be readily applied to the interpenetrational-plus-compressional domain. Unlike the interpenetrational domain, however, $\hat{\rho}_d$ no longer equals $\hat{\rho}_\infty$ and so the intramolecular segmental interpenetration that occurs on compression of the chains contributes to the total mixing interaction. In this domain, the mixing free energy change is supplemented by an elastic contribution arising from the decrease in configurational entropy of the stabilizing moieties on compression. This elastic contribution differs from the mixing term in that its sign is invariant. The elastic free energy change is always positive and so favours repulsion, irrespective of the quality of the solvency of the dispersion medium. It can be calculated from

$$\Delta G^{\mathrm{EL}} = kT \ln\{W(\infty)/W(d)\} \tag{8.8.13}$$

where $W(d)$ is the number of configurations accessible to the stabilizing moieties at a separation d. The elastic repulsive energy increases abruptly as the separation is decreased and for many purposes can be approximated by a hard sphere potential in both good and poor solvents.

8.8.4 *Potential energy diagrams*

Typical potential energy diagrams for the close approach of two sterically stabilized particles are displayed in Fig. 8.8.4. In dispersion media that are good solvents for the stabilizing moieties, the mixing free energy change is positive in both the interpenetrational and interpenetrational-plus-compressional domains. The elastic contribution is also positive in the latter domain so that the steric free energy rises monotonically as the interparticle distance is reduced (see Fig. 8.8.4(a)). Usually the van der Waals attraction between the colloidal particles is relatively small. As a result, the potential energy diagram for sterically stabilized particles does not display the characteristic maximum and primary minimum exhibited by the diagrams for electrostatically stabilized dispersions (see inset of

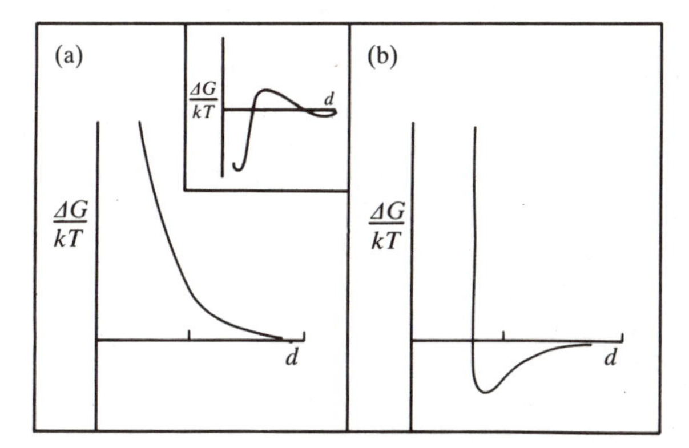

FIG. 8.8.4. Characteristic potential energy diagrams for the close approach of two sterically stabilized particles: (a) a good solvent; (b) a poor solvent. The inset shows the characteristic potential energy diagram for two electrostatically stabilized particles.

Fig. 8.8.4(a)). This means that many sterically stabilized dispersions can exhibit genuine thermodynamic stability. This contrasts sharply with the metastability (i.e. kinetic stability) displayed by electrostatically stabilized dispersions (see Chapter 7). In dispersion media that are worse solvents than θ-solvents, the potential energy diagram displays a *pseudo-secondary minimum* ('pseudo' because there is no primary minimum; see Fig. 8.8.4(b)). It is flocculation in this pseudo-secondary minimum that is responsible for the observation of aggregation at the critical flocculation point. The depth of this minimum is usually so large that the contribution to the total free energy of the system by the configurational entropy of the particles can be ignored. As a result, the particle concentration dependence of the critical flocculation point is negligible under these conditions.

Exercises. 8.8.1 Show that eqn (8.8.3) can be derived from eqn (8.8.2).

8.8.2 Use eqn (8.8.6) to derive eqn (8.8.7).

8.8.3 Derive eqns (8.8.9) and (8.8.10).

8.8.4 Use the Deryaguin integration procedure and eqn (8.8.10) to verify the validity of eqn (8.8.12).

8.8.5 Construct the potential energy vs. distance diagram in the interpenetrational domain for two polystyrene latex particles of radius 100 nm coated by poly(oxyethylene) stabilizing moieties of molecular weight 10 000 in water at 25 °C. Take the steric barrier layer thickness as 10 nm and the partial specific volume of poly(oxyethylene) in water as $0.84\,\text{cm}^3\,\text{g}^{-1}$. The interaction parameter under these conditions is $\chi_1 = 0.45$. Assume that the mass of stabilizing moieties per unit area is $5\,\text{mg}\,\text{m}^{-2}$. Construct the curve for a worse than θ-solvent with $\chi_1 = 0.52$. Also sketch the van der Waals attraction between the particles in this domain. Assume that the effective Hamaker constant for polystyrene in water is 5×10^{-21} J (ignore the effects of steric layers on the van der Waals attraction).

8.9 Dispersions displaying deviant aggregation behaviour

The behaviour described in the preceding sections whereby a correlation is observed between the critical flocculation point and the θ-point of the stabilizing moieties, is exhibited by colloidal particles sterically stabilized by relatively thick steric layers. Large colloidal particles stabilized by very thin steric layers display quite a different pattern of behaviour: coagulation occurs in dispersion media that are significantly better solvents than θ-solvents for the stabilizing moieties (Vincent and Whittington 1981). This is a consequence of the van der Waals attraction between the core particles becoming dominant. The presence of repulsive steric layers ensures that the van der Waals attraction is relatively weak so that the free energy minimum is shallow. The critical coagulation point of such dispersions then displays a weak particle concentration dependence. This arises from the configurational entropy of the colloidal particles which, although relatively small, contributes to the total free energy of the system. As a result, the more dilute dispersions appear to be more stable. Coagulation in better than θ-solvents is also observed if the stabilizing moieties are poorly anchored and/or if the surface is not fully coated. In both of these cases, however, coagulation occurs as a result of the displacement of the stabilizing moieties from the interaction zone under the stress generated in a Brownian collision. This allows the van der Waals attraction between the core particles to come into play.

Surprisingly, it is also possible to observe stability in significantly worse than θ-solvents. This is termed *enhanced steric stabilization*. It occurs if the stabilizing moieties are attached at many points along the chain (*multipoint anchoring*). One example of a system that can display enhanced steric stabilization is a dispersion of polystyrene latex particles, (whose surfaces contain un-ionized carboxylic acid groups), which are stabilized by poly(oxyethylene). Hydrogen bonding between the surface carboxylic acid groups and the oxygen atoms in the poly(oxyethylene) chains leads to multipoint anchoring and to stability in worse than θ-solvents. Precise details of the mechanistic origins of enhanced steric stabilization have yet to be unravelled (Napper 1983; Tadros 1982).

8.10 Effects of free polymer

Up to this point, we have considered solely the effects on colloid stability of polymer chains that are attached in some fashion to the colloidal particles. It may perhaps come as a surprise to find that not only can attached macromolecules influence colloid stability but so also can polymer chains that are free in solution. At the practical level, it is important to recall that free polymer is often present in biological dispersions and also for example in a paint film that is undergoing drying. The presence of free polymer can certainly induce the aggregation of colloidal particles. The possibility also arises that the free polymer chains may impart colloid stability but this is more speculative (Feigin and Napper 1980; Fleer *et al.* 1984).

8.10.1 *Depletion flocculation*

Flocculation induced by polymer that is in free solution is called depletion flocculation. The origin of the force that is responsible for this type of flocculation was first recognized by Asakura and Oosawa (1954). This force arises whenever the colloidal particles approach so closely as to exclude polymer chains from the interparticle region (see Fig. 8.10.1). Crudely, this depletion of the polymer molecules occurs if the interparticle distance is less than the 'diameter' of the polymer molecules (for

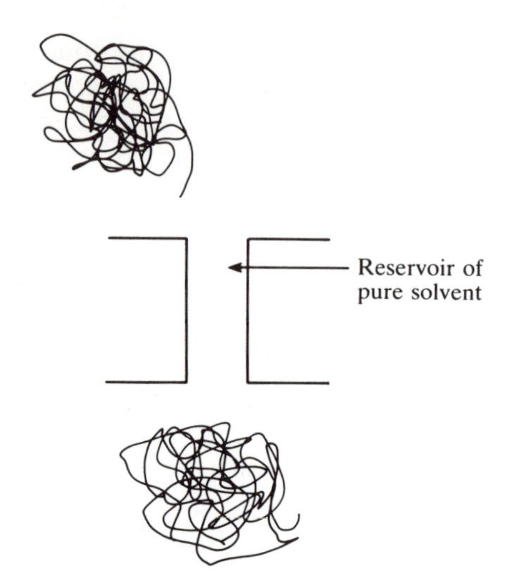

FIG. 8.10.1. Schematic representation of the origins of depletion flocculation.

linear polymer molecules the 'diameter' is roughly equal to the r.m.s. end-to-end distance of the chains). At such separations, polymer chains can only fit between the particles by undergoing a significant decrease in spatial extension. This compression is accompanied by a loss of configurational entropy, which is thermodynamically unfavourable. The depletion of the polymer chains from the space between the particles means that microreservoirs of essentially pure solvent are generated in the interparticle zones. Closer approach of the particles is then favoured in a good solvent because it leads to a reduction in free energy as the pure solvent is pushed from between the particles and mixes spontaneously with the bulk polymer solution.

An alternative, but entirely equivalent, way of viewing the origins of depletion flocculation is to consider the bulk solution as exerting a compressive osmotic pressure on the colloidal particles, one that is not counterbalanced by the microreservoirs of pure solvent between the particles. This unbalanced force causes the particles to aggregate. According to this viewpoint, the two particles function effectively as a semi-permeable membrane. This prevents the passage of the polymer chains into the space between the particles and so generates a compressive osmotic interaction.

8.10.2 *Depletion stabilization*

The preceding discussion of depletion flocculation began at a point where the colloidal particles had already approached so closely that the polymer chains were excluded from the space between them. No consideration was given to the free energy changes that must be involved in forming the microreservoirs of essentially pure solvent in the volume elements between the particles. It is obvious, however, that the polymer-depleted regions can only be formed by demixing polymer coils and solvent. In good solvents, this demixing is thermodynamically unfavourable. It follows that work must be done in order to bring the particles from infinite separation to an interparticle distance somewhat less than the 'diameter' of the polymer chains. This work requirement merely reflects the existence of a repulsive potential energy barrier that opposes the close approach of the particles, a repulsion that can be sufficient to impart stability. It is this repulsion which, if sufficiently large, gives rise to the possibility of free polymer imparting colloid stability (Feigin and Napper 1980). The experimental confirmation of this phenomenon is still open to some question.

8.10.3 *The polymer depletion layers*

The time averaged segmental concentration in a bulk polymer solution is everywhere uniform. Near to an interface, however, this uniformity is lost. Molecules whose centres of mass attempt to approach closer than half the 'diameter' (D) of the polymer possess configurations that would transgress the interface (see Fig. 8.10.2). Such configurations are forbidden. As a result, the concentration of polymer molecules close to an impenetrable interface is depleted. A gradient in the segmental concentration is established in the vicinity of the interface. If the surface is inert, the segmental concentration must be zero at the interface, rising to its bulk value over a characteristic distance comparable to the 'diameter' of the polymer chains (see Fig. 8.10.2). The latter represents the maximum distance normal to the interface at which there would be segmental contributions from the depleted molecules if the interface were notionally absent. This sets the outer limit to the thickness of the depletion layer.

8.10.4 *The three depletion domains of close approach*

For two parallel flat plates immersed in a polymer solution, three domains of close approach may be usefully distinguished (see Fig. 8.10.3).

The first domain occurs if the plates are separated by a distance (d) greater than twice the 'diameter' (D) of the polymer chains. The depletion layers then have sufficient distance to build up to their bulk segmental concentration value. Closer approach of the plates merely

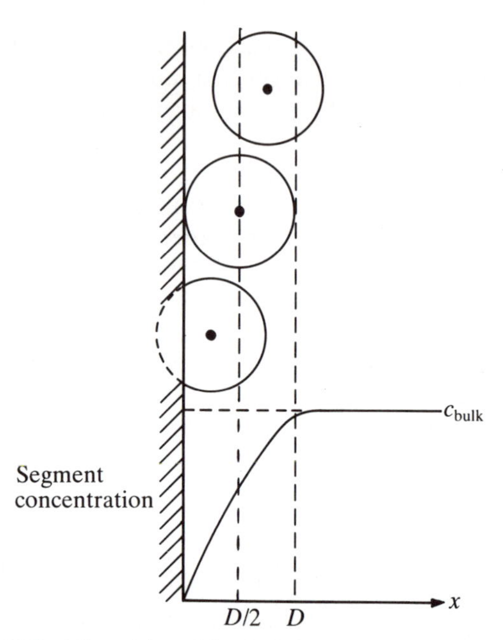

FIG. 8.10.2. The origins of the depletion layers at an interface.

excludes bulk polymer solution from between the particles and forces it into the bulk solution. No free energy change occurs.

The second domain occurs when the distance between the plates is reduced to between one and two polymer 'diameters'. Neither depletion layer now has sufficient distance to rise to the bulk segmental concentration. Closer approach of the plates therefore demands that the dilute polymer solution between the plates be depleted even further. This requires a demixing of polymer segments and solvent which, in a good solvent, inevitably increases the free energy of the system.

The third domain arises on even closer approach so that the distance between the plates is less than the 'diameter' of the polymer chains. The polymer segment concentration between the plates is then reduced effectively to zero. As discussed previously for depletion flocculation, even closer approach of the particles leads to the mixing of essentially pure solvent with bulk polymer solution. This lowers the free energy of the system.

8.10.5 *Potential energy diagram for colloidal particles in free polymer solution*

The foregoing discussion enables the qualitative features of the potential energy diagram for two colloidal particles immersed in a polymer solution

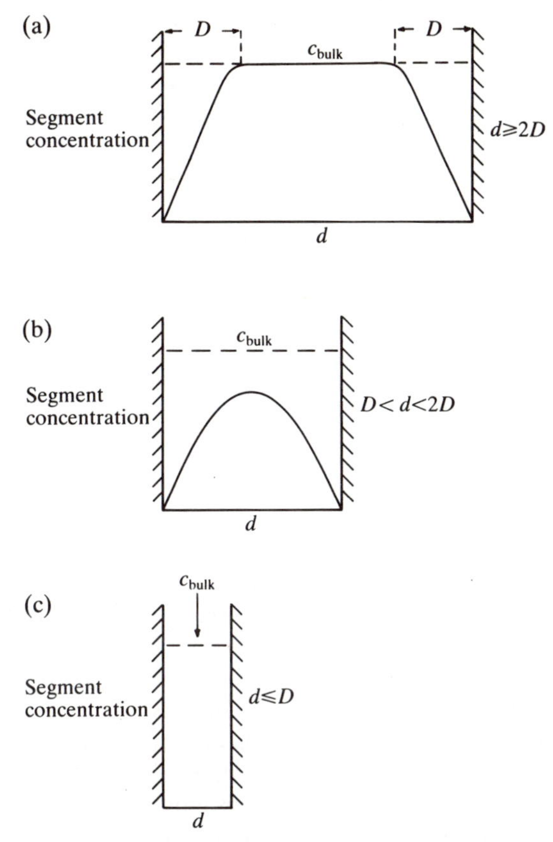

FIG. 8.10.3. The three domains of close approach of flat plates immersed in a free polymer solution.

to be readily discerned. One such diagram is shown in Fig. 8.10.4. Note that the polymer molecules have been assumed not to adsorb on the colloidal particles and further that the van der Waals attraction between the colloidal particles has been omitted.

It is immediately apparent that the shape of the potential energy diagram for colloidal particles immersed in a free polymer solution is very different from that for two sterically stabilized particles (see Fig. 8.8.4). Interestingly, the shape resembles more closely the potential energy diagram for two electrostatically stabilized particles whose van der Waals attraction is operative. Inspection of Fig. 8.10.4 reveals that the free energy of the system increases when the particles are brought from infinite separation to a surface separation comparable to the 'diameter' of the polymer molecules. Even closer approach leads to a decrease in the free energy of the system.

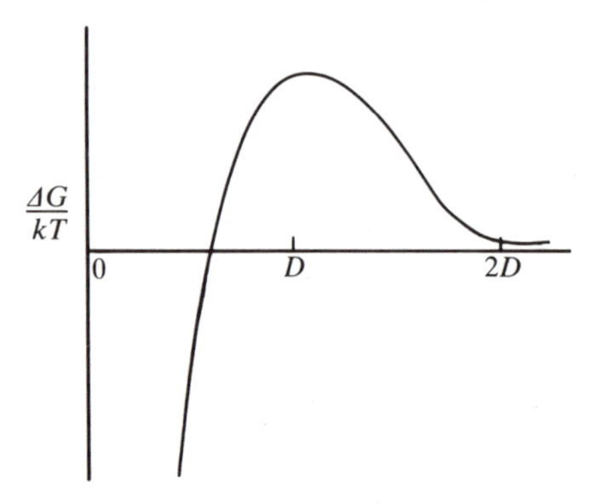

FIG. 8.10.4. Characteristic potential energy diagram for the close approach of two colloidal particles immersed in a free polymer solution.

8.10.6 *Flocculation effects induced by free polymer*

The shape of the potential energy vs. distance of separation curve determines the effects observed when free polymer is added to a colloidal dispersion. For the purposes at hand, it is convenient to continue to ignore the van der Waals attraction between the particles and to assume that the particles constitute a stable dispersion. In practice, this can be crudely simulated by sterically stabilizing the particles using low molecular weight polymer of the same composition as the higher molecular weight free polymer. Low concentrations of free polymer are found experimentally to have no apparent effect on stability. Increasing the free polymer concentration, however, ultimately results in the onset of flocculation. This occurs when the depth of the primary minimum in the potential energy curve is sufficiently great (of the order of several kT). The height of the primary maximum, which is comparable in magnitude to the depth of the minimum, is insufficient at these polymer concentrations to prevent passage into the primary minimum at an observable rate.

Exercises. 8.10.1 It can be shown that the force F between two parallel flat plates immersed in a non-adsorbing polymer solution is given by

$$\mathsf{F} = 2\int_{-\infty}^{\mu_2^*} \left(\frac{\partial \Gamma_{2,1}}{\partial h}\right)_{\mu_2} d\mu_2$$

where μ_2 = chemical potential of the polymer in solution, the * denotes pure polymer, and $\Gamma_{2,1}$ = surface excess of the polymer relative to the solvent (section 5.4). Show that this formula predicts the possible occurrence of depletion flocculation.

8.11 Polymeric flocculants

There exists a class of polymers that are widely exploited industrially as flocculants of colloidal dispersions (Gregory 1983). These *polymeric flocculants* may be anionic, cationic or even non-ionic in character (section 2.6.2). The vast majority of commerical anionic polymers are based on partially hydrolysed polyacrylamide:

$$\begin{array}{c} -\!\!(CH_2\text{—}\underset{\underset{\displaystyle CONH_2}{|}}{CH}\text{—}CH_2\text{—}\underset{\underset{\displaystyle COO^-Na^+}{|}}{CH}\!\!)\!-_n \end{array}$$

so that they are, in effect, statistical copolymers of acrylamide and acrylic acid. Polyethyleneimine $-\!(NH\text{–}CH_2\text{–}CH_2)\!-_n$ would be representative of the type of polymer which, at suitable pH, functions as a cationic flocculant. Note that both of the examples given here are normally polyelectrolytes when functioning as polymeric flocculants. The reasons for this will be set forth later. The charge densities of polyelectrolytes with weakly acidic or basic groups, such as carboxylic acid groups or secondary or tertiary amines, may depend strongly upon the solution pH. In contrast, polyelectrolytes with strongly ionized groups, such as sodium polystyrene sulphonate, display charge densities that are essentially independent of pH.

Although the molecular weights of available polymeric flocculants range from a few thousand to many millions, those of higher molecular weight (say, a few millions) are the more important. They act as effective flocculants at relatively low concentrations, typically of the order of several parts per million. Stated another way, the weight of polymeric flocculant required for floc formation is often only 10^{-4}–10^{-3} of that of the solids present.

8.12 Mechanisms of flocculation

8.12.1 *Bridging flocculation*

Ruehrwein and Ward (1952) were the first to propose that polymeric flocculants function by a bridging mechanism. Bridging is considered to be a consequence of the adsorption of the segments of individual polymeric flocculant molecules onto the surfaces of more than one particle (see Fig. 8.12.1). Such bridging links the particles together into loose aggregates, which often sediment rapidly or are easily removed by filtration. The reasons why high molecular weight polyelectrolytes are such effective flocculants can be readily discerned from this postulated mode of action.

Consider two electrostatically stabilized colloidal particles. In order for

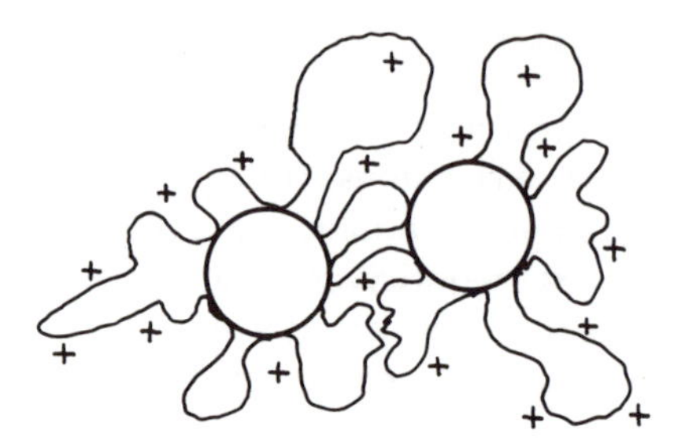

FIG. 8.12.1. Bridging flocculation of colloidal particles by polymeric flocculants.

a polymeric flocculant to induce bridging aggregation, it is necessary for the macromolecules not only to adsorb onto the surface of the particles, but also for loops of the adsorbed chains to extend further into the dispersion medium than the minimum distance of close approach between the particles. This requires the spatial extension of the loops of the polymeric flocculants to span at least the distance over which the electrostatic repulsion between particles is operative. This distance is of an order equal to the sum of the thicknesses of the electrostatic double layers surrounding the approaching colloidal particles, which is usually appreciable. The spatial extension of polymer molecules increases not only with molecular weight but also with increasing electrostatic charge density. Satisfaction of this demand for significant extension in space explains why high molecular weight polyelectrolytes can behave as extremely effective polymeric flocculants. Of course, as the ionic strength of the dispersion medium increases, so the thickness of the double layers decreases; this permits progressively lower molecular weight polymers to act as polymeric flocculants, even though, for a given molar mass (i.e. molecular weight) the extension of the polyelectrolyte is reduced somewhat by increasing ionic strength.

Note that for bridging flocculation to occur the segments of the polymeric flocculant must attach themselves to the surfaces of more than one colloidal particle. If the particles and the flocculant have opposite charges (e.g. positively charged particles and anionic polymer), Coulombic attraction clearly promotes segmental adsorption. However, it is found that anionic polymers are also able to flocculate negatively charged particles in some instances. There is compelling evidence that, in these systems, counterions, such as Ca^{2+}, are required to facilitate the binding of the anionic polymer segments onto the negatively charged surface. Such counterions thus act as 'bridges-with-bridges' to induce flocculation. Of course, hydrophobic interactions (see section 10.1), as well as hydrogen bonding, may also promote segmental adsorption, irrespective of the charges involved. These last-mentioned adsorption mechanisms are highly specific and lead to the possibility of selective polymeric flocculation of, for example, mixed mineral dispersions.

The foregoing discussion of bridging flocculation explains why it is found experimentally that the optimum flocculation occurs at flocculant dosages corresponding to a particle surface coverage that is significantly less than complete. Incomplete surface coverage ensures that there is sufficient unoccupied surface available on each particle to allow the additional adsorption during Brownian collisions of segments of one or more chains attached to other particles. It might be expected (Exercise 8.12.1) that optimum flocculation would occur at a surface coverage of one half; experimentally, however, optimum flocculation is often observed at significantly lower surface coverages.

At higher polymer concentrations than the optimum for flocculation, the particles become fully coated by segments of chains attached solely to the one particle. Steric (or electrosteric) stabilization is then observed.

8.12.2 *Charge neutralization*

In certain practical applications, such as water and effluent treatment, it is sometimes found that the only effective polymeric flocculants are those whose charge is opposite in sign to that of the particles. Since most naturally occurring particles are negatively charged, cationic polymers are commonly used in this context. Bridging flocculation may still be operative in such systems but, in certain instances, the polyelectrolytes may also induce flocculation through simple charge neutralization. In such systems, it is found that flocculation corresponds quite closely to the dosage of polymer required to neutralize the surface charge of the particles. The latter is detected experimentally by the reduction of the measured electrophoretic mobility of the particles to zero (Hunter *et al.* 1983). It has also been postulated that, even when the overall neutralization of the surface charge of the particles has been achieved, patches of electrostatic charge, both positive and negative, could well remain on the surfaces of the particles. Coulombic attraction between such mosaics of charge may then promote aggregation during a Brownian collision. This is a type of bridging flocculation, although it differs from that discussed in the preceding section in that significant extensions of loops into the continuous phase are improbable in these systems.

Exercises. 8.12.1 Show that the probability of forming a polymer bridge between two particles is expected to be proportional to $\theta(1-\theta)$ where θ is the fraction of surface covered by polymer. Hence show that bridging should be a maximum where $\theta = \frac{1}{2}$.

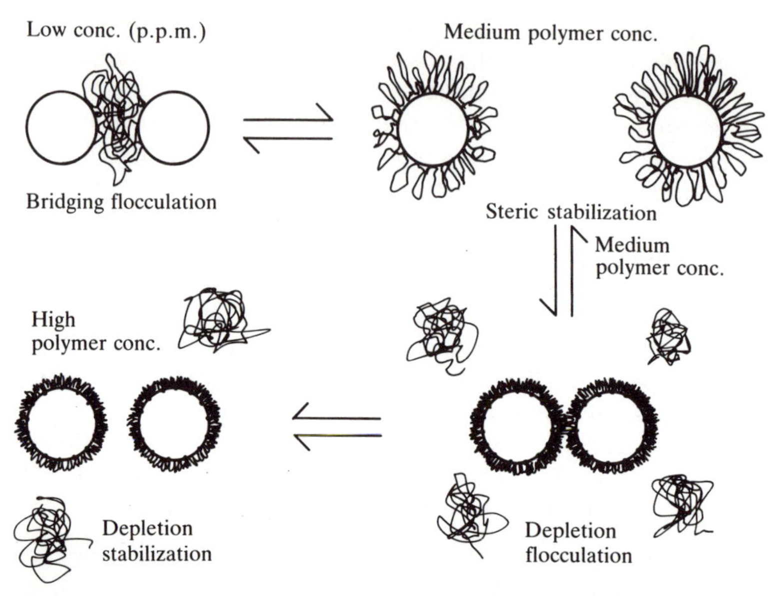

FIG. 8.13.1. Effects of polymer chains on colloidal dispersions. Bridging flocculation (section 8.12.1) can occur at very low polymer concentrations. Steric stabilization (sections 8.4–8.8) requires moderate polymer concentrations and the polymer must be adsorbed and anchored. Free polymer effects occur only at moderate to high polymer concentrations; depletion flocculation (section 8.10.1) is essentially an osmotic effect, whereas depletion stabilization (section 8.10.2) can be understood in terms of the work required to create a polymer free region between two approaching particles.

8.13 Summary of the effects of polymer on colloid stability

The effects of polymer, whether free or attached, on the stability of colloidal dispersions are represented schematically in a greatly simplified manner in Fig. 8.13.1. At very low free polymer concentrations (say a few parts per million), adsorption of the polymer onto the surface of more than one particle can induce bridging flocculation. At higher polymer concentrations, steric stabilization may be imparted. Increasing the free polymer concentration even further may result in depletion flocculation, although stabilization may be observed at even higher polymer concentrations. This diagram emphasizes the complexity of the effects that polymers exert on colloid stability. An extensive review of this entire field has recently appeared (Napper 1983) and the interested reader is referred to that work for further references to the recent literature.

References

Asakura, S. and Oosawa, F. (1954). *J. Chem. Phys.* **22,** 1255.
Barrett, K. E. J. (1975). *Dispersion polymerization in organic media.* Wiley, London.
Croucher, M. D. and Hair, M. L. (1980). *Colloids Surfaces* **1,** 349.
Croucher, M. D. and Hair, M. L. (1981). *J. Colloid interface Sci.* **81,** 257.
Derham, K. W., Goldsbrough, J., and Gordon, M. (1974). *Pure appl. Chem.* **38,** 97.
Deryaguin, B. V. (1934). *Kolloid Z.* **69,** 155.
Evans, R. and Napper, D. H. (1975). *J. Colloid interface Sci.* **52,** 260.
Evans, R., Davison, J. B., and Napper, D. H. (1972). *J. Polymer Sci.* **B10,** 449.
Faraday, M. (1857). *Phil. Trans.* **147,** 145.
Feigin, R. I. and Napper, D. H. (1980). *J. Colloid interface Sci.* **75,** 525.
Fleer, G. J., Scheutjens, H. M. H., and Vincent, B. (1984). In *Polymer adsorption and dispersion stability* (ed. E. D. Goddard and B. Vincent). ACS Symposium Series No. 240, p. 245. American Chemical Society, Washington D.C.
Flory, P. J. (1941). *J. Chem. Phys.* **9,** 660.
Flory, P. J. (1953). *Principles of polymer chemistry.* Cornell University Press, Ithaca, N.Y.
Flory, P. J. (1969). *Statistical mechanics of chain molecules.* Interscience, New York.
Flory, P. J. (1970). *Discuss. Faraday Soc.* **49,** 7.
Gregory, J. (1983). In *Chemistry and technology of water soluble polymers* (ed. C. A. Finch) pp. 307–20. Plenum, New York.
Griffin, W. C. (1954). *J. Soc. cosmet. Chem.* **5,** 249.
Heller, W. and Pugh, T. L. (1954). *J. Chem. Phys.* **22,** 1778.
Huggins, M. L. (1941). *J. Chem. Phys.* **9,** 440.
Hunter, R. J., Matarese, R., and Napper, D. H. (1983). *Colloids Surfaces* **7,** 1–13.
Meier, D. J. (1967). *J. phys. Chem.* **71,** 1861.
Napper, D. H. (1970). *J. Colloid interface Sci.* **32,** 106.
Napper, D. H. (1983). *Polymeric stabilization of colloidal dispersions.* Academic Press, London.
Ottewill, R. H. (1973) in *Colloid science* (ed. D. H. Everett) Vol. 1, Chapter 5. Chemical Society, London.
Patterson, D. (1969). *Macromolecules* **2,** 672.
Ruehrwein, R. A. and Ward, D. W. (1952). *Soil Sci.* **73,** 485.
Tadros, Th.F. (1982). *The effects of polymers on dispersion properties.* Academic Press, London.
Vincent, B. and Whittington, S. (1981) *Surface and colloid science,* Vol. 12, p. 1. Plenum Press, New York.
Zsigmondy, R. (1901). *Z. analyt. Chem.* **40,** 697.

9

TRANSPORT PROPERTIES OF SUSPENSIONS

9.1 Introduction

In this chapter we will describe the calculation of three of the transport properties introduced in Chapter 2: the sedimentation coefficient, the Brownian diffusivity, and the effective viscosity of a colloidal dispersion.

For simplicity we will concentrate on the case of a suspension of rigid, force-free particles; this is an idealization that can be approximated in practice by a suspension in which the double-layer thickness is much smaller than the particle radius, and the repulsive forces are large enough to prevent coagulation. For most separations the colloidal forces are then unimportant. In the final section of the chapter we study a number of 'electrokinetic' transport phenomena that arise specifically from the presence of charges at the solid–liquid interface.

To calculate colloidal transport properties it is necessary to determine the way in which the solvent flows around the suspended particles. In preparation for this analysis we will devote the first portion of the chapter to a discussion of the relevant aspects of fluid mechanics. Although earlier colloidal texts were able to get by with a very elementary treatment of this subject, research in the field of transport properties has now reached a level where it is no longer possible, even in a qualitative sense, to understand the current work without a sound basis in fluid mechanics. It is hoped that the first seven sections of this chapter will provide such a basis. In this description it will be assumed that the reader is familiar with the notation and methods of vector calculus; a few revision notes are provided on this subject in Appendix 3.

9.2 The mass conservation equation

In our study of fluid motion the molecular nature of the fluid will be neglected and it will be treated as a continuum. For this approximation to be valid, attention must be restricted to regions of the fluid which contain many molecules. On the assumption that colloidal particles are much larger than the solvent molecules, we will use the terms 'fluid particle' and 'point in a fluid' to refer to regions that are much smaller than the colloidal particles, but much larger than the intermolecular spacing.

In general, the velocity $\boldsymbol{v}$ at a point in the fluid will depend on the position of the point and on the time. To emphasize this dependence the velocity is written as $\boldsymbol{v}(x_1, x_2, x_3, t)$ or $\boldsymbol{v}(\boldsymbol{x}, t)$, where x_1, x_2, and x_3 are the Cartesian coordinates of the point, and $\boldsymbol{x}$ is the 'position vector' from the origin to the point. These quantities are illustrated in Fig. 9.2.1. The components of $\boldsymbol{v}$ in the direction of the coordinate axes will be denoted by v_1, v_2, and v_3, respectively. The aim in this and the following sections is to set up the differential equations that must be solved in order to determine the flow around a colloidal particle. As the form of these equations does not depend on the presence of the colloidal particles, we will for the moment neglect the suspended particles and concentrate on the case of a pure solvent, since this simplifies the derivation.

The density of water changes by only 0.01 per cent if the pressure is

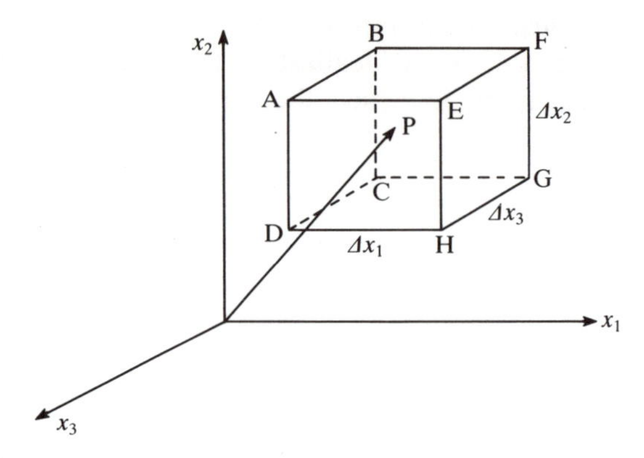

FIG. 9.2.1. The small rectangular volume referred to in the derivation of the continuity eqn (9.2.3). P has coordinates (x_1, x_2, x_3) and is in the centre of the cube.

increased from 1 to 2 atm; we can, therefore, treat water (and most other solvents for that matter) as incompressible. This incompressibility property, together with the principle of mass conservation, can be used to set up the first of the differential equations for $\boldsymbol{v}$. The mass of liquid that flows into a volume $\mathcal{V}$ over any period of time must be balanced by the amount flowing out. The required differential equation is obtained by writing this statement in mathematical terms.

Consider first the problem of calculating the mass of fluid that passes across a small portion ΔA of the surface of $\mathcal{V}$ during a small time interval Δt. On the assumption that the dimensions of the area element, ΔA are much smaller than the local radii of curvature of the surface we take ΔA to be approximately flat. If ΔA is also much smaller than the length scale over which the velocity $\boldsymbol{v}(\boldsymbol{x}, t)$ changes significantly, we can take $\boldsymbol{v}$ to be uniform in the neighbourhood of ΔA. During the interval Δt, the fluid particles in this region will move by $\boldsymbol{v}\Delta t$. If ΔA is rectangular, the volume of fluid that passes through the area element during Δt will lie, at the end of the interval, in the parallelepiped shown in Fig. 9.2.2. The volume of this parallelepiped is simply the base times the perpendicular height, or

$$\boldsymbol{v} \cdot \hat{n} \Delta t \Delta A$$

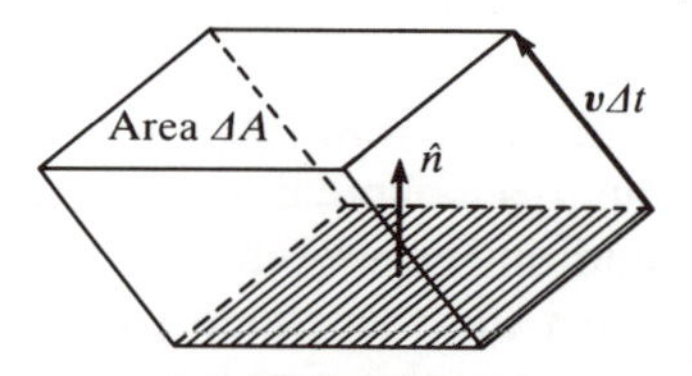

FIG. 9.2.2. A sketch of the parallelepiped-shape volume of fluid that passes across the shaded rectangular area element ΔA during a small time interval Δt.

where $\hat{n}$ is the 'unit vector normal to ΔA', that is, the unit vector perpendicular to ΔA, directed outwards from $\mathscr{V}$. Thus the rate of flow of mass out of $\mathscr{V}$ through ΔA is

$$\rho \boldsymbol{v} \cdot \hat{n} \Delta A \tag{9.2.1}$$

where ρ is the fluid density.

To write the principle of mass conservation in differential equation form it is convenient to take $\mathscr{V}$ to be a very small rectangular block with edges parallel to the coordinate axes, as shown in Fig. 9.2.1. The sides of the block have length Δx_1, Δx_2 and Δx_3 respectively. From eqn (9.2.1) it follows that the rate at which mass flows out of the volume through the face ABCD is given by (Exercise 9.2.1):

$$\rho v_1\left(x_1 - \frac{\Delta x_1}{2}, x_2, x_3\right)\Delta x_2 \Delta x_3$$

where (x_1, x_2, x_3) are the coordinates of the centre of the block. A similar expression can be written for the mass flow out of the opposing face EFGH. Using the same procedure as that described in Appendix A3 we can then calculate the total flux or flow into or out of the volume element. Since that is zero for an incompressible fluid we have

$$\sum_{i=1}^{3} \frac{\partial v_i}{\partial x_i} = 0. \tag{9.2.2}$$

In vector notation this equation is written (Appendix A3):

$$\nabla \cdot \boldsymbol{v} = 0. \tag{9.2.3}$$

This is one of the governing equations of fluid flow, referred to as the *mass conservation*, or *continuity equation*.

Exercises. 9.2.1 Justify the expression for the rate of mass flow through ABCD of Fig. 9.2.1 using eqn (9.2.1). Hence establish eqn (9.2.2).

9.3 Stress in a moving fluid

The remaining equations for the fluid velocity are obtained by applying Newton's Second Law of motion to a small block of fluid. In preparation for this step we first discuss the nature of the forces that act in a fluid, and the way these forces are related to the local velocity field.

In general the forces can be labelled as either short-range or long-range, where the short-range forces are those that act over molecular distances and the long-range ones include colloidal dispersion forces and gravity. In this and the following two sections we concentrate on the short-range forces and their relationship to the local velocity field.

Consider the forces acting across an area element ΔA that is translating and rotating with the local fluid. The fluid that lies on the side of ΔA to which $\hat{n}$ points exerts a force $\Delta \boldsymbol{F}$ on the fluid on the other side of ΔA. For small ΔA it is assumed that the force per unit area

$$\frac{\Delta \boldsymbol{F}}{\Delta A}$$

is independent of the size and shape of ΔA. This ratio is called the *stress* and is denoted by

$$\mathscr{S}(\boldsymbol{x}, \hat{n}),$$

where $\boldsymbol{x}$ is the position vector to the centre of the area element. As well as depending on the position and orientation of the area element, the stress may also depend on time. To keep the notation compact, however, we will not show this time-dependence explicitly.

The stresses on different planes passing through a point are related; in particular if we know the stress at $\boldsymbol{x}$ on any three mutually orthogonal planes we can calculate the stress at this point for a plane of arbitrary orientation. This relationship is obtained from an analysis of the forces that act on the small tetrahedral block of fluid shown in Fig. 9.3.1. The sloping face, with unit normal $\hat{n}$, coincides with the surface on which we seek to calculate the stress, and the other three faces are parallel to the coordinate planes. The unit normals to these surfaces will therefore be $-\hat{e}_1$, $-\hat{e}_2$, and $-\hat{e}_3$, respectively, where $\hat{e}_i$ denotes the unit vector in the direction of the x_i axis. If the block is very small, the stress will be approximately uniform over each face. Thus the short-range force on the sloping face, of area ΔA is:

$$\mathscr{S}(\boldsymbol{x}, \hat{n})\Delta A. \tag{9.3.1}$$

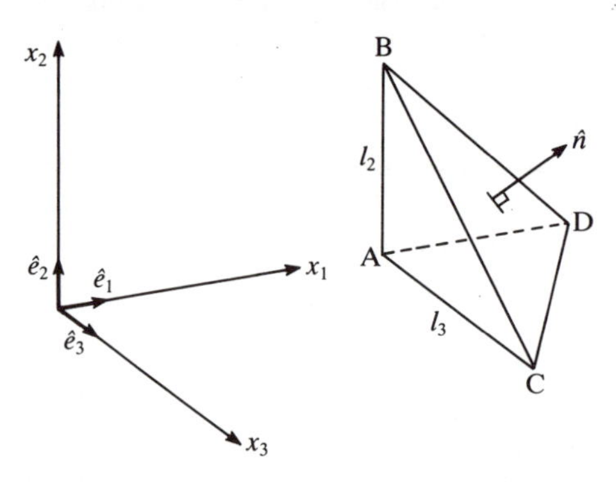

FIG. 9.3.1. The small block of fluid used in the derivation of the formula (9.3.5) for the stress on a plane of arbitrary orientation.

Since we are neglecting the spatial variation of stress over the block we may take $\boldsymbol{x}$ in this expression to be the position vector to any point in the block. Similarly, the force due to the stresses on the other faces is given by

$$\mathcal{S}(\boldsymbol{x}, -\hat{e}_j)\Delta A_j, \tag{9.3.2}$$

for $j = 1, 2, 3$. Each of these forces is proportional to the area of the corresponding face. The long-range forces, on the other hand, will be proportional to the volume of the block; for example the gravitational force is

$$\rho \boldsymbol{g} \Delta \mathcal{V},$$

where $\boldsymbol{g}$ is the gravitational acceleration vector, directed vertically downwards, and $\Delta\mathcal{V}$ is the volume. Thus as the block is reduced in size, the long-range forces will diminish at a faster rate than the stress forces, which will eventually dominate.

By Newton's Second Law, the net force on the block is equal to the mass times the acceleration. However, since the mass of the block is proportional to the volume, the inertia term, like the long-range forces, will be negligible in comparison with the stress forces. Thus for a very small block, Newton's Second Law reduces to the force balance equation

$$\mathcal{S}(\boldsymbol{x}, \hat{n})\Delta A + \sum_{j=1}^{3} \mathcal{S}(\boldsymbol{x}, -\hat{e}_j)\Delta A_j = 0. \tag{9.3.3}$$

This equation can be simplified by using the expression (see Exercise 9.3.1):

$$\Delta A_j = \Delta A \hat{e}_j \cdot \hat{n}, \tag{9.3.4}$$

together with the identity

$$\mathcal{S}(\boldsymbol{x}, \hat{e}_j) = -\mathcal{S}(\boldsymbol{x}, -\hat{e}_j),$$

which follows from the principle of action and reaction. Combining these results with eqn (9.3.3), we get

$$\mathcal{S}(\boldsymbol{x}, \hat{n}) = \sum_{j=1}^{3} \mathcal{S}(\boldsymbol{x}, \hat{e}_j)\hat{e}_j \cdot \hat{n}. \tag{9.3.5}$$

This is the required formula, which enables us to calculate the stress on an arbitrary plane through $\boldsymbol{x}$, given the stress on the three planes with unit normals $\hat{e}_1$, $\hat{e}_2$, and $\hat{e}_3$.

It is customary to write the components of the stresses $\mathcal{S}(\boldsymbol{x}, \hat{e}_i)$ on these three orthogonal planes as σ_{ij}, where

$$\mathcal{S}(\boldsymbol{x}, \hat{e}_j) = \sum_{i=1}^{3} \sigma_{ij}(\boldsymbol{x})\hat{e}_i. \tag{9.3.6}$$

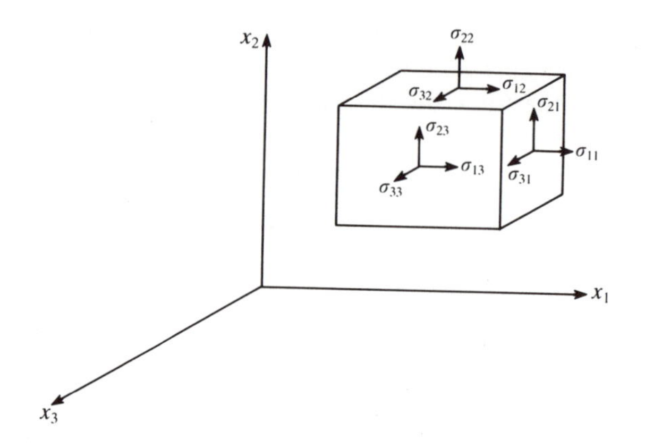

FIG. 9.3.2. An illustration of the stresses acting on a small rectangular block with faces parallel to the coordinate planes. The shear stress, $\mathscr{S}$, depicted in Fig. 2.4.2 would correspond to σ_{12}. (Note that the origin of the stresses labelled σ_{i3} lies in the plane of the front face of the block.)

σ_{ij} is the ith component of the stress on the plane with unit normal $\hat{e}_j$. This convention is illustrated in Fig. 9.3.2. In terms of the quantities σ_{ij}, the formula (9.3.5) for the stress on a plane of arbitrary orientation with unit normal $\hat{n}$ becomes

$$\mathscr{S}(\boldsymbol{x}, \hat{n}) = \sum_{j=1}^{3} \sum_{i=1}^{3} \sigma_{ij}(\boldsymbol{x})\hat{e}_i(\hat{e}_j \cdot \hat{n}). \tag{9.3.7}$$

The nine quantities $\sigma_{ij}(\boldsymbol{x})$ completely characterize the state of stress at $\boldsymbol{x}$, for once these quantities are known it is possible to calculate the stress on any plane through the point. These quantities depend on the orientation of the coordinate axes in the same sort of way as the components of a vector depend on the orientation of the axes, for as this orientation varies, so the direction of the unit vectors $\hat{e}_i$ and the orientation of the three orthogonal planes will alter. It is convenient to think of these nine quantities as the components of a *stress tensor*†, denoted by $\boldsymbol{\sigma}$. The expression (9.3.5) is formally written in terms of $\boldsymbol{\sigma}$ as

$$\mathscr{S}(\boldsymbol{x}, \hat{n}) = \boldsymbol{\sigma}(\boldsymbol{x}) \cdot \hat{n}. \tag{9.3.8}$$

It is customary to write the components of the stress tensor in the

† In mathematical terminology, $\boldsymbol{\sigma}$ is called a 'second-order Cartesian tensor'. Not every set of nine quantities can be regarded as the components of a tensor; they must transform on rotation of axes in a particular way. The quantities σ_{ij} *do* transform in the right manner, but we will not go into the details here.

matrix form

$$\begin{bmatrix} \sigma_{11} & \sigma_{12} & \sigma_{13} \\ \sigma_{21} & \sigma_{22} & \sigma_{23} \\ \sigma_{31} & \sigma_{32} & \sigma_{33} \end{bmatrix} \tag{9.3.9}$$

Since students often encounter difficulties with tensors, we should emphasize that this is simply a convenient piece of notation. The important thing to bear in mind about the stress tensor is its physical significance; that is that the components σ_{ij} are the components of the stresses on three orthogonal planes through the point.

The net (turning) moment due to the stresses on any small block of fluid is approximately zero† so, (from Fig. 9.3.2), it is clear that:

$$\sigma_{ij} = \sigma_{ji} \tag{9.3.10}$$

for all i and j. For this reason the stress tensor is said to be 'symmetric'.

Finally, for future reference we note that the net force due to the stresses over a macroscopic surface A is obtained by adding the contributions from the surface elements. On taking the limit of very small area elements and using the formula (9.3.8) we find that the net stress force is

$$\int_A \boldsymbol{\sigma} \cdot \hat{n}\, \mathrm{d}A. \tag{9.3.11}$$

Exercises. 9.3.1. Consider the problem of deriving the formula (9.3.4) for the area of the orthogonal faces of the block shown in Fig. 9.3.1. By the definition of the cross product,

$$\Delta A\hat{n} = \tfrac{1}{2}(\overrightarrow{\mathrm{BC}} \times \overrightarrow{\mathrm{BD}}),$$

where ΔA is the area of the sloping face, and $\overrightarrow{\mathrm{BC}}$ and $\overrightarrow{\mathrm{BD}}$ are vectors along the edges of that face.

(a) Show that

$$\overrightarrow{\mathrm{BC}} = l_3\hat{e}_3 - l_2\hat{e}_2,$$

and

$$\overrightarrow{\mathrm{BD}} = l_1\hat{e}_1 - l_2\hat{e}_2,$$

where l_i is the length of the block edge parallel to the x_i axis.

(b) By calculating the cross-product of these vectors and taking the dot product of the result with the unit vectors $\hat{e}_i$, verify eqn (9.3.4).

† As in eqn (9.3.3), we are assuming here that the block is so small that inertia terms are negligible.

9.4 Stress and velocity field in a fluid in thermodynamic equilibrium

By thermodynamic arguments it can be shown (Landau and Lifshitz 1969, section 12) that the stress in a fluid in equilibrium is given simply by

$$\mathscr{S}(\boldsymbol{x}, \hat{n}) = -p(\boldsymbol{x})\hat{n}, \tag{9.4.1}$$

where p is the pressure, the minus sign being included to indicate the compressive nature of the stress.

Recalling the definition (9.3.8) of the components of the stress tensor, we see that

$$\boldsymbol{\sigma} = -p\boldsymbol{I} \tag{9.4.2}$$

in a fluid in equilibrium, where $\boldsymbol{I}$ is the 'unit tensor', which has components

$$\begin{bmatrix} 1 & 0 & 0 \\ 0 & 1 & 0 \\ 0 & 0 & 1 \end{bmatrix}$$

If a fluid in equilibrium is in motion, it must move as a rigid body, for this is the motion that maximizes the entropy (Landau and Lifshitz 1969, section 10). It can be shown (Meriam 1966, section 30) that the velocity field for an arbitrary rigid body motion must have the form

$$\boldsymbol{v}(\boldsymbol{x}) = \boldsymbol{V} + \boldsymbol{\Omega} \times \boldsymbol{x}, \tag{9.4.3}$$

where $\boldsymbol{V}$ and $\boldsymbol{\Omega}$ are independent of $\boldsymbol{x}$, and are the *translational* and *angular velocities* respectively.

Those readers who are unfamiliar with this formula may find it helpful to consider the example of a glass of liquid placed on the centre of a steadily rotating turntable, as shown in Fig. 9.4.1. In a frame of reference

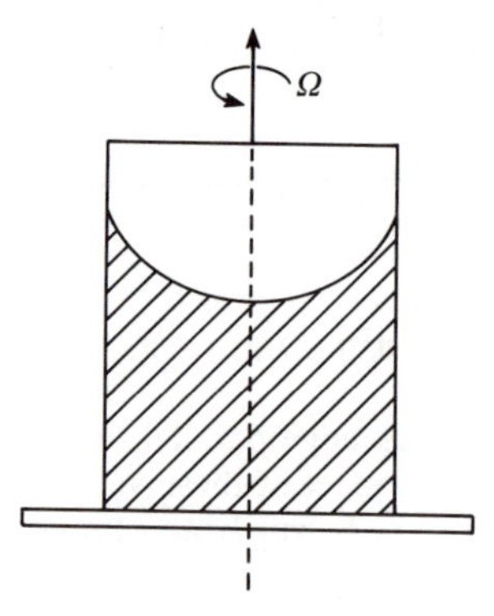

FIG. 9.4.1. An illustration of a fluid that is in equilibrium and in motion: a glass of liquid on the centre of a rotating turntable.

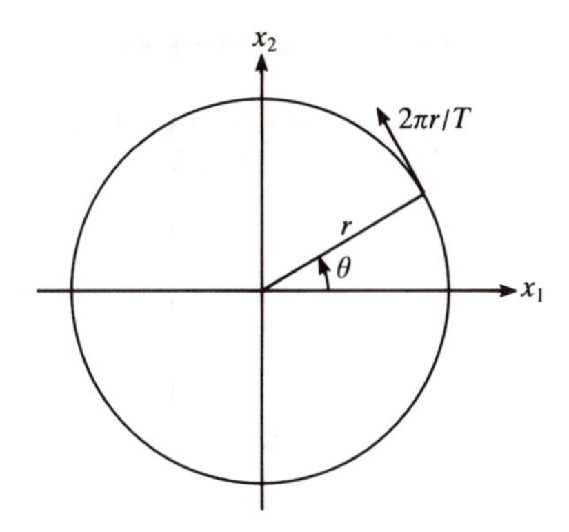

FIG. 9.4.2. A top view of the path traced out by a particle in the rotating glass in Fig. 9.4.1. The x_3 axis coincides with the axis of rotation.

that moves with the turntable the fluid appears to be at rest. In the laboratory frame the fluid particles will move steadily around circles centred on the axis of rotation with a speed of $2\pi r/T$, where T is the time for each rotation and r is the distance of the particle from the axis of rotation.

From Fig. 9.4.2 it can be seen that the components of the fluid velocity in the laboratory frame of reference are therefore given by

$$v_1 = \frac{-2\pi r}{T}\sin\theta = \frac{-2\pi x_2}{T},$$

$$v_2 = \frac{2\pi r}{T}\cos\theta = \frac{2\pi x_1}{T},$$

and

$$v_3 = 0.$$

These component expressions can be written in the vector form (Appendix A3):

$$\boldsymbol{v} = \frac{2\pi\hat{\boldsymbol{e}}_3}{T}\times\boldsymbol{x}. \tag{9.4.4}$$

Comparing this result with the general formula (9.4.3) for rigid body motion, we see that the translational velocity is zero, while the angular velocity is given by

$$\boldsymbol{\Omega} = \frac{2\pi\hat{\boldsymbol{e}}_3}{T}$$

in this case.

Exercises. 9.4.1. Establish eqn (9.4.4).

9.5 Relationship between the stress tensor and the velocity field

In a fluid that is undergoing deformation, frictional stresses are set up which tend to retard the deforming motion. Our aim in this section is to find the relationship between these frictional stresses and the deforming motion.

Although the equilibrium formula (eqn (9.4.2)) for the stress tensor is not valid here, it is convenient to define the pressure in a deforming fluid by

$$p = -\tfrac{1}{3}(\sigma_{11} + \sigma_{22} + \sigma_{33}), \tag{9.5.1}$$

and to write the stress tensor as

$$\boldsymbol{\sigma} = -p\boldsymbol{I} + \boldsymbol{\sigma}^{\mathrm{D}}. \tag{9.5.2}$$

The quantity $\boldsymbol{\sigma}^{\mathrm{D}}$ is known as the 'deviatoric stress tensor', since it represents the deviation of the stress tensor from the equilibrium form (eqn 9.4.2)). $\boldsymbol{\sigma}^{\mathrm{D}}$ will depend on the history of the motion of the fluid, being zero for a fluid which has been in steady rigid body motion for a sufficiently long time.

Since the stresses arise from short-range forces in the liquid, the deviatoric stress tensor at a point will only depend on the history of the motion of the fluid in the neighbourhood of that point. Furthermore, since the fluid molecules jiggle around and rearrange themselves very rapidly, the effect of past motions will soon fade. For most liquids the time for this molecular rearrangement is much smaller than the time required for the macroscopic velocity $\boldsymbol{v}(\boldsymbol{x}, t)$ to change significantly. Thus, to a good approximation, $\boldsymbol{\sigma}^{\mathrm{D}}(\boldsymbol{x}, t)$ will only depend on the *instantaneous* velocity field in the neighbourhood of $\boldsymbol{x}$, since this is the quantity that characterizes the recent deformation history.

Presumably $\boldsymbol{\sigma}^{\mathrm{D}}$ will not depend on the absolute velocity of the neighbouring fluid particles, but only on their velocity relative to the particle at $\boldsymbol{x}$. If the distance $\Delta\boldsymbol{x}$ between neighbouring fluid particles is sufficiently small we may approximate this relative velocity with the aid of the formula for a total differential, viz.

$$\boldsymbol{v}(\boldsymbol{x} + \Delta\boldsymbol{x}, t) - \boldsymbol{v}(\boldsymbol{x}, t) \approx \sum_{j=1}^{3} \left(\frac{\partial \boldsymbol{v}}{\partial x_j}\right) \Delta x_j, \tag{9.5.3}$$

where the derivatives are evaluated at $(\boldsymbol{x}, t)$. Thus the local relative velocity field is determined by the three derivatives $\partial \boldsymbol{v}/\partial x_j$, and therefore $\boldsymbol{\sigma}^{\mathrm{D}}(\boldsymbol{x}, t)$ will be a function of these three vector quantities, or equivalently, a function of the nine scalar quantities $\partial v_i/\partial x_j$.

Although we cannot predict the exact relationship between $\boldsymbol{\sigma}^{\mathrm{D}}$ and the velocity gradients for any particular liquid, we can make a reasonable assumption about the form of this relationship in the case of small velocity gradients.

In preparation for this step, the reader may find it helpful to recall that a function of a single variable $f(\xi)$ can be approximated in the neighbourhood of a point ξ_0 by the tangent to f at ξ_0, viz

$$f(\xi) \approx f(\xi_0) + f'(\xi_0)(\xi - \xi_0). \tag{9.5.4}$$

In this case we have a function σ_{ij}^{D} that depends on the nine variables $\partial v_k/\partial x_l$, and we seek an approximation valid in the neighbourhood of $\partial v_k/\partial x_l = 0$. When the velocity gradients are all zero, $\sigma_{ij}^{\mathrm{D}} = 0$, and the appropriate generalization of eqn (9.5.4) is

$$\sigma_{ij}^{\mathrm{D}} = \sum_{m=1}^{3} \sum_{n=1}^{3} A_{ijmn} \frac{\partial v_m}{\partial x_n}, \tag{9.5.5}$$

where the 81 parameters A_{ijmn} are properties of the fluid. Since the stress tensor is symmetric (see eqn (9.3.10)) it follows that

$$A_{ijmn} = A_{jimn}. \tag{9.5.6}$$

This still leaves 54 unknown constants. Fortunately, however, most of these are zero for commonly occurring liquids such as water, while the remaining constants are all equal. This follows from the fact that the simple liquids, like water, are *isotropic,* i.e. their properties are the same in all directions. For such fluids, the relation (9.5.5) must have a form which is unaffected by a rotation of the coordinate axes. By using this fact, together with the symmetry relations (eqn (9.5.6)) it can be shown (Batchelor 1970, section 3.3) that there is only one independent constant in eqn (9.5.5), denoted by η and that this relationship reduces to

$$\sigma_{ij}^{\mathrm{D}} = \eta\left(\frac{\partial v_i}{\partial x_j} + \frac{\partial v_j}{\partial x_i}\right). \tag{9.5.7}$$

For the simple shear flow given by (Fig. 2.4.2)

$$v_1 = \dot{\gamma} x_2, \qquad v_2 = 0, \qquad v_3 = 0,$$

eqn (9.5.7) yields

$$\sigma_{12} = \eta \dot{\gamma}$$

where $\dot{\gamma}$ is the shear rate. Comparing with the shear stress formula (eqn (2.4.3)) we see that η is the familiar shear viscosity.

9.5.1 *The rate of strain tensor* $\boldsymbol{e}$

The expression (eqn (9.5.7)) for the stress tensor is usually written in the form

$$\sigma_{ij}^{\mathrm{D}} = 2\eta e_{ij}, \tag{9.5.8}$$

where

$$e_{ij} = \frac{1}{2}\left(\frac{\partial v_i}{\partial x_j} + \frac{\partial v_j}{\partial x_i}\right). \tag{9.5.9}$$

The nine quantities e_{ij} can be regarded as the components of a tensor called the 'rate of strain tensor', denoted by $\boldsymbol{e}$. In this notation eqn (9.5.8) becomes

$$\boldsymbol{\sigma}^{\mathrm{D}} = 2\eta\boldsymbol{e}.$$

Combining this formula with the expression (9.5.2) for the total stress tensor, we get

$$\boldsymbol{\sigma} = -p\boldsymbol{I} + 2\eta\boldsymbol{e}, \tag{9.5.10}$$

or

$$\sigma_{ij} = -p\delta_{ij} + 2\eta e_{ij}, \tag{9.5.11}$$

where δ_{ij} is the 'Krönecker delta', defined by

$$\delta_{ij} = \begin{cases} 1 & \text{if} \quad i = j \\ 0 & \text{otherwise.} \end{cases} \tag{9.5.12}$$

Unfortunately there is no direct way of testing to see if a particular fluid does in fact satisfy the above relations, for it is not possible to set up a flow field in which all the components of the rate of strain tensor can be varied independently. It is possible, however, to test the validity of these formulae *indirectly* by comparing measurements made in various flow fields with predicted values obtained from solutions of the differential equations for the velocity and pressure fields. Since these equations, which will be set out in the following section, are derived using the formula (9.5.11), the measurements provide a test of the formula. For many commonly occurring liquids it is found that eqn (9.5.11) is valid over the entire range of strain rates encountered in normal practice. Such liquids are said to be 'Newtonian'. The viscosities of some of these liquids are listed in Table 9.1. Since viscosity decreases rather rapidly with

Table 9.1

The viscosities of commonly occurring liquids at 15 °C *and* 1 *atm. The unit in the SI system is* 1 N m^{-2} s = 1 *Pascal second* = 10 *poise.* (1 *poise* = 1 *dyne* cm^{-2} s = 1 g cm^{-1} s^{-1})

	Water	Mercury	Ethyl alcohol	Carbon tetrachloride	Olive oil	Glycerol
η(Pa s)	0.00114	0.00158	0.00134	0.00104	0.099	2.33

Table 9.2
The viscosity of water over a range of temperatures

Temperature (°C)	0	5	10	15	20	25
η(mPa s)	1.787	1.514	1.304	1.137	1.002	0.891

increase in temperature it is important to note the temperature at which the measurements were made; in this case 15 °C. Table 9.2 gives the viscosities of water for a range of temperatures. Some experimental techniques for measuring viscosities will be described in section 9.7.

9.5.2 *Physical significance of* $\boldsymbol{e}$

Before proceeding to the derivation of the differential equations for the velocity field referred to above, we pause briefly to discuss the physical significance of the rate of strain tensor $\boldsymbol{e}$. With the aid of the formulae (9.5.9) for the components of $\boldsymbol{e}$ we can write the expression (9.5.3) for the relative velocity in the component form

$$v_i(\boldsymbol{x}+\Delta\boldsymbol{x}) = v_i(\boldsymbol{x}) + \tfrac{1}{2}\sum_{j=1}^{3}\left(\frac{\partial v_i}{\partial x_j} - \frac{\partial v_j}{\partial x_i}\right)\Delta x_j + \sum_{j=1}^{3} e_{ij}\Delta x_j, \quad (9.5.13)$$

where e_{ij} is evaluated at the point $\boldsymbol{x}$. The first two terms on the right-hand side of this formula are the components of the vector expression

$$\boldsymbol{v}(\boldsymbol{x}) + \boldsymbol{\Omega}(\boldsymbol{x}) \times \Delta\boldsymbol{x}, \quad (9.5.14)$$

where the components of the 'local angular velocity' $\boldsymbol{\Omega}(\boldsymbol{x})$ are related to the quantities $\frac{1}{2}[\partial v_i/\partial x_j - \partial v_j/\partial x_i]$ (see Exercise 9.5.1).

On comparing eqn (9.5.13) with the general formula (eqn (9.4.3)) for rigid body motion, we see that the first two terms on the right-hand side of eqn (9.5.13) correspond to a translation and a rigid body rotation of the fluid. Thus the local rate of deformation or 'strain' of the fluid in the neighbourhood of $\boldsymbol{x}$ is entirely determined by the last term in eqn (9.5.13), which in turn depends only on $\boldsymbol{e}(\boldsymbol{x})$. This is the reason why $\boldsymbol{e}$ is known as the 'rate of strain tensor'.

9.5.3 *Relationship between stress and strain rate in suspensions*

Although we will be studying the rheology of suspensions in detail in section 9.10, it is appropriate at this point to describe the limitations of the above arguments when applied to suspensions.

In section 2.4.3 we noted that most suspensions behave, from the macroscopic point of view, like non-Newtonian liquids, even if the suspending liquid is Newtonian. To understand why this is so it is important to appreciate the fact that the macroscopic stress tensor represents an average of the stress over regions containing a large number of colloidal particles, and that this average therefore depends on the particle configuration. For the case of a dilute suspension of rod-like particles for example, this configuration is characterized by the particle orientation distribution.

In general the macroscopic flow tends to give the particles a 'preferred' configuration, as for example in the case of a shear flow, where rod-like particles tend to be aligned with the streamlines (section 9.10.4). This 'ordering' tendency is opposed by the Brownian motion, and the final particle configuration is determined by the balance between these two opposing forces. In the limit of weak strain rates the Brownian motion dominates, and the configuration becomes statistically isotropic; thus in this limit the Newtonian form (eqn (9.5.7)) applies; at higher strain rates this is not generally the case, and as a result the rheological behaviour is characterized not by one, but by a number of viscosity coefficients, and these in turn usually depend on the strain rates.

For unsteady flows the situation can be even more complicated, because a change in the imposed strain rates leads to a change in particle configuration, a change that may take a significant time; in a suspension of rod-like particles this 'relaxation time' is determined by the time required for Brownian motion to reorient the particles, which for a 1 μm long particle is of the order of 1 s. If the imposed strain rates change significantly over this period, the assumption that the stress depends on the instantaneous rate of strain breaks down, and to determine the stress it is necessary to look at the recent history of strain rates. Clearly the combination of rigid particles and Newtonian liquid can lead to some formidable complications!

Exercises. 9.5.1 By writing out the components of the expression (9.5.14) for the local velocity field explicitly, and comparing the result with the first two terms of eqn (9.5.13), show that the components of the angular velocity $\mathbf{\Omega}$ are given by

$$\Omega_1 = \frac{1}{2}\left(\frac{\partial v_3}{\partial x_2} - \frac{\partial v_2}{\partial x_3}\right),$$

$$\Omega_2 = \frac{1}{2}\left(\frac{\partial v_1}{\partial x_3} - \frac{\partial v_3}{\partial x_1}\right),$$

and

$$\Omega_3 = \frac{1}{2}\left(\frac{\partial v_2}{\partial x_1} - \frac{\partial v_1}{\partial x_2}\right).$$

9.6 The Navier–Stokes equations

In this section we derive the remaining differential equation for the velocity field in a Newtonian liquid. This equation is obtained by applying Newton's Second Law of motion to a small rectangular block of liquid centred on the point $\boldsymbol{x}$. The edges of the block are assumed to be instantaneously parallel to the coordinate axes and of length Δx_1, Δx_2, and Δx_3, respectively (Fig. 9.2.1).

In section 9.3 we showed that the stress on an area element with unit normal $\hat{n}$ is given by

$$\mathscr{S}(\boldsymbol{x}, \hat{n}) = \sum_{i=1}^{3} \sum_{j=1}^{3} \sigma_{ij}(\boldsymbol{x}) \hat{e}_i (\hat{e}_j \cdot \hat{n}), \qquad [9.3.7]$$

where $\boldsymbol{x}$ denotes the centre of the area element. If the block in Fig. 9.2.1 is sufficiently small we can treat each face of the block as an area element. Then the force on the face with unit normal $\hat{e}_1$ is

$$\sum_{i=1}^{3} \sigma_{i1}(\boldsymbol{x} + \tfrac{1}{2}\Delta x_1 \hat{e}_1) \hat{e}_i \Delta x_2 \Delta x_3,$$

a result obtained by replacing $\hat{n}$ in eqn (9.3.7) by $\hat{e}_1$ and using the fact that the unit vectors $\hat{e}_1$, $\hat{e}_2$, and $\hat{e}_3$ are orthogonal (so $\hat{e}_i \cdot \hat{e}_j = 0$, $i \neq j$).

Subtracting the force on the opposing face of the block we find that the net force on the two faces is

$$\sum_{i=1}^{3} [\sigma_{i1}(\boldsymbol{x} + \tfrac{1}{2}\Delta x_1 \hat{e}_1) - \sigma_{i1}(\boldsymbol{x} - \tfrac{1}{2}\Delta x_1 \hat{e}_1)] \hat{e}_i \Delta x_2 \Delta x_3,$$

which for small Δx_1 can be approximated by

$$\sum_{i=1}^{3} \frac{\partial \sigma_{i1}}{\partial x_1} \hat{e}_i \Delta \mathscr{V},$$

where $\Delta \mathscr{V}$ is the volume of the block and the derivatives are evaluated at $\boldsymbol{x}$.

The forces on the remaining faces are obtained similarly; hence the net force per unit volume due to the stresses on the surface of the block is:

$$\sum_{j=1}^{3} \sum_{i=1}^{3} \frac{\partial \sigma_{ij}}{\partial x_j} \hat{e}_i. \qquad (9.6.1)$$

With the aid of the formula (eqn (9.5.11)) for the stress in a Newtonian liquid we can write this expression in the form

$$-\sum_{i=1}^{3} \frac{\partial p}{\partial x_i} \hat{e}_i + 2\eta \sum_{j=1}^{3} \sum_{i=1}^{3} \frac{\partial e_{ij}}{\partial x_j} \hat{e}_i. \qquad (9.6.2)$$

From the definition (eqn (9.5.9)) of the rate of strain tensor it follows that

$$\sum_{j=1}^{3} \frac{\partial e_{ij}}{\partial x_j} = \frac{1}{2} \sum_{j=1}^{3} \left(\frac{\partial^2 v_i}{\partial x_j \partial x_j} + \frac{\partial^2 v_j}{\partial x_j \partial x_i} \right). \tag{9.6.3}$$

Now

$$\sum_{j=1}^{3} \frac{\partial^2 v_j}{\partial x_j \partial x_i} = \frac{\partial}{\partial x_i} \sum_{j=1}^{3} \left(\frac{\partial v_j}{\partial x_j} \right) = \frac{\partial}{\partial x_i} (\nabla \cdot \boldsymbol{v}),$$

which is zero for an incompressible fluid (eqn (9.2.3)). Thus eqn (9.6.3) reduces to

$$\sum_{j=1}^{3} \frac{\partial e_{ij}}{\partial x_j} = \tfrac{1}{2} \nabla^2 v_i,$$

and the formula (9.6.2) for the stress force per unit volume becomes

$$-\nabla p + \eta \nabla^2 \boldsymbol{v}. \tag{9.6.4}$$

In order to make the physical significance of the viscous force term $\eta \nabla^2 \boldsymbol{v}$ a little clearer, we will rederive this term for a unidirectional flow, starting with the Newtonian formula (eqn (2.4.3)) for shear stress.

The flow is assumed to be everywhere parallel to the x_1 axis with the velocity depending only on x_2. In this case the shear force on the top of the block is

$$\eta \left(\frac{\partial v_1}{\partial x_2} \right)_{x_2 + \Delta x_2} \hat{e}_1 \Delta x_1 \Delta x_3 .$$

Evaluating the force on the bottom of the block and then subtracting, we find that the net force per unit volume is

$$\eta \frac{\partial^2 v_1}{\partial x_2^2} \hat{e}_1,$$

a result which is consistent with eqn (9.6.4) for this type of flow.

If v_1 also depends on x_3 there will be additional shear forces on the faces with unit normals $\pm \hat{e}_3$, leading to an extra force

$$\eta \left(\frac{\partial^2 v_1}{\partial x_3^2} \right) \hat{e}_1.$$

Finally, from eqn (9.6.2) we see that in the general case, where v_1 also depends on x_1, and the components v_2 and v_3 are non-zero, there will be additional *normal* stresses on the $\pm \hat{e}_1$ faces, which lead to a force per unit volume

$$2\eta \left(\frac{\partial^2 v_1}{\partial x_1^2} \right) \hat{e}_1.$$

This force, which *cannot* be obtained from the simple Newtonian expression (eqn (2.4.3)), arises not from shearing of the liquid but from stretching in the x_1 direction. Additional shear stresses in the x_1 direction due to the velocity gradients $\partial v_2/\partial x_1$ and $\partial v_3/\partial x_1$ lead to a force per unit volume

$$\eta\left(\frac{\partial^2 v_2}{\partial x_2\,\partial x_1}+\frac{\partial^2 v_3}{\partial x_3\,\partial x_1}\right)\hat{e}_1$$

which by the incompressibility constraint, can be written as

$$-\eta\frac{\partial^2 v_1}{\partial x_1^2}\hat{e}_1.$$

Thus on summing these results, we find that the net force in the x_1 direction is, per unit volume:

$$\eta\nabla^2 v_1\hat{e}_1,$$

with similar results for the x_2 and x_3 directions.

In addition to the pressure gradient and the viscous force there will also be a contribution from the long-range forces. We let $\boldsymbol{F}$ denote the total long-range force per unit volume. If gravity is the only such force:

$$\boldsymbol{F}=\rho\boldsymbol{g}, \tag{9.6.5}$$

where, as before, $\boldsymbol{g}$ is the gravitational acceleration vector, directed vertically down. The total force per unit volume on a Newtonian liquid is therefore given by the sum

$$\boldsymbol{F}-\nabla p+\eta\nabla^2\boldsymbol{v}. \tag{9.6.6}$$

The required differential equation for $\boldsymbol{v}$ is obtained by equating this term to the mass times the acceleration per unit volume of the block and writing the acceleration in terms of the velocity field.

During a time Δt the velocity of a particle originally at $\boldsymbol{x}$ changes by

$$\boldsymbol{v}(\boldsymbol{x}+\Delta\boldsymbol{x},t+\Delta t)-\boldsymbol{v}(\boldsymbol{x},t).$$

If Δt is sufficiently small we can approximate this difference by the total differential formula

$$\frac{\partial\boldsymbol{v}}{\partial t}\Delta t+\sum_{i=1}^{3}\frac{\partial\boldsymbol{v}}{\partial x_i}\Delta x_i$$

and hence obtain for the acceleration (Exercise 9.6.2):

$$\frac{\partial\boldsymbol{v}}{\partial t}+\boldsymbol{v}\cdot\nabla\boldsymbol{v}. \tag{9.6.7}$$

Although there will be some slight variation in the acceleration over the block, we can to first approximation take the acceleration to be uniform. By combining the above result with the expression (9.6.6) for the force per unit volume on the block, we obtain

$$\rho\left(\frac{\partial \boldsymbol{v}}{\partial t}+\boldsymbol{v}\cdot\nabla\boldsymbol{v}\right)=\boldsymbol{F}-\nabla p+\eta\nabla^2\boldsymbol{v}. \qquad (9.6.8)$$

This equation and the incompressibility constraint

$$\nabla\cdot\boldsymbol{v}=0, \qquad (9.6.9)$$

are known as the Navier–Stokes equations. These are the equations that govern most of the commonly occurring fluid flows, from the large scale flows in the oceans down to the small scale flows of interest here.

In most applications the long-range force distribution $\boldsymbol{F}(\boldsymbol{x})$ is given† and the unknowns are the velocity and pressure fields. To determine these quantities for any situation it is necessary to solve the Navier–Stokes equations subject to conditions on the boundary of the fluid. These conditions depend on the nature of the boundaries, which in our case are the solid particle surfaces. Experimental observations indicate that the layer of fluid next to the surface sticks to that surface; thus the particle and the fluid adjacent to the particle move with the same local velocity. This is referred to as the 'no-slip' boundary condition.

Exercises. 9.6.1 Establish eqn (9.6.7) using the fact that $\Delta x_i = v_i \Delta t$ if Δt is small.

9.7 Methods for measuring the viscosity

In this section we will study some simple devices for measuring viscosity, namely the Couette, the Ostwald, and the cone/plate viscometers. We will not attempt to present a detailed description of the experimental procedure, but will concentrate instead on the theory behind these devices. This discussion will serve to illustrate the principles of viscometric measurement, and give some familiarity with the results of applying the Navier–Stokes equations before we turn to the more complicated colloidal flow problems.

9.7.1 *The Couette (cylinder-in-cylinder) viscometer*

The Couette viscometer consists of two coaxial cylinders, as shown in Fig. 9.7.1. The space between the cylinders is filled with liquid. One of the cylinders (preferably the outer one) is rotated and the other is held in

† In flows around charged colloidal particles, $\boldsymbol{F}$ depends on the distribution of ions in the liquid; in this case the Navier–Stokes equations must be supplemented by equations for the ion distribution and electrical potential. See section 9.11 below and Chapter 13 of Vol. II.

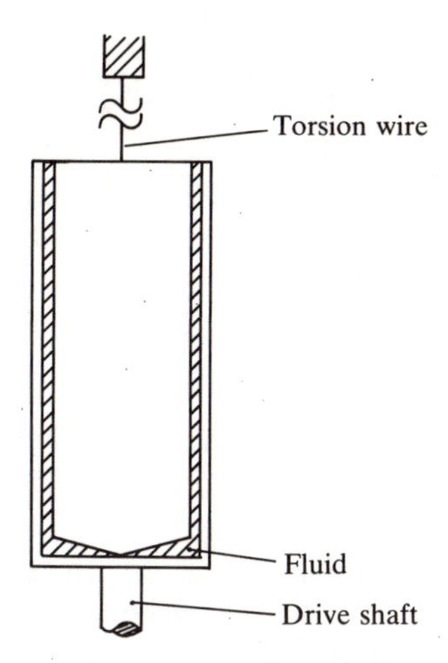

FIG. 9.7.1. A cross-sectional sketch of a Couette viscometer. The liquid occupies the shaded region between the cylinders.

place by a torsion wire. The viscous drag of the fluid against the surface of this (inner) cylinder causes it to twist against the torsion wire and it comes to rest when the torque imparted to it by the moving fluid is equal to the restoring torque in the wire.

With the aid of the Navier–Stokes equations we can determine the relationship between the torque and the speed of the moving cylinder. However, instead of working directly from the Navier–Stokes equations, we will in effect, rederive these equations for this particular flow, once again with the aim of making the meaning of the various terms clearer.

Consider the forces which act on the cylinder of material lying within an arbitrary distance r from the axis of the apparatus (see Fig. 9.7.2). If

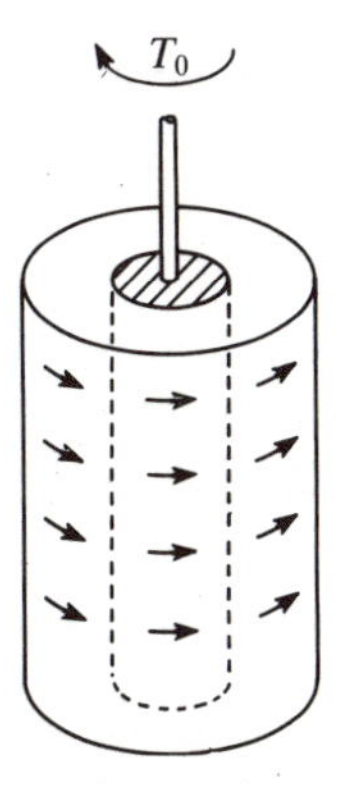

FIG. 9.7.2. A sketch of the forces acting on the inner cylinder and the liquid lying within a distance r of the axis of the Couette device. The outer cylinder is taken to be rotating in the counterclockwise direction when viewed from above.

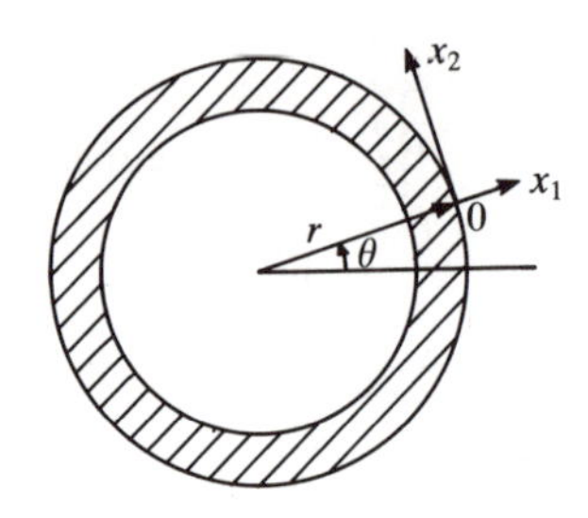

FIG. 9.7.3. A cross-sectional slice of the volume shown in the previous figure, illustrating the local Cartesian coordinates at a point 0 on the surface.

the flow is steady (that is, independent of t) the net torque on the volume will be zero. Thus the torque due to the torsion wire T_0 must be balanced by the torque due to shear stresses on the curved surface of the cylinder, which is assumed to be in the fluid. In Fig. 9.7.2 the direction of these shear stresses is indicated by the arrows, which will point in the direction of rotation of the outer cylinder.

To calculate the shear stress at an arbitrary point 0 on the curved surface of the volume we must first determine the rate of strain tensor $\boldsymbol{e}$ at that point. In preparation for this step we set up local Cartesian coordinates with the origin at the point 0, as shown in Fig. 9.7.3.

It is assumed that the fluid velocity is independent of distance along the cylinder axis, and that the fluid particles move in circles around the axis, that is

$$\boldsymbol{v} = v\hat{\theta} \tag{9.7.1}$$

where $\hat{\theta}$ is the unit vector in the direction of increasing θ. By symmetry, v must be independent of angle θ.

The velocity of any point in the neighbourhood of 0 can be written as $\boldsymbol{v}_0 + \Delta\boldsymbol{v}$, where $\boldsymbol{v}_0$ is the velocity at 0, and, by the chain rule

$$\Delta\boldsymbol{v} \approx \left(\frac{\mathrm{d}v}{\mathrm{d}r}\right)_0 \Delta r\hat{\theta} + v_0\frac{\mathrm{d}\hat{\theta}}{\mathrm{d}\theta}\Delta\theta; \tag{9.7.2}$$

where Δr and $\Delta\theta$ denote the small changes in the polar coordinates of Fig. 9.7.3 in moving from 0 to the point in question. With the same order of accuracy, we can put

$$\Delta r = x_1, \quad \text{and} \quad r\Delta\theta = x_2,$$

where x_1 and x_2 are the local Cartesian coordinates of the point. Combining this result with eqn (9.7.2), and using the fact that

$$\hat{\theta} \approx \hat{e}_2,$$

and

$$\frac{\partial \hat{\theta}}{\partial \theta} = -\hat{r} \approx -\hat{e}_1$$

(see Exercise 9.7.1), we find the local Cartesian components of $\Delta \boldsymbol{v}$ are given by

$$\Delta v_1 = -\frac{v_0}{r} x_2$$

and

$$\Delta v_2 = \left(\frac{\mathrm{d}v}{\mathrm{d}r}\right)_0 x_1.$$

From the formula (9.5.9) for the rate of strain tensor, we see that the only non-zero components of $\boldsymbol{e}$ at 0 are e_{12} and e_{21}, both of which are given by

$$\frac{1}{2}\left(\frac{\mathrm{d}v}{\mathrm{d}r} - \frac{v}{r}\right)_0. \tag{9.7.3}$$

With the aid of the Newtonian expression (eqn (9.5.11)) for the Cartesian components of the stress tensor we see that the stress on the curved surface of the cylindrical volume is composed of a pressure, normal to the surface, and a shear stress, of magnitude given by

$$\mathcal{S} = \eta\left(\frac{\mathrm{d}v}{\mathrm{d}r} - \frac{v}{r}\right). \tag{9.7.4}$$

This formula differs from the Newtonian expression (eqn (2.4.3)) for $\mathcal{S}$ in a simple shear flow, for in addition to the velocity gradient term there is a term associated with the curvature of the streamlines.

The shear stresses on the outer surface of the cylinder lead to a torque, given by

$$2\pi r^2 L \mathcal{S},$$

where L is the length of the cylinder. Replacing $\mathcal{S}$ by the formula (9.7.4) and equating the result to the torque T_0 of the torsion wire, we get

$$2\pi r^2 \eta\left(\frac{\mathrm{d}v}{\mathrm{d}r} - \frac{v}{r}\right) L = T_0. \tag{9.7.5}$$

Integrating with respect to r, and using the fact that $v = 0$ on the fixed inner cylinder $r = R_1$, we obtain (Exercise 9.7.2):

$$v = \frac{T_0 r}{4\pi\eta L}\left\{\frac{1}{R_1^2} - \frac{1}{r^2}\right\}. \tag{9.7.6}$$

On the outer cylinder v is equal to $R_2\Omega$ where R_2 is the radius and Ω is the angular velocity of the cylinder. Substituting these values for r and v

in the above expression and rearranging, we find that torque, T_0, is given by

$$T_0 = \left[\frac{4\pi L R_1^2 R_2^2}{R_2^2 - R_1^2}\right] \eta \Omega. \tag{9.7.7}$$

This is the required expression, relating measured torque to angular velocity. The quantity in square brackets and the relationship between torque and twist of the wire are instrument constants that can be evaluated by making measurements on a liquid of known viscosity. It can usually be assumed that, even for quite large angles of twist (up to $\sim\pi$), the restoring torque in the wire is directly proportional to the angle of twist of the wire.

Using the formula (9.7.6) for v, we find that the shear rate, given by eqn (9.7.3), is

$$T_0/4\pi\eta L r^2 \tag{9.7.8}$$

in the Couette device. Thus if $R_2 = 2R_1$ for example, the strain rate at the surface of the outer cylinder will be one quarter of the value at the inner cylinder. Such a device would not be well suited for use with non-Newtonian liquids since the shear viscosity generally varies with shear rate, making the above analysis invalid. Although it is possible to modify the analysis to take into account the shear rate dependence of the viscosity (Krieger and Maron 1952, 1954), the problem is usually overcome by making the gap width $R_2 - R_1$ much smaller than the cylinder radius; in this case the strain rate is approximately uniform across the gap (Exercise 9.7.4) and so the Newtonian analysis can be applied.

The above results can, as we mentioned earlier, be derived directly from the Navier–Stokes equations; in fact by combining the Newtonian expression (eqn (9.5.11)) for the stress, with the torque balance constraint in the above analysis we have, in effect retraced the derivation of part of the Navier–Stokes equations for this flow, namely the equation associated with force balance in the $\hat{\theta}$ direction. While this procedure will hopefully provide some insight into the application of the Navier–Stokes equations to this flow it is obviously inefficient to rederive the equations for each application. Normally, a problem such as this would be solved by first writing the Navier–Stokes equations in cylindrical coordinates in order to take advantage of the symmetry of the problem. The appropriate forms of these equations in cylindrical and other curvilinear coordinate systems are given in standard fluid texts (see for example Appendix 2 of Batchelor 1967). The assumption that the flow is circumferential then leads to three differential equations. The first equation is equivalent to eqn (9.7.5), while the remaining two have the form (Landau and Lifshitz

1959, section 18)

$$\frac{\partial p}{\partial r} = \rho \frac{v^2}{r} \tag{9.7.9}$$

and

$$\frac{\partial p}{\partial z} = \rho g.$$

The last equation simply represents the hydrostatic variation of pressure with depth z, assuming the axis of the apparatus to be vertical, while eqn (9.7.9) represents the balance between centrifugal forces and radial pressure gradients in the liquid. For the case of a rotating inner cylinder, this balance breaks down at high speed and a steady radial flow pattern develops. The speed at which this flow begins can be predicted from stability analysis of the Navier–Stokes equation. For the fixed inner cylinder device the circumferential flow pattern breaks down at much higher speeds; in this case the breakdown leads to a turbulent flow. The speed at which the turbulence begins cannot as yet be predicted theoretically. Since the measured torques can only be interpreted theoretically for the *circumferential* flow regime it is important to test any device to ensure that torque and speed are linearly related over the experimental range of speeds; this will ensure that there is no radial flow. Clearly, Couette devices with a rotating *outer* cylinder are preferable, since the circumferential flow assumption breaks down at higher speeds for these devices.

One final point on the Couette viscometer. The flow regime in the bottom of the cylinder is different from that in the annulus and an end correction may, therefore, be necessary. It can be assessed (Alexander and Johnson 1949) by filling the cylinder to two different depths, L_1 and L_2, and determining the difference in deflection $(\theta_2 - \theta_1)$ for a given rotational speed (Exercise 9.7.6). Hunter and Nicol (1968) in their experiments shaped the bottom of the two cylinders in the form of cones (Fig. 9.7.1) calculated to give approximately the same shear rate as that in the annulus. This seems to eliminate the end effect quite satisfactorily. (See Van Wazer *et al.* 1963, pp. 68–72 for details.)

9.7.2 *The Ostwald viscometer*

Another device which is commonly used by colloid chemists is the *Ostwald viscometer,* shown in Fig. 9.7.4. This viscometer consists essentially of two reservoirs linked by a fine capillary tube. Fluid is drawn up to the top mark of the left-hand tube in the diagram and allowed to drain out while the tube is held vertical. The time required for the level

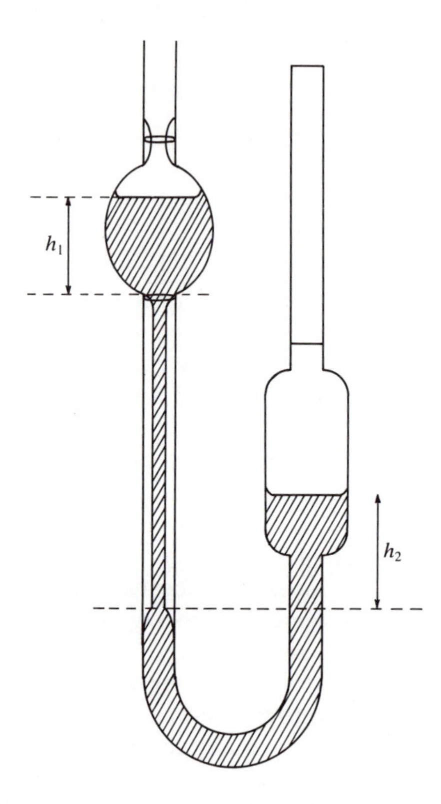

FIG. 9.7.4. The Ostwald viscometer. The volume of liquid used is such that the height in the right-hand reservoir moves symmetrically about the centre of that reservoir as the height in the left-hand arm falls from the upper to the lower mark. The constriction makes timing more accurate.

to drop to the lower mark is related to the viscosity; the more viscous the fluid the longer it takes to drain.

In order to find the precise form of this relationship we must determine the flow field in the capillary. We take the x_1 axis of our coordinate system to coincide with the centre line of the capillary tube, with x_1 increasing down the tube. It is assumed that the fluid flows in the x_1 direction only. The continuity equation therefore reduces to

$$\frac{\partial v_1}{\partial x_1} = 0,$$

implying that the flow profile is independent of distance down the tube.

Writing out each component of the remaining Navier–Stokes equation

(9.6.8) separately, we get

$$\rho \frac{\partial v_1}{\partial t} = \rho g - \frac{\partial p}{\partial x_1} + \eta \left(\frac{\partial^2 v_1}{\partial x_2^2} + \frac{\partial^2 v_1}{\partial x_3^2} \right) \tag{9.7.10}$$

and

$$\left. \begin{aligned} \frac{\partial p}{\partial x_2} &= 0, \\ \frac{\partial p}{\partial x_3} &= 0. \end{aligned} \right\} \tag{9.7.11}$$

For most viscometers of this type, the inertia term in eqn (9.7.10) may be neglected, and we may treat the flow as quasi-steady. In order to find out when such an approximation is valid, we must estimate the magnitude of the various terms in the equation. Since the arguments used to obtain these estimates will be used a number of times in this chapter we will describe them here in some detail.

The fluid velocity will vary over the tube cross-section, from zero at the wall to a maximum value V, which will presumably occur somewhere near the centre of the tube. Thus the magnitude of the velocity gradients in the tube should be of order V/a, while the second derivatives will be of order V/a^2 where a is the tube radius. Similarly the magnitude of the inertia term $\rho(\partial \boldsymbol{v}/\partial t)$ is expected to be of order $\rho V/t_0$, where t_0 is the time required for the velocity in the tube to vary significantly. Since, as we shall see, the velocity is related to the height of the fluid in the reservoirs, t_0 is the time for the liquid level to drop by an appreciable amount. Thus the ratio

$$\left| \rho \frac{\partial \boldsymbol{v}}{\partial t} \right| \Big/ |\eta \nabla^2 v_1|$$

of the inertia to the viscous term in eqn (9.7.10) is of the order of $\rho a^2/\eta t_0$. For a typical Ostwald device a is a millimetre or so, and t_0—which we may take to be the time required for the liquid to drain between the two marks—is around 100 s. The ratio in this case is 0.01 and the inertia terms may therefore be neglected. Arguments such as this are commonly used in fluid mechanics to simplify the Navier–Stokes equations. (See section 9.8.1.)

From eqn (9.7.11) it follows that the pressure is uniform over the cross-section of the tube. Thus the eqn (9.7.10) contains a term, $\partial p/\partial x_1$, which is independent of x_2 and x_3, while the remaining terms are independent of x_1. It follows that

$$\eta \left(\frac{\partial^2 v_1}{\partial x_2^2} + \frac{\partial^2 v_1}{\partial x_3^2} \right) = \frac{\mathrm{d}p}{\mathrm{d}x_1} - \rho g = G, \tag{9.7.12}$$

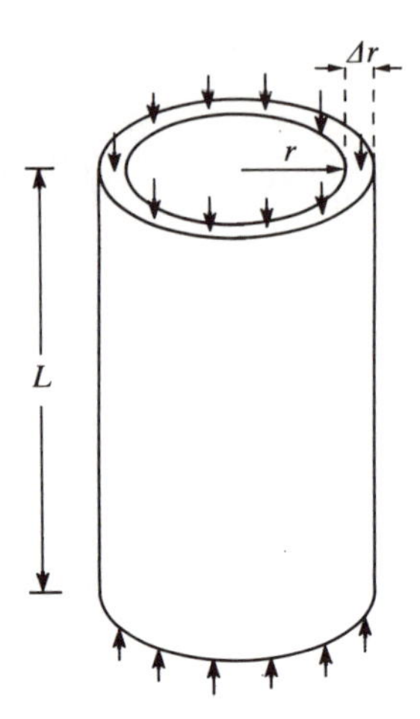

FIG. 9.7.5. The hollow cylinder of liquid used in the alternative derivation of the eqn (9.7.15) for the flow down a capillary.

where G is a constant. On using the fact that the pressure gradient is uniform down the tube, we find that

$$G = \frac{p_2 - p_1}{L} - \rho g \tag{9.7.13}$$

where p_1 and p_2 are the pressures at the top and bottom of the capillary and L is the tube length.

Since the tube has a circular cross-section, v_1 can only depend on distance r from the centre line. Thus eqn (9.7.10) reduces to the ordinary differential equation (see Exercise 9.7.7)

$$\frac{1}{r}\frac{\mathrm{d}}{\mathrm{d}r}\left(r\frac{\mathrm{d}v_1}{\mathrm{d}r}\right) = \frac{G}{\eta}. \tag{9.7.14}$$

As in the previous section, this differential equation for the velocity can also be obtained by combining the Newtonian expression for the stress tensor with a force balance on a suitably chosen volume of fluid. In this case the appropriate volume is the thin-walled cylinder shown in Fig. 9.7.5. With the aid of eqns (9.5.9) and (9.5.11) it can be shown that the shear stress on the inner surface acts in the direction of the flow, and is given by

$$\eta\left(\frac{\mathrm{d}v}{\mathrm{d}r}\right)_r;$$

exactly what you would expect from the formula (eqn (2.4.3)) for stress in a simple shear flow.

Since the inertia terms are negligible, the sum of the forces on the volume must be zero, that is

$$(p_1 - p_2) \cdot 2\pi r \Delta r - 2\pi\eta L\left\{\left(r\frac{\mathrm{d}v}{\mathrm{d}r}\right)_{r+\Delta r} - \left(r\frac{\mathrm{d}v}{\mathrm{d}r}\right)_r\right\} + 2\pi r \rho g L \Delta r = 0,$$

where Δr is the thickness of the cylinder wall. The required differential equation (9.7.14) is obtained by dividing by $2\pi r L \Delta r$ and taking the limit as $\Delta r \rightarrow 0$.

The solution of eqn (9.7.14) is readily found to be (Exercise 9.7.8)

$$v_1 = -\frac{G}{4\eta}(a^2 - r^2). \tag{9.7.15}$$

This parabolic flow profile is referred to as 'Poiseuille flow'. It applies not just to capillaries but to flow down any circular tube, provided the flow rate is not greater than the value at which turbulence sets in; for capillaries this is not usually a problem.

Integration of the velocity over the tube cross-section leads to the Poiseuille expression for volumetric flow rate, viz. (Exercise 9.7.8):

$$Q = \int_0^a 2\pi r v_1 \, \mathrm{d}r = -\frac{\pi G a^4}{8\eta}. \tag{9.7.16}$$

In order to find the relationship between G and the level of the liquid in the reservoirs we must solve the Navier–Stokes equations over the remaining portion of the viscometer. Fortunately, this is not so daunting a task as it sounds, for the flow in these regions is relatively slow and as a result the pressure distribution is approximately hydrostatic, viz.

$$p \approx p_0 + \rho g x_1 \tag{9.7.17}$$

where p_0 is a constant. The error involved in using this approximation is estimated in Exercise 9.7.11.

From eqn (9.7.17) it follows that

$$p_1 - p_2 = \rho g(h_1 - h_2) \tag{9.7.18}$$

where h_1 and h_2 are shown in Fig. 9.7.4. Combining this estimate with the formulae (9.7.13) and (9.7.16) for G and Q, we find that the flow rate is given by

$$Q = \frac{\pi \rho g a^4}{8\eta}\left\{1 + \frac{h_1 - h_2}{L}\right\}. \tag{9.7.19}$$

If the liquid levels h_1 and h_2 only change by a small fraction of L during the course of the experiment, the flow rate Q will be approximately constant; in this case Q is simply equal to the volume between the two markers on the left-hand reservoir, divided by the drainage time and from the measurement of Q, η can be determined using eqn (9.7.19). The various quantities in this expression that depend on the apparatus can be determined from a measurement of the flow rate for a liquid of known viscosity and density.

The Ostwald viscometer is very useful for obtaining accurate measurements of the viscosity of Newtonian fluids. Its major limitation in applications to colloidal systems is that the shear rate varies so widely across the capillary. It is necessarily zero at the axis and at the wall it can be as high as $2000\,\mathrm{s}^{-1}$. Obviously it is unwise to use such a device for measuring the viscosity of a non-Newtonian liquid, although some correction procedures are available. (See, for example, Maron *et al.* 1954.)

9.7.3 *The cone and plate viscometer*

Another interesting device for measuring the flow behaviour of non-Newtonian fluids is the cone and plate viscometer (Fig. 9.7.6). It is obvious from the figure that this instrument cannot be used for liquids of very low viscosity because the only method of restraining the fluid is through the surface tension generating a Laplace pressure (section 5.7.1) across the curved meniscus around the edges. The analysis of the flow in this device is simplified by the fact that the angle α between the cone and the plate is of the order of a degree or so, and thus the opposing surfaces are nearly parallel. By using order of magnitude estimates of the various terms in the Navier–Stokes equations for this situation it can be shown that the flow is locally the same as that between two infinite parallel plates, separated by the local gap thickness h. This is a result which is not just limited to cone and plate flow, but applies to any flow between

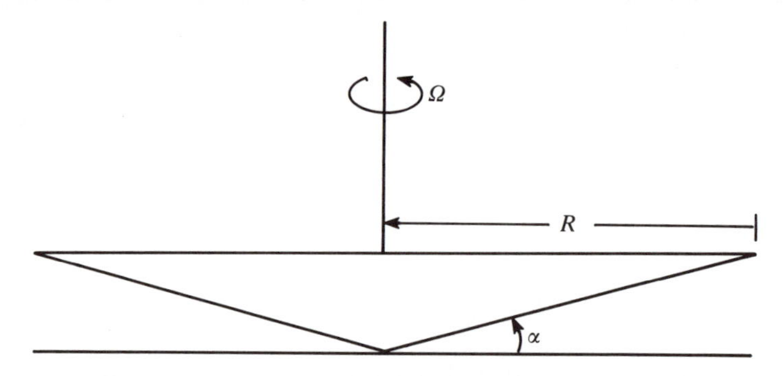

FIG. 9.7.6. The cone and plate viscometer. The angle α is typically very small ($<3°$ and often $<1°$). The cone is usually rotated but this not always the case. Rotating the plate allows the cone to be suspended on a torsion wire.

closely spaced nearly parallel solid surfaces and forms the basis of most studies on lubricant flow (Batchelor 1967, section 4.8). For this reason this parallel-plate approximation is called the lubrication approximation.

In this case one of the parallel plates is fixed while the other moves tangentially. Thus the flow is locally a simple shear flow with shear rate V/h, where V is the local cone or plate velocity, depending on which of the surfaces is fixed. By using the fact that

$$V = r\Omega,$$

and

$$h \approx r\alpha$$

where r is the distance from the axis of rotation and Ω is the angular velocity, we see that the local shear rate is

$$\Omega/\alpha.$$

Thus the shear rate is uniform in this device, and it is therefore well suited to the study of non-Newtonian liquids.

The moment M due to the uniform shear stresses on the cone or plate is readily found to be

$$M = \left(\tfrac{2}{3}\pi R^3 \frac{\Omega}{\alpha}\right)\eta \tag{9.7.20}$$

where R is the cone/plate radius.

Exercises. 9.7.1 Derive the result

$$\frac{\partial \hat{\theta}}{\partial \theta} = -\hat{r}$$

used in setting up the strain rate tensor for Couette flow. [Hint: First show that

$$\hat{\theta} = -\sin\theta\hat{\imath} + \cos\theta\hat{\jmath}$$

where $\hat{\imath}$ and $\hat{\jmath}$ are the fixed unit vectors shown in Fig. 9.7.7.] The required result can then be obtained by differentiating with respect to θ and using the representation of $\hat{r}$ in terms of $\hat{\imath}$ and $\hat{\jmath}$.

9.7.2 Derive eqns (9.7.6) and (9.7.7). [Hint: Consider $r(\mathrm{d}(v/r)/\mathrm{d}r)$.]

9.7.3 By solving eqn (9.7.5) show that the velocity field due to a rotating cylinder of radius R_1 in an infinite liquid is given by

$$v = \frac{\Omega R_1^2}{r}$$

where Ω is the angular velocity of the cylinder. Calculate the torque per unit length required to rotate the cylinder.

9.7.4 Show that the strain rate, (from eqns (9.7.3 and 4)) in the gap of a

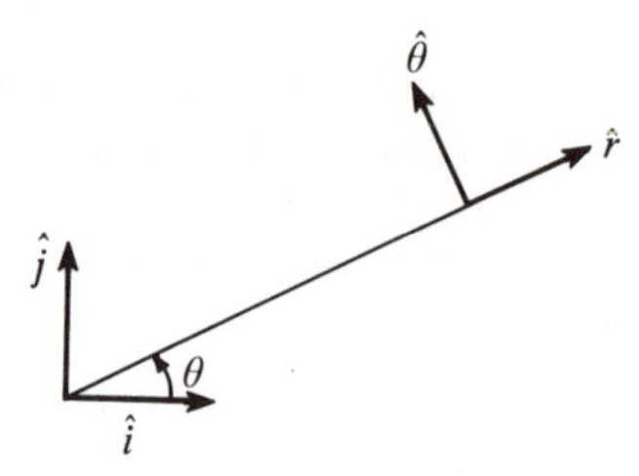

FIG. 9.7.7. An illustration of the unit vectors $\hat{i}$ and $\hat{j}$ used in Exercise 9.7.1.

Couette viscometer is equal to

$$\dot{\gamma} = \frac{\mathrm{d}v}{\mathrm{d}r} - \frac{v}{r} = 2\Omega\left(\frac{1}{r^2}\right)\left[\frac{1}{R_1^2} - \frac{1}{R_2^2}\right]^{-1}$$

where Ω is the angular velocity of the outer cylinder. Show that for very small gap widths, d, this can be reduced to strain rate $= \Omega(R_1/d)$, a value which can be obtained by treating the flow locally as a simple shear.

9.7.5 If the radius of the cylinders in a Couette viscometer is large compared to the gap width it is possible to approximate the flow in the gap by assuming it occurs between two parallel flat plates (see Fig. 9.7.8).

(a) Show that in that case the continuity equation gives $(\partial v_1/\partial x_1) = 0$, $v_2 = v_3 = 0$ and that the Navier–Stokes equation reduces to

$$-\frac{\partial p}{\partial x_1} + \frac{\mathrm{d}^2 v_1}{\mathrm{d}x_2^2} = 0; \qquad \frac{\partial p}{\partial x_2} = 0; \qquad \frac{\partial p}{\partial x_3} = \rho g.$$

(b) Hence show that $(\mathrm{d}^2 v_1/\mathrm{d}x_2^2) = 0$ and so $v_1(x_2) = (\Omega R_2 x_2)/d$.

(c) Calculate the torque on the inner cylinder and verify that the result is in agreement with the general formula (eqn (9.7.7)) in this limit.

9.7.6 Show that if the cylinders in a Couette viscometer are filled to two different depths, L_1 and L_2 then even in the presence of 'end effects'

$$\phi_2 - \phi_1 = K'(L_2 - L_1)\eta\Omega,$$

where ϕ_1 and ϕ_2 are the corresponding angles of twist of the torsion wire, and K' is an instrument constant.

9.7.7 Assuming that the velocity v in eqn (9.7.12) depends only on the distance

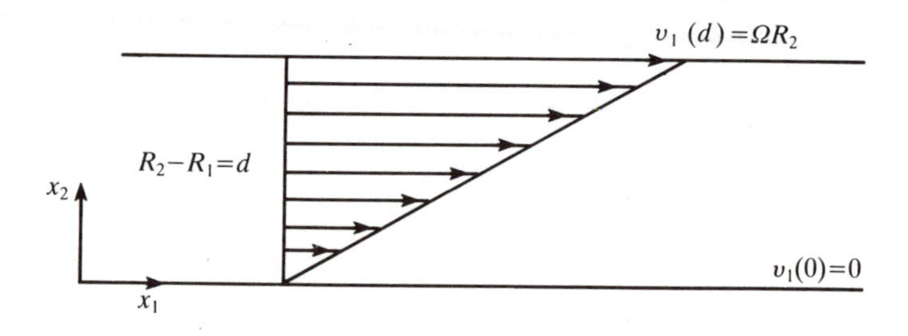

FIG. 9.7.8. Refer to Exercise 9.7.5.

r from the centreline of the capillary tube, show that this equation reduces to the form (9.7.14). (Hint: first show that

$$\frac{\partial^2 v}{\partial x_2^2}=\frac{\mathrm{d}^2 v}{\mathrm{d}r^2}\left(\frac{\partial r}{\partial x_2}\right)^2+\frac{\mathrm{d}v}{\mathrm{d}r}\frac{\partial^2 r}{\partial x_2^2} \quad \text{and} \quad \partial r/\partial x_2 = x_2/r.)$$

9.7.8 Establish the results (9.7.15) and (9.7.16) for Poiseuille flow.

9.7.9 A certain Ostwald viscometer has a capillary tube of length 5 cm and diameter 0.5 mm. Calculate the time required for the level of the water in the top reservoir to drop by 1 cm, assuming both reservoirs have a circular cross-section of diameter 1 cm. The temperature of the water is 20 °C. You may assume that the term

$$(h_1 - h_2)/L \ll 1$$

in the eqn (9.7.19) for the flow rate.

9.7.10 Calculate the shear rate at the wall of the capillary tube for the viscometer described in the previous question.

9.7.11 In applying the hydrostatic formula (eqn (9.7.17)) for the pressure in the Ostwald viscometer, we have, in effect, neglected the $\eta\nabla^2\boldsymbol{v}$ term in the equation of motion in the portion of the apparatus beyond the capillary tube. The aim of this exercise is to estimate the resulting error in the formula (9.7.19) for the flow rate.

The order of magnitude of the neglected $\eta\nabla^2\boldsymbol{v}$ term in this region is $\eta V/(a')^2$, where V and a' are the typical velocity and radius of the tube. These viscous forces must be balanced by an extra pressure gradient.

With the aid of this estimate, show that the relative error in the formula (9.7.19) for Q is of the order of

$$\left(\frac{a}{a'}\right)^4\left(\frac{L'}{L}\right),$$

where L' is the length of the viscometer tube beyond the capillary.

9.7.12 Hiemenz (1977) gives the following expression for the torque on the cone of a cone/plate viscometer.

$$M=\frac{\frac{4}{3}\pi R^3\eta\Omega\cos\alpha}{\tan\alpha+\frac{1}{2}\{\ln[(1+\sin\alpha)/(1-\sin\alpha)]\}\cos\alpha}.$$

Show that this reduces to eqn (9.7.20) for small α. (You will need the approximation $\ln(1+x)\approx x$ for small x.)

9.8 Sedimentation of a suspension

9.8.1 *The Stokes equations*

To determine the transport properties of a suspension, it is necessary to solve the Navier–Stokes equations for the flow field around the individual particles. This problem is greatly simplified by the fact that the flow field in this case has a very small length-scale, and as a result it is usually possible to neglect the inertia terms in the equation of motion. In this section we will justify this statement by estimating the order of magnitude of the various terms in the Navier–Stokes equations for the case of a sedimenting spherical particle.

In the neighbourhood of the particle the fluid velocity will presumably vary with distance on a length-scale of the order of the particle radius a. The velocity gradients should therefore be of order V/a, where V is the particle velocity; hence the magnitude of the inertia term $\rho \boldsymbol{v} \cdot \nabla \boldsymbol{v}$ in the equations of motion will be of the order of $\rho(V^2/a)$. By similar arguments the magnitude of the viscous force term $\eta \nabla^2 \boldsymbol{v}$ is estimated to be $\eta V/a^2$, and thus the ratio

$$|\rho \boldsymbol{v} \cdot \nabla \boldsymbol{v}|/|\eta \nabla^2 \boldsymbol{v}| = \rho \frac{Va}{\eta}. \tag{9.8.1}$$

This non-dimensional quantity is known as the Reynolds number; for most macroscopic flows, such as the flow around a tennis ball, the Reynolds number is very large; for colloidal flows however it is usually very small, thanks to the small particle radius and velocity; for example for a particle of 0.5 μm radius moving with a typical velocity of 10^{-6} m s^{-1} in water at 20 °C the Reynolds number is 5×10^{-7}. Thus for colloidal scale flows we may neglect the $\rho \boldsymbol{v} \cdot \nabla \boldsymbol{v}$ term in the equation of motion (eqn (9.6.8)), which reduces to

$$\rho \frac{\partial \boldsymbol{v}}{\partial t} = \boldsymbol{F} - \nabla p + \eta \nabla^2 \boldsymbol{v}. \tag{9.8.2}$$

Unlike the original equation, this equation is *linear,* and hence the sum of two solutions is also a solution (see Exercise 9.8.1). We can therefore analyse a combination of effects such as Brownian motion and sedimentation, by first analysing each component in isolation and then superposing to obtain the combined effect.

In this section we will begin by studying sedimentation in the absence of Brownian motion. In a frame of reference which moves with the steadily descending particle, the fluid velocity field is independent of time, and thus eqn (9.8.2) reduces to

$$\eta \nabla^2 \boldsymbol{v} = \nabla p - \boldsymbol{F}. \tag{9.8.3}$$

This equation and the incompressibility constraint

$$\nabla \cdot \boldsymbol{v} = 0, \qquad [9.6.9]$$

are known as the Stokes equations. These are the equations that must be solved for the calculation of the sedimentation coefficient, and, as we shall see later, for the calculation of the viscosity of a suspension.

9.8.2 *Sedimentation coefficient of a spherical colloidal particle*

In the sedimentation experiment, the body force $\boldsymbol{F}$ is approximately uniform over the neighbourhood of the particle, and the Stokes equation (9.8.3) can thus be written as (Exercise 9.8.3):

$$\eta \nabla^2 \boldsymbol{v} = \nabla p' \tag{9.8.4}$$

where

$$p' = p - \boldsymbol{F} \cdot \boldsymbol{r}$$

is a 'modified' pressure, and $\boldsymbol{r}$ is the position vector from the particle centre to the point in question. The calculation of the flow field around a sedimenting particle involves the solution of this equation and eqn (9.6.9) subject to the boundary conditions that

$$\boldsymbol{v} = \boldsymbol{0} \tag{9.8.5}$$

on the particle surface $r = a$, and

$$\boldsymbol{v} \to -\boldsymbol{V} \tag{9.8.6}$$

far from the particle. The sedimentation velocity $\boldsymbol{V}$ is determined from the constraint that the net force on the particle is zero.

The solution to this problem, which is described in standard texts (see for example section 4.9 of Batchelor 1970 or section 20 of Landau and Lifshitz 1959) is given by

$$\left.\begin{aligned} \boldsymbol{v} &= \left(-1 + \frac{3}{2}\frac{a}{r} - \frac{1}{2}\frac{a^3}{r^3}\right) V \cos\theta \hat{\boldsymbol{r}} + \left(1 - \frac{3}{4}\frac{a}{r} - \frac{1}{4}\frac{a^3}{r^3}\right) V \sin\theta \hat{\boldsymbol{\theta}} \\ &\text{and} \\ p' &= -\frac{3}{2}\eta\frac{a}{r^2} V \cos\theta, \end{aligned}\right\} \tag{9.8.7}$$

where θ is the angle between the direction of particle motion and the radius vector to the point in question; see Fig. 9.8.1. Note that the disturbance velocity $\boldsymbol{v} + \boldsymbol{V}$ due to the presence of the particle decays at large distances like r^{-1} (see Exercise 9.8.4). As we shall see, this relatively slow drop-off leads to significant particle interactions which limit the above analysis to very dilute suspensions.

The force on the particle due to the fluid is given by

$$\int_A -(\boldsymbol{F} \cdot \boldsymbol{r})\hat{\boldsymbol{n}}\, \mathrm{d}A + \int_A (-p' + 2\eta \boldsymbol{e}) \cdot \hat{\boldsymbol{n}}\, \mathrm{d}A$$

where A denotes the particle surface. With the aid of eqns (9.8.7) for $\boldsymbol{v}$

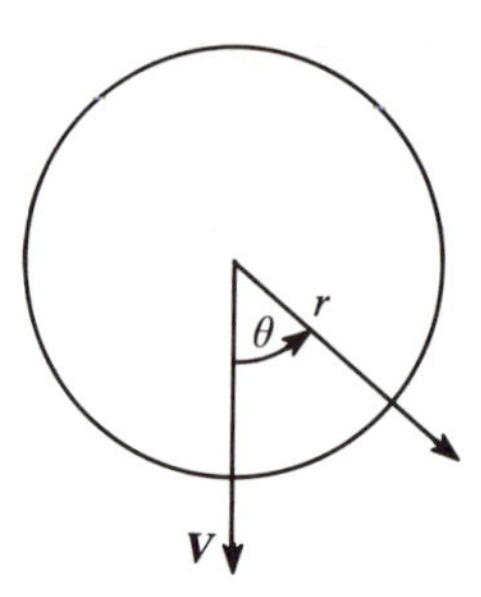

FIG. 9.8.1. An illustration of the coordinates r and θ, which appear in the formulae (9.8.7) for the velocity and pressure field around a sedimenting particle.

and p' it can be shown that the second integral, which represents the force due to the motion, yields the Stokes formula

$$-6\pi\eta a \boldsymbol{V}$$

while the first integral gives the buoyancy force

$$-\boldsymbol{F}\frac{4}{3}\pi a^3.$$

Thus on setting the sum of the hydrodynamic and sedimentation forces to zero, we find that (compare eqn (3.5.3)):

$$\boldsymbol{V} = \frac{2}{9}\frac{a^2}{\eta}\left(\frac{\rho_p}{\rho} - 1\right)\boldsymbol{F}, \tag{9.8.8}$$

where it is assumed that the body force on the particle is $(\rho_p/\rho)\boldsymbol{F}$, ρ_p being the particle density.

The Stokes equations have also been solved for ellipsoidal particles; in this case the sedimentation velocity depends on the particle orientation. Perrin (1934) has calculated the average sedimentation velocity for a dilute suspension of ellipsoids on the assumption that the Brownian motion has given the particles a uniform orientation distribution.

9.8.3 *Sedimentation in a concentrated suspension*

In a concentrated suspension the sedimentation velocity of a particle is affected by its hydrodynamic interaction with neighbouring particles. Since the disturbance velocity due to an isolated particle drops off on a length-scale of the order of the particle radius, the hydrodynamic interaction between a pair of particles will only be significant if their separation is of the order of a or less. Thus at low particle volume fractions ϕ, the fraction of interacting particles should be proportional to

ϕ, and of these the vast majority will be interacting with only one neighbour, since the probability of finding two (or more) particles within one or two radii of a given particle is proportional to ϕ^2 (see Exercise 9.8.2).

The average sedimentation velocity $\langle \boldsymbol{V} \rangle$ for such a suspension will therefore be given by an expression of the form

$$\langle \boldsymbol{V} \rangle = \boldsymbol{V}_0(1 + \alpha\phi + \beta\phi^2 + \ldots), \tag{9.8.9}$$

where $\boldsymbol{V}_0$ is the sedimentation velocity of an isolated particle, and the coefficient α can be obtained from the solution of Stokes' equations for a pair of interacting particles, averaged over all possible separations. The difficulties involved in the solution of this Stokes flow problem are so great that at present it is only possible to calculate α for spherical particles. For the important case of force-free spheres for which Brownian motion has made all separations equally likely it has been found that $\alpha = -6.55$ (Batchelor 1972).

In Fig. 9.8.2 the measured sedimentation velocities are shown for latex suspensions over a range of volume fractions. The broken line represents

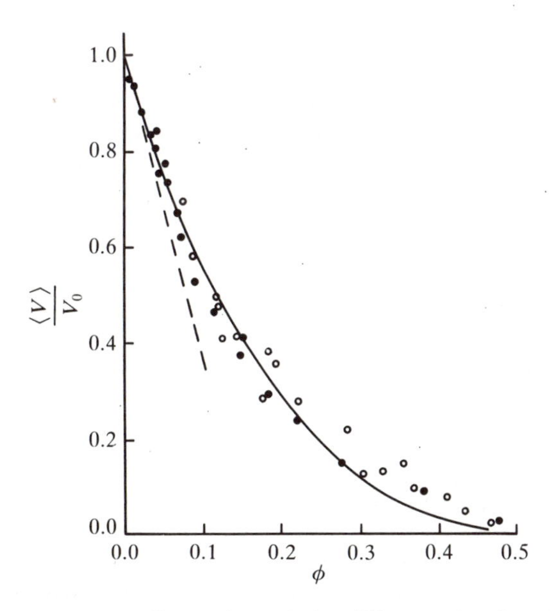

FIG. 9.8.2. The average sedimentation velocity $\langle \boldsymbol{V} \rangle$ as a function of particle volume fraction. The points were obtained from measurements on latex suspensions and the curve represents eqn (9.8.11) with $k_1 = 5.4$ and $p = 0.585$. (From Buscall *et al.* 1982, with permission).

the approximate formula

$$\langle V \rangle = \langle V_0 \rangle (1 - 6.55\phi) \tag{9.8.10}$$

obtained by truncating the series eqn (9.8.9) at the ϕ term, and using Batchelor's value for α. From the figure it can be seen that this approximation is accurate to about 5 per cent if $\phi < 0.05$. At higher volume fractions the $0(\phi^2)$ term in eqn (9.8.9) becomes significant, reflecting interactions between groups of three or more particles; the fact that these interactions are significant at such a low volume fraction can be attributed to the slow $1/r$ drop-off in the velocity field of a sedimentary particle.

The unbroken line in Fig. 9.8.2 represents the formula (Ekdawi and Hunter 1985)

$$\langle V \rangle = V_0 \left\{ 1 - \frac{\phi}{p} \right\}^{k_1 p} \tag{9.8.11}$$

with the parameters k_1 and p set equal to 5.4 and 0.585 respectively, in order to fit the data. The origins of this type of representation are discussed briefly at the end of section 9.10.

The above formulae for average sedimentation velocity are only valid over regions in which the suspension is macroscopically homogeneous, because in an inhomogeneous suspension the Brownian motion, which we have so far neglected, can lead to a flux of particles in addition to that due to the sedimentation force. The problem of calculating this diffusive flux in a dilute suspension has been discussed in section 2.1.4. In the following section we will look at the case of concentrated suspensions.

Exercises. 9.8.1 Let $\boldsymbol{v}_1$, p_1, and $\boldsymbol{v}_2$, p_2 be two solutions to the continuity equation and eqn (9.8.2). Verify that the sum $\boldsymbol{v}_1 + \boldsymbol{v}_2$, $p_1 + p_2$ also satisfies the equations (this is a consequence of the linearity of the equations). Show that this result is *not* true if eqn (9.8.2) is replaced by the full Navier–Stokes equation (eqn (9.6.8)).

9.8.2 In section 9.8.3 we stated that the fraction of interacting particles in a dilute suspension is proportional to ϕ, the volume fraction, while the fraction that interact with two or more neighbours is proportional to ϕ^2. This can be verified by the following steps:

(a) show that in a homogeneous suspension the probability of finding at least one particle in a volume V is nV, where n is the particle number density;

(b) assuming for simplicity that the probability of finding a particle at any point is independent of the fact that there are nearby particles, show that the probability of finding a particle within several radii of a reference particle is proportional to ϕ;

(c) again assuming the probabilities are independent show that the probability of finding two particles within several radii of the reference particle is proportional to ϕ^2.

9.8.3 Suppose that in eqn (9.8.3) the body force $\boldsymbol{F}$ is gravity ($=\rho\boldsymbol{g}$). How do you interpret the term $\boldsymbol{F}\cdot\boldsymbol{r}$ that appears in eqn (9.8.4)? At which points is the pressure unaffected by the presence of the particle? What is the pressure at those points? Where is the pressure most affected? Show that the maximum value of the modified pressure p' is equal to the Stokes force divided by the area of the particle.

9.8.4 Satisfy yourself that the disturbance velocity $\boldsymbol{v}+\boldsymbol{V}=\boldsymbol{v}+V(\cos\theta\hat{r}-\sin\theta\hat{\theta})$ where $\hat{r}$ and $\hat{\theta}$ are unit vectors (Fig. 9.7.7). Hence show that, for large distances:

$$\boldsymbol{v}+\boldsymbol{V}=\frac{3a}{2}\left[V+\frac{V}{2}\sin\theta\hat{\theta}\right]\frac{1}{r}.$$

9.9 Brownian motion revisited

9.9.1 *Gradient diffusion in a concentrated suspension*

Gradient diffusion refers to the motion of particles as a consequence of a concentration gradient. The diffusive flux in an inhomogeneous concentrated suspension can be calculated by an extension of the argument used by Einstein in his original study of dilute suspensions (see Einstein 1956). The argument centres on the case of a suspension in equilibrium under an applied field. The field is assumed to act only on the particles, giving rise to a force $\boldsymbol{f}(\boldsymbol{x})$, where $\boldsymbol{x}$ denotes the position of the particle centre.

As a result of this field the equilibrium particle density $n(\boldsymbol{x})$ will be non-uniform. There will therefore be a flux due to Brownian motion, which will be balanced by the sedimentation flux. As mentioned in the previous section, these two fluxes can be calculated separately and superposed.

The sedimentation flux is obtained by multiplying the concentration by the velocity (recall Fig. 2.1.3):

$$n(\boldsymbol{x})\frac{K(\phi)}{6\pi\eta a}\boldsymbol{f} \qquad (9.9.1)$$

where n is the particle number density, and $K(\phi)/6\pi\eta a$ is the average sedimentation velocity in a suspension in which the particles are acted on by a unit force. Comparing with the formula (9.8.10) for the sedimentation velocity for rigid spheres in non-concentrated suspensions, we find

$$K(\phi)\left(=\frac{|V|}{|V_0|}\right)=1-6.55\phi. \qquad (9.9.2)$$

For more concentrated suspensions ($\phi>0.05$), the empirical expression (9.8.11) can be used for calculating K.

The Brownian flux will presumably depend on the local variations in particle density, variations that are characterized by the quantity ∇n. By similar arguments to those that lead to the Newtonian formula (eqn (9.5.10)) between stress and rate of strain it can be shown that the flux density due to Brownian motion has the form

$$-\mathscr{D}\nabla n \tag{9.9.3}$$

for a locally isotropic suspension. This is identical to Fick's first law (eqn (2.1.18)), but now we must allow for the fact that $\mathscr{D}$ depends on the local particle volume fraction.

In equilibrium, the net particle flux is zero, so

$$\mathscr{D}(\phi)\nabla n = \frac{-n(\boldsymbol{x})K(\phi)\boldsymbol{f}}{6\pi\eta a}. \tag{9.9.4}$$

We also have the constraint that the net body force $-n(\boldsymbol{x})\boldsymbol{f}$ per unit volume of suspension must be balanced by the surface forces on the volume†. Since the suspension is in equilibrium the surface forces must take the form of an osmotic pressure Π, acting normal to the surface. The force balance equation therefore takes the form

$$-n(\boldsymbol{x})\boldsymbol{f} = \nabla\Pi,$$

or, since the local osmotic pressure depends only on particle density in this case,

$$-n(\boldsymbol{x})\boldsymbol{f} = \frac{\mathrm{d}\Pi}{\mathrm{d}n}\nabla n. \tag{9.9.5}$$

Using this equation to eliminate the $n\boldsymbol{f}$ term in eqn (9.9.4) we find that the diffusivity is given by

$$\mathscr{D}(\phi) = \frac{K(\phi)}{6\pi\eta a}\frac{\mathrm{d}\Pi}{\mathrm{d}n}. \tag{9.9.6}$$

Thus the diffusivity can be calculated from sedimentation and osmotic pressure data for the suspension. On replacing $K(\phi)$ in the above expression by the form (9.9.2) and using the approximate result

$$\frac{\mathrm{d}\Pi}{\mathrm{d}n} = kT(1+8\phi), \tag{9.9.7}$$

(see Batchelor 1976) we find that

$$\frac{\mathscr{D}}{\mathscr{D}_0} = 1 + 1.45\phi, \tag{9.9.8}$$

† We are speaking here of a volume containing many particles.

where

$$\mathscr{D}_0 = \frac{kT}{6\pi\eta a}$$

is the Einstein formula for the diffusivity in a dilute suspension, and terms of order ϕ^2 have been neglected. It can be seen that the effects of particle interaction on sedimentation velocity and osmotic pressure nearly cancel out, leaving a relatively weak diffusion–concentration dependence. The experimental verification of eqn (9.9.8) is described in Russel's (1981) review article.

In a non-equilibrium suspension, differences between the sedimentation and diffusion fluxes lead to variations in the concentration n, variations that can be calculated using the equation

$$\frac{\partial n}{\partial t} = \nabla\left\{\frac{K(\phi)}{6\pi\eta a}\left[\frac{\mathrm{d}\Pi}{\mathrm{d}n}\nabla n - n\boldsymbol{F}\right]\right\} \tag{9.9.9}$$

where $\boldsymbol{F}$ is the net sedimentation force on the particle. The derivation of this equation follows similar lines to that of eqn (2.1.21) for one-dimensional diffusion in a dilute suspension.

For suspensions of non-spherical particles, and for suspensions of spheres at high concentrations, the quantities $K(\phi)$ and $\mathrm{d}\Pi/\mathrm{d}n$ must be determined experimentally. Once these quantities are known, eqn (9.9.9) can be used for the prediction of sedimentation behaviour in any situation. This approach has been successfully applied to a number of colloids by Philip and Smiles (1982).

9.9.2 *Self-diffusion in a concentrated suspension*

The formula

$$\langle x^2 \rangle^{1/2} = \sqrt{(2\mathscr{D}t)} \qquad [2.1.26]$$

for the root-mean-squared displacement of an isolated spherical particle can also be applied to concentrated isotropic suspensions, with the proviso that the quantity $\mathscr{D}$ is not in general equal to the gradient diffusivity studied in the previous section. The quantity $\mathscr{D}$ in the above expression is usually referred to as the 'tracer' or 'self' diffusion coefficient. It measures the movement of an individual particle surrounded by a (uniform) collection of like particles. To calculate this quantity it is necessary to solve the Stokes equations for the velocity of the tracer particle moving under the steady force in the presence of force-free neighbouring particles, and then average over all particle configurations. As in the case of gradient diffusion, the exact theoretical analysis is limited to the low concentration range where pair interactions dominate. The formula for the tracer diffusion coefficient in a suspension of spheres is found to be (Batchelor 1976)

$$\mathscr{D} = \frac{kT}{6\pi\eta a}(1 - 1.83\phi). \tag{9.9.10}$$

Not surprisingly, the Brownian motion of a particle is hindered by interactions with its neighbours. The small amount of experimental evidence currently available appears to be in agreement with the above result (Russel 1981, p. 437).

9.9.3 *The Langevin equation*

In Chapter two the Brownian motion of an isolated particle was treated as a random walk, with the direction of each step being independent of previous steps. This assumption of independence places a lower limit on the step time τ, for τ must be large enough to allow the particle to 'forget' its previous movements.

In order to estimate this lower limit for τ, and to obtain other details of the dynamics of the Brownian motion, it is customary to use the Langevin equation

$$m\frac{\mathrm{d}\boldsymbol{V}}{\mathrm{d}t} + 6\pi\eta a\boldsymbol{V} = \boldsymbol{F}(t) \tag{9.9.11}$$

for the velocity $\boldsymbol{V}$ of an isolated spherical particle; this is simply Newton's second law for a particle of mass m acted upon by a fluctuating Brownian force $\boldsymbol{F}(t)$, with the liquid drag represented by the Stokes formula $-6\pi\eta a\boldsymbol{V}$. Since this formula was derived in section 9.8.2 for a particle in steady motion, its application to this problem can only be justified if the inertia term in the equation of motion of the liquid (eqn (9.8.2)) is negligible. After we have obtained our solution to the Langevin equation, we must therefore check for consistency by estimating the size of this neglected inertia term.

Since the particle is much more massive than the surrounding molecules, its response time will be much greater than the time scale of the fluctuating force $\boldsymbol{F}(t)$, which will presumably be of the order of the relaxation time for water molecules ($\sim 10^{-13}$ s). Thus the velocity of a particle at any time is determined not just by the instantaneous force on the particle, but by the history of that force over several particle relaxation times. In order to estimate this relaxation time, we take the average of the Langevin equation over all those particles that have a given velocity $\boldsymbol{V}_0$ at $t = 0$. By the above argument, we expect that the particle velocity and the force will be statistically independent; hence the average of $\boldsymbol{F}(t)$ over this group of particles will be zero, and the average of eqn (9.9.11) becomes

$$m\frac{\mathrm{d}\langle \boldsymbol{V}\rangle}{\mathrm{d}t}+6\pi\eta a\langle \boldsymbol{V}\rangle=0.$$

The solution to this equation is (recall Exercise 7.8.4):

$$\langle \boldsymbol{V}\rangle = V_0\exp(-t/\tau_\rho), \tag{9.9.12}$$

where

$$\tau_\rho=\frac{m}{6\pi\eta a}. \tag{9.9.13}$$

From eqn (9.9.12) it can be seen that $\langle \boldsymbol{V}\rangle\approx\mathbf{0}$ if t is large compared with τ_ρ; since the average velocity of a *random* sample of particles is zero, τ_ρ provides a measure of the time required for a particle to 'forget' its initial velocity. For a 0.1 μm radius particle of density $2\,\mathrm{g\,cm^{-3}}$ in water at 25 °C, $\tau_\rho\approx5\times10^{-9}$ s. For the random walk analogue of Chapter two to be valid, the step times τ must be much greater than this relaxation time.

We are now in a position to test the validity of the Langevin equation by estimating the size of the neglected 'inertia' term $\rho(\partial\boldsymbol{v}/\partial t)$ in the equation of motion (eqn (9.8.2)) of the liquid. Since the particle velocity decays from its initial value V_0 over a time τ_ρ, this inertia term is of order $\rho V_0/\tau_\rho$. The viscous term $\eta\nabla^2\boldsymbol{v}$ will be of order $\eta V_0/a^2$, and thus the ratio

$$\left|\rho\frac{\partial\boldsymbol{v}}{\partial t}\right|\Big/\left|\eta\nabla^2\boldsymbol{v}\right|=\frac{\rho a^2}{\eta\tau_\rho}. \tag{9.9.14}$$

From eqn (9.9.13) we see that this ratio is of order ρ/ρ_p, where ρ_p is the particle density. So for cases of practical interest, in which particle density is not large compared to the fluid density, the Langevin equation is invalid!

The correct analysis must be based on the equation of motion (eqn (9.8.2)), which takes into account the inertial forces in the liquid, together with a fluctuating 'body force' $\boldsymbol{F}$ representing the effects of Brownian motion. As in the Langevin equation, this fluctuating term can be removed by averaging. The solution of these equations for the average velocity $\langle \boldsymbol{V}\rangle$ yields a similar estimate for τ_ρ, but the decay in $\langle \boldsymbol{V}\rangle$ at large t is found to have a $t^{-3/2}$ form, instead of the exponential form found earlier (Russel 1981, p. 428). Thus while the Langevin equation may not be strictly valid, it provides qualitatively correct results, and serves as a simple model for illustrating the techniques used in the analysis of the complete equations describing Brownian motion dynamics.

9.9.4 *The Brownian motion of non-spherical particles*

In this section we will study the effect of Brownian motion on the orientation and position of spheroidal particles (Fig. 3.1.1) in a dilute suspension.

The rotational diffusion of these particles can be analysed along similar lines to the translational diffusion discussed in Chapter 2. In particular, the motion can be treated as a random walk, with the result that (compare eqn (2.1.26)):

$$\langle \theta^2(t) \rangle = 4\mathcal{D}_r t, \qquad (9.9.15)$$

where θ is the change in the particle orientation during time t, and $\mathcal{D}_r$ is termed the rotational diffusion coefficient. This result is valid provided the change in orientation is small (see Exercise 9.9.1).

The effect of Brownian motion on the particle orientation distribution in a dilute suspension is described by the differential equation (McQuarrie 1976, p. 398)

$$\frac{\partial R}{\partial t} = \mathcal{D}_r \left\{ \frac{1}{\sin\theta} \frac{\partial}{\partial \theta} \left(\sin\theta \frac{\partial R}{\partial \theta} \right) + \frac{1}{\sin^2\theta} \frac{\partial^2 R}{\partial \phi^2} \right\}, \qquad (9.9.16)$$

where the angles θ and ϕ used to specify the orientation of the particle axis are illustrated in Fig. 9.9.1. By definition, the quantity $R(\theta, \phi, t) \sin\theta \Delta\theta \Delta\phi$ is equal to the fraction of particles whose orientation lies in the range $(\theta \pm (\Delta\theta/2), \phi \pm \Delta\phi/2))$.

By multiplying eqn (9.9.16) by $\cos\theta$ and integrating over all orientations it can be shown that

$$\langle \cos\theta \rangle = e^{-2\mathcal{D}_r t} \qquad (9.9.17)$$

for a suspension in which all the particles are initially aligned in the $\theta = 0$ direction. Equation (9.9.17) refers to what is called Debye relaxation, since it was used by Peter Debye (1929) to describe the relaxation behaviour of molecular dipoles. From eqn (9.9.17) it can be seen that the

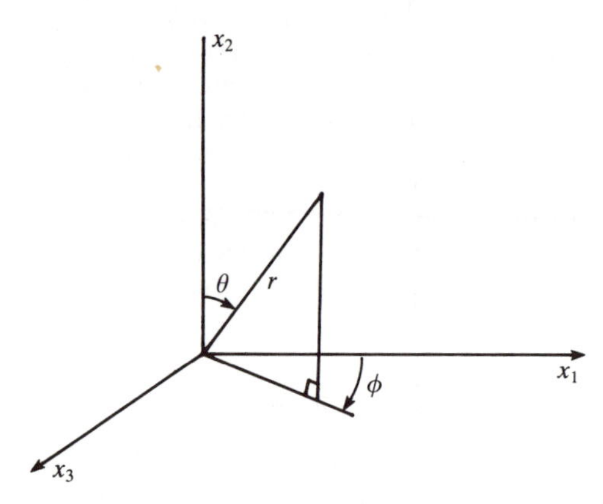

FIG. 9.9.1. The angles θ and ϕ used to specify the orientation of a particle, relative to a fixed set of Cartesian axes.

time required for the particle distribution to become isotropic (i.e. $\langle \cos\theta \rangle \approx 0$) is of order $1/\mathscr{D}_r$.

The quantity $\mathscr{D}_r$ can be calculated using a variation of the Einstein argument described in section 9.9.1, that is by considering a suspension in equilibrium under the action of an applied field which in this case tends to align the particles. The applied field leads to an extra term in eqn (9.9.16) corresponding to the rotation of the particles in the absence of Brownian motion. The solution of eqn (9.9.16) for this equilibrium problem yields the formula

$$\mathscr{D}_r = \frac{kT}{B_r}, \tag{9.9.18}$$

where B_r is the friction factor for particle rotation, equal in magnitude to the torque required to rotate the particle with unit angular velocity, in the absence of Brownian motion. From the solution of the Stokes equations around a rotating spheroid it is found that (Perrin 1934)

$$\mathscr{D}_r = \frac{kT}{8\pi\eta ab^2}\left\{3q^2(2-q^2)(1-q^2)^{-1/2}\ln\left\{\frac{1+(1-q^2)^{1/2}}{q}\right\} - 3q^2\right\}\Big/ 2(1-q^4) \tag{9.9.19}$$

for a cigar-shaped (prolate) spheroid, where $q = b/a$ is the ratio of the length of the minor axis to the major axis (the axis of rotation of the particle). The formula for disc-shaped spheroids is given in Exercise 9.9.4.

The problem of translational diffusion can be treated in a similar manner: for a suspension with a uniform orientation distribution, the translational diffusion coefficient is given by

$$\mathscr{D} = kT\left\{\frac{1}{3B_{\parallel}} + \frac{2}{3B_{\perp}}\right\}$$

where $B_{\parallel}$ and $B_{\perp}$ are the friction coefficients for motion parallel and perpendicular to the particle axis respectively. The formulae for these quantities are given in Perrin's (1934) paper.

With the aid of these formulae and the expression for the rotational diffusivity it is possible to estimate the particle size and shape from combined translational and rotational diffusivity measurements on dilute suspensions. The results relate of course to the size of the particle in solution, and this may differ from that of the dry particle as a result of solvent absorption or adsorption. This method has mainly been applied to protein and virus particles, but some work has also been done on montmorillonite suspensions (Shah 1963).

Exercises. 9.9.1 By using the small θ approximation

$$\cos\theta \approx 1 - \frac{\theta^2}{2},$$

in eqn (9.9.17), derive eqn (9.9.15) for the mean squared orientation change of a particle during a small period t.

9.9.2 Calculate the value of $\mathscr{D}_r$ for values of a/b from 1 to 20 and $a = 10^{-7}$ m using eqn (9.9.19). Show that for large a/b, this formula takes the approximate form

$$\mathscr{D}_r \approx \frac{3kT}{32\pi\eta a^3}\left\{2\ln\left(\frac{2a}{b}\right) - 1\right\}.$$

Compare values obtained with this formula with the exact value obtained earlier.

9.9.3 Alexander and Johnson (1949; p. 400) quote a value of $\mathscr{D}_r = 7\ \mathrm{s}^{-1}$ obtained by Edsall for the rotary diffusion constant of rabbit myosin at $T = 276$ K. Taking the viscosity of water as 1.6×10^{-3} Pa s at this temperature and a high value (say $a/b \approx 100$) for the axial ratio, estimate the length of the myosin molecule. (Independent measurements of the particle volume would permit a further refinement of this result.)

9.9.4 For an oblate (disc-shaped) particle the rotational diffusivity $\mathscr{D}_r$ is given by

$$\mathscr{D}_r = \frac{kT}{8\pi\eta ab^2}\left\{3\frac{q^2(2-q^2)}{(q^2-1)^{1/2}}\tan^{-1}(q^2-1)^{1/2} + 3\right\}\Big/ 2(1-q^4).$$

Show that for $q \gg 1$ (i.e. a flat disc)

$$\mathscr{D}_r \approx \frac{3kT}{32\eta b^3}.$$

9.10 The flow properties of suspensions

One of the most important transport properties of a colloidal dispersion is its viscosity. The problem of estimating the influence of the particles on the macroscopically observed viscosity has attracted much attention from some eminent scientists. Undoubtedly the most famous result is that derived by Einstein for the effect of volume fraction of particles on viscosity (eqn (9.10.2) below). It is a deceptively simple expression that represents quite well the limiting behaviour of smooth spheres at low concentrations in a Newtonian fluid. We will not follow Einstein's method here, but a rather more general procedure that proves to be more productive in the long run.

9.10.1 *The macroscopic flow field*

In our analysis of fluid motion we neglected the molecular nature of the fluid, with the proviso that attention be restricted to volumes of fluid containing large numbers of molecules. By the same token, a suspension can be treated as a one-component continuum, provided we only consider volumes containing many particles. 'Local properties' such as the velocity and density for this continuum can be defined as averages over volumes containing large numbers of particles. For example, the average velocity is defined as

$$\langle \boldsymbol{v}(\boldsymbol{x}) \rangle = \frac{1}{\mathcal{V}} \int_{\mathcal{V}} \boldsymbol{v} \, d\mathcal{V}, \tag{9.10.1}$$

where the sample volume $\mathcal{V}$ is centred on the point $\boldsymbol{x}$. In a flowing suspension this average velocity will vary on a macroscopic length scale L determined by the apparatus; in flow down a pipe for example, L will be of the order of the pipe radius. Thus in order to give formula (9.10.1) an unambiguous meaning we must specify that the sample volume $\mathcal{V}$ be much smaller than L^3, while still being large enough to contain many particles.

Differential equations for the macroscopic velocity field $\langle \boldsymbol{v}(\boldsymbol{x}) \rangle$ can be derived in a similar manner to the Navier–Stokes equations for the local velocity field; that is by applying the Principle of Conservation of Mass and Newton's Second Law to a rectangular block of fluid. However, whereas in the earlier sections we took this block to be much larger than the molecules and much smaller than the particles, the block in this case is taken to have the dimensions of a sample volume, as defined above.

The most difficult step in the derivation of these differential equations for the macroscopic velocity field is the determination of the relationship between the short-range forces that act on the surface of the block and the macroscopic velocity field; that is, the analogue of the Newtonian expression (eqn (9.5.10)) for the microscopic stress tensor in a liquid. The remainder of this section is concerned with the derivation of this relationship.

9.10.2 *The macroscopic stress tensor*

Let ΔA denote an imaginary flat surface that lies in the suspension and translates with the local macroscopic velocity field. The dimensions of ΔA are of the order of $\mathcal{V}^{2/3}$, where $\mathcal{V}$ is the sample volume. The short-range forces that act across ΔA are made up of the microscopic forces, characterized by the local stress tensor $\boldsymbol{\sigma}$, together with interparticle forces such as dispersion and electrical forces.

As in earlier sections we will limit our attention to the case of rigid force-free particles. In such suspensions the short-range force per unit area acting across ΔA is simply

$$\left[\frac{1}{\Delta A}\int_{\Delta A} \boldsymbol{\sigma}\, \mathrm{d}A\right]\cdot \hat{n},$$

where $\hat{n}$ is the unit normal to ΔA. The quantity in brackets represents an average of the local stress tensor over ΔA. We will denote this average, or 'macroscopic' stress tensor by $\langle \boldsymbol{\sigma} \rangle$.

Our aim is to calculate the relationship between $\langle \boldsymbol{\sigma} \rangle$ and the macroscopic velocity field. Although $\langle \boldsymbol{\sigma} \rangle$ has been defined as an area average, it is convenient to calculate this quantity using the volume average formula

$$\langle \boldsymbol{\sigma} \rangle = \frac{1}{\mathcal{V}}\int_{\mathcal{V}} \boldsymbol{\sigma}\, \mathrm{d}\mathcal{V}. \tag{9.10.2}$$

The area average and volume average are equivalent, since they both represent averages over the same statistical population (see Batchelor 1970 for more details on this point).

As in the case of the microscopic stress tensor, we write (cf. eqn (9.5.2)

$$\langle \boldsymbol{\sigma} \rangle = -P\boldsymbol{I} + \langle \boldsymbol{\sigma}^{\mathrm{D}} \rangle,$$

where P is an average pressure. $\langle \boldsymbol{\sigma}^{\mathrm{D}} \rangle$ is the part of $\langle \boldsymbol{\sigma} \rangle$ that depends on the motion. By using the fact that (eqn (9.5.8))

$$\boldsymbol{\sigma}^{\mathrm{D}} = 2\eta \boldsymbol{e}$$

in the liquid, we write eqn (9.10.2) in the form

$$\langle \boldsymbol{\sigma} \rangle = -P\boldsymbol{I} + \frac{2\eta}{\mathcal{V}}\int_{\mathcal{V}_l} \boldsymbol{e}\, \mathrm{d}\mathcal{V} + \frac{1}{\mathcal{V}}\int_{\mathcal{V}_{\mathrm{p}}} \boldsymbol{\sigma}^{\mathrm{D}}\, \mathrm{d}\mathcal{V}, \tag{9.10.3}$$

where $\mathcal{V}_{\mathrm{p}}$ is the volume occupied by the particles in $\mathcal{V}$, and $\mathcal{V}_l$ is the liquid volume.

The rate of strain tensor $\boldsymbol{e}$ is zero inside the rigid particles (see Exercise 9.10.1), and so the integral over $\mathcal{V}_l$ in eqn (9.10.3) can be formally extended over the particle volume as well, resulting in

$$\langle \boldsymbol{\sigma} \rangle = -P\boldsymbol{I} + 2\eta \langle \boldsymbol{e} \rangle + n\langle \boldsymbol{S} \rangle, \tag{9.10.4}$$

where

$$\langle \boldsymbol{e} \rangle = \frac{1}{\mathcal{V}}\int_{\mathcal{V}} \boldsymbol{e}\, \mathrm{d}\mathcal{V} \tag{9.10.5}$$

is the average rate of strain tensor; as before, n denotes the particle number density, and $\langle \boldsymbol{S} \rangle$ denotes an average, over the particles in $\mathcal{V}$, of

$$\int_{\mathcal{V}^i} \boldsymbol{\sigma}^{\mathrm{D}}\, \mathrm{d}\mathcal{V}, \tag{9.10.6}$$

where $\mathcal{V}^i$ is the volume of the ith particle in $\mathcal{V}$. $\langle \boldsymbol{S} \rangle$ is referred to as the average particle dipole strength. The term 'dipole strength' arises from the application of this volume averaging technique to the electrical transport properties of suspensions. There, each uncharged particle behaves, when viewed from a distance, like an electric dipole with a dipole strength given by an integral form analogous to eqn (9.10.6). The average dipole strength in eqn (9.10.4) represents the particle contribution to the stress. To calculate this term we must solve the Stokes equations around the particles. In the following section we will outline the solution to this problem for a dilute suspension of spheres.

9.10.3 *Effective viscosity of a dilute suspension of spheres*

For a dilute suspension we can treat each particle as being alone in an infinite liquid. To determine the dipole strength in this case, in the absence of Brownian motion, we must solve the Stokes equations

$$\left.\begin{aligned} \eta \nabla^2 \boldsymbol{v} &= \nabla p' \\ \text{and} \qquad \nabla \cdot \boldsymbol{v} &= 0, \end{aligned}\right\} \tag{9.10.7}$$

for the velocity and pressure field around the isolated particle, where p' is the modified pressure, defined in section 9.8.2.

At large distances the disturbance in the velocity due to the particle will approach zero, that is

$$\boldsymbol{v}(\boldsymbol{x}) \to \boldsymbol{v}^{\mathrm{a}}(\boldsymbol{x}) \tag{9.10.8}$$

where $\boldsymbol{v}^{\mathrm{a}}$ is the ambient velocity field that would exist in the volume in the absence of the particles. On the surface of the particle we have the no-slip boundary condition (cf. eqn (9.4.3))

$$\boldsymbol{v}(\boldsymbol{x}) = \boldsymbol{V} + \boldsymbol{\Omega} \times \boldsymbol{x}. \tag{9.10.9}$$

where $\boldsymbol{V}$ and $\boldsymbol{\Omega}$ are the translational and angular velocity of the particle, to be determined from the condition that the net force and torque on the particle are zero.

The ambient velocity field will presumably vary on the macroscopic length-scale L. Thus in the region where the velocity field is disturbed by the particle we can approximate $\boldsymbol{v}^{\mathrm{a}}$ by the formula:

$$v_i^{\mathrm{a}}(\boldsymbol{x}) = v_i^{\mathrm{a}}(\mathbf{0}) + \tfrac{1}{2}\sum_{j=1}^{3}\left(\frac{\partial v_i^{\mathrm{a}}}{\partial x_j} - \frac{\partial v_j^{\mathrm{a}}}{\partial x_i}\right)x_j + \sum_{j=1}^{3} e_{ij}^{\mathrm{a}} x_j, \qquad (9.10.10)$$

where the derivatives are evaluated at the origin, which is assumed to coincide instantaneously with the centre of the sphere. This formula is obtained in the same manner as the expression (eqn (9.5.13)) for the local velocity field in a liquid, with Δx_j replaced by x_j.

By similar arguments to those that follow eqn (9.5.13) it can be shown that the first two terms on the right-hand side of eqn (9.10.10) represent a rigid body translation and rotation. Since the problem is linear, we can write $\boldsymbol{S}$ as the sum of a dipole strength due to the rigid body velocity field, plus a dipole strength due to the portion of the ambient field

$$\sum_{j=1}^{3} e_{ij}^{\mathrm{a}} x_j. \qquad (9.10.11)$$

Since there is no force or torque on the particle, it will simply translate and rotate with the rigid body motion. The stress tensor in this case will have the equilibrium form (eqn (9.4.2)), and the corresponding dipole strength will be zero.

Thus the entire contribution to the dipole strength is due to the portion (eqn (9.10.11)) of the ambient velocity field involving e_{ij}^{a}; this is the part of the velocity field that tends to distort the particle (see Exercise 9.10.2). For the dilute suspension considered here e_{ij}^{a} is approximately equal to $\langle e_{ij}\rangle$, for since $\boldsymbol{v}^{\mathrm{a}}$ varies on the macroscopic length scale, the formula (9.10.10) is valid over the *entire* sample volume, and as $\boldsymbol{v} - \boldsymbol{v}^{\mathrm{a}}$ is only non-zero in the neighbourhood of the suspended particles, it follows that

$$\langle \boldsymbol{e}\rangle = \boldsymbol{e}^{\mathrm{a}},$$

to within a relative error of order ϕ, where, as usual, ϕ is the particle volume fraction.

To determine the particle dipole strength we must therefore solve the Stokes equations subject to the no-slip condition (eqn (9.10.9)) on the surface of the particle and the outer condition (eqn (9.10.8)), where v_i^{a} is given by the formula (9.10.11). From the symmetry of the problem in this case it can be shown that the sphere will remain at rest in the ambient field (eqn (9.10.11)) (see Exercise 9.10.2) and hence the boundary condition (eqn (9.10.9)) reduces to

$$\boldsymbol{v}(\boldsymbol{x}) = \mathbf{0}, \quad \text{on} \quad r = a.$$

The mathematical details of the solution of this problem for $\boldsymbol{v}$ are given in section 4.11 of Batchelor (1967). It is found that at large distances, the disturbance in the velocity field due to the body decays like $1/r^2$, in contrast to the $1/r$ decay found for sedimenting particles. Thus particles in the viscosity problem will exhibit less interaction than in the sedimentation problem, and our dilute approximation should, therefore, be valid for a greater concentration range than in the sedimentation problem.

For a rigid particle, the expression (9.10.6) for the dipole strength can be converted into an integral over the particle surface, which can then be evaluated directly from the flow field (Landau and Lifshitz 1959) yielding

$$\boldsymbol{S} = \frac{20}{3}\pi a^3 \eta \langle \boldsymbol{e} \rangle.$$

In the absence of particle interaction, the dipole strength is the same for each sphere in the sample volume, and the formula (9.10.4) becomes

$$\langle \boldsymbol{\sigma} \rangle = -P\boldsymbol{I} + 2\eta\left(1 + \frac{5\phi}{2}\right)\langle \boldsymbol{e} \rangle.$$

This has the same form as the expression (eqn (9.5.10)) for the stress in a Newtonian liquid. Thus from the macroscopic point of view, a dilute suspension of spheres behaves as a Newtonian liquid with viscosity η^*, where

$$\eta^* = \eta(1 + \tfrac{5}{2}\phi). \tag{9.10.12}$$

This result was obtained by Einstein from a calculation of the energy dissipation in the suspension. This dissipation method suffers from the disadvantage that it yields only a single number, namely the rate of energy dissipation. While this may be adequate for isotropic suspensions, which can be characterized by a single viscosity, it fails when applied to a non-isotropic suspension such as suspensions of rod-like particles aligned by a flow. Such suspensions require a number of 'viscosities' to describe their flow properties. For this reason, the recent papers on this subject have adopted the volume average approach described here, an approach first devised by Landau and Lifshitz (1959).

9.10.4 *Dilute suspensions of spheroidal particles*

The calculation of the flow behaviour of dilute suspension of spheroids is both more complicated and more interesting than the problem for spheres, for as we shall see, these suspensions exhibit non-Newtonian effects that are common to a large class of suspensions and liquids; by analysing these effects for dilute suspensions of this kind we can, therefore, gain some insight into the factors that are responsible for non-Newtonian behaviour in more complicated suspensions.

Although the mathematical details of the calculation of $\langle \boldsymbol{\sigma} \rangle$ for a dilute suspension of spheroids are beyond the scope of this volume, we have established a sufficiently solid foundation in this chapter to enable us to give a fairly detailed description of the methods used, and the results that have so far been obtained.

The formula (9.10.4) relating the macroscopic stress to the average dipole strength is valid for these suspensions, and since the suspensions are dilute, we can again calculate the dipole strength of each particle as if it were alone in an infinite liquid. This calculation involves the solution of the Stokes equations subject to the same boundary conditions (eqns (9.10.8) and (9.10.9)) as for the spherical particle. As before, the dipole strength due to the rigid body component of the ambient field (eqn (9.10.10)) is zero and the particle simply translates and rotates with this field. Unlike the spherical particle however, the dipole strength of the spheroid in the ambient straining field (eqn (9.10.11)) is found to depend on its orientation relative to the field, and furthermore this straining field induces an angular velocity in the particle that depends on particle orientation. This rotation is superposed on the rigid body rotation described earlier. Thus spheroidal particles do not simply rotate with a uniform angular velocity like spheres, but move in a more complicated fashion that depends on the type of flow; for example, in a shear flow it is found (Goodwin 1975) that cigar-shaped particles rotate in a periodic fashion, with the minimum rate of rotation occurring when the particle is most nearly aligned with the streamlines. In other flows such as the extensional flow of a stream of liquid from a hole in the bottom of a container, there is a preferred orientation (also along the streamlines in this case) at which the particles have *zero* angular velocity; such flows tend to align the particles in the preferred direction. Since the average dipole strength depends on the orientation distribution $R(\theta, \phi)$, the same suspension may behave quite differently for these different flows.

To calculate the distribution function R for any flow it is necessary to solve eqn (9.9.16) with an extra term due to the rotation of the particles by the ambient flow. Since the rotation rate is non-uniform, the ambient flow tends to make the distribution function non-uniform; for example in the shear flow case described above, the particles move more slowly when they are aligned with the streamlines, and so R will be larger for the aligned orientation. This aligning tendency is opposed by the Brownian motion term in eqn (9.9.16), and the final form of the distribution function in a steady flow is determined by the relative magnitude of these two opposing terms. In a shear flow this quantity is characterized by a 'Peclet number' $\dot{\gamma}/\mathscr{D}_r$; when this ratio is small the

Brownian motion dominates and the orientation distribution is approximately uniform. At large Peclet numbers the distribution is quite non-uniform, but rather surprisingly the Brownian motion still plays an important part in the shear flow case, for in the absence of Brownian motion the particles simply rotate in a periodic fashion leading to an orientation distribution R that varies periodically. In such a situation the effect of a small amount of Brownian motion over a long period of time results in a steady-state orientation distribution.

The orientation distribution has been calculated for a number of limiting cases such as nearly spherical particles at low or high Peclet numbers, for a range of flows (Leal and Hinch 1973). Once the orientation distribution is known, the average dipole strength can be calculated directly, but even at this stage there is an unexpected complication, for in addition to the dipole strength obtained from the solution of the Stokes equations in the ambient flow (eqn (9.10.10)) there is a direct 'contribution' from the Brownian motion. The dipole strength of a particle undergoing Brownian motion will, like its angular velocity, depend on the history of the fluctuating torque on the particle over the previous few relaxation periods. In a suspension in which the orientation distribution is non-uniform, the particles that instantaneously have a given orientation represent a biased sample, since more will have come (via Brownian rotations) from neighbouring orientations where R is large than from those where R is small. Thus on average the particles will have experienced more torques in one direction than in another, and as a result their average dipole strength due to these torques is non-zero. Although this extra dipole strength is the consequence of fluctuating torques it can in fact be calculated from the solution of Stokes equations around a spheroid rotating under the action of a steady 'thermal torque' analogous to the thermal force introduced in section 2.1.3 (see Hinch and Leal 1972).

Since the orientation distribution depends on the type of flow, it is difficult to make general statements about the flow properties of these suspensions; consequently we will limit our attention to the results obtained for the important case of a steady shear flow.

Even in this restricted case it is not possible to describe the flow properties in terms of a single viscosity, for when a suspension such as this is sheared, stresses arise in addition to those encountered in a Newtonian liquid; for example if the suspension is subjected to the idealized shear flow of Fig. 2.4.1, the top plate would experience both a shear force *and* a normal force, both depending on the shear rate. We will concentrate here on the calculation of the 'shear viscosity', which represents the ratio of shear stress to shear strain rate.

It is found that the shear viscosity of the suspension, η^*, decreases

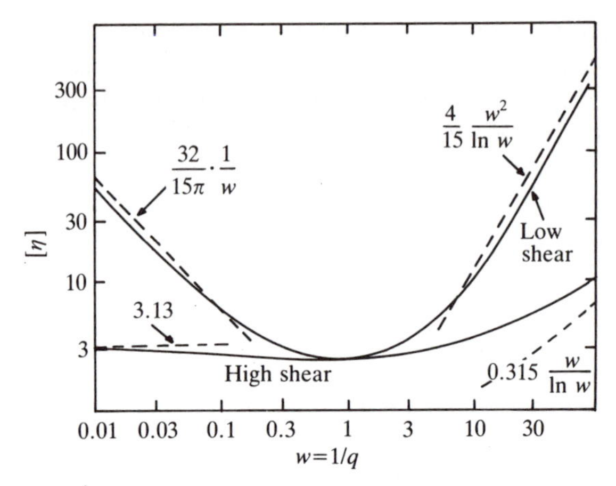

FIG. 9.10.1. The high and low shear rate intrinsic viscosities of a suspension of spheroidal particles. From Figs 5 and 8 of Hinch and Leal's (1972) paper.

with increasing shear rate. Although this viscosity dependence cannot be calculated for arbitrary particle shapes, the limiting viscosities at small and large Peclet numbers *have* been calculated for arbitrary aspect ratios $w(=a/b)$ (Hinch and Leal 1972). These results are illustrated in Fig. 9.10.1, when $[\eta]$ is the 'intrinsic viscosity' defined by

$$[\eta] = \lim_{\phi \to 0} \left(\frac{\eta^*}{\eta} - 1\right) \Big/ \phi. \tag{9.10.13}$$

The broken lines represent approximate formulae valid for disc-shaped ($w \ll 1$), and rod-like ($w \gg 1$) particles. With the aid of these results it should be possible to obtain information about particle shape from measurements of the high and low shear rate viscosities.

9.10.5 *Concentrated suspensions*

As with the other transport properties, theoretical studies in this area have been limited to the case of suspensions of rigid spheres, and attention has centred on the case of semi-dilute suspensions, in which pair interactions are dominant. For such suspensions, the average dipole strength can be approximated by the formula

$$\langle \boldsymbol{S} \rangle = \boldsymbol{S}_0 + \phi \boldsymbol{S}_1,$$

where $\boldsymbol{S}_0$ is the dipole strength of an isolated sphere, and $\boldsymbol{S}_1$ is related to the dipole strength of one of a pair of particles in the ambient flow, averaged over all pair configurations.

To calculate this average it is necessary to determine the pair

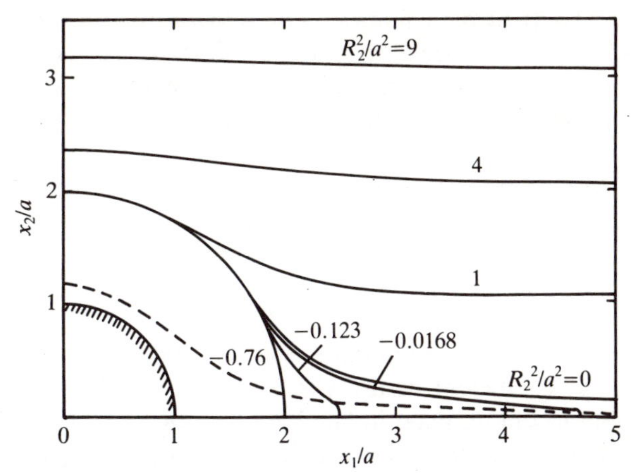

FIG. 9.10.2. The trajectories of one member of a pair of spheres in a shear flow, relative to the other sphere, for pairs lying in the plane of the flow. The circle of radius 2 is the trajectory for touching particles. (From Batchelor and Green 1972a, with permission.)

probability function $p(\boldsymbol{r})$, where $p(\boldsymbol{r})\Delta\mathcal{V}$ is the probability of finding a particle with its centre in the small volume $\Delta\mathcal{V}$ around the point $\boldsymbol{r}$, given that there is a sphere at the origin. $p(\boldsymbol{r})$ is the analogue of the orientation distribution $R(\theta, \phi)$ for the dilute spheroid problem, and like the orientation distribution, $p(\boldsymbol{r})$ depends on the type of flow.

In Fig. 9.10.2 we show the relative trajectories of pairs of spheres in a shear flow (Batchelor and Green 1972a); the lines represent the paths traced out by the centre of one member of the pair relative to the other for the case of pairs lying in the plane of the flow. Since the path lines are symmetrical about the x_2 and x_1 axes, only one quadrant has been shown in the figure. If the ambient shear rate $\partial v_1^a/\partial x_2$ is negative, the second particle will move from right to left along the path lines in the diagram. The quantity R_2 represents the x_2 coordinate of the particle far upstream, where the path lines are parallel to the x_1 axis. From the diagram it can be seen that the flow tends to bring sphere pairs together. For the case $R_2/a<1$, the pairs are brought into such proximity that even a slight attractive force may be enough to bring about 'shear-induced' coagulation. Those trajectories that lie below the $R_2=0$ line represent pairs in closed orbits; in the absence of Brownian motion the pair distribution function in this region will vary periodically with time, like the orientation distribution for the spheroid problem. (For such particles $R_2^2/a^2<0$ so there are no (real) values of R_2 far upstream of the particle.) In extensional flows there is also a tendency for pairs to be brought together, but there are no closed orbits. These flows lead to an increase in the probability $p(\boldsymbol{r})$ at small separations, a tendency that is of course opposed by the Brownian motion. For a given flow the form of $p(\boldsymbol{r})$ is determined by the relative magnitude of these two opposing terms, characterized by the Peclet number $ea^2/\mathscr{D}$, where e is a typical macroscopic strain rate and $\mathscr{D}$ is the translational diffusivity of an isolated sphere.

At low Peclet numbers the Brownian motion dominates and $p(\boldsymbol{r})$ is approximately uniform. In this case it is found that the suspension behaves as a Newtonian liquid with viscosity

$$\eta^* = \eta(1 + 2.5\phi + 6.2\phi^2)$$

(Batchelor 1977). The experimental accuracy required for the determination of the ϕ^2 coefficient of the viscosity is so great that it has not yet been possible to obtain experimental verification of this result.

At high Peclet numbers the problem of determining $p(\boldsymbol{r})$ in a shear flow is complicated by the closed trajectories referred to earlier, for in the region of closed trajectories a small amount of Brownian motion can have a significant effect on $p(\boldsymbol{r})$, in the same way that the Brownian motion has a significant effect in the spheroid case at high shear rates. In the extensional flow, where there are no closed trajectories, the effect of the Brownian motion is unimportant. The calculation of the bulk stress in this flow is described by Batchelor and Green (1972b).

For more concentrated suspensions there are a number of semi-empirical formulae that can be used for calculating the shear viscosity. The data for the limiting shear viscosities at high and low shear rates (that is, high and low Peclet numbers) can be adequately represented by the Dougherty–Krieger formula

$$\frac{\eta^*}{\eta} = \left(1 - \frac{\phi}{p}\right)^{-[\eta]p}, \tag{9.10.14}$$

where $[\eta]$ is the intrinsic viscosity (eqn (9.10.13)) and p is an adjustable parameter. This formula can be derived by considering the change in the viscosity $\delta\eta^*$ caused by an increase in the volume fraction $\delta\phi$. If the suspension before the addition of particles is treated as a Newtonian liquid with viscosity η^*, then by the Einstein formula

$$\delta\eta^* = \eta^* \tfrac{5}{2} \delta\phi.$$

In order to take into account the fact that the suspension is rigid when the particles are closely packed, $\delta\phi$ is replaced in the above formula by $\delta\phi/(1 - K\phi)$, where $1 - K\phi$ is the volume available for the added particles. Integration of the resulting formula yields (Exercise 9.10.3):

$$\frac{\eta^*}{\eta} = \left(1 - \frac{\phi}{p}\right)^{-2.5p}$$

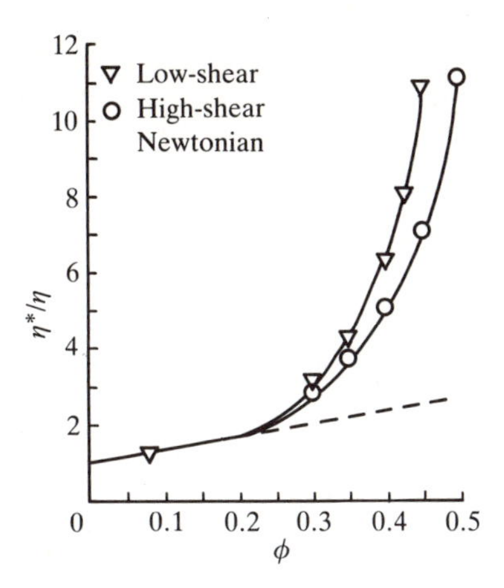

FIG. 9.10.3. Measured high and low shear-rate viscosities of concentrated latex suspensions. The lines represent eqn (9.10.14). (Reproduced from Krieger 1972, with permission).

where $p = 1/K$ is expected to be approximately equal to the volume fraction for close packing. In practice, p is treated as an adjustable parameter, and the 2.5 is replaced by the measured intrinsic viscosity $[\eta]$, which can differ from the Einstein value due to effects such as non-sphericity and the presence of small numbers of permanent doublets.

From Fig. 9.10.3 it can be seen that this formula provides an excellent fit of high and low shear rate viscosity measurements for latex suspensions; in both cases $[\eta] = 2.67$, while $p = 0.57$ for the low shear and 0.68 for the high shear viscosity.

For such suspensions, the viscosity at intermediate shear rates can be calculated using an equation of the form (Krieger 1972)

$$\frac{\eta^* - \eta_1^*}{\eta_2^* - \eta_1^*} = \left(1 + \frac{|\mathcal{S}|}{\mathcal{S}_i}\right)^{-1}, \tag{9.10.15}$$

where η_1^* and η_2^* are the high and low shear limiting viscosities respectively. $\mathcal{S}$ is the shear stress, and $\mathcal{S}_i$ is a characteristic shear stress related to the particle diffusivity (Exercise 9.10.4):

$$\mathcal{S}_i = \frac{kT}{\alpha a^3}$$

where α is an adjustable parameter, independent of ϕ and strain rate. Considering the fact that there are only two adjustable parameters in these formulae, viz. p and α, the fit of the data is most impressive. In Fig. 9.10.4 we show the measured shear viscosities for latex suspensions

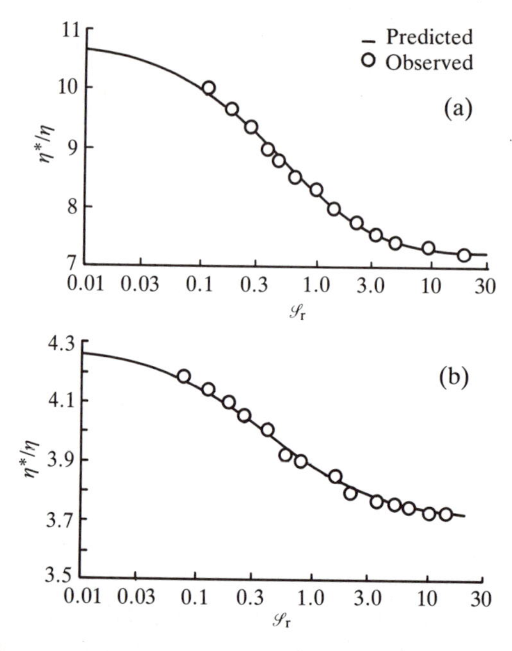

FIG. 9.10.4. Measured shear viscosities of concentrated latex suspensions as a function of non-dimensional shear stress

$$\mathscr{S}_r = \frac{\mathscr{S}a^3}{kT}.$$

The volume fraction ϕ is 0.35 for (a) and 0.45 for (b). The lines represent eqn (9.10.15) with $\alpha = 0.431$. (From Krieger 1972, with permission.)

at $\phi = 0.35$ and 0.45. The lines represent the Dougherty–Krieger formula with $\alpha = 0.431$, with the $[\eta]$ and p values given earlier.

There is an obvious similarity between eqn (9.10.14) and eqn (9.8.11) for the sedimentation velocity of a concentrated suspension. The latter equation gives a very good description of the sedimentation velocity at concentrations far in excess of those for which the more exact theory is applicable (Fig. 9.8.2). It can be rationalized by treating the sedimentation of each particle as though it were obeying eqn (9.8.8) but assuming that the surrounding medium has the average density and the average shear viscosity of the *suspension* (not the liquid).

Exercises. 9.10.1 In deriving the formula (9.10.4) for the macroscopic stress tensor in a suspension of rigid particles, we used the fact that $\boldsymbol{e} = \boldsymbol{0}$ inside the particles. Establish this result by calculating the strain tensor for the rigid body velocity field $\boldsymbol{V} + \boldsymbol{\Omega} \times \boldsymbol{x}$.

9.10.2. In section 9.10.3 we discussed the solution for the flow field around a

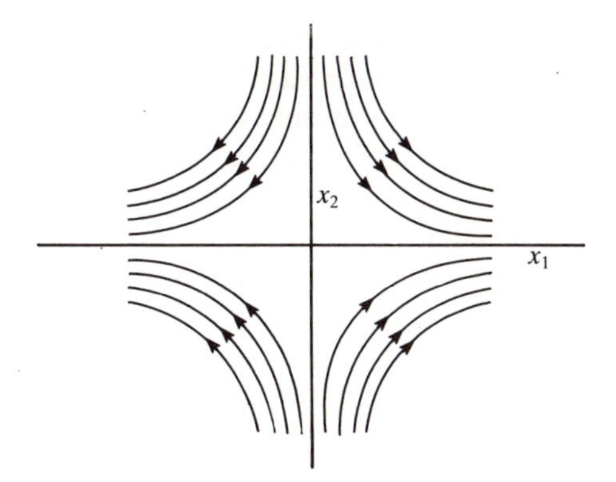

FIG. 9.10.5. Streamlines for a flow field of the form (9.10.11). See Exercise 9.10.2.

spherical particle, placed in the ambient velocity field

$$v_i^a = \sum_{j=1}^{3} e_{ij}^a x_j.$$

In order to visualize the distorting effect of this flow on the particle it is useful to sketch the streamlines of the undisturbed flow, for the case when $e_{11}^a = -e_{22}^a = e$, and all other components are zero. By putting

$$v_1^a = \frac{dx_1}{dt}, \quad \text{and} \quad v_2^a = \frac{dx_2}{dt}$$

in the above expression solve the resulting equations for the trajectories of fluid particles in this flow, and hence show that the fluid particles move along hyperbolae, as shown in Fig. 9.10.5.

If a sphere were now placed at the centre of this flow, there would be a tendency to stretch the sphere along the x_1 axis and compress it along the x_2 axis; as a result, there are set up on the surface of the sphere, stresses which oppose this distortion and thereby increase the effective viscosity.

Can you see why a sphere will not move when it is placed at the centre of the above flow?

9.10.3 Establish eqn (9.10.14) using the suggested integration procedure.

9.10.4 Show that $\mathcal{S}_i$ in eqn (9.10.15) is proportional to $\mathcal{D}\eta/a^2$ where $\mathcal{D}$ is the particle diffusion coefficient.

9.10.5 Calculate $\mathcal{S}_i$ for the suspensions shown in Fig. 9.10.4 assuming the particle radius is 100 nm and temperature is 25 °C. At what shear rate is the suspension viscosity halfway between η_1^* and η_2^*? What do η_1^* and η_2^* represent?

9.11 Electrokinetic effects

9.11.1 *Introduction*

Until now we have neglected the effect of the electrical charges that are inevitably present at the particle surface and in the surrounding liquid.

While this assumption is valid in certain circumstances—for example in the calculation of the viscosity of suspensions with thin double-layers—there are many instances in which the electric charge plays a vital role. In this section we will study some simple examples of these 'electrokinetic effects'; a more thorough discussion will be given in Vol. II (Chapter 13).

The charges in the liquid give rise to an electrical body force, which appears as an extra term in the Stokes equations:

$$\eta \nabla^2 \boldsymbol{v} - \nabla p = \rho_e \nabla \psi \tag{9.11.1}$$

and

$$\nabla \cdot \boldsymbol{v} = 0. \tag{9.11.2}$$

The quantity $-\rho_e \nabla \psi$ is the electrical force per unit volume of liquid, where ρ_e is the electrical charge density, due to the ions in the liquid, and ψ is the electrical potential. As before, inertia forces are assumed to be negligible.

In general the ion density ρ_e is unknown, and eqns (9.11.1) and (9.11.2) must be supplemented by a set of ion-conservation equations, one for each species of ion. In certain circumstances however the disturbing influence (e.g. an applied electric field) does not affect the ion density and thus ρ_e can be approximated in eqn (9.11.1) by its equilibrium value, obtained from the solution of the Poisson–Boltzmann equation (section 6.3). In this section we will concentrate on problems of this type.

9.11.2 *Electro-osmosis*

This term refers to the motion of liquid induced by an applied electric field. Such a motion occurs when an electric field is applied across a porous plug, but we begin with the much simpler case of the flow induced in a capillary tube by an electric field $\boldsymbol{E}$ parallel to the tube axis.

This flow results from the presence of a double layer at the tube wall. The application of the electric field causes the ions in the double layer to move towards one electrode or the other. Since the ions are predominantly of one sign their motion gives rise to a body force on the liquid in the double layer, and it is this body force which sets the liquid in motion.

In most cases, the tube radius is much larger than the double-layer thickness, and we can analyse the flow in the double layer on the assumption that the surface is locally flat.

Since the applied field is parallel to the tube surface, the resulting ion migration will not affect the charge density ρ_e. Thus the body force on the liquid may be written as

$$-\varepsilon_0 \varepsilon_r \nabla^2 \psi_e \boldsymbol{E} \tag{9.11.3}$$

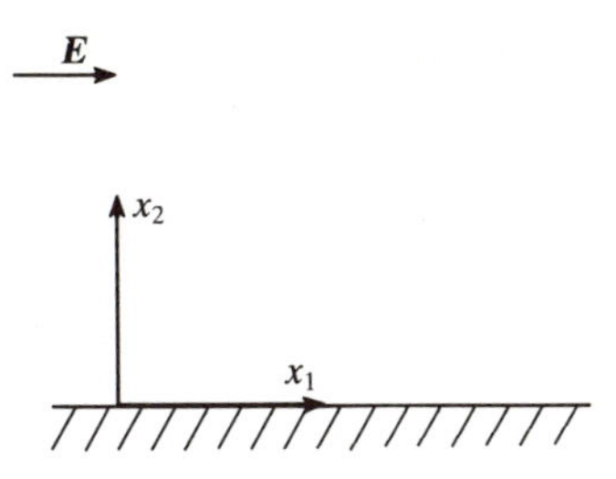

FIG. 9.11.1. The local Cartesian coordinate system used in the analysis of electro-osmotic flow near a solid boundary.

where we have used Poisson's equation (eqn (6.3.4)) relating ρ_e to the equilibrium potential ψ_e.

In analysing the flow we will use the Cartesian coordinates shown in Fig. 9.11.1, where the x_1 axis is parallel to the applied field, and the x_2 axis is normal to the local tube surface. In this coordinate system the force–balance equation (9.11.1) becomes

$$\eta \frac{\mathrm{d}^2 v_1}{\mathrm{d}x_2^2} - \frac{\partial p}{\partial x_1} = -\rho_e E, \tag{9.11.4}$$

and

$$-\frac{\partial p}{\partial x_2} = \rho_e \frac{\mathrm{d}\psi_e}{\mathrm{d}x_2}. \tag{9.11.5}$$

In setting up these equations we have assumed that the fluid flows parallel to the tube; from the continuity equation it then follows that v_1 is independent of x_1; it is further assumed that none of the quantities will depend on x_3, and for this reason we have only shown the x_1 and x_2 components of the force balance equation.

In the absence of an applied pressure gradient, p will be independent of x_1 and thus eqn (9.11.4) reduces to

$$\eta \frac{\mathrm{d}^2 v_1}{\mathrm{d}x_2^2} = \varepsilon_0 \varepsilon_r \frac{\mathrm{d}^2 \psi_e}{\mathrm{d}x_2^2} E$$

where we have replaced the electrical force by the formula (9.11.3). Integrating this equation twice with respect to x_2 and using the no-slip boundary condition, we obtain (Exercise 9.11.1):

$$\eta v_1 = \varepsilon_0 \varepsilon_r (\psi_e - \zeta) E + A_1 x_2 \tag{9.11.6}$$

where A_1 is a constant of integration, and ζ is the equilibrium potential at the 'plane of shear', which represents the effective location of the solid–liquid interface. Since the equilibrium potential may be a rapidly varying function of position near the particle surface, the value that we assign to ζ will be very sensitive to the position of this plane of shear. The plane will presumably be displaced out from the tube surface by a distance of the order of the thickness of the adsorbed ion layer on the surface, so that $\zeta \approx \psi_d$, the diffuse double layer potential, but the exact position of the plane is still the subject of some debate (see for example, Hunter 1981).

The quantity A_1 in eqn (9.11.6) is related to the velocity gradient beyond the double-layer (where the gradient in ψ_e is negligible). In the absence of an applied pressure gradient field A_1 must be zero, for the velocity gradients and the associated shear stresses in this region only arise in order to balance the pressure forces on the liquid. Thus from eqn (9.11.6) we see that the fluid velocity rises from zero at the plane of shear to a limiting value v_s beyond the double layer, where

$$v_s = -\frac{\varepsilon_0 \varepsilon_r \zeta E}{\eta} = -\frac{\varepsilon \zeta E}{\eta}. \tag{9.11.7}$$

From the macroscopic point of view the fluid appears to slip past the surface with this velocity, hence the subscript *s*.

Since only a small fraction of the total fluid volume lies in the double layer, the total flow rate is approximately given by $v_s A$, where A is the cross-sectional area of the tube. Thus with the aid of the formula (9.11.7) it is possible to calculate the ζ potential from measurements of the electro-osmotic flow rate. Such measurements provide useful information about the charging process at the glass–solution interface, and they will be discussed further in Chapter 12.

Equation (9.11.7) was first obtained by Smoluchowski in 1921. Since that time the analysis has been extended to the case of capillaries in which the radius is not large compared with κ^{-1} (Hunter 1981, section 3.5) and, more importantly it has also been extended to the case of porous plugs with thin double-layers; in the latter case the local analysis of the velocity field given here is still valid, but the local applied field $\boldsymbol{E}$ is distorted by the presence of the particles. Overbeek (1952) has shown that the velocity is given by eqn (9.11.7) everywhere in the pores beyond the double layer, provided $\boldsymbol{E}$ represents the *local* electric field.

The justification for this surprising observation is quite straightforward: substituting the form

$$\boldsymbol{v} = -\frac{\varepsilon_0 \varepsilon_r \zeta}{\eta} \boldsymbol{E} \tag{9.11.8}$$

in the Stokes equations, and using the fact that

$$\nabla \cdot \boldsymbol{E} = 0$$

we see that the continuity equation (9.11.2) is automatically satisfied while the force balance equation (9.11.1) is satisfied beyond the double layer with zero pressure gradient. Thus the formula (9.11.8) is the required solution to Stokes' equations with zero applied pressure gradient.

The macroscopic electro-osmotic velocity in a porous plug is therefore given by

$$\langle \boldsymbol{v} \rangle = -\frac{\varepsilon_0 \varepsilon_r}{\eta} \frac{1}{\mathcal{V}} \int_{\mathcal{V}_l} \boldsymbol{E} \, d\mathcal{V}, \tag{9.11.9}$$

where the integral extends over the liquid in the sample volume $\mathcal{V}$ and the contribution from the double layer is assumed to be negligible. The integral in this expression can be evaluated by measuring the macroscopic electric current density, given by

$$\langle \boldsymbol{i} \rangle = \frac{K_e}{\mathcal{V}} \int_{\mathcal{V}_l} \boldsymbol{E} \, d\mathcal{V} \tag{9.11.10}$$

where K_e is the electrolyte conductivity. Combining these equations we get

$$\frac{|\langle \boldsymbol{v} \rangle|}{|\langle \boldsymbol{i} \rangle|} = \frac{\varepsilon \zeta}{\eta K_e}, \tag{9.11.11}$$

an equation which can be used for the calculation of ζ-potentials in porous plugs, provided that the contribution to the conductivity from the double layer is negligible.

9.11.3 *Streaming potential*

When the liquid in a capillary tube or porous plug is set in motion by an applied pressure gradient, the double-layer charge moves with the surrounding liquid, giving rise to an electric current. The resulting transfer of charge leads to electric fields that tend to reduce the current until, after a very short time the current due to the pressure gradient is balanced by the current due to the induced electric field. The potential drop associated with this electric field is called the 'streaming potential'.

The calculation of the streaming potential for a circular capillary tube is quite straightforward: the velocity due to the applied pressure difference is given by the Poiseuille formula (eqn (9.7.15)), which in this case takes the form

$$v = \frac{\Delta p}{4\eta L}(a^2 - r^2), \tag{9.11.12}$$

where Δp is the pressure difference across the tube, L is the tube length and a is the radius. The electric current due to convection of the charge with the flow is

$$I_1 = \int_0^a 2\pi r \rho_e(r) v_l(r)\, dr. \tag{9.11.13}$$

This integral is dominated by the contribution from the double layer where ρ_e is non-zero. In this region the formula (9.11.12) for the velocity can be approximated by (Exercise 9.11.2)

$$v_l \approx \frac{\Delta p \cdot a}{2\eta L}(a - r), \tag{9.11.14}$$

and the formula (9.11.13) therefore reduces to

$$I_1 = -\frac{\Delta p a^2 \pi}{\eta L} \int_0^a \rho_e(y) y\, dy,$$

where

$$y = a - r.$$

Using Poisson's equation (eqn (6.3.4)) to replace ρ_e by

$$-\varepsilon_0 \varepsilon_r \frac{d^2 \psi_e}{dy^2},$$

and integrating by parts we find (Exercise 9.11.2):

$$I_1 = -\varepsilon_0 \frac{\varepsilon_r \zeta}{\eta} \frac{\pi a^2}{L} \Delta p. \tag{9.11.15}$$

This current is balanced by the current due to the induced field E_s, a current that is approximately given by

$$I_2 = K_e \pi a^2 E_s,$$

in the case when the double-layer contribution can be neglected. This assumption is valid if (Exercise 9.11.3)

$$\frac{\exp(ze\zeta/2kT)}{\kappa a} \ll 1 \tag{9.11.16}$$

for a symmetrical electrolyte. Assuming that this constraint is satisfied, we find that the condition of zero net current yields

$$E_s = \frac{\varepsilon_0 \varepsilon_r \zeta}{\eta K_e} \frac{\Delta p}{L},$$

and hence the potential difference across the tube is

$$\Delta E = \frac{\varepsilon_0 \varepsilon_r \zeta}{\eta K_e} \Delta p. \qquad (9.11.17)$$

Overbeek (1952, p. 204) has shown that this equation can be extended to porous plugs, provided the constraint (9.11.16) is satisfied, where a is the particle radius.

9.11.4 *Electrophoresis: the Smoluchowski formula*

The term electrophoresis refers to the motion of suspended particles in an applied electric field. Experimentally it is found that the particle velocity is proportional to the applied field strength; for spherical particles this relationship takes the form

$$\boldsymbol{v} = \mu_E \boldsymbol{E}, \qquad (9.11.18)$$

where μ_E is called the electrophoretic mobility of the particle. In this section we will discuss the link between electrophoretic mobility and ζ-potential. In all cases it will be assumed that the particle can be treated as being alone in an infinite liquid.

The earliest solution to this problem was given by Smoluchowski (1921) for the $\kappa a \gg 1$ (thin double-layer) case. Smoluchowski reasoned that the problem of determining the local flow in the double layer in this case is the same as the electro-osmosis problem described in section 9.11.2, provided we take a frame of reference which moves with the particle. Although this is a reasonable assumption for particles with a dielectric constant which is much less than that of water, it is hard to justify in the general case when the local electric field can have a component directed into the surface. As it turns out, this objection is unimportant, for the electrophoretic mobility has been shown to be independent of particle dielectric constant (O'Brien and White 1978), and thus results obtained using Smoluchowski's argument for the zero dielectric constant case can be extended to the general case.

Having made the connection between electrophoresis and electro-osmosis, Smoluchowski concluded that the relationship between particle velocity and electric field should have the form

$$\boldsymbol{v} = \frac{\varepsilon_0 \varepsilon_r \zeta}{\eta} \boldsymbol{E},$$

and hence that the electrophoretic mobility is given by

$$\mu_E = \frac{\varepsilon_0 \varepsilon_r}{\eta} \zeta = \frac{\varepsilon \zeta}{\eta}. \qquad (9.11.19)$$

The justification for this step, which was given by Overbeek in 1952 (p. 207) follows very similar lines to the analysis of the electro-osmotic flow described in section 9.11.2.

The formula (9.11.19) can be applied to a particle of arbitrary shape provided the particle dimensions are much greater than the double-layer thickness and that the constraint (eqn (9.11.16)) is satisfied (O'Brien 1983). Since the double layer is thin, this constraint can only be violated at high ζ-potentials (Exercise 9.11.4).

9.11.5 *The Henry formula*

To study the effect of varying double-layer thickness on mobility, Henry (1931) calculated μ_E for a spherical particle with arbitrary double-layer thickness on the assumption that the charge density is unaffected by the applied field. This assumption is valid provided the ζ-potential is sufficiently low.

To justify this statement, consider first the case of an uncharged particle with zero dielectric constant. The instant after the field has been switched on the charge density will still be zero and the local electric field $\boldsymbol{E}$ will satisfy

$$\nabla \cdot \boldsymbol{E} = 0 \tag{9.11.20}$$

at each point in the electrolyte. This field will give rise to a local current density $K_e\boldsymbol{E}$. The net flux of charge into any volume element ΔV in the electrolyte will be (see Exercise 9.11.5)

$$\nabla \cdot (K_e\boldsymbol{E})\Delta V,$$

which is zero by eqn (9.11.20). Thus there is no tendency for the field to produce a charge build up in the liquid, and since

$$\boldsymbol{E} \cdot \hat{n} = 0$$

at the surface of the (zero dielectric constant) particle, there will be no accumulation of charge at the surface. Thus $\rho_e = 0$ for an uncharged particle in an electric field, and the electrical body forces are zero.

Now consider the body force term $\rho_e \nabla \psi$ in the case of a particle with low ζ-potential. We write:

$$\rho_e = \rho_e^0 + \delta\rho_e$$

and

$$\psi = \psi^0 + \delta\psi,$$

where the superscript 0 refers to the equilibrium value, in the absence of the applied field. Since the electrophoretic velocity is a linear function of field strength $\boldsymbol{E}$ we can calculate it, in the limit as $E \to 0$, neglecting $0(E^2)$ terms; thus we can treat the perturbations $\delta\rho_e$ and $\delta\psi$ as small quantities, and we can approximate the force term by

$$\rho_e^0 \nabla \psi^0 + \rho_e^0 \nabla \delta\psi + \delta\rho_e \nabla \psi^0 \qquad (9.11.21)$$

where the product of the small quantities $\delta\rho_e$ and $\delta\psi$ has been neglected. This is an approximation which is made in nearly every theoretical study of electrophoresis.

The first term in eqn (9.11.21) is of little interest, since it is balanced by the equilibrium osmotic pressure. Assuming that each of the remaining quantities in the equation can be expanded in a power series in ζ, we find that $\rho_e^0 \nabla \delta\psi$ is proportional to ζ for small ζ, while $\delta\rho_e \nabla \psi^0$ is proportional to ζ^2 since both $\delta\rho_e$ and $\nabla \psi^0$ approach zero as $\zeta \to 0$. Thus the dominant term in the small ζ limit is

$$\rho_e^0 \nabla \delta\psi \qquad (9.11.22)$$

where $\nabla \delta\psi$ is the field around an uncharged sphere. The double-layer distortion, which is embodied in the $\delta\rho_e$ terms, can therefore be neglected if the ζ-potential is sufficiently small; this simplification provides the basis for Henry's analysis.

To determine the electrophoretic mobility of a particle with low ζ potential it is therefore necessary to solve the Stokes equations (9.11.1) and (9.11.2) with the body force term (eqn (9.11.22)), subject to the constraints that the velocity tends to zero far from the particle, and that the net force on the particle is zero.

The techniques for solving problems of this sort will be described in Volume II. The resulting formula for the electrophoretic mobility is

$$\mu_E = 2\frac{\varepsilon_r \varepsilon_0}{3\eta} \zeta f_1(\kappa a), \qquad (9.11.23)$$

where $f_1(\kappa a)$ is a monotonically varying function which increases from 1.0 at $\kappa a = 0$ to 1.50 at $\kappa a = \infty$, in which case eqn (9.11.23) reduces to the Smoluchowski formula. A graph of $f_1(\kappa a)$ is given in Fig. 9.11.2.

In order to give an idea of the range of validity of Henry's formula we have included some curves of mobility versus ζ-potential obtained from a computer solution of the exact equations for the flow and the ion densities around a sphere (O'Brien and White 1978).

From these curves, in Fig. 9.11.3, it can be seen that the Henry formula, which represents the tangent at the origin, is valid for a range of ζ potentials that increases with κa, from a value of about 25 mV at $\kappa a = 1$ to the value indicated by the constraint (9.11.16) at large κa values.

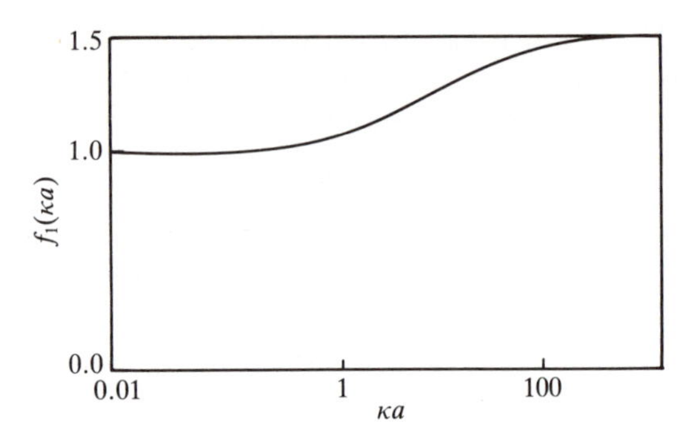

FIG. 9.11.2. The function f_1 (κa) that appears in the Henry formula eqn (9.11.23) for electrophoretic mobility.

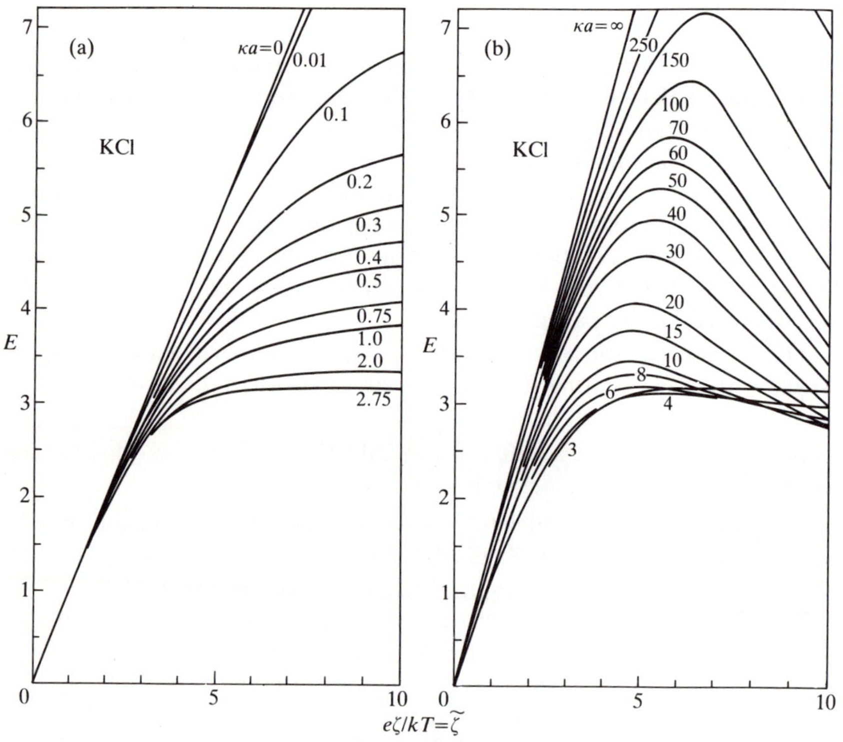

FIG. 9.11.3. Computed electrophoretic mobilities for a spherical particle in a KCl electrolyte. The quantity E is here a non-dimensional mobility, given by

$$E=\frac{3\eta e}{2\varepsilon_0\varepsilon_r kT}\mu_E.$$

(a) Small κa; (b) large κa. (Reproduced from O'Brien and White 1978, with permission.)

This completes our discussion of the elementary electrokinetic effects. The details of the recent work in this area, including the calculation of the electrophoresis curves of Fig. 9.11.3, will be set out in Volume II; for this reason we will not attempt to review the literature as we did for the other transport properties; interested readers are referred to Chapters 3 to 5 of Hunter's (1981) book.

Exercises. 9.11.1 Establish eqn (9.11.6). Calculate the electro-osmotic velocity for a capillary tube using eqn (9.11.7), for a ζ-potential of 25 mV, and an applied field of 1000 V m^{-1}. The water temperature is 20 °C.

9.11.2 By putting

$$(a^2 - r^2) = (a - r)(a + r)$$

in the formula (9.11.12) for Poiseuille flow, obtain the approximation (9.11.14) for flow near the tube wall ($r \approx a$). Establish eqn (9.11.15).

9.11.3 In deriving eqn (9.11.17) for the streaming potential in a capillary tube, we neglected the contribution of the electrical double layer to the current I_2 caused by the streaming field E_s. For a symmetrical electrolyte in which both species of ion have approximately the same mobility, the exact formula for I_2 is

$$I_2 = \left\{\int_A K_e \frac{\{n_+ + n_-\}}{2n^0} \, dA\right\} E_s + \int_A \rho_e v_1 \, dA$$

where A is the cross-sectional area of the tube, n_+ and n_- are the equilibrium ion densities and v_1 is the osmotic velocity due to E_s, given by eqn (9.11.6) (with $A_1 = 0$). By using the Boltzmann formula (eqn (6.3.5)) for the ion densities and eqn (6.3.23) for ψ, evaluate this integral assuming that the double layer is locally flat. Hence verify that

$$I_2 \approx K_e A E_s$$

provided the constraint (9.11.16) is satisfied.

9.11.4 For $\kappa a = 50$, determine the value of ζ at which

$$\frac{\exp\left(\dfrac{e\zeta}{2kT}\right)}{\kappa a} = 1;$$

the Smoluchowski formula is invalid for such ζ potentials (see section 9.11.4)

9.11.5 In section 9.2 we showed that the net volume flux of fluid into a small block-shaped volume element $\Delta\mathscr{V}$ is

$$(\nabla \cdot \boldsymbol{v})\Delta\mathscr{V}.$$

Show, by similar arguments that the net flux of electric charge into $\Delta\mathscr{V}$ is

$$(\nabla \cdot \boldsymbol{i})\Delta\mathscr{V}$$

where $\boldsymbol{i}$ is the local current density; this is a result used in the discussion of electrophoresis in section 9.11.5.

9.11.6 *Reciprocity relations*

Comparison of eqns (9.11.11) and (9.11.17) reveals that the phenomena of electro-osmosis and streaming potential are connected by the relation

$$\left(\frac{|\langle \boldsymbol{v} \rangle|}{|\langle \boldsymbol{i} \rangle|}\right)_{\Delta p=0} = \left(\frac{\mathcal{V}_{\mathrm{T}}}{i}\right)_{\Delta p=0} = \left(\frac{\Delta E}{\Delta p}\right)_{i=0} \tag{9.11.24}$$

where Δp is the applied pressure difference, $\mathcal{V}_{\mathrm{T}}$ is the total volume flow per unit time and i is the current carried. This is known as Saxén's relation and it has been recognized for a long time (Saxén 1892). It is one of a general class of relations that can be derived from Onsager's reciprocity relations, using the thermodynamics of irreversible processes. Similar relations can be established between, for example, electrophoretic mobility and the sedimentation potential. Mazur and Overbeek (1951) showed that for porous plugs as well as single capillaries eqn (9.11.24) can be supplemented by:

$$\left(\frac{i}{\Delta p}\right)_{\Delta E=0} = \left(\frac{\mathcal{V}_{\mathrm{T}}}{\Delta E}\right)_{\Delta p=0} \tag{9.11.25}$$

and

$$\left(\frac{i}{\Delta p}\right)_{\mathcal{V}_{\mathrm{T}}=0} = \left(\frac{\mathcal{V}_{\mathrm{T}}}{\Delta E}\right)_{i=0} \tag{9.11.26}$$

and

$$\left(\frac{i}{\mathcal{V}_{\mathrm{T}}}\right)_{\Delta E=0} = -\left(\frac{\Delta p}{\Delta E}\right)_{\mathcal{V}_{\mathrm{T}}=0}. \tag{9.11.27}$$

These equations apply even in systems involving surface conductance or double-layer overlap when the simple relations like eqns (9.11.11) and (9.11.17) will certainly fail. They demonstrate that agreement between the various experimental procedures (e.g. electro-osmosis and streaming potential) is bound to be observed for a given surface system even if the equations for the zeta potential are quite erroneous.

The theory of irreversible processes provides some important unifying equations for electrokinetics but it must be admitted that its application beyond the sort of relation set out above (eqns (9.11.24)–(9.11.27)) has not so far proved very profitable.

References

Alexander, A. E. and Johnson, P. (1949). *Colloid science.* Oxford University Press, Oxford.

Batchelor, G. K. (1967). *An introduction to fluid dynamics*. Cambridge University Press, Cambridge.
Batchelor, G. K. (1970). *J. Fluid Mech.* **41**(3), 545.
Batchelor, G. K., (1972). *J. Fluid Mech.* **52,** 245.
Batchelor, G. K., (1976). *J. Fluid Mech.* **74,** 1.
Batchelor, G. K., (1977). *J. Fluid Mech.* **83,** 97.
Batchelor, G. K. and Green, J. T., (1972a), *J. Fluid Mech.* **56,** 375.
Batchelor, G. K. and Green, J. T. (1972b). *J. Fluid Mech.* **56,** 401.
Buscall, R., Goodwin, J. W., Ottewill, R. H., and Tadros, Th. F., (1982). *J. Colloid interface Sci.* **85,** 78.
Debye, P. (1929). *Polar molecules*. Reinhold, New York.
Einstein, A., (1956). *Investigations on the theory of the Brownian movement*. Dover, New York.
Ekdawi, N. and Hunter, R. J. (1985). *Colloids and surfaces* **15,** 147–59.
Goodwin, J. W. (1975). The rheology of dispersions. In *Colloid science* (ed. D. H. Everett), Vol. 2, pp. 246–293. Chemical Society, London.
Henry, D. C. (1931). *Proc. R. Soc. Lond.* **A133,** 106.
Hiemenz, P. C., (1977). *Principles of colloid and surface chemistry*. Marcel Dekker, New York.
Hinch, E. J. and Leal, L. G. (1972). *J. Fluid Mech.* **52,** 683–712.
Hunter, R. J. (1981). *Zeta potential in colloid science*. Academic Press, London.
Hunter, R. J. and Nicol, S. K. (1968). *J. Colloid interface Sci.* **28,** 250.
Krieger, I. M. (1972). *Adv. Colloid interface Sci.* **3,** 111.
Krieger, I. M. and Maron, S. (1952). *J. appl. Phys.* **23,** 147–9.
Krieger, I. M. and Maron, S. (1954). *J. appl. Phys.* **25,** 72–5.
Landau, L. D., and Lifshitz, E. M. (1959). *Fluid mechanics*. Pergamon Press, Oxford.
Landau, L. D. and Lifshitz, E. M. (1969). *Statistical physics*. Pergamon Press, Oxford.
Leal, L. G. and Hinch, E. J. (1973). *Rheol. Acta* **12,** 127.
Maron, S. H., Krieger, I. M. and Sisko, A. W. (1954). *J. appl. Phys.* **25,** 971–6.
Mazur, P. and Overbeek, J. Th. G. (1951). *Recl. Trav. Chim.* **70,** 83.
Meriam, J. L. (1966). *Dynamics*. Wiley, New York.
McQuarrie, D. A. (1976). *Statistical mechanics*. Harper and Row, New York.
O'Brien, R. W. and White, L. R. (1978). *J. chem. Soc. Faraday Trans. 1* **74,** 1607.
O'Brien, R. W. (1983). *J. Colloid interface Sci.* **92,** 204.
Overbeek, J. Th. G. (1952). In *Colloid science* (ed. H. R. Kruyt) Vol. 1, p. 197. Elsevier, Amsterdam.
Perrin, F. (1934). *J. Phys. Radium* **5**(7) 497.
Philip, J. R. and Smiles, D. E. (1982). *Adv. Colloid interface Sci.* **17,** 83.
Russel, W. B. (1981). *Ann. Rev. Fluid Mech.* **13,** 425–55.
Saxén, U. (1892). *Wied. Ann.* **47,** 46.
Shah, M. J. (1963). *J. phys. Chem.* **67,** 2215–9.
Smoluchowski, M. von (1921). In *Handbuch der Electrizität und des Magnetismus* (Graetz) Vol. II, p. 366. Barth, Leipzig.
Van Wazer, J. R., Lyons, J. W., Kim, K. Y., and Colwell, R. E. (1963). *Viscosity and flow measurement*. Interscience, New York.

10

ASSOCIATION COLLOIDS

10.1 The critical micellization concentration

We noted in section 1.5.3 that certain molecules (called amphiphiles) are able to form aggregates called *micelles* in aqueous solution, provided their concentration is sufficiently high. The concentration at which this micelle formation occurs is usually fairly sharply defined and it can be identified by observing the behaviour of any one of a number of equilibrium or transport properties of the solution, (Fig. 10.1.1) each of which undergoes a rather abrupt change in concentration dependence at much the same point (called the critical micellization concentration or c.m.c.). We do not propose to review the many methods which have been used to detect the onset of micellization although some reference will be made to the recent spectroscopic procedures in section 10.6.1.

Close examination reveals that, in some cases, different methods of measurement would yield c.m.c. values varying by almost 50 per cent (see e.g. Kresheck 1975, Fig. 1) and the same method in different hands can produce a similar spread. Some of the variation may be due to the presence of impurities, small amounts of which are known to have a significant effect on the c.m.c. Some variation can be traced to uncertainties in the extrapolation procedures used to define the c.m.c. (The value derived from the *same* conductance data can vary significantly, depending on whether a plot is made of the molar conductance against $C^{1/2}$ or the specific conductivity (conductance) against $\log C$.) Despite

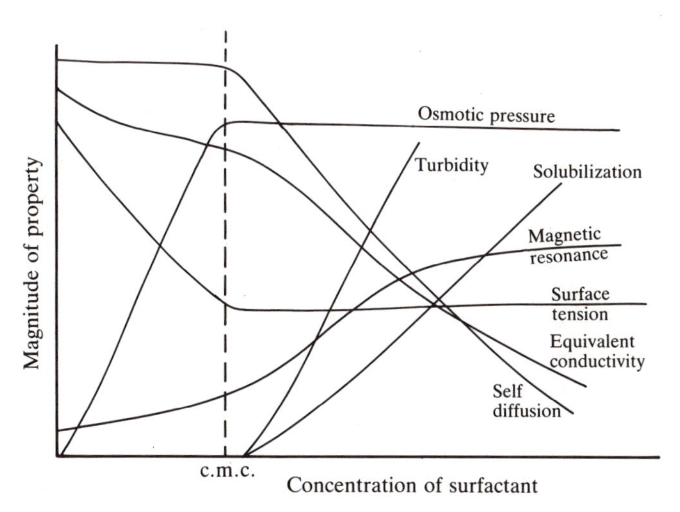

FIG. 10.1.1. Schematic representation of the concentration dependence of some physical properties for solutions of micelle-forming amphiphiles. (After Lindman and Wennerström 1980, with permission.)

these limitations the concept of critical micellization concentration remains an important one. It can be defined in terms of one or other of the properties suggested by Fig. 10.1.1 but a more general definition is (Phillips 1955):

$$\left(\frac{d^3\phi}{dC_T^3}\right)_{C_T=\mathrm{cmc}} = 0 \qquad (10.1.1)$$

where ϕ is any one of the properties (Exercise 10.1.1) and C_T is the total concentration of the amphiphile or surfactant.

Micelle formation is only one of a number of characteristic aggregation phenomena which amphiphilic molecules undergo and one might well ask what causes this behaviour pattern, intermediate as it is between true solution and a separation of the components into two distinct phases. After all, it does not occur with long chain alcohols, amides, or amines which also have a polar head group and a lipophilic or hydrophobic chain attached. It seems to be necessary to have either a charged head group (carboxylate, sulphate, sulphonate, or quaternary ammonium), a zwitterionic group, or a rather bulky oxygen-containing hydrophilic group (polyoxyethylene, phosphine oxide, amine oxide, or a sugar residue) able to undergo significant hydrogen bonding as well as dipolar interaction with water. (We concentrate mainly on the formation of micelles in aqueous solution because the characteristics of micelles formed in non-aqueous solutions are less well established. For strongly hydrogen bonded solvents (e.g. formamide and hydrazine), however, the behaviour is similar in many respects to water (Kresheck 1975). Evidently, the balance of forces which leads to micelle formation is a subtle one. If the hydrophilic effect is sufficiently strong the molecule can enjoy complete solution; if it is too weak the substance is merely insoluble (e.g. octadecanol).

The structure of micelles has been a subject of controversy for many years, since the pioneering works of Hartley (1936) and McBain (1950) arguing the cases for spherical and lamellar structures respectively. The spherical (or near-spherical) form was generally accepted as the dominant species in dilute aqueous solutions up until recent times. Tanford (1980) in his extensive review of the mechanism of micelle formation prefers the description 'disc-like' since the weight of evidence from transport and equilibrium properties suggests that the structures are better described as oblate spheroids (Fig. 3.1.1). This view is not, however, universally accepted and the most elaborate theoretical calculations (to be discussed in section 10.10) are based on a spherical shape, at least for the dodecyl sulphate micelle near its c.m.c. Throughout the following discussion it would be wise to bear in mind the fact that the forces controlling micellar structure are delicately balanced so that distortions of the shape (either due to shearing processes or fluctuations) may occur fairly easily (but see section 10.10).

Furthermore, any attempt to represent the structure can at best be a statement of the average shape over some discrete time interval. The exchange that occurs between monomer molecules and those in the micelle occupies a time-scale of the order of 1–10 μs (Aniansson 1978) and the entry or exit of a monomer presumably occurs by a series of diffusive steps, one methylene group at a time. The surface must, therefore, be somewhat rough on the nanosecond time scale (section 10.7).

More detailed thermodynamic analysis (section 10.4) reveals some, at first sight, quite surprising results. For one thing, the aggregation process in water at room temperature is accompanied by a significant *increase* in entropy, which is the main contribution to the negative $\Delta G^{\ominus}$ value for micellization. The $\Delta H^{\ominus}$ value is usually small, at least for systems with small degrees of aggregation (50–100), and often slightly positive. The traditional view of the mechanism of micelle formation has been based on the study of the (very slight) solubility of hydrocarbons in water, and what has come to be known as the *hydrophobic effect* (Tanford 1980). It seems that, at room temperature, the presence of a hydrocarbon molecule in water causes a significant *decrease* in the (partial molar) entropy of the water, suggesting that it induces an increase in the degree of structuring of the water molecules. The isolated hydrocarbon molecule forms a cavity in the water structure and the walls of that cavity are lined with water molecules with a bonding pattern that differs, on average, from the bulk pattern and that, furthermore, varies in a complex and subtle way with change in temperature. The predominant effect of a hydrocarbon molecule is to increase the degree of structure in the immediately surrounding water and this is one of the main features of the hydrophobic effect. The other major effect is to disrupt the extensive hydrogen bonding pattern in the water. Evidently the entropy increase associated with this latter process is outweighed by the energy increase involved and its contribution to $\Delta G^{\ominus}$ is again positive (Frank and Evans 1945).

When hydrocarbon residues aggregate in aqueous solution to produce a micelle, the reverse process occurs: the hydrogen bonding structure in the water is, to a large extent, restored. For this process, both enthalpy and entropy changes are negative. The 'melting' of the cavities that surrounded the hydrocarbon chains gives rise to an entropy increase in the water that more than compensates for the decreased randomness of the hydrocarbon chains as they enter the micelles. This view of the micellization process, in which the principal driving force is the partial molar entropy increase of the water, has been strongly challenged recently by some studies of aqueous systems at high temperatures (up to 166 °C) and micellization in pure hydrazine solutions (Ramadan *et al.* 1983) and we will return to this discussion when we have established the basis for a thermodynamic analysis (section 10.4).

Since the head group remains surrounded by water its contribution to the energetics of micellization is much less but its role is essential. It is the nature of the head group and the interactions that occur between head groups that, in principle, determine the size and shape of the aggregate structure (Israelachvili *et al.* 1976). We will however, defer more detailed discussion of these matters until later (section 10.9). We will first look at the phenomenology of micelle formation—what effect factors like temperature, pressure, and salt content have on the c.m.c. We will then examine some of the simple mass-action models of micelle formation and discuss in more detail the thermodynamics of the process.

For ionic surfactants, one of the most important determinants of behaviour is the extent to which the counterions are bound tightly to the head groups, i.e. the extent of dissociation. That question is examined in section 10.5, using a number of experimental techniques. We then consider some of the methods (especially spectroscopic procedures) that have been used to study other features of micellar structure and dynamics. The chapter concludes with a brief description of the current ideas about the geometric factors that determine aggregate shapes and the statistical mechanical description of a typical micelle. More detailed discussion of these matters will be deferred to Volume II, when the theory of liquid structure and scattering behaviour can be brought to bear on the problem. A brief coda on micellization in mixed surfactant systems is included. Such mixed systems give rise to a variety of aggregation phenomena including the formation of bilayers and biological membrane analogues. A more detailed treatment of such systems must also be deferred to Volume II.

Exercises. 10.1.1 Show, using appropriate sketches, that eqn (10.1.1) is a general statement of the fact that the property $\phi(C_T)$ undergoes a sharp change in slope at $C_T =$ c.m.c.

10.1.2 Use eqns (5.10.1) and (5.10.2), together with the data in Exercise 5.11.2, to show that the work of cohesion between two hydrocarbon phases is almost exactly equal to the work of adhesion between hydrocarbon and water.

10.2 Factors affecting the c.m.c.

A very large and valuable collection of data on the c.m.c. of various surfactants has been compiled by Mukerjee and Mysels (1971) from which a number of important generalizations emerge.

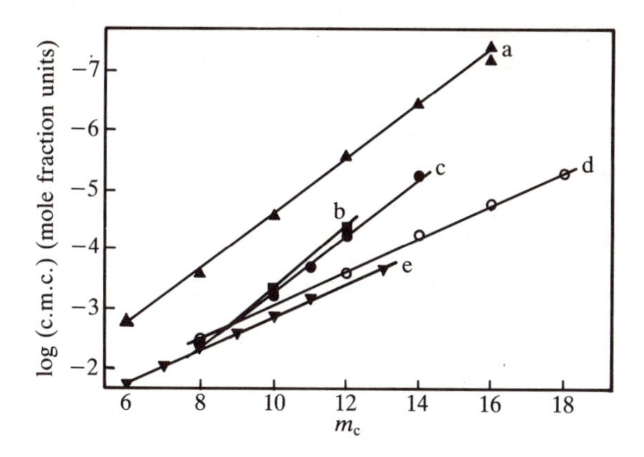

FIG. 10.2.1. Plots of $\log_{10}$ c.m.c. (in mole fraction units) versus m_c, the number of carbon atoms in the alkyl chain. (Temperature in general 25 °C.) (a) Alkyl hexaoxyethylene glycol monoethers. (b) Alkyl trimethylammonium bromides in 0.5 M NaBr. (c) *N*-alkyl betaines. (d) Sodium alkyl sulphates (40 °C). (e) Sodium alkylcarboxylates. (From Lindman and Wennerström 1980, with permission.)

10.2.1 *Effect of head group and chain length*

For surfactants with a single straight hydrocarbon chain, the c.m.c. is related to the number of carbon atoms in the chain (m_c) by:

$$\log_{10} \text{c.m.c.} = b_0 - b_1 m_c \tag{10.2.1}$$

where b_0 and b_1 are constants. Figure 10.2.1 illustrates this point for a number of typical non-ionic and ionic surfactants. Some of the reasons for the differences in b_0 and b_1 values revealed by this figure and Table 10.1, will be discussed below (section 10.4.2). It is hardly surprising that

Table 10.1

Values of b_0 *and* b_1 *in eqn* (10.2.1) *for various surfactants.* (*Selected from Kresheck* 1975)

Surfactant	Temperature (°C)	b_0	b_1
Na carboxylates	20	2.41	0.341
K carboxylates	25	1.92	0.290
Alkane sulphonates	40	1.59	0.294
Alkyl sulphates	45	1.42	0.295
Alkylammonium chlorides	25	1.25	0.265
Alkyltrioxyethylene-glycol monoether	25	2.32	0.554
Alkyldimethylamine oxide	27	3.3†	0.5†

† These values are likely to be pH dependent because this amphiphile becomes cationic at low pH.

the nature of the head group should affect the value of b_0, but it is also apparent that it profoundly affects b_1 as well.

It should also be noted that the non-ionics usually have much lower c.m.c.s than the ionics despite their generally larger b_0 values. Most ionics of given chain length show very similar values for c.m.c. Lindman and Wennerström (1980) quote, for the straight chain C_{12} (dodecyl) surfactants the following values (with the counterion in brackets):

$$8\,\text{mM for } -\text{O.SO}_3^-(\text{Na}^+) \quad 10\,\text{mM for } -\text{SO}_3^-(\text{Na}^+)$$
$$15\,\text{mM for } -\text{NH}_3^+(\text{Cl}^-) \quad 12\,\text{mM for } -\text{CO}_2^-(\text{K}^+)$$
$$20\,\text{mM for } -\text{N}^+(\text{CH}_3)_3(\text{Cl}^-).$$

Modifications to the hydrocarbon chain (such as introducing branching, or double bonds, or polar functional groups along the chain) usually lead to increases in the c.m.c., but the introduction of a benzene ring is equivalent to adding about 3.5 methylene groups to the chain length. Introducing fluorine in place of hydrogen has a quite marked effect, at first increasing and then decreasing the c.m.c. (ultimately to less than 10 per cent of the hydrocarbon value), as the proportion of fluorine is increased towards saturation (Mukerjee and Mysels 1975). We will return to this point in section 10.7.4.

10.2.2 *Effect of counterion*

It should come as no surprise to learn that counterion valency has a strong effect on c.m.c. and for various ions of the same valency, the lyotropic series (section 7.7.4) has a role to play in the explanation of smaller variations. The values for b_0 and b_1 quoted in Table 10.1 for the sodium and potassium salts of carboxylic acids are, however, rather more difficult to rationalize. Small differences in b_0 can be attributed to differences in degree of hydration and cation binding (section 10.5), but it seems surprising that the b_1 value should vary so much in this case.

10.2.3 *Effect of temperature and pressure*

One of the most surprising things about micellization is the very weak temperature and pressure dependence of the c.m.c., considering that it is an association process (Lindman and Wennerström 1980). This is a reflection, of course, of the very subtle changes in bonding, heat capacity, and volume that accompany the micellization process. It seems likely that if a wide enough temperature range were accessible, all amphiphile systems would show a temperature at which the c.m.c. was a minimum (Kresheck 1975).

Raising the temperature has a quite different effect on ionic and non-ionic surfactants. For ionics, there is a temperature (called the *Krafft point*) below which the solubility is quite low and the solution appears to contain no micelles. Above the Krafft temperature, micelle formation evidently becomes possible and there is a rapid increase in solubility of the surfactant. It is significant that surfactants are usually much less effective (as, for example, detergents) below the Krafft point. Non-ionic surfactants tend to behave in the opposite manner. As the temperature is raised, a point may be reached at which large aggregates of the non-ionic separate out into a distinct phase and the temperature at which this occurs is referred to as the *cloud point*. It is usually rather less sharp than the Krafft point (Leja 1982).

We will postpone further discussion of temperature and pressure effects until the thermodynamics of micellization has been examined in more detail.

10.2.4 *Effect of added salt*

Adding an indifferent electrolyte (section 2.5) to an amphiphile/water system has a pronounced effect on the c.m.c., especially for ionics. For non-ionics the effect is smaller but still significant and the difference between the two is dramatically demonstrated by the difference in the functional dependence of c.m.c. on salt concentration, C:

$$\log(\text{c.m.c.}) = b_2 + b_3 C \quad \text{(non-ionic)} \tag{10.2.2}$$

and

$$\log(\text{c.m.c.}) = b_4 + b_5 \log C \quad \text{(ionic)}. \tag{10.2.3}$$

The constants, b_i, depend upon the nature of the electrolyte (Fig. 10.2.2). For the ionics, the principal effect of the salt is to partially screen the electrostatic repulsion between the head groups and so lower the c.m.c. Values of b_5 of from -0.6 to -1.2 are given (Kresheck 1975) for the influence of sodium salts on sodium carboxylates. The more subtle influences of salts of the same valence type are again usually discussed in terms of the lyotropic series (section 7.7.4).

For the non-ionics the concentrations of salts required to produce significant effects are much higher and the discussion of such behaviour introduces the notion of 'salting in' and 'salting out' of non-electrolytes by the electrolyte (Ray and Nemethy 1971). This amounts to a description of the competition between the surfactant (chiefly the head group) and the electrolyte for the opportunity to associate with the water. To put it more precisely, the activity coefficient of the monomer surfactant changes as the electrolyte concentration *and type* alters. If the

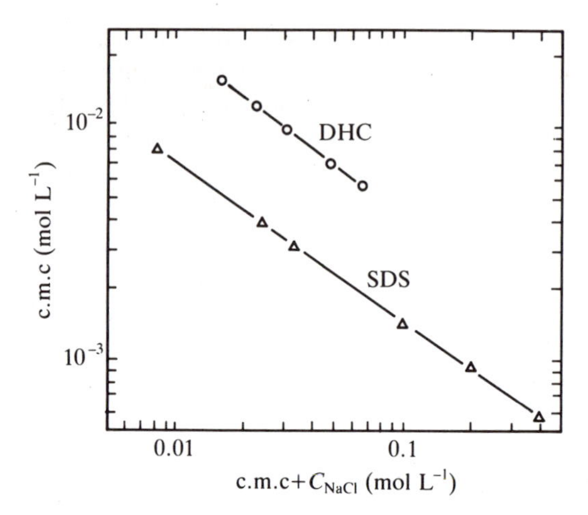

FIG. 10.2.2. The effect of added salt on the c.m.c. of SDS and dodecylamine hydrochloride (DHC). (From Stigter 1975a, with permission.)

monomer is salted out by electrolyte then micellization is thermodynamically favoured and the c.m.c. is reduced. The reverse situation applies if the monomer is salted in.

10.2.5 *Effect of organic molecules*

Quite small amounts of organic material can have a significant influence on the c.m.c., and the properties of micellar solutions. Recall Fig. 5.4.2, where the presence of small amounts of a certain impurity caused a minimum in the plot of surface tension against surfactant concentration in the neighbourhood of the c.m.c. The classical example of this behaviour occurs in aqueous solutions of sodium dodecyl sulphate (SDS), where the presence of dodecanol (a hydrolysis product) causes a minimum in the surface tension due to the competing effects of adsorption of dodecanol at the aqueous solution–air interface and solubilization in the SDS micelles. Such anomalous behaviour is to be expected between supposedly similar industrial surfactants, as a consequence of the presence of impurities or manufacturing by-products.

It is an important aspect of the behaviour of micelles that they are able to act as sites for the dissolution of lipophilic (i.e. fat-soluble) molecules. The use of surfactants as detergents, stabilizers and dispersing agents depends on this property, known as *solubilization*. It is characterized by a dramatic increase in the solubility of the lipophilic material at concentrations of the surfactant above the c.m.c. It is, in fact, used as a means of detecting the onset of micellization. One of the problems with this method, however, is that the lipophilic material itself may influence the value of the c.m.c. by contributing to or opposing the aggregation forces.

It is common practice to divide organic materials into two main groups, depending on their mode of action in influencing the c.m.c. Group A is composed of molecules (like alcohols with moderate to long hydrocarbon chains) that appear to be adsorbed in the outer regions of the micelle, forming a palisade (i.e. fence-like) structure with the surfactant molecules. This lowers the free energy of micellization to more negative values and so reduces the c.m.c.; such molecules can also influence the micelle shape. Straight chain molecules have the most marked effect, the latter reaching a maximum when the length of the hydrophobic chain of the additive is about the same as that of the surfactant. A decreased electrostatic repulsion between ionized head groups, and reduction in steric hindrance for non-ionic surfactants have been proposed as likely explanations for these observed effects. Group A compounds are generally effective at quite low bulk concentrations. They behave in an analogous fashion in other areas such as in the addition of small quantities of non-ionic molecules to flotation pulps (Fuerstenau 1976), in the penetration of insoluble charged monolayers by compounds such as hexadecanol (Gaines 1966) and in enhancing foam stability (Kitchener 1964).

Group B materials alter the c.m.c. at substantially higher bulk concentrations and probably exert their influence through modification of the bulk water structure†. The effect is usually discussed in terms of whether the additive is a (water) *structure maker* or a *structure breaker*. Typical 'structure makers' are xylose and fructose (Schwuger 1971) and 'structure breakers' are urea and formamide (Schick 1967).

Structure breakers increase the c.m.c. of surfactants in aqueous solution, exerting their strongest influence on non-ionic surfactants of the polyethyleneoxide type. Presumably the presence of a structure breaker reduces the amount of water structure that the hydrophobic residues of the surfactant can induce. The entropy increase on micelle formation is thus reduced and so the c.m.c. is raised. The concept is not, however, a very straightforward one to apply. Even where the solute is able to interact very strongly with water its effect may be overall structure breaking, firstly because it has to pull water from its existing structure and secondly, because the resulting entity may substantially disrupt the remaining water structure. For a description of these complexities, see the recent treatment by Franks, (1983, Chapters 9 and 10). It cannot be emphasized too strongly that the interactions between organic molecules

† The short-chain alcohols like ethanol and methyl propanol-2 can act as both group A and B materials.

and water are subtle and complex. The enthalpic and entropic contributions are often finely balanced so that the free energy of solution suggests a simple pattern that cannot be sustained on deeper analysis. Furthermore, the variations of these thermodynamic parameters with temperature and with composition are often so complex and difficult to explain that many researchers in the field would reject the dichotomy into 'makers' and 'breakers' of structure as being too simplistic. It remains, however a useful notion in the extreme case to distinguish the increased interactions that occur *between water molecules* because a hydrocarbon moiety is present and those that occur *between solute and water* (usually hydrogen bonding) that disturb the normal (very strong) hydrogen bonding of the water itself.

Exercises. 10.2.1 Estimate the c.m.c.s for the sodium salts of the C_8, C_{10}, C_{12}, and C_{14} straight-chain alkyl sulphates from the data in Table 10.1. The experimental values at 25 °C are:

130.3 33.0 8.08 and 2.05 mM

and at 40 °C are:

136 33.5 8.7 and 2.21 mM, respectively.

Comment on the relation between the results.

10.3 Equilibrium constant treatment of micelle formation

As Tanford (1977) has pointed out, the formation of micelles can be treated in a formally rigorous way in terms of all of the possible equilibria:

$$Z + Z \overset{K_2}{\rightleftarrows} Z_2 + Z \overset{K_3}{\rightleftarrows} Z_3 \ldots \overset{K_n}{\rightleftarrows} Z_n + Z \rightleftarrows \ldots \qquad (10.3.1)$$

with equilibrium constants K_n for $n = 2$–∞. The various thermodynamic parameters ($\Delta G^{\ominus}$, $\Delta H^{\ominus}$, $\Delta S^{\ominus}$) for the aggregation process could then be expressed in terms of the K_n. Unfortunately, it is not possible to measure the individual equilibrium constants, and recourse must be had to one of a number of models to simplify the situation. We will describe here two of the simplest possible models, each of which finds some applications. They are the closed association model (section 10.3.1) and the multiple equilibria model (section 10.3.2) in its three manifestations:

(a) dimers dominate;
(b) all K_n of equal size; and
(c) one K_n much larger than the rest.

The last of these is an improvement on the closed association model.

10.3.1 *The closed association model*

Observations of the size of more or less spherical micelles in the neighbourhood of the c.m.c. (such as those of sodium dodecyl sulphate (SDS)) suggest that the size range is very limited. The simplest assumption to make in treating eqn (10.3.1) is, therefore, that only one of the K_n values is important. (For SDS it would be about K_{60} at 25 °C.) In that case the micelle formation is represented as:

$$n \text{ monomers} \rightleftarrows \text{micelle}$$

or

$$n\mathrm{Z} \rightleftarrows \mathrm{M}$$

for which the equilibrium constant, K, is:

$$K = \frac{[\text{micelles}]}{[\text{monomers}]^n} = \frac{C_\mathrm{m}}{C_\mathrm{s}^n}. \tag{10.3.2}$$

The inherent assumption here is that the activities may be replaced by concentrations. For the monomer this amounts to assuming that the only departure from ideal behaviour is the aggregation process. It could, in principle, be removed by estimating other activity corrections from solution theory. Assuming ideal behaviour for the micelles is more problematical because of the large size difference between monomers and micelles. The micelles will also interact strongly; for ionics the interaction will become very significant as soon as the mean separation is less than about $(8\text{–}10)/\kappa$, where κ is the Debye–Hückel parameter (section 6.3), and that occurs at surfactant concentrations not far above the c.m.c. (Exercise 10.3.1). It must also be noted that when ionic micelles are formed there is a strong tendency for the counterions to be associated closely with the head groups, because of the high electrostatic potential in that region. This is a further source of non-ideality, which is discussed in section 10.5 (Evans and Ninham 1983).

From eqn (10.3.2) we have (Exercise 10.3.4):

$$\Delta G^\ominus = -RT \ln K \tag{10.3.3}$$

$$= -RT \ln C_\mathrm{m} + nRT \ln C_\mathrm{s} \tag{10.3.4}$$

and

$$\frac{-\Delta G^\ominus}{n} = -\overline{\Delta G}^\ominus = \frac{RT}{n} \ln C_\mathrm{m} - RT \ln C_\mathrm{s}. \tag{10.3.5}$$

At the c.m.c., we set

$$C_\mathrm{s} = C_0 \tag{10.3.6}$$

and the total surfactant concentration, C_T, above this point is given by:

$$C_T = C_0 + nC_m. \qquad (10.3.7)$$

From eqns (10.3.2) and (10.3.7):

$$K = \frac{C_m}{(C_T - nC_m)^n}. \qquad (10.3.8)$$

Mukerjee (1975) shows how eqn (10.3.8), with a K value of unity and $n = 100$ gives rise to a sharp transition from a system in which all of the surfactant is present as monomer to one in which the monomer concentration remains essentially constant above the c.m.c. and all additional surfactant goes into micelle formation (Exercise 10.3.2). As an alternative approach (Exercise 10.3.3), eqn (10.3.8) can be differentiated to obtain:

$$\frac{dC_m}{dC_T} = \frac{K^{1/n}}{\left[nK^{1/n} + \frac{1}{n} C_m^{(1-n)/n}\right]}. \qquad (10.3.9)$$

The plot of dC_m/dC_T against concentration for $K = 1$ and various values of n is shown in Fig. 10.3.1.

When it is realized that for most of the commonly used surfactants the value of n is at least 50 it becomes clear that the concept of a critical micellization concentration (being the concentration in the neighbourhood of which, micelle formation begins) is a reasonable one.

As $n \to \infty$ the transition becomes sharper and ultimately approaches the behaviour expected of a first-order phase transition. In this extreme case,

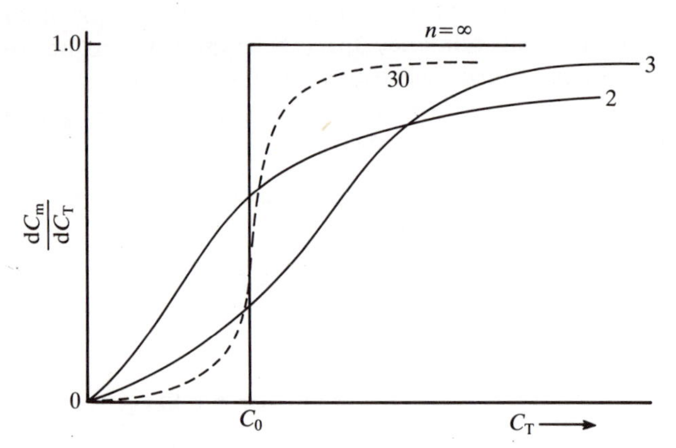

FIG. 10.3.1. Variation of dC_m/dC_T with total surfactant concentration for different values of the aggregation number, n. C_0 is the critical micellization concentration and C_m the concentration of micelles.

it is possible to treat the micelle formation as a phase separation. For smaller values of n there are problems involved in the phase separation model (Hall and Pethica 1967) and these can only be properly resolved by resorting to the formalism of small system thermodynamics (Hill 1963).

10.3.2 *Multiple equilibrium models*

The closed association model (section 10.3.1) is not physically appealing. If, for example, $n = 50$, it is difficult to see why the addition of one extra monomer to the aggregate of 49 should drastically reduce the free energy of the aggregate. And why should it be difficult, or impossible, to add an additional monomer? It may be argued that a certain number of monomers is required to build a complete structure and certainly one can see that some minimal number is required to produce a structure in which the head groups can effectively shield the hydrocarbon residues from the water. If the aggregate were crystalline there might be geometric reasons for some rather closely specified aggregation number, but there is abundant evidence (see e.g. Phillips 1955; Fisher and Oakenfull 1977) that the interior of most micelles (at least those formed from long-chain surfactants) is liquid-like. To obtain a physically reasonable model, it is therefore necessary to write out the full equilibrium between monomer and micelles of all sizes and then to seek a physically reasonable basis for defining a relationship between the equilibrium constants. Such a relationship should predict the observed facts of micelle size (or size distribution) and the thermodynamic parameters, while still having only a small number of adjustable parameters. The treatment here follows that of Mukerjee (1975) with some modifications.

Consider again the equilibria in eqn (10.3.1). For any n-mer, the stepwise association constant is:

$$x_{n-1} + x_1 \overset{K_n}{\rightleftarrows} x_n. \qquad (10.3.10)$$

The overall association constant, then, for the formation of x_n from x_1 (i.e. $nx_1 \rightleftarrows x_n$) is:

$$^*K_n = \frac{[x_n]}{[x_1]^n} \qquad (10.3.11)$$

where

$$^*K_n = \prod_2^n K_n \qquad (10.3.12)$$

(i.e. $^*K_n$ is the product of all stepwise association constants K_2, K_3 . . . up to K_n).

Again in defining $^*K_n$ we replace activities with concentrations with all of the uncertainties previously referred to.

The concentration of all solute species, S, (in mol dm^{-3}) is:

$$S = \sum [x_n] \tag{10.3.13}$$

and, expressed in terms of the total monomer concentration, M_1, this becomes:

$$M_1 = \sum n[x_n]. \tag{10.3.14}$$

Then the number average degree of association $\bar{N}_n$, of all species (including the monomer) is (see section 3.3.2):

$$\bar{N}_n = \frac{\sum n[x_n]}{\sum [x_n]} = \frac{M_1}{S}. \tag{10.3.15}$$

The mass average, $\bar{N}_w$, is given by:

$$\bar{N}_w = \frac{\sum n^2[x_n]}{\sum n[x_n]} = \frac{Z}{M_1} \tag{10.3.16}$$

where

$$Z \stackrel{\text{def}}{=} \sum n^2[x_n]. \tag{10.3.17}$$

We normally concern ourselves only with associated species and exclude the monomer. In these terms, the corresponding values of number and mass average are:

$$^*\bar{N}_n = \frac{\sum_2 n[x_n]}{\sum_2 [x_n]} = \frac{M_1 - [x_1]}{S - [x_1]} \tag{10.3.18}$$

and

$$^*\bar{N}_w = \frac{\sum_2 n^2[x_n]}{\sum_2 n[x_n]} = \frac{Z - [x_1]}{M_1 - [x_1]}. \tag{10.3.19}$$

It is evident from eqn (10.3.11) that, as $[x_1]$ increases, the concentration of each associated species increases. The value of n influences each associated species and the percentage increase of each increases with n. From eqns (10.3.16)–(10.3.19) both number average and mass average aggregation numbers must increase as the total concentration, M_1, increases. On the other hand, with increasing dilution $M_1 \to 0$, $\bar{N}_n \to 1$, $\bar{N}_w \to 1$ and $M_1 \approx [x_1]$.

The individual K_n values define the concentrations at which particular n-mers become most important and, hence, they control the values of $\bar{N}_n$ and $\bar{N}_w$ at any concentration. The products of association and their size distribution depend on concentration and on $K_n(n)$, i.e. how K_n depends upon n, which is a reflection of the molecular architecture of the associated species.

Three main types of association behaviour may be identified:

(a) simple dimerization (K_2 dominant);

(b) formation of micelles with a wide range of aggregation numbers (all values of K_n approximately equal); and

(c) formation of micelles with a narrow size range (strong dependence of K_n on n). The first two cases are rather trivial, while the third covers most of the interesting micelle-forming compounds and will, therefore, be discussed in more detail.

(a) *Dimerization.* The formation of dimers will of course take place in all self-associating systems. Whether or not the process is limited to dimer formation or continues predominantly on to multimers will be decided by the value of K_2 in comparison with other K values. Dimerization appears to be restricted to dilute aqueous solutions of some flexible chain surfactants, such as carboxylic acids (Oakenfull and Fenwick 1974) and to solutions of some bile salts, notably sodium cholate (Small 1968; Oakenfull and Fisher 1977). Both of these are rather special cases, in which the structure of the interacting molecules favours formation of a 'closed' dimer. The carboxylic acids for example, show cooperative hydrogen bonding to form a cyclic structure:

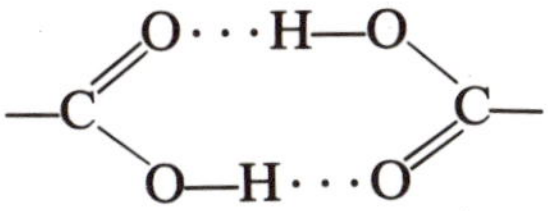

$$\begin{array}{c} \mathrm{O\cdots H{-}O} \\ -\mathrm{C}\!\!\!\quad\quad\quad\quad\mathrm{C}- \\ \mathrm{O{-}H\cdots O} \end{array}$$

Bile salts may show a similar interaction, with three cooperative hydrogen bonds, although the evidence for this has been disputed (Zana 1978; Oakenfull and Fisher 1978).

(b) *K_n values of similar magnitude.* For the equilibrium between monomers and micelles (eqn. (10.3.1)), let

$$K_2 = K_3 \ldots K_n = K \ldots \tag{10.3.20}$$

The total concentration, S, can now be directly related to the equilibrium monomer concentration $[x_1]$. Defining

$$X = K[x_1] \tag{10.3.21}$$

then

$$\begin{aligned} S &= [x_1] + [x_2] + [x_3] + \ldots [x_n] \ldots \\ &= [x_1]\{1 + K[x_1] + K^2[x_1]^2 \ldots K^{n-1}[x_1]^{n-1}\} \\ &= [x_1]\{1 + X + X^2 + X^3 + \ldots X^{n-1}\} \\ &= \frac{[x_1]}{1-X} \quad \text{for large } n, \end{aligned} \tag{10.3.22}$$

if $X < 1$, which, of course, it must be in real systems that conform to the association scheme outlined above. M_1, the total monomer concentration, now becomes (Exercise 10.3.5):

$$M_1 = \frac{[x_1]}{(1-X)^2} \tag{10.3.23}$$

and thus

$$\left(\frac{[x_1]}{M_1}\right)^{1/2} = 1 - K[x_1] \tag{10.3.24}$$

so that if $[x_1]$ is measured experimentally (Mukerjee and Ghosh 1970), K may be evaluated.

$\bar{N}_n$, the number average degree of association of all species is:

$$\bar{N}_n = \frac{M_1}{S} = \frac{1}{1-X} \tag{10.3.25}$$

and the mass average, $\bar{N}_w$, is (Exercise 10.3.5):

$$\bar{N}_w = \frac{Z}{M_1} = \left(\frac{1+X}{1-X}\right) \tag{10.3.26}$$

It is not difficult to establish the following expressions (Exercise 10.3.6):

$$\frac{S-[x_1]}{[x_1]} = \frac{X}{1-X}; \qquad \frac{M_1-[x_1]}{[x_1]} = \frac{X(2-X)}{(1-X)^2}$$

and

$$\frac{Z-[x_1]}{[x_1]} = \frac{X(4-3X+X^2)}{(1-X)^3} \tag{10.3.27}$$

from which the modified number and mass average degrees of association (eqns (10.3.18) and (10.3.19)) follow (Exercise 10.3.6):

$$^*\bar{N}_n = 1 + \frac{1}{1-X} \tag{10.3.28}$$

and

$$*\bar{N}_w = 1 + \frac{2}{(2-X)(1-X)}. \tag{10.3.29}$$

The index of polydispersity is given by (recall section 3.3.6)

$$\frac{*\bar{N}_w}{*\bar{N}_n} = 1 + \frac{X}{(2-X)^2} \tag{10.3.30}$$

which approaches 1 for small X and 2 as X approaches 1. The breadth of the size distribution thus increases with the degree of association. It is not always broad as is suggested by Mukerjee (1975).

This model often gives a good description of the behaviour of molecules that are rigid and flat, with faces of approximately equal hydrophobicity. Such molecules associate by a simple stacking arrangement, as has been demonstrated with the methylene blue† system (Mukerjee and Ghosh 1970). As the size of the stack increases, charge repulsion builds up and K_n gradually decreases at higher concentrations. This association model is also a fairly good representation of the behaviour of many nucleosides and of the stability of double stranded DNA (Ts'o 1968). In both cases, stacking interactions occur between the organic bases.

(c) *Strong dependence of K_n on n.* This is the situation that was crudely examined in the closed association model. The delicate force balance involved in micelle formation leads, not to an increase or decrease in K_n with n, but rather to a value of n for which K_n is a maximum. The values of K_{n-1} and K_{n+1} will be of comparable magnitude but one will expect to find the resulting size distribution to be fairly narrow.

As noted earlier (section 10.1), the existence of a preferred micelle size is easily explained on the basis of the competing requirements for bringing the hydrocarbon chains into intimate contact, away from the aqueous environment, while maintaining the head groups as far apart as possible. There is very little energy change in separating the hydrocarbon chains from the water (Exercise 10.1.2) and the negative free energy associated with that process stems largely from the concomitant entropy increase in the water. In the case of an ionic surfactant, as each additional monomer is added to the micelle, the contribution to the free energy change becomes less negative because the developing micellar charge causes an increasing (positive) free energy change, reflecting the repulsion between the head groups.

Even when n is large, a relatively broad maximum in the values of K_n

† Methylene blue is a cationic dye.

may produce a narrow size distribution. The free energy change involved in the formation of the n-mer from the monomer is, from eqns (10.3.3) and (10.3.11):

$$\Delta G_n^{\ominus} = -RT \ln {}^*K_n \tag{10.3.31}$$

and only a slight minimum in the plot of $\Delta G_n^{\ominus}/n$ against n is required to produce narrow size distributions of micelles, as Stigter and Overbeek (1957) have shown. Mukerjee (1975) gives a striking illustration of this point. Using the following empirical expression for $\ln {}^*K_n$:

$$\ln {}^*K_n = 2(n-1)\ln(n-1) - 0.02(n-1)^2 + 2.7896(n-1) \tag{10.3.32}$$

which exhibits a very broad maximum (Fig. 10.3.2) Mukerjee calculated the values of $n[x_n]$ from eqn (10.3.11) (i.e. the concentration of n-mers expressed in terms of monomer concentration), for a monomer concentration of $4.11 \times 10^{-5}\ \text{mol L}^{-1}$ (i.e. $\ln[x_1] = -10$). The result is a narrow size distribution (Fig. 10.3.2) peaking at $n = 97$ and with a half-width of less than 10, even though $\Delta G_n^{\ominus}/n$ changes by less than 2 per cent over the whole range from 70–120 for n.

This figure illustrates clearly why the calculation of the anticipated size

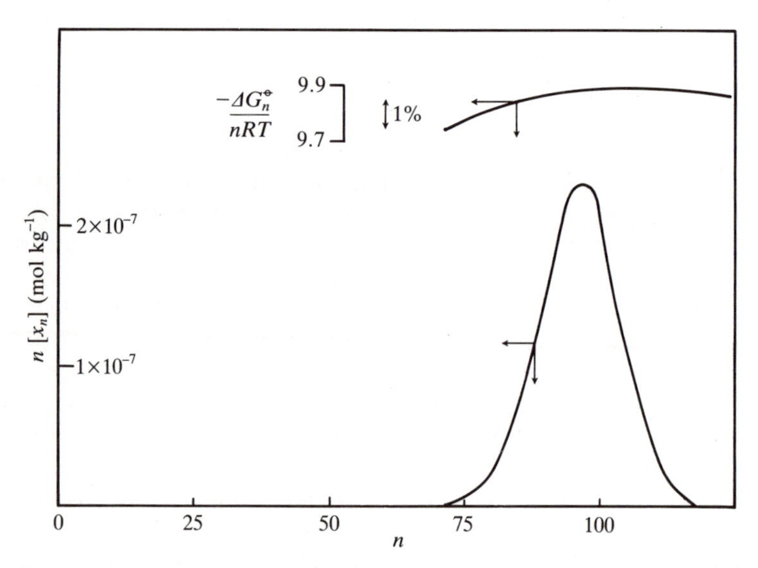

FIG. 10.3.2. Variation in the concentration of monomers existing in the form of micelles ($n[x_n]$), as a function of the number of monomers per micelle, n, for an assumed free energy profile, $\Delta G_n^{\ominus}/n$. (From Mukerjee 1975, with permission.)

distribution in a particular case requires accurately measured $^*K_n$ values near the maximum or good estimates of $\Delta G_n^\ominus$. We will examine the extent to which $\Delta G_n^\ominus$ can be estimated in the next section.

When the monomer chains are very long, very large polydisperse aggregates apparently form, even in dilute solution (Debye and Anacker 1951). These micelles are thought to be flexible cylinders and may be described by a self-association model similar to that discussed above (Mukerjee 1974). The molecular requirements for micelles of various geometries will be discussed in section 10.9.

Our immediate aim is to estimate the value of $\Delta G_n^\ominus$ as accurately as possible and, from this, to determine the micelle size distribution and other properties.

Exercises. 10.3.1 Estimate the mean centre-to-centre distance between the micelles of sodium dodecyl sulphate (SDS) when the surfactant concentration is 5×10^{-2} M. Assume that the c.m.c. is 8 mM. Show that this corresponds to about $4/\kappa$ when κ is calculated on the basis of the residual monomer concentration. (Note that the minimum micelle radius is about 2.5 nm, corresponding to a stretched C_{12} alkyl chain).

10.3.2 Take $K = 1$ and $n = 100$ in eqn (10.3.8) and discuss the variation of $C_s(=C_T - nC_m)$ and nC_m with C_T/C_0 in the range $0 \leq C_T/C_0 \leq 3$. (Assume that in this formulation the concentration of micelles is expressed in moles of micelles per litre.)

10.3.3 Establish eqn (10.3.9).

10.3.4 Integrate eqn (A5.13) to establish eqn (10.3.3) for the case where only P, V work is involved. (Use the fact that $\Delta G = 0$ at equilibrium if P and T are constant.)

10.3.5 Establish eqn (10.3.23) using the series expansion for $(1 - X)^{-2}$. Establish eqns (10.3.25) and (10.3.26) by showing that $Z = [x_1](1 + X)/(1 - X)^3$.

10.3.6 Establish the expressions (10.3.27) and use them to derive (10.3.28) and (10.3.29).

10.4 Thermodynamics of micelle formation

Careful analyses of the thermodynamics of micelle formation have been given by a number of authors, including Hall and Pethica (1967) and Tanford (1980). We will follow the latter treatment, with some modifications. Our aim is to relate the chemical potential of an amphiphile or surfactant in free solution with that of the same molecule in a micelle of arbitrary size. In an isothermal system at equilibrium this quantity must be constant throughout the system.

Tanford distinguishes what he calls the '*cratic*' contribution to the chemical potential from the intrinsic contribution, which is due to local (chemical and physical) interactions. The cratic part is that due to the entropy of mixing and so for any particular size of micelle, is equal to:

$$RT \ln(\text{mole fraction of micelles of size } n).$$

This would give the contribution per mole of micelles of size n, *assuming ideal behaviour*. (The use of mole fractions is connected with the most appropriate choice of standard state and will be discussed in more detail in section 10.4.3). Again (cf. Fig. 10.3.2) it is more convenient to express this contribution in terms of the concentration of the monomeric surfactant: $RT\ln\left(\frac{X_n}{n}\right)$ where X_n is the mole fraction of monomer in micelles of size n. The cratic contribution *per mole of monomeric amphiphile is* $1/n$ of this and so:

$$\mu_{\text{mic},n} = \mu^{\ominus}_{\text{mic},n} + \frac{RT}{n}\ln(X_n/n). \tag{10.4.1}$$

[Do not confuse X_n with $[x_n]$ as used in section 10.3; the latter is the concentration expressed in terms of moles of n-mers.] Note that in this formulation, each of the micellar sizes is treated as a separate component, with its own standard state chemical potential. Equating $\mu_{\text{mic},n}$ with the value for the free surfactant gives:

$$\mu^{\ominus}_{\text{mic},n} - \mu^{\ominus}_1 = RT\ln a_1 - \frac{RT}{n}\ln\frac{X_n}{n} \tag{10.4.2}$$

or

$$\ln X_n = \frac{-n(\mu^{\ominus}_{\text{mic},n} - \mu^{\ominus}_1)}{RT} + n\ln a_1 + \ln n \tag{10.4.3}$$

where a_1 is the activity of the monomer.

Equation (10.4.3) gives explicitly the distribution function for micelles of different size, in terms of the quantity $(n\mu^{\ominus}_{\text{mic},n} - n\mu^{\ominus}_1)$, which is the value of $\Delta G^{\ominus}_n$ for the reaction in which an n-mer is formed from monomers. An optimal size n^* can be defined as that value of n for which X_n is a maximum at the particular surfactant activity:

$$(\partial\ln X_n/\partial n)_{a_1} = 0 \quad (n = n^*) \tag{10.4.4}$$

and Tanford (1980) points out that if the size distribution is reasonably narrow, the value of n^* is experimentally indistinguishable from the number average or mass average micellar size. All can then be set approximately equal to a mean size, $\bar{n}$. At this level of approximation, all micelles are treated as the same and lumped together with a standard chemical potential, $\mu^{\ominus}_{\text{mic}}$.

$$\ln X_{\text{mic}} = \frac{-\bar{n}(\mu^{\ominus}_{\text{mic}} - \mu^{\ominus}_1)}{RT} + \bar{n}\ln a_1 + \ln\bar{n}. \tag{10.4.5}$$

The activity of the free surfactant a_1 is, of course, given by y_1X_1, where y_1

is the activity coefficient and it is tempting to assume that $y_1 \approx 1$, since X_1 is usually fairly small. (Even above the c.m.c. the concentration of the free surfactant remains close to the c.m.c. value.) The problem with this procedure is that even at the concentrations normally encountered, ionic surfactants exhibit activity coefficient effects due to interactions other than micelle formation. Some account could be taken of such effects using, say, the extended Debye–Hückel theory but the experimental procedure of extrapolating to infinite dilution breaks down in this case because the presence of the micelles may itself influence the interactions between the free surfactant molecules.

This effect may not be as serious as Tanford (1980) suggests, however, since the free surfactant ions will tend to be excluded from the double layer regions surrounding the micelles. For non-ionic surfactants the activity coefficient correction can probably be dispensed with altogether with negligible error (Desnoyers *et al.* 1983).

The relation between the standard chemical potential change and the c.m.c. (X_0) can be obtained from eqn (10.4.5) by introducing the ratio $\sigma = X_{mic}/X_0$ and recognizing that, at the c.m.c., $X_1 = X_0 - X_{mic}$. We then have (Exercise 10.4.2):

$$(\mu^{\ominus}_{mic} - \mu^{\ominus}_1)/RT = [(\bar{n} - 1)/\bar{n}]\ln X_0 + \ln y_1 + \ln(1 - \sigma) + (1/\bar{n})\ln(\bar{n}/\sigma). \qquad (10.4.6)$$

To estimate values of $(\mu^{\ominus}_{mic} - \mu^{\ominus}_1)$ thus requires a knowledge of the c.m.c. (X_0) and the mean aggregation number, $\bar{n}$. The value of σ can be taken to be anywhere from 0.01 to 0.1 with negligible effect on the result but, of course, one has to either assume $y_1 = 1$ or make some correction for it. This latter correction should be relatively unimportant if one wishes only to evaluate the change in $(\mu^{\ominus}_{mic} - \mu^{\ominus}_1)$ as the chain length changes. Indeed, for many purposes, the following approximation, valid for large $\bar{n}$ and small σ (Exercise 10.4.3) is sufficiently accurate:

$$\mu^{\ominus}_{mic} - \mu^{\ominus}_1 = \overline{\Delta G^{\ominus}} = RT \ln X_0. \qquad (10.4.7)$$

Note that this is identical to eqn. (10.3.5) if n is very large and the surfactant concentration is expressed in mole fraction terms. It cannot be applied in this form to ionic surfactants, for which a further correction is essential. (See section 10.5 below.) Further discussion of the thermodynamic determinants of micellar size and shape is given by Missel *et al.* (1980). See also Ekwall *et al.* (1971).

10.4.1 *Estimation of* $\overline{\Delta G^{\ominus}}$

It seems reasonable to assume that the free energy change associated with the transfer of one mole of monomer from free solution into a micelle:

$$\Delta G_n^\ominus / n = \overline{\Delta G^\ominus} = \mu_{\text{mic}}^\ominus - \mu_1^\ominus \quad (10.4.8)$$

could be broken into various contributions:

$$\begin{aligned}\overline{\Delta G^\ominus} &= \overline{\Delta G^\ominus}(\mathrm{CH_3}) + (m-1)\overline{\Delta G^\ominus}(\mathrm{CH_2}) + \overline{\Delta G^\ominus}(\mathscr{H}) \\ &= \overline{\Delta G_{\text{hc}}^\ominus} + \overline{\Delta G^\ominus}(\mathscr{H}) \end{aligned} \quad (10.4.9)$$

where $\mathscr{H}$ represents a (hydrophilic) head group.

This kind of breakdown is suggested by the solubility and vapour pressure behaviour of homologous series of organic molecules, in which a constant increment or decrement is noted for each additional $-CH_2$ group. Studies of the solubilities of alkanes in water suggest values for $\overline{\Delta G^\ominus}(CH_3)$ of ~ -8.8 kJ mol^{-1} or $-3.5\,RT$, whilst the free energy of transfer of a mole of methylene groups from water to a hydrocarbon environment (obtained from studies of adsorption at the oil–water interface) is about $-1.4\,RT$. Tanford (1980) quotes the work of Swarbrick and Daruwala (1969, 1970) on the *N*-alkyl betaines ($\mathscr{R}N(CH_3)_2^+CH_2COO^-$) for which $\bar{n}$ data are available, so that $\overline{\Delta G^\ominus}$ can be estimated from eqns (10.4.6) or (10.4.7). The change in $\overline{\Delta G^\ominus}$ for each additional CH_2 in the group $\mathscr{R}$ is -3.06 kJ mol^{-1} or $-1.23\,RT$, in reasonable agreement with the value quoted above for $\overline{\Delta G^\ominus}(CH_2)$. Tanford (1980) argues that the first CH_2 group, next to the head group, is not able to entirely escape interaction with the water (refer to Fig. 1.5.5) and this accounts for the fact that $\overline{\Delta G^\ominus}(CH_2)$ for micellization is some 10 per cent lower than the value estimated for complete transfer from water to hydrocarbon. He proposes a more elaborate expression to take account of this effect and of the variation in the freedom of motion of the CH_2 groups down the chain:

$$\overline{\Delta G_{\text{hc}}^\ominus}(J) = -8800 - 2900(m-2) + 10{,}500(A_{\text{hm}} - 0.21) - Q \quad (10.4.10)$$

where the first term on the right is the contribution of the terminal $-CH_3$ group and Q is a constant (<2900), which represents the contribution of the $-CH_2$ group next to the head group. A_{hm} is the area (in nm^2) per emerging hydrocarbon chain for a micelle of aggregation number $\bar{n}$; this quantity must be estimated from geometrical considerations. We will return to these matters in section 10.9. It should be understood, however, that these empirical correlations should not be taken too seriously. The statistical mechanical model developed in section 10.10 shows that each of the CH_2 groups (and even the terminal CH_3) has some access to the water so that arguments that ignore this fact can at best only lead to semi-quantitative conclusions.

10.4.2 *Estimation of* $\overline{\Delta G^\ominus}(\mathscr{H})$

The contribution of the hydrophilic head group $\overline{\Delta G^\ominus}(\mathscr{H})$) to the free energy of micellization is invariably positive and so opposes the process. Very little progress has been made on the calculation of $\overline{\Delta G^\ominus}(\mathscr{H})$ for non-ionic molecules. It is assumed to arise from steric interactions as the large head groups crowd the surface but beyond that there is little that can be said from a theoretical point of view; the contributions of different non-ionic head groups can, however, be evaluated from experimental data.

By contrast, more progress has been made on the calculation of $\overline{\Delta G^\ominus}(\mathscr{H})$ for ionic surfactants, assuming that it is dominated by the electrostatic effects predictable from the Gouy–Chapman theory of the double layer (section 6.3). The procedure is analogous to that used in the estimation of the electrical part of the free energy of other colloidal particles (section 7.2), except that instead of working in terms of unit area we consider the energy change involved in establishing an n-mer (of charge $q = ne$). The electrical contribution per mole of monomer is then:

$$\overline{\Delta G^\ominus}(\mathscr{H}) = \overline{\Delta G^\ominus_{\mathrm{elec}}} \approx \overline{\Delta F}_{\mathrm{elec}} = \frac{1}{n}\int_0^{ne} \psi_0'\,\mathrm{d}q' \qquad (10.4.11)$$

where q' and ψ_0' refer to the charge and potential on the micelle surface during the charging process. The problem is that, for spherical particles, there is no explicit relation between the surface charge, q, and the surface potential. Only at low potentials (strictly $\psi_0 < 25\,\mathrm{mV}$) can the Debye–Hückel relation (eqn (6.11.6)) be invoked:

$$q' = 4\pi\varepsilon\psi_0' a(1 + \kappa a) \qquad (10.4.12)$$

where a is the micelle radius and then (Exercise 10.4.4):

$$\overline{\Delta G^\ominus}(\mathscr{H}) = \psi_0 e/2 = ne^2/8\pi\varepsilon a(1 + \kappa a) \qquad (10.4.13)$$

where e is the proton charge. Unfortunately, this approximation is rarely, if ever, valid in micellar systems. Stigter (1975a) has shown, however, that a more exact calculation, using the computer-calculated relation between ψ_0 and q for spherical particles, can satisfactorily account for the variation of c.m.c. with salt concentration in SDS solutions at 25 °C (Table 10.2) over the range from 0 to 0.2 M sodium chloride. In this region, the micellar aggregation number changes over less than a factor of two (from 65–110) and the concentration of free surfactant remains approximately equal to the c.m.c.

Using eqns (10.4.7) and (10.4.9) we can then write

$$\delta \ln X_0 = \delta \ln C_0 = \delta[\overline{\Delta G^\ominus}(\mathscr{H})/RT] \qquad (10.4.14)$$

Table 10.2

Calculation of the electrical contribution to the free energy of sodium dodecyl sulphate (SDS) micelles at 25 °C. (From Stigter 1975a, with permission)

C_{NaCl} (mol L^{-1})	c.m.c. (mol L^{-1} × 10^3)	$\overline{\Delta G^{\ominus}_{elec}}/RT$ (D–H)	$\overline{\Delta G^{\ominus}_{elec}}/RT$ (G–C)	δ ln(c.m.c.)	$\delta(\overline{\Delta G^{\ominus}_{elec}}/RT)$ (G–C)
0	8.12	7.96	4.85	0	0
0.01	5.29	7.28	4.39	0.43	0.46
0.03	3.13	6.04	3.81	0.95	1.04
0.05	2.27	5.70	3.62	1.27	1.23
0.1	1.46	4.80	3.21	1.72	1.64
0.2	0.92	4.15	2.85	2.18	2.00

Column 3 uses the Debye–Hückel approximation (eqn (10.4.13)) while column 4 uses the 'exact' $\psi_0' - q'$ relation from Gouy–Chapman (G–C) theory.

where δ measures the change compared to the value in the absence of salt. Equation (10.4.14) assumes that C_0(mol L^{-1}) $\propto X_0$ at low concentrations. The agreement between columns 5 and 6 is excellent, suggesting that double-layer theory gives a good account of the effect of added salt, at least up to 0.2 M. The remaining discrepancy is of the order of the activity coefficient correction for unassociated ions and this has not been included. The Debye–Hückel procedure would obviously give a very poor representation, being some 50–65 per cent higher at all concentrations. Note that in this approach, no attempt was made to relate the absolute magnitude of $\overline{\Delta G^{\ominus}_{elec}}$ to the experimental estimates of $\overline{\Delta G^{\ominus}}(\mathscr{H})$. That would require an assumption about the degree of ion binding and a choice of the appropriate value of ψ_0 (or ψ_d) to characterize the surface. More elaborate calculations of this nature are given by Stigter (1975b). In Table 10.2 it is assumed that ion binding is not affected over the concentration range involved. A more recent calculation of $\overline{\Delta G^{\ominus}_{elec}}$ along similar lines is given by Evans and Ninham (1983).

At somewhat higher salt concentrations (above about 0.45 M) the aggregation number of SDS increases dramatically to over 1000 and a more elaborate model is required (see e.g. Mazer *et al.* 1977 and Gunnarson *et al.* 1980).

Rather than attempt a theoretical estimate of $\overline{\Delta G^{\ominus}}(\mathscr{H})$, it is possible to estimate it from the behaviour of surfactant films at the oil–water interface (Tanford 1980). The work of compression of a monolayer of material adsorbed at that interface is a direct measure of the repulsive interaction between the head groups, and its measurement in different electrolyte systems gives an estimate of $\overline{\Delta G^{\ominus}}(\mathscr{H})$ with allowance for ion binding automatically taken into account. Tanford (1980, p. 74) gives an expression of the form:

$$B_1/A_{rm} \pm B_2/A_{rm}^2 - B_3/A_{rm}^3 \qquad (10.4.15)$$

for $\overline{\Delta G^{\ominus}}(\mathscr{H})$ where the B_i are empirical constants and A_{rm} is the area occupied by the head group, which again, must be estimated from a suitable choice of the effective position of the surface.

The ionic strength effect (Table 10.2) has another important consequence that was noted in connection with Table 10.1. The constant b_1, which measures the effect of each additional CH_2– group on the c.m.c., is much smaller for most ionics than it is for non-ionics. This may be partly due to the much better shielding of the hydrocarbon chains from the water in the case of non-ionics. In the ionics, the repulsion forces prevent the head groups from packing close together so that there is always a significant hydrocarbon–water interface (Fig. 1.5.5). Tanford (1980), however, claims that the difference in b_1 values can be almost entirely accounted for by the ionic strength effect. As the number of carbon atoms decreases, the c.m.c. tends to increase but for ionics this is partly offset by the increasing ionic strength due to the surfactant itself. The resulting tendency to lower c.m.c. (Table 10.2) thus opposes the effect of shortening the chain. This explanation is supported by the behaviour of the *n*-alkyltrimethylammonium bromides in 0.5 M salt solution. (Exercise 10.4.6).

10.4.3 *Choice of standard state and concentration units*

For a detailed discussion of this question the reader is referred to specialized treatments such as those of Shinoda (1963, 1978), Phillips (1955), Anacker (1970) and Kishimoto and Sumida (1974). The choice of mole fraction as the concentration unit implies that the standard state for the surfactant is the pure material but physically it is more reasonable to interpret it in terms of a totally hydrated state.

Consider the following schematic diagram:

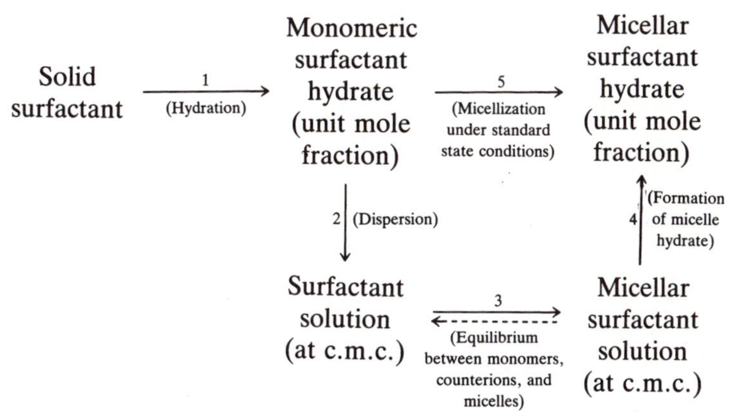

ΔG_5 may be identified with the standard free energy change in terms of unit mole fraction, so that ΔG_5 becomes $\overline{\Delta G^\ominus}$ and may be calculated through eqn (10.4.2) (with $\overline{\Delta G^\ominus} = \mu^\ominus_{\text{mic}} - \mu^\ominus_1$). The condition under which micelles are formed spontaneously is given by:

$$\Delta G_1 + \Delta G_2 \leqslant 0. \tag{10.4.16}$$

For compounds such as alcohols, amides, and like substances, the free energy of solution, ΔG_1, is positive to the extent that $\Delta G_1 + \Delta G_2 > 0$ and no micelles are formed. ΔG_1 is reduced by the presence of an ionic head group, or of a strongly polar head group of the ethylene oxide variety and micelles will begin to form when $\Delta G_1 + \Delta G_2 = 0$. As the temperature is lowered, the positive entropy of micelle formation means that at some point (called the Krafft temperature) the overall free energy change for micelle formation is no longer negative and the solubility of the surfactant decreases dramatically. (See section 10.2.3.)

10.4.4 *Enthalpy and entropy of micelle formation*

From eqn (10.4.5) we can write, for the free energy change on micelle formation:

$$\overline{\Delta G^\ominus} = \mu^\ominus_{\text{mic}} - \mu^\ominus_1 = RT \ln a_1 - \frac{RT}{n} \ln X_{\text{mic}}$$

$$\approx RT \ln(\text{c.m.c.}) - \frac{RT}{n} \ln X_{\text{mic}} \tag{10.4.17}$$

neglecting the $(n^{-1} \ln n)$ term. The temperature and pressure derivatives of eqn (10.4.17) give the standard enthalpy change $\overline{\Delta H^\ominus}$ and volume change $\overline{\Delta V^\ominus}$, per mole of monomer (Kresheck 1975; Exercise 10.4.5):

$$\overline{\Delta H^\ominus} = -RT^2\left(\frac{\partial \ln(\text{c.m.c.})}{\partial T}\right)_P + \frac{RT^2}{n}\left(\frac{\partial \ln X_{\text{mic}}}{\partial T}\right)_P \tag{10.4.18}$$

and

$$\overline{\Delta V^\ominus} = RT\left(\frac{\partial \ln(\text{c.m.c.})}{\partial P}\right)_T - \frac{RT}{n}\left(\frac{\partial \ln X_{\text{mic}}}{\partial P}\right)_T. \tag{10.4.19}$$

$\overline{\Delta S^\ominus}$ can be obtained from:

$$\overline{\Delta S^\ominus} = (\overline{\Delta H^\ominus} - \overline{\Delta G^\ominus})/T. \tag{10.4.20}$$

In the neighbourhood of the c.m.c. the value of X_{mic} is small, and since n is not always known, it is common practice to neglect the second term on the right of each of these expressions and to estimate $\overline{\Delta G^\ominus}$ from the approximate eqn (10.4.7). Although Kresheck (1975) defends this

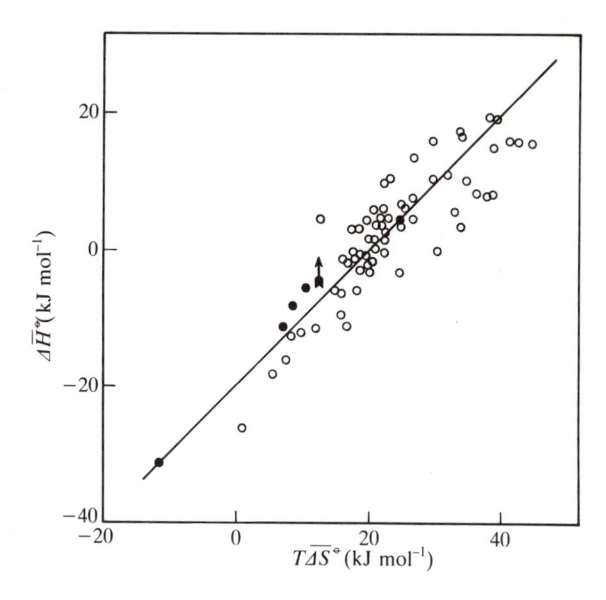

FIG. 10.4.1. Compensation plot of data for a variety of ionic and non-ionic surfactants in various liquids. Open circles are for water, with or without additives. Filled circles are for benzene and filled triangles for formamide. (Modified from Kresheck 1975, Fig. 9, with permission.)

procedure, chiefly on the grounds that the $\overline{\Delta H^{\ominus}}$ data obtained agree with the calorimetric estimates, Muller (1977) has called it into question, on account of the large changes in aggregation number that can occur with temperature.

Despite some reservations then, we will examine the resulting $\overline{\Delta H^{\ominus}}$ and $\overline{\Delta S^{\ominus}}$ data because they have been used to develop a deeper insight into the micelle formation process. The most striking feature is the relation between $\overline{\Delta H^{\ominus}}$ and $T\overline{\Delta S^{\ominus}}$, which is shown in Fig. 10.4.1. Taking $T = 300$ K gives a slope of unity, so that this is called the *compensation temperature*. It is very significant that this compensation (making $\overline{\Delta G^{\ominus}} = 0$) should occur so close to room temperature, emphasizing how delicate the balance between energy and entropy must be in the micellization process.

As noted earlier it has, until very recently, been assumed that the major factor driving the surfactant molecules into aggregation in water is a positive entropy change, presumably associated with breakdown of the structured water which surrounds the hydrocarbon chain in the unassociated species. Such an interpretation carries over readily to the formamide system in which some structuring by dissolved hydrocarbon also occurs. The recent more extensive studies of Evans *et al.* (1984) on the alkyltrimethyl ammonium bromides in water (from 25 °C to 166 °C) and in hydrazine (Ramadan *et al.* 1983) suggest that this interpretation is erroneous or, at any rate, misleading. At high temperatures (>90 °C) water loses most of its peculiar structural properties and the formation of structured water in the walls of the hydrocarbon cavities is no longer possible. Neither is it expected to occur in hydrazine and yet both of these systems exhibit micellization phenomena. The difference is that $\Delta S^\ominus$ for the process is now negative, as is also $\Delta H^\ominus$. Evans *et al.* argue that, in such circumstances, it is not sensible to attribute the micellization at room temperature to the positive entropy change. That change is made up of two parts: a large positive part due to removal of water from around the hydrocarbon and a (smaller) negative part due to transfer of the hydrocarbon (and counterions) into the micelle. Since the first water structure part is not present in hydrazine or hot water it cannot be the general driving force for micellization. Rather it is the second part (for which $\Delta H^\ominus$ is also negative, due to some extent to the re-establishment of the hydrogen bonds in the solvent) that must be the usual driving force. Evans and Ninham (1986) have also applied this analysis to changes in protein conformation, to vesicle formation and fusion and to other biochemically important self-assembly processes.

The limited amount of data on other solvents also presents a complex picture. Why should $T\overline{\Delta S^\ominus}$ be positive (although smaller than the water value) for the formation of (presumably inverse) micelles of dodecylammonium alkanoates in benzene? The problem of interpretation is highlighted by the comparison of the behaviour of dodecylammonium octanoate in benzene and cyclohexanes (Kresheck 1975, taken from Kitahara 1967):

	$\overline{\Delta H^\ominus}$ kJ mol^{-1}	$T\overline{\Delta S^\ominus}$ kJ mol^{-1}
In benzene	−5.4	+10.0
In cyclohexane	−32	−12.1

Although these data were derived at different temperatures (299 K and 313 K, respectively) it is obvious that no naïve interpretation will 'explain' such a difference. This could be a situation in which the changes in aggregation number over the temperature range are impossible to ignore, as Muller (1977) has argued. Certainly it is important to recognize that these molecules, when dissolved in benzene at low concentrations (before they aggregate), are present as *ion pairs,* so the starting point for the aggregation processes is very different from that in aqueous solution. Indeed, Kertes (1977) has called into question the whole concept of micelle formation as applied to solutions in aprotic non-aqueous media (like benzene). It should also be noted that the presence of trace amounts of water has a profound effect on the micellization process.

One thing is clear from the minimal amount of data considered here: the very simple behaviour of the free energy function often masks a very complex shift in ΔS and ΔH values as temperature changes. Sometimes the study of these detailed shifts provides deeper insights into the process. In the present case it may well have led to an overemphasis of the role of 'structured water' in the phenomenon referred to as 'the hydrophobic bond'.

Exercises. 10.4.1 Note that as $\bar{n} \rightarrow \infty$ eqn (10.4.5) reduces to $\mu^{\ominus}_{\text{mic}} = \mu^{\ominus}_1 + RT \ln a_1$. How would this be interpreted in terms of a phase separation model?

10.4.2 Establish eqn (10.4.6).

10.4.3 Show that a crude estimate of $\overline{\Delta G^{\ominus}}$ for the micellization process can be obtained from:

$$\overline{\Delta G^{\ominus}} \doteqdot RT \ln X_0$$

where $X_0 = $ c.m.c. Show that, in water:

$$\ln C_0 \doteqdot \frac{\overline{\Delta G^{\ominus}}}{RT} + \ln 55.5$$

where C_0 is the c.m.c. expressed in mol L^{-1}, provided the c.m.c. occurs at low concentration.

10.4.4 Establish eqn (10.4.13).

10.4.5 Show that $T(\partial(\Delta G/T)/\partial T)_P = -\Delta H/T$. Hence show that, when applied to the components of a reaction,

$$d(\Delta G^{\ominus}/T)/dT = -\Delta H^{\ominus}/T^2.$$

Use this to establish eqn (10.4.18). Also establish eqn (10.4.19).

10.4.6 Discuss the relation between the b_1 values obtained for alkyltrimethylammonium bromides in the presence of salt (Fig. 10.2.1) and those for the corresponding chlorides in the absence of salt (Table 10.1).

10.5 Ion binding

It was noted several times in the preceding discussion that the repulsive interaction between the head groups in an ionic micelle is moderated to a considerable extent by the fact that a large proportion of the counterions (~50–80 per cent) are strongly bound to the surface. The effective charge on a micelle can be measured by a variety of techniques, involving equilibrium, transport or kinetic properties, and all lead to substantially the same conclusion, though there are some quantitative variations, depending on the method used. Ion binding of this sort is exactly analogous to the formation of a Stern layer (section 6.3.4) and its assessment by various techniques will give slightly different results depending on what fraction of the diffuse and compact double layers contributes to the measured effect.

The aggregation process can then be represented, for an anionic surfactant, as:

$$\underset{\text{Surfactant}}{nS^-} + \underset{\text{Counterions}}{(n-p)Na^+} \rightleftharpoons \underset{\text{micelles}}{M^{p^-}} \tag{10.5.1}$$

where n = aggregation number, p = effective charge on the micelle and $(n-p)$ is the number of ions bound in the Stern layer.

Using, for simplicity, the closed association model, we obtain for the equilibrium constant:

$$K = \frac{C_m}{C_s^n C_{Na^+}^{n-p}} \tag{10.5.2}$$

which introduces two new variables, (C_{Na^+} and p) compared to eqn (10.3.2). In order to estimate the equilibrium constant it is necessary to know, in addition to n and p, the concentrations of both the monomeric surfactant ions (C_s) and the micelles (C_m), at some total surfactant ion concentration C_T. Because n is so large (often ~80) these concentrations are very difficult if not impossible to acquire experimentally, but we can obtain estimates of p/n, for example, by measuring the electrical conductance or electrophoresis (section 9.11.4) of the micelles and obtaining a separate estimate of n from light scattering of the micelles. The c.m.c. can be identified by the Phillips criterion (eqn (10.1.1)) and the property, ϕ, can be assumed to obey a relation of the form:

$$\phi = AC_s + BC_m \tag{10.5.3}$$

where A and B are constants. We have the following mass and charge balance relationships:

$$C_T = C_s + nC_m \tag{10.5.4}$$

and

$$C_{Na^+} = x + C_s + pC_m \tag{10.5.5}$$

where x is the concentration of added electrolyte. (Note that in eqns (10.5.4) and (10.5.5), C_m is again expressed as moles of micelles, which is a change from the procedure used in eqns (10.3.7), (10.3.8), and (10.3.9).

In the neighbourhood of the c.m.c., the concentration of micelles is very small and so eqn (10.5.5) can be approximated as:

$$C_{Na^+} \approx x + C_s \tag{10.5.6}$$

and one can then solve eqns (10.1.1) and (10.5.2), (3), (4), and (6) simultaneously to obtain K as a function of n, p, and the c.m.c. This is a lengthy and not very instructive procedure but there are two limiting results that are of some value: (i) the case where no salt is added ($x = 0$) and (ii) the case where a swamping excess of electrolyte is involved ($C_{Na^+} \approx x$).

10.5.1 *Approximate relations for the equilibrium constant*

(a) *No added salt.* It is not too difficult to show (Exercises 10.5.1 and 2) that in this case the equilibrium constant is given by:

$$\frac{1}{K} \simeq 4n^2 C_0^{2n-p-1} \tag{10.5.7}$$

where C_0 is the c.m.c., expressed in terms of moles of monomeric surfactant. The ratio of monomer concentration to micelle concentration at the c.m.c. is given by (Exercise 10.5.1):

$$\frac{C_s}{C_m} = \frac{nC_s}{C_0 - C_s} = \frac{n(2n-p)(4n-2p-1)}{2n-p-2} \simeq 4n^2 \tag{10.5.8}$$

so the main approximation involved in deriving eqn (10.5.7) is amply justified.

(b) *In the presence of a swamping excess of salt.* In this case we again use eqn (10.5.6), but in the form

$$C_{Na^+} \simeq x \tag{10.5.9}$$

and a similar analysis (Exercises 10.5.3 and 4) gives:

$$\frac{1}{K} \simeq 2n^2 C_0^{n-1} x^{n-p}. \tag{10.5.10}$$

At the c.m.c. (Exercise 10.5.4), we have:

$$\frac{C_s}{C_m} = \frac{n^2(2n-1)}{n-2} \approx 2n^2. \tag{10.5.11}$$

For practical purposes, K is given over the entire range of electrolyte concentrations by the approximate expression:

$$\frac{1}{K} = 3n^2 C_0^{n-1}[x + C_0]^{n-p}. \tag{10.5.12}$$

Although this equation has a maximum error in K of ±33 per cent this only results in very small errors in the estimation of standard free energy of formation of the micelles.

10.5.2 *Extent of ion binding*

The standard free energy change per mole, $\overline{\Delta G^{\ominus}}$, is related to the equilibrium constant by:

$$\overline{\Delta G^{\ominus}} = \frac{\Delta G_n^{\ominus}}{n} = -\frac{RT}{n} \ln K \tag{10.5.13}$$

and substituting from eqn (10.5.12):

$$\begin{aligned}\frac{\overline{\Delta G^{\ominus}}}{RT} &= \frac{\ln 3 + 2 \ln n}{n} + \frac{n-1}{n} \ln C_0 + \frac{n-p}{n} \ln[x + C_0] \\ &= \mathscr{L}(n) + (1 - 1/n) \ln C_0 + (1 - p/n) \ln[x + C_0].\end{aligned} \tag{10.5.14}$$

Obviously, the function $\mathscr{L}(n)$ tends to zero for large n and we have, for the case where x is small compared to C_0:

$$\frac{\overline{\Delta G^{\ominus}}}{RT} \simeq (2 - p/n) \ln C_0. \tag{10.5.15}$$

Note the relation between this expression and eqn (10.4.7) for a non-ionic surfactant. When $p = 0$ (complete binding of the counterions), the equilibrium equation (eqn (10.5.1)) corresponds to the simple association of a 1 : 1 electrolyte:

$$n\text{S}^- + n\text{Na}^+ \rightleftarrows \text{Micelle}$$

and the appropriate form of eqn (10.4.7) would be:

$$\begin{aligned}\frac{\overline{\Delta G^{\ominus}}}{RT} &= RT \ln a_0 \simeq RT \ln C_{\pm}^2 \\ &= 2RT \ln C_0\end{aligned} \tag{10.5.16}$$

where a_0 is the activity of the surfactant at the c.m.c. When $p = n$ (complete dissociation) eqn (10.3.5) is again recovered (assuming $C_m \ll C_s = C_0$). The factor p/n is, thus, the anticipated correction for partial dissociation.

When x is very large, eqn (10.5.14) becomes, for large n:

$$\frac{\overline{\Delta G^{\ominus}}}{RT} = \ln C_0 + (1 - p/n) \ln[x + C_0]. \tag{10.5.17}$$

Equation (10.5.14) can be used to estimate the value of $\overline{\Delta G^{\ominus}}$ under various conditions, if an estimate can be made of the aggregation number, n, and the effective charge, p, on the micelles. As noted above, the aggregation number can be obtained from the result of light scattering experiments, which measure the effective molar mass of the micelle (section 3.7.1) and the effective charge can be estimated from electrophoretic measurements or by analysing the effect of added salt on the c.m.c. (see Vold and Vold 1983, p. 599). Nuclear magnetic resonance (n.m.r.) spectroscopy is also used in the study of ion binding (section 10.7.4). The results of such estimations are shown in Table 10.3.

Note that between 75 and 90 per cent of the micellar charge is balanced by the counterions bound to the surface. The presence of so many oppositely charged ions significantly lowers the repulsive contribution of the head groups and, hence, the c.m.c. Comparison of the data for decyl- and dodecyltrimethylammonium bromide suggests a value of $\overline{\Delta G^{\ominus}}(CH_2)$ (eqn (10.4.9)) of about $\frac{3}{2}RT$, very similar to the value obtained above (section 10.4.1) for the alkyl betaines. The differences between the halides of the C_{12} pyridinium ion at a given concentration reflect the differences in polarizability and hydration between those ions, and are another example of the Hofmeister or lyotropic series effect (section 7.7.4).

A cautionary note is required at this point. Implicit in the discussion above is the assumption that the counterions behave ideally and that their activity coefficient is unity. This is a normal assumption in light scattering but it may well be incorrect. Gilanyi (1973) has shown that experimental estimates of p/n can vary widely and has selected a range of experimental data for sodium dodecyl sulphate to illustrate this point. Ultracentrifugation, e.m.f., light scattering and conductance measurements all yield $p/n = 0.18 \pm 0.02$. In contrast, electrophoretic mobility data give ~ 0.5 and other measurements give p/n values ranging from 0.28 to 0.5. Thus the convenient assumption of ideal behaviour *may* lead to incorrect data and erroneous conclusions. One of the thorny problems in dealing with micelles and associated models is to obtain *accurate* activity coefficients. Research in this area is sorely needed.

10.5.3 *Enthalpy and entropy contributions to $\overline{\Delta G^{\ominus}}$ in bound ion systems*

If it is assumed that p and n remain constant over the small temperature range involved, it is possible to extend the analysis of section 10.4.4 and obtain an expression of the form:

$$\overline{\Delta H^{\ominus}} = (2 - p/n)RT^2\left(\frac{\partial \ln(\text{c.m.c.})}{\partial T}\right)_p \tag{10.5.18}$$

for the case where there is little or no added electrolyte. Estimates of

Table 10.3

Critical micelle concentrations, aggregation numbers, effective degree of ionization of micelle (p/n) and free energies of micelle formation for various ionic surfactants. (*Phillips* 1955; *Ford et al.* 1966). (*Bracketed n values* (*for SDS*) *are more recent values from Kratohvil* 1980.)

	Material	Solvent	c.m.c. (M)	n	p/n	$\overline{\Delta G^{\ominus}}/RT$
A.	Sodium dodecyl sulphate	Water	8.1×10^{-3}	80(58)	0.18	−15.8
		0.02 M NaCl	3.82×10^{-3}	94	0.14	−16.0
		0.03 M NaCl	3.09×10^{-3}	100	0.13	−16.2
		0.10 M NaCl	1.39×10^{-3}	112(91)	0.12	−15.9
		0.20 M NaCl	8.3×10^{-4}	118(105)	0.14	−15.8
		0.40 M NaCl	5.2×10^{-4}	126(~129)	0.13	−15.7
	Dodecylamine hydrochloride	Water	1.31×10^{-2}	56	0.14	−15.2
		0.0157 M NaCl	1.04×10^{-2}	93	0.13	−15.1
		0.0237 M NaCl	9.25×10^{-3}	101	0.12	−15.0
		0.0460 M NaCl	7.23×10^{-3}	142	0.09	−15.2
	Decyltrimethyl ammonium bromide	Water	6.80×10^{-2}	36	0.25	−11.3
		0.013 M NaCl	6.34×10^{-2}	38	0.26	−11.3
	Dodecyl trimethyl ammonium bromide	Water	1.53×10^{-2}	50	0.21	−14.3
		0.013 M NaCl	1.07×10^{-2}	56	0.17	−14.6
B.	Dodecyl pyridinium chloride	Water	1.47×10^{-2}	20†	0.22	−13.8
		0.02 M KCl	1.13×10^{-2}	37†	0.25	−13.6
		0.05 M KCl	8.46×10^{-3}	48†	0.22	−13.7
		0.08 M KCl	6.88×10^{-3}	49†	0.23	−13.6
	Dodecyl pyridinium bromide	Water	1.16×10^{-2}	58	0.20	−14.9
		0.02 M KBr	7.32×10^{-3}	80	0.19	−14.7
		0.04 M KBr	4.88×10^{-3}	95	0.15	−15.1
		0.06 M LiBr	3.96×10^{-3}	87	0.17	—
		0.06 M KBr	3.96×10^{-3}	95	0.18	−14.9
		0.06 M RbBr	3.35×10^{-3}	98	0.18	—
		0.08 M KBr	3.36×10^{-3}	95	0.21	−14.6
		0.10 M KBr	2.74×10^{-3}	139	0.19	−14.8
	Dodecyl pyridinium iodide	Water	5.60×10^{-3}	87	0.13	−16.7
		0.0025 M KI	4.53×10^{-3}	90	0.11	−17.2
		0.0050 M KI	3.87×10^{-3}	94	0.09	−17.2
		0.0100 M KI	2.94×10^{-3}	124	~0	−18.0

† These data (from Ford *et al.* 1966) seem to be on the low side.

Table 10.4

$\overline{\Delta G^{\ominus}}$, $\overline{\Delta S^{\ominus}}$, and $\overline{\Delta H^{\ominus}}$ contributions to micellization (Kishimoto and Sumida 1974; *Tori and Nakagawa* 1963; *Corkill et al.* 1964)

Compound	Solvent	Temperature °C	$\overline{\Delta G^{\ominus}}$	$\overline{\Delta H^{\ominus}}$ (kJ mol^{-1})	$T\overline{\Delta S^{\ominus}}$
$CH_3(CH_2)_{11}O.SO_3^-Na^+$	H_2O	20	−38.6	+1.7	+40.3
$C_{12}H_{25}CH(COO^-N^+(CH_3)_3$	H_2O	10–60	−26.7	+2.3	+29.0
$C_{12}H_{25}(OC_2H_4)_6OH$	H_2O	25	−33.0	+16.3	+49.3

$\overline{\Delta H^{\ominus}}$ obtained in this way can be combined with values of $\overline{\Delta G^{\ominus}}$ to determine the entropic contribution as shown in Table 10.4.

Again, it appears from these results that the driving force for the micellization process in water at room temperature is the increase in entropy that is involved (but see section 10.4.4 above).

Exercises. 10.5.1 Use eqns (10.5.2), (3), (4), and (6) together with the condition $x = 0$ to show that:

$$\frac{d\phi}{dC_T} = \frac{d\phi}{dC_s} \cdot \frac{dC_s}{dC_T} = \frac{A + B(2n-p)KC_s^{2n-p-1}}{1 + n(2n-p)KC_s^{2n-p-1}}.$$

Hence, using eqn (10.1.1), show that

$$\frac{1}{K} = \frac{n(2n-p)(4n-2p-1)}{2n-p-2} C_s^{2n-p-1}$$

where C_s is the concentration of monomeric surfactant at the c.m.c., where $C_T = C_0$. (Hint: you can simplify the algebra a little by introducing a function $f = (2n-p)KC_s^{2n-p-1}$).

10.5.2 Writing eqn (10.5.2) in the form

$$\frac{1}{K} = \frac{nC_s^{2n-p}}{C_0 - C_s}$$

show that

$$C_s = \frac{(2n-p)(4n-2p-1)C_0}{2(2n-p+1)(2n-p-1)}$$

and hence, for $n \gg p \gg 1$

$$\frac{1}{K} \simeq 4n^2 C_0^{2n-p-1}.$$

10.5.3 Use the same procedure as in Exercises (10.5.1) and (10.5.2) above to show that for the case of swamping electrolyte, $d^3\phi/dC_T^3 = 0$ when

$f=(n-2)/n(2n-1)$, where $f=nKx^{n-p}C_s^{n-1}$ and $dC_T/dC_s=1+nf$. Hence show that

$$\frac{1}{K}=\frac{n^2(2n-1)}{n-2}C_s^{n-1}x^{n-p}.$$

10.5.4 Show that, for the case of swamping electrolyte:

$$\frac{C_s}{C_0}=\frac{n(2n-1)}{2(n^2-1)}\simeq 1 \quad \text{for large } n,$$

and hence obtain eqn (10.5.10).

10.6 Spectroscopic techniques for investigating micelle structure

Spectroscopic techniques based either on optical absorption or emission of light from some 'probe' molecule are now well established for investigating a wide range of physical properties of micellar solutions. A brief outline of several of these techniques is given in the following discussion. For a general description of the principles involved see Turro *et al.* (1977).

10.6.1 *Methods for determining the c.m.c.*

(a) *Solubilization of additives.* The fact that micelles can solubilize relatively large amounts of sparingly water-soluble compounds has been used to measure the onset of micelle formation (Mukerjee and Mysels 1971). The method is to measure the concentration of a chosen sparingly water-soluble substance, possessing a convenient UV-visible absorbing chromophore, in the presence of increasing amounts of surfactant. Below the c.m.c., the concentration of the solubilizate in solution is the same as in aqueous solution in the absence of surfactant. Above the c.m.c., the total amount of the additive in solution increases sharply as the total micelle concentration increases.

(b) *Spectral change of additives.* Some dyes such as Pinacyanol and Rhodamine 6G show changes in their absorption spectrum when solubilized by micelles (Mukerjee and Mysels 1971). Other aromatic organic molecules such as naphthalene, anthracene and, in particular, pyrene (Almgren *et al.* 1979a) show similar changes but also show changes in their fluorescence spectra when associated with micelles. These spectral changes have been used to monitor the c.m.c. in the same way as the solubilization procedure (see Fig. 10.6.1(a) and (b)).

An important feature of the fluorescence method is that since it is far more sensitive than optical absorption, lower concentrations of probe molecules can be used. This avoids the problems of the additive

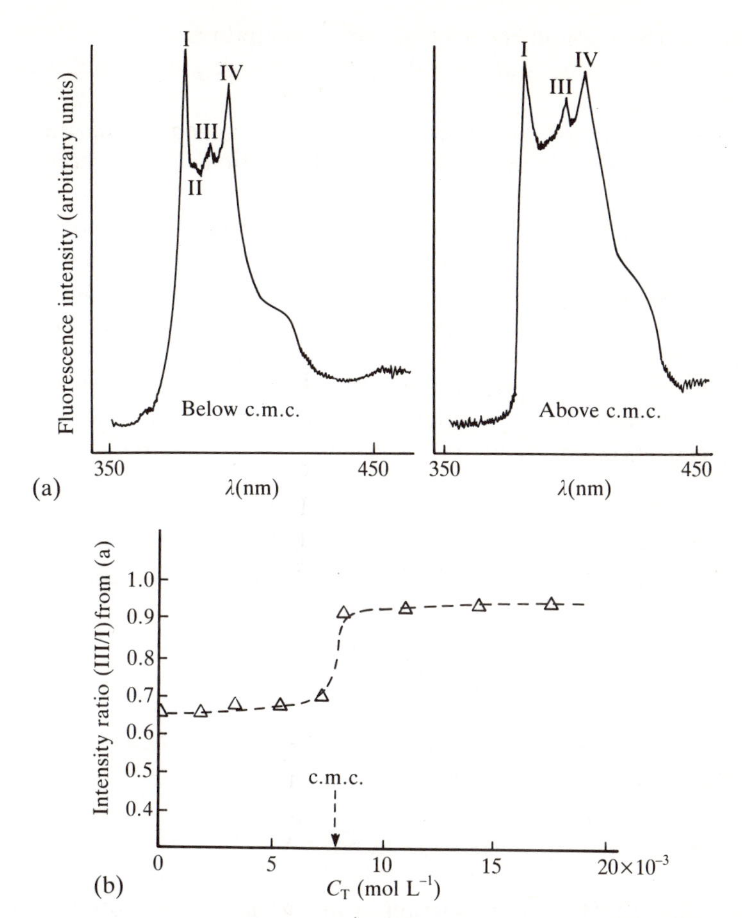

FIG. 10.6.1. (a) Pyrene monomer fluorescence in aqueous sodium dodecyl (lauryl) sulphate (SDS) solutions, at SDS concentrations below and above the c.m.c. (After Kalyanasundaram and Thomas 1977, with permission. Copyright American Chemical Society 1977.)
(b) Variation of the ratio of intensity of peaks III and I from Fig. 10.6.1 (a) as a function of the SDS concentration.

influencing the c.m.c. of the surfactant, which can occur at high probe concentrations.

10.6.2 *Fluorescent methods for determining aggregation number*

(a) *Steady-state emission quenching method.* In 1978 Turro and Yekta presented an extremely simple method for determining the number average aggregation number, $\bar{n}$, of micelles. The method is based on the quenching of a luminescent probe by a known amount of quencher molecules. The method relies on several assumptions:

(a) both the probe and quencher are completely associated with the micelles;

(b) both the probe and quencher remain attached to the micelle at times much longer than the unquenched lifetime of the luminescent probe;

(c) the quenching in or on a micelle containing both a probe and a quencher molecule is much faster than the emission lifetime of the probe, so that emission is observed only from micelles without quenchers;

(d) the distribution of the probe and quencher among the micelles is known; in practice, a Poisson distribution is assumed. (Exercise 10.7.1).

If the above assumptions are valid, the relative intensity of the fluorescence of a probe is related to the quencher concentration and micelle concentration by the relationship,

$$I = I_0 \exp[-C_q/C_m] \tag{10.6.1}$$

where I_0 is the emission intensity in the absence of quencher ($C_q = 0$) and I the emission in the presence of quencher. The micelle concentration, C_m, is given by

$$C_m = \frac{C_T - \text{c.m.c.}}{\bar{n}} \tag{10.6.2}$$

where C_T is the total surfactant concentration. Rearranging eqns (10.6.1) and (10.6.2) yields (Exercise 10.7.2):

$$\ln(I_0/I) = \frac{C_q \cdot \bar{n}}{C_T - \text{c.m.c.}}. \tag{10.6.3}$$

Thus a plot of $\ln(I_0/I)$ as a function of the quencher concentration allows $\bar{n}$ to be determined.

For the probe [$Ru(bipy)_3^{2+}$] and quencher [9-methyl anthracene] used by Turro and Yekta for SDS micelles, the range of validity of equation (10.6.3) has been investigated by Almgren and Löfroth (1980). Using a time-resolved procedure (discussed in the next section) they showed that the method is quite accurate up to aggregation numbers of about 120. Beyond this size the micelles become sufficiently large that condition (c) above is no longer valid. However, the method would still be applicable if a longer lived probe, such as a phosphorescent molecule, were to replace the ruthenium tris-(bipyridyl) ion.

(b) *Time-resolved emission quenching method.* If a probe molecule in a micellar solution is excited by a short radiation pulse (from a laser, say) then, in the presence of quencher molecules, its emission intensity will decay with time according to an expression of the form (see e.g. Almgren and Löfroth 1980):

$$I(t) = I(0)\exp\left[-t/\tau_0 + \frac{C_q}{C_m}(e^{-k_q t} - 1)\right] \tag{10.6.4}$$

or in the more convenient log form:

$$\ln\left(\frac{I(t)}{I(0)}\right) = -\frac{t}{\tau_0} + \frac{C_q}{C_m}(\exp(-k_q t) - 1) \tag{10.6.5}$$

where k_q (s^{-1}) is the quenching rate constant in the micelle, and τ_0 the emission lifetime of the solubilized excited probe in the absence of a quencher.

It is instructive to consider the form of eqn (10.6.5) at long and short times. For $t \to \infty$ we have

$$\ln\frac{I(t)}{I(0)} = -\frac{t}{\tau_0} - \frac{C_q}{C_m} \tag{10.6.6}$$

and for $t \to 0$ when $\exp(-k_q t) \approx 1 - k_q t$:

$$\ln\frac{I(t)}{I(0)} = -\frac{t}{\tau_0} - \frac{C_q k_q t}{C_m}. \tag{10.6.7}$$

Thus, the logarithmic emission decay curves for different C_q/C_m (changing C_q rather than C_m), will be a family of curves that have the same slope at long times equal to $-1/\tau_0$. However, at short times the slopes depend on C_q according to eqn (10.6.7). An example of such a set of emission–time quenching curves is presented in Fig. 10.6.2.

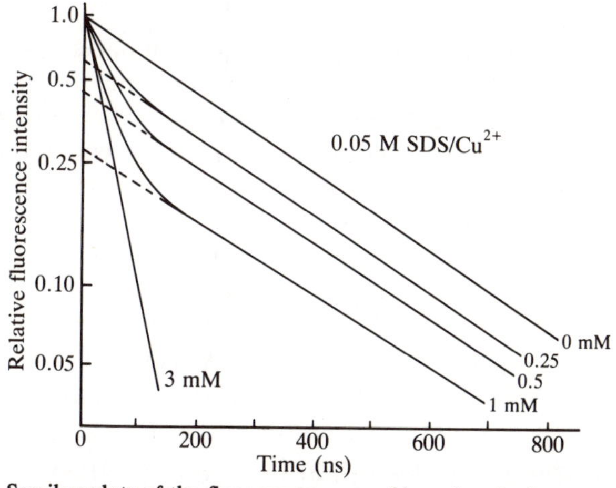

FIG. 10.6.2. Semilog plots of the fluorescence quenching of excited pyrene in 0.05 M SDS as a function of Cu^{2+} concentration. (From Grieser and Tausch-Treml 1980, with permission. Copyright American Chemical Society 1980.)

Computer fitting routines are generally used to determine k_q and C_q/C_m (and, subsequently, $\bar{n}$) from the emission decay curves. A simple method of determining $\bar{n}$ is to extrapolate the long-time slopes to $t = 0$, and, if the family of curves has been standardized, the intercept gives C_q/C_m from eqn (10.6.6) and $\bar{n}$ follows from eqn (10.6.2) at the known C_q.

The advantage of the time-resolved method over the steady-state method for determining $\bar{n}$ is that it can be used over a much wider range of micelle sizes.

10.6.3 *Interfacial electrostatic potentials of micelles using solubilized pH indicators*

To determine the electrical potential at the interface of a charged micelle, such as SDS or cetyltrimethylammonium bromide (CTAB), use can be made of pH indicators that are bound to the micelle surface. This method was originally explored by Hartley and Roe (1940). They observed a shift in the pK of the indicator in micellar solutions compared to pure aqueous solutions. This shift they attributed to a change in the 'local interfacial' proton concentration H_s^+ at the surface of the micelles compared to that in bulk solution, H_b^+. The relation between the proton concentration at the interface and that in the bulk solution is given by the Boltzmann equation (Section 6.3):

$$[H^+]_s = [H^+]_b \exp(-e\psi/kT) \qquad (10.6.8)$$

where ψ is the interfacial potential and the other constants have their usual meaning.

However, the shift in the pK of the indicator in micellar solutions may be caused not only by the electrostatic potential but also by a different local environment at the micellar surface, e.g. by a lower dielectric constant as compared to bulk water. Mukerjee and Banerjee (1964) pointed out that to measure the electrostatic contribution to the pK shift, the intrinsic interfacial pK, pK^i, must be known. The 'apparent' pK(pK^a) is related to pK^i by:

$$pK^a - pK^i = -e\psi/2.3kT. \qquad (10.6.9)$$

Fernandez and Fromherz (1977) in an excellent study on the surface potential of micelles, titrated alkyl coumarins (the fluorescence intensity of these molecules is pH-sensitive) in charged and neutral micelles. Figure 10.6.3 shows the characteristic titration curves obtained. To calculate an electrostatic potential for the charged system, the apparent pK in the neutral micelles was taken as the intrinsic interfacial pK^i. Their

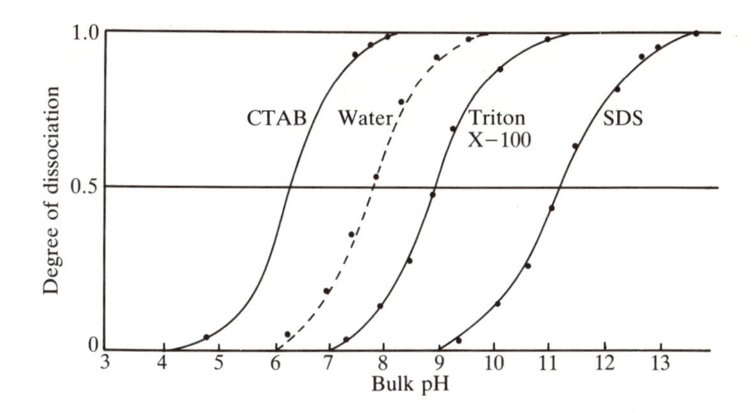

FIG. 10.6.3. Degree of dissociation of acid pH indicator versus bulk pH. The figure compares the titration of hydroxycoumarin chromophore, hydrophobically bound to positively (CTAB) and negatively (SDS) charged micelles in 24 mM surfactant concentration in water. (From Fernandez and Fromherz 1977, with permission. Copyright American Chemical Society 1977.)

assumption that the intrinsic pK is similar in the charged and uncharged interface was supported by other experimental data.

The extent of the pH shift due to the zero-charge surface environment can be seen in Fig. 10.6.3, where the titration curve of the coumarins in water is shown alongside the titration curve in neutral Triton X-100 micelles.

The surface potentials calculated using eqn (10.6.9) for CTAB and SDS were +148 mV and −134 mV respectively. The potentials are sensitive to the total ionic strength of the solutions and the background electrolyte, but these values are very reasonable estimates of the surface potential, ψ_0, expected for spherical micelles of the size suggested in Table 10.3. (Exercise 10.7.3). It appears, therefore, that the probe molecule is in this case sampling the potential in the plane of the head groups and is not merely being affected by the diffuse layer potential, which would be much lower in this case.

10.6.4 *Polarity of the micelle–water interface*

As already indicated in the previous section, molecules located at the micelle–water interface appear to sense an environment which is neither completely water-like nor completely hydrocarbon-like. An exact description of the environment is impossible, but it is possible to relate the characteristics to an apparent dielectric permittivity. This approach has been taken in several investigations of the micelle–water interface.

Molecules that show wavelength changes in their spectral absorption or emission bands are commonly used. The method is to measure the position of the spectral feature as a function of the known dielectric permittivity of a solvent. Usually these are made up of dioxane/water, or alcohol/water mixtures. The position of the spectral band of the probe molecule in the micelle is compared to its position in the calibration solvent, and an effective permittivity is obtained. Various molecules have been used as dielectric probes including pyrene carboxaldehyde, pyridinium *N*-phenol betaine, dodecylpyridinium iodide, and benzophenone. The most extensive study of the effective interfacial permittivity was made by Zachariasse *et al.* (1981). Some of the permittivities they determined for micelles using pyridinium –*N*-phenol betaine as a probe are listed in Table 10.5.

Although there are some differences evident in Table 10.5, depending on the reference solvent, the results show a remarkable degree of consistency. It should be noted, however, that the value obtained is crucially dependent on the structure of the probe molecule, which will presumably tend to sample the environment that is most energetically favourable, and that will depend on the disposition of its own hydrophilic and lipophilic parts. For further discussion see Mukerjee *et al.* (1977).

Table 10.5

Effective permittivity at the surface of various micelles

Micelles	ε_r^a	ε_r^b	ε_r^c
C_{12} trimethylammonium chloride (DTAC)	40	36	30
C_{12} TA bromide	35	33	29
Cetyl (C_{16}) TA chloride (CTAC)	31	31	28
C_{16} TA bromide (CTAB)	30	30	27
Sodium decyl sulphate (SDeS)	51[d]	55	
Sodium dodecyl sulphate (SDS)	51[d]	55	
Triton X-100	30	30	27
Brij 35	28	29	27
C_{12} (ethylene oxide)$_8$	28	29	27

Effective permittivity based on [a] ethanol/water mixtures; [b] dioxane/water mixtures; [c] *n*-alcohols; [d] methanol/water mixtures.

10.7 Micellar dynamics

As mentioned earlier in this chapter micelles are dynamic units, constantly forming and dissociating on a time-scale in the microsecond to millisecond range. The kinetics of micelle formation and breakdown have mainly been studied by fast relaxation methods (temperature-jump, pressure-jump, ultrasonic absorption, shock-tube methods, etc; see Aniansson *et al.* 1976 and Muller 1977). In these methods a system is subjected to a sudden perturbation and it is then monitored as it returns to equilibrium or departs from it. Such measurements reveal the presence of two† relaxation processes in the perturbed system—a fast step (nanosecond to microsecond time-scale), related to the exchange of monomers between the bulk aqueous phase and micelles of different sizes, and a slower relaxation (microsecond to millisecond) related to the formation or break-up of micelles.

The types of equilibria that exist in micellar solutions can be divided into the following (Muller 1977):

A. Ionization

$$M_nX_p \rightleftharpoons M_nX_{p-1} + X \tag{10.7.1}$$

Here M_n is a micelle with n monomers, and p counterions, X, of either positive or negative charge.

B. Monomer exchange

$$M_{n-1} + S \underset{k^-}{\overset{k^+}{\rightleftharpoons}} M_n \tag{10.7.2}$$

where S is the surfactant monomer.

C. Formation/dissolution process

$$nS \rightleftharpoons M_n. \tag{10.7.3}$$

D. Partial breakdown and reformation

$$M_n \rightleftharpoons M_a + M_{n-a} \tag{10.7.4}$$

(a is an integer not greatly different from $n/2$).

E. Size redistribution (number of micelles unchanged)

$$M_n + M_m \rightleftharpoons M_{n+b} + M_{m-b} \tag{10.7.5}$$

(b is small such that M_{m-b} is still considered a micelle).

F. Size redistribution (number of micelles changed)

$$(n+1)M_n \rightleftharpoons nM_{n+1}. \tag{10.7.6}$$

† In some cases three relaxation processes have been observed for ionic surfactant solutions—a very fast process (*ca.* 50 ns) has been attributed to counter-ion relaxation in the system.

It is interesting to note that although a non-equilibrated system can relax by all of the above processes, only two well-defined relaxation times are consistently observed. This is partly a consequence of the widely varying rates involved in the above reaction steps. As already mentioned, process A is very rapid, and generally outside the time regime studied. Process B is the only other process that can occur rapidly and in a single step, and hence it is assigned to the fast relaxation time. (An independent pulse radiolysis study (Almgren *et al.* 1979b) has confirmed that the fast reaction is due to monomer exchange, at least in the case of SDS.)

Process C is actually a shorthand representation of a mechanism with $(n-1)$ steps, i.e.

$$\begin{aligned} 2S &\rightleftharpoons M_2 \\ M_2 + S &\rightleftharpoons M_3 \\ &\vdots \\ M_{n-1} + S &\rightleftharpoons M_n. \end{aligned} \tag{10.7.7}$$

This sequence has been identified with the slower relaxation event. It has been argued (Muller 1977) that although reactions (10.7.4) and (10.7.6) may also be slow, because they must also proceed through a sequence such as (10.7.7), the relative concentration of species involved is small and therefore not easily recognized.

10.7.1 *Kinetics of micelle formation*

Although there are some differences of opinion on the most appropriate theoretical treatment of micelle formation (see e.g. Muller 1977; Kahlweit 1981), the most widely accepted approach is that of Aniansson and Wall (1974). Their analysis is based on the consideration that there is a distribution of aggregates and micelle sizes in any micellar system. A schematic form of such a distribution is given in Fig. 10.7.1 (after Kahlweit (1981)).

In the relaxation treatment of Aniansson and Wall the fast relaxation time constant τ_1 is given as

$$\frac{1}{\tau_1} = \frac{k^-}{\sigma^2} + k^-(C_T - C_e)/\bar{n}C_e \tag{10.7.8}$$

where k^- is the reverse rate constant for reaction (10.7.2), and C_e is the monomer concentration at equilibrium following the perturbation. This result is in accord with the observed linear rise of $1/\tau_1$ with C_T, the total surfactant concentration, noted in most experiments. A brief discussion of the derivation of this kind of relaxation is given by Vold and Vold (1983, pp. 612–4).

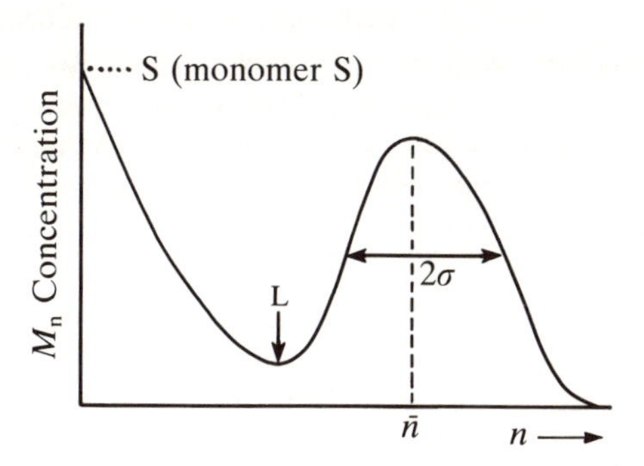

FIG. 10.7.1. Micelle size distribution. M_n is the number of aggregates of size n. The aggregates on the left side of the minimum (L) are called submicellar, those on the right-hand side (proper) micelles with mean size of $\bar{n}$, and the width of their size distribution is given as σ.

The derivation of the slow relaxation process involves a number of assumptions and approximations. The reason for this can be readily appreciated when it is remembered that n steps are required, the rate constants of the individual steps are not all equal, and the concentrations of the intermediate sized aggregates need to be specified. Based on a model of mass flow, the simplifications made by Aniansson and Wall allowed a single relaxation time to be derived and it is given by,

$$\frac{1}{\tau_2} = (\bar{n}^2/C_e R_1) \Big/ \left(1 + \sigma^2 \frac{(C_0 - C_e)}{C_e \bar{n}}\right). \tag{10.7.9}$$

R_1 in the above equation is a function related to the restrictions on flow of monomers from the aggregates based on the mass flow model. R_1 itself is dependent on C_0 in a complex way, part of which is due to the fact that the theory does not take into account the redistribution of free counterions, i.e. the theory is a better description for non-ionic surfactants. Advances on the Aniansson and Wall model by Chan, Kahlweit and co-workers (1977) resulted in a theory specifically designed for ionic surfactants. Although there are weaknesses in the model, it predicts that, for ionic surfactants, a plot of $1/\tau_2$ against C_0 should exhibit a maximum, while for non-ionics the relaxation rate should be a monotonically increasing function of C_0, in good agreement with observed results.

Some examples of the relaxation times observed for alkyl sulphates are given in Table 10.6, showing the time scales involved in the fast and slow processes.

The relaxation times are, of course, related to the rate constants for association (k^+) and dissociation (k^-) as represented by reaction (10.7.2) earlier. The rate constants for a series of sodium alkyl sulphates are given

Table 10.6
Relaxation times τ_1 (μs) and τ_2 (ms) for some sodium alkyl sulphates

Surfactant	Temperature (°C)	Concentration (M)	τ_1 (μs)	τ_2 (ms)
SHS (C_{16})	30	1×10^{-3}	760	350
STS (C_{14})	25	2.1×10^{-3}	320	41
	30	2.1×10^{-3}	245	19
	35	2.1×10^{-3}	155	7
	25	3×10^{-3}	125	34
SDS (C_{12})	20	1×10^{-2}	—	1.8
	20	5×10^{-2}	—	50

in Table 10.7. The obvious pattern shown in Table 10.7 is that, as the alkyl chain becomes larger, the exit rate of monomer becomes appreciably slower, as one would expect with increasing hydrophobic character of the monomer. The association rate constant also decreases with increasing chain length (by about a factor of six from C_6 to C_{14}), although only slightly, reflecting basically a diffusion controlled rate step with some electrostatic repulsion involved between the charged micelle and the anionic monomer (Aniansson *et al.* 1976).

In summary it may be said that although some uncertainties remain in understanding the results of the relaxation processes in micellar systems, the time ranges of the dynamic processes are well defined, and a reasonably good picture of the equilibria has been established.

Table 10.7
Kinetic parameters of association and dissociation of alkyl sulphates from their micelles (from Aniansson et al. 1976)

Surfactant	$\bar{n}$	c.m.c. (M)	k^- (s^{-1})	k^+ ($M^{-1}s^{-1}$)	K
NaC_6SO_4	17	0.42	1.32×10^9	3.2×10^9	40.5
NaC_7SO_4	22	0.22	7.3×10^8	3.3×10^9	100
NaC_8SO_4	27	0.13	1×10^8	7.7×10^8	207
NaC_9SO_4	33	6×10^{-2}	1.4×10^8	2.3×10^9	550
$NaC_{11}SO_4$	52	1.6×10^{-2}	4×10^7	2.6×10^9	3250
$NaC_{12}SO_4$	64	8.2×10^{-3}	1×10^7	1.2×10^9	7800
$NaC_{14}SO_4$	80	2.05×10^{-3}	9.6×10^5	4.7×10^8	3.9×10^4

Note: The equilibrium constant $K=\dfrac{k^+}{k^-/\bar{n}}\simeq\dfrac{\bar{n}}{\text{c.m.c.}}$; $\bar{n}$ is the average aggregation number.

Table 10.8
Residence times in micelles (μs)

Probe molecule	SDS	CTAB	Solubility in water ($mol\ L^{-1}$)
Anthracene	59	303	2.2×10^{-7}
Pyrene	243	588	6×10^{-7}
Biphenyl	10	62	4.1×10^{-5}
Naphthalene	4	13	2.2×10^{-4}
Benzene	0.23	1.3	2.3×10^{-2}

10.7.2 *Residence times of probe molecules in micelles*

Like the surfactant monomer the micelle-solubilized additive is not rigidly fixed in the micelles. Not only can it move about within the micelle, but it is in constant dynamic equilibrium with the bulk aqueous phase. The residence time of small molecular substances, such as those commonly used as probes, is an important consideration when interpreting dynamic results such as fluorescence quenching behaviour. Since most hydrophobic probes have residence times that are longer than their fluorescence lifetimes, fluorescence quenching techniques are unsuitable for determining probe residence lifetimes in micelles. However, it was shown by Almgren *et al.* (1979b) that if phosphorescence is used, residence times can be determined. In Table 10.8 several probe molecules are listed with their residence times† and solubility in water.

As can be seen from Table 10.8, apart from anthracene the residence times of the various probes decreases with increase in their solubility in water. Also, the residence times are dependent on the micelle type, which in part may be due to weak complexes forming between the quaternary ammonium head group of CTAB and the solubilized additive.

The existence of finite residence times means that over long periods of time the additive is uniformly distributed among the micelles in solution. However, at time intervals less than their residence time the distribution of the additives will be statistical, having a form given by the Poisson distribution (Exercise 10.7.1). Virtually all kinetic studies to date have arrived at this conclusion.

10.7.3 *Determination of microfluidity using fluorescence probes*

The most common method for studying fluidity of micelles has been fluorescence depolarization, which measures the resistance of a probe molecule to rotational or reorientational motion in its local environment.

† This is the time taken for a fraction $(1 - e^{-1})$, i.e. 63 per cent, of the probe molecules to escape.

The fluorescence emission intensity of a solubilized probe is measured at crossed ($I_\perp$) and parallel ($I_\parallel$), positions of the polarizers. The degree of polarization is given by,

$$d = \frac{I_\parallel - I_\perp}{I_\parallel + I_\perp}. \tag{10.7.10}$$

The Perrin (1932) equation relates the degree of polarization d to other parameters of the system by:

$$\frac{d^{-1} - \frac{1}{3}}{d_0^{-1} - \frac{1}{3}} = 1 + \frac{\tau_F RT}{V_0 \eta} \tag{10.7.11}$$

where d_0 is the degree of polarization in an extremely viscous solvent, τ_F is the average lifetime of the fluorescent molecule, V_0 is the effective volume of the molecule, and η is the viscosity of the environment. The more viscous the environment, the larger the value of d.

An alternative method (Zachariasse 1978) for measuring microfluidity is based on the formation of intramolecular excimers of molecules with two chromophores (such as dipyrenyl alkanes). The relative yield of excimer to monomer (obtained from their fluorescence emission) is viscosity dependent, so by comparing this ratio in micellar solutions with a reference solvent of known viscosity the microfluidity can be found.

The results obtained using different methods and different probes are, however, highly variable (ranging from 4–50 centipoise for SDS), which is probably a reflection of the fact that the effective viscosity depends on position in the micelle—varying from high fluidity in the interior to more viscous at the interface. Such variations with position in the micelle can be monitored by n.m.r. spectroscopy (see next section).

10.7.4 *Other spectroscopic techniques*

The most important technique remaining is that of n.m.r. spectroscopy, which can be used in a variety of modes and with a number of probe nuclei (hydrogen, the halides, the alkali metals) to study the phenomena of hydration, ion binding, and the mobility of segments of the hydrocarbon chain. The interpretation of results is, however, rather too complicated to go into here and the reader is referred to more specialized reviews (e.g. Lindman *et al.* 1977; Wennerström and Lindman 1979).

Unfortunately, one of the earliest applications of n.m.r. (using fluorine-substituted hydrocarbon chains) led to the conclusion that there must be significant penetration of water into the core (Muller and Birkhahn 1967; Muller and Platko 1971), a result that is now generally regarded as erroneous (Mukerjee and Mysels 1975). That work did, however, ultimately point up the anomalous nature of the fluoro-hydrocarbon mixing behaviour as indicated in section 10.2.1; evidently a terminal $-CF_3$ group on a hydrocarbon chain spends much more time near the surface of the micelle than does the corresponding $-CH_3$ terminal group.

Studies of the alkali metal ions by n.m.r. generally reveal that, even when bound to the head groups, they retain their primary hydration sheath. This is consistent with the usual picture of the Stern layer.

The n.m.r. data also confirm the notion, mentioned above, that the rotational freedom of the segments of the hydrocarbon chain increases as one moves away from the head group. The 'microviscosity' likewise tends to be higher in the neighbourhood of the head groups. The most convincing evidence for a liquid-like structure of the hydrocarbon core also comes from the n.m.r. studies. Wennerström *et al.* (1979) used T_1 relaxation times of the ^{13}C nuclei along the chain to obtain an estimate of the (rotational) correlation time. Values of around 10 ps were obtained, in close agreement with measurements on liquid hydrocarbons.

Electron spin resonance (e.s.r.) spectroscopy has also been used in micellar studies by Stilbs and Lindman (1974). They investigated the mobility of the vanadyl ion (VO^{2+}) as a counterion on the micellar surface, and found its mobility to be very high: the rotational correlation time was $\sim$70 ps, consistent with a fairly loose association with the head groups. Substitution of a nitrosyl group (–NO) along the hydrocarbon chain could also provide an unpaired electron, from which the e.s.r. signal would give information on the local environment in the core. Such procedures, however, can lead to equivocal results because the presence of the probe group can modify the mobility of the chain to which it is attached.

Exercises. 10.7.1 The probability of finding a micelle with i solute species is given by the Poisson relationship,

$$P_i = \frac{\bar{S}^i \exp(-\bar{S})}{i!}$$

where

$$\bar{S} = \frac{[\text{solute}]}{[\text{micelles}]}.$$

Given that $[S] = 10^{-3}$ M and $[\text{micelles}] = 5 \times 10^{-4}$, 10^{-3} M, and 2×10^{-3} M, calculate the probability of micelles containing 0, 1, 2, 3, and 4 solute molecules. Plot a graph of probability vs. number of solute molecules for each micelle concentration; note the relative proportion of molecules per micelle. Is it misleading to think that if [solute] = [micelle] each micelle contains one solute molecule?

10.7.2 Establish eqn (10.6.3) assuming that both probe and quencher are spread amongst the micelles in a Poisson distribution.

10.7.3 Calculate the charge, Q, on a micelle of radius 2.5 nm in 24 mM solution if the surface potential is -138 mV, using the semi-empirical Gouy–Chapman expression (eqn (6.11.10)). Compare your result with the approximate value from eqn (10.4.12). (Note that eqn (6.11.10) is quite accurate for such large potentials and small κa values. The error, compared to the exact computer solution for $\psi = 150$ mV and $\kappa a \approx 1$ is only about 1 per cent (Loeb *et al.* 1961).)

10.7.4 Addition of indifferent electrolyte usually lowers the estimated surface potential of a micelle. Fernandez and Fromherz (1977) found that, for 1 : 1 electrolyte, $d\psi/d \log_{10} C \simeq -60$ mV. Show that this is the expected result for a system where the diffuse layer charge remains constant. (Use the equations of Section 6.3 for a flat double layer.)

10.8 Micellization of surfactant mixtures

In aqueous solutions containing more than one surfactant component, a common situation when dealing with commercial non-ionic surfactants (e.g. Triton X-100), the formation of mixed micelles is found to be related to the c.m.c. of the individual components and their relative amount by (Clint 1975):

$$\frac{1}{C_M} = \sum_i (\alpha_i / C_{M_i}) \tag{10.8.1}$$

where C_M is the c.m.c. of the mixed system, α_i the mole fraction of component i, and C_{M_i} the c.m.c. of pure i.

Clint (1975) and others (see, for example, Mysels and Otter 1961) have used relatively simple treatments to calculate the variation of the concentration of the individual components in both the aqueous phase and in the mixed micelle.

Below the c.m.c. of the mixture (i.e. $C_T < C_M$) the concentration of the individual components is given by

$$C_i = \alpha_i C_T \tag{10.8.2}$$

where C_T is the total surfactant concentration in solution.

Using the phase separation model for micelle formation it can be shown (Warr *et al.* 1983) that the concentration of component i as monomer above the c.m.c. is given by:†

$$C_i = \{[(C_T + C_{M_i} - C_{\mu_i})^2 + 4\alpha_i C_T(C_{\mu_i} - C_{M_i})]^{1/2} - C_T - C_{M_i} + C_{\mu_i}\}/\{2[C_{\mu_i}/C_{M_i} - 1]\} \tag{10.8.3}$$

where C_{μ_i}, defined as the c.m.c. of the mixed system containing all

† This equation is misquoted in the original.

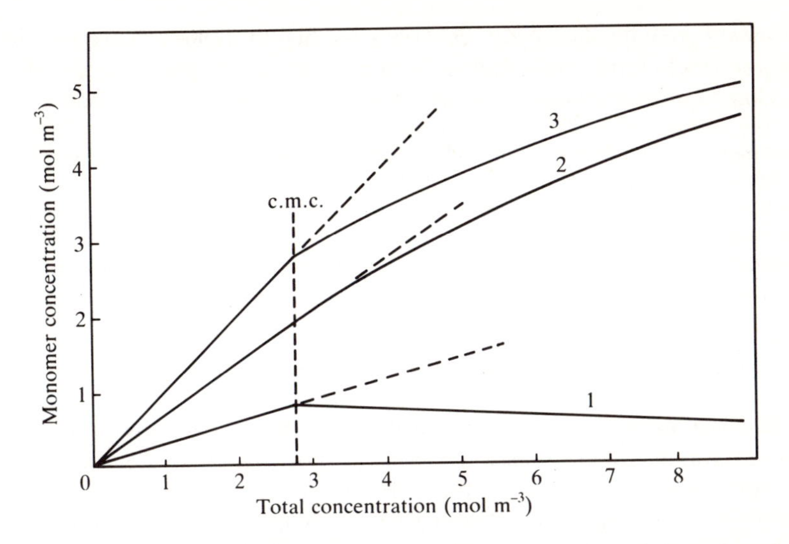

FIG. 10.8.1. Monomer concentrations in a mixed micellar solution plotted against total concentration. Curve 1, monomer concentration of component 1 ($c_{M_1} = 1\,\text{mol m}^{-3}$). Curve 2, monomer concentration of component 2 ($c_{M_2} = 10\,\text{mol m}^{-3}$). Curve 3, total monomer concentration. (From Clint 1975, with permission.)

species except i, is

$$(1-\alpha_i)C_{\mu_i}^{-1} = \left(\frac{1}{C_M} - \alpha_i/C_{M_i}\right). \tag{10.8.4}$$

The mole fraction of i in the micelle phase is given by,

$$x_i = (\alpha_i C_T - C_i)\Big/\Big(C_T - \sum_j C^j_{\text{mon}}\Big) \tag{10.8.5}$$

where $\sum_j C^j_{\text{mon}}$ is the total concentration of monomer in equilibrium with micelles.

Figure 10.8.1 shows the concentration behaviour below and above the c.m.c. of a two-component system. The main observations are that the free monomer concentration keeps increasing above the c.m.c. (compare this to a one-component system where the monomer concentration is virtually constant above the c.m.c.), and that the more hydrophobic component (the one with the low c.m.c.) decreases its concentration in free solution. The consequence of the latter observation is that the micelles become enriched with the more hydrophobic species as C_T increases, as described by eqn (10.8.5).

The increase in the monomer concentration above the c.m.c. for mixed surfactant systems has been observed by Schott (1964) using an ultra-filtration technique, and confirmed more recently by Warr *et al.* (1983), in good support of the predictions of the phase separation model.

Exercises. 10.8.1 Calculate the individual monomer concentrations for a three-component system, given that $\alpha_1 = 0.2$, $\alpha_2 = 0.4$, and $\alpha_3 = 0.4$ with c.m.c.$_{,1} = 2 \times 10^{-4}$ M, c.m.c.$_{,2} = 5 \times 10^{-4}$ M and c.m.c.$_{,3} = 10^{-3}$ M over a concentration range of 0 to 4 times the C_M.

10.8.2 Calculate, using eqn (10.8.5), the variation in the composition of the mixed micelle in Exercise (10.8.1).

10.9 Molecular packing and its effect on aggregate formation

So far, we have dealt in detail with both thermodynamic and kinetic considerations, but have assumed for the most part that the micelles are roughly spherical in shape. Little allowance has been made for the existence of rod-like micelles, vesicles or bilayers. Recently Tanford (1972), Mitchell and Ninham (1981) and Israelachvili *et al.* (1976, 1977, 1980) have revived an idea originally proposed by Hartley (1941), viz. that molecular packing plays a crucial role in determining allowed structures, at least in the case of dilute surfactant solutions.

Aggregation is, as outlined previously, described by either the mass action or phase models. Typically, chemical potentials of the monomer in the aggregate and in solution at equilibrium are equated and the normal relationships are derived. In this section we wish to examine the contributions to the standard chemical potential or molar free energy per surfactant molecule in the aggregate, $\mu_m^\ominus$:

$$\mu_m^\ominus = \mu_m^B + \mu_m^s + \mu_m^c + \text{molecular packing term}. \qquad (10.9.1)$$

These terms may be further described, referring to Fig. 10.9.1 for clarification. The bulk term, μ_m^B, is a constant and is a measure of the free energy change involved in removing hydrophobic tails from an aqueous environment into the micelle interior. The latter is assumed to be liquid-like which, as we have already seen, is a sound assumption (Vikingstad and Hoiland 1978). The surface term, μ_m^s, includes a quantity γA to account for the fact that the hydrophobic tails have some residual contact with the aqueous phase. A is the area per surfactant head group and γ is the hydrocarbon–water interfacial tension. Head group interactions, which may be due to steric, hydration, electrostatic, and other forces contribute a repulsion energy. Their quantitative description is still elusive; however, for electrostatic repulsion, an energy contribution varying inversely with A would be expected. μ_m^s then takes the form:

$$\mu_m^s = \gamma\left(A + \frac{A_0^2}{A}\right) \qquad (10.9.2)$$

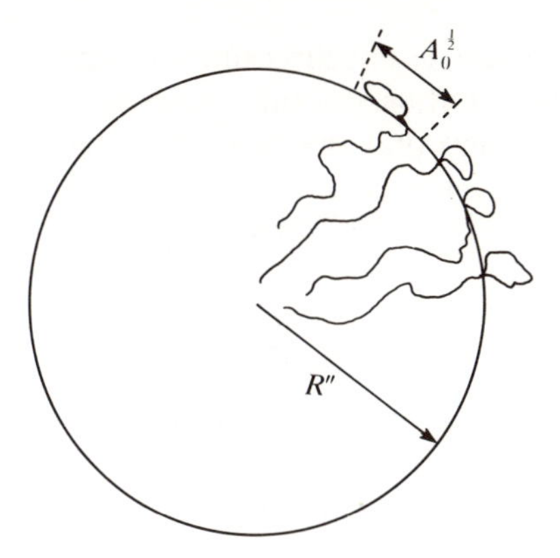

FIG. 10.9.1. Schematic representation of a model spherical micelle of core radius R''. (After Mitchell and Ninham 1981, with permission.)

reaching a maximum at $2\gamma A_0$ where A_0 is a limiting or optimal area per head group.

The curvature term μ_m^c accounts for reductions in the effective surface tension and alterations in the electrostatic energy when a spherical surface is formed, rather than a planar one. Generally speaking these corrections may be ignored, except for very special cases (Israelachvili *et al.* 1976).

Lastly the packing term needs to be considered. One needs to account for the fact that the interior of the aggregate is fluid-like *and* incompressible. The aggregate radius or radii, hydrophobic chain volume, and surface area per head group are of paramount importance here, as discussed below.

In more concentrated systems, where other interactions take place, eqn (10.9.2) will contain additional contributions. For the present, however, the discussion is only concerned with dilute solutions. Ninham (1981) and Israelachvili and Ninham (1977) have shown that when $\mu_m^\ominus$ is decomposed into the first three terms, the correct dependence of c.m.c. on hydrocarbon chain length is predicted as are the effects of temperature and ionic strength. However the existence of rod-like micelles or bilayers suggests certain geometric constraints and it is with these that we shall now concern ourselves, i.e. the nature of the molecular packing term.

For simplicity, consider firstly a spherical micelle (Figure 10.9.1). The radius, R'', surface area per head group A and hydrophobic chain volume V are linked by

$$\frac{V}{A}=\frac{R''}{3}. \qquad (10.9.3)$$

The radius of a spherical micelle cannot be greater than a specific critical length l_c, slightly less than the fully extended length of the hydrocarbon chain, if it is assumed that the head group never moves into the core. Clearly, then, when $V/A_0l_c > \frac{1}{3}$ spherical micelles will not form unless $A > A_0$. The critical condition for the formation of spheres is

$$\frac{V}{A_0 l_c} = \tfrac{1}{3}. \tag{10.9.4}$$

Similarly, for cylindrical micelles, it may be shown that

$$\frac{V}{A_0 l_c} = \tfrac{1}{2} \tag{10.9.5}$$

and for planar bilayers

$$\frac{V}{A_0 l_c} = 1. \tag{10.9.6}$$

Any aggregated structure must satisfy two basic criteria:

(a) no point within the aggregate can be farther from the surface of tension than l_c;

(b) the total hydrocarbon core volume, $\mathcal{V}$, and the total surface area, $\mathcal{A}$, must approximately satisfy $\mathcal{V}/\bar{V} = \mathcal{A}/A_0 = n$, the aggregation number. Here $\bar{V}$ is the volume occupied by a single hydrocarbon molecule of the appropriate chain length in the liquid state. This criterion is only an approximate one since the average surface area per surfactant head group is assumed to be equal to A_0.

Between a sphere and a cylinder, one might reasonably expect to find a variety of transition shapes. Thus if we consider a surfactant (Fig. 10.9.2) with a head group of surface area A and liquid hydrocarbon core volume V in a micelle or bilayer vesicle where the local radii of curvature are R_1 and R_2, the following equation results:

$$\frac{V}{A} = l\left[1 - \frac{l}{2}\left(\frac{1}{R_1} + \frac{1}{R_2}\right) + \frac{l^2}{3R_1R_2}\right] \tag{10.9.7}$$

where l is the length of the hydrocarbon region of the amphiphile. Equation (10.9.7) is exact for spheres ($R_1 = R_2$) cylinders ($R_2 = \infty$) and planar surfaces ($R_1 = R_2 = \infty$) and is accurate to within one per cent for other cases (Israelachvili *et al.* 1976). It may then be predicted that bilayers or vesicles exist when

$$\tfrac{1}{2} < \frac{V}{A_0 l_c} < 1 \tag{10.9.8}$$

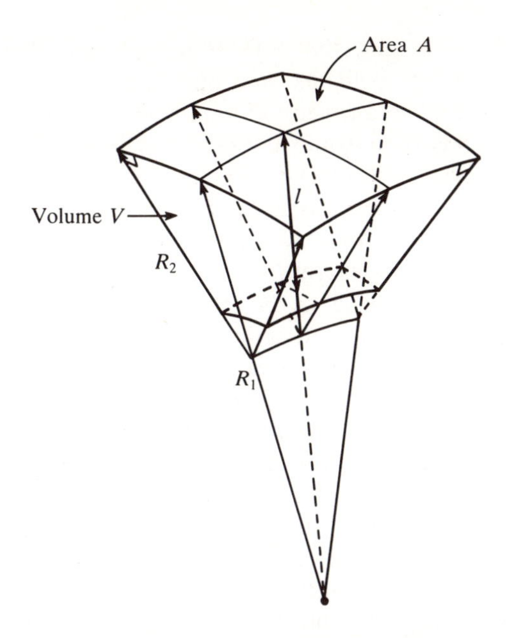

FIG. 10.9.2. Geometric packing of a hydrocarbon region of volume V and surface area A, at a surface with two radii of curvature, R_1 and R_2. Eqn (10.9.7) gives the relation between V, A, R_1 and R_2 and the length, l, of the hydrocarbon. For both R_1, $R_2 > l$ a void region is formed behind the hydrocarbon region. (After Israelachvili *et al.* 1976, with permission.)

and inverted structures occur when

$$\frac{V}{A_0 l_c} > 1. \tag{10.9.9}$$

This deceptively simple packing model allows many physical properties of micelles and vesicles such as size, shape, polydispersity, etc. to be predicted. The interested reader should consult the original references for more detailed discussions.

10.10 Statistical thermodynamics of chain packing in micelles

In order to model, theoretically, the hydrocarbon chains in micelles, it is first necessary to understand several points. Fully saturated alkyl chains are very flexible. Each C–C bond in a fully extended alkyl chain exists in a *trans* state but each bond can also exist in two other conformations—the *gauche*$^+$ and *gauche*$^-$ states, which can each be formed at an energy of $\sim 0.8kT$ at room temperature. Therefore, in a fully saturated alkyl chain of length 12 (a dodecylsulphate chain for instance) there are a very large number of fairly low energy states of the chain.

It is possible to imagine that fully saturated alkyl chains can pack in a frozen array of all-*trans* chains, and indeed this is the packing found in solid bulk *n*-alkane. However, without considerable (energetically unfavourable) hydrocarbon–water contact, it is impossible to see how such an array could exist in a micelle. Elementary geometrical considerations (section 10.9) and the experimental evidence discussed in section 10.6 suggest that the interior of a micelle is liquid, and that the alkyl chains are conformationally disordered.

In a non-polar liquid (such as the micellar interior), the packing density is determined by the combined action of very short-range intermolecular repulsive forces (arising from overlap of electron clouds) and longer range van der Waals attractions. Since the chains are chemically identical to *n*-alkane chains, we should expect a packing density almost identical to that found for bulk liquid *n*-alkane, as noted in section 10.9.

There is little doubt that, for an ionic surfactant, the head groups are almost completely excluded from the hydrophobic core of a micelle. There was, however, some disagreement about the extent of water penetration in the core, as noted in section 10.7.4. Experiments in which probe molecules chemically bonded to different parts of the alkyl chains were observed to behave as if they sat in a partly hydrophilic environment have, in the past, been interpreted as implying extensive water penetration in the core (see e.g. Menger *et al.* 1978). We have seen earlier, however, that free energies of micellization are comparable to free energies of transfer of alkyl chains from water to bulk *n*-alkane, which implies that the core of a micelle must be almost devoid of water. Recently, n.m.r. relaxation experiments (Halle and Carström 1981) and neutron scattering experiments (Bendedouch *et al.* 1983; Cabane *et al.* 1983) have given unequivocal evidence that water penetration in the core is minimal and that the hydrocarbon–water interface is smooth, with an average roughness of the order of the diameter of a water molecule.

A theoretical model has been developed in the light of all this evidence by Gruen (1981). The model assumes the existence of a hydrophobic core that neither head groups nor water can enter. The amphiphile chains are allowed to exist in all their possible conformations (*trans, gauche*$^+$ and *gauche*$^-$ states for each C–C bond). On average, the chains are constrained to pack into the hydrophobic core of the micelle at liquid alkane density throughout. To a limited extent they may also exist outside the hydrophobic core, but they are then subject to an increased free energy. Each chain conformation is assigned its appropriate Boltzmann factor and thermodynamic averages over all conformations are evaluated.

Several illuminating conclusions emerge from the study. They may be illustrated by considering a spherical micelle formed from a surfactant with an alkyl chain of length 12 (e.g. a dodecylsulphate). If the hydrophobic core of the micelle has a radius equal to the length of a fully extended chain (1.67 nm), it will contain approximately 56 chains. The model predicts that the free energy cost of packing the chains into this structure is less than $0.5kT$ per chain. By comparison, the free energy gained when a C_{12} chain is transferred from water to bulk *n*-alkane is $\sim 20kT$ (recall section 10.4 above).

In the micelle one or two chains must be completely straight (in their all-*trans* state) in order to fill the volume at the centre of the micelle, but *no more* than one or two. On average, the model predicts a loss of only 0.2 *gauche* bonds per chain on transfer from a bulk *n*-alkane environment to the micelle.

The fact that half the volume of the hydrophobic core occurs within 0.34 nm of its surface has important consequences. The mean position of each segment in the chain is nearer to the surface of the aggregate than to the micelle centre. The terminal CH_3 group, although on average closer to the centre than any other group, sits a mean distance of 1.04 nm from the centre (and only 0.63 nm from the core surface). Because of the liquid-like micelle interior, and the flexibility of the chains, all segments sample the surface of the micelle. Even the terminal CH_3 group is in contact with the surface (and hence in contact with a partly hydrophilic environment) approximately 20 per cent of the time. (This observation explains why probes attached to any part of the chain behave as though they were in a partly hydrophilic environment.)

The model suggests that the hydrocarbon–water interface is fairly sharp. 0.2 nm beyond the core surface, the average hydrocarbon volume fraction is 0.02 and falling fast. It may seem extraordinary that a structure as dynamic as a micelle (with monomers being associated with the micelle for only 10^{-6}–10^{-5} s) could have such a smooth surface. It occurs because the hydrophobic effect is very strong and so the vast majority of time that any monomer is associated with the micelle it sits with almost all of its chain inside the hydrophobic core.

It is this picture of the SDS micelle that was used in Fig. 1.5.5 for the spherical micelle structure. Each of the five spherical shells contains approximately the correct number of chain segments to ensure an even packing density throughout. Note particularly the large fraction that is in the outermost shell. This model is capable of reconciling a considerable body of experimental evidence but there remain some disagreements over details. Hayter and Penfold (1981), for example, argue that the neutron scattering results suggest a slightly rougher hydrocarbon–water interface, though the difference is not great. We will return to this discussion in Volume II after the theory of scattering has been treated more exactly.

References

Almgren, M., Grieser, F., and Thomas, J. K. (1979a). *J. Am. chem. Soc.* **101,** 279.
Almgren, M., Grieser, F., and Thomas, J. K. (1979b). *J. chem. Soc. Faraday Trans. 1* **75,** 1674.
Almgren, M. and Löfroth, J. E. (1980). *J. Colloid interface Sci.* **81,** 486.
Anacker, E. W. (1970). Micelle formation of cationic surfactants in aqueous media. In *Cationic surfactants* (ed. E. Jungermann). Marcel Dekker, New York.
Aniansson, E. G. (1978). *J. phys. Chem.* **82,** 2805.
Aniansson, E. G. and Wall, S. N. (1974). *J. phys. Chem.* **78,** 1024.
Aniansson, E. G., Wall, S. N., Almgren, M., Hoffmann, H., Kielmann, I., Ulbricht, W., Zana, R., Lang, J., and Tondre, C. (1976). *J. phys. Chem.* **80,** 905.
Bendedouch, D., Chen, S.-H., and Koehler, W. C. (1983). *J. phys. Chem.* **87,** 153.
Cabane, B., Duplessix, R., and Zemb, T. (1983). In *Surfactants in solution* (ed. K. L. Mittal and B. Lindman). Plenum Press, New York.
Chan, S.-K. and Kahlweit, M. (1977). *Ber. Buns. phys. Chem.* **81,** 1294.
Chan, S.-K., Herrman, U., Ostner, W., and Kahlweit, M. (1977). *Ber. Buns. phys. Chem.* **81,** 60 and 396.
Chan, S.-K., Herrman, U., Ostner, W., and Kahlweit, M. (1978). *Ber. Buns. phys. Chem.* **82,** 380.
Clint, J. H. (1975). *J. chem. Soc. Faraday Trans. 1* **71,** 1327.
Corkill, J. M., Goodman, J. F., and Harrold, S. P. (1964). *Trans. Faraday Soc.* **60,** 202.
Debye, P. and Anacker, E. W. (1951). *J. phys. Colloid Chem.* **55,** 644.
Desnoyers, J. E., Caron, G., DeLisi, R., Roberts, D., Roux, A., and Perron, G. (1983). *J. phys. Chem.* **87,** 1397–1406.
Ekwall, P., Mandell, L., and Solyom, P. (1971). *J. Colloid interface Sci.* **36,** 519–28.
Evans, D. F., and Ninham, B. W. (1983). *J. phys. Chem.* **87,** 5025–32.
Evans, D. F., and Ninham, B. W. (1986). *J. phys. Chem.* **90,** 226–34.
Evans, D. F., Allen, M., Ninham, B. W., and Fouda, A. (1984). *J. Solution Chem.* **13,** 87–101.
Evans, D. F., Yamauchi, A., Roman, R., and Casassa, E. Z. (1982). *J. Colloid interface Sci.* **88,** 89–96.
Fernandez, M. S. and Fromherz, P. (1977). *J. phys. Chem.* **81,** 1755.
Fisher, L. R. and Oakenfull, D. G. (1977). *Chem. Soc. Rev.* **6,** 25.
Ford, W. P. J., Ottewill, R. H., and Parreira, H. C. (1966). *J. Colloid interface Sci.* **21,** 522.
Frank, H. S. and Evans, M. W. (1945). *J. Chem. Phys.* **13,** 507.
Franks, F. (1983). *Water.* Royal Society of Chemistry, London.
Fuerstenau, M. C. (1976). *Flotation,* Vols 1 and 2. Am. Inst. MMPE, New York.
Gaines, G. L. (1966). *Insoluble monolayers at liquid–gas interfaces.* Interscience, New York.
Gilanyi, T. (1973). *Acta chim. scand.* **27,** 729.
Grieser, F., and Tausch-Treml, R. (1980). *J. Am. chem. Soc.* **102,** 7258.
Gruen, D. W. R. (1981). *J. Colloid interface Sci.* **84,** 281.
Gunnarson, G., Jönsson, B., and Wennerström, H. (1980). *J. phys. Chem.* **84,** 3114–21.

Hall, D. G., and Pethica, B. A., (1967). Thermodynamics of micelle formation. In *Nonionic surfactants* (ed. M. Schick) Chapter 16. Marcel Dekker, New York.
Halle, B. and Carlström, G. (1981). *J. phys. Chem.* **85,** 2142.
Hartley, G. S. (1936). *Aqueous solutions of paraffin chain salts.* Herman et Cie, Paris.
Hartley, G. S. (1941). *Trans. Faraday Soc.* **37,** 130.
Hartley, G. S. and Roe, J. W. (1940). *Trans. Faraday Soc.* **36,** 101.
Hayter, J. B. and Penford, J. (1981). *J. chem. Soc. Faraday Trans. 1* **77,** 1851–63.
Hill, T. L. (1963). *Thermodynamics of small systems,* Vol. 1. Benjamin, New York.
Israelachvili, J. N., and Ninham, B. W. (1977). *J. Colloid interface Sci.* **58,** 14–25.
Israelachvili, J. N., Mitchell, D. J., and Ninham, B. W. (1976). *J. chem. Soc. Faraday Trans. 2* **72,** 1525.
Israelachvili, J. N., Mitchell, D. J., and Ninham, B. W. (1977). *Biochim. Biophys. Acta* **470,** 185.
Israelachvili, J. N., Marcelja, S., and Horn, R. G. (1980). *Q. Rev. Biophys.* **13,** 121.
Kahlweit, M. (1981). *Pure appl. Chem.* **53,** 2069.
Kalyanasundaram, K. and Thomas, J. K. (1977). *J. Am. chem. Soc.* **99,** 2039–44.
Kertes, A. S. (1977). Aggregation of surfactants in hydrocarbons. In *Micellization, solubilization and microemulsions* (ed. K. L. Mittal) Vol. 1, pp. 445–54. Plenum Press, New York.
Kishimoto, J. and Sumida, K. (1974). *Chem. pharm. Bull.* (*Japan*) **22,** 1108.
Kitahara, A. (1967). In *Nonionic surfactants* (ed. M. J. Schick) p. 289. Marcel Dekker, New York.
Kitchener, J. A. (1964). In *Recent progress in surface science* (ed. J. F. Danielli, K. Pankhurst, and A. C. Riddiford) Vol. 1. Academic Press, New York.
Kratohvil, J. (1980). *J. Colloid interface Sci.* **75,** 271–5.
Kresheck, G. C. (1975). Surfactants. In *Water—a comprehensive treatise* (ed. F. Franks) Chapter 2, pp. 95–167. Plenum Press, New York.
Leja, J. (1982). *Surface chemistry of froth flotation,* pp. 284–6. Plenum Press, New York.
Lindman, B., Lindblom, G., Wennerström, H., and Gustavsson, H. (1977). Ionic interactions in amphiphilic systems studied by n.m.r. In *Micellization, solubilization and microemulsions* (ed. K. L. Mittal) pp. 195–227. Plenum Press, New York.
Lindman, B. and Wennerström, H. (1980). *Topics in current chemistry* Vol. 87, pp. 1–83. Springer, Berlin.
Loeb, A. L., Wiersema, P. H., and Overbeek, J. Th. G. (1961). *The electrical double layer around a spherical colloid particle,* p. 37. MIT Press, Cambridge, Mass.
Mazer, N. A., Carey, M. C., and Benedek, G. B. (1977). The size, shape and thermodynamics of sodium dodecyl sulphate (SDS) micelles using quasielastic light-scattering spectroscopy. In *Micellization, solubilization and microemulsions* (ed. K. L. Mittal) Vol. 1, pp. 359–81. Plenum Press, New York.
McBain, J. W. (1950). *Colloid science.* D. C. Heath, Boston.
Menger, F. M., Jerkumica, J. M., and Johnson, J. C. (1978). *J. Am. chem. Soc.* **100,** 4676.
Missel, P. J., Mazer, N. A., Benedek, G. B., Young, C. Y., and Carey, M. C. (1980). *J. phys. Chem.* **84,** 1044–57.

Mitchell, D. J., and Ninham, B. W. (1981). *J. chem. Soc. Faraday Trans. 2* **77,** 601.
Mukerjee, P. (1974). *J. pharm. Sci.* **63,** 972.
Mukerjee, P. (1975). Differing patterns of self-association and micelle formation. In *Physical chemistry, enriching topics from colloid and surface science* (ed. H. van Olphen and K. J. Mysels) Chapter 9; IUPAC Commission I.6. Theorex, La Jolla, California.
Mukerjee, P. and Banerjee, K. (1964). *J. phys. Chem.* **68,** 3567.
Mukerjee, P. and Ghosh, A.-K. (1970). *J. Am. chem. Soc.* **92,** 6419.
Mukerjee, P. and Mysels, K. J. (1971). *Critical micelle concentrations of aqueous surfactant systems.* NSRDS–NBS **36,** National Bureau of Standards. US Government Printing Office, Washington, D.C.
Mukerjee, P. and Mysels, K. J. (1975). Anomalies of partially fluorinated surfactant micelles. *A.C.S. Symposia* (ed. K. L. Mittal) Series 9, p. 239. American Chemical Society, Washington, D.C.
Mukerjee, P., Cardinal, J. R., and Desai, N. R. (1977). The nature of the local microenvironments in aqueous micellar systems. In *Micellization, solubilization and microemulsions* (ed. K. L. Mittal) pp. 241–61. Plenum Press, New York.
Muller, N. (1977). Errors in micellization enthalpies from temperature dependence of c.m.c.s. In *Micellization, solubilization and microemulsions.* (ed. K. L. Mittal) Vol. 1, pp. 229–39. Plenum Press, New York.
Muller, N., and Birkhahn, R. H. (1967). *J. phys. Chem.* **71,** 957.
Muller, N., and Platko, F. E. (1971). *J. phys. Chem.* **75,** 547.
Mysels, K. J., and Otter, R. J. (1961). *J. Colloid Sci.* **16,** 462 and 474.
Ninham, B. W. (1981). *Pure app. Chem.* **53,** 2135.
Oakenfull, D. G. and Fenwick, D. E. (1974). *J. phys. Chem.* **78,** 1759.
Oakenfull, D. G. and Fisher, L. R. (1977). *J. phys. Chem.* **81,** 1838.
Oakenfull, D. G., and Fisher, L. R. (1978). *J. phys. Chem.* **82,** 2443–5.
Perrin, F. (1932). *Ann. Phys. Paris.* **17,** 283.
Phillips, J. N. (1955). *Trans. Faraday Soc.* **51,** 561.
Ramadan, M. S., Evans, D. F., and Lumry, R. (1983). *J. phys. Chem.* **87,** 4538–43.
Ray, A. and Nemethy, G. (1971). *J. Am. chem. Soc.* **93,** 6787.
Schick, M. J., ed. (1967). *Nonionic surfactants.* Marcel Dekker, New York.
Schott, H. (1964). *J. phys. Chem.* **68,** 3612.
Schwuger, M. (1971). *Ber. Buns. phys. Chem.* **75,** 167.
Shinoda, K. (1963). *Colloidal surfactants: some physico-chemical properties.* Academic Press, New York.
Shinoda, K. (1978). *Principles of solution and solubility.* Marcel Dekker, New York.
Small, D. M. (1968). *Adv. Chem. Ser.* **84,** 31.
Stigter, D., and Overbeek, J. T. G. (1957). *Proceedings of the Second International Congress on Surface Activity* Vol. I, p. 311.
Stigter, D. (1975a). Electrostatic interactions in aqueous environments. In *Physical chemistry: enriching topics from colloid and surface science* (ed. H. van Olphen and K. J. Mysels) Chapter 12; IUPAC Commission I.6. Theorex, La Jolla, California.
Stigter, D. (1975b). *J. phys. Chem.* **79,** 1015–22.
Stilbs, P., and Lindman, B. (1974). *J. Colloid interface Sci.* **46,** 177.
Swarbrick, J., and Daruwala, J. (1969). *J. phys. Chem.* **73,** 2627.
Swarbrick, J., and Daruwala, J. (1970). *J. phys. Chem.* **74,** 1293.

Tanford, C. (1972). *J. phys. Chem.* **76,** 3020.
Tanford, C. (1977). Thermodynamics of micellization of simple amphiphiles in aqueous media. In *Micellization, solubilization and microemulsions.* (ed. K. L. Mittal) pp. 119–32. Plenum Press, New York.
Tanford, C. (1980). *The hydrophobic effect. Formation of micelles and biological membranes* (2nd edn). Wiley, New York.
Tori, K., and Nakagawa, T. (1963). *Kolloid-Z. Z. Polym.* **188,** 47; **189,** 50.
Ts'o, P.O.P. (1968). In *Molecular associations in biology* (ed. B. Pullman). Academic Press, New York.
Turro, N. J., Geiger, M. W., Hautala, R. R., and Schore, N. E. (1977). Fluorescent probes for micellar systems. In *Micellization, solubilization and microemulsions* (ed. K. L. Mittal) pp. 75–86. Plenum Press, New York.
Turro, N. J., and Yekta, A. (1978). *J. Am. chem. Soc.* **100,** 5951.
Vikingstad, E., and Hoiland, H. (1978). *J. Colloid interface Sci.* **64,** 510.
Vold, R. D., and Vold, M. J. (1983). *Colloid and interface chemistry.* Addison-Wesley, Reading, Mass.
Warr, G. G., Grieser, F., and Healy, T. W. (1983). *J. phys. Chem.* **87,** 1220.
Wennerstöm, H., and Lindman, B. (1979). *Phys. Rep,* **I,** 1.
Wennerstöm, H., Lindman, B., Söderman, O., Drakenberg, T., and Rosenholm, J. B. (1979). *J. Am. chem. Soc.* **101,** 6860.
Zachariasse, K. A. (1978). *Chem. Phys. Lett.* **57,** 429.
Zachariasse, K. A., Nguyen van Phuc, and Kozankiewicz, B. (1981). *J. phys. Chem.* **85,** 2676–83.
Zana, R. (1978). *J. phys. Chem.* **82,** 2440–3.

APPENDICES

Appendix A1. London theory of dispersion forces

The instantaneous configuration of a molecule consisting of a set of nuclei and electrons with charges q_i $(i = 1, 2, \ldots)$ can be designated by specifying the positions $\boldsymbol{r}_i$ $(i = 1, 2, \ldots)$ of each charge relative to the centre of mass of the molecule. In general the net charge on the molecule is:

$$Q = \sum_i q_i \tag{A1.1}$$

and the instantaneous electric dipole moment is:

$$\boldsymbol{p} = \sum_i q_i \boldsymbol{r}_i. \tag{A1.2}$$

In principle at least, Schrödinger's equation for the isolated molecule can be solved to yield a set of eigenfunctions $|n\rangle$ and corresponding energy levels E_n. The state $|0\rangle$ corresponding to the lowest energy level E_0 is the ground state of the system. If the molecule in its ground state is neutral and non-polar then:

$$Q = 0 \tag{A1.3}$$

and

$$\langle \boldsymbol{p} \rangle_0 = \langle 0 | \boldsymbol{p} | 0 \rangle = 0 \tag{A1.4}$$

(We use here the compact bra$\langle|$ and ket$|\rangle$ notation instead of the more familiar wave function form: $\langle n|\boldsymbol{p}|m\rangle = \int \psi_n^* \boldsymbol{p} \psi_m \, \mathrm{d}\tau$).
Consider two such molecules, A and B, in their respective ground states and originally an infinite distance apart, brought to a separation distance R. In general, the charge distribution that is molecule A will interact with that of molecule B to change the energy of the system from $E_0^A + E_0^B$ to some new value $E^{AB}(R)$. The interaction energy $V_{int}(R)$ is defined by

$$V_{int}(R) = E^{AB}(R) - E_0^A - E_0^B. \tag{A1.5}$$

To solve the quantum mechanical problem of two neutral, non-polar molecules A and B at a fixed separation distance R, we need to specify the interaction of the molecular charge distributions to obtain the interaction energy operator $\mathscr{V}_{int}$ in the total Hamiltonian for the system. We write:

$$\mathscr{V}_{int} = \frac{1}{4\pi\varepsilon_0} \sum_i \sum_j \frac{q_i^A q_j^B}{|\boldsymbol{R} + \boldsymbol{r}_j^B - \boldsymbol{r}_i^A|} \tag{A1.6}$$

where $\boldsymbol{R}$ is the position of the centre of mass of molecule B relative to the centre of mass of molecule A and ε_0 is the dielectric permittivity of free space. The quantity $|\boldsymbol{R} + \boldsymbol{r}_j^B - \boldsymbol{r}_i^A|$ is just the distance between charge q_i^A in molecule A and charge q_j^B in molecule B (see Fig. A1.1). Thus eqn (A1.6) is the Coulombic interaction energy of the instantaneous charge distribution that is molecule A with that of molecule B. Note that we have not included any spin–spin interactions and, more importantly, that we have used an *electrostatic* approximation for the electromagnetic interaction energy. We discuss this point in section 4.5.

If the intermolecular distance R is greater than $|\boldsymbol{r}_i^A|$ and $|\boldsymbol{r}_j^B|$ then eqn (A1.6) can be expanded in powers of $1/R$. The $0(1/R)$ term of eqn (A1.6) is just

$$\left(\sum_i q_i^A\right)\left(\sum_j q_j^B\right) \Big/ (4\pi\varepsilon_0 R)$$

which vanishes since we are concerned here with neutral molecules (eqn (A1.3)).

The $0(1/R^2)$ term vanishes for the same reason. Thus we may write to leading order in $(1/R)$:

$$\mathscr{V}_{int} \simeq \frac{1}{4\pi\varepsilon_0} (\boldsymbol{p}^A \cdot \boldsymbol{p}^B / R^3 - 3(\boldsymbol{p}^A \cdot \boldsymbol{R})(\boldsymbol{p}^B \cdot \boldsymbol{R})/R^5) \tag{A1.7}$$

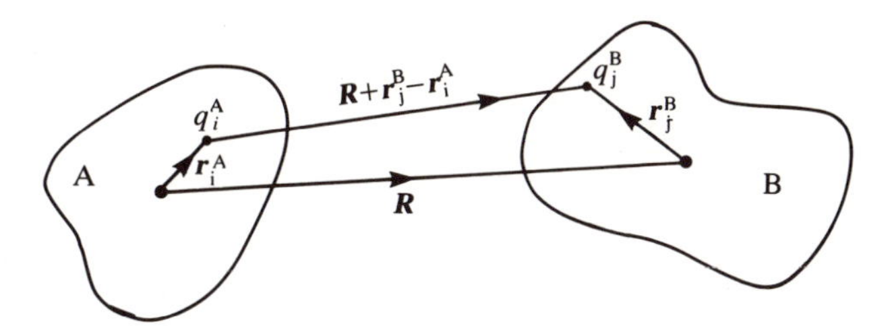

FIG. A1.1. Interaction between atoms in body A and body B.

where $\boldsymbol{p}^{\mathrm{A}}$ and $\boldsymbol{p}^{\mathrm{B}}$ are the instantaneous dipole moments of molecules A and B as defined in eqn (A1.2). It is sufficient for our purposes to retain only the instantaneous dipole–dipole interaction term in $\mathcal{V}_{\mathrm{int}}$. Higher-order terms in this expansion for $\mathcal{V}_{\mathrm{int}}$ represent instantaneous dipole–quadrupole, quadrupole–quadrupole etc. interactions and are usually negligible. For a more complete discussion of the expansion see Dalgarno and Davison (1966) or Hirschfelder *et al.* (1954).

Even with the simplified expression eqn (A1.7) for $\mathcal{V}_{\mathrm{int}}$ the general solution of Schrödinger's equation for the two-molecule system is intractable and we must resort to perturbation theory to obtain $E^{\mathrm{AB}}(R)$ (Davydov 1965). In first-order perturbation theory

$$E^{\mathrm{AB}}(R) = E_0^{\mathrm{A}} + E_0^{\mathrm{B}} + \langle 0, 0|\, \mathcal{V}_{\mathrm{int}}\, |0, 0\rangle \tag{A1.8}$$

where $|0, 0\rangle$ represents the unperturbed ($\mathcal{V}_{\mathrm{int}} = 0$) state of the AB system and is given by

$$|0, 0\rangle = |0^{\mathrm{A}}\rangle\, |0^{\mathrm{B}}\rangle \tag{A1.9}$$

where $|0^{\mathrm{A}}\rangle$ is the ground state wave function of the isolated molecule A. Thus, in first order theory

$$V_{\mathrm{int}}^{(1)}(R) = \frac{1}{4\pi\varepsilon_0}\{(\langle \boldsymbol{p}^{\mathrm{A}}\rangle_0 \cdot \langle \boldsymbol{p}^{\mathrm{B}}\rangle_0)/R^3 - 3(\langle \boldsymbol{p}^{\mathrm{A}}\rangle_0 \cdot \boldsymbol{R})(\langle \boldsymbol{p}^{\mathrm{B}}\rangle_0 \cdot \boldsymbol{R})/R^5\} \tag{A1.10}$$

$$= 0 \tag{A1.11}$$

for non-polar molecules by virtue of eqn (A1.4). Note that if we were concerned with polar molecules,

$$\langle \boldsymbol{p}\rangle_0 = \boldsymbol{\mu} \tag{A1.12}$$

where $\boldsymbol{\mu}$ is the permanent electric dipole moment of the molecule. In this case eqn (A1.10) would be just the permanent dipole–dipole interaction energy. Therefore, first order theory will produce a non-vanishing interaction energy only when the individual molecules possess non-vanishing expectation values for charge, dipole moment, quadrupole moment etc.

To achieve a non-zero $V_{\mathrm{int}}(R)$ in the absence of permanent multipole moments, it is necessary to calculate $E^{\mathrm{AB}}(R)$ to second order in the perturbation $\mathcal{V}_{\mathrm{int}}$ i.e. to consider how $\mathcal{V}_{\mathrm{int}}$ causes the internal state of each molecule to change. To this order of approximation

$$V_{\mathrm{int}}^{(2)}(R) = -\sum_{\substack{m,n\\(\neq 0)}} \frac{|\langle 0, 0|\, \mathcal{V}_{\mathrm{int}}\, |m, n\rangle|^2}{E_m^{\mathrm{A}} + E_n^{\mathrm{B}} - E_0^{\mathrm{A}} - E_0^{\mathrm{B}}} \tag{A1.13}$$

where

$$|m, n\rangle = |m^{\mathrm{A}}\rangle\, |n^{\mathrm{B}}\rangle. \tag{A1.14}$$

Note that $V_{\mathrm{int}}^{(2)}(R)$ is always negative since $E_{m,n} > E_{0,0}$. This is a direct consequence of the original molecules being in their ground states and does not hold in general if the interacting molecules are in excited states initially.

To simplify eqn (A1.13), we will assume that the molecules A and B are symmetrical. Clearly such molecules will automatically satisfy eqn (A1.4).

Substituting eqn (A1.7) into (A1.13) we obtain the result:

$$V_{\text{int}}^{(2)}(R) = -\frac{C_{\text{AB}}}{R^6} \tag{A1.15}$$

where the constant C^{AB} is given by

$$C_{\text{AB}} = \frac{3e^4\hbar}{2m_e^2(4\pi\varepsilon_0)^2} \sum_{\substack{m,n \\ (\neq 0)}} \frac{f_{0m}^{\text{A}} f_{0n}^{\text{B}}}{\omega_{0m}^{\text{A}} \omega_{0n}^{\text{B}} (\omega_{0m}^{\text{A}} + \omega_{0n}^{\text{B}})} \tag{A1.16}$$

and

$$\omega_{0m} = (E_m - E_0)/\hbar \tag{A1.17}$$

is the frequency (rad s^{-1}) of electromagnetic radiation that would cause the transition from the ground state $|0\rangle$ to the excited state $|m\rangle$ in the isolated molecule. The corresponding oscillator strengths f_{0m} defined by

$$f_{0m} = \frac{2m_e\omega_{0m}}{\hbar e^2} |\langle 0| p_z |m\rangle|^2 \tag{A1.18}$$

(e and m_e are the electron charge and mass respectively) involve only the expectation value of one of the Cartesian components of the vector dipole moment operator $\boldsymbol{p} = (p_x, p_y, p_z)$, since the sphericity assumption means that

$$|\langle 0| p_x |m\rangle| = |\langle 0| p_y |m\rangle| = |\langle 0| p_z |m\rangle|. \tag{A1.19}$$

If the molecules A and B are not symmetric the expression (A1.16) must be modified to include the orientation of molecules A and B relative to the line of centres (Hirschfelder *et al.* 1954).

It should be noted that in the approximation leading to eqn (A1.7), the expression (A1.15) is only the first term in an infinite series of the form

$$V_{\text{int}}(\text{R}) = -\sum_{l_1=1} \sum_{l_2=1} \frac{C_{l_1 l_2}}{R^{2(l_1+l_2+1)}}. \tag{A1.20}$$

In colloid science applications, it is an article of faith that only the $l_1 = l_2 = 1$ term is important in this sum. For a discussion of the magnitude of the higher order terms, the reader is referred to Hirschfelder *et al.* (1954).

References

Dalgarno, A. and Davison, W. D. (1966). *Adv. Atom. mol. Phys.* **2,** 1.
Davydov, A. S. (1965). *Quantum mechanics*. Pergamon Press, Oxford.
Hirschfelder, J. O., Curtis, C. F., and Bird, R. B. (1954). *Molecular theory of gases and liquids*. Wiley, New York.

Appendix A2. Evaluation of the sum of the roots of the dispersion relation

Cauchy's integral theorem states that the value of a function $f(\omega)$ of a complex variable ω at a point (say $a = \eta + i\xi$) can be obtained from its values on the curve

C using the expression

$$f(a) = \frac{1}{2\pi i} \oint_C \frac{f(\omega)\,\mathrm{d}\omega}{\omega - a} \tag{A2.1}$$

where C is any closed curve surrounding the point a, and $f(\omega)$ is analytic in the region containing C.

The solutions, ω_j, of the dispersion equation (4.7.7) must satisfy:

$$\mathfrak{D}(\omega) = \prod_j (\omega - \omega_j) = 0 \tag{A2.2}$$

which may be written

$$\ln \mathfrak{D}(\omega) = \sum_j \ln(\omega - \omega_j). \tag{A2.3}$$

Differentiating with respect to ω gives:

$$\frac{\mathfrak{D}'(\omega)}{\mathfrak{D}(\omega)} = \sum_j \frac{1}{\omega - \omega_j}. \tag{A2.4}$$

Now putting $f(\omega) = \omega$ and $a = \omega_j$ in eqn (A2.1) gives:

$$f(a) = \omega_j = \frac{1}{2\pi i} \oint_C \frac{\omega\,\mathrm{d}\omega}{\omega - \omega_j}$$

and, hence:

$$\sum \omega_j = \frac{1}{2\pi i} \oint_C \omega \sum \left(\frac{1}{\omega - \omega_j}\right) \mathrm{d}\omega = \frac{1}{2\pi i} \oint_C \omega \frac{\mathfrak{D}'(\omega)}{\mathfrak{D}(\omega)} \mathrm{d}\omega \tag{A2.5}$$

using eqn (A2.4). The closed curve (or contour) C must exclude any poles in the function $\mathfrak{D}(\omega)$. The most convenient choice is a semicircle in the complex ω plane with centre at the origin and radius R. As R increases to infinity (Fig. A2.1), the semicircle encompasses all of the zeros of $\mathfrak{D}(\omega)$ on the real axis and

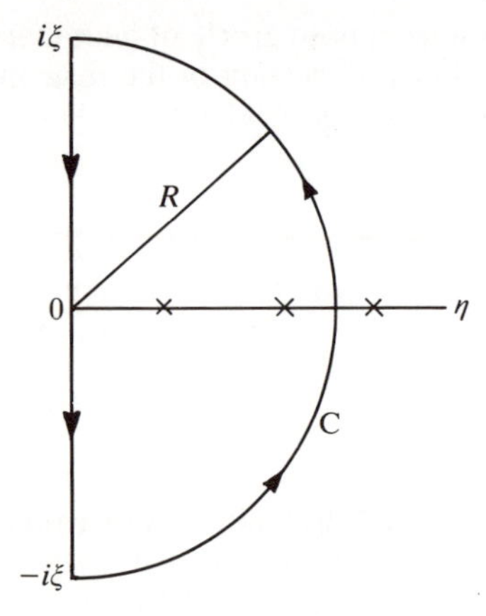

FIG. A2.1. The contour around which integration yields the sum of the zeros of the dispersion relation.

the contribution to the integral from the circular part of the contour decreases to zero. What remains then is the line integral along the imaginary axis:

$$U_d = \tfrac{1}{2}\sum \hbar\omega_j(d) = \frac{-\hbar}{2(2\pi i)}\int_{-i\infty}^{i\infty} \frac{\omega\mathfrak{D}'(\omega)}{\mathfrak{D}(\omega)}\,d\omega. \tag{A2.6}$$

Introducing the value of U for infinite separation and integrating by parts leads to (Ninham and Parsegian 1970):

$$U_d - U_\infty = \frac{\hbar}{4\pi i}\int_{-i\infty}^{i\infty} \ln\frac{\mathfrak{D}(\omega)}{\mathfrak{D}_\infty(\omega)}\,d\omega. \tag{A2.7}$$

This expression must then be summed over all possible values of the wave vector Q, which gives:

$$V_A = \frac{\hbar}{8\pi^2 i}\int_0^\infty Q\,dQ\int_{-i\infty}^{i\infty} \ln\left[1 - \left\{\frac{\varepsilon(\omega)-1}{\varepsilon(\omega)+1}\right\}^2\right] e^{-2QL}\,d\omega. \tag{A2.8}$$

Substituting $\omega = i\xi$ and $x = 2QL$ then gives:

$$V_A = \frac{\hbar}{32\pi^2 L^2}\int_0^\infty x\,dx\int_{-\infty}^{\infty} \ln\left[1 - \left\{\frac{\varepsilon(i\xi)-1}{\varepsilon(i\xi)+1}\right\}^2\right] e^{-x}\,d\xi.$$

This expression has many of the features of the more exact solution (eqn (4.7.4)) but it lacks the temperature-dependence that is introduced by calculating the Helmholtz free energy change $F_d - F_\infty$ rather than the internal energy change. There are added mathematical complications in that case because the sinh function (eqn (4.7.27)) has branch points along the imaginary axis and these must be circumvented.

(The above treatment is adapted from some lectures given by Dr. Sam Levine in the University of Sydney in 1972.)

References

Hunter, R. J. (1975). Electrochemical aspects of colloid chemistry. In *Modern aspects of electrochemistry* (ed. B. E. Conway and J. O'M. Bockris) No. 11, Chapter 2. Plenum Press, New York.

Ninham, B. W. and Parsegian, V. A. (1970). *J. Chem. Phys.* **52,** 4578 and **53,** 3398.

Appendix A3. Vector calculus and Poisson's equation

The basic equations of electrostatic potential theory (and of hydrodynamics) are couched in the language of vector calculus. In this appendix we review a few of the salient features of that subject in order to interpret the Poisson equation. In Chapter 9 a more elaborate discussion is required to permit the proper statement of the equations of hydrodynamics. Much of the material in this appendix is taken from Hunter (1981).

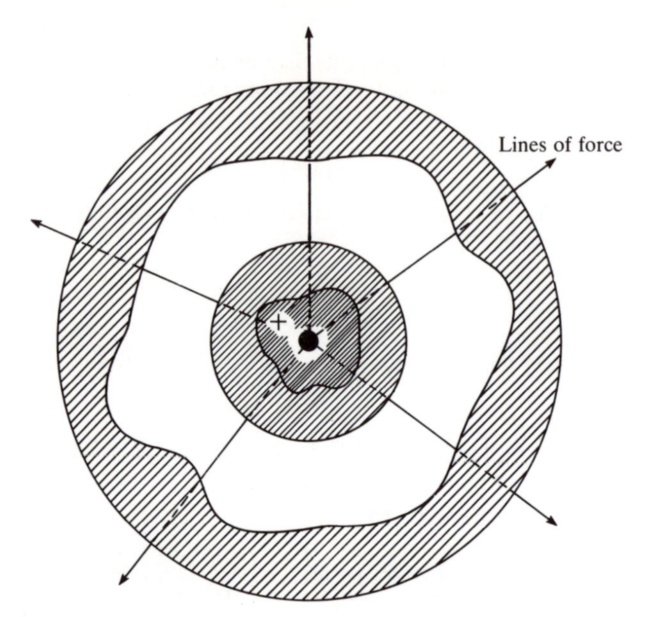

FIG. A3.1. Equipotential surfaces surrounding a point charge.

A3.1 *Scalar and vector fields*

The electrostatic potential at any point in space can be represented by a number, with suitable units attached. In general it will vary with position and can be represented by a function $\psi(x_1, x_2, x_3)$ or in polar coordinates $\psi(r, \theta, \phi)$. The values of the potential are said to form a *scalar field.* The temperature of a body is another example of a scalar field. On the other hand, the values of the velocity at each point in a fluid must be specified in terms of a number or magnitude *and* a direction. They, therefore, form a *vector field*.

A3.2 *The gradient of a scalar field*

Consider the electrostatic potential in the region around a point charge. It is the same at all points on a sphere centred on the charge (Fig. A3.1). Concentric spheres can be drawn around the central charge and each will be an *equipotential surface,* with the potential decreasing as the sphere increases in size. The lines of force emanating from the charge always pass through these equipotential surfaces at right angles. For an arbitrary charge or potential distribution it is still possible to imagine equipotential surfaces, of more exotic shape, obtained by linking together points with the same potential.

The *gradient* of the potential measures how rapidly the potential changes when we move from one equipotential surface to a nearby one *along the normal* to the equipotential surface. It is a vector quantity, which is written:

$$\text{grad } \psi = \nabla \psi$$

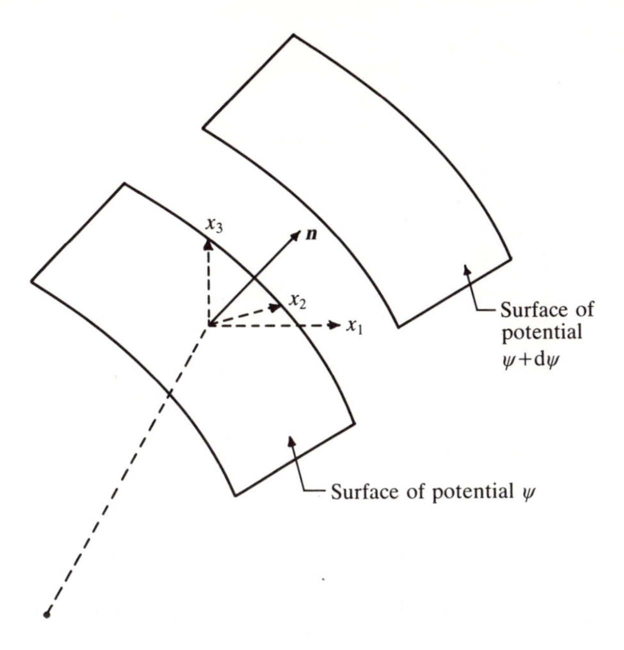

FIG. A3.2. The gradient of the potential.

and in Cartesian coordinates (Fig. A3.2):

$$\text{grad}\ \psi = \left(\frac{\partial \psi}{\partial x_1}, \frac{\partial \psi}{\partial x_2}, \frac{\partial \psi}{\partial x_3}\right). \tag{A3.1}$$

The symbol ∇ (called 'del') represents the operator

$$\left(\frac{\partial}{\partial x_1}, \frac{\partial}{\partial x_2}, \frac{\partial}{\partial x_3}\right)$$

which in this case operates on a scalar function to produce a vector. It is a very useful concept because it can be manipulated very easily to produce results of great importance. In many respects it behaves formally like a vector with the components shown. It can, for example, be combined with another vector in a scalar product:

$$\nabla \cdot \boldsymbol{A} = \left(\frac{\partial A_1}{\partial x_1} + \frac{\partial A_2}{\partial x_2} + \frac{\partial A_3}{\partial x_3}\right)$$

which is analogous to the scalar (or inner or dot) product (obtained by multiplying together the x_1, the x_2, and the x_3 components of the two vectors and summing the result). Note that this produces a *scalar* quantity:

$$\boldsymbol{B} \cdot \boldsymbol{A} = B_1A_1 + B_2A_2 + B_3A_3.$$

Later on (section A3.5) we will see that the operator ∇ can also form a cross (or vector) product.

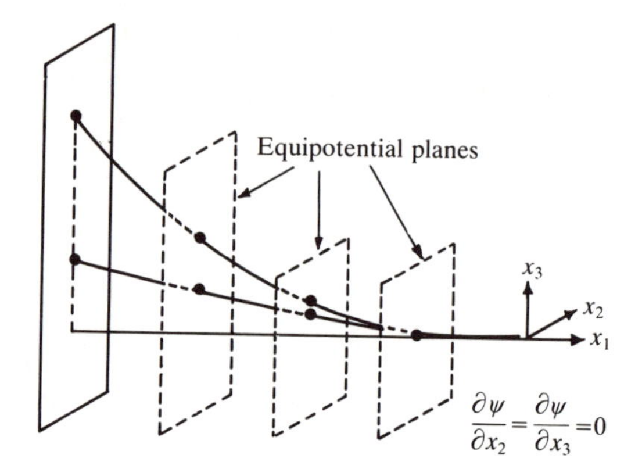

FIG. A3.3. Potential in the neighbourhood of a flat plate.

The negative of *grad* ψ measures the electric field strength $\boldsymbol{E}$ at each point and the values of $\boldsymbol{E}$, in general, vary from place to place, forming a vector field. The direction of the field is indicated by the lines of force which, as noted above, are always normal to the equipotential surfaces.

In the special case of an infinitely large, uniformly charged plate immersed in an electrolyte, ψ is constant in any plane parallel to the plate (Fig. A3.3) so that $\partial\psi/\partial x_2 = \partial\psi/\partial x_3 = 0$ and then

$$\boldsymbol{E} = (E_1, E_2, E_3) = -\text{grad}\ \psi = -\frac{\mathrm{d}\psi}{\mathrm{d}x_1} \tag{A3.2}$$

In this case the lines of force run out perpendicular to the plate. The strength of the field can be indicated by the number of lines of force passing through unit area of each plane and in this case it diminishes steadily to zero in the bulk electrolyte. The reason for this is discussed in the next section. (Actually eqn (A3.2) must be slightly modified in the SI system of units, even *in vacuo,* to include the permittivity, ε, and this point is also discussed in more detail below.)

A3.3 *Divergence of a vector field*

Consider an infinitesimal cube of material at the point $\boldsymbol{x} = (x_1, x_2, x_3)$ (Fig. A3.4). If the field strength on the side ABCD is E_1, then on the opposite side it will be

$$E_1 + \left[\frac{\partial E_1}{\partial x_1}\right] \mathrm{d}x_1$$

and the net change in E_1 in the x_1 direction is

$$\left[\frac{\partial E_1}{\partial x_1}\right] \mathrm{d}x_1.$$

This will, therefore, be equal to the difference in the number of lines of force per

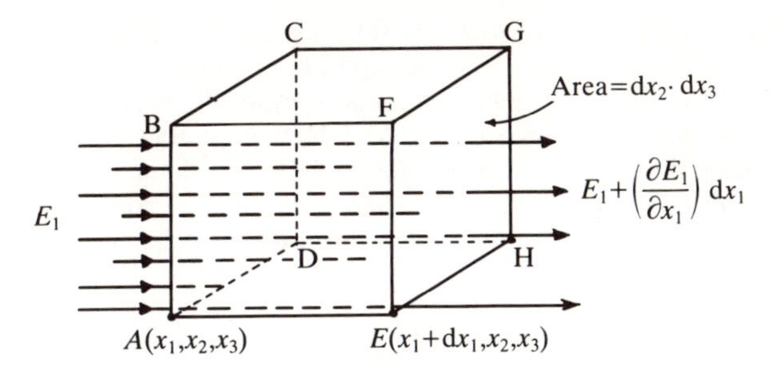

FIG. A3.4. The flux of a vector field.

unit area emanating from the side EFGH compared to the number which went into side ABCD. Multiplying by the area $dx_2 \cdot dx_3$ we obtain the *net flux* (or flow) of lines of force out of the box in the x_1 direction.

The same calculation can be made in the other two directions so that the total flux out of this small volume is

$$\left(\frac{\partial E_1}{\partial x_1}+\frac{\partial E_2}{\partial x_2}+\frac{\partial E_3}{\partial x_3}\right) dx_1\, dx_2\, dx_3.$$

The total flux *per unit volume* at the point $\boldsymbol{x}$ is then

$$\frac{\partial E_1}{\partial x_1}+\frac{\partial E_2}{\partial x_2}+\frac{\partial E_3}{\partial x_3} = \operatorname{div} \boldsymbol{E} = \nabla \cdot \boldsymbol{E}. \tag{A3.3}$$

The flux per unit volume of a vector field is called the *divergence* or *div* for short. A very obvious application of the divergence operator is the continuity equation (section 9.2):

$$\operatorname{div} \boldsymbol{v} = \nabla \cdot \boldsymbol{v} = 0 \tag{A3.4}$$

where $\boldsymbol{v}$ is the velocity at a point in an incompressible fluid. Equation (A3.4) expresses the fact that what flows into a volume element must also flow out (so that the net flux of material is zero). For a charge-free region of space $\operatorname{div} \boldsymbol{E}$ is also zero because the number of lines of force representing $\boldsymbol{E}$ is affected only by the presence of electric charges.

When a vector field can be represented as the gradient of a scalar potential, as is the case with the electric field vector $\boldsymbol{E}$, then we can write:

$$\begin{aligned}\operatorname{div} \boldsymbol{E} &= \operatorname{div}(-\operatorname{grad} \psi)\\ &= -\left(\frac{\partial}{\partial x_1}\left(\frac{\partial \psi}{\partial x_1}\right)+\frac{\partial}{\partial x_2}\left(\frac{\partial \psi}{\partial x_2}\right)+\frac{\partial}{\partial x_3}\left(\frac{\partial \psi}{\partial x_3}\right)\right)\\ &= -\left[\frac{\partial^2 \psi}{\partial x_1^2}+\frac{\partial^2 \psi}{\partial x_2^2}+\frac{\partial^2 \psi}{\partial x_3^2}\right]\\ &= -\nabla \cdot \nabla \psi = -\nabla^2 \psi. \end{aligned} \tag{A3.5}$$

Notice that ∇ is here involved in a scalar product with the vector $\nabla \psi$ but the effect is the same as if ∇ is regarded as forming a scalar product with itself to produce a new operator: $\nabla \cdot \nabla \equiv \nabla^2$. This operator, called the Laplace operator, occurs so frequently that it is given a special symbol $\boldsymbol{\Delta}$, which means *div grad.* (Do not confuse this with the symbol Δ, which is simply a difference.) The advantage of using $\boldsymbol{\Delta}$ is that it is independent of the coordinate system being used whereas the symbol ∇^2 implies that Cartesian (rectangular) coordinates are being used (Kemmer, 1977)—a distinction that is not always made.

In a charge-free region of space

$$\boldsymbol{\Delta}\psi = \nabla^2\psi = 0 \tag{A3.6}$$

which is Laplace's equation.

A3.4 *Poisson's equation*

When electric charges are immersed in a dielectric medium (like water) the strength of the electric field is significantly reduced because the molecular dipoles tend to align themselves in such a way as to cancel part of the field. The capacity of a substance to affect the electric field strength is measured by its *permittivity,* ε. To correct for this effect we introduce a new vector called the *dielectric displacement,* $\boldsymbol{D}$, defined by (section 2.2):

$$\boldsymbol{D} = \varepsilon \boldsymbol{E}. \tag{A3.7}$$

For water ε is about 80 times larger than its value *in vacuo,* so $\boldsymbol{D}$ is much larger than $\boldsymbol{E}$. For a charge distribution *in vacuo* Poisson's equation used to read:

$$\operatorname{div} \boldsymbol{D} = \operatorname{div} \boldsymbol{E} = \nabla \cdot \boldsymbol{E} = \nabla \cdot (-\nabla\psi) = \rho$$

or

$$\boldsymbol{\Delta}\psi = \nabla^2\psi = -\rho \tag{A3.8}$$

because the permittivity of a vacuum was taken as unity. Equation (A3.8) says, in effect, that the flux of the field into any element of volume is determined by the amount of charge in that volume. To put it another way, the lines of force of the electric field begin on elements of positive charge and terminate on elements of negative charge.

In the modern system (SI) of units, consistency in units is achieved at the price of assigning a value of $\varepsilon_0 = 8.85 \times 10^{-12}\,\mathrm{F\,m^{-1}}$ for the permittivity of a vacuum. Poisson's equation is then, *in vacuo*:

$$\boldsymbol{\Delta}\psi = \nabla^2\psi = -\frac{\rho}{\varepsilon_0}. \tag{A3.9}$$

The corresponding expression for charges immersed in a dielectric medium is:

$$\operatorname{div} \boldsymbol{D} = \operatorname{div}(-\varepsilon \operatorname{grad} \psi) = \rho. \tag{A3.10}$$

If ε can be regarded as a constant, independent of position, then eqn (A3.10) becomes:

$$\operatorname{div} \operatorname{grad} \psi = \boldsymbol{\Delta}\psi = -\frac{\rho}{\varepsilon}. \tag{A3.11}$$

In the case of the (positively charged) plate immersed in an electrolyte referred to above (section A3.2), the excess of negative ions means that $\rho < 0$ so that div $\boldsymbol{D}$ (and hence div $\boldsymbol{E}$) is also negative (from eqn (A3.10)), i.e. more lines of force go into each volume element than come out. The others are swallowed up by the negative ions.

A3.5 *Vector (cross) product and the curl of a vector field*

The cross product of two vectors $\boldsymbol{b} \times \boldsymbol{a}$ can be represented

$$\boldsymbol{b} \times \boldsymbol{a} = \begin{vmatrix} \hat{e}_1 & \hat{e}_2 & \hat{e}_3 \\ b_1 & b_2 & b_3 \\ a_1 & a_2 & a_3 \end{vmatrix}$$

$$= (b_2a_3 - b_3a_2)\hat{e}_1 + (b_3a_1 - b_1a_3)\hat{e}_2 + (b_1a_2 - b_2a_1)\hat{e}_3 \qquad \text{(A3.12)}$$

where $\hat{e}_1$, $\hat{e}_2$, $\hat{e}_3$ are unit vectors along the three coordinate directions.

The curl of a vector field is defined as

$$\text{curl}\, \boldsymbol{v} = \nabla \times \boldsymbol{v} = \begin{vmatrix} \hat{e}_1 & \hat{e}_2 & \hat{e}_3 \\ \dfrac{\partial}{\partial x_1} & \dfrac{\partial}{\partial x_2} & \dfrac{\partial}{\partial x_3} \\ v_1 & v_2 & v_3 \end{vmatrix}$$

$$= \left[\left(\frac{\partial v_3}{\partial x_2} - \frac{\partial v_2}{\partial x_3}\right), \left(\frac{\partial v_1}{\partial x_3} - \frac{\partial v_3}{\partial x_1}\right), \left(\frac{\partial v_2}{\partial x_1} - \frac{\partial v_1}{\partial x_2}\right)\right] \qquad \text{(A3.13)}$$

where the bracketed terms are the components of a new vector field, which may itself have a curl. Note the similarity between these components and those of $\boldsymbol{\Omega}$ in eqn (9.5.14). The curl of a vector field $\boldsymbol{v}$ (sometimes written rot $\boldsymbol{v}$ in older texts) measures what is called the *circulation* of the field. In the case of the vector describing the velocity of a fluid, the circulation has a simple interpretation: it measures at each point the extent to which the field is causing the liquid to circulate around that point, i.e. the local rotational component of the flow. An extreme situation would be that of a whirlpool or vortex (though non-zero circulation is not always so obvious; see Fig. A3.5).

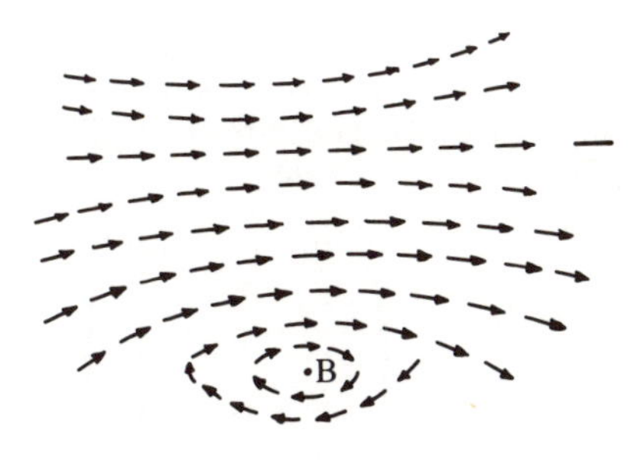

FIG. A3.5. The direction of a fluid velocity field showing a large circulation around the point, B (i.e. curl $\boldsymbol{v}$ is large in the neighbourhood of B.)

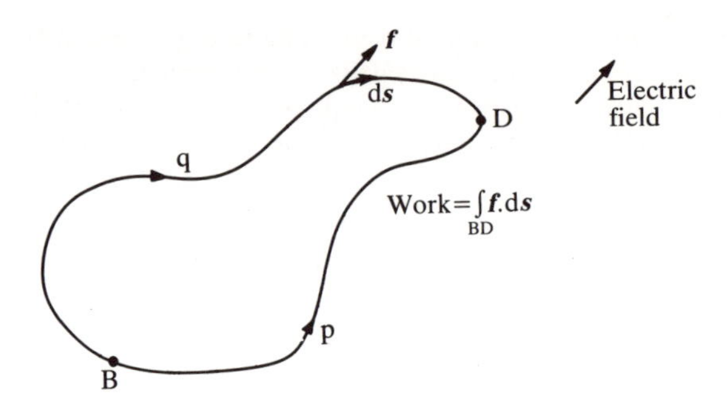

FIG. A3.6. Work done in going from B to D along p is the same as from B to D along q so the integral around the closed curve B→D→B is zero.

The line integral. If the vector being considered is a force then the line integral of that force along any curve (or path) measures the work done in moving through the field. The line integral may be called a *work integral.* We can convert the electrical field vector into a force at each point by determining the force caused by the field on a unit charge. The integral of this electrical force along any closed curve is then zero (Fig. A3.6) because the work done in going from point B to point D is independent of the path. The electric field is said to be a *conservative field*.

The notion of a work integral can be generalized to include any vector field. The line integral of the vector field around a closed curve surrounding a point in the field then measures the circulation of the field around the point. As noted above, this is measured by the *curl* of the field at that point (via another theorem due to Stokes). The electrical field is a particular case of a field for which the curl is zero everywhere (Exercise A3.1). Any vector field that can be represented as the gradient of a (scalar) potential has this property (i.e. curl grad $\phi = 0$ for any scalar ϕ). Such a field is said to be *irrotational.* On the other hand the velocity in a viscous fluid will in general be described by a rotational field (i.e. curl $\boldsymbol{v} \neq 0$).

A3.6 *Gradient of a vector*

To specify the gradient of a vector quantity $\boldsymbol{v}$ (assuming it is a function of three position coordinates) we need to specify how each of its components is changing in each of the three different coordinate directions. To do so requires a 3×3 matrix of coefficients (called a *tensor*). Thus

$$\operatorname{grad} \boldsymbol{v} = \nabla \boldsymbol{v} = \begin{pmatrix} \dfrac{\partial v_1}{\partial x_1} & \dfrac{\partial v_2}{\partial x_1} & \dfrac{\partial v_3}{\partial x_1} \\ \dfrac{\partial v_1}{\partial x_2} & \dfrac{\partial v_2}{\partial x_2} & \dfrac{\partial v_3}{\partial x_2} \\ \dfrac{\partial v_1}{\partial x_3} & \dfrac{\partial v_2}{\partial x_3} & \dfrac{\partial v_3}{\partial x_3} \end{pmatrix} \tag{A3.14}$$

The divergence of this quantity $\nabla \cdot \nabla \boldsymbol{v}$ is again a vector arrived at by applying the same principle as is used in eqn (A3.5). The first component of that vector is then

$$((\partial^2 v_1/\partial x_1^2) + (\partial^2 v_1/\partial x_2^2) + (\partial^2 v_1/\partial x_3^2)).$$

The operator $\nabla \cdot \nabla$ applied to a vector thus has the effect of applying the Laplace operator ∇^2 to each of the three components of the original vector in turn to produce a new vector.

Exercises. A3.1 Show that if vector $\boldsymbol{a} = \text{grad } \phi$ where ϕ is a scalar, then curl $\boldsymbol{a} = 0$.

A3.2 Show that $\nabla^2 \boldsymbol{v} = \nabla(\nabla \cdot \boldsymbol{v}) - \nabla \times \nabla \times \boldsymbol{v}$ and hence show that for an incompressible fluid $\eta \nabla^2 \boldsymbol{v}$ can be written $-\eta$ curlcurl $\boldsymbol{v}$. (This form is often used in statements of the Navier–Stokes equation.)

References

Hunter, R. J. (1981). *Zeta potential in colloid science,* Appendix A1. Academic Press, London.

Kemmer, N. (1977). *Vector analysis—A physicist's guide to the mathematics of fields in three dimensions.* Cambridge University Press, Cambridge.

Appendix A4. Electrical units

To reconcile eqn (A3.11) with the form of Poisson's equation used in the older literature we shall initially write it in the form:

$$\Delta \psi = \nabla^2 \psi = \frac{-1}{4\pi\varepsilon_0} \cdot \frac{4\pi\rho}{\varepsilon_r} \tag{A4.1}$$

where $\varepsilon = \varepsilon_0 \varepsilon_r$ and ε_r is the (dimensionless) dielectric constant, often symbolized by D or K, which measures the relative permittivity. The factor 4π appears because the earlier unit systems were 'unrationalized'. 'Rationalization' is achieved by defining the magnitude of the force between two charged particles in a dielectric as:

$$f = \frac{q_1 q_2}{4\pi\varepsilon_0\varepsilon_r r^2} = \frac{1}{4\pi} \cdot \frac{q_1 q_2}{\varepsilon r^2}. \tag{A4.2}$$

In this way the factor 4π appears in formulae involving spherical symmetry and disappears from those involving flat plates. Whenever ε_r, K, or D appears in the old formulae we replace it by $4\pi\varepsilon_0\varepsilon_r$ with the following results, e.g:

1. The potential of a charged sphere immersed in a dielectric used to be written:

$$\psi = \frac{Q}{Kr} \tag{A4.3a}$$

but now becomes:

$$\psi = \frac{Q}{4\pi\varepsilon_0\varepsilon_r r} = \frac{1}{4\pi} \frac{Q}{\varepsilon r} \tag{A4.3b}$$

where Q is the total charge on the sphere and r is the radius.

2. The capacity per unit area, C, of a parallel plate condenser used to be written:

$$C = \frac{\sigma}{\psi} = \frac{K}{4\pi d} \tag{A4.4a}$$

which now becomes:

$$C = \frac{\sigma}{\psi} = \frac{4\pi\varepsilon_0\varepsilon_r}{4\pi d} = \frac{\varepsilon}{d} \tag{A4.4b}$$

where σ is the charge per unit area, ψ is the potential difference between the plates and d is the plate separation.

We use the symbol ε_r to represent the relative permittivity to conform to the common usage of physicists. The same quantity is still often written K in order to emphasize the fact that K has *no* units and is, therefore, a different kind of quantity to ε_0, the permittivity of a vacuum (cf. Atkins 1978). Since the introduction of the rationalized system of units simplifies the electrokinetic equations (just like eqn (A4.4b) above), we shall frequently cancel the factor 4π and often replace the composite $\varepsilon_0\varepsilon_r$ by the symbol ε, which must be understood to carry its units with it. The formulae that appear throughout the text (especially those in Chapters 6 and 9) should then be used with SI units. To regenerate the older formulae it is necessary to replace ε, wherever it appears, by $4\pi\varepsilon_0 K/4\pi$ and then to set the quantity $(4\pi\varepsilon_0)$ equal to unity with appropriate units attached. For example, to use the CGS system, we see from eqn (A4.2) that the units to be assigned to $(4\pi\varepsilon_0)$ must be (statcoulomb)^2 cm^{-2} dyne^{-1} or (statcoulomb)^2 cm^{-1} erg^{-1}, since an erg is a dyne cm. But 1 erg = 1 statcoulomb $\times$ 1 statvolt, so the units of $4\pi\varepsilon_0$ can also be represented as statcoulomb (statvolt)$^{-1}$ cm^{-1} or statfarad cm^{-1}.

For example, in eqn (A4.4b), if $d = 10$ Å $= 10^{-7}$ cm and $K = \varepsilon_r = 80$, then:

$$C = \frac{1 \times 80}{4\pi \times 10^{-7}} \text{ statfarad cm}^{-2}$$
$$= 6.37 \times 10^7 \text{ statfarad cm}^{-2}.$$

Since 1 statcoulomb (=1 e.s.u. of charge) = $1/(3 \times 10^9)$ C and 1 statvolt (=1 e.s.u. of voltage) = 300 V we see that 1 statfarad = $10^{-11}/9$ F.

The capacity of this condenser in practical units is, therefore, 7.08×10^{-5} F cm^{-2} = 70.8 μF cm^{-2} or 0.708 F m^{-2}.

The same calculation in SI units requires us to use the value $\varepsilon_0 = 8.854 \times 10^{-12}$ F m^{-1} and then:

$$C = \frac{8.854 \times 10^{-12} \times 80}{10^{-9}} = 0.708 \text{ F m}^{-2}.$$

The application of these procedures to electrokinetic problems is shown in the following example.

Smoluchowski's equation for the electrophoretic mobility, eqn (9.11.19) in older texts is written:

$$\mu_E = \frac{\text{velocity}}{\text{field strength}} = \frac{K\zeta}{4\pi\eta} = \frac{\varepsilon_r\zeta}{4\pi\eta}$$

which becomes:

$$\mu_E = (4\pi\varepsilon_0)\frac{\varepsilon_r\zeta}{4\pi\eta} = \frac{\varepsilon\zeta}{\eta}.$$

For water at 25 °C, $\eta = 8.95 \times 10^{-3}$ poise $= 8.95 \times 10^{-4}$ SI units (Pa s) and $\varepsilon = \varepsilon_0\varepsilon_r = 8.854 \times 10^{-12} \times 79$ F m^{-1}.

$$\mu_E = \frac{\text{velocity (m s}^{-1})}{\text{field (V m}^{-1})} = \frac{8.854 \times 10^{-12} \times 79}{8.95 \times 10^{-4}} \times \zeta(\text{V})$$
$$= 7.815 \times 10^{-7} \times \zeta(\text{V}).$$

But the mobility is usually expressed as a velocity in μm s^{-1} per unit field strength (in V cm^{-1}). We have

$$\zeta \text{ (in mV)} = \frac{10^3}{7.815 \times 10^{-7}} \mu_E \text{ (m s}^{-1} \text{ per V m}^{-1})$$
$$= 12.8 \times \mu_E \text{ (in μm cm V}^{-1} \text{ s}^{-1}).$$

Reference

Atkins, P. W. (1978). *Physical chemistry*, pp. 316, 338. Oxford University Press, Oxford.

Appendix A5. Elementary thermodynamic relationships in the absence of surface contributions

The fundamental equation for the internal energy, U, of a one-phase system of many components is

$$dU = T\,dS - P\,dV + \sum_i \left(\frac{\partial U}{\partial n_i}\right)_{S,V,n_j} dn_i \quad \text{(A5.1)}$$

$$= T\,dS - P\,dV + \sum_i \mu_i\,dn_i \quad \text{(A5.2)}$$

where μ_i is the chemical potential of i. The first term represents the heat flow, the second term is the mechanical work and the last term is a 'chemical work' term.

Introducing the definitions of Helmholtz free energy, F, and Gibbs free energy, G:

$$F = U - TS; \qquad G = H - TS \quad \text{(A5.3)}$$

$$= U + PV - TS \quad \text{(A5.4)}$$

we have

$$dF = -P\,dV - S\,dT + \sum \mu_i\,dn_i \quad \text{(A5.5)}$$

$$dG = V\,dP - S\,dT + \sum \mu_i\,dn_i. \quad \text{(A5.6)}$$

The chemical potential can thus be defined in terms of the internal energy, U, using eqns (A5.1) and (A5.2) but is more usefully obtained from eqns (A5.5) and (A5.6)

$$\mu_i = \left(\frac{\partial F}{\partial n_i}\right)_{V,T,n_j} = \left(\frac{\partial G}{\partial n_i}\right)_{P,T,n_j} = \bar{G}_i. \tag{A5.7}$$

The significance of μ_i lies in the fact (eqn (5.3.21)) that at equilibrium, in a constant temperature system, its value for each component must be the same in every phase to which the particular component has access.

It is important to note that the condition for equilibrium of one component i is independent of that of any other component. This is particularly important in membrane phenomena where the membrane may be permeable to only one component. The hydrostatic pressure may also be different on the two sides of the membrane but μ_i will still be the same on both sides of the membrane *for any component to which the membrane is permeable.*

A5.1 *Capacity of a system to do work*

If work w_0 other than P,V work is permitted then it, too, must be included and, for a closed system, we have:

$$\mathrm{d}F = -P\,\mathrm{d}V - S\,\mathrm{d}T + \mathrm{d}w_0 \tag{A5.8}$$

and

$$\mathrm{d}G = V\,\mathrm{d}P - S\,\mathrm{d}T + \mathrm{d}w_0. \tag{A5.9}$$

At constant temperature and pressure

$$\mathrm{d}G = \mathrm{d}w_0 \quad (P,\ T \text{ constant}) \tag{A5.10}$$

so that

$$\Delta G = w_0 \quad (P,\ T \text{ constant}).$$

Thus ΔG measures the capacity of the system to do work (other than P,V work) under conditions of constant temperature and pressure. If w_0, the reversible work done on the system, is positive then ΔG is positive and the capacity of the system to do work is increased. If the system does work on its surroundings then ΔG is negative and the capacity of the system to do further work is diminished. It is in this sense that the G function, under constant pressure and temperature conditions, behaves as the counterpart of a potential energy in a simple mechanical system.

If the system represented by eqn (A5.6) is able to undergo chemical reaction, then the $\mathrm{d}n_i$ terms are no longer zero and we have, for a reversible process, at constant temperature and pressure,

$$\mathrm{d}G = \mathrm{d}w_0 = \sum \mu_i\,\mathrm{d}n_i. \tag{A5.11}$$

This equation gives the (maximum) amount of useful (i.e. non P,V) work which the system would be able to do on its surroundings as a consequence of the chemical changes taking place within it. Note that the quantity $\sum \mu_i \, dn_i$ would have to be negative for the chemical process to proceed spontaneously since only then can the system do work on its surroundings ($w_0 < 0$).

In a chemical reaction, the dn_i values are not independent since they must obey the stoichiometry of the reaction. A general reaction can be represented:

$$\nu_{1,R}A + \nu_{2,R}B + \ldots \rightleftarrows \nu_{1,P}L + \nu_{2,P}M + \ldots$$

where R refers to reactants and P to products and the ν_i are stoichiometric coefficients. Adopting the IUPAC convention that the $\nu_{i,R}$ are negative and the $\nu_{i,P}$ are positive, we can define a parameter ξ which measures the extent of reaction:

$$d\xi = \frac{dn_{i,R}}{\nu_{i,R}} = \frac{dn_{i,P}}{\nu_{i,P}}. \quad (A5.12)$$

Then at constant T and P

$$dG = dw_0 = \sum \mu_i \nu_{i,P} \, d\xi + \sum \mu_i \nu_{i,R} \, d\xi. \quad (A5.13)$$

If the reactants are converted reversibly into products, the maximum work (other than P,V work) which can be extracted from the system is

$$\begin{aligned} w_0 &= \int_0^1 \sum (\mu_i \nu_i) \, d\xi \\ &= \sum_{\substack{\text{reactants} \\ \text{and} \\ \text{products}}} \mu_i \nu_i = \sum_{\text{products}} \mu_i \nu_i - \sum_{\text{reactants}} \mu_i |\nu_i| \quad (A5.14) \\ &= \Delta G(\text{reaction}). \end{aligned}$$

When $\Delta G = 0 = w_0$, the system is in equilibrium and $\sum \mu_i \nu_i = 0$, from which it can be shown that $\Delta G^{\ominus} = -RT \ln \Pi(a_i^{\nu_i})_{\text{equil}}$ where Π indicates the product of the $a_i^{\nu_i}$ terms, i.e.

$$\Delta G^{\ominus} = -RT \ln K_a$$

where K_a is the thermodynamic equilibrium constant.

One application of (A5.14) in which we will be particularly interested is that where the chemical system is made to do electrical work. Suppose electrodes are immersed in the system so that some charge q can flow from one electrode to the other during the course of the reaction. If the potential difference between the electrodes is ΔE (>0) then the total (reversible) work done $w_0 = -q\Delta E$, and if z is the number of electrons associated with the reaction of $|\nu_{i,R}|$ moles of reactants to form $\nu_{i,P}$ moles of product then $q = z\mathscr{F}$ where $\mathscr{F}$ is the Faraday ($= 96485$ C). We then have

$$w_0 = -z\mathscr{F}\Delta E = \sum \mu_i \nu_i. \quad (A5.16)$$

The terms in the summation can be written in the form

$$\nu_i \mu_i^{\ominus} + RT \ln a_i^{\nu_i}$$

for each of the reactants and products and then

$$\Delta E = \Delta E^{\ominus} - \frac{RT}{z\mathcal{F}} \ln\left[\frac{\Pi(a_i^{\nu_i})_{\text{prod}}}{\Pi(a_i^{|\nu_i|})_{\text{react}}}\right]$$

$$= \Delta E^{\ominus} - \frac{RT}{z\mathcal{F}} \ln Q_{\text{R}} \tag{A5.16}$$

where Q_{R} is called the reaction quotient, and $\Delta E^{\ominus}$ $(= -\Delta G^{\ominus}/z\mathcal{F})$, the standard electrode potential difference of the reaction, depends only on the standard state values of the chemical potentials. Equation (A5.16) is the *Nernst equation.*

A5.2 *Situations involving external fields*

We encountered in section 2.1 a problem involving a gravitational field and, in Chapter 6, situations involving the thermodynamics of electrically charged particles and electric fields.

The characteristic feature of a field in this context is that the individual components of a system may do work, or have work done on them, as they move about in the field. That work can be incorporated into eqns (A5.5) and (A5.6) by modifying the chemical potential (as we did in section 2.1) because the work done is related to the numbers of moles of each component. Consider, for example, the case of an external gravitational field. It is characterized by a gravitational potential Φ $(= gh$ in the case of the Earth's gravitational field) and the work done in transferring a small mass $\mathrm{d}m_i$ of component i from a point where $\Phi = \Phi^{\alpha}$ to one where $\Phi = \Phi^{\beta}$ is

$$\mathrm{d}w_0 = (\Phi^{\beta} - \Phi^{\alpha})\, \mathrm{d}m_i$$

$$= (\Phi^{\beta} - \Phi^{\alpha}) M_i\, \mathrm{d}n_i \tag{A5.17}$$

where M_i is the molar mass of component i. If α and β refer to separate phases then $\mathrm{d}n_i^{\alpha} = -\mathrm{d}n_i^{\beta} = -\mathrm{d}n_i$ (where $\mathrm{d}n_i$ is a positive quantity in this case). The work done, $\mathrm{d}w_0$, will then be positive if the magnitude of the potentials is such that $\Phi^{\beta} > \Phi^{\alpha}$. If this is the only form of work (other than P,V work) permitted then eqns (A5.5) and (A5.6) become, for each phase

$$\mathrm{d}F = -P\, \mathrm{d}V - S\, \mathrm{d}T + \sum (\mu_i + M_i\Phi)\, \mathrm{d}n_i \tag{A5.18}$$

and

$$\mathrm{d}G = V\, \mathrm{d}P - S\, \mathrm{d}T + \sum (\mu_i + M_i\Phi)\, \mathrm{d}n_i. \tag{A5.19}$$

The new potential $\tilde{\mu}_i = (\mu_i + M_i\Phi)$ is now the quantity which must be the same in every phase to which component i has access for equilibrium to be established at constant temperature (compare eqn (5.3.21)). A similar argument gives rise to the electrochemical potential function:

$$\tilde{\mu}_i = \mu_i + z_i \mathscr{F} \phi \tag{A5.20}$$

where ϕ is the electrostatic potential (section 6.2) and z_i is the valency of an electrically charged species.

A5.3 *The Gibbs–Duhem equation*

Consider a system of several components in a single phase. If the size of the system is increased in the ratio $(1 + \mathrm{d}\xi):1$, *whilst keeping T and P and the composition unchanged*, then for every component, $\mathrm{d}n_i = n_i\,\mathrm{d}\xi$ and the magnitude of each of the extensive variables is increased in the same proportion: $\mathrm{d}G = G\,\mathrm{d}\xi$, $\mathrm{d}U = U\,\mathrm{d}\xi$ etc. Substituting in eqn (A5.2) we have

$$\mathrm{d}U = U\,\mathrm{d}\xi = TS\,\mathrm{d}\xi - PV\,\mathrm{d}\xi + \sum \mu_i n_i\,\mathrm{d}\xi$$

and integrating this expression from $\xi = 0$ to 1 (which corresponds to doubling the size of the system) we have:

$$U = TS - PV + \sum \mu_i n_i. \tag{A5.21}$$

(This procedure is used very frequently and gives rise to a number of important results.) If, in particular, we take the complete differential of eqn (A5.21):

$$\mathrm{d}U = T\,\mathrm{d}S + S\,\mathrm{d}T - P\,\mathrm{d}V - V\,\mathrm{d}P + \sum \mu_i\,\mathrm{d}n_i + \sum n_i\,\mathrm{d}\mu_i \tag{A5.22}$$

and compare eqn (A5.22) with (A5.2), we have

$$S\,\mathrm{d}T - V\,\mathrm{d}P + \sum n_i\,\mathrm{d}\mu_i = 0. \tag{A5.23}$$

Equation (A5.23) is called the Gibbs–Duhem equation. It describes the fact that only $m + 1$ of the variables T, P, $\mu_1, \mu_2 \ldots \mu_m$ are independent. Its most important application is in the study of two-component systems at constant temperature and pressure, for then:

$$n_1\,\mathrm{d}\mu_1 + n_2\,\mathrm{d}\mu_2 = 0 \tag{A5.24}$$

and so

$$\mu_2 = -\int \frac{n_1}{n_2}\,\mathrm{d}\mu_1. \tag{A5.25}$$

Equation (A5.25) suggests that the chemical potential of one component can be obtained from a knowledge of the chemical potential of the other component. It is particularly useful for the study of solutions of a non-volatile solute. The vapour pressure of the solvent is a direct measure of its chemical potential and so can be used to measure the chemical potential of the solute.

A5.4 *Partial molal quantities*

Equation (A5.6) can be integrated under constant pressure and temperature conditions (with only P, V work permitted) to yield:

$$G = \sum \mu_i n_i = \sum n_i \bar{G}_i \tag{A5.26}$$

at constant T and P where $\bar{G}_i$ is given by eqn (A5.7). This equation expresses the fact that the partial molal Gibbs free energy of component i measures the contribution that i makes (per mole) to the total Gibbs free energy. This is, of course, true of any partial molal quantity and is the reason for the significance of such quantities.

The concept of a partial molal quantity is a little easier to understand if applied to a quantity like the volume. You are aware of the fact that when two liquids are mixed the total volume V may not be equal to the sum of the initial volumes. It is still true, however, that

$$V = n_1 \bar{V}_1 + n_2 \bar{V}_2 = \sum n_i \bar{V}_i \tag{A5.27}$$

if $\bar{V}_i = (\partial V / \partial n_i)_{P,T,n_j}$ is the partial molal volume of component i. It must be recognized, of course, that $\bar{V}_i$ is a function of the concentration of component i (and may be influenced by the presence of other components).

Partial molal quantities are intensive properties which can be derived from any extensive thermodynamic property, B:

$$\bar{B}_i = \left(\frac{\partial B}{\partial n_i}\right)_{P,T,n_j} \tag{A5.28}$$

and in any mixture the total value of B for the whole system will be given by (at constant T and P):

$$B = \sum n_i \bar{B}_i. \tag{A5.29}$$

All of the thermodynamic equations that are derived for a single pure substance can, in general, be applied to the $\bar{B}_i$ values of the individual components of a mixture. For example:

$$\mathrm{d}\bar{G}_i = \mathrm{d}\mu_i = \bar{V}_i \,\mathrm{d}P - \bar{S}_i \,\mathrm{d}T. \tag{A5.30}$$

Appendix A6. Osmotic pressure

As an illustration of the application of some of the basic concepts of thermodynamics we will examine the phenomenon of osmotic pressure, and in particular its application to colloidal suspensions.

Consider the system illustrated in Fig. A6.1, consisting of a cylinder, C, containing a membrane, M, which is permeable to water but not to the solute. The pistons are able to apply a direct (hydrostatic) pressure to the liquids on either side of M and M is assumed to be sufficiently rigid to be able to support the

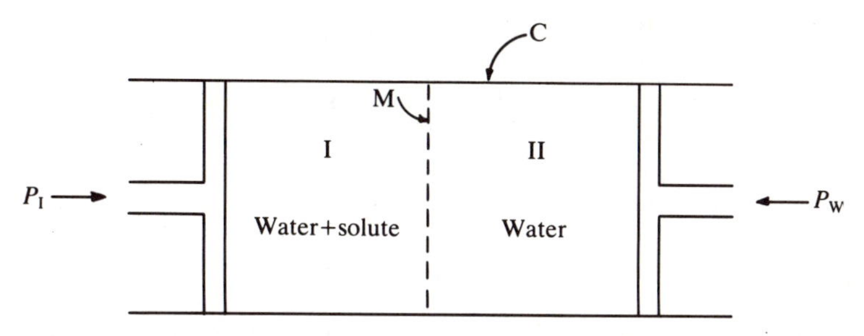

FIG. A6.1. Schematic arrangement of an osmotic pressure cell. The cylinder, C, is fitted with a semipermeable membrane M. The liquids I and II can be subjected to hydrostatic pressures P_I and P_w through the pistons.

resulting pressure difference across it. At constant temperature the equilibrium condition is:

$$\mu_w^I = \mu_w^{II} \tag{[5.3.21]}$$

where μ_w is the chemical potential of the water, since it has access to both phases I and II. The chemical potential of water in a solution is given by:

$$\mu_w = \mu_w^\ominus + RT \ln a_w \tag{A6.1}$$

where $\mu_w^\ominus$ is the chemical potential of pure water at 1 atmosphere pressure and temperature T, and a_w is the activity of the water.

If $P_I = P_w$ then water will tend to diffuse from phase II into the solution because its chemical potential is lower there ($a_w < 1$). In order to prevent it from doing so it is necessary to apply an excess pressure to phase I which is sufficient to raise the chemical potential of the water in that phase up to the value in the pure water phase at pressure P_w. The chemical potential of the water in phase II is obtained from eqn (A5.30), which at constant temperature gives:

$$\mu_w^{II} = \mu_w^\ominus + \int_{1\,\text{atm}}^{P_w} \bar{V}_w^{II}\, dP \tag{A6.2}$$

$$= \mu_w^\ominus + \bar{V}_w^{II}(P_w - 1) \tag{A6.3}$$

assuming that the molar volume of the water is unaffected by the pressure. The chemical potential of the water in phase I at pressure P_w is calculated by adding the effects of solute and of the pressure change:

$$\mu_w^I = \mu_w^\ominus + RT \ln a_w^I + \bar{V}_w^I(P_w - 1).$$

The excess pressure $(P_I^{eq} - P_w)$ that must be applied to phase I to bring the chemical potential up to the value in phase II is given by

$$\int_{P_w}^{P_I^{eq}} \bar{V}_w^I\, dP = \bar{V}_w^I(P_I^{eq} - P_w)$$

$$= (\mu_w^{II} - \mu_w^I) \quad \text{at pressure } P_w$$

$$= \bar{V}_w^{II}(P_w - 1) - RT \ln a_w^I - \bar{V}_w^I(P_w - 1). \tag{A6.4}$$

The osmotic pressure Π is defined as $(P_{\mathrm{I}}^{\mathrm{eq}} - P_{\mathrm{w}})$ and so:

$$\Pi = \frac{-RT}{\bar{V}_{\mathrm{w}}} \ln a_{\mathrm{w}}^{\mathrm{I}} \tag{A6.5}$$

if we assume that $\bar{V}_{\mathrm{w}}^{\mathrm{I}} = \bar{V}_{\mathrm{w}}^{\mathrm{II}} = \bar{V}_{\mathrm{w}}$.

From the Gibbs–Duhem eqn (A5.24) we have:

$$\frac{1000}{18} \mathrm{d}\mu_{\mathrm{w}} + m_{\mathrm{s}} \, \mathrm{d}\mu_{\mathrm{s}} = 0 \tag{A6.6}$$

where m_{s} is the molality of the solute (mol kg^{-1}) or $55.5 \, \mathrm{d} \ln a_{\mathrm{w}} + m_{\mathrm{s}} \, \mathrm{d} \ln a_{\mathrm{s}} = 0$ and so

$$\ln a_{\mathrm{w}} = -\frac{1}{55.5} \int_0^{m_{\mathrm{s}}} m_{\mathrm{s}} \, \mathrm{d} \ln a_{\mathrm{s}}. \tag{A6.7}$$

Substituting in eqn (A6.5):

$$\Pi = \frac{RT}{55.5 \bar{V}_{\mathrm{w}}} \int_0^{m_{\mathrm{s}}} m_{\mathrm{s}} \, \mathrm{d} \ln a_{\mathrm{s}}. \tag{A6.8}$$

In dilute solution $a_{\mathrm{s}} \simeq m_{\mathrm{s}}$ and $55.5 \bar{V}_{\mathrm{w}} = 1$ and so

$$\Pi = m_{\mathrm{s}} RT. \tag{A6.9}$$

Babcock (1963), on whose analysis the above is based, asserts that 'The similarity between eqn (A6.9) and the ideal gas law is wholly without significance'. That is, perhaps, an extreme view but it is certainly true that the use of crude kinetic theory ideas to 'explain' osmotic pressure often leads to difficulties.

In dilute electrolyte solutions the value of m_{s} in eqn (A6.9) is replaced by $\sum m_i$, where the summation is over all of the ions present. Osmotic pressure is, therefore, referred to as a colligative property since it depends on the number of particles of solute present.

Equation (A6.9) can be applied to systems containing colloidal particles if they are at sufficiently high concentration. It is often applied to polymer (including protein) solutions as a method of obtaining molar masses and the procedures involved are treated by Hiemenz (1977, Chapter 4). Since the osmotic pressure depends on the number of particles and not their size it gives a measure of number average molar mass. Its application to the study of interaction between colloidal particles is examined in section 7.3.

Reference

Babcock, K. L. (1963). *Hilgardia* **34,** 417–542.

Hiemenz, P. C. (1977). *Principles of colloid and surface chemistry*. Marcel Dekker, New York.

INDEX

Entries in bold face refer to major sections devoted to the topic. Entries in italics refer to minor sections. The letters 'f' and 't' refer to figures and tables, respectively. Multiple entry has been used in preference to cross-entry unless there are many listings under one heading.